Preliminary Edition for

Introductory Algebra

Models, Concepts, Skills

Preliminary Edition for
Introductory Algebra
Models, Concepts, Skills

Douglas F. Robertson
Scott A. Storla
University of Minnesota

BROOKS/COLE PUBLISHING COMPANY

I(T)P®   An International Thomson Publishing Company

Pacific Grove • Albany •Belmont • Boston • Cincinnati • Detroit • Johannesburg • London
Madrid • Melbourne • Mexico·City • New York • Scottsdale • Singapore • Tokyo • Toronto

A Robert W. Pirtle Book

Publisher: *Bob Pirtle*
Editorial Assistant: *Erin Wickersham*
Marketing Manager: *Leah Thomson*
Marketing Associate: *Debra Johnston*
Production Editor: *Mary Vezilich*
Manufacturing Buyer: *Vena Dyer*
Cover Design: *Vernon T. Boes*
Printing and Binding: *Globus Printing Co.*

COPYRIGHT© 1999 by Brooks/Cole Publishing Company
A division of International Thomson Publishing Inc.

I(T)P The ITP logo is a registered trademark under license.

For more information, contact:

BROOKS/COLE PUBLISHING
511 Forest Lodge Road
Pacific Grove, CA 93950
USA

International Thomson Editores
Seneca 53
Col. Polanco
11560 México, D. F., México

International Thomson Publishing Europe
Berkshire House 168-173
High Holborn
London WC1V 7AA
England

International Thomson Publishing Japan
Palaceside Building, 5F
1-1-1 Hitotsubashi
Chiyoda-ku, Tokyo 100-0003
Japan

Thomas Nelson Australia
102 Dodds Street
South Melbourne, 3205
Victoria, Australia

International Thomson Publishing Asia
60 Albert Street
#15-01 Albert Complex
Singapore 189969

Nelson Canada
1120 Birchmount Road
Scarborough, Ontario
Canada M1K 5G4

International Thomson Publishing GmbH
Königswinterer Strasse
53227 Bonn 418
Germany

Printed in the United States of America

5   4   3   2   1

ISBN 0-534-36894-8

Contents

Chapter 7 Factoring Algebraic Expressions

Chapter 8 Rational Expressions and Equations

Chapter 9 Radical Expressions and Equations

Answers to Selected Exercises 615

Index

Preface

We designed this book to fit the needs of self–paced, laboratory–based courses as well as traditional lecture–based formats. The content includes the standard curriculum for introductory algebra but the material is geared for students who previously have had difficulty in mathematics. We have been classroom testing the material for the past two years at Minneapolis Community and Technical College and in the General College of the University of Minnesota.

Writing this book has been an evolutionary process for us that began in 1986. At that time, we began to design a series of developmental mathematics courses based on modeling and problem solving. For these new courses, the instructor's role was largely as a "mediator" or "facilitator" and the preferred classroom activity was group work with little lecture. The general environment of the classroom was one of student activity where we could observe and influence student thought processes rather than as a place for the transfer of knowledge from expert to novice.

Over the last 13 years we have found that some of our original ideas were wrong, such as it is bad to have students memorize procedures, while other ideas were correct, such as the importance of developing and sharing a common in-class vocabulary.

The textbook we have written includes many of the ideas we have developed, adapted, and refined while working in this alternative environment. Many of the ideas have been derived from what other mathematics educators have discovered about student learning at the developmental level, from analyzing the way students construct their own versions of mathematics, and from the students themselves through our use of student interviews and the monitoring of student to student conversations. Student input has been particularly helpful to us:

- Student interviews — Many authors are greatly influenced by focus groups consisting of mathematics instructors. We don't feel this is a bad idea, it's just incomplete because the primary purpose of the textbook is to help students learn mathematics rather than to help instructors present material. Therefore, we feel it's important for the text to present topics in a language that novices can easily understand. For this reason our in-book dialogue is heavily influenced by interviews conducted with our students. Each term we have interviewed a subset of students to determine the depth and breadth of understanding of various topics. This has often lead to a revision of the way the textbook explanations are presented as it became apparent that what seemed unambiguous to us as experts was often not unambiguous to many of our students.

- Student conversations — Students in our classes engage in content-specific group work (that is, group work which is centered on the skills to be learned that day). This affords us the opportunity to listen to mathematical exchanges between students and to hear first hand how they are constructing (and misconstructing) their own versions of mathematics. Student misconceptions generally are not random, nor are they necessarily the result of poor thinking; misconceptions tend to occur for specific reasons and an analysis of student conversations has often led us to consider alternative ways of interpreting and explaining various topics.

We have done a great deal of training of undergraduate and graduate teaching assistants who are experienced at mathematics but who have had little teacher training. It is apparent to us that for subject-matter experts one of the most difficult parts of instruction is remembering what it was like to be a novice. Analysis of student thought processes through interviews and listening to conversations has helped us write materials that will make sense to the novice math student.

While we have tried to keep a good deal of traditional introductory algebra intact, we have devoted a large amount of space to topics we have found to be especially difficult for novices:

- Word problems — Rather than treating word problems as a small section in a chapter on equations and inequalities, we have devoted an entire chapter to them. In Chapter 3 (and in later sections throughout the book) we have tried to illustrate a procedure used by experts in many different fields to solve non-routine problems. We have tried to shift the focus from memorizing a solution process for some subset of common algebra problems (e. g., coin, distance, mixture) to the connection and interaction between the English and algebraic representations of problems. We have stressed the importance of units in both setting up the problems and for verifying intermediary steps and final results.

  Interestingly, a common feedback from our former students is that they found themselves using the general procedure for solving word problems to solve non-mathematical problems in their work outside of school.

- Modeling — When we began to introduce linear modeling into our classrooms many years ago we did not realize the gulf that exists between experts, who have been initiated into the process of thinking mathematically, and novices. We are not just referring to the scientific method but to such basic elements as the concepts of input and output and the ways we represent input and output in data tables, graphs, and equations, Over time, we developed Chapter 4 Mathematical Models in an attempt to properly present to students ideas that were automatic to us but which made little intuitive sense to them. This is why we spend so much time building the foundations for modeling in chapters three and four rather than simply including a few "real world" problems at the end of a chapter on the linear equation.

  We have found that to help students learn how to model mathematically the four types of representations, verbal, symbolic, tabular, and graphic, need to be carefully and fully explained and practiced with emphasis on their interconnectedness. We use the following graphic to illustrate this idea throughout Chapter 4:

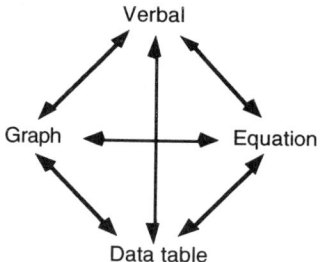

- Variable — As experts in mathematics we often unconsciously and easily shift between the different meanings for variable. Our interviews with students showed that meanings which we as experts took for granted our students had not even considered. This is why we make explicit the difference between the Chapter 3 concept of variable (a symbol which represents an unknown number) and the Chapter 4 concept of variable (a symbol which represents a set of values). This is helpful to students who must make sense of something like $x = 3$ within different contexts (for example, does this help them find the solution to an equation such as $2x = 6$ or is it the equation of a vertical line?).

- Rate — Like variable, rate has several definitions and uses which novices find confusing. In Chapter 3 we explicitly define rate and use it as a conversion factor (as in miles per hour). In Chapter 4 we expand the definition to describe rate as a rate of change (as in the slope of a line). As experts we can easily switch between the two definitions depending on the context; however, novices have to be carefully guided in doing this.

- Procedures and skills — We are concerned with current thinking by some that time spent becoming automatic at skills or memorizing procedures is either wasted or harmful to students. Understanding concepts is necessary but so is some memorization. Without a knowledge base of facts and procedures it is impossible to make sense of concepts and ideas. We encourage students to copy procedures to note cards and to develop an internal dialogue based on these procedures as they work through their homework problems. Just as a student who needs to focus his or her attention on decoding sentences will have great difficulty understanding the meaning of a novel, a student who is not automatic at the simplification of integer expressions will find it difficult to understand the meaning of slope.

- Traditional content — We believe it's important for students to be exposed to mathematics for its own sake, as a way of knowing and thinking. Obviously, our technological society is dependent on some people knowing and understanding mathematics at a very high level. However, our English teacher colleagues use algebra about as often as we use our knowledge of Shakespeare. We don't think students should take mathematics only because it might be directly useful to them in their future lives; mathematics is an important field of study in its own right, and like literature, art, humanities, social science, and natural science, mathematics is something that all educated people should be exposed to and have some knowledge of. Therefore, we make no apologies for including in our book most of the traditional topics of introductory algebra. The beauty of mathematics lies not only in its usefulness but in its logic and structure.

We hope you and your students find this book effective. If you would like to send us comments or ask questions, we can be reached at droberts@tc.umn.edu or jsstorla@worldnet.att.net

Chapter 1
The Real Number System

Section 1.1 Introduction to Algebra

Although many people think of algebra as a collection of abstract rules and procedures it's really a language, much like English, which has its own vocabulary and rules of grammar. Like other languages, algebra will help you clarify thoughts, communicate ideas, and solve problems. For example, suppose you are given the problem:

If a can of juice costs 50 cents, how much would 2 cans cost?

Your intuition probably says $1.00, which is correct. Now, suppose you are given the problem:

A can of juice costs 50 cents. The juice alone costs 40 cents more than the can. How much does the can alone cost?

Your intuition probably says 10 cents, which turns out to be incorrect. (We use algebra to solve this problem in Chapter 3.) Intuition can help solve many problems, but when intuition fails, algebra is an excellent problem solving tool.

This book is filled with concepts and skills that will help you use algebra effectively. Our first step is to discuss some of the vocabulary and notation of algebra.

Topic 1.1 A — Sets

A **set** is a collection of distinct things that are to be treated as an entity. The things, which may be numbers, objects, etc., are called **elements** or **members** of the set. A set can be described by listing the elements inside braces { }. For example, the days of the week that begin with the letter S can be written as the set {Saturday, Sunday}. If there are too many elements to list, a description of the elements may be written inside braces. Some sets of numbers are used so often that mathematicians have given them special names.

Natural Numbers are the numbers 1, 2, 3, etc., which we use to count things. We cannot list all the elements of this set so we will write the set as {1, 2, 3, 4, ...}. The three dots ... mean the sequence of numbers goes on forever in the same pattern. In this case, the ... means 5, 6, 7, 8 and so on.

Natural numbers are also known as **counting numbers**., because they are used to count things, such as 15 books, 38 shares of stock, or 49 computers.

Whole Numbers are the natural numbers with the number zero included. This set can be written as {0, 1, 2, 3, 4, ...}.

Topic 1.1 B — Variables and Constants

When we wish to represent a number whose value we currently don't know, or whose value changes from situation to situation, we use a letter. We refer to the letter as a **variable**. For example, we might let x represent your current age.

We sometimes use letters to represent quantities that always have the same value. In that case, the letter is referred to as a **constant**. For example, we use the Greek letter π (pronounced *pi*) to represent the ratio of the circumference to the diameter of a circle. Pi always represents a value that is approximately 3.14. The Arabic numerals we use to represent numbers, such as 2 for the number two, are also referred to as constants.

Topic 1.1 C — Mathematical Operations

A mathematical **operation**, such as addition, is a process that transforms one or more numbers. The symbol we use to indicate the operation, such as +, is called an **operator**. For example, if your current age is 18, and we wish to know your age a year from now, we use the operation of addition to find that you will be 19 years old. We may use the operator, +, to write the operation of addition symbolically as 18 + 1.

In algebra, when we want to discuss general mathematical operations, we use variables as well as constants. In the following definitions we will use a and b to represent any numbers, such as 2 or 5:

Addition: When we add two numbers we have a **sum**. We use + to show the operation of addition. Algebraically, the sum of any two numbers is written $a + b$, where a represents the value of the first number and b represents the value of the second. We refer to the quantities being added, a and b in this case, as **terms**. Terms may be constants, variables, or the product of constants and variables. For example, in the statement "2 + 3 is 5", 2 is a term and 3 is a term; 2 + 3 is a sum and 5 is a sum.

Subtraction: When we subtract two numbers we have a **difference**. We use – to show the operation of subtraction. Algebraically, the difference of any two numbers is written $a - b$.

Multiplication: When we multiply two numbers we have a **product**. We use $\times$, $\cdot$, parentheses, or no symbol at all to show the operation of multiplication. For example,

$$\text{multiply 2 and 5} \Rightarrow \quad 2 \times 5 \quad 2 \cdot 5 \quad 2(5) \quad (2)5 \quad (2)(5)$$

$$\text{multiply the value of } a \text{ and the value of } b \Rightarrow \quad a \times b \quad a \cdot b \quad a(b) \quad (a)b \quad (a)(b) \quad ab$$

Note the following:

- The product of a constant and a variable usually is written without an operator. For example, $4a$ means multiply 4 by the value of a. This is read "the product of four and a", "four times a", or simply "four a". We refer to 4 as the **coefficient** of the variable a.

- A variable by itself literally means 1 times the value of the variable. For example, a means $1a$. We say that 1 is the coefficient of a.

- Each number being multiplied is called a **factor**. For example, in the product ab, a is a factor and b is a factor; in the product $2a$, 2 is a factor and a is a factor (and 2 is the coefficient of a).

- The symbol $\times$ is not the letter x, even though it looks like it when we write it long hand. The $\times$ is called *St. Andrew's cross*.

Division: When we divide two numbers we have a **quotient**. To show the operation of division use $\div$, /, or a **fraction bar** (the line that separates the numerator and denominator in a fraction, as in $\frac{2}{3}$). We say that one whole number is **divisible** by another if the division results in a remainder of 0. Algebraically, the quotient of any two numbers may be written as $a \div b$, a/b, or $\frac{a}{b}$.

As you will see throughout the book, being able to distinguish terms, sums, factors, and products is especially important.

Example 1.1.1   Identify the sums, terms, products, and factors.

a) $5 + 1$

•SOLUTION•
- 5 is a term, because terms are added.
- 1 is a term.
- $5 + 1$ is a sum.

— *Study Tip* — ***How to use the*** ***worked-out examples***	This book contains many examples with worked-out solutions. The examples usually further explain or apply the ideas you just read about. To get the most from the examples, you should read the question and then try to answer it yourself, before you read our solution. In most cases this will require you to do several steps so have a pencil, paper, and your calculator at hand. If your answer is correct, you should still read through our solution to see if your answer is correct for the right reasons. The idea is to get your brain out of a passive reading mode and into an active thinking mode and to help you analyze the correctness of your thought processes.

b) $3 \cdot 2$

•SOLUTION•
- 3 is a factor, because factors are multiplied.
- 2 is a factor.
- $3 \cdot 2$ is a product.

c) $3y$

•SOLUTION•
- 3 is a factor.
- y is a factor.
- $3y$ is a product.

d) $2l + 2w$

•SOLUTION•
- $2l$ is a term, because terms are added, and $2l$ is also a product.
- $2w$ is a term and it is also a product.
- $2l + 2w$ is a sum.
- Within the term $2l$, both 2 and l are factors.
- Within the term $2w$, both 2 and w are factors.

Practice 1.1.1 Identify the sums, terms, products, and factors.

1. $25 + 4$ 2. $5w$ 3. $6x + 2$ 4. $2m + 3n$ 5. $5abc$

Topic 1.1 D — The Equal Sign

In previous math classes, you may have used the **equal sign**, =, to write a problem and its solution on a single line. For example, $2 \cdot 3 + 5 = 6 + 5 = 11$. In general, the quantities on the left and right of = represent two different ways of saying the same thing. For example, $1 + 8 = 9$ means that $1 + 8$ and 9 are two ways of representing the number nine. We might read $1 + 8 = 9$ as

- "One plus eight simplifies to nine" meaning that if we begin with $1 + 8$ and carry out the addition we end up with 9.

- Or, we might say "the number one, increased by eight, produces the number nine".

- Or, we might say "wherever we see $1 + 8$ we can use 9 instead".

In algebra, the equal sign sometimes has a different meaning. We will discuss the new way of interpreting = in Chapter 2.

When we want to indicate that two things are not equal we draw a slash through the = to get ≠. We refer to ≠ as the **not equal sign** For example, we might write "the sum of two and three is not equal to six" as $2 + 3 \neq 6$; or, we might write $x \neq 0$, meaning that x can represent any value except 0.

Topic 1.1 E — Exponential Notation

To make it easier to write repeated multiplication, such as $5 \cdot 5 \cdot 5 \cdot 5$, we use **exponential notation**. For example, using exponential notation we would write $5 \cdot 5 \cdot 5 \cdot 5$ as 5^4 (read "five to the fourth power").

- 5 is called the **base**. It tells us what we are multiplying.

- 4 is called the **exponent**. It tells us how many factors of the base we are multiplying. The exponent is always written a "half–step" above the base.

Exponents of 2 and 3 have special names. An exponent of 2 is read "squared" (because a square has two dimensions). An exponent of 3 is read "cubed" (because a cube has 3 dimensions). For example,

- 7^2 can be read "seven to the second power" or "seven squared". It means $7 \cdot 7$.

- 4^3 can be read "four to the third power" or "four cubed". It means $4 \cdot 4 \cdot 4$.

Exponential notation has the same meaning when the base contains letters. For example, x^2 is read "x squared" and means multiply 2 factors of x. This can be written $x \cdot x$. In x^2, the base is x and the exponent is 2.

We usually don't write exponents of 1. For example, rather than writing 2^1 we usually write 2; rather than writing x^1 we usually write x.

Topic 1.1 F — Simplifying Expressions and the Order of Operations

An **expression** is a collection of numerals, operators, letters, and grouping symbols. To **simplify** an expression means to perform all the allowable operations. For example, to simplify $1 + 2 + 3$ we would do the additions to get 6. When we simplify an expression we transform it into an equivalent expression; that is, we transform it into a different expression which represents the same value. For example, the expressions $2 + 3$ and $10 \div 2$ are equivalent because they represent the same number, 5.

When an expression contains two or more operations we have to decide which operation to do first. For example, does $1 + 2 \cdot 3$ simplify to 9 or 7? If we do the addition first we have $3 \cdot 3$, which is 9; if we do the multiplication first we have $1 + 6$, which is 7. A practical example might help us decide which way to do the operations. If we buy a quart of milk for \$1 and two boxes of doughnuts for \$3 each, we have spent \$7. That is, $1 + 2 \cdot 3$ is 7. This implies that we must do the multiplication before the addition.

Grouping symbols allow us to change the order in which mathematical operations are carried out. **Grouping symbols** include parentheses ( ), brackets [ ], the fraction bar, and the radical symbol, $\sqrt{}$ †. The operations inside grouping symbols are done before other operations. For example, $2(5 + 1)$ says to first add 5 and 1 and then to multiply this sum and 2.

In general, we use the following order when doing mathematical operations:

— Procedure — Simplifying expressions using the proper order of operations	
Step 1	Simplify expressions inside grouping symbols.
Step 2	Simplify exponents and square roots.
Step 3	Simplify multiplication and division, working left to right.
Step 4	Simplify addition and subtraction, working left to right.

— Study Tip — Making procedure study cards	
	Throughout this book we have listed many procedures like the one shown above. To help you learn the procedures, it would be a good idea to make a study card for each one as you encounter it.
	To make procedure study cards:
	• On one side of a 3 by 5 index card write the procedure's name and the textbook page number where it was discussed.
	• On the reverse side write the steps.
	To use procedure study cards:
	• When you work on your homework exercises, have the appropriate cards in front of you and refer to them as often as necessary.
	• Before a test or quiz, review the cards.
	• Before you begin your next class in mathematics, review the cards.

Example 1.1.2   Simplify (that is, do all the allowable operations).

a) $4 + 3 \cdot 2$

•SOLUTION•

$$4 + 6 \quad \Leftarrow \text{ Multiplication is done before addition.}$$

$$10 \quad \Leftarrow \text{ Added.}$$

---

† The radical symbol is used to indicate numbers like square roots, which we discuss in much detail in Chapter 9. An example of a square root is $\sqrt{9}$, which is "the principal square root of 9". The principal square root of a positive number is a number, which when multiplied by itself, produces the original number. Therefore, $\sqrt{9}$ simplifies to 3 because 3×3 is 9.

— *Study Tip* —	Many of our multi–step solutions include an explanation of each step, written to the right of an arrow, $\Leftarrow$. Be sure to read explanations, even if you know how to do the problem correctly by yourself. For example, in the above problem you should
How to use the	• Read the problem, $4 + 3 \cdot 2$.
explanations in the	• Think about how you would solve the problem and write your solution in the margin or on another sheet of paper.
worked examples	• Read the explanation to the right of the $\Leftarrow$.
	• Check your solution. Does your solution match ours? Do you understand the explanation? If not, you may want to go back and read through the instructional material again. If you are still confused, ask a classmate, a tutor, or the instructor, or write a note to yourself to ask about the problem at the next class session.
	Keep in mind that the answer is secondary to your understanding the process.

b) $2 + 6 - 2 + 1$

•SOLUTION•

$8 - 2 + 1 \quad \Leftarrow$ Addition and subtraction are done left to right, so we added $2 + 6$.

$6 + 1 \quad \Leftarrow$ Subtracted.

$7 \quad \Leftarrow$ Added.

— *Caution* —	Remember that we do addition and subtraction working from left to right. In this example, if we do all the additions before the subtraction we get a different (incorrect) answer:
Working from left to right	$2 + 6 - 2 + 1$
	$8 \quad - \quad 3 \quad$ **WRONG (see correct solution above)**
	5

c) $36 \div 4(4 + 2)$

•SOLUTION•

$36 \div 4(6) \quad \Leftarrow$ Simplified inside parentheses first.

$9(6) \quad \Leftarrow$ There are no exponents so we did the multiplication and division working left to right. In this instance, division comes before multiplication.

$54 \quad \Leftarrow$ Multiplied.

Practice 1.1.2 Simplify.

1. $8 + 12 \div 3 \cdot 4$

2. $5^2 - 2^2$

3. $(5 - 2)^2$

4. $(4 - 1)(12 - 7)$

5. $24 \div (4)(16 - 10)$

6. $8 - \dfrac{2 + 5 \cdot 4}{15 - 2 \cdot 2}$

When one set of grouping symbols is contained within another we say the grouping symbols are **nested**. When the grouping symbols are nested, the calculations within the innermost pair are performed first.

Example 1.1.3   Simplify $12 - [(5 - 2)^2 + 1]$

•SOLUTION• This expression contains nested grouping symbols so we simplify inside the innermost grouping symbols first. Then, we simplify inside the brackets [ ].

$12 - [(5 - 2)^2 + 1]$ $\Leftarrow$ Expression to simplify.

$12 - [(\mathbf{3})^2 + 1]$ $\Leftarrow$ Subtracted $5 - 2$ inside the parentheses.

$12 - [\mathbf{9} + 1]$ $\Leftarrow$ Multiplied 2 factors of 3 to get 9.

$12 - [\mathbf{10}]$ $\Leftarrow$ Added inside [ ].

2 $\Leftarrow$ Subtracted.

Practice 1.1.3   Simplify.

1. $14 - [(5 - 3)^2 + 6]$ 2. $9 - (12 - [2^2 + 4])$ 3. $60 - 3(2 + 3[2 + 3])$ 4. $15 - [(2 + 4) \cdot 2 - (12 - 8)]$

— *Study Tip* — ***Writing out steps***	These problems can get complicated! It will help if you keep a detailed written record as you work out the solutions. There are three compelling reasons for writing out the steps: • The process of writing helps you focus on what you are doing. • With a written record, it's easier to find errors. If you did all the work in your head and made an error, you will have to begin all over from scratch and you might make the same mistake again. • Without a written record, your teacher cannot see what you did and help you improve; on the job, your colleagues or boss cannot check what you have done. A good rule of thumb is to use one line per operation. It's usually better to write more than less.

Topic 1.1 G — Evaluating Expressions

To **evaluate an expression** means to substitute specific numbers for the letters and then simplify. *It's best to place parentheses around the numbers when doing the substitution.* For example, to evaluate $5x$ when x is 2, we write $5(2)$ and then simplify to get 10.

Example 1.1.4   a)   Evaluate $x - y$ if x is 6 and y is 2.

•SOLUTION•

$(\mathbf{6}) - (\mathbf{2})$ $\Leftarrow$ Substituted 6 for x and 2 for y.

4 $\Leftarrow$ Subtracted.

b)   Evaluate $2l + 2w$ if l is 3 and w is 4.

•SOLUTION• Remember, $2l$ means multiply 2 by the value of l and $2w$ means multiply 2 by the value of w.

$2l + 2w$ $\Leftarrow$ Expression to evaluate.

$2(\mathbf{3}) + 2(\mathbf{4})$ $\Leftarrow$ Substituted 3 for l and 4 for w.

$6 + 8$ $\Leftarrow$ Multiplied.

14 $\Leftarrow$ Added.

c) Evaluate πd if d is 10.

•SOLUTION• We are not given the value of π and so we assume it's approximately 3.14. Be sure to memorize this value for π.

$\pi d \Leftarrow$ Expression to evaluate.

$(3.14)(10) \Leftarrow$ Substituted 3.14 for π and 10 for d.

$31.4 \Leftarrow$ Multiplied.

Practice 1.1.4 Evaluate the expressions for the given values of the variables.

1. $4t + 3$ if t is 7
2. πr^2 if r is 3
3. $5x + 25(x + 30)$ if x is 9
4. lwh if l is 8, w is 2 and h is 3
5. $a^2 + b^2$ if a is 9 and b is 4
6. $(a + b)^2$ if a is 9 and b is 4.

Topic 1.1 H — The Importance of Vocabulary

At the beginning of this section we mentioned that algebra is a language with its own vocabulary and rules of grammar. In order to talk to each other about algebra we need to share a common and well defined vocabulary. When you see words highlighted like **this** it's critical that you learn their meanings and how they are used.

As your vocabulary improves you will find that this book makes more sense, your instructors discussions will be easier to follow, and most importantly, when you are away from the classroom and thinking about algebra your self–discussions will be productive.

— Study Tip — *Making vocabulary study cards*	A good way to learn vocabulary is by creating and using vocabulary study cards. *To make vocabulary study cards:* • After you have read a section, go through it again and select the most important words and definitions. These usually are in **bold** type or inside boxes. • On one side of a 3 by 5 index card write a word or phrase that identifies the idea and the page in this book where the idea was discussed. • On the reverse side write the corresponding definition. Write one or two examples to help clarify the words. *To use vocabulary study cards:* • Every day, shuffle the cards and quiz yourself randomly on a few cards. Since mathematics is always using past ideas to explain present work this review helps you recall important information you may have forgotten. • Before a test, review the cards. • Before you begin your next class in mathematics, review the cards.

Example 1.1.5 Make a study card for the word *expression*.

•SOLUTION• One side of the card has the phrase *expression*. The reverse side has the definition, an example, and a page reference.

Expression	Def (pg. 107): a collection of numerals, operators, letters, and grouping symbols. Ex: $3x + 2$ y -8

Practice 1.1.5 Make study cards for the following words:

1. Variable
2. Simplify
3. Factor
4. Exponent

Exercise Set 1.1 The answers to the odd numbered exercises are at the back of the book.

Identify the sums, terms, products, and factors.

1. $35 + 21$
2. $2.3 + 0.5$
3. $(3.5) \cdot 2$
4. $6 \cdot 3$
5. $4u$
6. $7p$
7. $5v + 3w$
8. $3j + 9k$

Simplify.

9. $31 - 8 \cdot 3$
10. $60 - 9 \cdot 5$
11. $18 - 22/11 \cdot (5)$
12. $63 - 45/9 \cdot (8)$
13. $(2 \cdot 3)^3$
14. $(5 \cdot 2)^2$
15. $2^3 \cdot 3^3$
16. $5^2 \cdot 2^2$
17. $2 + 3(2 + 1)$
18. $5(8) + 10(11 - 8)$
19. $15 - 3 - 6 + 4 \cdot 2$
20. $18 - 3 + 2 - 4 \cdot 3$
21. $[12 - (12 - 6)] + 8$
22. $5 + 4[3 + 2(1 + 6)]$
23. $[3 + (10 - 7)^2]^2$
24. $[36 - (8 - 5)^3]^2$
25. $12 + \dfrac{12 - 2 \cdot 3}{10 - 3^2}$
26. $30 - \dfrac{20 + 2 \cdot 8}{8 - 2^2}$
27. $15 - [(8 - 5)^2 - 2]$
28. $20 - [(12 - 8)^2 + 2^2]$
29. $5 + 4(5 + 4[5 - 2])$
30. $18 - 2[8 - 3(9 - 8) + 1]$
31. $23 - [(8 - 2) \cdot 3 - (9 - 5)]$
32. $21 - [5(8 - 3) - 3^2]$

Evaluate the expressions for the given values of the variables.

33. $a^2 - b^2$ if a is 8 and b is 6
34. $x^3 - y^3$ if x is 3 and y is 2
35. $b^2 - 4ac$ if a is 3, b is 9 and c is 2
36. $b^2 - 4ac$ if a is 2, b is 6 and c is 2
37. $x^2 + 2xy + y^2$ if x is 6 and y is 5
38. $x^2 - 2xy + y^2$ if x is 8 and y is 3
39. $0.05(x) + 0.10(30 - x)$ if x is 15
40. $0.45(x) + 0.55(9 - x)$ if x is 6

Section 1.2 Fractions

A **fraction** is a number that can be written in the form $\frac{a}{b}$, where a and b are real numbers and $b \neq 0$. In arithmetic, fractions can be used to indicate parts of a whole. For example, the fraction $\frac{2}{3}$ can mean that we have divided something into 3 equal parts and that we have 2 of those parts. The top number, the numerator, tells us how many parts we have and the bottom number, the denominator, tells us how many equal parts are needed to make the whole.

In algebra, we will use fractions just as you did in arithmetic, but we will extend the concept in several important ways. Because many of the problems in this book involve fractions, and because many of the processes you learned in arithmetic will be generalized so that we can deal with algebraic fractions, it's important to review some of the basics of fractions now. You must be automatic at reducing, multiplying, dividing, adding, and subtracting fractions in order to do much of the work that follows in this book.

— Caution —	Being able to identify the numerator and denominator is important so it generally is not a good idea to use a slant bar to indicate a fraction. For example, a fraction with 1 in the numerator and $3x$ in the denominator, $\dfrac{1}{3x}$, may NOT be written as $1/3x$.
Identifying numerator *and denominator*	Because we do multiplication and division working left to right, $1/3x$ means $\dfrac{1}{3}\,x$ and not $\dfrac{1}{3x}$. If you choose to use a slant bar (not a good choice) put the denominator inside parentheses, as in $1/(3x)$.

Topic 1.2 A — Prime Numbers, Composite Numbers, and Prime Factoring

Prime numbers are useful when working with fractions. A **prime number** is a whole number, greater than 1, which is divisible only by 1 and itself. The first fifteen prime numbers are 2, 3, 5, 7, 11, 13, 17, 19, 23, 29, 31, 37, 41, 43, and 47.

A **composite number** is a whole number, greater than 1, which is not prime. Every composite number can be written as a unique product of prime numbers. For example, 6 is composite and can be written as 2 • 3.

To **prime factor** a composite number means to write it as a product of prime numbers. For example we can prime factor 12 as 2 • 2 • 3. We call 2 • 2 • 3 the prime factorization of 12 since the product is 12 and the factors, 2 and 3, are prime; 2 and 3 are called the prime factors of 12. We do not consider 2 • 6 the prime factorization of 12 because 6 is not prime.

There are several ways to prime factor a number. Small numbers, like 15, are easily done in your head (3 • 5). When the numbers are larger, the following procedure works well:

— Procedure —	**Step 1**	Divide the given number by the smallest prime number, 2. If a remainder results, go to Step 3.
Prime factoring	**Step 2**	Divide the quotient from the previous step by 2. Keep dividing the quotients by 2 until 2 no longer works (that is, until the division yields a remainder other than 0).
	Step 3	Divide the quotient from the previous step by the next prime number, 3. Keep dividing the quotients by 3 until it no longer works.
	Step 4	Continue the process of dividing by prime numbers (5, 7, 11, ...) until the quotient is a prime number.
	Step 5	Write the prime factorization as the product of all the divisors and the final quotient.

Example 1.2.1 a) Prime factor 90.

•SOLUTION• **Step 1** Divide 90 by 2 to get 45.

Step 2 Divide 45 by 2 to get 22.5. Since the quotient has a decimal part (that is, the remainder was not 0) we say that 2 no longer works.

Step 3 Divide 45 by 3 to get 15.

Step 4 Divide 15 by 3 to get 5. Since 5 is a prime number we can stop.

Step 5 Write the final answer as the product of all the divisors and the final quotient. That is, write 90 as 2 • 3 • 3 • 5, or $2 \cdot 3^2 \cdot 5$.

We can write out the entire division process compactly as follows:

$$
\begin{array}{r}
5 \\
3\,)\overline{15} \\
3\,)\overline{45} \\
2\,)\overline{90}
\end{array}
$$

b)   Prime factor 539.

•SOLUTION•   Start by dividing by 2. Since 2 does not work, try 3 (which does not work), then 5 (which does not work), then 7 (it works) and so on.

$$
\begin{array}{r}
11 \\
7\overline{\smash{)}77} \\
7\ \diagdown\ \diagdown\ \diagdown\overline{\smash{)}539}
\end{array}
$$

The prime factorization of 539 is $7 \cdot 7 \cdot 11$, or $7^2 \cdot 11$.

Practice 1.2.1   Prime factor.

1.   18   2.   54   3.   385   4.   119   5.   105   6.   60

Topic 1.2 B — Reducing Fractions

A fraction is reduced to lowest terms when the numerator and denominator have no common whole number factors, other than 1. For example, the fraction $\frac{3}{4}$ is reduced to lowest terms because 3 and 4 have no common whole number factors, other than 1. However, $\frac{8}{12}$ is not reduced to lowest terms because 8 and 12 have a common factor of 4. Since 4 is the largest factor that is common to the numerator and denominator we say that 4 is the **Greatest Common Factor** (**GCF**) of the fraction. To reduce the fraction we can do the following:

$\dfrac{8}{12}$ $\Leftarrow$ Original fraction

$\dfrac{2 \cdot 4}{3 \cdot 4}$ $\Leftarrow$ Wrote numerator and denominator as a product where one of the factors is the GCF, 4.

$\dfrac{2}{3} \cdot \dfrac{4}{4}$ $\Leftarrow$ Broke up the fraction into two separate fractions.

$\dfrac{2}{3} \cdot 1$ $\Leftarrow$ Simplified $\dfrac{4}{4}$. Note that factors which reduce completely (the 4's in this case) are replaced by 1.

$\dfrac{2}{3}$ $\Leftarrow$ Multiplied by 1.

In practice we don't show all these steps. We simply divide both numerator and denominator by 4 (the GCF) and say that we *reduced by a factor of 4*.

If you cannot determine the GCF by inspection (that is, by just looking at it) you may reduce the fraction by following the procedure given below.

— *Procedure* —	**Step 1** Prime factor the numerator and the denominator.
Reducing fractions	**Step 2** Reduce all factors common to the numerator and the denominator.

Example 1.2.2   Reduce to lowest terms.

a) $\dfrac{18}{24}$

•**SOLUTION**• If you can see that 6 is the GCF, just divide both numerator and denominator by 6 to get $\dfrac{3}{4}$. Otherwise, use the steps.

Step 1   Prime factor numerator and denominator to get $\dfrac{2 \cdot 3 \cdot 3}{2 \cdot 2 \cdot 2 \cdot 3}$.

Step 2   Reduce common factors $\dfrac{\cancel{2} \cdot \cancel{3} \cdot 3}{\cancel{2} \cdot 2 \cdot 2 \cdot \cancel{3}}$ to get $\dfrac{3}{4}$.

b) $\dfrac{95}{114}$

•**SOLUTION**• You probably need to follow the procedure for this one.

Step 1   Prime factor numerator and denominator to get $\dfrac{5 \cdot 19}{2 \cdot 3 \cdot 19}$.

Step 2   Reduce common factors $\dfrac{5 \cdot \cancel{19}}{2 \cdot 3 \cdot \cancel{19}}$ to get $\dfrac{5}{6}$.

Practice 1.2.2 Reduce to lowest terms.

1. $\dfrac{18}{20}$ 2. $\dfrac{24}{64}$ 3. $\dfrac{72}{40}$ 4. $\dfrac{104}{156}$ 5. $\dfrac{95}{152}$ 6. $\dfrac{84}{120}$

Topic 1.2 C — Multiplying Fractions

Multiplying fractions involves the same skills as reducing fractions.

— Procedure —	
Multiplying fractions	*Step 1 Reduce* Reduce factors common to the numerators and denominators.
	Step 2 Multiply Multiply the remaining factors in the numerators and multiply the remaining factors in the denominators.

Example 1.2.3   Simplify.

a) $\dfrac{21}{15} \cdot \dfrac{60}{56}$

•**SOLUTION**• *Step 1 Reduce*   We can do this by inspection or by using prime factoring. If you can see the common factors of 7, 15, and 4 then just divide out these factors:

$$\dfrac{\overset{3}{\cancel{21}}}{\underset{1}{\cancel{15}}} \cdot \dfrac{\overset{1}{\underset{}{\cancel{60}}^{\,4}}}{\cancel{56}^{\,8}_{\,2}} \quad \Leftarrow \text{ Reduced common factors.}$$

$$\dfrac{3}{1} \cdot \dfrac{1}{2} \quad \Leftarrow \text{ Fractions to multiply.}$$

The above is quick but it is not very clear as to what was reduced with what. The prime factoring method takes a few more seconds but is easier to follow:

$$\frac{21}{15} \cdot \frac{60}{56} \quad \Leftarrow \text{ Expression to simplify.}$$

$$\frac{3 \cdot 7}{3 \cdot 5} \cdot \frac{2 \cdot 2 \cdot 3 \cdot 5}{2 \cdot 2 \cdot 2 \cdot 7} \quad \Leftarrow \text{ Prime factored each numerator and denominator.}$$

$$\frac{\cancel{3} \cdot \cancel{7}}{\cancel{3} \cdot \cancel{5}} \cdot \frac{\cancel{2} \cdot \cancel{2} \cdot 3 \cdot \cancel{5}}{\cancel{2} \cdot \cancel{2} \cdot 2 \cdot \cancel{7}} \quad \Leftarrow \text{ Reduced common factors.}$$

Step 2 Multiply We multiply numerators and multiply denominators to get $\frac{3}{2}$.

b) $\dfrac{68}{171} \cdot \dfrac{95}{34}$

•**SOLUTION**• This one would be difficult to do by inspection unless your know your 17 and 19 times tables.

Step 1 Reduce

$$\frac{\cancel{2} \cdot 2 \cdot \cancel{17}}{3 \cdot 3 \cdot \cancel{19}} \cdot \frac{5 \cdot \cancel{19}}{\cancel{2} \cdot \cancel{17}} \quad \Leftarrow \text{ Reduced common factors.}$$

Step 2 Multiply We multiply numerators and denominators to get $\frac{10}{9}$.

Practice 1.2.3 Simplify.

1. $\dfrac{12}{25} \cdot \dfrac{10}{27}$ 2. $\dfrac{20}{3} \cdot \dfrac{15}{20}$ 3. $\dfrac{4}{42} \cdot \dfrac{14}{6}$ 4. $\dfrac{18}{25} \cdot \dfrac{35}{12}$ 5. $\dfrac{92}{57} \cdot \dfrac{190}{115}$ 6. $\dfrac{42}{98} \cdot \dfrac{175}{165}$

Topic 1.2 D — Reciprocals

Two numbers whose product is 1 are called **reciprocals** or **multiplicative inverses**. For example, $\frac{2}{3}$ and $\frac{3}{2}$ are reciprocals because $\frac{2}{3} \cdot \frac{3}{2}$ simplifies to 1. Likewise, 5 and $\frac{1}{5}$ are reciprocals because $5 \cdot \left(\frac{1}{5}\right)$ simplifies to 1. Note that 0 has no reciprocal because there is no number we can multiply 0 by to get 1.

To find the reciprocal of a number (other than 0) simply exchange its numerator and denominator.

Example 1.2.4 Find the reciprocals.

a) $\dfrac{2}{7}$

•**SOLUTION**• Exchange the numerator and denominator to get $\frac{7}{2}$.

b) 3

•**SOLUTION**• Think of 3 as $\frac{3}{1}$. Then, exchange the numerator and denominator to get $\frac{1}{3}$.

Practice 1.2.4 Find the reciprocals.

1. $\dfrac{32}{33}$ 2. 6 3. $\dfrac{1}{x}, x \neq 0^{\dagger}$ 4. $y, y \neq 0$

$\dagger$ Remember, $x \neq 0$ means that x can take on any value except 0.

We may use reciprocals to write fractions as products. For example, we may write $\frac{3}{4}$ as the product of 3 and the reciprocal of 4. That is, $\frac{3}{4}$ is equivalent to $3 \cdot \frac{1}{4}$. We could also write this as $\frac{1}{4} \cdot 3$.

— Notation — Writing quotients as products	English: A fraction may be written as the product of the reciprocal of its denominator and its numerator. Example: $\frac{x}{2}$ may be written as $\frac{1}{2}x$ Algebra: Assuming $b \neq 0, \frac{a}{b} = \frac{1}{b} \cdot a$

Example 1.2.5   Rewrite as a product.

a) $\frac{6}{7}$

•SOLUTION• We can write this as $6 \cdot \frac{1}{7}$ or $\frac{1}{7} \cdot 6$.

b) $\frac{y}{5}$

•SOLUTION• Since we usually write the numeric coefficient to the left of the variable, we will write this as $\frac{1}{5}y$.

c) $\frac{2x}{3}$

•SOLUTION• We could write this as $\frac{1}{3}(2x)$, but we usually keep the constants together so we will write this as $\frac{2}{3}x$.

Practice 1.2.5 Rewrite as a product.

1. $\frac{5}{3}$ 2. $\frac{x}{3}$ 3. w 4. $\frac{5p}{4}$ 5. $\frac{3c}{5}$

Topic 1.2 E — Dividing Fractions

We divide fractions by transforming the division to multiplication.

— Procedure — Dividing fractions	*Step 1 Reciprocal* Rewrite the division as the multiplication of the numerator by the reciprocal of the denominator. *Step 2 Reduce* Reduce factors common to the numerators and denominators. *Step 3 Multiply* Multiply the remaining factors in the numerators and multiply the remaining factors in the denominators.

Example 1.2.6   Simplify.

a) $\dfrac{\dfrac{3}{8}}{\dfrac{7}{24}}$

•SOLUTION• ***Step 1 Reciprocal*** $\dfrac{3}{8} \cdot \dfrac{\mathbf{24}}{\mathbf{7}}$

   Step 2 Reduce   We reduce the common factor of 8 to get $\dfrac{3}{1} \cdot \dfrac{3}{7}$

   Step 3 Multiply $\dfrac{9}{7}$

b) $\dfrac{10}{21} \div \dfrac{5}{14}$

•SOLUTION• The $\dfrac{5}{14}$ is considered to be in the denominator so we multiply by its reciprocal.

   Step 1 Reciprocal $\dfrac{10}{21} \cdot \dfrac{\mathbf{14}}{\mathbf{5}}$

   Step 2 Reduce   We reduce the common factors of 5 and 7 to get $\dfrac{2}{3} \cdot \dfrac{2}{1}$

   Step 3 Multiply $\dfrac{4}{3}$

c) $\dfrac{5}{\dfrac{25}{3}}$

•SOLUTION• Think of 5 as $\dfrac{5}{1}$ so the fraction becomes $\dfrac{\dfrac{5}{1}}{\dfrac{25}{3}}$

   Step 1 Reciprocal $\dfrac{5}{1} \cdot \dfrac{\mathbf{3}}{\mathbf{25}}$

   Step 2 Reduce   We reduce the common factor of 5 to get $\dfrac{1}{1} \cdot \dfrac{3}{5}$

   Step 3 Multiply $\dfrac{3}{5}$

d) $\dfrac{5}{12} \div 15$

•SOLUTION• Think of 15 as $\dfrac{15}{1}$

   Step 1 Reciprocal $\dfrac{5}{12} \cdot \dfrac{\mathbf{1}}{\mathbf{15}}$

   Step 2 Reduce   We reduce the common factor of 5 to get $\dfrac{1}{12} \cdot \dfrac{1}{3}$

   Step 3 Multiply $\dfrac{1}{36}$

Practice 1.2.6   Simplify.

1. $\dfrac{3}{4} \div \dfrac{9}{16}$

2. $\dfrac{\dfrac{5}{9}}{\dfrac{35}{27}}$

3. $\dfrac{10}{8} \div 5$

4. $\dfrac{7}{\dfrac{14}{20}}$

5. $\dfrac{51}{76} \div \dfrac{102}{95}$

6. $\dfrac{48}{56} \div 12$

7. $\dfrac{\dfrac{15}{21}}{14}$

8. $\dfrac{\dfrac{70}{3}}{\dfrac{175}{15}}$

Topic 1.2 F — Adding and Subtracting Fractions That Have the Same Denominator

When adding (or subtracting) fractions that have the same denominators we simply add (or subtract) the numerators and then reduce, if possible.

Example 1.2.7　　Simplify: $\dfrac{5}{7} + \dfrac{6}{7} - \dfrac{3}{7}$

•SOLUTION• Since the denominators are the same, we combine the numerators as indicated.

$$\dfrac{5}{7} + \dfrac{6}{7} - \dfrac{3}{7} \quad \Leftarrow \text{ Expression to simplify.}$$

$$\dfrac{11}{7} - \dfrac{3}{7} \quad \Leftarrow \text{ Added numerators of first two fractions.}$$

$$\dfrac{8}{7} \quad \Leftarrow \text{ Subtracted numerators.}$$

The fraction $\dfrac{8}{7}$ cannot be reduced because there are no factors common to the numerator and denominator (other than 1). However, we could write it as the mixed number $1\dfrac{1}{7}$, but that is not necessary.

Practice 1.2.7　Simplify.

1. $\dfrac{5}{8} + \dfrac{7}{8}$

2. $\dfrac{1}{12} + \dfrac{5}{12}$

3. $\dfrac{4}{15} + \dfrac{1}{15} - \dfrac{2}{15}$

4. $\dfrac{7}{20} - \dfrac{3}{20} + \dfrac{1}{20}$

Topic 1.2 G — The Least Common Denominator

If the fractions to be added have different denominators we must convert the fractions into equivalent fractions that have the same denominator before adding the numerators. We call this new denominator the Least Common Denominator or LCD.

The LCD of a set of fractions is the smallest number that is divisible by all the denominators in the set. For example, the LCD of $\dfrac{1}{6}$ and $\dfrac{1}{9}$ is 18 because 18 is the smallest number that is divisible by both 6 and 9.

When the denominators are small, as with $\dfrac{1}{3}$ and $\dfrac{1}{2}$, the LCD may be obvious (it is 6). With larger denominators, the following procedure is helpful:

— *Procedure* — *Finding the LCD*	*Step 1 Prime factor*　Prime factor each denominator and write it in exponential form.
	Step 2 Select　For each prime factor, select the one with the largest power.
	Step 3 Multiply　Multiply the selected factors. This is the LCD.

Example 1.2.8   Find the LCD of $\frac{5}{24}$ and $\frac{7}{60}$.

•SOLUTION• *Step 1 Prime factor*

24 factors as $2 \cdot 2 \cdot 2 \cdot 3$ or $2^3 3^1$

60 factors as $2 \cdot 2 \cdot 3 \cdot 5$ or $2^2 3^1 5^1$

Step 2 Select

Largest power of 2 is 2^3 (from 24)

Largest power of 3 is 3^1 (from either 24 or 60)

Largest power of 5 is 5^1 (from 60)

Step 3 Multiply   The LCD is $2^3 3^1 5^1$, which is 120. That is, 120 is the smallest number which is divisible by 24 and 60.

— Note — LCD contains all denominators	You can check your LCD by seeing if it contains all the denominators of the given set of fractions. For example, 120 contains the factors $2 \cdot 2 \cdot 2 \cdot 3$ (which is 24) and the factors $2 \cdot 2 \cdot 3 \cdot 5$ (which is 60). 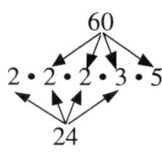

Practice 1.2.8   Find the LCD.

1. $\frac{7}{9}, \frac{5}{12}$ 2. $\frac{17}{18}, \frac{13}{24}$ 3. $\frac{3}{8}, \frac{4}{9}, \frac{7}{10}$ 4. $\frac{9}{34}, \frac{25}{51}, \frac{2}{85}$

Topic 1.2 H — Equivalent Fractions

Equivalent fractions are fractions that have different denominators but which represent the same quantity. For example $\frac{6}{8}$ and $\frac{3}{4}$ have different denominators but they are equivalent since they both represent three-fourths. We *reduced* fractions, such as $\frac{6}{8}$, by removing the same factor from the numerator and denominator. We can *build* equivalent fractions by inserting the same factor into the numerator and the denominator.

— Procedure — Building equivalent fractions	*Step 1 Find Needed Factors* Determine what the denominator of the original fraction must be multiplied by to transform it into the desired number. *Step 2 Multiply* Multiply BOTH the numerator and the denominator of the original fraction by the factors found in Step 1.

Example 1.2.9   a)   Build a fraction equivalent to $\frac{7}{12}$ but which has a denominator of 24.

•SOLUTION• *Step 1 Find Needed Factors*   To transform 12 into 24 we multiply by 2.

Step 2 Multiply $\frac{7 \cdot 2}{12 \cdot 2}$, or $\frac{14}{24}$.

b) Build a fraction equivalent to 3 but which has a denominator of 15.

•SOLUTION• *Step 1 Find Needed Factors* First, we write 3 as the fraction $\frac{3}{1}$ so we can see its denominator. Then, we notice that to transform 1 into 15 we multiply by 15.

Step 2 Multiply $\frac{3 \cdot 15}{1 \cdot 15}$, or $\frac{45}{15}$.

c) Build a fraction equivalent to $\frac{5}{28}$ but which has a denominator of 476.

•SOLUTION• *Step 1 Find Needed Factors* It's hard to "see" what factor we need to multiply 28 by to get 476. To find the needed factor we divide 476 by 28 to get 17. Therefore, to transform 28 into 476 we multiply by 17.

Step 2 Multiply $\frac{5 \cdot 17}{28 \cdot 17}$, or $\frac{85}{476}$.

Practice 1.2.9 Write each fraction as an equivalent fraction but with the given denominator.

1. $\frac{4}{5}$ with a denominator of 40

2. $\frac{5}{16}$ with a denominator of 48

3. 4 with a denominator of 18

4. 3 with a denominator of 54

Topic 1.2 I — Adding and Subtracting Fractions That Have Different Denominators

Before we can add or subtract fractions that have different denominators, we must first convert them to equivalent fractions with the same denominator.

— Procedure —	
Adding and subtracting fractions with different denominators	*Step 1 Build LCD* Build the Least Common Denominator (LCD). *Step 2 Build Equivalent Fractions* Build equivalent fractions that contain the LCD. *Step 3 Combine Numerators* Add or subtract the numerators and use the LCD as the denominator. *Step 4 Reduce* Reduce if possible.

Example 1.2.10 Simplify: $\frac{5}{12} + \frac{7}{18}$

•SOLUTION• *Step 1 Build LCD* You may be able to find the LCD by inspection, but we will use the factoring method for practice.

- 12 factors as $2 \cdot 2 \cdot 3$ or $2^2 3^1$

- 18 factors as $2 \cdot 3 \cdot 3$ or $2^1 3^2$

- Largest power of 2 is 2^2

- Largest power of 3 is 3^2

LCD is $2^2 3^2$ or 36.

Step 2 Build Equivalent Fractions

- First term: When we compare 12 with the LCD we see that 12 needs one more factor of 3. Therefore, we multiply the numerator and denominator by 3.

$$\frac{5}{12} \quad \Leftarrow \text{ Original fraction.}$$

$$\frac{5 \cdot 3}{12 \cdot 3} \quad \Leftarrow \text{ Multiplied numerator and denominator by 3.}$$

$$\frac{15}{36} \quad \Leftarrow \text{ Multiplied.}$$

- Second term: 18 needs one more factor of 2.

$$\frac{7}{18} \quad \Leftarrow \text{ Original fraction}$$

$$\frac{7 \cdot \mathbf{2}}{18 \cdot \mathbf{2}} \quad \Leftarrow \text{ Multiplied numerator and denominator by 2.}$$

$$\frac{14}{36} \quad \Leftarrow \text{ Multiplied.}$$

Step 3 Combine Numerators

$$\frac{15}{36} + \frac{14}{36}, \text{ which simplifies to } \frac{29}{36}$$

Step 4 Reduce There is no obvious factor common to 29 and 36. To be sure, we will factor $\frac{29}{36}$ to get $\frac{29}{2 \cdot 2 \cdot 3 \cdot 3}$. The numerator and denominator have no common factors, other than 1, so this fraction cannot be reduced.

Practice 1.2.10 Simplify.

1. $\frac{5}{12} + \frac{5}{18}$
2. $\frac{7}{8} - \frac{17}{24}$
3. $\frac{9}{10} - \frac{1}{9} + \frac{14}{15}$
4. $\frac{4}{9} + \frac{4}{7} - \frac{1}{14}$
5. $\frac{7}{15} + \frac{5}{12} + \frac{7}{10}$
6. $\frac{3}{4} + \frac{1}{9} - \frac{13}{36}$

Topic 1.2 J — Order of Operations and Fractions

We use the standard order of operations for expressions involving fractions. Note that the fraction bar is considered a grouping symbol, like parentheses. For example, the fraction $\frac{2 + 3}{4 + 6}$ may be thought of as $(2 + 3) \div (4 + 6)$. The work inside the grouping symbols is done first to give $\frac{5}{10}$, which reduces to $\frac{1}{2}$.

— Procedure — Simplifying expressions using the proper order of operations	
	Step 1 Simplify expressions inside grouping symbols.
	Step 2 Simplify exponents and square roots.
	Step 3 Simplify multiplication and division, working left to right.
	Step 4 Simplify addition and subtraction, working left to right.

Example 1.2.11 Simplify.

a) $\dfrac{2^2 + 10}{2 \cdot 3 + 1}$

• SOLUTION •

$$\frac{4 + 10}{2 \cdot 3 + 1} \quad \Leftarrow \text{ In numerator, multiplied 2 factors of 2 to get 4.}$$

$$\frac{14}{2 \cdot 3 + 1} \quad \Leftarrow \text{ Added in numerator.}$$

$$\frac{14}{6 + 1} \quad \Leftarrow \text{ Multiplied in denominator.}$$

$$\frac{14}{7} \quad \Leftarrow \text{ Added in denominator.}$$

$$2 \quad \Leftarrow \text{ Divided.}$$

b) $2 - \dfrac{\dfrac{1}{2} \cdot 8}{7 - \dfrac{5}{3}}$

•SOLUTION•

$2 - \dfrac{4}{7 - \dfrac{5}{3}}$ ⇐ Multiplied in numerator.

$2 - \dfrac{4}{\dfrac{16}{3}}$ ⇐ Subtracted $\dfrac{21}{3} - \dfrac{5}{3}$ in denominator.

$2 - 4 \cdot \dfrac{3}{16}$ ⇐ Inverted denominator and multiplied.

$2 - \dfrac{3}{4}$ ⇐ Multiplied.

$\dfrac{5}{4}$ ⇐ Subtracted $\dfrac{8}{4} - \dfrac{3}{4}$

c) $\dfrac{2}{3} - \dfrac{2}{3} \cdot \left(\dfrac{1}{2}\right)^2$

•SOLUTION•

$\dfrac{2}{3} - \dfrac{2}{3} \cdot \dfrac{1}{4}$ ⇐ Multiplied 2 factors of $\dfrac{1}{2}$ to get $\dfrac{1}{4}$.

$\dfrac{2}{3} - \dfrac{1}{6}$ ⇐ Multiplied.

$\dfrac{3}{6}$ ⇐ Subtracted $\dfrac{4}{6} - \dfrac{1}{6}$.

$\dfrac{1}{2}$ ⇐ Reduced.

Practice 1.2.11 Simplify.

1. $10 - \dfrac{4(3)}{2 + 4}$

2. $5 - \dfrac{2 \cdot 6}{4}$

3. $\dfrac{1}{2} + \dfrac{1}{3}\left(\dfrac{5}{4} - \dfrac{1}{6}\right)$

4. $\dfrac{2}{3} + \dfrac{2}{7}\left(\dfrac{1}{4} + \dfrac{1}{3}\right)$

5. $\left(\dfrac{4}{5} + \dfrac{2}{10}\right)^2 - \dfrac{3}{4} + \dfrac{1}{4}$

6. $\dfrac{5 + 2^3}{6 - 2} \div \dfrac{24 + 2}{3^2 + 5}$

— *Study Tip* — *Making problem study* *cards*	When you find a problem that's difficult for you it's important to make a study card for that problem so that you can review it several times.

To make problem study cards:

- On one side of a 3 by 5 index card write the problem and the textbook page number where it was discussed.

- On the reverse side write the solution.

To use problem study cards:

- Every day before starting the new homework, quiz yourself on a few randomly chosen problem cards. This will get you into a mathematical frame of mind and will review past ideas which often turn up in new homework.

- Four or five times a week shuffle the cards, deal ten to twenty cards off the top, and give yourself two minutes to do each card. Make sure you time yourself. This helps you get ready for tests by having you perform under time pressure.

- At least twice a week go through the cards and discuss with yourself the procedure needed to solve the problem. You don't have to actually solve the problem, just say to yourself "this is what I would do" and then check to see if the problem is actually solved that way.

- Before a test or quiz, review the cards.

Types of problems that are good candidates for study cards:

- Homework problems you found difficult.

- Situations where you make the same mistake over and over.

- Questions you miss on a test or quiz.

- If your instructor says something like, "now here's a problem students often miss", make a study card.

Exercise Set 1.2   The answers to the odd numbered exercises are at the back of the book.

Prime factor.

1. 96	2. 98	3. 378	4. 68
5. 825	6. 392	7. 484	8. 2205

Reduce to lowest terms.

9. $\dfrac{6}{20}$ 10. $\dfrac{63}{42}$ 11. $\dfrac{36}{90}$ 12. $\dfrac{56}{98}$ 13. $\dfrac{135}{126}$ 14. $\dfrac{210}{1050}$

Simplify.

15. $\dfrac{8}{15} \cdot \dfrac{20}{12}$ 16. $\dfrac{6}{35} \cdot \dfrac{14}{18}$ 17. $\dfrac{30}{42} \cdot \dfrac{8}{60}$ 18. $\dfrac{15}{63} \cdot \dfrac{70}{100}$

19. $\dfrac{110}{30} \cdot \dfrac{21}{33}$ 20. $\dfrac{30}{28} \cdot \dfrac{245}{150}$ 21. $\dfrac{98}{50} \cdot \dfrac{60}{315}$ 22. $\dfrac{36}{150} \cdot \dfrac{105}{84}$

Find the reciprocals.

23. 8   24. 15   25. $\dfrac{5}{6}$ 26. $\dfrac{18}{11}$

27. $\dfrac{7}{a}, a \neq 0$ 28. $\dfrac{9}{c}, c \neq 0$ 29. $t, t \neq 0$ 30. $z, z \neq 0$

Rewrite as a product.

31. $\dfrac{13}{15}$ 32. $\dfrac{3}{4}$ 33. $\dfrac{x}{2}$ 34. $\dfrac{m}{7}$ 35. $\dfrac{3t}{8}$ 36. $\dfrac{12d}{7}$

Simplify.

37. $\dfrac{4}{9} \div \dfrac{12}{27}$

38. $\dfrac{\frac{18}{16}}{\frac{126}{48}}$

39. $\dfrac{9}{\frac{18}{5}}$

40. $\dfrac{21}{\frac{14}{6}}$

41. $\dfrac{135}{210} \div 18$

42. $8 \div \dfrac{32}{35}$

43. $\dfrac{14}{15} \div \dfrac{60}{90}$

44. $\dfrac{66}{75} \div \dfrac{121}{60}$

45. $\dfrac{\frac{3}{4}}{\frac{1}{8}} \div 6$

46. $\dfrac{\frac{10}{9}}{\frac{8}{27}} \div \dfrac{3}{2}$

Simplify.

47. $\dfrac{1}{15} + \dfrac{11}{15}$

48. $\dfrac{3}{16} + \dfrac{5}{16}$

49. $\dfrac{13}{9} - \dfrac{2}{9} + \dfrac{10}{9}$

50. $\dfrac{7}{4} - \dfrac{2}{4} + \dfrac{3}{4}$

51. $\dfrac{19}{54} - \dfrac{5}{54} - \dfrac{5}{54}$

52. $\dfrac{32}{41} - \dfrac{18}{41} - \dfrac{11}{41}$

Find the LCD.

53. $\dfrac{5}{8}, \dfrac{9}{14}$

54. $\dfrac{11}{36}, \dfrac{57}{45}$

55. $\dfrac{1}{9}, \dfrac{14}{15}, \dfrac{8}{35}$

56. $\dfrac{11}{10}, \dfrac{13}{12}, \dfrac{15}{14}$

57. $\dfrac{8}{27}, \dfrac{1}{36}, \dfrac{5}{48}$

58. $\dfrac{1}{5}, \dfrac{1}{9}, \dfrac{1}{20}, \dfrac{1}{30}$

Write each fraction as an equivalent fraction but with the given denominator.

59. $\dfrac{1}{6}$ with a denominator of 18

60. $\dfrac{5}{8}$ with a denominator of 40

61. $\dfrac{1}{36}$ with a denominator of 180

62. $\dfrac{2}{15}$ with a denominator of 225

63. 3 with a denominator of 91

64. 11 with a denominator of 33

Simplify.

65. $\dfrac{2}{9} + \dfrac{5}{24}$

66. $\dfrac{4}{15} - \dfrac{1}{6}$

67. $\dfrac{6}{8} - \dfrac{3}{20}$

68. $\dfrac{8}{35} + \dfrac{4}{56}$

69. $\dfrac{5}{6} + \dfrac{1}{2} + \dfrac{8}{3}$

70. $\dfrac{5}{12} + \dfrac{3}{8} + \dfrac{1}{18}$

71. $\dfrac{9}{28} + \dfrac{10}{21} - \dfrac{7}{9}$

72. $\dfrac{17}{18} - \dfrac{7}{24} - \dfrac{29}{60}$

73. $\left(\dfrac{2}{3}\right)^2 + \left(\dfrac{1}{4}\right)^2$

74. $\left(\dfrac{1}{2}\right)^2 - \left(\dfrac{1}{4}\right)^2$

75. $\left(\dfrac{3}{7}\right)\left(\dfrac{1}{10} + \dfrac{5}{6}\right)$

76. $\left(\dfrac{9}{11}\right)\left(\dfrac{3}{8} + \dfrac{7}{18}\right)$

77. $2 + \dfrac{3+5}{3(2)}$

78. $3 + \dfrac{4 + 3(6)}{4(3) + 6}$

79. $\left(\dfrac{4}{15} + \dfrac{1}{5}\right)\left(\dfrac{7}{6} + \dfrac{4}{3}\right)$

80. $\left(\dfrac{1}{2} + \dfrac{9}{22}\right)\left(\dfrac{2}{3} + \dfrac{5}{9}\right)$

81. $\dfrac{3}{8} \div \dfrac{9}{4} + \dfrac{1}{9} \div \dfrac{1}{3}$

82. $\dfrac{6}{5} \div 3 - \dfrac{8}{9} \div 4$

83. $\dfrac{1}{3} + \dfrac{2}{3}\left(\dfrac{1}{2} + \dfrac{1}{7}\right)$

84. $\dfrac{1}{2} + \dfrac{1}{2}\left(\dfrac{15}{6} - \dfrac{1}{8}\right)$

Review Exercises   The answers to all of these exercises are at the back of the book.

85. Identify the sums, terms, products, and factors in $-2r + 5s$

Simplify.

86. $18/3 \cdot 2 + 3^2$

87. $3 \cdot 2^2 - 3^2$

88. $8 + 24 \div 2 \cdot 3$

89. $5 + \dfrac{8 - 2 \cdot 2^2}{4 + 3^2}$

90. $(3^2 - 2 \cdot 4)(20 - 12/2 \cdot 3)$

91. $4 + 2(109 - 5(2 + 6^2/2))$

92. $18 - 3(8 - 3(2^2 - 3))$

93. $6 \div 2 \cdot \sqrt{9}$

94. Evaluate $8c - c^2$ if c is 3

95. Evaluate $2a(a^2 \div 2 \cdot 3 - 50)$ if a is 6

Section 1.3 Real Numbers

Topic 1.3 A — Positive and Negative Numbers

Numbers greater than 0 are **positive**. We indicate positive numbers by placing a positive sign, +, to the left of the number. Numbers written without the + are assumed to be positive. For example, +5 and 5 both represent the number *positive five*.

Numbers less than 0 are **negative**. We show that a number is negative by placing a negative sign, –, to its left. For example, –5 (pronounced "negative five") represents a number that is 5 less than 0. You may have already used negative numbers to represent things like:

*   temperatures below zero: on a cold day the temperature in White Bear Lake, Minnesota can reach –30 degrees (that is, 30 degrees below zero).

*   a debt: if you have overdrawn your bank account by $10 then your bank balance is –10 dollars.

*   a loss: Jim lost $100 in the stock market so his earnings were –100 dollars.

*   elevations below sea level: Sea level is defined as having an elevation of 0 feet. Therefore, the elevation of Death Valley, which is 282 feet below sea level, is –282 feet.

The – symbol is also used for the operation of subtraction. Because of this, we sometimes place negative numbers in parentheses like this (–5) to make it clear that we are talking about a negative number rather than the operation of subtraction.

Positive and negative numbers are called **signed numbers** because they have either a + sign or a – sign associated with them. The number 0 is neither negative nor positive. For this reason we say 0 has no sign. We use the term **non-zero** to indicate any number *except* zero.

Topic 1.3 B — Special Sets of Numbers

In Section 1.1 we identified the set of natural numbers, {1, 2, 3, 4, ...} and the set of whole numbers, {0, 1, 2, 3, 4, ...}. There are 4 additional sets of numbers that we will need in the study of algebra:

Integers (pronounced in-tea-jers) This set consists of the natural numbers, the negatives of the natural numbers, and 0. We can indicate this set as

$$\{ \ldots -3, -2, -1, 0, 1, 2, 3, \ldots \}$$

The ... implies that this set goes on forever in both directions. You can think of this set as having 3 parts:

*   **positive integers**, 1, 2, 3, and so on. These are the natural numbers.
*   zero, which is neither positive nor negative.
*   **negative integers**, –1, –2, –3, and so on. These are the negatives of the natural numbers.

Rational Numbers These are numbers that can be written as the ratio of two integers, where the denominator is not 0. These include the fractions and decimals you worked with in arithmetic. Note that whole numbers are also considered rational numbers because they can be written as the ratio of integers. For example, since the whole number 2 can be written as $\frac{2}{1}$ we say that 2 is a rational number.

Irrational Numbers These are numbers that cannot be written as the ratio of integers. The decimal representations of these numbers go on forever without repeating.

Numbers such as π and $\sqrt{2}$ (the square root of 2) are examples of irrational numbers. We discuss square roots in detail in Chapter 9.

Real Numbers These are the rational and irrational numbers combined.

Below is a diagram showing the relation between the different sets of numbers.

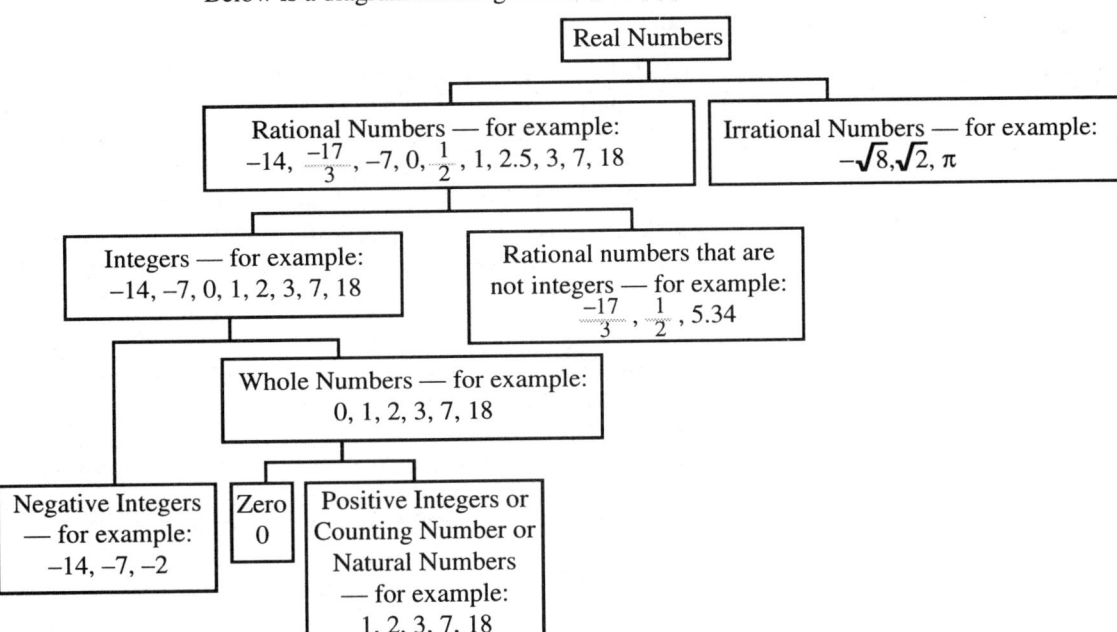

Topic 1.3 C — The Number Line

Sometimes it's helpful to represent a set of numbers using a picture. We can visualize sets of numbers by associating each number in the set with a point on a line, called a **number line**. A number line looks like a thermometer laid on its side with the positive numbers (temperatures above 0) on the right and the negative numbers (temperatures below 0) on the left.

Thermometer

Colder   Hotter

Number line

We designate 0 as the "center" or **origin** of the number line. The distance between each integer on the number line is called the **unit length** or simply *unit*.

On a thermometer, every temperature corresponds to a point on the scale. Likewise, every real number corresponds to a point on the number line. For example, the number 4 corresponds to the point on the number line that is 4 units to the right of 0. We can show this by placing a dot on the number line just above 4.

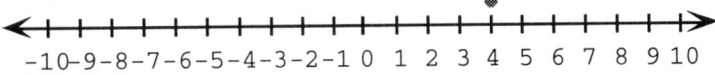

Locating a point on the number line is called **plotting** the point. The location of the point is called its **coordinate**. So, plotting 4 on the number line means to place a dot at 4. The coordinate of that point is 4.

Don't be mislead by the picture of the number line we have drawn. Even though we have only labeled integer values on the number line, numbers like $2\frac{1}{2}$ are also on the number line. We think of $2\frac{1}{2}$ as located half-way between 2 and 3. Likewise, to plot -3.25 we locate the point between -4 and -3, but closer to -3 (we can only approximate this).

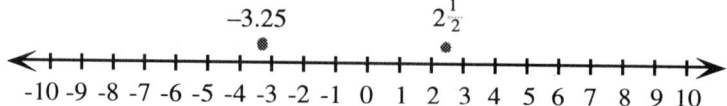

The picture that results when numbers are plotted on a number line is referred to as a graph. A **graph** is a visual representation of a set of numbers. Instead of saying "plot these numbers on a number line" we might say "graph these data on a number line".

Example 1.3.1   Plot on a number line: $\left\{-5, 2, 0, -7.6, -\frac{5}{3}, 2\pi\right\}$.

•SOLUTION•   Find the location on the number line that corresponds to each number in the set and then place a dot there. We have to approximate locations for numbers that are not integers. To plot 2π, we multiply $(2) \bullet (3.14)$ and then plot 6.28.

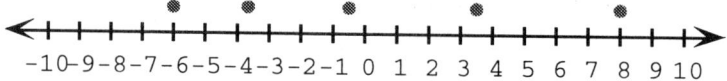

Practice 1.3.1   Plot on a number line.

1. $\left\{-10, 9, 1.1, -4.7, \frac{25}{4}, \pi\right\}$ 2. $\left\{-\frac{33}{34}, 2\frac{1}{5}, 3.4, -8.5, -7, 5.3\right\}$

Example 1.3.2   Find the approximate coordinate for each point shown on the number line.

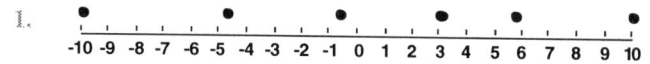

•SOLUTION•   We have to approximate some of the coordinates because they are not integer values. From left to right, the coordinates of the points are -6, -3.8 (between -4 and -3, but closer to -4), -0.5 (midway between -1 and 0), 3.5 (midway between 3 and 4), and 8.

Practice 1.3.2   Find the approximate coordinate for each point shown on the number line.

1.   2.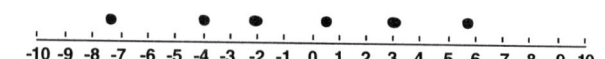

To locate large or small numbers on the number line we expand or contract the scale so that everything can fit in a reasonable space.

Example 1.3.3   a)   Draw a number line from -100 to 400 using an interval of 50. Then, plot 300.

•SOLUTION•

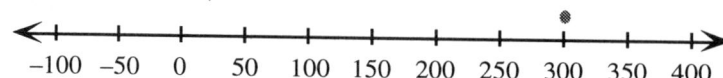

We chose to show multiples of 50 because that makes the number line easy to construct and read. We could have used intervals of 72 or 37 if we had wanted. It's not necessary to show 0 on the number line, but it usually is a good idea because 0 serves as a good point of reference.

b) Draw a number line from –200 to 500 using an interval of 100. Then, plot 300.

•SOLUTION•

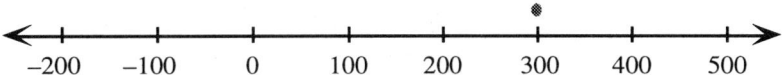

This graph and the one above show the same point plotted but use different scales. The information is essentially the same but the way it's presented looks different.

c) Graph –28 on a number line. Use an interval of 5.

•SOLUTION•

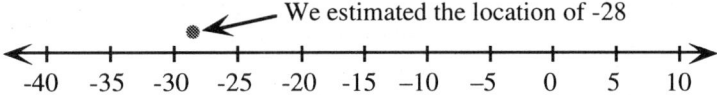

d) Graph 0.3 on a number line. Use an interval of 0.1.

•SOLUTION• This number is small so we expand the scale on the number line. Instead of labeling every unit we label every 0.1 unit.

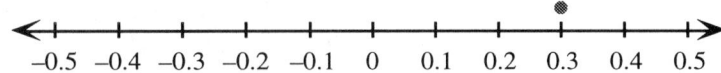

Practice 1.3.3 Draw a number line in the given interval and plot the given point.

1. –300 to 100 in steps of 50. Plot –5.
2. –200 to 800 in steps of 100. Plot –50
3. 20 to 80 in steps of 10. Plot 65.
4. 5 to 45 in steps of 5. Plot 17.
5. –0.5 to 1.5 in steps of 0.1. Plot 0.2
6. –2.5 to 0.5 in steps of 0.5. Plot –1.5

Topic 1.3 D — Order and Inequalities

When two numbers are located on a number line, the number on the left is **less than** the number on the right. For example, 5 lies to the left of 8 on the number line so we say that 5 is less than 8. We use the symbol < to mean "less than". We would write *5 is less than 8* as 5 < 8.

Another way of stating the relation between 8 and 5 is to say that 8 is **greater than** 5. This means that 8 is to the *right* of 5 on the number line. We would use the "greater than" symbol > to write this as 8 > 5.

This order relation holds for all real numbers. For example, we write –3 < –2 because –3 lies to the left of –2 on the number line.

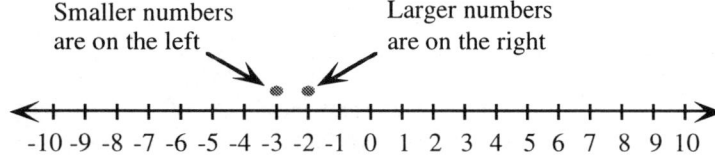

We can also show this relation by writing –2 > –3 (that is, –2 is greater than –3) since –2 is to the right of –3 on the number line.

We refer to statements like 5 < 8 and 8 > 5 as inequalities. An **inequality** is a statement that says one expression is less than (or greater than) another. Note that an inequality refers to relative sizes ("less than" or "greater than"), rather than "not equal". If we want to say that 5 is not equal to 8 we would use a "not equal to sign" to write 5 ≠ 8.

When interpreting inequality symbols > and <, remember the smaller part of the symbol always points to the *smaller* number. For example, in the inequality –8 < –5, the inequality symbol < points to –8 because it's the smaller number.

Example 1.3.4   Replace the word "and" with > or < to show the relation between each pair of numbers.

a)   3 and 5

> •SOLUTION•   On the number line, 3 lies to the left of 5, so 3 is smaller. Thus, we write $3 < 5$.

b)   –4 and –8

> •SOLUTION•   –8 lies to the left of –4, so –8 is smaller. The inequality symbol always points to the smaller number so we write $-4 > -8$ (–4 is greater than –8).

c) $\dfrac{3}{10}$ and $\dfrac{1}{4}$

> •SOLUTION• $\dfrac{1}{4}$ lies to the left of $\dfrac{3}{10}$ so $\dfrac{1}{4}$ is smaller. Therefore we write $\dfrac{3}{10} > \dfrac{1}{4}$.

Practice 1.3.4   Replace the word "and" with > or < to show the relation between each pair of numbers.

1.   7 and 12   2. $\dfrac{1}{2}$ and $\dfrac{3}{4}$ 3.   –12 and –22   4.   5 and –12

The inequality symbol ≤ means **less than or equal to**. The statement $3 \le 5$ is true because 3 is less than OR equal to 5. The statement $3 \le 3$ is also true because 3 is less than OR equal to 3. However, the statement $3 < 3$ is false because 3 is not less than 3.

The inequality symbol ≥ means **greater than or equal to**. Thus, $5 \ge 2$ is true, as is $5 \ge 5$.

A number that is greater than or equal to 0 is called **non-negative**. In terms of symbols, we can say that the letter a represents a non–negative number by writing $a \ge 0$. Here, a could have a value of 5 or $\dfrac{8}{3}$ or 27 or 0. But, the value of a could not be –2 or –5.3.

Topic 1.3 E — Opposites

Except for 0, each real number has two fundamental attributes: its magnitude (size), and its sign. For example, –2 has a magnitude of 2 and a negative sign. Both attributes tell us something about the number. The magnitude tells us how far the number is from 0 and the sign tells us on which side of 0 the number lies. For example, if the temperature is –60 degrees, the 60 tells us we are 60 degrees away from 0 and the – tells us we are below 0 (cold). A temperature of +60 degrees tells us we are also 60 degrees away from 0 but the + tells us we are above 0 (warm). The numbers +60 and –60 are called opposites. The **opposite** of a number has the same magnitude but a different sign. Opposites are also called **additive inverses** because they "undo" each other's effects.

The opposite of a real number is obtained by changing its sign. For example, to obtain the opposite of 5, we change its (understood) + sign to – . Thus, the opposite of 5 is –5. Similarly, to obtain the opposite of –5, we change its – sign to + and write +5 (or just 5). Thus, the opposite of –5 is 5.

Note that because 0 has no sign, we cannot "change its sign" to find its opposite. We say that 0 has no opposite.

Mathematically, we will write the opposite of a number a as $-a$. Using this same idea, the opposite of $-a$ would be written as $-(-a)$. But, we know that the opposite of $-a$ is also equal to a. Therefore, it must be true that $-(-a)$ can be written as a.

Note that the *reciprocal* of a number and the *opposite* of the number are different.

Reciprocal (multiplicative inverse)	*Opposite (additive inverse)*
Reciprocals have the **same** sign.	Opposites have **different** signs.
• the reciprocal of +5 is $+\frac{1}{5}$.	• the opposite of +5 is –5.
• the reciprocal of $-\frac{1}{2}$ is $-\frac{2}{1}$ (or –2).	• the opposite of $-\frac{1}{2}$ is $+\frac{1}{2}$.
The **product** of reciprocals is **1**.	The **sum** of opposites is **0**.
• $(+5)\left(+\frac{1}{5}\right)$ can be written as $\frac{5}{5}$, which reduces to 1.	• +5 + (–5) simplifies to 0.
• $\left(-\frac{1}{2}\right)(–2)$ can be written as $\frac{-2}{-2}$, which reduces to 1.	• $-\frac{1}{2}+\frac{1}{2}$ simplifies to 0.

Example 1.3.5   Write the opposite.

a)   4

•**SOLUTION**• To find the opposite, change the sign. The opposite of 4 is –4.

b)   –3

•**SOLUTION**• 3 (or +3).

c)   8y

•**SOLUTION**• –8y

Practice 1.3.5 Write the opposite.

1. 36   2. –45   3. –2w   4. 7t   5. –(–2)

— Note — *Uses of the "dash"*	We have used the dash "–" in three different ways: • To show the operation of subtraction, as in 6 – 2. This is read "six minus two." • To indicate a negative number, as in –3. This is read "negative three". • To indicate the opposite of a number, as in "the opposite of 5" is –5. Mathematics, like most other languages, sometimes has different meanings for the same symbol. The meaning depends on the context in which the symbol is used. For example, we read –(–7) – 2 as "the opposite of negative seven minus 2".

Topic 1.3 F — Absolute Value

We mentioned that all real numbers, except 0, have both a magnitude and a sign. Sometimes we don't care about the sign. For example, a machinist might need to make a part that is within a tenth of an inch of the value specified on a blueprint. The machinist cannot make the part exactly the right size (no measurement in the real world is exact) but as long as it's within 0.1 inch of the proper value it will work. If the part is off by 0.2 inch, it really doesn't matter if it's too big (that is, the part is off by +0.2 inch) or too small (the part is off by –0.2 inch); in either case, the part will not fit properly. The sign of the error does not matter, only the magnitude is important.

Mathematically, if we want to consider only the magnitude of a number and disregard its sign, we use the idea of absolute value. The **absolute value** of a number is defined as its distance from 0 on the number line, regardless of whether the number is greater than 0 (positive) or less than 0 (negative). We use vertical lines $|\ \ |$ to indicate absolute value. For example, $|-2|$ means "absolute value of –2". To simplify $|-2|$ we find the distance of –2 from 0 on a number line. Since this distance is 2 units, we say $|-2|$ is 2.

Example 1.3.6 Simplify.

a) $|5|$

•SOLUTION• 5, because the distance of 5 from 0 on the number line is 5.

b) $|-4|$

•SOLUTION• 4, because the distance of –4 from 0 on the number line is 4.

c) $-|-2|$

•SOLUTION• This is the opposite of $|-2|$. The absolute value of –2 is 2, and the opposite of that is –2. Therefore, $-|-2|$ is –2.

d) $|0|$

•SOLUTION• 0, because the distance of 0 from 0 on the number line is 0.

Practice 1.3.6 Simplify.

1. $|12|$ 2. $|-57|$ 3. $\left|-\dfrac{44}{45}\right|$ 4. $-|179|$ 5. $-|-59|$

Notice that the absolute value has no effect on positive numbers. For example, $|6|$ is 6. But, it transforms negative numbers into their opposites. For example, $|-6|$ is 6.

Topic 1.3 G — Absolute Value and the Order of Operations

Absolute value is considered an operation because it takes one number, which can be negative, zero, or positive and transforms it into a number which must be zero or positive. In the order of operations, absolute value is considered a grouping symbol and absolute values should be evaluated in Step 2 along with exponents and square roots.

— Procedure — *Simplifying expressions using the proper order of operations*	
	Step 1 Simplify expressions inside grouping symbols. Grouping symbols include parentheses ( ), brackets [ ], the fraction bar, **absolute value bars** $\| \ \|$, and the radical symbol, $\sqrt{}$.
	Step 2 Simplify **absolute values**, exponents, and square roots.
	Step 3 Simplify multiplication and division, working left to right.
	Step 4 Simplify addition and subtraction, working left to right.

— Study Tip — **GAREMDAS**	
	An easy way to remember the order of operations is to make an acronym out of the first letter of each important word:
	Grouping symbols
	Absolute value, **R**oots, **E**xponents
	Multiplication, **D**ivision
	Addition, **S**ubtraction
	Put together, the letters spell **GAREMDAS** (pronounced gar–em'–das).
	If you use this memory device, be sure to remember that multiplication and division are at the same level and must be done from left to right (in other words, sometimes multiplication is done before division, sometimes not). The same is true for addition and subtraction.

Example 1.3.7   Simplify.

a) $|2| + 7 \cdot |-5|$

•SOLUTION• Simplify the absolute values before doing other operations.

$|2| + 7 \cdot |-5|$ ⇐ Expression to simplify.

2 + 7•5 ⇐ Simplified absolute values.

2 + 35 ⇐ Multiplied.

37 ⇐ Added.

b) $8 + |-3| - 2$

•SOLUTION•

$8 + 3 - 2$ ⇐ Simplified absolute value.

11 − 2 ⇐ Added.

9 ⇐ Subtracted.

c) $3 - |-2| + |-12| \div 4$

•SOLUTION•

$3 - 2 + 12 \div 4$ ⇐ Simplified absolute values.

$3 - 2 + 3$ ⇐ Divided.

1 + 3 ⇐ Subtracted.

4 ⇐ Added.

Practice 1.3.7 Simplify.

1. $15 + |-12|$

2. $25 - |7|$

3. $|-17| + 2 \cdot |21|$

4. $5 + |-45| \div 5 + |-7|$

5. $22 - |-10| - |3|$

6. $|-32| + |-15| \cdot |3|$

Exercise Set 1.3   The answers to the odd numbered exercises are at the back of the book.

Plot on a number line.

1. $\left\{ -4.5, 3, -6, -0.5, \dfrac{9}{2}, 8 \right\}$

2. $\left\{ -\dfrac{7}{8}, -4.1 \, -7.75, 3.4, 0.25, 7 \right\}$

Find the approximate coordinate for each point shown on the number line.

3.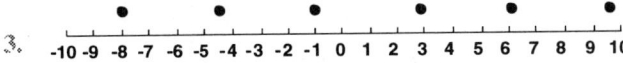

4.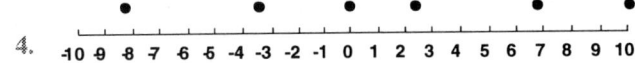

Draw a number line in the given interval and plot the given point.

5. 1940 to 2000 in steps of 10. Plot 1980.

6. −100 to 100 in steps of 20. Plot −20.

7. 0 to 40 in steps of 5. Plot 28.

8. 0 to 80 in steps of 10. Plot 55.

9. −2 to 2 in steps of 0.5. Plot −1.75.

10. −0.5 to 0.5 in steps of 0.1. Plot 0.1.

Replace the word "and" with > or < to show the relation between each pair of numbers.

11. −3 and −4

12. −9 and −6

13. $-\dfrac{1}{2}$ and $-\dfrac{1}{3}$

14. $-1\dfrac{1}{3}$ and $-2\dfrac{1}{5}$

15. 5 and –12

16. 188 and –25

17. 0 and $-\dfrac{7}{9}$

18. 4.75 and 0

19. $\dfrac{5}{12}$ and $-\dfrac{6}{7}$

20. $\dfrac{15}{4}$ and $\dfrac{-7}{9}$

Write the opposite of the number.

21. 12

22. –1.6

23. –5x

24. 3m

25. –(–7)

26. –(–3)

Simplify.

27. $\left|-577\right|$

28. $\left|\dfrac{-2}{5}\right|$

29. $-\left|-7.5\right|$

30. $-\left|-0.04\right|$

31. $8-\left|-5\right|$

32. $2+\left|-9\right|$

33. $\left|15-3\right|\div\left|5-3\right|$

34. $\left|6+2\right|\div\left|6-2\right|$

35. $-\left(8+\left|-6\right|\right)$

36. $-\left(\left|-6\right|+12\right)$

37. $\left|8-2\right|\div\left(6-\left|6\right|\right)$

38. $\left|5+4\right|\left((8)\div\left|-2\right|\right)$

Review Exercises   The answers to all of these exercises are at the back of the book.

39. Identify the sums, terms, products, and factors: 8 + 2

40. Reduce to lowest terms: $\dfrac{98}{343}$

41. Find the reciprocal of $\dfrac{x}{2}$

42. Rewrite as a product: $\dfrac{7}{3}$

43. Simplify: $16/2 \cdot (13-2(5-1)-2)$

Simplify.

44. $8+5\cdot 2$

45. $8-\dfrac{\frac{1}{2}\cdot 14}{\frac{1}{3}+2}$

46. $\dfrac{\frac{32}{25}}{\frac{24}{35}}$

47. $\dfrac{\frac{84}{15}}{\frac{4}{25}}$

48. $\dfrac{4}{15}+\dfrac{7}{25}$

49. $\dfrac{3}{4}+\dfrac{7}{10}+\dfrac{1}{5}$

50. $\dfrac{6^2-2\cdot 3}{15-5\cdot 2}$

51. $\dfrac{9}{24}\cdot\dfrac{60}{72}$

Section 1.4 Multiplying and Dividing Real Numbers

Topic 1.4 A — Basic Ideas of Multiplying Real Numbers

From arithmetic, we know how to multiply positive numbers. To develop a rule for multiplying real numbers, which include negative as well as positive numbers, we will write several multiplications and try to discover a pattern. Let's start with the product 3 • 3, which we know is 9; next, we find the product 2 • 3, which is 6; next, we multiply 1 • 3 to get 3; and so on. The process looks like this:

$$(3)\cdot(3)=9$$
$$(2)\cdot(3)=6$$
$$(1)\cdot(3)=3$$
$$(0)\cdot(3)=0$$

We know these products from arithmetic.

$$(-1)\cdot(3)=\square$$
$$(-2)\cdot(3)=\square$$
$$(-3)\cdot(3)=\square$$

Looking at the pattern, what should the boxes be?

Notice that as the first factor is decreased by 1, the product is decreased by 3. To fill in the first box we need three less than zero or –3; the second box will be three less than –3 or –6; and so on.

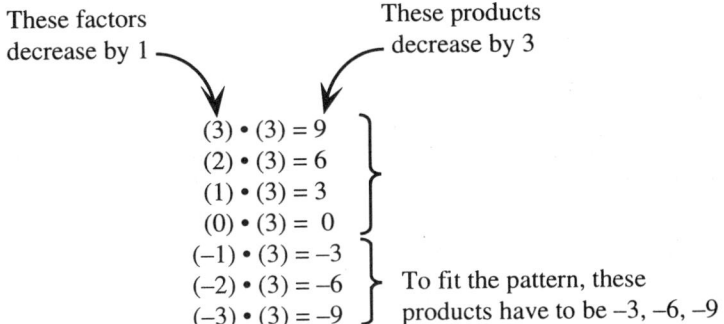

These factors decrease by 1

These products decrease by 3

$$(3) \cdot (3) = 9$$
$$(2) \cdot (3) = 6$$
$$(1) \cdot (3) = 3$$
$$(0) \cdot (3) = 0$$
$$(-1) \cdot (3) = -3$$
$$(-2) \cdot (3) = -6$$
$$(-3) \cdot (3) = -9$$

To fit the pattern, these products have to be –3, –6, –9

This suggests, correctly, that if two factors have different signs their product is negative.

To determine the sign of the product of two negative numbers we will again look for patterns. Let's multiply (–2) by decreasing integers (3 then 2, then 1, and so on) and see if a pattern emerges:

$$(3) \cdot (-2) = -6$$
$$(2) \cdot (-2) = -4$$
$$(1) \cdot (-2) = -2$$
$$(0) \cdot (-2) = 0$$

We know these products from the previous discussion.

$$(-1) \cdot (-2) = \square$$
$$(-2) \cdot (-2) = \square$$
$$(-3) \cdot (-2) = \square$$

Looking at the pattern, what should these be?

Following the pattern, what should be the product (–1) • (–2)? Each time the first factor decreases by 1 the product *increases* by 2.

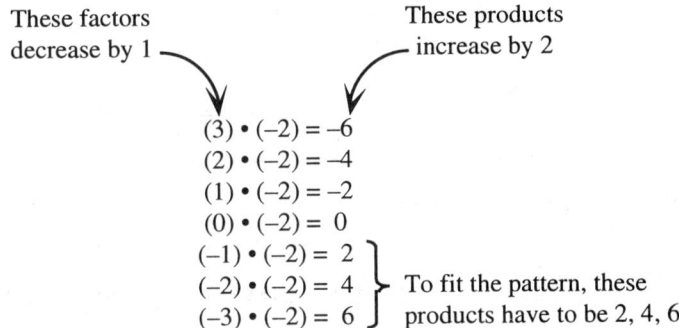

These factors decrease by 1

These products increase by 2

$$(3) \cdot (-2) = -6$$
$$(2) \cdot (-2) = -4$$
$$(1) \cdot (-2) = -2$$
$$(0) \cdot (-2) = 0$$
$$(-1) \cdot (-2) = 2$$
$$(-2) \cdot (-2) = 4$$
$$(-3) \cdot (-2) = 6$$

To fit the pattern, these products have to be 2, 4, 6

This suggests, correctly, that the product of two negative numbers is a positive number.

We can summarize the results for multiplication as follows:

- When we multiply two factors that have the different signs, the product is negative.

- When we multiply two factors that have the same sign, the product is positive.

This leads us to the following procedure for multiplying real numbers:

— Procedure — Multiplying two real numbers	**Step 1** Multiply the absolute values.
	Step 2 Attach the proper sign:
	• If the factors had different signs, the product is negative.
	• If the factors had the same sign, the product is positive.

Example 1.4.1   Simplify.

a)   5(–4)

•**SOLUTION**• We multiply the absolute values, 5 • 4, to get 20. Because the original factors had different signs, we attach a negative sign to get –20.

b) $(-3)(-4)$

• SOLUTION • We multiply the absolute values, $3 \cdot 4$, to get 12. Because the original factors had the same sign, we attach a positive sign to get $+12$, or simply 12.

c) $(-8)(2)(-3)$

• SOLUTION • Because this expression has 3 factors, we multiply the first pair to get -16. Then, we multiply -16 by -3 to get $+48$.

Practice 1.4.1 Simplify.

1. $6(-5)$

2. $(-6)(-9)$

3. $(-12)(5)$

4. $(8)\left(\dfrac{-3}{4}\right)$

5. $(-12)\left(-\dfrac{2}{3}\right)$

6. $(-2)(5)(4)$

7. $(-2)(-7)(-4)$

What happens if one of the factors is 0? We know from arithmetic that the product of 0 and any positive number is just 0. This is also true for the product of 0 and any negative number.

Topic 1.4 B — Evaluating Expressions

As we did before, we evaluate expressions by replacing the variables with numbers and then simplifying. When substituting, be sure to use parentheses around the numbers to help keep the signs straight.

Example 1.4.2 a) Evaluate $-2x$ if x is -6.

• SOLUTION •

$$-2(-6) \quad \Leftarrow \text{ Substituted } -6 \text{ for } x.$$

$$12 \quad \Leftarrow \text{ Multiplied (signs are the same so product is positive).}$$

b) Evaluate xy if x is -6 and y is 3.

• SOLUTION •

$$(-6)(3) \quad \Leftarrow \text{ Substituted } -6 \text{ for } x \text{ and } 3 \text{ for } y.$$

$$-18 \quad \Leftarrow \text{ Multiplied (signs are different so product is negative).}$$

c) Evaluate x^2z if x is -6 and z is $-\frac{1}{2}$.

• SOLUTION •

$$(-6)^2\left(-\frac{1}{2}\right) \quad \Leftarrow \text{ Substituted } -6 \text{ for } x \text{ and } -\frac{1}{2} \text{ for } z.$$

$$36\left(-\frac{1}{2}\right) \quad \Leftarrow \text{ Multiplied 2 factors of } -6 \text{ to get 36.}$$

$$-18 \quad \Leftarrow \text{ Multiplied (signs are different so product is negative).}$$

Practice 1.4.2 Evaluate the following if m is -5 and n is -4.

1. $-7m$

2. $5m$

3. n^2m

4. m^3n

5. $-\dfrac{3}{16}n$

6. $(mn)^3$

Recall that a variable by itself may be interpreted as 1 times the variable. For example, we may think of x as $1 \cdot x$. Similarly, the opposite of a variable may be interpreted as -1 times the variable. For example, $-x$ may be written as $-1 \cdot x$. Explicitly writing coefficients of 1 and -1 may be helpful when evaluating expressions

Example 1.4.3   a)   Evaluate $-x$ when x is -6.

     •SOLUTION•

$$-1 \cdot x \quad \Leftarrow \text{ Explicitly wrote the coefficient of } x.$$

$$-1 \cdot (-6) \quad \Leftarrow \text{ Substituted } -6 \text{ for } x.$$

$$6 \quad \Leftarrow \text{ Signs are the same so product is positive.}$$

In practice, we probably would not write out these steps. We probably would interpret $-x$ as *the opposite of* x so when -6 is substituted for x, we have the opposite of -6, which is 6.

     b)   Evaluate $-x^2$ when x is -6.

     •SOLUTION•

$$-1 \cdot x^2 \quad \Leftarrow \text{ Explicitly wrote the coefficient of } -x.$$

$$-1 \cdot (-6)^2 \quad \Leftarrow \text{ Substituted } -6 \text{ for } x.$$

$$-1 \cdot (36) \quad \Leftarrow \text{ Multiplied 2 factors of } -6 \text{ to get 36.}$$

$$-36 \quad \Leftarrow \text{ Multiplied.}$$

Here, again, we might think of $-x^2$ as *the opposite of* x^2. When -6 is substituted for x, we have the opposite of $(-6)^2$, which is -36.

Practice 1.4.3   Evaluate the expression when m is -8 and n is 3.

1. $-m$ 2. $-n^2$ 3. $(-n)^2$ 4. $-(-m)(-n)$

— *Note* — ***The sign of a variable***	A variable that has a negative sign in front of it can be misleading. For example, does $-x$ represent a positive or a negative number? The answer is, it could be either or neither! x is a variable and so it may represent many different values:
	• If the value of x happens to be -5, then $-x$ is $-(-5)$, which is 5, a positive number.
	• If the value of x happens to be 5, then $-x$ is $-(5)$, which is -5, a negative number.
	• If the value of x happens to be 0, then $-x$ is neither negative nor positive because 0 has no sign.

Topic 1.4 C — Mathematical Models and Translations Involving Multiplication

When we translate from the English words of an applied problem to mathematical symbols we say we are creating a mathematical model. In everyday usage, one meaning of the word model is *a representation of an existing object displayed in a form that is different from the original.* For example, a model airplane is a miniature representation of a real airplane. A **mathematical model** is a representation that uses mathematics to describe the original. For example, the English phrase *twice the number of days in a week* may be modeled by the expression $2 \cdot 7$. We refer to the English phrase as the *verbal model* and the mathematical expression as the *mathematical model.*

There are several key words that indicate the operation of multiplication in applied problems.

Verbal Model	Key Word(s)	Mathematical Model
The product of 8 and –3	product	$(8)(-3)$
–6 multiplied by 3	multiplied by	$(-6)(3)$
6 squared	squared	6^2
2 cubed	cubed	2^3
twice 6	twice	$(2)(6)$
double 6	double	$(2)(6)$
three times the number of minutes in an hour	times	$(3)(60)$
two–thirds of 12	of	$\frac{2}{3} \cdot 12$
18 percent of 20	percent of	$(18\%) \cdot (20)$

Example 1.4.4   Translate to a mathematical model and simplify.

a)   The product of –2 and 8.

   •**SOLUTION**• The word "product" tell us to multiply $(-2)(8)$, which simplifies to –16.

b)   Double $\frac{5}{7}$.

   •**SOLUTION**• The word "double" tell us to multiply $(2)\left(\frac{5}{7}\right)$, which simplifies to $\frac{10}{7}$.

c)   Six percent of 5.

   •**SOLUTION**• The word "of" tells us to multiply. We usually write a percent as a decimal, so we will write this as $(0.06)(5)$, which simplifies to 0.3.

d)   One–half of 18.

   •**SOLUTION**• "Of" tells us to multiply $\left(\frac{1}{2}\right)(18)$, which simplifies to 9.

Practice 1.4.4 Translate to a mathematical model and simplify.

1.   The cube of –2   2.   The product of 5 and –1   3.   Twice –10   4.   Two thirds of 9

5.   Five percent of 8   6.   15 multiplied by –4

Example 1.4.5   Translate to an English phrase.

a) $\frac{2}{3}(9)$

   •Solution• We may translate this in several ways:

   •   The product of $\frac{2}{3}$ and 9.

   • $\frac{2}{3}$ multiplied by 9.

   • $\frac{2}{3}$ of 9.

b) 5^2

•SOLUTION• We may translate this is several ways:

- 5 squared.

- 5 to the second power.

Note that 5^2 is NOT *twice 5*. Twice 5 is $2 \cdot 5$, which is 10.

Practice 1.4.5 Translate to an English phrase.

1. $-2(4)$ 2. $7(8)$ 3. $\frac{1}{2} \cdot 5$ 4. 6^3

Topic 1.4 D — Basic Ideas of Dividing Real Numbers

Division may be viewed as multiplication by a fraction. For example, $\frac{6}{2}$ can be written as

$$\left(\frac{6}{1}\right) \cdot \left(\frac{1}{2}\right) \text{ or } (6) \cdot \left(\frac{1}{2}\right)$$

For this reason, the procedure for finding a quotient is the same as for finding a product.

— Procedure — *Dividing two real numbers*	*Step 1* Divide the absolute values. *Step 2* Attach the proper sign (same as for multiplication): • If the numbers had different signs, the quotient is negative. • If the numbers had the same sign, the quotient is positive.

Division involving 0 is a special case. Dividing a number into 0 results in 0. Algebraically, we write this as $\frac{0}{a} = 0$, assuming $a \neq 0$.

We have a different situation when we try to divide a number *by* 0. In that case, the quotient is said to be **undefined**. In this sense, undefined means the quotient is not a real number. For example, suppose we want to divide $6 among 0 people. How much does each person get? The question makes no sense because we are trying to break up 6 things into pieces of size 0.

Example 1.4.6 Simplify.

a) $-42 \div 6$

•SOLUTION• -7, because the quotient of the absolute values is 7 and the signs are different.

b) $\frac{15}{-3}$

•SOLUTION• -5, because the quotient of the absolute values is 5 and the signs are different.

c) $(-16) / (-2)$

•SOLUTION• 8, because the quotient of the absolute values is 8 and the signs are the same.

Practice 1.4.6 Simplify.

1. $-20/4$ 2. $\frac{49}{-7}$ 3. $(-54) \div (0)$ 4. $0/(-9)$ 5. $(-1.5)/(-0.3)$

Since division can be represented in several ways including $\div$, /, and the fraction bar, it's important to understand how the negative sign is interpreted using these different notations.

- When using $\div$ or $/$ we need parentheses when the divisors are negative numbers. For example, dividing 24 by -6 should be written as $24 \div (-6)$ or $24/(-6)$.

- Fractions that contain negative signs can be written in several ways. For example, we can write the fraction $\dfrac{24}{-6}$ as $\dfrac{-24}{6}$ or $-\dfrac{24}{6}$ because they all are equivalent to -4. This situation is true for all real numbers.

— *Notation* — *Notation for fractions with one negative sign*	English: Suppose a fraction contains a single term in both the numerator and denominator and only one of those terms has a negative sign. The negative sign can be written in the numerator or in the denominator or to the left of the fraction bar. Example: $\dfrac{-2}{3}$ and $\dfrac{2}{-3}$ and $-\dfrac{2}{3}$ all represent the same number. Algebra: If $b \neq 0$, then $\dfrac{-a}{b} = \dfrac{a}{-b} = -\dfrac{a}{b}$

Likewise, we can rewrite fractions with two negative signs. For example, we can write $\dfrac{-24}{-6}$ as $\dfrac{24}{6}$ because each of these fractions is equivalent to 4. Again, this will be true in general:

— *Notation* — *Notation for fractions with two negative signs*	English: Suppose a fraction contains a single term in both the numerator and denominator and both of those terms have a negative sign. The fraction can be written without the negative signs. Example: $\dfrac{-8}{-5}$ and $\dfrac{8}{5}$ represent the same number. Algebra: If $b \neq 0$, then $\dfrac{-a}{-b} = \dfrac{a}{b}$

Example 1.4.7   Simplify $\dfrac{\frac{7}{9}}{\frac{-21}{18}}$

•**SOLUTION**• We invert the denominator and multiply to get $\dfrac{7}{9} \cdot \dfrac{18}{-21}$, which simplifies to $\dfrac{2}{-3}$. Even though we may leave the answer in this form, we usually will write the negative sign in front of the fraction like this $-\dfrac{2}{3}$.

Practice 1.4.7   Simplify.

1. $\dfrac{\frac{5}{6}}{\frac{-10}{9}}$ 2. $(24) \div \left(\dfrac{1}{-3} \right)$ 3. $\dfrac{\frac{-4}{5}}{-6}$

Topic 1.4 E — Evaluating Expressions

To evaluate expressions, we replace the variables with the given values and simplify.

Example 1.4.8   a)   Evaluate $\dfrac{x}{y}$ when x is -12 and y is 4.

•SOLUTION•

$$\dfrac{(-12)}{(4)} \quad \Leftarrow \text{ Substituted } -12 \text{ for } x \text{ and } 4 \text{ for } y.$$

$$-3 \quad \Leftarrow \text{ Signs are different so quotient is negative.}$$

b)   Evaluate $\left(\dfrac{y}{x}\right)^3$ when x is -6 and y is 3.

•SOLUTION•

$$\left(\dfrac{3}{-6}\right)^3 \quad \Leftarrow \text{ Substituted } -6 \text{ for } x \text{ and } 3 \text{ for } y.$$

$$\left(\dfrac{1}{-2}\right)^3 \quad \Leftarrow \text{ Reduced the fraction.}$$

$$\left(\dfrac{1}{-2}\right)\left(\dfrac{1}{-2}\right)\left(\dfrac{1}{-2}\right) \quad \Leftarrow \text{ Applied definition of exponent.}$$

$$\dfrac{1}{-8} \quad \Leftarrow \text{ Multiplied.}$$

We could write $\dfrac{1}{-8}$ as $\dfrac{-1}{8}$ or $-\dfrac{1}{8}$. By tradition, $-\dfrac{1}{8}$ or $\dfrac{-1}{8}$ are preferred. We rarely write a fraction with a negative denominator.

c)   Evaluate $x\dfrac{y}{x^2}$ when x is -5 and y is -2.

•SOLUTION•

$$(-5)\left(\dfrac{-2}{(-5)^2}\right) \quad \Leftarrow \text{ Substituted } -5 \text{ for } x \text{ and } -2 \text{ for } y.$$

$$(-5)\left(\dfrac{-2}{25}\right) \quad \Leftarrow \text{ Multiplied 2 factors of } -5 \text{ to get 25.}$$

$$\dfrac{2}{5} \quad \Leftarrow \text{ Reduced common factor of 5.}$$

Practice 1.4.8   Evaluate when x is -5, y is -3 and z is $\dfrac{1}{4}$.

1. $\dfrac{y}{x}$
2. $\dfrac{z}{y}$
3. $\dfrac{-xy^2}{z}$
4. $\dfrac{\dfrac{x^2}{y}}{\dfrac{y^2}{x}}$

Topic 1.4 F — Translations Involving Division

There are a number of key words that indicate the operation of division in applied problems.

Verbal Model	Key Word(s)	Mathematical Model
the quotient of 8 and -3	quotient	$8 \div (-3)$ or $\dfrac{8}{-3}$
-6 divided by 3	divided by	$-6 \div 3$ or $\dfrac{-6}{3}$
the reciprocal of $\dfrac{2}{3}$	reciprocal	$\dfrac{1}{\frac{2}{3}}$ or $\dfrac{3}{2}$

Example 1.4.9   Translate to a mathematical model and simplify.

a)   The quotient of 15 and 3.

     •SOLUTION• The word "quotient" tells us to divide $15 \div 3$, which simplifies to 5.

b)   24 divided by -8.

     •SOLUTION• The words "divided by" tell us to divide $24 \div (-8)$, which simplifies to -3.

Practice 1.4.9   Translate to a mathematical model and simplify.

1.   The quotient of -8 and 4
2.   The quotient of -24 and $\dfrac{4}{3}$
3. $\dfrac{8}{9}$ divided by -3
4. -36 divided by -4
5.   The reciprocal of 2
6.   The reciprocal of 4 less than 6.

Example 1.4.10   Translate to an English phrase.

a) $17 \div 3$

     •SOLUTION• We may translate this in several ways:

- The quotient of 17 and 3.
- 17 divided by 3.

b) $\dfrac{\frac{1}{2}}{3}$

     •SOLUTION• We may translate this is several ways:

- The quotient of $\dfrac{1}{2}$ and 3.
- $\dfrac{1}{2}$ divided by 3.

Practice 1.4.10   Translate to an English phrase.

1. $18 \div 4$
2. $39 / (-5)$
3. $\dfrac{1}{5+3}$
4. $\dfrac{1}{2(-4)}$

Exercise Set 1.4   The answers to the odd numbered exercises are at the back of the book.

Simplify.

1. $-15(-5)$ 2. $-7(-12)$ 3. $(-2)(-5)(-3)$ 4. $(-6)(-6)(-6)$

5. $\left(\dfrac{-6}{5}\right)(-5)$ 6. $\left(\dfrac{-9}{4}\right)(-12)$ 7. $\left(\dfrac{15}{18}\right)\left(-\dfrac{6}{5}\right)$ 8. $\left(\dfrac{12}{8}\right)\left(-\dfrac{4}{3}\right)$

Evaluate the following if x is -2, y is -3 and z is 4.

9. $2x$ 10. $5y$ 11. $-6y$ 12. $-4x$ 13. $-xy$ 14. $-xz$

15. $-x^2$ 16. $-y^2$ 17. $(yz)^2$ 18. $(xz)^2$ 19. xy^2 20. yx^2

21. $-\left(\dfrac{5}{4}\right)x^2$ 22. $-\left(\dfrac{4}{9}\right)y^2$

Translate to a mathematical model and simplify.

23. The square of -5 24. The product of 2 and -7 25. Double -8 26. Four times 12

27. Six–sevenths of -14 28. $\dfrac{2}{3}$ of -12 29. 12% of 80   30. 8% of -40

Translate to an English phrase.

31. $6(3)$ 32. $7(-8)$ 33. $\dfrac{3}{4}(12)$ 34. $\dfrac{1}{3}(6)$

Simplify.

35. $-45/9$ 36. $-32/8$ 37. $\dfrac{-27}{-3}$ 38. $\dfrac{-48}{-12}$ 39. $54 \div (-9)$ 40. $33 \div (-11)$

41. $\dfrac{15}{\frac{-3}{4}}$ 42. $\dfrac{8}{\frac{-4}{9}}$ 43. $\dfrac{-5}{8} \div \left(-\dfrac{15}{16}\right)$ 44. $\dfrac{-7}{6} \div \left(\dfrac{2}{3}\right)$ 45. $\dfrac{2}{3} \div (-6)$ 46. $\dfrac{12}{35} \div (-3)$

47. $\dfrac{\frac{18}{-25}}{\frac{8}{-15}}$ 48. $\dfrac{\frac{-20}{9}}{\frac{15}{-63}}$ 49. $\dfrac{\frac{24}{-5}}{-8}$ 50. $\dfrac{\frac{-36}{25}}{9}$ 51. $\dfrac{-2}{\frac{-3}{-4}}$ 52. $\dfrac{-2}{\frac{-1}{8}}$

Evaluate when p is -4, q is -2 and r is $-\dfrac{1}{3}$.

53. $\dfrac{p}{r}$ 54. $\dfrac{q}{r}$ 55. $\dfrac{q}{p \cdot q}$ 56. $\dfrac{r}{r \cdot q}$

57. $\left(\dfrac{q}{p^2}\right)^2$ 58. $\left(\dfrac{p^2}{q^2}\right)^2$ 59. $r(p \div q)$ 60. $q(r \div p)$

Translate to a numeric expression and simplify.

61. 50 divided by -150.   62. -72 divided by 9.   63. The reciprocal of 5 less than 12.

64. The reciprocal of 6 more than 3.   65. $\dfrac{4}{5}$ divided by -2.   66. 16 divided by $\dfrac{4}{3}$.

67. The quotient of $\dfrac{6}{15}$ and $\dfrac{-2}{5}$.   68. The quotient of $-\dfrac{13}{12}$ and $\dfrac{26}{24}$.

Translate to an English phrase.

69. $64 \div (-8)$ 70. $55 \div (-11)$ 71. $\dfrac{2}{\frac{1}{3}}$ 72. $\dfrac{\frac{-5}{6}}{4}$ 73. $\dfrac{1}{3} \div \dfrac{2}{9}$ 74. $\dfrac{4}{7} \div \dfrac{6}{5}$

Review Exercises   The answers to all of these exercises are at the back of the book.

75. Identify the sums, terms, products, and factors: $12 + 5$

76. Prime factor: 300

77. Find the reciprocal of $\frac{3}{8}$

78. Rewrite as a product: $\frac{x}{9}$

79. Reduce to lowest terms: $\frac{16}{20}$

80. Write the opposite: 8

81. Graph 46 on a number line. Use an interval from -10 to 55 in steps of 5.

Simplify.

82. $6 \cdot 2 + 3$

83. $8 - [(2 + 3)^2 - 18]$

84. $\frac{12}{24} \cdot \frac{6}{30}$

85. $\frac{18}{12} \div \frac{6}{15}$

86. $\frac{1}{2} + \frac{5}{6}$

87. $\frac{24 - 2 \cdot 3^2}{3 \cdot 2^2 - 5 \cdot 2}$

88. $\left| -2 \right| + 5$

89. $\left| -3 \right| \cdot 18 \div (-9 \cdot 2)$

Section 1.5 Adding and Subtracting Real Numbers

Topic 1.5 A — Basic Ideas of Adding Real Numbers

To develop a general procedure for adding real numbers we will discuss an example that involves money in a checking account. We think of money that you can use, such as money deposited or money already in the account, as positive. We think of money you have already spent, such as checks you wrote or the account being overdrawn, as negative. Let's look at the following four situations:

- The account begins with a balance of $30 and you deposit $20. The account now is worth $50. Mathematically, we write this as

$$30 + 20 = 50$$

This is the way we add positive numbers in arithmetic.

- The account is already overdrawn by $30, which we will represent as -30. You write a check for $20, which we can think of as -20. The account now is overdrawn by $50, which we will write as -50. Mathematically, we may write this as

$$-30 + (-20) = -50$$

This says we added two negative numbers and obtained a sum that is a negative number.

- The account begins with a balance of $30 and you write a check for $20. The new account balance is $10. If we represent the beginning balance as 30 and the check as -20 we have

$$30 + (-20) = 10$$

We added a positive and a negative number and obtained a positive number. The "size" or absolute value of the answer is the *difference* of the absolute values of the terms $(30 - 20)$. Note that the answer is positive because the amount in the account was more "positive" than the check was "negative".

- The account is overdrawn by $30 and you make a deposit of $20. The new balance is still overdrawn by $10. If we use -30 to represent the amount the account was originally overdrawn by, and if we represent the deposit by 20, we have

$$-30 + 20 = -10$$

We added a negative and a positive number and obtained a negative number. Like the previous example, the "size" or absolute value of the answer is the *difference* of the absolute values of the terms $(30 - 20)$. In this case, however, the answer is negative because the original debt was more "negative" than the deposit was "positive".

These four examples illustrate the possible situations for adding real numbers. The procedure for adding real numbers is summarized below:

— *Procedure* — **Adding two real numbers**	• If the numbers have the same sign,
	     Step 1 Add the absolute values.
	     Step 2 Attach the common sign to the sum.
	
	 • If the numbers have different signs,
	     Step 1 Subtract the smaller absolute value from the larger.
	     Step 2 Attach the sign of the number that has the larger absolute value.

Example 1.5.1   Simplify.

   a) $-7 + (-8)$

     •SOLUTION• The signs are the same so we add the absolute values and attach the common sign.

       Step 1   *Add absolute values:* $|-7| + |-8|$ simplifies to $7 + 8$, which is 15.

       Step 2   *Attach common sign:* We attach the common sign, –, to get –15.

   b) $-1 + 3$

     •SOLUTION• The signs are different so we subtract the absolute values and attach the sign of the larger absolute value.

       Step 1   *Subtract absolute values:* $|3| - |-1|$ simplifies to $3 - 1$, which is 2.

       Step 2   *Attach sign of larger absolute value:* Since $|3|$ is larger than $|-1|$ we attach a + to get +2.

   c) $4 + (-9)$

     •SOLUTION• The signs are different so we subtract the absolute values and attach the sign of the larger absolute value.

       Step 1   *Subtract absolute values:* $|-9| - |4|$ simplifies to $9 - 4$, which is 5.

       Step 2   *Attach sign of larger absolute value:* Since $|-9|$ is larger than $|4|$ we attach a – to get –5.

   d) $-2 + 3 + (-7)$

     •SOLUTION• To add three numbers we find the sum of the first two and then add the third.

       $-2 + 3 + (-7)$ ⇐ Expression to simplify.

       $1 + (-7)$ ⇐ Added first two numbers (different signs so subtracted absolute values and used sign of larger absolute value).

       -6 ⇐ Added (different signs so subtracted absolute values and used sign of larger absolute value).

— *Note* — *Doing things mentally*	Rather than use a lot of words as we did above, it's more efficient to keep most of the work in your head and only write an abbreviated version of the calculation. Write as much as you need to make things clear to you and to anyone else who is reading your work. And practice, practice, practice until you become automatic at doing this type of problem.

Practice 1.5.1 Simplify.

1. $(-8) + (-10)$ 2. $(-25) + (-16)$ 3. $9 + (-36)$ 4. $(-15) + 18$

5. $(-25) + (-10) + 14$ 6. $8 + (-31) + 34$ 7. $21 + (-27) + (-14)$ 8. $(-12) + (-26) + (-13)$

9. $\left(-\dfrac{1}{2}\right) + \left(-\dfrac{3}{4}\right)$ 10. $\left(-\dfrac{5}{6}\right) + \left(\dfrac{3}{4}\right)$ 11. $-\dfrac{3}{8} + \left(-3\dfrac{1}{2}\right)$ 12. $-2\dfrac{1}{4} + \dfrac{3}{2}$

Topic 1.5 B — Adding Opposites

The sum of two opposites is always 0 because the magnitude of each number is the same but the numbers are going in opposite directions. For example, a gain of $100 in the stock market, represented by +100, is the opposite of a loss of $100, represented by −100. The sum $100 + (−100)$ is 0.

Example 1.5.2   Fill in the box so the sum is 0.

  a) $5 + \square$

   •SOLUTION• The $\square$ is −5, because $5 + (−5)$ is 0

  b) $−3 + \square$

   •SOLUTION• The $\square$ is 3, because $−3 + 3$ is 0

Practice 1.5.2   Fill in the box so the sum is 0.

1. $11 + \square$ 2. $\square + (−22)$ 3. $\square + \left(-\dfrac{5}{6}\right)$ 4. $\square + 0.002$

Topic 1.5 C — Translations Involving Addition

There are a number of words that indicate the operation of addition. Knowing these words will help when you have to create mathematical models for applied problems.

Verbal Model	*Key Word(s)*	*Mathematical Model*
The sum of 8 and −3	sum	$8 + (−3)$
−6 increased by 3	increased by	$−6 + 3$
8 more than −6	more than	$−6 + 8$
4 added to −3	added to	$−3 + 4$

Example 1.5.3   Translate to a mathematical model and simplify.

  a)   −3 increased by 12

   •SOLUTION• The words "increased by" tell us to add.

   −3 increased by 12   ⇐ Phrase to translate.

   −3   +   12   ⇐ Translated.

         9   ⇐ Signs are different so we subtracted the smaller absolute value from the larger and attached the sign of the larger absolute value.

  b)   the sum of −4 and −2

   •SOLUTION• The word "sum" means to add while "and" tells us where to place the symbol for addition.

   the sum of −4 and −2   ⇐ Phrase to translate.

   −4 + −2 ⇐ Translated.

         −6 ⇐ Signs are the same so added and used common sign.

c) 0.25 more than –5

•**SOLUTION**• The words "more than" mean to add.

$$0.25 \text{ more than } {-5} \quad \Leftarrow \text{ Phrase to translate.}$$

$${-5} \quad + \quad 0.25 \quad \Leftarrow \text{ Translated.}$$

$${-4.75} \quad \Leftarrow \text{ Signs are different so we subtracted the smaller absolute value from the larger and attached the sign of the larger absolute value.}$$

Practice 1.5.3 Translate to a mathematical model and simplify.

1. –18 increased by 10

2. –8.6 increased by 15

3. $\frac{3}{7}$ more than $-\frac{5}{7}$

4. 1 more than $-\frac{7}{8}$

5. –10 added to –16

6. The sum of 25 and –36

We may also have to translate from algebra to English.

Example 1.5.4 Translate to an English phrase: –4 + 7

•**SOLUTION**• We can translate this expression in several ways.

- The sum of –4 and 7.
- –4 increased by 7.
- 7 more than –4.
- 7 added to –4.

Practice 1.5.4 Translate to an English phrase.

1. –12 + 7

2. 15.9 + 25

Many times applied problems involve addition but the key words are not present.

Example 1.5.5

a) The current temperature is –20 °F and it warms up 7 °F. What's the new temperature?

•**SOLUTION**• The new temperature is given by the sum –20 + 7. The signs are different so subtract the absolute values and attach the sign of the larger absolute value. The new temperature is –13 °F.

b) Your checking account is overdrawn by $35 and you deposit $65. What is the new balance?

•**SOLUTION**• We may say the balance of an account overdrawn by $35 is –35. If you deposit $65 in the account you are adding 65 to the –35. Thus, we must find the sum –35 + 65. The signs are different so subtract the absolute values and attach the sign of the larger absolute value. The new balance is $30.

c) A stock traded at \$40 per share yesterday and today its value changed by $-2\frac{1}{4}$ dollars per share. What is the new share price?

- **SOLUTION** - The new share price is given by the sum $40 + \left(-2\frac{1}{4}\right)$. First, we will write both numbers as improper fractions and then combine the fractions following the rules for addition:

$$40 + \left(-2\frac{1}{4}\right) \quad \Leftarrow \text{ Expression to simplify.}$$

$$40 + \left(-\frac{9}{4}\right) \quad \Leftarrow \text{ Converted } -2\frac{1}{4} \text{ to an improper fraction.}$$

$$\left.\begin{array}{c} \dfrac{40}{1} \cdot \dfrac{4}{4} + \left(-\dfrac{9}{4}\right) \\[2mm] \dfrac{160}{4} + \left(-\dfrac{9}{4}\right) \end{array}\right\} \quad \Leftarrow \text{ Converted 40 to 4ths, the LCD.}$$

$$\left.\begin{array}{c} \dfrac{160 + (-9)}{4} \\[2mm] \dfrac{151}{4} \end{array}\right\} \quad \Leftarrow \begin{array}{l}\text{Signs are different so we subtracted the smaller} \\ \text{absolute value from the larger and attached} \\ \text{the sign of the larger absolute value.}\end{array}$$

$$37\frac{3}{4} \quad \Leftarrow \text{ Converted to a mixed number.}$$

The new price of the stock is $37\frac{3}{4}$. This corresponds to \$37.75 per share.

Practice 1.5.5

1. The current temperature is –40 degrees and it warms up 3 degrees. What is the new temperature?

2. You begin on a hill whose elevation is 650 feet above sea level and you descend 780 feet from the top of the hill into a valley. What is your new elevation?

3. Your checking account is overdrawn by \$36 and you deposit \$15. What is the new balance?

4. Your checking account has a balance of \$45 and you write a check for \$60. What is the new balance?

5. A stock traded at \$136 per share yesterday and today its value changed by $-2\frac{1}{2}$ dollars per share. What is the new price?

6. Today a stock changed its value by $-3\frac{1}{4}$ dollars per share. If its price yesterday was \$255 per share what is its new price?

Topic 1.5 D — Basic Ideas of Subtracting Real Numbers

We know from arithmetic that $30 - 20$ is 10. In the last section we found that $30 + (-20)$ is also 10. In this case, subtraction of +20 gives the same answer as addition of –20. Let's see if this pattern continues:

Subtraction problem	Answer to subtraction problem	Addition problem	Answer to addition problem
$6 - 2$	4	$6 + (-2)$	4
$12 - 3$	9	$12 + (-3)$	9
$2.85 - 1.62$	1.23	$2.85 + (-1.62)$	1.23

For each of the above calculations it was possible to write the subtraction problem as an equivalent addition problem. It turns out that this is true whether or not the numbers being subtracted are positive or negative. Algebraically, we write this as

$$a - b = a + (-b)$$

where a and b are any real numbers.

We can use this idea to develop a procedure for subtracting real numbers.

| — *Procedure* — *Subtracting two real numbers* | **Step 1** Change the operation of subtraction into the operation of addition AND change the number being subtracted to its opposite. |
| | **Step 2** Follow the procedure for addition. |

This way of doing subtraction does not conflict with what you have learned in the past. It's simply a different way of subtracting that helps us calculate differences which involve negative numbers.

Example 1.5.6   a) Simplify $9 - 17$

•SOLUTION• We are subtracting $+17$ so we will change it to adding -17.

$9 - 17$ $\Leftarrow$ Expression to simplify.

$9 - (+17)$ $\Leftarrow$ Explicitly wrote sign of 17.

$9 + (-17)$ $\Leftarrow$ Changed subtraction to addition of the opposite.

-8 $\Leftarrow$ Followed rules for addition.

b) Simplify $-10 - 5$

•SOLUTION• We are subtracting $+5$ so we will change it to adding -5.

$-10 - 5$ $\Leftarrow$ Expression to simplify.

$-10 - (+5)$ $\Leftarrow$ Explicitly wrote sign of 5.

$-10 + (-5)$ $\Leftarrow$ Changed subtraction to addition of the opposite.

-15 $\Leftarrow$ Followed rules for addition.

c) Simplify $7 - (-3)$

•SOLUTION•

$7 + (+3)$ $\Leftarrow$ Changed subtraction to addition of the opposite.

$+10$ $\Leftarrow$ Followed rules for addition.

Notice how the subtraction of a negative is equivalent to the addition of a positive. [†]

d) Simplify $-8 - (-1)$

•SOLUTION•

$-8 + (+1)$ $\Leftarrow$ Changed subtraction to addition of the opposite.

-7 $\Leftarrow$ Followed rules for addition.

Practice 1.5.6 Simplify.

1. $-8 - 10$

2. $-9 - 12$

3. $42 - (-27)$

4. $4 - (-36)$

5. $-10 - (-4)$

6. $-7 - (-3)$

7. $18 - 27$

8. $11 - 32$

9. $-25 - (-10) - 14$

10. $-8 - (-31) - 34$

11. $\left(-\frac{1}{2}\right) - \left(\frac{3}{4}\right)$

12. $\left(-\frac{3}{8}\right) - \left(\frac{3}{5}\right)$

[†]   This is analogous to the double negative in English. If I say *It is not not hot*, what is it, hot or cold? *Not hot* means it is cold. Therefore, *not not hot* means not cold, which is hot. We do not use double negatives in English because they are confusing. In mathematics, we have to learn to deal with them.

— *Caution* — ***The dash and subtraction***	Students usually find subtracting real numbers more difficult than adding, multiplying, or dividing. This may be true because we use the same symbol, –, for both the operation of subtraction and for the sign of a negative number. After changing subtraction to addition of the opposite the – symbol will always represent the sign of a number.
	Be sure you know the signs of the numbers you are working on. For example,
	• 12 – 8 means (+12) – (+8), so we are subtracting +8 from +12.
	• –10 – 5 means (–10) – (+5), so we are subtracting +5 from –10.
	• 7 – (–3) means (+7) – (–3), so we are subtracting –3 from +7.

— *Caution* — ***The dash and your*** ***calculator***	Calculators usually have different keys to represent the operation of subtraction and the sign of a number. Therefore, you cannot use the subtraction key to display a number negative.
	• The operation of subtraction usually appears as a dash on a key that is near the addition $\boxed{+}$, multiplication $\boxed{\times}$, and division $\boxed{\div}$ keys.
	• On some calculators the key for the sign of a number is labeled $\boxed{+/\_}$. This is called the *change sign* key. Tapping it changes the sign of the number on the display. For example, to calculate –7 – 3 you would tap $\boxed{7}$ $\boxed{+/\_}$ $\boxed{-}$ $\boxed{3}$ $\boxed{=}$. On some calculators you tap the change sign key before you tap the number. For example, to calculate –7 – 3 you would tap $\boxed{+/\_}$ $\boxed{7}$ $\boxed{-}$ $\boxed{3}$ $\boxed{=}$. The display will read –10
	• On some calculators the key for the sign of a number is labeled with the dash inside parentheses, $\boxed{(-)}$. On such calculators, to display a negative number you tap the key before you tap the number. For example, to calculate –7 – 3 you would tap $\boxed{(-)}$ $\boxed{7}$ $\boxed{-}$ $\boxed{3}$ $\boxed{\text{ENTER}}$. The display will read –10
	The moral is ***GET TO KNOW HOW YOUR CALCULATOR WORKS***.

Topic 1.5 E — Subtraction and Grouping Symbols

Grouping symbols are used to indicate operations that must be done first in a calculation.

Example 1.5.7   a)   Simplify –12 – (5 + 3)

• SOLUTION •

$$-12 - (\mathbf{8}) \quad \Leftarrow \text{ Did addition inside parentheses.}$$

$$-12 + (\mathbf{-8}) \quad \Leftarrow \text{ Changed subtraction to addition of the opposite.}$$

$$-20 \quad \Leftarrow \text{ Added.}$$

b)   Simplify –5 – (3 – 7)

• SOLUTION •

$$-5 - (\mathbf{-4}) \quad \Leftarrow \text{ Did subtraction inside parentheses.}$$

$$-5 + \mathbf{4} \quad \Leftarrow \text{ Changed subtraction to addition of the opposite.}$$

$$-1 \quad \Leftarrow \text{ Added.}$$

c) Simplify $(-5 - 4) - (-12 + 5)$

•SOLUTION•

$(-9) - (-7)$ ⇐ Simplified inside each set of parentheses.

$-9 + 7$ ⇐ Changed subtraction to addition of the opposite.

-2 ⇐ Added.

Practice 1.5.7 Simplify.

1. $7 - (6 - 9)$	2. $-7 - (4 + 2)$	3. $25 - (-10 + 13)$	4. $19 - (25 - 36)$
5. $25 - (-10 - 13)$	6. $(19 - 25) - 36$	7. $(-8 - 2) - (-5 + 4)$	8. $(-5 - 1) - (-3 + 8)$

— *Study Tip* — *Practice*	In order to do mathematics well and learn new material quickly you must become automatic with the four basic operations on real numbers. By *automatic* we mean that you must practice enough so that you are able to do calculations quickly in your head without having to think about the rules and procedures. Just as it is impractical to do computations such as $8 - 5$ on a calculator, you should not have to rely on a calculator to do calculations like $-3 - 8$.

Topic 1.5 F — Absolute Value Versus Parentheses as Grouping Symbols

When simplifying numeric expressions, it's important to keep in mind the differences and similarities between absolute value symbols and parentheses.

- Both are considered grouping symbols. That is, we must do operations inside them before operations outside them. For example:

Example with *Parentheses*	*Example with* *Absolute Value*	
$5 \cdot (3 - 8)$	$5 \cdot \lvert 3 - 8 \rvert$	⇐ Expression to simplify.
$5 \cdot (-5)$	$5 \cdot \lvert -5 \rvert$	⇐ Simplified inside grouping symbols first.
	$5 \cdot (-5)$	⇐ Simplified absolute value.
25	-25	⇐ Multiplied.

- Both can imply multiplication. For example, we could have written the previous two problems without explicitly writing the multiplication symbol • .

   $5 \cdot (3 - 8)$ means the same thing as $5(3 - 8)$

   $5 \cdot \lvert 3 - 8 \rvert$ means the same thing as $5 \lvert 3 - 8 \rvert$

- Parentheses group expressions while the absolute value symbol groups expressions AND tells us to do something. For example,

   $-(3 - 8)$ means the opposite of the difference of 3 and 8, which is 5.

   $-\lvert 3 - 8 \rvert$ means the opposite of the absolute value of the difference of 3 and 8. The difference of 3 and 8 is -5; the absolute value of -5 is 5; and the opposite of 5 is -5.

Topic 1.5 G — Evaluating Expressions

When evaluating expressions that involve subtraction, special care must be taken with the signs. Parentheses will help keep the operation and sign separate.

Example 1.5.8 a) Evaluate $x - y$ if x is –6 and y is 3.

•SOLUTION• Replace the variables with the given values and simplify. Remember to place both the numeral and its sign inside parentheses when you do the substitutions.

$x - y$ ⟸ Expression to evaluate.

$(-6) - (3)$ ⟸ Substituted –6 for x and 3 for y.

$-6 + (-3)$ ⟸ Changed subtraction to addition of the opposite.

-9 ⟸ Followed rules for addition.

b) Evaluate $x - (x - y)$ if x is –6 and y is 3.

•SOLUTION•

$(-6) - [(-6) - (3)]$ ⟸ Substituted –6 for x and 3 for y. We changed the original parentheses ( ) to brackets [ ] to make the expression easier to read. This does not change the algebraic meaning of the problem.

$-6 - (-9)$ ⟸ Simplified inside parentheses.

$-6 + 9$ ⟸ Changed subtraction to addition of the opposite.

3 ⟸ Followed rules for addition.

Practice 1.5.8 Evaluate the expression for the given values of the variables.

1. $x - y$ if x is –8 and y is 7

2. $-3p - q$ if p is 10 and q is –4

3. $w - (v - w)$ if w is –32 and v is –25

4. $p - 2(q - 3p)$ if p is –5 and q is 3

Topic 1.5 H — Translations Involving Subtraction

Knowing which words indicate subtraction will help you solve applied problems. Note that unlike addition, the order in which we do the subtraction will make a difference so we have to be extra careful when doing the translations.

Verbal Model	Key Word(s)	Mathematical Model
The difference of 8 and –3	difference	$8 - (-3)$
The difference of –3 and 8		$-3 - 8$
–6 decreased by 3	decreased by	$-6 - 3$
8 less than –6	less than	$-6 - 8$ Note that terms are in reverse order from the English.
8 less 5	less	$8 - 5$
5 less 8		$5 - 8$
4 subtracted from –3	subtracted from	$-3 - 4$
–3 subtracted from 4		$4 - (-3)$ Note that terms are in reverse order from the English.

Example 1.5.9   Translate to a mathematical model and simplify.

    a) -3 decreased by 12

       •**SOLUTION**• The words "decreased by" tell us to subtract.

$$-3 \text{ decreased by } 12 \quad \Leftarrow \text{ Phrase to translate.}$$

$$-3 \quad - \quad 12 \quad \Leftarrow \text{ Translated.}$$

$$-15 \quad \Leftarrow \text{ Simplified.}$$

    b)   the difference of -4 and -2

       •**SOLUTION**• The word "difference" tells us to subtract while "and" shows us where the symbol for subtraction should be placed.

the difference of -4 and -2 $\Leftarrow$ Phrase to translate.

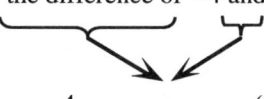

$$-4 \quad - \quad (-2) \quad \Leftarrow \text{ Translated.}$$

$$-4 + (\textbf{+2}) \quad \Leftarrow \text{ Changed subtraction to addition of the opposite.}$$

$$-2 \quad \Leftarrow \text{ Signs are different so subtracted and used sign of larger absolute value.}$$

    c)   0.25 less than -5

       •**SOLUTION**• The words "less than" tell us to subtract but in the reverse order of the way in which the numbers are written.

$$0.25 \text{ less than } -5 \quad \Leftarrow \text{ Phrase to translate.}$$

$$-5 \quad - \quad 0.25 \quad \Leftarrow \text{ Translated (rearranged order to match English meaning).}$$

$$-5 + (\textbf{--0.25}) \quad \Leftarrow \text{ Changed subtraction to addition of the opposite.}$$

$$-5.25 \quad \Leftarrow \text{ Signs are the same so added and used sign present.}$$

    d) -1 subtracted from -5

       •**SOLUTION**• The words "subtracted from" tell us to subtract but in the reverse order of the way in which the numbers are written.

$$-1 \text{ subtracted from } -5 \quad \Leftarrow \text{ Phrase to translate.}$$

$$-5 - (-1) \quad \Leftarrow \text{ Translated (rearranged order to match English meaning).}$$

$$-5 + (\textbf{+1}) \quad \Leftarrow \text{ Changed subtraction to addition of the opposite.}$$

$$-4 \quad \Leftarrow \text{ Signs are different so subtracted and used sign of larger.}$$

Practice 1.5.9   Translate to a mathematical model and simplify.

1. -18 decreased by 10   2.   15 decreased by 9   3.   1 less than $-\dfrac{7}{8}$

4. -10 subtracted from -16 5. -13 subtracted from 28   6.   The difference of 5 and $-\dfrac{3}{4}$

We must also be able to do the translation from algebra to English.

Example 1.5.10   Translate to an English phrase.

a) $5 - 12$

•SOLUTION• We may translate this in several ways. Notice the order of the numbers in the English translation is sometimes different from the order given in the algebraic expression.

- The difference of 5 and 12.
- 5 decreased by 12.
- 12 less than 5.
- 12 subtracted from 5.

b) $-3 - 2$

•SOLUTION•

- The difference of -3 and 2.
- -3 decreased by 2.
- 2 less than -3.
- 2 subtracted from -3.

Practice 1.5.10   Translate to an English phrase.

1. $-12 - 7$ 2. $-1 - (-3)$

Exercise Set 1.5   The answers to the odd numbered exercises are at the back of the book.

Simplify.

1. $-5 + (-9)$ 2. $-2 + (-11)$ 3. $24 + (-26)$ 4. $-11 + (45)$

5. $-39 + 54 + (-7)$ 6. $-21 + 75 + (-14)$ 7. $36 + (-45) + (-22) + 73$ 8. $27 + (-14) + (-54) + (-18)$

9. $-\dfrac{5}{12} + \dfrac{3}{4}$ 10. $-\dfrac{6}{7} + \dfrac{4}{5}$ 11. $-1 + \dfrac{1}{8} + \left(-\dfrac{5}{6}\right)$ 12. $\dfrac{3}{5} + (-3) + \dfrac{2}{15}$

Fill in the box so the sum is 0.

13. $29 + \square$ 14. $18 + \square$ 15. $\square + \left(-\dfrac{13}{15}\right)$ 16. $\square + \dfrac{4}{11}$ 17. $-24.9 + \square$ 18. $26.7 + \square$

Translate to a mathematical model and simplify.

19. -32 increased by 24   20.   5 increased by -8 21. -22 added to 13   22. -6 added to -4

23.   The sum of $\dfrac{5}{8}$ and $\dfrac{-1}{6}$ 24.   The sum of $-\dfrac{5}{12}$ and $\dfrac{3}{8}$ 25.   5 more than $\dfrac{-2}{3}$ 26.   2 more than $\dfrac{-1}{2}$

Translate to English.

27. $-36 + 11$ 28. $-\dfrac{7}{6} + \dfrac{1}{8}$ 29. $-5 + (-2)$ 30. $-7 + (-1)$

Solve the following problems:

31.   Your checking account is overdrawn by $25 and you deposit $40. What is the new balance?

32.   What is the new balance of your checking account if it is overdrawn by $72 and you deposit $56?

33.   What is the current price per share of a stock that traded at $25 per share yesterday and today its value changed by $-\dfrac{3}{4}$ dollars per share?

34.   A stock traded at $125 per share yesterday and today its value changed by $-5\dfrac{3}{4}$ dollars per share. What is the new share price?

Simplify.

35. $-5 - 9$
36. $-3 - 18$
37. $-24 - (-26)$
38. $-11 - (-45)$

39. $-36 - (-22) - 18$
40. $(-15) - (-24) - 19$
41. $11 - 18 - (-30)$
42. $20 - 42 - (-16)$

43. $36 - (-45) - (-22) - 73$
44. $27 - (-14) - (-54) - 18$
45. $\dfrac{-3}{14} - \dfrac{6}{7}$
46. $\dfrac{-8}{5} - \dfrac{9}{10}$

47. $12 - (4 - 8)$
48. $8 - (9 - 12)$
49. $(12 - 4) - 8$
50. $(8 - 9) - 12$

51. $(-10 - 7) - 2$
52. $(10 - 12) - 2$
53. $(36 - 45) - (6 - 8)$
54. $(3.8 - 7.2) - (4.3 - 6.1)$

Evaluate the expression for the given values of the variables.

55. $x - y$ if x is -28 and y is 15
56. $p - q$ if p is -12 and q is 34
57. $m - (m - n)$ if m is -2 and n is -5

58. $a - (2b - a)$ if a is -3 and b is -4
59. $3x - 5y$ if x is 4 and y is 9
60. $4x - 2z$ if x is -3.2 and y is -4.2

Translate to a numeric expression and simplify.

61. -32 decreased by 24
62. -8 decreased by 5
63. -22 subtracted from 13

64. -8 subtracted from -12
65. the difference of $-\dfrac{1}{3}$ and 3
66. the difference of $-\dfrac{3}{8}$ and -8

67. 5 less than -24
68. 8 less than -3

Translate to an English phrase.

69. $-36 - 11$
70. $-9 - 14$
71. $\dfrac{4}{9} - 4$
72. $\dfrac{4}{6} - 15$

Review Exercises The answers to all of these exercises are at the back of the book.

73. Prime factor: 150
74. Find the reciprocal of x
75. Identify the sums, terms, products, and factors: $4 \cdot 2$

76. Rewrite as a product: $\dfrac{2t}{3}$
77. Write the opposite: 0.23
78. Reduce to lowest terms: $\dfrac{24}{36}$

Simplify.

79. $8 + \dfrac{2 \cdot 3 + 6}{12 - 2 \cdot 3}$
80. $\dfrac{18}{35} \cdot \dfrac{21}{24}$
81. $\dfrac{5}{8} + \dfrac{2}{3}$
82. $\dfrac{2^4 - 5 \cdot 2}{5^2 - 2 \cdot 3^2 - 1}$

83. $8 - 2(8 + 2(8 - 2) - 19)$
84. $2 \cdot \left| -3 \right| - 4$
85. $6 \cdot \left(-\dfrac{1}{2} \right)$
86. $-0.5 \cdot 2 \div (-0.4)$

87. Translate to a mathematical model: The product of -5 and 12.

88. Translate to a mathematical model: The quotient of a number and 7.

Section 1.6 Properties of Real Numbers

A property is a characteristic trait. A property of water is that it can flow; a property of fire is that it's hot. Numbers and operations have properties too. A property of whole numbers is that they are non–negative; a property of multiplication is that the order of the factors does not affect the product. You can use properties to save time by classifying new things as an example of something you already know.

Some of the properties listed below will be familiar to you from your study of arithmetic. Even if you know a property, you should learn its formal algebraic definition so that you can become experienced using the symbolism of algebra.

Topic 1.6 A — Properties of 0 and 1

The sum of any number and 0 is identical to the number. For this reason, we refer to 0 as the **additive identity**.

— *Property* — ***Addition Property of 0***	English: The sum of 0 and any number is identical to that number. Example: $\quad 0 + 8 = 8$ $\quad x + 0 = x$ Algebra: $0 + a = a$

The product of any number and 0 is 0.

— *Property* — ***Multiplication Property of 0***	English: The product of 0 and any number is 0. Example: $\quad 0 \cdot 8 = 0$ $\quad x \cdot 0 = 0$ Algebra: $0 \cdot a = 0$

The product of any number and 1 is just that number. For this reason, we refer to 1 as the **multiplicative identity**.

— *Property* — ***Multiplication Property of 1***	English: The product of 1 and any number is identical to that number. Example: $\quad 8 \cdot 1 = 8$ $\quad 1 \cdot x = x$ Algebra: $1 \cdot a = a$

Example 1.6.1   Make each sentence true by replacing the box with a number.

a) $\square + (-8) = -8$

•SOLUTION• The $\square$ is 0 because adding 0 to –8 will leave it unchanged.

b) $5 \cdot \square = 5$

•SOLUTION• The $\square$ is 1 because multiplying 5 by 1 will leave it unchanged.

c) $\square + 3 + 2x = 2x$

•SOLUTION• The $\square$ is –3 because adding –3 and 3 will make 0 and adding 0 to $2x$ will leave $2x$ unchanged.

d) $(\square)\left(\dfrac{1}{3}\right)(5) = 5$

•SOLUTION• The $\square$ is 3 because the product of 3 and $\dfrac{1}{3}$ is 1 and multiplying 1 by 5 will leave 5 unchanged.

Practice 1.6.1   Make each sentence true by replacing the box with a number.

1. $\square + (-36) = -36$

2. $-12 \cdot \square = -12$

3. $\square + (-7) + 2z = 2z$

4. $\square + 33 + 8n = 8n$

5. $(\square)\left(\dfrac{2}{3}\right)(9) = 9$

6. $(\square)(23)(11) = 11$

— *Caution* — *Keep vocabulary straight*	Be sure to keep the vocabulary straight. • Additive identity is another name for 0 because the sum of 0 and a number is the number. • Additive inverse of a number is the opposite of that number. For example, the additive inverse of 8 is –8. • Multiplicative identity is another name for 1 because the product of 1 and a number is that number. • Multiplicative inverse of a number is the reciprocal of the number. For example, the multiplicative inverse of 8 is $\frac{1}{8}$.

Topic 1.6 B — Commutative Properties of Addition and Multiplication

The order in which we add terms does not affect their sum. For example, $3 + 5$ simplifies to the same number as $5 + 3$. This property holds true for all real numbers and is called the **Commutative Property of Addition**.

— *Property* — *Commutative Property of Addition*	English: The order in which we add terms does not affect their sum. Examples: $3 + 8 = 8 + 3$ $-2 + x = x + (-2)$ Algebra: $a + b = b + a$

One use of the commutative property of addition is to help us simplify expressions more quickly. For example, suppose we are asked to simplify $-17 + 35 + 17$. Following the order of operations, we first have to add $-17 + 35$. But, if we use the commutative property to rearrange the second addition, we first add $-17 + 17$, which is easier.

Another use of the property is to arrange terms in a standard way to make them easier to understand. For example, we usually write expressions with the terms in decreasing powers of the variable. Therefore, if we have an expression like $3x + 5x^3 + 2x^2$ we would, by tradition, write it as $5x^3 + 2x^2 + 3x$. The expressions are equivalent but the second expression is easier to work with.

Note that we must be careful when we rearrange terms where subtraction is involved. We defined *terms* as constants, variables, or the product of constants and variables. We said that terms either stand alone, such as $2w$, or they are separated from each other by addition as in $x + 3$, where x is a term and 3 is a term. To identify terms in an expression that involves subtraction we first think of the subtraction as addition of the opposite. For example, to identify the terms of $2x - 5$ we mentally visualize this as $2x + (-5)$. Now, the terms are easily identifies as $2x$ and -5.

Example 1.6.2   Use the Commutative Property of Addition to rewrite each expression.

a) $5 + 3$

     •SOLUTION• $3 + 5$

b) $2x + 3x^4 + x^2$

     •SOLUTION•   By tradition, we write expressions like this with the terms in decreasing powers of x as $3x^4 + x^2 + 2x$

c) $5 - 2x$

     •SOLUTION•   This is a subtraction, which is not commutative. To use the Commutative Property of Addition we must first write the expression as an addition.

         $5 + (-2x)$ $\Leftarrow$ Rewrote subtraction as addition of the opposite.

         $-2x + 5$ $\Leftarrow$ Applied Commutative Property of Addition.

Practice 1.6.2 Use the Commutative Property of Addition to rewrite each expression. Where possible, write the terms in decreasing powers of the variable.

1. $5 + 22$

2. $8 - 23$

3. $3y + 5y^2 + 2$

4. $w - 2w^2 + 5w^3$

5. $-12 - 8x$

6. $p - 5 - p^2$

7. $a^2 - 3 - 2a$

The Commutative Property is also true for multiplication. We can multiply factors in any order and obtain the same product. For example, $2 \cdot 3$ and $3 \cdot 2$ both simplify to 6.

— *Property* — *Commutative Property of Multiplication*	English: The order in which we multiply factors does not affect their product. Examples: $3 \cdot 8$ is equivalent to $8 \cdot 3$ $-2x$ is equivalent to $x(-2)$ Algebra: $a \cdot b = b \cdot a$

The **Commutative Property of Multiplication** can be used to make calculations easier. For example, multiplying in your head $17 \cdot 5 \cdot 5 \cdot 2 \cdot 2$ is fairly difficult. The problem is much easier if you rearrange the factors like this $2 \cdot 5 \cdot 2 \cdot 5 \cdot 17$.

We can also use the Commutative Property of Multiplication to make expressions easier to read by writing factors in alphabetical order. For example, even though the Commutative Property says expressions such as *ab* and *ba* have the same meaning, we will choose to write the product with the letters in alphabetical order as *ab*. Also, the product of a constant and a variable is usually written with the constant first. Thus, even though $5x$ and $x5$ are equivalent, $5x$ is the preferred way of writing this expression.

Example 1.6.3 Use the Commutative Property of Multiplication to rewrite the expression.

a) $x3$

   •SOLUTION• $3x$

b) $(x + 1)5$

   •SOLUTION• The parentheses tell us to think of the sum $x + 1$ as a single factor. Therefore, we may write $(x + 1)5$ as $5(x + 1)$

Practice 1.6.3 Use the Commutative Property of Multiplication to rewrite the expression.

1. $4 \cdot 5$

2. $w5$

3. $(y - 8)12$

4. $8(d - 3)$

5. $(x - 12)(x + 2)$

6. ca^2b

While addition and multiplication are commutative, subtraction and division are not commutative. The order in which we subtract or divide is important.

— *Caution* — *Subtraction and division are not commutative*	Subtraction and division are NOT commutative. For example, $7 - 3 \neq 3 - 7$ because $4 \neq -4$. Likewise, $\dfrac{8}{2} \neq \dfrac{2}{8}$ because $4 \neq \dfrac{1}{4}$

Topic 1.6 C — Associative Properties of Addition and Multiplication

An association is a grouping. The **Associative Property of Addition** says the way we group terms does not affect their sum. For example, $(2 + 3) + 8$ produces the same sum as $2 + (3 + 8)$. Notice we have not changed the order of the terms — we have only changed the way they are grouped. We can check that this regrouping does not affect the sum if we simplify both expressions using the standard order of operations, which says that we must do the work inside parentheses first:

Original expression	*Regrouped expression*
$(2 + 3) + 8$	$2 + (3 + 8)$
5 + 8	**2 + 11**
13	13

— Property — *Associative Property of* *Addition*	English: The way in which terms are grouped does not affect their sum. Examples: $\quad 3 + (8 + 1)$ is equivalent to $(3 + 8) + 1$ $\quad (3 + 2x) + 9x$ is equivalent to $3 + (2x + 9x)$ Algebra: $a + (b + c) = (a + b) + c$

Example 1.6.4 Use the Associative Property of Addition to rewrite the expressions.

a) $\dfrac{1}{3} + \left(\dfrac{2}{3} + 8\right)$

> •SOLUTION• Keep the terms in the same physical order and rearrange the parentheses to get $\left(\dfrac{1}{3} + \dfrac{2}{3}\right) + 8$

b) $(5 + y) + 4y$

> •SOLUTION• Keep the terms in the same physical order and rearrange the parentheses to get $5 + (y + 4y)$.

Practice 1.6.4 Use the Associative Property of Addition to rewrite the expressions.

1. $\dfrac{5}{6} + \left(\dfrac{1}{6} + 9\right)$ 2. $\left(-10 + \dfrac{1}{3}\right) + \dfrac{2}{3}$ 3. $(6 + 7x) + 3x$ 4. $3x + (4x + 12)$

Multiplication is also associative.

— Property — *Associative Property of* *Multiplication*	English: The way in which factors are grouped does not affect their product. Examples: $\quad 3 \cdot (8 \cdot 1)$ is equivalent to $(3 \cdot 8) \cdot 1$ $\quad 2x \cdot (9x \cdot 3)$ is equivalent to $(2x \cdot 9x) \cdot 3$ Algebra: $a\,(bc) = (ab)c$

Example 1.6.5   Use the Associative Property of Multiplication to rewrite the expressions.

a) $(3 \cdot 2) \cdot 5$

•SOLUTION• Keep the factors in the same physical order and rearrange the parentheses to get $3 \cdot (2 \cdot 5)$. Notice that the original expression and the rearranged expression simplify to the same value:

Original	Rearranged
$(3 \cdot 2) \cdot 5$	$3 \cdot (2 \cdot 5)$
$(6) \cdot 5$	$3 \cdot (10)$
30	30

b) $\dfrac{1}{17}(17x)$

•SOLUTION• Keep the factors in the same physical order and rearrange the parentheses to get $\left(\dfrac{1}{17} \cdot 17\right)x$.

c) $5\left(\dfrac{x}{5}\right)$

•SOLUTION• Remember that $\dfrac{x}{5}$ can be written as $\dfrac{1}{5}x$.

$$5\left(\frac{x}{5}\right) \quad \Leftarrow \text{ Original expression.}$$

$$5\left(\frac{1}{5}x\right) \quad \Leftarrow \text{ Rewrote } \frac{x}{5} \text{ as } \frac{1}{5}x.$$

$$\left(5 \cdot \frac{1}{5}\right)x \quad \Leftarrow \text{ Kept the factors in the same physical order and rearranged the parentheses.}$$

Practice 1.6.5   Use the Associative Property of Multiplication to rewrite the expressions.

1. $(7 \cdot 3) \cdot 5$ 2. $(-4 \cdot 2) \cdot (-8)$ 3. $4 \cdot (5 \cdot 3)$ 4. $\dfrac{2}{3}(3x)$ 5. $12\left(\dfrac{1}{12}y\right)$ 6. $8\left(\dfrac{z}{8}\right)$

— *Note* —
 	
 Distinguishing Between the Commutative and Associative Properties
	Students sometimes have difficulty keeping these properties straight.
	
	 • The Commutative Property allows us to change the *physical* order of the numbers:
	
	     $a + b$ becomes $b + a$
	
	     $a \cdot b$ becomes $b \cdot a$
	
	 • The Associative Property allows us to change the way the numbers are grouped, but the physical order remains the same:
	
	     $a + (b + c)$ becomes $(a + b) + c$
	
	     $a \cdot (b \cdot c)$ becomes $(a \cdot b) \cdot c$
	
	 Remember: The Commutative Property commutes (interchanges) the physical order of the numbers while the Associative Property changes the associations (groupings) of the numbers.

Topic 1.6 D — Distributive Property of Multiplication Over Addition

In arithmetic, we defined multiplication as repeated addition. Another relation between addition and multiplication can be seen in the Distributive Property of Multiplication Over Addition:

— *Property* — *Distributive Property of* *Multiplication Over* *Addition*	English: The product of a single factor and the sum of two terms can be rewritten as the sum of two products. Example: $2(y + 4)$ is equivalent to $2 \cdot y + 2 \cdot 4$ Algebra: $a(b + c) = ab + ac$

Example 1.6.6 Use the Distributive Property of Multiplication Over Addition to rewrite each expression. Then simplify, if possible.

a) $5(x + 4)$

•SOLUTION•

$5 \cdot x + 5 \cdot 4$ ⟸ Distributed 5 to terms inside parentheses.

$5x + 20$ ⟸ Simplified.

b) $6(x + y + 5)$

•SOLUTION•

$6 \cdot x + 6 \cdot y + 6 \cdot 5$ ⟸ Distributed 6 to terms inside parentheses.

$6x + 6y + 30$ ⟸ Simplified.

c) $-6(x - 2)$

•SOLUTION• If the number inside the parentheses includes subtraction, you may find it helpful to rewrite the subtraction as addition of the opposite.

$-6(x - 2)$ ⟸ Expression to rewrite.

$-6(x + (-2))$ ⟸ Rewrote subtraction as addition of the opposite. We usually do this step mentally.

$-6 \cdot x + (-6) \cdot (-2)$ ⟸ Distributed 6 to terms inside parentheses.

$-6x + 12$ ⟸ Simplified.

Practice 1.6.6 Use the Distributive Property of Multiplication Over Addition to rewrite each expression.

1. $2(y + 7)$ 2. $-4(x + 2)$ 3. $6(r + 3s + 2)$ 4. $-2(5x + y + 4)$ 5. $-7(y - 4)$ 6. $-4(-2x - 3)$

— *Caution* — *We cannot distribute over* *multiplication*	We can distribute multiplication over addition but NOT over multiplication. For example, we cannot distribute the 2 in $2(5 \cdot 3)$. **INCORRECT:** $2(5 \cdot 3)$ ⟸ Expression to simplify. $(2 \cdot 5)(2 \cdot 3)$ ⟸ ***INCORRECTLY*** distributed the 2. 60 ⟸ Multiplied. **CORRECT:** $2(5 \cdot 3)$ ⟸ Expression to simplify. $2(15)$ ⟸ Simplified inside parentheses. 30 ⟸ Multiplied.

This property is called the *distributive* property because in the definition

$$a(b + c) = ab + ac$$

multiplication by the variable a is "distributed" to b and c inside the parentheses. This property has two important uses in algebra:

- It allows us to remove parentheses from an expression without having to do the operations inside the parentheses. We will use this fact to simplify algebraic expressions later.

- If we use the property in reverse, it allows us to convert a sum, such as $ab + ac$, into the product $a(b + c)$. We call this process factoring.

Topic 1.6 E — Factoring

We can use the Distributive Property in reverse to convert a sum into a product. For example, we can convert $4x + 5x$ into a product by "undistributing" the common factor of x. That is, $4x + 5x$ may be written as $x(4 + 5)$. We say that the x is "factored out" of $4x + 5x$. Rewriting an expression as a product is called **factoring**.

Example 1.6.7   Factor. That is, use the Distributive Property to rewrite the sum as a product.

a) $5x + 5y$

  •SOLUTION• $5(x + y)$ We "undistributed" 5 from each term.

b) $3y - 7y$

  •SOLUTION• $y(3 - 7)$ We "undistributed y from each term.

c) $mx + nx$

  •SOLUTION• $x(m + n)$ We "undistributed x from each term.

Practice 1.6.7   Factor. That is, use the Distributive Property to rewrite the sum as a product.

1. $2y + 12y$ 2. $5y + 8y + 11y$ 3. $3x - 5x$ 4. $xy + wy$ 5. $nm + mp$

Topic 1.6 F — More on Exponential Notation

In Section 1.1, we discussed exponential notation as a way of indicating repeated multiplication. For example, the expression x^4 means $x \bullet x \bullet x \bullet x$:

- x is called the *base* and indicates the factor to be multiplied.

- 4 is called the *exponent* and tells us how many factors of the base are to be multiplied.

In some expressions, such as 2^4, it's easy to identify the base (2). In other expressions, such as -2^4, the base is not so obvious (it is NOT -2). We read -2^4 as "the opposite of 2 to the fourth power" rather than "negative 2 to the fourth power". By convention, when we write a number raised to a power the base does not include the negative sign unless the number and sign are inside parentheses. Study the following examples:

Expression	Base	*English translation*	*Meaning*	*Equivalent expression*
2^4	2	two to the fourth power	$2 \bullet 2 \bullet 2 \bullet 2$	16
$(-2)^4$	-2	the quantity negative two to the fourth power	$(-2) \bullet (-2) \bullet (-2) \bullet (-2)$	16
$-(2^4)$	2	the opposite of the quantity two to the fourth power	$-(2 \bullet 2 \bullet 2 \bullet 2)$	-16
-2^4	2	the opposite of two to the fourth power	$-2 \bullet 2 \bullet 2 \bullet 2$	-16

In the above examples, the only expression that has –2 as a base is $(-2)^4$ because that is the only expression that has the negative sign inside parentheses.

Example 1.6.8   Identify each base and its exponent, and then simplify.

a) 5^3

•SOLUTION•

- base is 5.
- exponent is 3.
- 5^3 says to multiply 3 factors of 5, which is $5 \cdot 5 \cdot 5$, which is 125.

b) $5 \cdot 2^3$

•SOLUTION•

- base of the first factor is 5, exponent is 1 (remember, if we don't write an exponent it's assumed to be 1).
- base of the second factor is 2, exponent is 3. The exponent 3 does not apply to the 5 because the order of operations says we must simplify exponents before multiplying.
- we simplify the exponents first and then multiply the results.

$$5 \cdot 2^3 \quad \Leftarrow \text{ Expression to simplify.}$$

$$5 \cdot \mathbf{8} \quad \Leftarrow \text{ Multiplied 3 factors of 2 to get 8.}$$

$$40 \quad \Leftarrow \text{ Multiplied.}$$

c) $(5 \cdot 2)^3$

•SOLUTION•

- base is $(5 \cdot 2)$.
- exponent is 3.
- do the work inside parentheses first, and then simplify the exponent.

$$(5 \cdot 2)^3 \quad \Leftarrow \text{ Expression to simplify.}$$

$$(\mathbf{10})^3 \quad \Leftarrow \text{ Simplified inside parentheses.}$$

$$1000 \quad \Leftarrow \text{ Multiplied 3 factors of 10 to get 1000.}$$

d) -3^2

•SOLUTION•

- base is 3 (it is NOT –3).
- exponent is 2.
- -3^2 means the opposite of 3^2. We must simplify the exponent before taking the opposite to get –9.

Practice 1.6.8 Identify each base and its exponent, and then simplify.

1. 6^3 2. 4^5 3. $5 \cdot 4^2$ 4. $(3 \cdot 2)^3$

5. $(-7)^2$ 6. $-(8)^2$ 7. -7^2 8. $-(-8)^2$

When we write a product as a repeated multiplication we say it's written in **expanded form**. For example, $x \cdot x \cdot x \cdot x$ is written in expanded form. When we use an exponent to write a product we say it's written in **exponential notation**. The product x^4 is written in exponential notation.

Example 1.6.9   Identify each base and its exponent, and then write in expanded form.

a) $2b^4$

•SOLUTION•

- the base of the first factor is 2, and its exponent is 1.
- the base of the second factor is b, and its exponent is 4.
- written in expanded form $2b^4$ is $2 \cdot b \cdot b \cdot b \cdot b$.

b) $(2b)^4$

    • **SOLUTION** •

- the base is $2b$, and its exponent is 4.
- written in expanded form $(2b)^4$ is $(2b) \cdot (2b) \cdot (2b) \cdot (2b)$.

c) $2^4 b^4$

    • **SOLUTION** •

- the base of the first factor is 2, and its exponent is 4.
- the base of the second factor is b, and its exponent is 4.
- written in expanded form $2^4 b^4$ is $2 \cdot 2 \cdot 2 \cdot 2 \cdot b \cdot b \cdot b \cdot b$.

Practice 1.6.9   Identify each base and its exponent, and then write in expanded form.

1. y^7 2. $8z^5$ 3. $(4y)^3$ 4. $5^3 b^4$

Example 1.6.10   Identify each base and the number of factors, and then write in exponential form.

a) $xxyyy$

    • **SOLUTION** •

- one base is x, the other is y.
- there are 2 factors of x so its exponent is 2.
- there are 3 factors of y so its exponent is 3.
- written in exponential notation, we have $x^2 y^3$.

b) $(-3)(-3)(-3)(-3)aab$

    • **SOLUTION** •

- one base is -3, one base is a, one base is b.
- there are 4 factors of -3 so its exponent is 4.
- there are 2 factors of a so its exponent is 2.
- there is 1 factor of b so its exponent is 1 (but we usually don't write exponents of 1).
- written in exponential notation, we have $(-3)^4 a^2 b$.

Practice 1.6.10   Identify the base and the exponent, and then write using exponential notation

1. $3 \cdot 3 \cdot 3 \cdot 3$ 2. $xxyyyyy$ 3. $8 \cdot 8 \cdot 8 \cdot x \cdot x \cdot x \cdot x$

4. $2 \cdot 2 \cdot 2 \cdot 2 \cdot 2 \cdot 2 \cdot y$ 5. $(-3)(-3)(-3)(-3)mmmmmnnn$

Exercise Set 1.6   The answers to the odd numbered exercises are at the back of the book.

Make each sentence true by replacing the box with a number.

1. $\square + (-18) = -18$ 2. $2 + \square = 2$ 3. $8 \cdot \square = 8$ 4. $-24 \cdot \square = -24$

5. $\square + (-7.7) + 6z = 6z$ 6. $\square + 8.5 + 11w = 11w$ 7. $\square + \dfrac{4}{5} + 12n = 12n$ 8. $\square + \left(-\dfrac{7}{8}\right) + 10x = 10x$

9. $\square \cdot \left(\dfrac{5}{6}\right)(33) = 33$ 10. $\square \cdot \left(-\dfrac{7}{9}\right)(13) = 13$

Use the Commutative Property of Addition to rewrite the expression (write the terms in decreasing powers of the variable).

11. $7 - 13$ 12. $-4 - 18$ 13. $3x - 2x^2 + 5$ 14. $15 - 8x^2 - x^3$

Use the Commutative Property of Multiplication to rewrite the expression.

15. $12 \cdot 26$ 16. $8 \cdot 13$ 17. $r3$ 18. $u5$

19. $(x - 5)9$ 20. $(y + 7)3$ 21. $(2x + 3)(x + 9)$ 22. $(y - 5)(3y + 10)$

Use the Associative Property of Addition to rewrite the expressions.

23. $\frac{1}{16} + \left(\frac{15}{16} + 10\right)$ 24. $\left(-5 + \frac{1}{4}\right) + \frac{3}{4}$ 25. $(-2 + 6y) + 8y$ 26. $15x + (3x + 5)$

Use the Associative Property of Multiplication to rewrite the expressions.

27. $(6 \cdot 2) \cdot 3$ 28. $5 \cdot (7 \cdot 10)$ 29. $\frac{7}{9}(9y)$ 30. $22\left(\frac{1}{22}z\right)$

Use the Distributive Property of Multiplication Over Addition to rewrite each expression.

31. $12(y + 3)$ 32. $-5(x + 7)$ 33. $8(x - 2)$ 34. $6(x - 5)$ 35. $-3(y - 5x)$
36. $-5(x - 2y)$ 37. $-2(x + y - 6)$ 38. $-3(x - 5 - y)$ 39. $3(y - 2x + 6)$ 40. $8(x - y - 3)$

Factor. That is, use the Distributive Property to rewrite the sum as a product.

41. $3h + 6h$ 42. $4x + 8x$ 43. $8x - 2x$ 44. $-3w + 5w$
45. $2y - 3y + 5y$ 46. $6m - 5m - 11m$ 47. $rs - rt$ 48. $xy - wy$

Identify each base and its exponent, and then simplify.

49. 7^3 50. 2^6 51. $2 \cdot 4^3$ 52. $5 \cdot 3^3$
53. $(2 \cdot 4)^3$ 54. $(5 \cdot 3)^3$ 55. -6^2 56. -12^2

Identify the base and exponent, and then write using expanded form.

57. $9p^4$ 58. $4t^6$ 59. $(-8r)^3$ 60. $(-5s)^3$ 61. 4^2q^4
62. 7^3k^6 63. $-r^3$ 64. -5^2 65. $-(-t)^4$ 66. $-(-w)^6$

Identify each base and the number of its factors, and then write using exponential notation

67. $11 \cdot 11 \cdot 11$ 68. $13 \cdot 13 \cdot 13 \cdot 13 \cdot 13 \cdot 13$ 69. $pppppprr$ 70. $vvwwwwwww$
71. $(zzz)(zzzz)$ 72. $(ffffff)(f)$ 73. $-5 \cdot 5 \cdot 5 \cdot 5$ 74. $-2 \cdot 2 \cdot 2$
75. $(-3) \cdot (-3) \cdot (-3) \cdot (-3)$ 76. $(-8) \cdot (-8) \cdot (-8) \cdot (-8) \cdot (-8)$

Review Exercises   The answers to all of these exercises are at the back of the book.

77. Find the reciprocal: $-\frac{4}{5}$ 78. Rewrite as a product: $\frac{5x}{2}$ 79. Write the opposite: $-12x$

Translate to a mathematical model.

80. The quotient of a number and 6   81. A number decreased by 4.   82. Double $\frac{8}{9}$

Simplify.

83. $4[12 - (6 - 4)^2]$ 84. $\frac{64}{80} \cdot \frac{30}{15}$ 85. $\dfrac{\frac{15}{21}}{\frac{45}{49}}$ 86. $\frac{5}{6} + \frac{7}{8}$ 87. $12 - \left| -5 \right| + 2$

88. Evaluate $12 - y - xy$ when x is -2 and y is -3.   89. Reduce to lowest terms: $\frac{48}{12}$

90. Graph 0.2 on a number line. Use an interval from -0.5 to 0.5 in steps of 0.1.

Section 1.7 Simplifying Algebraic Expressions

At the end of a wonderful dinner, a father turned to his young son, sighed, and said "Gastronomic satiety admonishes me that I have reached the ultimate stage of deglutition consistent with dietetic integrity". "In other words", said the mother, "he's full." The father's and mother's verbal expressions are equivalent because they say the same thing, although one is quite a bit simpler and easier to understand than the other.

Like verbal expressions, by simplifying algebraic expressions we can sometimes make them easier to understand and use. Algebraic expressions that represent the same quantity for all possible values of the variables are called **equivalent expressions**.

In this section we will use the properties of real numbers to make expressions simpler. We begin with the idea of like terms.

Topic 1.7 A — Like Terms

If we add 5 apples to 3 apples we have 8 apples. If we add 5 hours to 3 hours we have 8 hours. If we add 5 apples to 3 hours we have ..., well we have 5 apples and 3 hours. The point is we can add things that are *like* each other but we cannot add things that are *unlike*. The same is true in algebra, but we need a precise definition of what *like* means. **Like terms** are terms [†] that have identical variable factors (letters and exponents). For example:

- $5q$ and $-3q$ are like terms because both have the variable q raised to the first power. It's true the constants 5 and -3 are different, but the definition is concerned only with variables and exponents.

- $6x^2y^3$ and $2x^2y^3$ are like terms because they both have factors of x squared and y cubed.

- $6x^2y$ and $2yx^2$ are also like terms because they have the same variables raised to the same powers. The definition does not mention the *order* of the factors.

- $3wt^2$ and $3w^2t$ are **not** like terms. They have identical variables but the t is squared in $3wt^2$ and it's not squared in $3w^2t$. Also, w is squared in the second term but not the first.

Example 1.7.1   Group the like terms: $2m^3n$, $2nm^2$, $-3nm^3$, $7m^2n^2$, $\dfrac{nm^2}{4}$, $0.75m^2n$

•**SOLUTION**• Like terms have the same variable factors. So,

- $2m^3n$ and $-3nm^3$ are like terms.

- $2nm^2$, $\dfrac{nm^2}{4}$, and $0.75m^2n$ are like terms.

- $7m^2n^2$ is not "like" any of the other terms.

Practice 1.7.1 Group the like terms

1. $2x^2$, $5x$, $3x^2$, $2y$

2. $5y$, $\dfrac{3}{2}y$, $3y^2$

3. $2x^2y^4$, $0.07y^3x^2$, $\dfrac{x^2y^4}{\pi}$, $-x^2y^3$, $7x^2y^4$, $-2x^3y^2$

4. $5x^5y^2$, $1.25y^2x^5$, $-9x^2y^5$, $1.1x^2y^5$, $10x^3y^2$

5. $-2mn^2$, $5m^2n$, $10m^2n^2$, $3.1m^2$, $4n^2$, $6mn$

6. $7r^2s$, $-4rs^2$, $15r^2s^2$, $6r^2$, $2s^2$, $8sr^2$

[†] Terms may be constants, variables, or products. Terms either stand alone, or they are separated from each other by the operation of addition.

Topic 1.7 B — Adding and Subtracting Like Terms

When we add like terms, as in $7x + 2x$, we simply add the coefficients to get $9x$. Mathematically, adding like terms is based on the Distributive Property, $a(b + c) = ab + ac$. For example,

$7x + 2x$ $\Leftarrow$ Expression to simplify.

$(7 + 2)x$ $\Leftarrow$ Used the Distributive Property to rewrite the sum as a product (that is, we factored out the x).

$9x$ $\Leftarrow$ Added.

To save time, we usually don't write out all these steps.

Note that since we can only add like terms it is not possible to simplify expressions such as $2x + 3y$.

— *Procedure* —	**Step 1** Decide which terms are like.
Adding Like Terms	**Step 2** Add the coefficients of the like terms.

Adding and subtracting like terms is also known as **combining like terms**.

Example 1.7.2   Simplify. That is, combine like terms to form the simplest expression possible.

a) $5x + 3x - 2x$

    •SOLUTION• The terms are $5x$, $3x$, and $-2x$. They are *like* because they have identical variable factors (x to the first power). We add the coefficients to get $6x$. Note that the expressions $6x$ and $5x + 3x - 2x$ are equivalent but it's simpler and faster to write $6x$.

b) $5m^4 - 3m^4 - m^4$

    •SOLUTION• The terms are $5m^4$, $-3m^4$, and $-1m^4$. All three terms have identical variable factors so we add their coefficients to get $1m^4$, which we would write as m^4.

c) $2x - x^2 + 6x + 4x^2$

    •SOLUTION• The terms are $2x$, $-1x^2$, $6x$, and $4x^2$. We have two groups of like terms involving x and x^2. We add the coefficients of the x's and then we add the coefficients of the x^2's to get $8x + 3x^2$. We will write the final answer with decreasing powers of x as $3x^2 + 8x$.

Practice 1.7.2   Simplify. That is, combine like terms to form the simplest expression possible.

1. $6m + 3m$ 2. $-6r + 8r$ 3. $7w - 15w$ 4. $4s - s$

5. $-8x^3 + 5x^3 - 12x^3$ 6. $-15a^2 + 7a^2 - 14a^2$ 7. $8x - 2y + 3y - x$ 8. $9xy^2 + 3x^2y - 7xy^2 + 2x^2y$

Topic 1.7 C — Simplifying Expressions Containing Grouping Symbols

When grouping symbols are present, we follow the order of operations and simplify inside the grouping symbols first. Then we remove the grouping symbols and simplify again.

Example 1.7.3   Simplify.

a) $2x + (3x - 7x)$

    •SOLUTION•

$2x + (\mathbf{-4x})$ $\Leftarrow$ Simplified inside parentheses by combining like terms.

$-2x$ $\Leftarrow$ Combined like terms.

b) $8p - (3p - 7p)$

   •SOLUTION•

$8p - (\mathbf{-4p})$ $\Leftarrow$ Simplified inside parentheses by combining like terms.

$8p + \mathbf{4p}$ $\Leftarrow$ Changed subtraction to addition of the opposite.

$12p$ $\Leftarrow$ Combined like terms.

c) $-(5y - 7y)$

   •SOLUTION•

$-(\mathbf{-2y})$ $\Leftarrow$ Simplified inside parentheses by combining like terms.

$2y$ $\Leftarrow$ Wrote the opposite of $-2y$.

Practice 1.7.3   Simplify.

1. $4x + (2x - 12x)$ 2. $-8y + (3y - 6y)$ 3. $5w - (4w - 6w)$ 4. $-7m - (4m - 3m)$

5. $(6x + 4x) - (3x + 7x)$ 6. $(5m + 7m) - (m - 8m)$ 7. $(5r - r) - (9r - 6r)$ 8. $(5n + 2n) - (6n - 15n)$

If the terms inside the grouping symbols cannot be combined we must use the Distributive Property of Multiplication Over Addition to remove the grouping symbols.

Example 1.7.4   Simplify.

a) $5w + 2(w + 4)$

   •SOLUTION• Distribute the 2 to the w and the 4 and then combine like terms.

$5w + 2(w + 4)$ $\Leftarrow$ Expression to simplify.

$5w + \mathbf{2}(w) + \mathbf{2}(4)$ $\Leftarrow$ Distributed the 2.

$5w + \mathbf{2w} + \mathbf{8}$ $\Leftarrow$ Multiplied.

$\mathbf{7w} + 8$ $\Leftarrow$ Combined like terms.

b) $-2(3x + 4)$

   •SOLUTION•

$(\mathbf{-2})(3x) + (\mathbf{-2})(4)$ $\Leftarrow$ Distributed the -2.

$\mathbf{-6}x + (\mathbf{-8})$ $\Leftarrow$ Multiplied.

$-6x - 8$ $\Leftarrow$ Wrote as a subtraction (optional).

c) $-4(5y - 3)$

   •SOLUTION•

$(\mathbf{-4})(5y) - (\mathbf{-4})(3)$ $\Leftarrow$ Distributed the -4.

$\mathbf{-20}y - (\mathbf{-12})$ $\Leftarrow$ Multiplied.

$-20y + 12$ $\Leftarrow$ Wrote as an addition.

| — Caution —
Distributing a negative | When distributing a negative factor be sure to keep the signs straight. The terms inside the parentheses change sign because they are being multiplied by a negative. Thus, $-4(5y - 3)$ may be written as $-20y + \mathbf{12}$ but NOT $-20y - \mathbf{12}$. |

d) $-2(3x(5))$

•SOLUTION• We can't use the distributive property here because there is no addition inside the parentheses. We follow the normal order of operations and do the work inside the parentheses first.

$-2(3x(5))$ ⟸ Expression to simplify.

$-2(\mathbf{15}x)$ ⟸ Multiplied inside the parentheses.

$-30x$ ⟸ Multiplied.

Practice 1.7.4 Simplify.

1. $5x + 3(x + 3)$ 2. $3m + 6(m + 6)$ 3. $4y + 3(2y - 7)$ 4. $2(3x)$ 5. $-12(3x + 8)$

6. $-5(4x + 4)$ 7. $-3(2x - 5)$ 8. $6x - 3(5x)$ 9. $1.5(6x + 10)$ 10. $\frac{1}{3}(6x - 9)$

A negative sign to the left of a set of parentheses can be interpreted as a –1 times the parentheses. For example,

- $-(-3x)$ can be written as $(\mathbf{-1})(-3x)$, which simplifies to $3x$. This confirms that the opposite of $-3x$ is $3x$.

- $-(2x + 5)$ can be written as $(\mathbf{-1})(2x + 5)$. We can distribute the -1 to each term inside the parentheses to get $-2x + (-5)$, which we write more compactly as $-2x - 5$.

The negative sign has the effect of changing the signs of each term inside the parentheses. This will always be the case, so we usually do this type of problem mentally.

Example 1.7.5 Simplify.

a) $-(-6y)$

•SOLUTION• Change the sign of the term inside the parentheses to get $6y$.

b) $-(3x^2 + 5x - 1)$

•SOLUTION• Change the sign of each term inside the parentheses to get $-3x^2 - 5x + 1$.

Practice 1.7.5 Simplify.

1. $-(-4x)$ 2. $-(6m^3 + 3m^2 - 4m)$ 3. $-(-3n^2 - 2n + 5)$ 4. $-(-2x^2 - 3x - 5)$

Special care must be taken removing grouping symbols when subtraction is involved. For example, $12x - 3(x - 2)$ can be simplified as follows:

$12x - 3(x - 2)$ ⟸ Expression to simplify.

$12x + (\mathbf{-3})(x + (\mathbf{-2}))$ ⟸ Converted subtraction into addition of the opposite.

$12x + (\mathbf{-3})(x) + (\mathbf{-3})(-2)$ ⟸ Distributed -3 to each term inside parentheses.

$12x + (\mathbf{-3}x) + (\mathbf{6})$ ⟸ Multiplied.

$9x + 6$ ⟸ Combined like terms.

In practice, we usually do most of the above steps in our heads. This is much faster but you run the risk of making an error with a sign. Practice this type of problem until you become automatic at the process.

Example 1.7.6   Simplify.

a) $4x - 2(3x + 8)$

•SOLUTION• Mentally, distribute –2 to the terms inside the parentheses and then simplify:

$$4x - 2(3x + 8) \quad \Leftarrow \text{ Expression to simplify.}$$

$$4x - \mathbf{6x} - \mathbf{16} \quad \Leftarrow \text{ Distributed } -2.$$

$$\mathbf{-2x} - 16 \quad \Leftarrow \text{ Combined like terms.}$$

b) $2w - 3(6 - 4w)$

•SOLUTION• Mentally, distribute –3 and then simplify:

$$2w - 3(6 - 4w) \quad \Leftarrow \text{ Expression to simplify.}$$

$$2w - \mathbf{18} + \mathbf{12w} \quad \Leftarrow \text{ Distributed } -3. \text{ Notice the 12 is positive since it is the product of two negatives, } (-3)(-4).$$

$$\mathbf{14w} - 18 \quad \Leftarrow \text{ Combined like terms.}$$

c) $8t - (4t^2 - 3t - 7)$

•SOLUTION• Mentally, distribute the negative sign (really a –1) and then simplify:

$$8t - (4t^2 - 3t - 7) \quad \Leftarrow \text{ Expression to simplify.}$$

$$8t - \mathbf{4t^2} + \mathbf{3t} + \mathbf{7} \quad \Leftarrow \text{ Distributed the negative sign.}$$

$$\mathbf{11t} - 4t^2 + 7 \quad \Leftarrow \text{ Combined like terms.}$$

$$-4t^2 + \mathbf{11t} + 7 \quad \Leftarrow \text{ Wrote terms in decreasing powers of the variable. We don't have to write the expression in decreasing powers of the variable, but we usually do.}$$

d) $-(2r + 3) - (5r - 9)$

•SOLUTION• Mentally, distribute both negative signs and then simplify:

$$-(2r + 3) - (5r - 9) \quad \Leftarrow \text{ Expression to simplify.}$$

$$\mathbf{-2r} - \mathbf{3} - \mathbf{5r} + \mathbf{9} \quad \Leftarrow \text{ Distributed negative signs. Be sure you understand where the signs of each term came from.}$$

$$\mathbf{-7r} + \mathbf{6} \quad \Leftarrow \text{ Combined like terms.}$$

Practice 1.7.6 Simplify.

1. $5x - 6(3x + 7)$ 2. $2n - 7(4n + 8)$ 3. $3r - 12(5 - 2r)$ 4. $5t - 3(7 - 4t)$

5. $4x - (2x^2 + 3x - 6)$ 6. $4n - (-3n^2 - 2n + 5)$ 7. $-(4x + 5) - (3x - 3)$ 8. $-(2y + 9) - (4y - 2)$

If we have one set of grouping symbols inside another we say the grouping symbols are nested. With nested grouping symbols we begin with the inner–most group and work outward.

Example 1.7.7   Simplify $x + 3[6x + 2(x + 4) - 5]$

$$x + 3[6x + \mathbf{2x} + \mathbf{8} - 5] \quad \Leftarrow \text{ Distributed 2 to remove inner–most grouping symbol.}$$

$$x + 3[\mathbf{8x} + \mathbf{3}] \quad \Leftarrow \text{ Combined like terms inside brackets.}$$

$$x + \mathbf{24x} + \mathbf{9} \quad \Leftarrow \text{ Distributed 3 to remove brackets.}$$

$$\mathbf{25x} + 9 \quad \Leftarrow \text{ Combined like terms.}$$

Practice 1.7.7 Simplify.

1. $x - 6[3x + (7x - 1) + 2]$ 2. $5x - 3[2 + 2(x + 1) + 4x]$ 3. $2x - [2x - (2x + 3) - 4]$ 4. $5x - [5x + 2(x + 3) + 3]$

5. $5x + [3x - (2x + 4x)]$ 6. $8w + [4w - (5w - 7w)]$ 7. $-3[2x + 4(x - 1)]$ 8. $-2[5x + 3(x - 4)]$

9. $5[3x - 4(2x + 1)]$ 10. $2[5x - 3(3x + 1)]$

Exercise Set 1.7   The answers to the odd numbered exercises are at the back of the book.

Simplify.

1. $-4m + 7m$
2. $-12x + 25x$
3. $-5mn + 2mn - 13mn$
4. $-5ab - 3ab + 11ab$

5. $1.2xy + 2.4xy - 5.7xy$
6. $-7.5ab + 3.4ab - 1.2ab$
7. $x - \frac{1}{2}x + \frac{3}{4}x$
8. $\frac{1}{3}x - x - \frac{2x}{5}$

9. $8x + (5x - 11x)$
10. $7r + (5r - 12r)$
11. $-2w + (4w - 5w)$
12. $4t + (5t - 7t)$

13. $(-5x + 3x) - (8x + 12x)$
14. $(-12y + 4y) - (7y + 3y)$
15. $5x + [3x - (2x + 4x)]$
16. $8w + [4w - (5w - 7w)]$

17. $-12m + 7(2m - 6)$
18. $-5y + 3(3y - 4)$
19. $-8(2y - 5)$
20. $-5(3m - 2)$

21. $2.5(4x + 6)$
22. $-1.2(5y + 2)$
23. $-\frac{1}{2}(6x - 2)$
24. $\frac{2}{3}(12 - 9x)$

25. $-(-12y)$
26. $-(-\frac{1}{2}y)$
27. $-(x^2 - 3x + 5)$
28. $-(-2n^2 + 3n - 6)$

29. $2w - 3(2w + 2)$
30. $15t - 3(3t + 1)$
31. $3r - (2r^2 + 5r - 3)$
32. $5t - (3t^2 - 6t - 9)$

33. $(2x + 5) - 3(x - 7)$
34. $(8x - 1) - 2(3x - 9)$
35. $(4t^2 + 3t - 8) - 2(3t^2 + 5t - 5)$

36. $(3x^2 - 2x + 7) - 4(x^2 - 5x - 6)$
37. $x + 4(2x + 6(x + 1) - 8)$

38. $y + 6(3y + 9(y + 3) - 12)$ 39. $x - 2(5x + 3(x - 8) + 7)$
40. $y - 4(2y + 5(y - 6) + 15)$

41. $3x - 2(4x - 3(x - 5) + 6)$ 42. $5y - 3(2y - 4(y + 8) - 2)$ 43. $-2[-4x - (1 - 4x)]$ 44. $-3[x - 2(5 - 3x)]$

Review Exercises   The answers to all of these exercises are at the back of the book.

45. Find the reciprocal of -6 46. Rewrite as a product: $\frac{w}{6}$ 47. Write the opposite: 0 48. Factor: $7x + 2x$

Simplify.

49. $2^3 \cdot 8 \div 4$
50. $39 - 2[3 + 2^2(5 - 1)]$
51. $\frac{42}{28} \cdot \frac{6}{24}$
52. $8 - |-6| \div 2 + 3 \cdot 4$

Translate to a mathematical model.

53. Three percent of 9.
54. A number divided by 3.
55. The difference of a number and 5.

56. Evaluate $3x - y$ when x is -2 and y is -3.
57. Name the property that justifies the statement: $5 + 2 = 2 + 5$

58. Identify each base and its exponent: $3w^5$

Chapter 1 Review

Vocabulary to Know

absolute value — the distance of a number from zero on the number line.

additive identity — the number 0.

additive inverse — same as opposite.

base — in exponential notation, the number being multiplied; in 3^2, 3 is the base.

coefficient — a constant multiplied by a variable; in $2x$, 2 is the coefficient of x.

combining like terms — adding and subtracting the coefficients of terms that have identical variable parts.

composite number — a whole number, greater than 1, which is not prime; examples include 4, 6, 8, 9.

coordinate — the location of a point on a number line.

Counting Numbers — the set of numbers 1, 2, 3, ...; same as natural numbers.

difference — the combination of two numbers using subtraction.

divisible — one whole number is divisible by another if the division results in a remainder of 0.

element — an object contained in a set; same as member; 8 is an element of the Whole Numbers.

equivalent expressions — expressions that represent the same quantity for all values of the variables; $2x + 3x$ and $5x$ are equivalent expressions.

evaluate an expression — to substitute specific numbers for letters in an expression and then simplify.

expanded form — form of a product where repeated bases are written explicitly — xxx is the expanded form of x^3.

exponent — in exponential notation, the number of bases being multiplied; in 3^2, 2 is the exponent and tells us to multiply $3 \cdot 3$.

exponential notation — a shorthand way of writing repeated multiplication, as in 3^2.

expression — a collection of numbers, operations, letters and grouping symbols such as x, $5y$, and $3x + 2$.

factor — a number being multiplied; in $2x$, both 2 and x are factors; in $5(3x - 7)$ both 5 and $(3x - 7)$ are factors.

factoring — rewriting an expression as a product; a factored form of 6 is $2 \cdot 3$.

fraction — a number that can be written in the form $\frac{a}{b}$, where a and b are real numbers and $b \neq 0$.

GCF — same as Greatest Common Factor.

graph — a visual representation of a set of numbers.

greatest common factor — largest factor common to a set of given numbers; same as GCF; the GCF of 4, 10, and 12 is 2.

grouping symbols — symbols used to group terms of an expression; grouping symbols include parentheses ( ), brackets [ ], the fraction bar, absolute value bars | |, and the radical symbol, $\sqrt{}$.

inequality — a statement that one expression is less than (<), less than or equal to (≤), greater than (>), or greater than or equal to (≥) another.

Integers — the set of numbers consisting of the negatives of the natural numbers, 0, and the natural numbers; examples include –3, –2, –1, 0, 1, 2, 3.

Irrational Numbers — the set of numbers consisting of numbers that cannot be written as rational numbers; examples include π and $\sqrt{2}$.

like terms — terms that have identical variable factors; examples are $5x$ and $\frac{2}{3}x$.

mathematical model — a representation that uses mathematics to describe an object, an operation, a procedure, a relation, etc..

member — an object contained in a set; same as element.

multiplicative identity — the number 1.

multiplicative inverse — same as reciprocal.

Natural Numbers — the set of numbers 1, 2, 3,...; same as counting numbers.

negative integers — the set of numbers consisting of –1, –2, –3, and so on.

negative numbers — numbers that are less than zero.

nested grouping symbols — multiple grouping symbols where one set is completely inside another as in $2[5 - (3 - 2)]$.

non-negative number — a number that is greater than or equal to 0.

non-zero — any number except 0.

number line — a horizontal line with unit markings used to illustrate the ordering of real numbers.

operation — a process, such as addition, that transforms one or more numbers.

operator — a symbol used to indicate an operation, such as + for addition.

opposite — The opposite of a number has the same absolute value as the number but a different sign; the opposite of 2 is –2; the sum of two opposites is 0; same as additive inverse.

origin — the location of 0 on a number line.

plotting — locating a point on a number line.

positive integers — the set of numbers consisting of 1, 2, 3, and so on.

positive numbers — numbers that are greater than zero.

prime factor — (verb) to write a composite number as a product of prime numbers.

prime factor — (noun) a factor which is prime.

prime number — a whole number, greater than 1, which is divisible only by 1 and itself; the first fifteen prime numbers are 2, 3, 5, 7, 11, 13, 17, 19, 23, 29, 31, 37, 41, 43, and 47.

product — the combination of two numbers using multiplication.

quotient — the combination of two numbers using division.

Rational Numbers — the set of numbers consisting of numbers that can be written as the ratio of two integers, where the denominator is not 0.

Real Numbers — the set of numbers consisting of the rational and irrational numbers.

reciprocals — two numbers whose product is 1, such as $\frac{3}{2}$ and $\frac{2}{3}$; to transform a number into its reciprocal exchange its numerator and denominator; same as multiplicative inverse.

set — a collection of objects.

signed numbers — numbers that are written with either a + or a − sign.

simplify — to perform some or all of the allowable operations in an expression.

sum — the combination of two numbers using addition.

symbols —

π Greek letter pi, represents the ratio of the circumference to the diameter of a circle (about 3.14).

{ } braces, indicate a set.

+ addition operator, indicates two numbers are to be added.

+ positive sign, indicates a number is greater than 0.

− subtraction operator, indicates two numbers are to be subtracted.

− negative sign, indicates a number is less than 0.

− opposite sign, indicates the opposite of a number.

− fraction bar, indicates a fraction or the operation of division (also a grouping symbol).

÷ division operator, indicates that two numbers are to be divided.

/ division operator, indicates that two numbers are to be divided.

/ slant fraction bar, indicates a fraction.

× St. Andrew's cross, indicates the operation of multiplication.

• dot, indicates the operation of multiplication.

= equal sign, indicates that two expressions represent the same value.

≠ not equal sign, indicates that two expressions do not represent the same value.

( ) parentheses, indicate multiplication as in 3(5).

( ) parentheses, indicate a grouping as in 2 − (1 + 5).

[ ] brackets, indicate a grouping.

> greater than, indicates the number on the left is larger than the number on the right as in 5 > 3.

≥ greater than or equal to, indicates the number on the left is larger than the number on the right or is equal to the number on the right, as in 5 ≥ 3.

< less than, indicates the number on the left is smaller than the number on the right as in 3 < 5.

≤ less than or equal to, indicates the number on the left is smaller than the number on the right or is equal to the number on the right, as in 3 ≤ 5.

| | absolute value symbol, indicates the distance of a number from 0 on the number line (also a grouping symbol).

undefined — the result of an operation that is not permitted (for example, division by 0 is undefined).

unit length — the distance on a number line between each integer.

variable — a symbol, usually a letter, that is used to represent something, usually an unknown quantity or a quantity that can take on more than one value.

Whole Numbers — the set of numbers 0, 1, 2, 3, ….

Properties, Definitions, Formulas, and Notation to Understand

Addition Property of 0 — The sum of 0 and any number is identical to that number.	$0 + a = a$
Multiplication Property of 0 — The product of 0 and any number is 0.	$0 \bullet a = 0$
Multiplication Property of 1 — The product of 1 and any number is identical to that number.	$1 \bullet a = a$
Commutative Property of Addition — The order in which we add terms does not affect their sum.	$a + b = b + a$
Commutative Property of Multiplication — The order in which we multiply factors does not affect their product.	$a \bullet b = b \bullet a$ $2x$ is equivalent to $x2$
Associative Property of Addition — The way in which terms are grouped does not affect their sum.	$a + (b + c) = (a + b) + c$ $(3 + 2x) + 9x$ is equivalent to $3 + (2x + 9x)$

Associative Property of Multiplication — The way in which factors are grouped does not affect their product.	$a(bc) = (ab)c$ $2x \cdot (9x \cdot 3)$ is equivalent to $(2x \cdot 9x) \cdot 3$
Distributive Property of Multiplication Over Addition — The product of a single factor and the sum of two terms can be rewritten as the sum of two products.	$a(b + c) = ab + ac$ $2(y + 4)$ is equivalent to $2 \cdot y + 2 \cdot 4$
reciprocal (multiplicative inverse) — two numbers are reciprocals if their product is 1.	If $a \neq 0$, a and $\dfrac{1}{a}$ are reciprocals the reciprocal of $-\dfrac{1}{2}$ is $-\dfrac{2}{1}$ (or -2).
opposite (additive inverse) — two numbers are opposites if their sum is 1.	a and $-a$ are opposites the opposite of $-a$, which may be written as $-(-a)$, is a the opposite of $-\dfrac{1}{2}$ is $\dfrac{1}{2}$
Notation for fractions with one negative sign	If $b \neq 0$, then $\dfrac{-a}{b} = \dfrac{a}{-b} = -\dfrac{a}{b}$ $\dfrac{-2}{3}$ and $\dfrac{2}{-3}$ and $-\dfrac{2}{3}$ all represent the same number
Notation for fractions with two negative signs	If $b \neq 0$, then $\dfrac{-a}{-b} = \dfrac{a}{b}$ $\dfrac{-8}{-5}$ and $\dfrac{8}{5}$ represent the same number

Key words for translating verbal models into mathematical models
 	
  Multiplication — see the table on page 136
  Division — see the table on page 140
  Addition — see the table on page 144
  Subtraction — see the table on page 150
	Double x may be written as $2x$
	
	The quotient of y and 3 may be written as $y \div 3$
	
	The sum of a number and 6 may be written as $x + 6$
	
	A number decreased by 2 may be written as $x - 2$

Procedures to Follow

Procedure for simplifying expressions using the proper order of operations	Simplify $6 - 2^2(15 - 12/2 \cdot 3)$	Simplify $8 + 72 \div 2 \cdot 3 - \left| -5 + 3 \right|$
Step 1 Simplify expressions inside grouping symbols.	$6 - 2^2(15 - \mathbf{6} \cdot 3)$ $6 - 2^2(15 - \mathbf{18})$ $6 - 2^2(\mathbf{-3})$	$8 + 72 \div 2 \cdot 3 - \left| \mathbf{-2} \right|$
Step 2 Simplify absolute values, exponents, and square roots.	$6 - \mathbf{4}(-3)$	$8 + 72 \div 2 \cdot 3 - \mathbf{2}$
Step 3 Simplify multiplication and division, working left to right.	$6 - (\mathbf{-12})$	$8 + \mathbf{36} \cdot 3 - 2$ $8 + \mathbf{108} - 2$
Step 4 Simplify addition and subtraction, working left to right.	18	114

Procedure for prime factoring	Prime factor 1144
Step 1 Divide the given number by the smallest prime number, 2. If a remainder results, go to Step 3.	$2\overline{)1144}$ with quotient 572
Step 2 Divide the quotient from the previous step by 2. Keep dividing the quotients by 2 until 2 no longer works (that is, until the division yields a remainder other than 0).	$2\overline{)286}$ $2\overline{)572}$ $2\overline{)1144}$
Step 3 Divide the quotient from the previous step by the next prime number, 3. Keep dividing the quotients by 3 until it no longer works.	$3\overline{)143}$ $2\overline{)286}$ $2\overline{)572}$ $2\overline{)1144}$
Step 4 Continue the process of dividing by prime numbers (5, 7, 11, ...) until the quotient is a prime number.	$11\ 3\ 3\ 3\overline{)143}$ quotient 13 $2\overline{)286}$ $2\overline{)572}$ $2\overline{)1144}$
Step 5 Write the prime factorization as the product of all the divisors and the final quotient.	The prime factorization of 1144 is $2 \cdot 2 \cdot 2 \cdot 11 \cdot 13$

Procedure for reducing fractions	Reduce $\dfrac{252}{210}$
Step 1 Prime factor Prime factor the numerator and the denominator.	$\dfrac{2 \cdot 2 \cdot 3 \cdot 3 \cdot 7}{2 \cdot 3 \cdot 5 \cdot 7}$
Step 2 Reduce all factors common to the numerator and the denominator.	$\dfrac{\cancel{2} \cdot 2 \cdot \cancel{3} \cdot 3 \cdot \cancel{7}}{\cancel{2} \cdot \cancel{3} \cdot 5 \cdot \cancel{7}}$, which simplifies to $\dfrac{6}{5}$

Procedure for multiplying fractions	Simplify $\dfrac{12}{15} \cdot \dfrac{25}{18}$
Step 1 Reduce Reduce factors common to the numerators and denominators.	$\dfrac{\cancel{2} \cdot 2 \cdot \cancel{3}}{3 \cdot \cancel{5}} \cdot \dfrac{\cancel{5} \cdot 5}{\cancel{2} \cdot \cancel{3} \cdot 3}$
Step 2 Multiply Multiply the remaining factors in the numerators and multiply the remaining factors in the denominators.	$\dfrac{10}{9}$

Procedure for dividing fractions	Simplify $\dfrac{4}{21} \div \dfrac{10}{45}$
Step 1 Reciprocal Rewrite the division as the multiplication of the numerator by the reciprocal of the denominator.	$\dfrac{4}{21} \cdot \dfrac{45}{10}$
Step 2 Reduce Reduce factors common to the numerators and denominators.	$\dfrac{\cancel{2} \cdot 2}{\cancel{3} \cdot 7} \cdot \dfrac{\cancel{3} \cdot 3 \cdot \cancel{5}}{\cancel{2} \cdot \cancel{5}}$
Step 3 Multiply Multiply the remaining factors in the numerators and multiply the remaining factors in the denominators.	$\dfrac{6}{7}$

Procedure for finding the LCD	Find the LCD of $\dfrac{5}{12}$ and $\dfrac{7}{30}$
Step 1 Prime factor Prime factor each denominator and write it in exponential form.	$12 = 2^2 \cdot 3^1$ $30 = 2^1 \cdot 3^1 \cdot 5^1$

Step 2 Select For each prime factor, select the one with the largest power.	largest power of 2 is 2^2 largest power of 3 is 3^1 largest power of 5 is 5^1
Step 3 Multiply Multiply the selected factors. This is the LCD	LCD is $2^2 \cdot 3^1 \cdot 5^1$, which is 60

Procedure for building equivalent fractions	Build a fraction equivalent to $\dfrac{5}{6}$ but which has a denominator of 84
Step 1 Find Needed Factors Determine what the denominator of the original fraction must be multiplied by to transform it into the desired number.	Divide 84 by 6 to get 14.
Step 2 Multiply Multiply BOTH the numerator and the denominator of the original fraction by the factors found in Step 1.	$\dfrac{5}{6} \cdot \dfrac{14}{14}$, which can be written as $\dfrac{70}{84}$

Procedure for adding and subtracting fractions with different denominators	Simplify $\dfrac{5}{12} + \dfrac{7}{30}$
Step 1 Build LCD Build the Least Common Denominator (LCD).	$12 = 2^2 \cdot 3^1$ $30 = 2^1 \cdot 3^1 \cdot 5^1$ LCD is $2^2 \cdot 3^1 \cdot 5^1$, which is 60
Step 2 Build Equivalent Fractions Build equivalent fractions that contain the LCD.	$\dfrac{5 \cdot \mathbf{5}}{12 \cdot \mathbf{5}} + \dfrac{7 \cdot \mathbf{2}}{30 \cdot \mathbf{2}}$ $\dfrac{25}{60} + \dfrac{14}{60}$
Step 3 Combine Numerators Add or subtract the numerators and use the LCD as the denominator.	$\dfrac{39}{60}$
Step 4 Reduce Reduce if possible.	Reduce common factor of 3 to get $\dfrac{13}{20}$

Procedure for multiplying two real numbers	Simplify $(-2)(-3)$	Simplify $(5)(-2)$
Step 1 Multiply the absolute values.	$\lvert -2 \rvert \cdot \lvert -3 \rvert$ $2 \cdot 3$ 6	$\lvert 5 \rvert \cdot \lvert -2 \rvert$ $5 \cdot 2$ 10
Step 2 Attach the proper sign: • If the factors had different signs, the product is negative. • If the factors had the same sign, the product is positive.	$+6$	-10

Procedure for dividing two real numbers	Simplify $12 \div (-2)$	Simplify $-8/(-4)$
Step 1 Divide the absolute values.	$\lvert 12 \rvert \div \lvert -2 \rvert$ $12 \div 2$ 6	$\lvert -8 \rvert / \lvert -4 \rvert$ $8 / 4$ 2
Step 2 Attach the proper sign (same as for multiplication)	-6	$+2$

Procedure for adding two real numbers that have the same sign	Simplify $-2 + (-3)$	
Step 1 Add the absolute values.	$\left\lvert -2 \right\rvert + \left\lvert -3 \right\rvert$ $2 + 3$ 5	
Step 2 Attach the common sign to the sum.	-5	

Procedure for adding two real numbers that have different signs	Simplify $-2 + 8$	Simplify $1 + (-6)$
Step 1 Subtract the smaller absolute value from the larger.	$\left\lvert 8 \right\rvert - \left\lvert -2 \right\rvert$ $8 - 2$ 6	$\left\lvert -6 \right\rvert - \left\lvert 1 \right\rvert$ $6 - 1$ 5
Step 2 Attach the sign of the original number that has the larger absolute value.	$+6$	-5

Procedure for subtracting two real numbers	Simplify $5 - 8$	Simplify $-3 - 5$	Simplify $-8 - (-2)$
Step 1 Change the operation of subtraction into the operation of addition AND change the number being subtracted to its opposite.	$5 + (-8)$	$-3 + (-5)$	$-8 + 2$
Step 2 Follow the procedure for addition.	-3	-8	-6

Procedure for adding like terms	Simplify $5x + 2y + 6x$
Step 1 Decide which terms are like.	$5x$ and $6x$ are like terms because they have the same variable parts.
Step 2 Add the coefficients of the like terms.	$\mathbf{11x + 2y}$

Chapter 1 Review Exercises   The answers to all the exercises are in the back of the book.

1. Identify sums, terms, products, factors: $-8rs$

2. Identify sums, terms, products, factors: $3w + 5p$

3. Prime factor: 392

4. Prime factor: 572

5. Reduce: $\dfrac{128}{312}$

6. Reduce: $\dfrac{118}{48}$

7. Write the reciprocal: $\dfrac{5}{7}$

8. Write the reciprocal: -2

9. Write the reciprocal: $\dfrac{3m}{5}$

10. Write as a product: $\dfrac{5b}{9}$

Simplify.

11. $(18 - 5 \cdot 2)^2$

12. $72/3^2 \cdot 5$

13. $10 - [(8 - 5)^2 - 12/(3 \cdot 4)]$

14. $8 + 2(35 - 2(24/6)^2 - 3]$

15. $\dfrac{40}{62} \cdot \dfrac{24}{30}$

16. $\dfrac{18}{70} \cdot \dfrac{60}{30}$

17. $\dfrac{90}{12} \div \dfrac{10}{15}$

18. $\dfrac{45}{56} \div \dfrac{27}{35}$

19. $\dfrac{9}{16} + \dfrac{5}{12}$

20. $\dfrac{6}{25} + \dfrac{7}{30} + \dfrac{8}{75}$

21. $\dfrac{4 \cdot 3^2 - 2^3}{2 \cdot 6 - 5}$

22. $12 - \dfrac{30 - \frac{1}{3} \cdot 12}{3 - \frac{2}{5}}$

23. $\dfrac{3}{2} + \dfrac{18}{5}\left(\dfrac{5}{6}\right)^2$

24. $\dfrac{30 + 5^2 \cdot 2}{3 \cdot 7 - 5}$

25. $12 - \left\lvert -8 \right\rvert \div 4 + 2 \cdot 5$

26. $\left\lvert 2 - 11 \right\rvert \cdot 2 - \left\lvert 12 \right\rvert \div 3$

27. -3^4

28. $-(x - 5x) + 3x$

29. $8x - 12x^2 - 13x + 19x^2$

30. $-(7x - x) + 3x$

31. $4x - (3x + 2)$

32. $-(3x - 2) - (5 - x)$

33. $z + 3[4z - 3(2z + 1)]$

34. $m + 4[5m - 2(5m + 1)]$

---

35. Plot on a number line: $\{-6, \frac{1}{2}, 4.3, 0, -2.7\}$

36. Write the opposite: -12.6

37. Write the opposite: $\frac{5}{8}$

38. Find the coordinate for each point shown:

    $\begin{array}{c} \text{-8 -7 -6 -5 -4 -3 -2 -1 0 1 2 3 4 5 6 7 8} \end{array}$

39. Graph -420 on a number line. Use an interval from -800 to 800 in steps of 100.

40. Graph 0.8 on a number line. Use an interval from -1 to 1 in steps of 0.2.

Translate to a mathematical model and simplify.

41. Eight percent of 12

42. Double a number

43. A number divided by -8.

44. The quotient of twice a number and 3

45. A number increased by 5

46. Evaluate $8 - (x - y)$ when x is -2 and y is -3

47. Evaluate $y - xy^2 - 3x$ when x is -2 and y is -3

48. Negative two subtracted from a number

49. The difference of a number and four.

50. Three more than half a number.

51. Name the property that justifies $5 - x = -x + 5$

52. Name the property that justifies $1x = x$

53. Factor: $2x + 7x$

54. Factor: $3a - a$

Chapter 1 Test   The answers to all the exercises are in the back of the book.

1. Identify the sums, terms, products, and factors: $8 + x$

2. Plot on a number line: $\{-7, -2, -4.5, -3\frac{1}{2}, 6\}$

3. Find the coordinate for each point shown on the number line:

    $\begin{array}{c} \text{-8 -7 -6 -5 -4 -3 -2 -1 0 1 2 3 4 5 6 7 8} \end{array}$

4. Graph 125 on a number line. Use an interval from -100 to 500 in steps of 75.

5. Find the reciprocal of -3

6. Write the opposite: $\frac{3}{2}$

7. Rewrite as a product: $\frac{w}{2}$

8. Factor: $5b + b$

9. Prime factor: 459

10. Translate to a mathematical model : Two more than one third of a number.

11. Reduce to lowest terms: $\frac{56}{112}$

12. Evaluate $2x^3 - xy^2$ when x is -2 and y is -3.

13. Name the property that justifies the statement: $3(x + 2) = 3(2 + x)$

14. Name the property that justifies the statement: $(6 + x) + 3 = 6 + (x + 3)$

Simplify:

15. $36/2^2 \cdot (5 - 2)$

16. $\frac{20}{39} \cdot \frac{51}{25}$

17. $\frac{21}{30} \div \frac{42}{20}$

18. -2^6

19. $\frac{8}{15} + \frac{3}{40}$

20. $20 - \dfrac{\frac{7}{5} \cdot 20}{2 - \frac{1}{4}}$

21. $9 - \left| -16 \right| \div 8 + 12/3$

22. $12x^3 - 3x^2 + 5x^3 - 4x^2$

23. $6x - (3x - 8x) + 2x$

24. $9x - 3(3x - 2)$

25. $6r - 2[r + 4(r - 3) + 1]$

26. $26 - 3(12 - (5 - 2)^2 + 36/12)$

Chapter 2
Equations and Inequalities

Section 2.1 Definitions and Solving Equations

Topic 2.1 A — Three Types of Equations

An **equation** is a statement that says two expressions are equal. Like an English sentence, an equation can be true, false, or conditionally true. For example:

True statements

- The sentence "January has 31 days." is true.

- The equation $3 = 1 + 2$ is true because the expressions on both sides of the equal sign represent the same number. An equation that is always true is called an **identity**.

False statements

- The sentence "February has 31 days." is false.

- The equation $3 = 1 + 1$ is false because the expressions on both sides of the equal sign do not represent the same number. An equation that is always false is called a **contradiction**.

Conditionally true statements

- The sentence "The current month has 31 days." is conditionally true. If the month we are talking about is January, March, May, July, August, October, or December, the sentence is true. If we are talking about any other month, the sentence is false.

- The equation $x = 3$ is conditionally true. If x represents the number 3 the equation is true. If x represents any other number, say 4, the equation is false. An equation that is true for some values of the variable and not others is called a **conditional equation**.

Topic 2.1 B — Solutions

To **solve** an equation means to find the values of the variables that make the equation true. One such value is called a **solution** or **root** of the equation. The collection of all solutions is called the **solution set** of the equation. Each value in the solution set is said to **satisfy** the equation.

For example, 3 is a solution of the equation $x = 3$ because if we replace x with 3 we have the true statement $3 = 3$. Four is not a solution of $x = 3$ because if we replace x with 4 we have the false statement $4 = 3$.

Equations may have any number of solutions. For example,

- The conditional equation $x + 2 = 7$ has one solution, 5.

- The identity $x + 1 = 1 + x$ has infinitely many solutions (this is just a statement of the Commutative Property of Addition).

- The contradiction $x + 1 = x$ has no solutions because the expression on the left is always one more than the expression on the right.

To see if a number is a solution of an equation, we replace each variable in the equation with the number and then simplify. If the expression on each side of the equation simplifies to the same value, the given number is a solution.

Example 2.1.1   a)   Is 5 a solution of the equation $6 - x = 1$?

• **SOLUTION** • We replace x with 5 and see if the expressions on either side of the equation simplify to the same value.

Left expression   ?=?   *Right expression*

$6 - x$	$=$	1	⇐ Equation to check.
$6 - (5)$		1	⇐ Substituted 5 for x.
1		1	⇐ Subtracted.

5 is a solution because the left and right expressions simplify to the same value.

— *Note* — ***Checking solutions***	The important point of the check is that both expressions simplify to the SAME number, but the particular number they simplify to has no significance. If both expressions had simplified to $\frac{5}{7}$, we would have drawn the same conclusion: the value we substituted for the variable (5) is probably the correct solution.

b)   Is 4 a solution of the equation $6 - t = 2(t - 8)$?

• **SOLUTION** • We replace every t with 4 and see if the expressions simplify to the same value.

Left expression   ?=?   *Right expression*

$6 - t$	$=$	$2(t - 8)$	⇐ Equation to check.
$6 - (4)$		$2((4) - 8)$	⇐ Substituted 4 for t.
2		$2(-4)$	⇐ Simplified.
2		-8	⇐ Simplified.

4 is *not* a solution because, after substituting and simplifying, the left and right expressions *do not* have the same value.

c)   Is $\frac{4}{3}$ a solution of the equation $\frac{x}{6} = \frac{2}{9}$?

• **SOLUTION** •

Left expression   ?=?   *Right expression*

$\dfrac{x}{6}$	$=$	$\dfrac{2}{9}$	⇐ Equation to check.
$\dfrac{\frac{4}{3}}{6}$		$\dfrac{2}{9}$	⇐ Substituted $\frac{4}{3}$ for x.
$\dfrac{4}{3} \cdot \dfrac{1}{6}$		$\dfrac{2}{9}$	⇐ Inverted the divisor and changed the division to multiplication.
$\dfrac{2}{9}$		$\dfrac{2}{9}$	⇐ Simplified fractions.

$\frac{4}{3}$ is a solution because both expressions simplified to the same value.

Practice 2.1.1

1. Is 8 a solution of $7 - y = -1$?   2. Is 6 a solution of $5 - n = 4(n - 3)$?   3. Is -4 a solution of $3(x - 2) = 4x - 2$?

4. Is 4 a solution of $3x - \left(\dfrac{x}{4} + 2\right) = x + 2$?   5. Is 16 a solution of $\dfrac{b}{12} + \dfrac{5}{3} = 3$?

Topic 2.1 C — Equivalent Equations

Equivalent equations are equations that have the same solution set. For example, the following equations are equivalent because each has –3 as its solution (that is, if we replace x with –3 each side of the equation will simplify to the same number).

$$\frac{x}{3} + 9 = 5 - x \qquad 2x + 1 = x - 2 \qquad x = -3$$

In this Chapter we will discuss how to transform an equation into a simpler, but equivalent, equation. Then, we can find the solution of the complex equation by finding the solution of the simple equivalent equation. For example, later we will show you that the equations $5(x - 2) = 3x - (x + 7)$ and $x = 1$ are equivalent equations. Since it's easy to see that 1 is the solution to $x = 1$, we can say that 1 is the solution to $5(x - 2) = 3x - (x + 7)$ because the equations are equivalent.

Example 2.1.2   Fill in the box so the first and second equations are equivalent. (These problems may seem confusing at first. Stick with them, the idea is important.)

a) $x = 3$ and $x + 1 = \square$

•SOLUTION• For the equations to be equivalent, the value of x that makes $x = 3$ true must be the same as the value that makes $x + 1 = \square$ true. Replacing x with 3 makes $x = 3$ true so x must be 3 in the second equation. This gives $3 + 1 = \square$. Therefore, $\square$ must be 4.

b) $-1 = x$ and $x + 1 = \square$

•SOLUTION• The first equation tells us that x is –1. If we put this value in for x in the second equation we get $-1 + 1 = \square$. Therefore, $\square$ is 0.

c) $y = 0$ and $\square = y - 6$

•SOLUTION• The first equation tells us that y is 0. Substituting 0 for y in the second equation gives $\square = 0 - 6$, therefore $\square$ is –6.

Practice 2.1.2   Fill in the box so the first and second equations are equivalent.

1. $x = 12$ and $x + 6 = \square$ 2. $z = -6$ and $z + 11 = \square$ 3. $m = -5$ and $6 - m = \square$ 4. $7 = s$ and $-3 - s = \square$

When writing an equation it's not important which side of the equal sign the variable is on. For example, $x = 3$ and $3 = x$ are equivalent equations. This idea is called the Symmetric Property of Equality.

— *Property* — ***Symmetric Property of Equality***	English: Exchanging the expressions on the left and right sides of an equation produces an equivalent equation. Example: $2x = 4$ and $4 = 2x$ are equivalent. Algebra: $a = b$ is equivalent to $b = a$

By tradition, we write the variable on the left so even though $2 = x$ is not wrong we will usually write it as $x = 2$.

Topic 2.1 D — Adding or Subtracting to Make Equivalent Equations

If we start with the true equation $3 = 3$ and make the expression on the left larger, say by adding 1, the new equation, $3 + 1 = 3$, is not true. If we make *both* expressions larger *in exactly the same way*, for example by adding 1 to both sides, the new equation, $3 + 1 = 3 + 1$, is true. This is an example of the addition property of equality.

— *Property* — *Addition Property of Equality*	English: Adding the same number (or expression) to both sides of an equation results in an equivalent equation.
	Examples:
	The equation $x - 5 = -12$ is equivalent to $x - 5 + \mathbf{5} = -12 + \mathbf{5}$
	The equation $8 - 3x = -2x$ is equivalent to $8 - 3x + \mathbf{3x} = -2x + \mathbf{3x}$
	Algebra: $a = b$ is equivalent to $a + c = b + c$

A similar situation occurs if we *subtract* the same number from both sides.

— *Property* — *Subtraction Property of Equality*	English: Subtracting the same number (or expression) from both sides of an equation results in an equivalent equation.
	Examples:
	The equation $x + 2 = 6$ is equivalent to $x + 2 - \mathbf{2} = 6 - \mathbf{2}$
	The equation $x + 5 = 2x$ is equivalent to $x + 5 - \mathbf{x} = 2x - \mathbf{x}$
	Algebra: $a = b$ is equivalent to $a - c = b - c$

We can use these properties to help us solve equations. The general idea is to rewrite the given equation as a simpler but equivalent equation with the variable isolated on one side and a constant on the other. For example, let's solve $2x - 7 = x - 20$. Since we want to isolate the variable, we will first use the Addition Property of Equality to "move" the -7 to the right side. Then, we will use the Subtraction Property of Equality to "move" the x to the left side. Here are the details of the solution:

$$2x - 7 = x - 20 \qquad \Leftarrow \text{Equation to solve.}$$

$$2x - 7 + \mathbf{7} = x - 20 + \mathbf{7} \qquad \Leftarrow \text{Added 7 to both sides to "move" } -7 \text{ to the right side.}$$

$$2x + 0 = x - 13 \qquad \Leftarrow \text{Combined like terms in each expression.}$$

$$2x = x - 13 \qquad \Leftarrow \text{Used the Addition Property of 0 to simplify the expression on the left.}$$

$$2x - \mathbf{x} = x - 13 - \mathbf{x} \qquad \Leftarrow \text{Subtracted } x \text{ from both sides to "move" } x \text{ to the left side.}$$

$$x = -13 + \mathbf{0} \qquad \Leftarrow \text{Combined like terms in each expression.}$$

$$x = -13 \qquad \Leftarrow \text{Used the Addition Property of 0 to simplify the right side.}$$

It is easy to see that the solution to $x = -13$ is -13. And, since $x = -13$ is equivalent to $2x - 7 = x - 20$, we conclude that -13 is the solution to this equation also. We can check that this is correct by substituting -13 for x in the original equation and then simplifying:

Left expression ?=? *Right expression*

$2x - 7$ =	$x - 20$	$\Leftarrow$ Equation to check.
$2(\mathbf{-13}) - 7$	$\mathbf{-13} - 20$	$\Leftarrow$ Substituted -13 for x.
$\mathbf{-26} - 7$	$\mathbf{-33}$	$\Leftarrow$ Simplified.
$\mathbf{-33}$		$\Leftarrow$ Simplified.

Since the left and right expressions simplify to the same number, we are confident that the solution is correct.

— Note — ***Writing out all the steps***	In the above example, we wrote out all the details to illustrate the process. As was the case with simplifying expressions, we usually don't write out everything because it takes too long. You should write out as many steps as you need to make what you are doing clear to yourself and to others.

Example 2.1.3   a)   Solve $x + 3 = -8$

• SOLUTION • We can isolate the variable by "undoing" the addition of 3 on the right side of the equation. To do this, we subtract 3 from both sides to get $x = -11$. Since $x = -11$ and $x + 3 = -8$ are equivalent equations, the solution to $x + 3 = -8$ is -11.

b)   Solve $3n + 5 = 1 + 2n$

• SOLUTION • To isolate n, we need to bring the terms with n to one side and the terms without n to the other side.

$$3n + 5 = 1 + 2n \qquad \Leftarrow \text{ Equation to solve.}$$

$$3n + 5 - \mathbf{2n} = 1 + 2n - \mathbf{2n} \qquad \Leftarrow \text{ Subtracted } 2n \text{ from both sides so the terms with } n \text{ are on the left side. }^{\dagger}$$

$$n + 5 = 1 \qquad \Leftarrow \text{ Combined like terms.}$$

$$n + 5 - \mathbf{5} = 1 - \mathbf{5} \qquad \Leftarrow \text{ Subtracted 5 from both sides so the terms without } n \text{ are on the right side.}$$

$$n = -4 \qquad \Leftarrow \text{ Combined like terms.}$$

The solution is -4. We check the solution by substituting -4 into the original equation.

Left expression   ?=?   Right expression

$$3n + 5 = 1 + 2n \qquad \Leftarrow \text{ Equation to check.}$$

$$3(\mathbf{-4}) + 5 \quad | \quad 1 + 2(\mathbf{-4}) \qquad \Leftarrow \text{ Substituted } -4 \text{ for } n.$$

$$\mathbf{-12} + 5 \quad | \quad 1 + (\mathbf{-8}) \qquad \Leftarrow \text{ Simplified.}$$

$$-7 \quad | \quad -7 \qquad \Leftarrow \text{ Simplified.}$$

Since both expressions simplified to the same number, we are confident that -4 is the solution.

— Note — ***Check the solution***	Be sure to check the solution in the **original** equation and not a simplified version of it. That way, you are more likely to catch an error you made during simplification.

c)   Solve $4(3 - p) = -3p$

• SOLUTION • Before moving terms, we need to simplify the expression on the left side. Then, we can isolate the variable.

$$4(3 - p) = -3p \qquad \Leftarrow \text{ Equation to solve.}$$

$$\mathbf{4}(3) - \mathbf{4}(p) = -3p \qquad \Leftarrow \text{ Distributed 4.}$$

$$\mathbf{12} - \mathbf{4p} = -3p \qquad \Leftarrow \text{ Multiplied.}$$

$$12 - 4p + \mathbf{4p} = -3p + \mathbf{4p} \qquad \Leftarrow \text{ Added } 4p \text{ to both sides. Note that we could have added } 3p \text{ to both sides but that would have resulted in a negative coefficient for } p, \text{ which is more difficult to work with}$$

$$12 = p \qquad \Leftarrow \text{ Combined like terms.}$$

---

† We could just as correctly subtract $3n$ from both sides. This would result in a negative coefficient of the n term. Some people prefer to avoid negative coefficients. The solution will be the same regardless of the technique you prefer.

The solution is 12. We check the solution by substituting 12 into the original equation.

Left expression	?=?	*Right expression*	
$4(3 - p)$	=	$-3p$	⇐ Equation to check.
$4(3 - \mathbf{12})$		$-3(\mathbf{12})$	⇐ Substituted 12 for p
$4(\mathbf{-9})$		$\mathbf{-36}$	⇐ Simplified
-36		$\mathbf{-36}$	⇐ Simplified

Since both expressions simplified to the same number, we are confident that 12 is the solution.

— Note — **Saving space**	In order to save space, from now on we usually will not write out the details of the check.

Practice 2.1.3 Solve.

1. $y - 8 = 17$
2. $-7 = v - 5$
3. $5y + 9 = 4y - 8$
4. $-14 + 2x = x - 3$
5. $5x - 36 = 4(x + 9)$
6. $-2x = -3(x - 4)$

Topic 2.1 E — Multiplication Property of Equality

If we start with the true equation $3 = 3$ and make the expression on the left larger, say by multiplying by 4, the new equation, $4(3) = 3$ is not true. If we make *both* expressions larger *in exactly the same way*, for example by multiplying *both* expressions by 4, the new equation $4(3) = 4(3)$ is true. This idea of using multiplication to make the expression on both sides of the equal sign larger (or smaller) in exactly the same way is an important equation–solving tool.

— *Property* — *Multiplication Property of* *Equality*	English: Multiplying both sides of an equation by the same *non–zero* number (or expression) results in an equivalent equation. Example: The equation $\dfrac{x}{3} = 7$ is equivalent to $3\left(\dfrac{x}{3}\right) = 3(7)$ Algebra: If $c \neq 0$, then $a = b$ is equivalent to $ca = cb$

The Multiplication Property of Equality says we may multiply by any number except 0. We can't multiply by 0 because doing so may or may not keep the equations equivalent. For example, if we begin with the false equation $4 = 5$ and multiply both sides by 0 we end up with the true equation $0 = 0$. But $4 = 5$ and $0 = 0$ cannot be equivalent equations because the contradiction $4 = 5$ cannot be equivalent to the identity $0 = 0$.

Example 2.1.4 a) Solve $\dfrac{y}{-8} = 5$

•**SOLUTION**• We wish to isolate the variable by "moving" the –8 to the right side of the equation. We can do this by multiplying both sides by –8 (essentially, we are "undoing" the division by –8).

$$(\mathbf{-8})\left(\frac{y}{-8}\right) = (\mathbf{-8})(5) \quad ⇐ \text{ Multiplied both sides by –8.}$$

$$y = -40 \quad ⇐ \text{ Simplified.}$$

The solution is –40. We leave the check to you.

b) Solve $2 = \dfrac{x-4}{3}$

•SOLUTION• First, we multiply both sides of the equation by 3 to "undo" the division by 3.

$$3(2) = 3\left(\frac{x-4}{3}\right) \quad \Leftarrow \text{ Multiplied both sides by 3.}$$

$$6 = x - 4 \quad \Leftarrow \text{ Simplified.}$$

$$10 = x \quad \Leftarrow \text{ Added 4 to both sides.}$$

The solution is 10. We leave the check to you.

c) Solve $-6 = \dfrac{2}{3}w$

•SOLUTION• We can isolate the variable by $\dfrac{2}{3}$ by multiplying both sides of the equation by the reciprocal of $\dfrac{2}{3}$, which is $\dfrac{3}{2}$.

$$\frac{3}{2}(6) = \frac{3}{2}\left(\frac{2}{3}w\right) \quad \Leftarrow \text{ Multiplied both sides by } \frac{3}{2}.$$

$$9 = w \quad \Leftarrow \text{ Simplified.}$$

The solution is 9. We leave the check to you.

Practice 2.1.4 Solve.

1. $\dfrac{x}{5} = -3$

2. $\dfrac{y-1}{2} = -5$

3. $-8 = \dfrac{2}{5}x$

Topic 2.1 F — Dividing to Make Equivalent Equations

If we start with the true equation $3 = 3$ and change the expression on the left, say by dividing by 3, the new equation, $\dfrac{3}{3} = 3$ is not true. If we change *both* expressions *in exactly the same way*, for example by dividing *both* expressions by 3, the new equation $\dfrac{3}{3} = \dfrac{3}{3}$ is true. This idea of using division to change the expressions on both sides of the equal sign in exactly the same way is an important equation–solving tool.

— *Property* — *Division Property of Equality*	English: Dividing both sides of an equation by the same non–zero number (or expression) results in an equivalent equation.
	Example: The equation $3x = 12$ is equivalent to $\dfrac{3x}{3} = \dfrac{12}{3}$
	Algebra: If $c \neq 0$, then $a = b$ is equivalent to $\dfrac{a}{c} = \dfrac{b}{c}$

Notice the Division Property of Equality states that the number we divide by cannot be 0. We exclude 0 because division by 0 is not possible.

Example 2.1.5   a)   Solve $4x + 6 = 34$

•SOLUTION•

$$4x + 6 - \textbf{6} = 34 - \textbf{6} \quad \Leftarrow \text{Subtracted 6 from both sides.}$$

$$4x = 28 \quad \Leftarrow \text{Combined like terms.}$$

$$\frac{4x}{\textbf{4}} = \frac{28}{\textbf{4}} \quad \Leftarrow \text{Divided both sides by 4.}$$

$$x = 7 \quad \Leftarrow \text{Simplified.}$$

The solution is 7. We leave the check to you.

b)   Solve $12 = 5 - (y - 1)$

•SOLUTION• First, we will simplify the right side and then isolate the variable.

$$12 = 5 - (y - 1) \quad \Leftarrow \text{Equation to solve.}$$

$$12 = 5 - y + 1 \quad \Leftarrow \text{Distributed the negative to the terms inside the parentheses.}$$

$$12 = -y + 6 \quad \Leftarrow \text{Combined like terms.}$$

$$12 - \textbf{6} = -y + 6 - \textbf{6} \quad \Leftarrow \text{Subtracted 6 from both sides.}$$

$$6 = -y \quad \Leftarrow \text{Combined like terms.}$$

$$\frac{6}{\textbf{-1}} = \frac{-y}{\textbf{-1}} \quad \Leftarrow \text{Divided both sides by } -1 \text{ (remember, } -y \text{ means } -1y\text{).}$$

$$-6 = y \quad \Leftarrow \text{Simplified.}$$

The solution is −6. We leave the check to you.

— Caution —	Be sure you understand the difference between expressions and equations.	
Expression versus	***Expressions***	***Equations***
Equation	• collection of numerals, operators, letters, and grouping symbols.	• statement of equality between two expressions.
	• we *simplify* expressions; for example, $2x + 3x$ simplifies to $5x$.	• we *solve* equations; for example, the solution to $2x + 3x = 5$ is 1.
	• expressions are like phrases in English, such as *a number increased by 3*.	• equations are like sentences in English, such as *a number increased by 3 is 5*.
	• expressions don't have equal signs.	• equations always have equal signs.

Practice 2.1.5 Solve.

1. $5x = 35$

2. $22 = -m$

3. $10 = -2(4y - 1)$

4. $4(y - 1) + 3 = -9$

5. $6(4 - a) + 5a = 3(6 - 1)$

Exercise Set 2.1   The answers to the odd numbered exercises are at the back of the book.

1. Is -2 a solution of $9 - w = 11$?

2. Is -6 a solution of $15 - y = 9$?

3. Is 300 a solution of $10x + 25(200 - x) = 450$?

4. Is 680 a solution of $5x + 10(400 - x) = 600$?

5. Is $\frac{7}{2}$ a solution of $2(5 - x) = x - \frac{1}{2}$?

6. Is $\frac{1}{2}$ a solution of $-3(6 + x) = 5x - 22$?

7. Is 5 a solution of $-2(6-x) = 4x - 2(x+1)$?

8. Is 0 a solution of $6(1-x) = -x - 3(x-2)$?

9. Is -40 a solution of $\frac{1}{4}m + \frac{1}{3}(2-m) = 4$?

10. Is 32 a solution of $7 = \frac{1}{2}r + \frac{1}{3}(5-r)$?

11. Is 21 a solution of $\frac{1}{3}x + \frac{1}{2}(1-x) = 4$?

12. Is 3 a solution of $\frac{5}{6}(x-2) + \frac{3}{2}(x+1) = 7$?

13. Is $\frac{1}{3}$ a solution of $\frac{3x-5}{2} = \frac{x}{3} - \frac{19}{9}$?

14. Is $\frac{1}{6}$ a solution of $\frac{2x-3}{4} = \frac{x}{3} + \frac{1}{4}$?

Fill in the box so the first and second equations are equivalent.

15. $d + 6 = \square$ and $d = -11$

16. $m + 12 = \square$ and $m = 3$

17. $z = -2.1$ and $z + 4.5 = \square$

18. $r = 0$ and $r + 6.7 = \square$

19. $z = \frac{1}{2}$ and $z + \frac{3}{4} = \square$

20. $d = \frac{2}{3}$ and $d + \frac{3}{4} = \square$

Solve.

21. $3y + 5 = 2y - 1$

22. $2x - 7 = 3 + x$

23. $24 - 9x = 3 - 8x$

24. $5 - 5w = 12 - 4w$

25. $\frac{x}{-6} = 4$

26. $\frac{z}{-3} = -5$

27. $\frac{w}{20} = \frac{1}{4}$

28. $\frac{1}{2} = \frac{t}{8}$

29. $\frac{1}{24}s = \frac{-1}{6}$

30. $\frac{1}{11}x = -\frac{5}{22}$

31. $\frac{x-3}{5} = -2$

32. $\frac{w-5}{8} = -1$

33. $\frac{5}{8}a = 15$

34. $21 = \frac{-3}{5}x$

35. $-3 = \frac{-t}{2}$

36. $\frac{-p}{5} = 6$

37. $-51 = 17x$

38. $57 = 19w$

39. $-36 = -18w$

40. $-48 = -4x$

41. $5x + 2 = 37$

42. $8 + 3x = 41$

43. $25 = 7 - 2x$

44. $-9 = 3 - x$

45. $9y + 9 = 4y + 9$

46. $12m - 13 = 4m - 13$

47. $4x - 24 = x - 3$

48. $5w - 27 = 2w + 12$

49. $-2(5 - 2n) = -3n$

50. $3(6 - 3y) = -8y$

51. $-5(3y - 7) = -10y$

52. $-2(7w - 8) = -10w$

Review Exercises The answers to all of these exercises are at the back of the book.

53. Find the reciprocal of $-2x$

54. Rewrite as a product: $\frac{5m}{8}$

55. Write the opposite: $\frac{1}{3}$

56. Identify the sums, terms, products, and factors: $-3y$

57. Reduce to lowest terms: $\frac{38}{20}$

58. Name the property that justifies the statement: $5(x + y) = 5x + 5y$

59. Factor: $3y + y$

Simplify.

60. $-(2x - 5x)$

61. $\frac{8}{15} + \frac{3}{20} + \frac{5}{12}$

62. $12 \div 3(5 - 2)$

63. $2x + 5y - 5x$

Section 2.2 More on Solving Equations

Topic 2.2 A — A General Procedure for Solving Equations

You may have noticed that the equations we have studied so far all have a single variable and the exponent of the variable was always 1. Such equations are called **linear equations in one variable**. Later in this book we will discuss literal equations (Section 2.3), linear equations in two variables (Section 4.1), and quadratic equations (Section 7.6).

As the one variable linear equations become more complicated it's useful to have a systematic procedure for solving them.

— Procedure — Solving equations in one variable	***Step 1 LCD*** If fractions are present, multiply each term by the LCD. ***Step 2 Simplify*** Simplify each side of the equation. ***Step 3 Move Terms*** Use addition or subtraction to move terms with variables to one side of the equation and terms without variables to the other side. Then, simplify each side again. ***Step 4 Divide*** Divide both sides of the equation by the coefficient of the variable. ***Step 5 Check*** Check your answer by substituting it into the original equation. If a true statement results, your answer is probably correct. If a false statement results, go back over your work to see if you made an error.

As long as you correctly apply the properties of algebra, don't make careless errors, and are patient, you should be successful using this general procedure. But remember that algebra is an art as well as a science. As you become more experienced you will find shortcuts and gain insights that will make you more efficient at solving equations.

— Study Tip — Writing procedures on note cards	Experts can solve equations automatically because they have repeatedly talked themselves through equations using a procedure like the one above. It would be best to copy the procedure above onto an index card and keep it next to you as you do your homework for the next few days. Consciously consider all five steps each time you solve an equation. In time, your mind will begin to consider each step automatically without you having to think about it. This will free up your attention so you can work with more interesting equations, like those arising from word problems.

Example 2.2.1   Solve $2(x + 1) = 6(x - 3) - 7x + 11$

•SOLUTION• ***Step 1 LCD*** There are no fractions so we skip this step.

Step 2 Simplify Both sides need to be simplified.

$$2(x + 1) = 6(x - 3) - 7x + 11 \quad \Leftarrow \text{ Equation to solve.}$$
$$2x + 2 = 6x - 18 - 7x + 11 \quad \Leftarrow \text{ Distributed.}$$
$$2x + 2 = -x - 7 \quad \Leftarrow \text{ Combined like terms.}$$

Step 3 Move Terms We will move the terms with variables to the left side of the equation and the constant terms to the right side.

$$2x + 2 + x = -x - 7 + x \quad \Leftarrow \text{ Added } x \text{ to both sides.}$$
$$3x + 2 = -7 \quad \Leftarrow \text{ Combined like terms.}$$
$$3x + 2 - 2 = -7 - 2 \quad \Leftarrow \text{ Subtracted 2 from both sides.}$$
$$3x = -9 \quad \Leftarrow \text{ Combined like terms.}$$

Step 4 Divide

$$\frac{3x}{3} = \frac{-9}{3} \quad \Leftarrow \text{ Divided both sides by 3.}$$
$$x = -3 \quad \Leftarrow \text{ Simplified.}$$

Step 5 Check The solution is –3. We leave the check to you.

Practice 2.2.1 Solve

1. $12 = 5(x + 2) - x - 10$ 2. $-4(3 - x) + 1 = 2x - 5$ 3. $a - 3(a - 3) + 15 = 5 - 3(2 - a)$

4. $7(y + 2) + 5y = -5 + 3y - (y + 1)$ 5. $3 - 3(2 - m) = 2(m + 1) - (m - 1)$

Topic 2.2 B — Equation Solving and LCD's

When an equation contains fractions it's usually a good idea to clear them using the Multiplication Property of Equality. For example to solve $\frac{x}{2} + 3 = \frac{5}{2}$ we begin by multiplying both sides of the equation by 2.

$$\frac{2}{1}\left(\frac{x}{2} + 3\right) = \frac{2}{1}\left(\frac{5}{2}\right) \qquad \Leftarrow \text{ Multiplied both sides by 2.}$$

$$\frac{2}{1}\left(\frac{x}{2}\right) + \frac{2}{1}(3) = \frac{2}{1}\left(\frac{5}{2}\right) \qquad \Leftarrow \text{ Distributed the 2 to each term.}$$

$$x + 6 = 5 \qquad \Leftarrow \text{ Reduced fractions and multiplied.}$$

As you can see, it's much easier to solve the equation once the fractions have been cleared.

— Note — *Multiplying both sides means multiplying each term*	When an equation has more than one term on a side, multiplying both sides by the LCD has the effect of multiplying *each term* by the LCD. To save space, from now on when we say we are multiplying both sides by the LCD we will directly multiply each term by the LCD. For example, we would combine the first two steps in the above problem into a single step.

When the fractions involve larger denominators, we need a procedure for finding the smallest factor to multiply by. For example, suppose we are asked to solve $\frac{x}{12} - \frac{7}{30} = \frac{1}{10}$.

If we multiply through by a number that's divisible by *all* the denominators then *all* the fractions would reduce to integers in one step. We could multiply by the product of all the denominators, 12 • 30 • 10, which is 3600. However, a much smaller number can be found that has 12 and 30 and 10 as factors. That number is the Least Common Denominator or LCD of the fractions. We discussed finding the LCD of a set of fractions in Section 1.2. In this case the LCD is 60.

Example 2.2.2   a)   Solve $2x + \frac{x}{3} = 7$

•SOLUTION•   *Step 1 LCD*   To clear the fraction, we multiply all the terms by the LCD, 3.

$$3(2x) + 3\left(\frac{x}{3}\right) = 3(7) \qquad \Leftarrow \text{ Multiplied every term by 3.}$$

Step 2 Simplify

$$6x + x = 21 \qquad \Leftarrow \text{ Multiplied and reduced.}$$

$$7x = 21 \qquad \Leftarrow \text{ Combined like terms.}$$

Step 3 Move Terms   No terms need to be moved.

Step 4 Divide

$$\frac{7x}{7} = \frac{21}{7} \qquad \Leftarrow \text{ Divided both sides by 7.}$$

$$x = 3 \qquad \Leftarrow \text{ Reduced.}$$

Step 5 Check   The solution is 3. We leave the check to you.

b) Solve $\frac{1}{3}x + \frac{1}{2}(1-x) = 4$

• SOLUTION •

Step 1 LCD   The LCD of 3 and 2 is 6.

$$6\left(\frac{1}{3}x\right) + 6\left(\frac{1}{2}\right)(1-x) = 6(4) \quad\Leftarrow \text{ Multiplied each term by the LCD, 6.}$$

— *Note* — *Multiplying 3 factors*	Be careful when multiplying $6\left(\frac{1}{2}\right)(1-x)$. We may multiply 6 by $\frac{1}{2}$ OR we may multiply 6 by $(1-x)$ but we may NOT do both (that is, we may NOT distribute across a multiplication). To reduce the fraction we multiply $6\left(\frac{1}{2}\right)$ to get 3, leaving the term $3(1-x)$.

Step 2 Simplify

$$2x + 3(1-x) = 24 \quad\Leftarrow \text{ Reduced fractions and multiplied.}$$
$$2x + 3 - 3x = 24 \quad\Leftarrow \text{ Distributed 3 to terms inside parentheses.}$$
$$3 - 1x = 24 \quad\Leftarrow \text{ Combined like terms.}$$

Step 3 Move Terms

$$3 - 1x - 3 = 24 - 3 \quad\Leftarrow \text{ Subtracted 3 from both sides.}$$
$$-1x = 21 \quad\Leftarrow \text{ Combined like terms.}$$

Step 4 Divide

$$\frac{-1x}{-1} = \frac{21}{-1} \quad\Leftarrow \text{ Divided both sides by } -1.$$
$$x = -21 \quad\Leftarrow \text{ Reduced fractions.}$$

Step 5 Check   The solution is –21. We leave the check to you.

c) Solve $\frac{5}{12}(y-2) + \frac{7}{18} = \frac{2}{9}(y+5)$

• SOLUTION • **Step 1 LCD**   To clear the fraction, we multiply all the terms by 36, the LCD.

$$36\left(\frac{5}{12}(y-2)\right) + 36\left(\frac{7}{18}\right) = 36\left(\frac{2}{9}(y+5)\right) \quad\Leftarrow \text{ Multiplied every term by 36.}$$

— *Note* — *Multiplying 3 factors*	We may multiply 36 by $\frac{5}{12}$ OR we may multiply 36 by $(y-2)$ but we may NOT do both. Multiplying $36 \cdot \frac{5}{12}$ will clear the fraction so we choose to do that.

Step 2 Simplify

$$15(y-2) + 14 = 8(y+5) \quad\Leftarrow \text{ Multiplied.}$$
$$15y - 30 + 14 = 8y + 40 \quad\Leftarrow \text{ Distributed.}$$
$$15y - 16 = 8y + 40 \quad\Leftarrow \text{ Combined like terms.}$$

Step 3 Move Terms

$$15y - 16 - 8y = 8y + 40 - 8y \quad \Leftarrow \text{Subtracted } 8y \text{ from both sides.}$$

$$7y - 16 = 40 \quad \Leftarrow \text{Combined like terms.}$$

$$7y - 16 + 16 = 40 + 16 \quad \Leftarrow \text{Added 16 to both sides.}$$

$$7y = 56 \quad \Leftarrow \text{Combined like terms.}$$

Step 4 Divide

$$\frac{7y}{7} = \frac{56}{7} \quad \Leftarrow \text{Divided both sides by 7.}$$

$$y = 8 \quad \Leftarrow \text{Reduced fractions.}$$

Step 5 Check   The solution is 8. We leave the check to you.

— Caution — ***Multiply each term by*** ***LCD***	When eliminating the fractions in an equation, be sure to multiply each term in the equation by the LCD. It's a common error to multiply only some of the terms by the LCD. If you fail to multiply all terms by the LCD the resulting equation will <u>not</u> be equivalent to the original and your solution will be incorrect.

Practice 2.2.2   Solve.

1. $\dfrac{x}{12} - \dfrac{7}{30} = \dfrac{1}{10}$

2. $\dfrac{1}{4} + \dfrac{4}{5}n = \dfrac{-11}{20}$

3. $\dfrac{x}{3} + 3 = -\dfrac{1}{3} + \dfrac{x}{5}$

4. $\dfrac{1}{3}b + 1 = \dfrac{b}{12} - 4$

5. $\dfrac{1}{5}(1 + 2x) + \dfrac{1}{3}(4 - x) = \dfrac{1}{15}$

Topic 2.2 C — Solving Equations Containing Decimals

Equations containing decimals are solved using the same general procedure.

Example 2.2.3   a)   Solve $55x + 45(x - 1.5) = 330$

•SOLUTION•   *Step 1 LCD*   There are no denominators to clear so we skip this step.

Step 2 Simplify

$$55x + 45(x) - 45(1.5) = 330 \quad \Leftarrow \text{Distributed the 45.}$$

$$55x + 45x - 67.5 = 330 \quad \Leftarrow \text{Multiplied.}$$

$$100x - 67.5 = 330 \quad \Leftarrow \text{Combined like terms.}$$

Step 3 Move Terms

$$100x - 67.5 + 67.5 = 330 + 67.5 \quad \Leftarrow \text{Added 67.5 to both sides.}$$

$$100x = 397.5 \quad \Leftarrow \text{Combined like terms.}$$

Step 4 Divide

$$\frac{100x}{100} = \frac{397.5}{100} \quad \Leftarrow \text{Divided both sides by 100.}$$

$$x = 3.975 \quad \Leftarrow \text{Reduced fractions.}$$

Step 5 Check   The solution is 3.975. We leave the check to you.

b) Solve $0.06(s) + 0.08(6000 - s) = 414$

•**SOLUTION**• ***Step 1 LCD*** There are no denominators to clear so we skip this step. Note that some people may prefer to clear the decimals by multiplying each term by 100.

Step 2 Simplify

$0.06(s) + \mathbf{0.08}(6000) - \mathbf{0.08}(s) = 414$	$\Leftarrow$	Distributed 0.08.
$0.06(s) + \mathbf{480} - \mathbf{0.08}s = 414$	$\Leftarrow$	Multiplied.
$\mathbf{-0.02}s + 480 = 414$	$\Leftarrow$	Combined like terms.

Step 3 Move Terms

$-0.02s + 480 - \mathbf{480} = 414 - \mathbf{480}$	$\Leftarrow$	Subtracted 480 from both sides.
$-0.02s = -66$	$\Leftarrow$	Combined like terms.

Step 4 Divide

$\dfrac{-0.02s}{\mathbf{-0.02}} = \dfrac{-66}{\mathbf{-0.02}}$	$\Leftarrow$	Divided both sides by −0.02.
$s = 3300$	$\Leftarrow$	Reduced fractions.

Step 5 Check The solution is 3300. We leave the check to you.

Practice 2.2.3 Solve.

1. $5(x + 3.50) + 20x = 105$

2. $0.30x + 0.45(7) = 0.35(x + 7)$

3. $0.06x + 0.05(15000 - x) = 830$

4. $0.10x + 0.11(12000 - x) = 1304.50$

Topic 2.2 D — Identities and Contradictions

In section 2.1 we said that an equation that is always true is called an identity and an equation that is always false is called a contradiction. For example the equation $x + 1 = 1 + x$ is an identity because no matter what value we replace x with, the resulting statement is true. Likewise, $x = x + 1$ is a contradiction because no matter what value we replace x with the resulting statement is false.

It may not be obvious that a given equation is an identity or a contradiction until we go through the process of rewriting it as a simpler, but equivalent, equation.

Example 2.2.4 a) Solve $6 + 4(x - 2) = 5x - 8 - x$

•**SOLUTION**• ***Step 1 LCD*** There are no fractions, so skip this step.

Step 2 Simplify

$6 + \mathbf{4x} - \mathbf{8} = 5x - 8 - x$	$\Leftarrow$	Distributed 4.
$4x - \mathbf{2} = \mathbf{4x} - 8$	$\Leftarrow$	Combined like terms.

Step 3 Move Terms

$4x - 2 - \mathbf{4x} = 4x - 8 - \mathbf{4x}$	$\Leftarrow$	Subtracted $4x$ from both sides.
$-2 = -8$	$\Leftarrow$	Combined like terms.

Where have all the x's gone? We began with an equation that contained unknowns, represented by the x's, and we have arrived at an equivalent equation that contains no x's. Because the final equation, $-2 = -8$, is a contradiction, and it's equivalent to the original equation, the original also must be a contradiction. In other words there is no value we can substitute for x that will make the original equation true. Thus, the answer is *no solution*.

b) Solve $2x - 6 = 5x - 3(x + 2)$

•SOLUTION• *Step 1 LCD*   There are no fractions, so skip this step.

Step 2 Simplify

$$2x - 6 = 5x - 3x - 6 \quad \Leftarrow \quad \text{Distributed } -3.$$

$$2x - 6 = 2x - 6 \quad \Leftarrow \quad \text{Combined like terms.}$$

Step 3 Move Terms

$$2x - 6 - 2x = 2x - 6 - 2x \quad \Leftarrow \quad \text{Subtracted } 2x \text{ from both sides.}$$

$$-6 = -6 \quad \Leftarrow \quad \text{Combined like terms.}$$

This situation is similar to the previous one, except that the resulting equation is an identity (it's always true) rather than a contradiction. We can use the same logic as before to find the solution of the original equation: Because $-6 = -6$ is an identity, and because it's equivalent to the original equation, the original equation must also be an identity. This means that we will get a true statement when we substitute any real number for x. So, the solution is *all real numbers*.

Step 4 Divide   Not applicable.

Step 5 Check   To check the solution, we obviously cannot try all real numbers to see if they work. Instead, let's try two values. If these work, we will be confident that any real number is a solution of the equation.

First, let's check 0:

Left expression	?=?	*Right expression*	
$2x - 6$	=	$5x - 3(x + 2)$	$\Leftarrow$ Equation to check.
$2(0) - 6$		$5(0) - 3((0) + 2)$	$\Leftarrow$ Substituted 0 for x.
$2(0) - 6$		$5(0) - 3(2)$	$\Leftarrow$ Simplified inside parentheses.
-6		-6	$\Leftarrow$ Simplified. Solution checks.

Next, let's check -1:

Left expression	?=?	*Right expression*	
$2x - 6$	=	$5x - 3(x + 2)$	$\Leftarrow$ Equation to check.
$2(-1) - 6$		$5(-1) - 3((-1) + 2)$	$\Leftarrow$ Substituted -1 for x.
$2(-1) - 6$		$5(-1) - 3(1)$	$\Leftarrow$ Simplified inside parentheses.
$-2 - 6$		$-5 - 3$	$\Leftarrow$ Multiplied.
-8		-8	$\Leftarrow$ Subtracted. Solution checks.

Because both of these values check, we are confident that the original equation is an identity.

We may generalize the reasoning of the last two examples as follows.

— *Note* — *Solutions to an equation where all the variables disappear*	When constructing equivalent equations, if all the variables disappear we have a contradiction or an identity: • if the equivalent equation is false, the original equation is a contradiction and the solution is *no solution*. • if the equivalent equation is true, the original equation is an identity and the solution is *all real numbers*.

It's important for you to distinguish between *no solution* and 0 as a solution, as the following example illustrates.

Example 2.2.5   Solve $5x = 3x$

•**SOLUTION**• You may be tempted to divide both sides of the equation by x and end up with the contradiction $5 = 3$. However, dividing by a variable is dangerous because we do not know if the variable represents 0 — if it does, then the division is not allowed. To solve this equation, follow the procedure:

Step 1 LCD   There are no fractions, so skip this step.

Step 2 Simplify   Both sides are already simplified.

Step 3 Move Terms

$$5x - 3x = 3x - 3x \qquad \Leftarrow \text{ Subtracted } 3x \text{ from both sides.}$$

$$2x = 0 \qquad \Leftarrow \text{ Combined like terms.}$$

Step 4 Divide

$$\frac{2x}{2} = \frac{0}{2} \qquad \Leftarrow \text{ Divided both sides by 2.}$$

$$x = 0 \qquad \Leftarrow \text{ Reduced fractions.}$$

The solution is 0. Note that if we had divided by x as a first step we really would have been dividing by 0, which is not allowed.

Step 5 Check   Substituting 0 for x in $5x = 3x$ gives $0 = 0$. Therefore, the solution checks.

Practice 2.2.5 Solve.

1. $5 + 4(x - 3) = 5x - 7 - x$

2. $3 - 3(y - 2) = 5y + 4 - 8y$

3. $7 + 5(x - 1) = 6x + 8 - x$

4. $8x - 2 = 2x - 4(x + 1) + 2$

5. $6 - 2(x + 5) = 4x - 4 - 6x$

6. $7x = -x$

Exercise Set 2.2   The answers to the odd numbered exercises are at the back of the book.

Solve.

1. $24 + 3x = 6(x + 7) - 42$

2. $-12 + 2y = 4(y - 6) + 24$

3. $35 + 9x = -4(x + 8) + 32$

4. $5 + 3(x - 2) = 4x + 7 - x$

5. $-4(x + 2) + 3x = -x + 7$

6. $3x - 5 = 5(x - 1) - 2x$

7. $-16 + 9x = -5(x + 2) + 10$

8. $5(a + 6) - 3(10 - 2a) = 22$

9. $-3(y - 4) + 6(y - 2) = 15$

10. $4(t + 6) - 6(t - 6) - 60 = 2$

11. $8(a - 5) + 4(5 - a) + 20 = -4$

12. $-3(x - 5(2x - 1)) = 39$

13. $16 = -4(y - (6y - 6))$

14. $-(h - 2(h - 1)) = -1$

15. $-(a - 5(a - 1)) = -3$

16. $3x + 4 = 5 - (x + 1)$

17. $-7 - (2x - 1) = 2x - 3(x + 2)$

18. $7(x + 9) - 6x = x + 4$

19. $2(m - 6) - 26 = -7 + 9(m - 4)$

20. $-8(p + 2) + 34 = -7 + 3(7 - p)$

21. $-6(k - 3) - 2 = -1 + 12(2 - k)$

22. $3(4 + 2(t + 1)) = 6(t + 7) - 4t$

23. $2(5 + 3(s - 2)) = 4(s - 1) - 6s$

24. $4x - 7(8 - x) = 3 + x - (57 + x)$

25. $-4n + 8(6 - n) = 54 + n - (12 + n)$

26. $-8p + 5(6 - p) = 11 - p - (7 - p)$

27. $15a + 7(-8 - a) = 4 - a - (12 - a)$

28. $\dfrac{x}{7} - \dfrac{2}{3} = \dfrac{1}{21}$

29. $\dfrac{x}{4} - \dfrac{2}{3} = \dfrac{-11}{12}$

30. $\dfrac{2}{3}y - \dfrac{1}{4} = \dfrac{13}{12}$

31. $\dfrac{1}{4} + \dfrac{4}{5}n = \dfrac{-11}{20}$

32. $\dfrac{x}{3} + 3 = -\dfrac{1}{3} + \dfrac{x}{5}$

33. $\dfrac{1}{3}b + 1 = \dfrac{b}{12} - 4$

34. $\dfrac{2}{3}y + \dfrac{1}{2}(8 - y) = 3$

35. $\dfrac{4}{5}x + \dfrac{2}{3}(3 - x) = 4$

36. $\dfrac{3}{4}z + \dfrac{2}{3}(5 - z) = 4$

37. $\dfrac{5}{6}r + \dfrac{3}{4}(4-r) = 2$

38. $3(x-4) - \dfrac{3}{4} = \dfrac{5x}{2} + 4$

39. $4(x-1) - \dfrac{2}{3} = \dfrac{3}{4}x + 2$

40. $\dfrac{5}{4}a + 1 = \dfrac{3}{2} + a$

41. $\dfrac{7}{3}a - 2 = -3 + 2a$

42. $\dfrac{3}{25}(x-4) + \dfrac{7}{10} = \dfrac{1}{5}(x+1)$

43. $\dfrac{3}{4}(x-2) = \dfrac{1}{5}(x+1) - \dfrac{3}{5}$

44. $\dfrac{x-2}{6} - \dfrac{1}{8}(x+2) = \dfrac{7-x}{12}$

45. $\dfrac{2x-1}{2} - \dfrac{x+1}{3} = \dfrac{1}{6}(x+4)$

46. $\dfrac{2}{3}(9-3x) - (6-3x) = \dfrac{1}{2}(x-4) + \dfrac{x}{3}$

47. $\dfrac{1}{5}(10-5x) - 2(1-x) = \dfrac{1}{4}(x-4) + \dfrac{x}{4}$

48. $45(x - 1.25) = 55x$

49. $5(x - 2.75) = 15x$

50. $0.08x = 0.09(x - 450)$

51. $0.03x = 0.07(x - 345)$

52. $0.12x + 0.18(x - 180) = 150$

53. $0.14x + 0.22(x - 175) = 51.5$

54. $35(y - 3.25) + 25(y + 0.05) = 300$

55. $15(n - 2.75) + 80(n + 1.25) = 130$

56. $2(x + 1.40) - 3(x - 2.25) = 112.75$

57. $5(x + 2.05) - 8(x - 0.01) = 28.48$

58. $0.25x + 0.10(325 - x) = 0.20(x - 7)$

59. $0.40t + 0.25(6 - t) = 0.30(t - 5)$

60. $3n - 2(5 - n) = -2(n + 1) + 27$

61. $2x - 3(7 + x) = -5(x + 2) - 3$

62. $3(y + 2) + 29 = 2 - 6(y - 6)$

63. $4 + \dfrac{1}{2}(x + 8) = -\dfrac{1}{4}x + 8 + \dfrac{3}{4}x$

64. $5 - \dfrac{1}{4}(x - 8) = -\dfrac{1}{2}x + 7 + \dfrac{1}{4}x$

65. $14x = 2x$

66. $7 - 2(5 - 3x) = 7x - 3 - x$

67. $0.23y + 0.31(y + 150) = 0.28(2y + 150)$

68. $0.05c + 0.08(c + 110) = 0.06(2c + 110)$

Review Exercises   The answers to all of these exercises are at the back of the book.

69. Reduce to lowest terms: $\dfrac{128}{90}$

70. Plot on a number line: $\{8, -2, 0.5, 6.5, -5\frac{2}{3}\}$

71. Write the opposite: $\dfrac{8}{5}$

72. Name the property that justifies the statement: $ab = ba$

73. Identify the base and exponent, and simplify: 6^2

74. Is 5 a solution of $3x - 7 = 12 - 2(x - 3)$?

75. Prime factor 120.

Simplify.

76. $12 + 6[3 + 4(12/4)^2 - 4]$

77. $\dfrac{12}{4} \div \dfrac{80}{24}$

78. $\dfrac{3}{25} + \dfrac{2}{15} + \dfrac{7}{10}$

79. $\dfrac{2 \cdot 3 \cdot 4^2 - 5 \cdot 12}{58 - 2 \cdot 3^3}$

80. $\dfrac{7}{3} - 8 \cdot \left(\dfrac{1}{2}\right)^2$

81. $\dfrac{7}{5} + \dfrac{\frac{2}{3} \cdot 18}{6 \cdot \frac{1}{4}}$

82. $-(5x - 2) - (x - 5)$

Section 2.3 Literal Equations and Formulas

Topic 2.3 A — Literal Equations

Equations that involve more than one variable are called **literal equations**. Literal equations that represent a rule, a fact, or a way of doing something are called **formulas**. Formulas generally have real–world applications. An example of a formula is the literal equation $p = 2l + 2w$. This formula from geometry shows the relation between the perimeter (p) of a rectangle and its length (l) and width (w). We sometimes say that the isolated variable is solved for "**in terms of**" the other variables. For example, we would say $p = 2l + 2w$ is solved for p in terms of l and w.

Example 2.3.1   Use "is solved for" and "in terms of" to discuss the literal equations.

a) $d = rt$

•SOLUTION• This is solved for d in terms of r and t.

b) $f = \frac{9}{5}c + 32$

> •SOLUTION• This is solved for f in terms of c.

c) $y = mx + b$ where m and b are constants.

> •SOLUTION• Even though this formula has four letters we are told that m and b are constants. Therefore, we would say this is solved for y in terms of x.

Practice 2.3.1 Use "is solved for" and "in terms of" to discuss the literal equations.

1. $a = p + prt$ 2. $A = \frac{1}{2}bh$ 3. $V = \frac{1}{3}\pi r^2 h$ 4. $y = ax^2 + bx + c$ where a, b, and c are constants.

Topic 2.3 B — Solving Literal Equations

The process used for solving a literal equation is the same as for solving equations that contain only one variable. To illustrate this, lets compare solving a one variable equation, $2x + 3 = 11$, with the literal equation, $ax + b = c$. Notice both equations have the same form.

Single Variable Equation		*Literal Equation*	
$2x + 3 = 11$	$\Leftarrow$ Equation to solve.	$ax + b = c$	$\Leftarrow$ Equation to solve.
$2x + 3 - 3 = 11 - 3$	$\Leftarrow$ Subtracted 3 from both sides.	$ax + b - b = c - b$	$\Leftarrow$ Subtracted b from both sides.
$2x = 8$	$\Leftarrow$ Simplified.	$ax = c - b$	$\Leftarrow$ Simplified.
$\dfrac{2x}{2} = \dfrac{8}{2}$	$\Leftarrow$ Divided both sides by 2.	$\dfrac{ax}{a} = \dfrac{c-b}{a}$	$\Leftarrow$ Divided both sides by a.
$x = 4$	$\Leftarrow$ Reduced fractions.	$x = \dfrac{c-b}{a}$	$\Leftarrow$ Reduced fractions.
The solution is 4. We say we have solved the equation.		The solution is $\dfrac{c-b}{a}$. We say we have solved the equation for x in terms of a, b, and c.	

The only difference in these procedures is that we could not actually carry out the arithmetic operations in the literal equation so the answer is the algebraic expression $\frac{c-b}{a}$, rather than the constant, 4.

Example 2.3.2   a)   Solve $d = rt$ for r

> •SOLUTION• To solve for r we must undo the multiplication by t.
>
> ***Step 1 LCD***   There are no fractions, so skip this step.
>
> ***Step 2 Simplify***   Each expression is in its simplest form.
>
> ***Step 3 Move Terms***   All terms that contain r are on one side of the equation.
>
> ***Step 4 Divide***
>
> $$\frac{d}{t} = \frac{rt}{t} \quad \Leftarrow \text{Divided both sides by } t.$$
>
> $$\frac{d}{t} = r \quad \Leftarrow \text{Reduced the common factor of } t.$$
>
> ***Step 5 Check***   The solution is $\frac{d}{t}$. We say this new equation is solved for r in terms of d and t. We will do the check later in this section.

b) Solve $y = mx + b$ for x

•**SOLUTION**• To solve for x we must undo the addition of b (so we subtract b) and the multiplication by m (so we divide by m).

Step 1 LCD   There are no fractions, so skip this step.

Step 2 Simplify   Each expression is in its simplest form.

Step 3 Move Terms

$y - b = mx + b - b$ ⇐ Subtracted b from both sides.

$y - b = mx$ ⇐ Combined like terms.

Step 4 Divide

$$\frac{y-b}{m} = \frac{mx}{m}$$ ⇐ Divided both sides by m.

$$\frac{y-b}{m} = x$$ ⇐ Reduced common factor of m.

Step 5 Check   The solution is $\frac{y-b}{m}$. We say this new equation is solved for x in terms of y, b, and m. We will do the check later in this section.

c) Barbara wants to fence in a rectangular garden that is to lie between two walkways 10 feet apart. She can buy rolls of fencing that are 25 feet, 50 feet, or 100 feet in length. How long would the garden be using the three different size rolls?

•**SOLUTION**• The perimeter formula, $p = 2l + 2w$ describes the relation between perimeter (the length of the fence) and the length and width of the enclosed rectangle. Since we are given the perimeter and width, it will be easier to solve the formula for the length before doing the calculations.

Step 1 LCD   There are no fractions, so skip this step.

Step 2 Simplify   Each expression is in its simplest form.

Step 3 Move Terms

$p - 2w = 2l + 2w - 2w$ ⇐ Subtracted $2w$ from both sides.

$p - 2w = 2l$ ⇐ Combined like terms.

Step 4 Divide

$$\frac{p-2w}{2} = \frac{2l}{2}$$ ⇐ Divided both sides by 2.

$$\frac{p-2w}{2} = l$$ ⇐ Reduced common factor of 2.

Step 5 Check   The solution is $\frac{p-2w}{2}$. We say this new equation is solved for l in terms of p and w. We will do the check later in this section.

Now that we have the formula in a more convenient form we can substitute 10 for w and the three different perimeters (25, 50, and 100) for p to find the corresponding lengths:

- if p is 25 the length is $\frac{25 - 2(10)}{2}$, which simplifies to 2.5 feet.

- if p is 50 the length is $\frac{50 - 2(10)}{2}$ which simplifies to 15 feet.

- if p is 100 the length is $\frac{100 - 2(10)}{2}$ which simplifies to 40 feet.

Practice 2.3.2 Solve the equation for the specified variable.

1. $i = prt$ for r

2. $PV = nRT$ for R

3. $M = \dfrac{a+b}{2}$ for b

4. $a = p + prt$ for t

5. $ax + by + c = 0$ for y

6. $F = \dfrac{9}{5}C + 32$ for C

Topic 2.3 C — Checking Literal Equations

We can check the solution of a literal equation in the same way we check the solution for a single variable equation. That is, we substitute the solution into the original equation. If this results in an identity, then the solution probably is correct.

Study the two checking processes described below, where we check to see if 4 is a solution of $2x + 3 = 11$ and we check to see if $\dfrac{c-b}{a}$ is a solution of $ax + b = c$.

Single Variable Equation			Literal Equation	
$2x + 3 = 11$	$\Leftarrow$ Equation to check.	$ax + b = c$	$\Leftarrow$ Equation to check.	
$2(4)+ 3 \mid 11$	$\Leftarrow$ Substituted 4 for x.	$a\left(\dfrac{c-b}{a}\right)+ b \mid c$	$\Leftarrow$ Substituted $\dfrac{c-b}{a}$ for x.	
$8 + 3 \mid 11$	$\Leftarrow$ Multiplied.	$c - b + b \mid c$	$\Leftarrow$ Reduced the common factor of a.	
$11 \mid 11$	$\Leftarrow$ Added.	$c \mid c$	$\Leftarrow$ Added.	

At the end of both checks, the left and right expressions are equivalent so we are confident that both solutions are correct.

Example 2.3.3 a) Show that $\dfrac{d}{t}$ a solution when $d = rt$ is solved for r.

• SOLUTION •

Left expression ?=? Right expression

$$d \;=\; rt \qquad\qquad \Leftarrow \text{Equation to check.}$$

$$d \;\Big|\; \left(\dfrac{d}{t}\right)t \qquad \Leftarrow \text{Substituted } \dfrac{d}{t} \text{ for } r$$

$$d \;\Big|\; d \qquad\qquad \Leftarrow \text{Reduced common factor of } t.$$

The final expressions are equivalent so the solution checks.

b) Show that $\dfrac{y-b}{m}$ a solution when $y = mx + b$ is solved for x.

• SOLUTION •

Left expression ?=? Right expression

$$y \;=\; mx + b \qquad\qquad \Leftarrow \text{Equation to check.}$$

$$y \;\Big|\; m\left(\dfrac{y-b}{m}\right)+ b \qquad \Leftarrow \text{Substituted } \dfrac{y-b}{m} \text{ for } x.$$

$$y \;\Big|\; y - b + b \qquad\qquad \Leftarrow \text{Reduced common factor of } m.$$

$$y \;\Big|\; y \qquad\qquad\qquad \Leftarrow \text{Combined like terms.}$$

The final expressions are equivalent so the solution checks.

c) Show that $\dfrac{p-2w}{2}$ is a solution when $p = 2l + 2w$ is solved for l.

•SOLUTION•

— Caution — **Canceling**	You may be tempted to make the check easier by "canceling" the 2's in $\dfrac{p-2w}{2}$ like this $\dfrac{p-\cancel{2}w}{\cancel{2}}$ to get $p-w$. WE CANNOT DO THIS because 2 is not a *factor* of the numerator and denominator. "Canceling" the 2's would be like "canceling" the 2's in $\dfrac{12}{32}$ to get $\dfrac{1}{3}$. "Canceling" is not a mathematical operation. To reduce a fraction we write the numerator and denominator in factored form and then reduce the common factors.

Left expression ?=? Right expression

$$p \quad = \quad 2l + 2w \qquad \Leftarrow \text{ Equation to check.}$$

$$p \quad \Big| \quad 2\left(\frac{p-2w}{2}\right) + 2w \qquad \Leftarrow \text{ Substituted } \frac{p-2w}{2} \text{ for } l.$$

$$p \quad \Big| \quad p - 2w + 2w \qquad \Leftarrow \text{ Reduced common factor of 2}$$

$$p \quad \Big| \quad p \qquad \Leftarrow \text{ Combined like terms.}$$

The final expressions are equivalent so the solution checks.

Practice 2.3.3

1. Show that $\dfrac{i}{pt}$ is a solution when $i = prt$ is solved for r.   2. Show that $\dfrac{c}{2\pi}$ is a solution when $c = 2\pi r$ is solved for r.

3. Show that $\dfrac{a-p}{pr}$ is a solution when $a = p + prt$ is solved for t.

4. Show that $\dfrac{-ax-c}{b}$ is a solution when $ax + by + c = 0$ is solved for y.

Exercise Set 2.3   The answers to the odd numbered exercises are at the back of the book.

Use "is solved for" and "in terms of" to discuss the literal equations.

1. $I = \dfrac{E}{R}$ 　　　　2. $z = \dfrac{x-5}{s}$ 　　　　3. $C = 180 - A - B$ 　　4. $H = 3x + 3y$

5. $A = mt - p$ (where m and p are constants) 　　　6. $y = ax + b$ (where a and b are constants)

7. $d = \sqrt{1.5}\,(t-n) + 0.5tn$ 　8. $S = \dfrac{3}{4}(W+H) - \dfrac{1}{4}WH$ 　　9. $t = 2x^2 + 3x - 5$ 　　10. $m = 3y^3 - 4y + 1$

Solve the equation for the specified variable.

11. $A = \dfrac{bh}{2}$ for h 　　　12. $A = \dfrac{bh}{2}$ for b 　　　13. $C = 180 - A - B$ for A 　14. $C = 180 - A - B$ for B

15. $A = 0.5h(b+d)$ for b 　16. $s = 180(n-2)$ for n 　17. $V = \dfrac{1}{3}\pi r^2 h$ for h 　18. $K = \dfrac{1}{2}mv^2$ for m

19. $A = \dfrac{a+b+c}{3}$ for b 　20. $z = \dfrac{x-m}{s}$ for x 　21. $h = vt - 16t^2$ for v 　22. $P = 9.337da - 299$ for a

23. $H = \dfrac{D^2 N}{2.5}$ for N 　24. $A = \dfrac{\pi r^2 s}{360}$ for s 　25. $s = 2\pi rh + 2\pi r^2$ for h 　26. $A = 3ab + 3bh$ for a

27. $F = \dfrac{kMm}{d^2}$ for M 　28. $F = \dfrac{kMm}{d^2}$ for k 　29. $W = \dfrac{LZ}{P}$ for P 　　30. $W = \dfrac{LZ}{P}$ for Z

31. Show that $B = 180 - A - C$ is a solution when $C = 180 - A - B$ is solved for B.

32. Show that $A = 180 - B - C$ is a solution when $C = 180 - A - B$ is solved for A.

33. Show that $n = \dfrac{s}{180} + 2$ is a solution when $s = 180(n - 2)$ is solved for n.

34. Show that $b = \dfrac{A}{0.5h} - d$ is a solution when $A = 0.5h(b + d)$ is solved for b.

35. Show that $m = \dfrac{2K}{v^2}$ is a solution when $K = \dfrac{1}{2}mv^2$ is solved for m.

36. Show that $h = \dfrac{3V}{\pi r^2}$ is a solution when $V = \dfrac{1}{3}\pi r^2 h$ is solved for h.

37. Show that $R = \dfrac{E}{I}$ is a solution when $I = \dfrac{E}{R}$ is solved for R.

38. Show that $h = \dfrac{2A}{b}$ is a solution when $A = \dfrac{bh}{2}$ is solved for h.

39. Show that $a = \dfrac{A - 3bh}{3b}$ is a solution when $A = 3ab + 3bh$ is solved for a.

40. Show that $h = \dfrac{s - 2\pi r^2}{2\pi r}$ is a solution when $s = 2\pi rh + 2\pi r^2$ is solved for h

---

Review Exercises   The answers to all of these exercises are at the back of the book.

41. Translate to a mathematical model: A number increased by 12.

42. Plot on a number line: $\{-\dfrac{3}{4}, 5, -8, -1\dfrac{2}{3}\}$

43. Find the coordinate for each point shown on the number line:

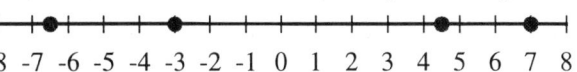

-8 -7 -6 -5 -4 -3 -2 -1 0 1 2 3 4 5 6 7 8

Simplify.

44. $4 \cdot 3^2 \div 6$

45. $\dfrac{56}{72} \cdot \dfrac{81}{26}$

46. $-\left| -3 \right| \cdot 2 + \left| -5 \right|$

47. $7x - 4[3x + 2(x + 1) + 7]$

Solve.

48. $2 - (x - 3) + 5x = 3x - 2$

49. $5 - \dfrac{x}{3} = -3 + x$

50. $0.05d - 0.1(d + 12) = 4.5$

51. $3(x - 5) + 2x = 5x - 15$

52. $12x = 3x$

53. $5x = 2(3 + x) + 3(x - 1)$

Section 2.4 Properties and Graphs of Inequalities

Topic 2.4 A — Inequalities, Solutions, and Solving

We defined an equation as a statement that two expressions are equal. Similarly, an **inequality** is a statement that one expression is greater than, or less than, another.

— *Misconception* — *Not equal is different* *from inequality*	The prefix *in* sometimes means *not*, as in the word *inflexible*. However, inequality does not mean *not equal*. For example, $3 = 5$ is not an inequality. It's an equation that happens to be a contradiction (because the statement is always false). Likewise, $3 \neq 5$ is not considered an inequality. Inequalities talk about whether one expression is larger (or smaller) than another, not about whether equations are *not equal*.

We use the following symbols when dealing with inequalities:

Symbol	Meaning	Example	English
$>$	greater than	$6 > 5$	six is greater than five
$\geq$	greater than or equal to	$6 \geq 5$	six is greater than or equal to five
$<$	less than	$5 < 6$	five is less than six
$\leq$	less than or equal to	$5 \leq 6$	five is less than or equal to six

— *Note*— *Changing the way a relation looks does not change it's meaning*	We can look at a greater than or less than relation in two different ways. For example if I am older than you, then you are younger then I. If six is greater than five ($6 > 5$) then five is less than six ($5 < 6$). Notice that even though the direction of the inequality symbol has changed the relation between the values has remained the same.

To solve an inequality means to find the values of the variables that make the inequality true. One such value is called a **solution** or **root** of the inequality. The collection of all solutions is called the **solution set** of the inequality. Each value in the solution set is said to **satisfy** the inequality.

For example, 8 is a solution of $x + 3 > 5$ because replacing x with 8 will make the statement true. Likewise, 1 is not a solution because replacing x with 1 results in a false statement.

Example 2.4.1 Determine whether the given value is a solution of the inequality.

a) Is -8 a solution of $-x - 4 \geq 1$?

•SOLUTION• We replace x with -8 and see if the left expression is greater than or equal to the right.

Left expression ?≥? Right expression

$$
\begin{array}{lll}
-x - 4 \ \geq \ 1 & & \Leftarrow \text{ Inequality to check.} \\
-(-8) - 4 \ \bigm| \ 1 & & \Leftarrow \text{ Substituted } -8 \text{ for } x. \\
\quad\quad 4 \ \bigm| \ 1 & & \Leftarrow \text{ Simplified.}
\end{array}
$$

-8 *is* a solution because, after substituting and simplifying, the resulting inequality is true (that is, 4 is greater than <u>or</u> equal to 1).

b) Is 2 a solution of $3x + 1 < 6$?

•SOLUTION• We replace x with 2 and simplify to see if the left expression is less than the right.

Left expression ?<? Right expression

$$
\begin{array}{lll}
3x + 1 \ < \ 6 & & \Leftarrow \text{ Inequality to check.} \\
3(2) + 1 \ \bigm| \ 6 & & \Leftarrow \text{ Substituted 2 for } x. \\
\quad\quad 7 \ \bigm| \ 6 & & \Leftarrow \text{ Simplified.}
\end{array}
$$

2 is *not* a solution because, after substituting and simplifying, we have the *false* statement 7 is less than 6.

c) Is -1 a solution of $5 + a > 4$?

•SOLUTION• We replace a with -1 and see if the left expression is greater than the right .

Left expression ?>? Right expression

$$
\begin{array}{lll}
5 + a \ > \ 4 & & \Leftarrow \text{ Inequality to check.} \\
5 + (-1) \ \bigm| \ 4 & & \Leftarrow \text{ Substituted} -1 \text{ for } x. \\
\quad\quad 4 \ \bigm| \ 4 & & \Leftarrow \text{ Simplified.}
\end{array}
$$

-1 is *not* a solution because, after substituting and simplifying, the resulting inequality is false (that is, 4 is not greater than 4).

Practice 2.4.1 Determine whether the given value is a solution of the inequality.

1. Is -3 a solution of $3x - 2 > 10$?   2. Is -4 a solution of $5x + 2 < 4$?   3. Is 4 a solution of $6 - y \geq -7$?

4. Is 9 a solution of $12 - m \geq 5$ 5. Is -8 a solution of $2 + x < -1$?   6. Is -1 a solution of $y + 3 > -4$?

Topic 2.4 B — Types of Inequalities

As was the case with equations, an inequality can be a conditional inequality, an identity, or a contradiction:

- An inequality that is sometimes true and sometimes false, depending upon the value used in place of the variable is called a **conditional inequality**. For example, $3x + 1 < 12$ is a conditional inequality because replacing x with some numbers, such as 2, results in a true statement but replacing x with other numbers, such as 8 results in a false statement.

- An inequality which is always true is called an **identity**. For example, $x + 1 > x$ is an identity because the left expression, $x + 1$, is always larger than the right expression, x, no matter what value we use in place of x.

- An inequality which is always false is called a **contradiction**. For example, $x + 1 < x$ is a contradiction because no matter what value we replace x with, the left expression will never be less than the right expression.

Example 2.4.2   Classify each inequality as a conditional inequality, an identity, or a contradiction.

a) $z - 3 > 2$

•**SOLUTION**• This is a conditional inequality because replacing z with some numbers, such as 8, results in a true statement while replacing z with other numbers, such as 0, results in a false statement.

b) $1 < -4$

•**SOLUTION**• This is a contradiction because it's always false.

c) $x \leq x + 5$

•**SOLUTION**• This is an identity because the expression on the left is less than the expression on the right no matter what value we use for x.

Practice 2.4.2 Classify each inequality as a conditional inequality, an identity, or a contradiction.

1. $x + 9 > 4$ 2. $m - 4 < m$ 3. $-11 > -4$ 4. $y - 8 < 7$ 5. $0 < -9$ 6. $y + 8 \geq y$

Topic 2.4 C — Graphing Inequalities

When we solved equations in sections 2.2 and 2.3 it wasn't hard to list all the solutions. For example, 3 is the only solution of $x = 3$. If we try to list all the solutions to the inequality $x > 3$ we would have to write forever. Inequalities usually have an infinite number of solutions. Even though we can't list all the solutions we still need a way to tell others what the solutions are. One way to do this is to draw a graph that represents all the solutions.

To draw a graph we start with the idea that the solutions will come from the set of all real numbers. We use the number line to represent all the real numbers.

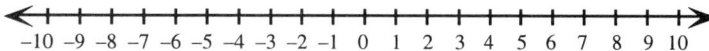

If we plotted a few numbers which made $x > 3$ true, for example 4, 5, 8, and 10, the picture would look like

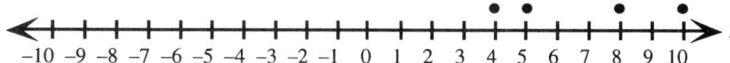

Notice that we now have labeled the number line with an *x* to show that we are displaying solutions to an inequality that involves the variable *x*.

Of course, numbers other than integers, such as 6.5, 3.75, and 5.213 would also be solutions to $x > 3$ so we would have to show these on the number line also.

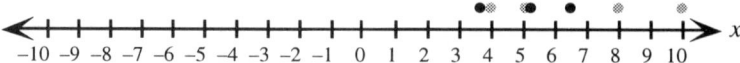

If we displayed another 50 solutions on the number line they would appear so close together that would blend into what looks like a line:

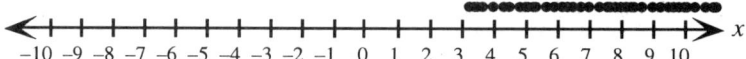

When we make a dot for *every* solution we get what looks like a straight line that begins at 3 and moves to the right. We represent this set of points as an arrow pointing to the right.

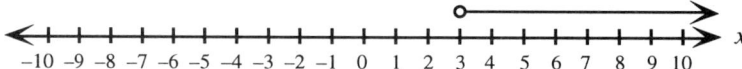

The number 3 is called the **critical point** of the graph because it separates the solutions from the non–solutions. We used an open circle **o** at 3 to tell us that 3 is not included in the solution set or the graph. This is because the inequality, $x > 3$ does not include 3 as a solution. If the inequality had been written using the greater than or equal to symbol, $\geq$, we would have used a closed circle **•** at 3 to indicate that 3 is part of the solution set and graph.

— Procedure — **Graphing an inequality**	***Step 1 Number line*** Draw a number line and label it with the appropriate variable. ***Step 2 Critical point*** Determine the critical point and plot it on the number line: • If the inequality symbol is $>$ or $<$, place an open circle at the critical point. • If the inequality symbol is $\geq$ or $\leq$, place a closed circle at the critical point. ***Step 3 Test point*** Select any other point on the number line to serve as a test point: • If the test point is a solution of the inequality, draw a line from the critical point through the test point and place an arrowhead on that end of the line. • If the test point is NOT a solution of the inequality, draw a line from the critical point away from the test point and place an arrowhead on that end of the line.

Example 2.4.3   a)   Graph $d \geq -8$

•SOLUTION• ***Step 1 Number line***   Draw a number line and label it with the appropriate variable, *d*.

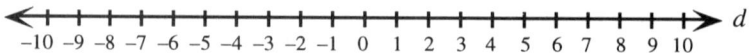

Step 2 Critical point   Determine the critical point and plot it on the number line. The critical point is –8. Since the inequality symbol is "greater than *or equal to*", we place a closed circle at –8 to indicate that –8 is a solution.

Plot critical point.

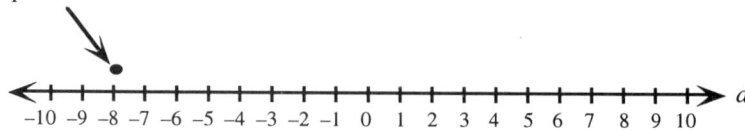

Step 3 Test point   Select a test point. We will choose 0. Since 0 is a solution of $d \geq -8$ we draw a line from -8 through 0 and place an arrow head on the right end of the line.

Plot critical point.   Test point.   Since test point, 0, is a solution of $d \geq -8$, draw arrow through test point.

$$\xleftarrow{\hspace{1cm}} \bullet \underset{\underset{-10\ -9\ -8\ -7\ -6\ -5\ -4\ -3\ -2\ -1\ \ 0\ \ 1\ \ 2\ \ 3\ \ 4\ \ 5\ \ 6\ \ 7\ \ 8\ \ 9\ \ 10}{}}{\longrightarrow} d$$

b)   Graph $x < 3$

   •SOLUTION•   ***Step 1 Number line***   See below.

   Step 2 Critical point   The critical point is 3. Since the inequality symbol is a less than (with no *or equal to*) we place an open circle at 3 to indicate that 3 is not part of the solution.

   Step 3 Test point   We will choose 0. Since 0 is a solution of $x < 3$ we draw a line from 3 through 0 and place an arrow head on the left end of the line.

Since test point, 0, is a solution to $x < 3$, draw arrow through test point.   Test point.   Plot critical point, 3.

$$\xleftarrow{\hspace{1cm}} \underset{\underset{-10\ -9\ -8\ -7\ -6\ -5\ -4\ -3\ -2\ -1\ \ 0\ \ 1\ \ 2\ \ 3\ \ 4\ \ 5\ \ 6\ \ 7\ \ 8\ \ 9\ \ 10}{}}{\circ} \longrightarrow x$$

c)   Graph $-1 > p$

   •SOLUTION•   ***Step 1 Number line***   See below.

   Step 2 Critical point   The critical point is -1. Since the inequality symbol is a greater than (with no *or equal to*) we place an open circle at -1 to indicate that -1 is not a solution.

   Step 3 Test point   We will choose 0. Since 0 is NOT a solution of $-1 > p$ we draw a line from -1 away from 0 and place an arrow head on the left end of the line.

Plot critical point, -1.

Since test point, 0, is NOT a solution to $-1 > p$, draw arrow away from test point.   Test point.

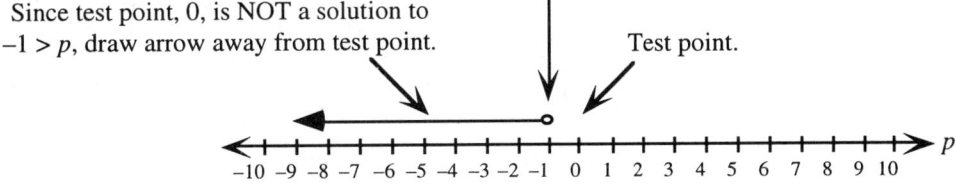

— Caution — ***deciding which way the*** ***arrow points***	Don't use the inequality symbol to decide which way the arrow on the graph should point. For example, the graph of $-1 > p$ points to the left while the graph of $d \geq -8$ points to the right. Both examples have "greater than" symbols and yet the arrows on the graphs point in different directions.

d) Graph $0 \leq w$

•SOLUTION• ***Step 1 Number line*** See below.

Step 2 Critical point The critical point is 0. Since the inequality symbol is "less than *or equal to*", we place a closed circle at 0 to indicate that 0 is a solution.

Step 3 Test point We cannot choose 0 as the test point because 0 is the critical point. So, we choose another point, say 3. Since 3 is a solution of $0 \leq w$ we draw a line from 0 through 3 and place an arrow head on the right end of the line.

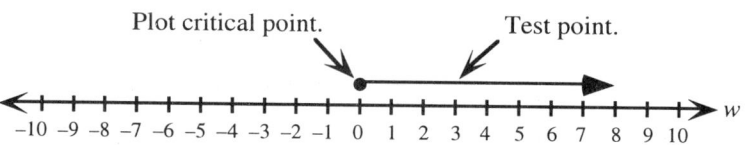

Plot critical point.   Test point.

Practice 2.4.3 Graph the inequality.

1. $m < 4$ 2. $m \geq -3$ 3. $10 \geq n$ 4. $t \geq 0$

5. $-1 > m$ 6. $r < -2$ 7. $-5 \geq p$ 8. $5 < t$

The Symmetric Property of Equality tells us that it does not matter on which side of the equal sign we place the variable. For example, $x = 5$ and $5 = x$ are equivalent. We have a similar situation with inequalities, but we have to keep the inequality symbol "pointing" in the correct direction, which is toward the smaller expression. For example, $x > 5$ means "numbers greater than 5"; the same set of numbers can be described using $5 < x$, which means "5 is less than the numbers".

Topic 2.4 D — Translations Involving Inequalities

Inequalities often appear in applied problems. There are a number of words that indicate an inequality.

Verbal Model	*Key Word(s)*	*Mathematical Model*
8 is greater than 5	greater than	$8 > 5$
-4 is less than -2	less than	$-4 < -2$
Bill's age is at most 18	at most	$b \leq 18$
a plumber's pay rate is at least $20 per hour	at least	$p \geq 20$
the maximum height of fence is 6 feet	maximum	$h \leq 6$
the minimum score on a test is 75%	minimum	$s \geq 75\%$
The numbers are over 7	over	$y > 7$
The values are under 3	under	$z < 3$
The temperatures were all above 110 °F.	above	$t > 110$
The temperatures were all below 104 °F.	below	$t < 104$

It's important to realize that there are other ways to indicate inequalities. The list above is only a few of the more common key words.

Example 2.4.4   Write a mathematical model for each sentence. Choose whatever variable you wish for the unknown.

   a)   Numbers that are at most 5.

   •SOLUTION• $x \le 5$ or $5 \ge x$

   b)   The minimum number of pets we care for is seven.

   •SOLUTION• $p \ge 7$ or $7 \le p$

   c)   All of the contestants were below five and a half feet tall.

   •SOLUTION• $c < 5.5$ or $5.5 > c$

   d) y was either 0.35 or had values less than 0.35.

   •SOLUTION• $y \le 0.35$ or $0.35 \ge y$

Practice 2.4.4   Write a mathematical model for each sentence. Choose whatever variable you wish for the unknown.

1.   At most twelve people can fit in the elevator.   2.   The numbers have a minimum value of –3.

3.   Five is greater than the other values.   4.   Three fish per day is the maximum.

5.   We can guarantee the commission will be under \$7,000.   6.   The value of m is greater than –4.

It's also important to be able to translate the math symbols back into English.

Example 2.4.5   Translate into English. There will be more than one way of doing this.

   a) $w > 4$

   •SOLUTION•

   • w is greater than four.
   • w is more than four.
   • w is above four.
   • w is over four.

   b)

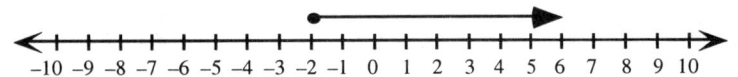

   •SOLUTION•

   • t is greater than or equal to negative two.
   • t is at least negative two.
   • t has a minimum value of negative two.

   c)

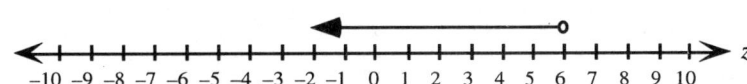

   •SOLUTION•

   • z is less than six.
   • z is under six.
   • z is below six.

Practice 2.4.5

1.   Translate into English: $y \le -6$ 2.   Translate into English: $q > 25$

3.   If p is the variable translate the graph into English:

4.   If t is the variable translate the graph into English:

Exercise Set 2.4   The answers to the odd numbered exercises are at the back of the book.

Determine whether the given value is a solution of the inequality.

1. Is -1.2 a solution to $-2x - 3.2 < 4.1$?

2. Is 2.5 a solution to $4 - m > 1.4$?

3. Is $\dfrac{2}{3}$ a solution to $3x + 7 \geq 2$?

4. Is $\dfrac{-1}{4}$ a solution to $5 - x < 12$?

5. Is -1 a solution to $-x < -4.1 + x$?

6. Is 0 a solution to $2(m + 1) > 2.4 + m$?

7. Is 5 a solution to $3(4 - x) \leq 1 - x$?

8. Is -2 a solution to $x + 12 > -2(13 - 2x)$?

Classify each inequality as conditional, identity, or contradiction.

9. $x \leq x$

10. $y + 1.8 < 7$

11. $y + 3 < y$

12. $x - 2 < x$

13. $x + 1 > 1$

14. $y + 1.8 < y$

15. $2y < y$

16. $x < x$

Graph the inequality

17. $n < 3.5$

18. $s > -2.5$

19. $-5.5 \leq p$

20. $-4.9 \geq t$

21. $a > 0.5$

22. $b > -0.5$

23. $7 \geq m$

24. $-3 < n$

25. $-7 < r$

26. $8 > w$

Write a mathematical model for each sentence. Choose whatever variable you wish for the unknown.

27. The price is at most $3.50.

28. The length is to be no more than 7.5 yards.

29. The maximum height of the tower is 100.75 feet

30. The minimum profit is $.50 for each unit.

31. The width is greater than $4\dfrac{3}{4}$ feet

32. The length is less than $2\dfrac{3}{4}$ yards.

33. The parking lot will not hold over 270 cars.

34. The number of entries must by under 25.

35. The values must exceed -275.

36. You must score above 90% to receive an A.

37. You must score at least 90% to receive an A.

38. If the water level remains below 14 feet no damage will occur.

Translate into English.

39. $y \geq -4.1$

40. $y > 3.7$

41. $s < -3.5$

42. $t \leq 11$

43. $y > -1.1$

44. $y \geq 23$

45.
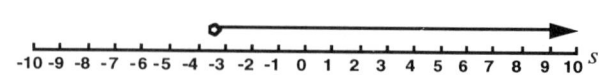

46.

47.

48.
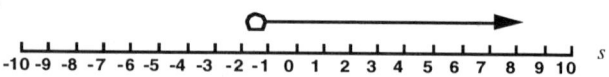

Review Exercises   The answers to all of these exercises are at the back of the book.

49. Find the coordinate for each point shown on the number line:
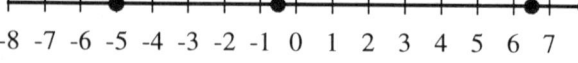

50. Name the property that justifies the statement: $5x - 10 = 5(x - 2)$

51. Solve $3x + 5y = 10$ for y

Simplify.

52. $8c - (c - 2) + 5c$

53. $\dfrac{3}{2} \cdot \dfrac{4}{15} + \dfrac{2}{3}\left(2 - \dfrac{1}{2}\right)$

54. $-(3x - x)$

55. $\dfrac{1}{2}(2 - 5w) - \dfrac{1}{3}w + \dfrac{w}{2}$

56. $x + 4[2x + 6(x + 1) - 8]$

57. $\dfrac{1}{2}x + 4 + 3(x - 2)$

Solve.

58. $0.1(10 - 3d) = 0.5d + 3$

59. $\dfrac{1}{2}x + 4 = 3(x - 2)$

60. $x - 5(2 - x) = 3(2x - 4) + 1$

Section 2.5 Solving Inequalities

In the last section we stated that to solve an inequality means to find all values of the variables that make the inequality true. Each value in the solution set is said to satisfy the inequality.

In this section we solve inequalities using techniques that are similar to those we use to solve equations.

Topic 2.5 A — Solving Inequalities Using Addition and Subtraction

We defined equivalent equations as equations that have the same solution set. We used this idea to solve complex equations by transforming them into simpler equivalent equations, which were easy to solve. Likewise, we define **equivalent inequalities** as inequalities that have the same solution set and we will use them to transform complicated inequalities into simple ones that are easier to solve.

With equations, we could add (or subtract) the same number to both sides of the equation and the result was an equivalent equation. This is also true for inequalities.

— *Property* — *Addition and subtraction* *properties of inequality*	English: • Adding the same number (or expression) to both sides of an inequality results in an equivalent inequality. • Subtracting the same number (or expression) from both sides of an inequality results in an equivalent inequality. Example: • The inequality $x - 5 > -12$ is equivalent to $x - 5 + \mathbf{5} > -12 + \mathbf{5}$ • The inequality $x + 8 > 1$ is equivalent to $x + 8 - \mathbf{8} > 1 - \mathbf{8}$ • The inequality $5 + 2x < 3x + 9$ is equivalent to $5 + 2x - \mathbf{2x} < 3x + 9 - \mathbf{2x}$ Algebra: • $a > b$ is equivalent to $a + c > b + c$ • $a > b$ is equivalent to $a - c > b - c$

We can use this property to solve inequalities.

Example 2.5.1   Solve and then graph the solution set.

a) $x - 2 < 5$

•SOLUTION•

$$x - 2 + \mathbf{2} \; < \; 5 + \mathbf{2} \qquad \Leftarrow \text{ Added 2 to both sides.}$$

$$x \; < \; 7 \qquad \Leftarrow \text{ Simplified.}$$

— Note —

Three ways of
representing the solution
of an inequality

Recall that the solution set of an inequality is the set of all values that make the inequality true. For the above example, we may write the solution set in several different ways:

- Verbally: Using English, we can represent the solution set as "all real numbers less than seven."

- Symbolically: Using **set builder notation** we can write the solution set as $\{x \mid x < 7\}$. This is read "The set of all real numbers represented by x, such that x is less than 7". This is a formal way of writing the solution set; however, to save space, we will write this using the abbreviated form $x < 7$.

- Graphically: We may use a picture to represent the solution set as follows:

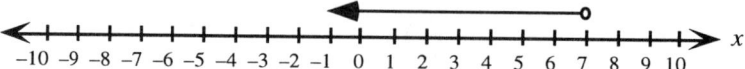

Remember, the open circle means that 7 is not included in the solution set. Also, the arrow pointing to the left is made up of an infinite number of dots, each of which represents a solution to the original inequality. The arrow head implies that the solutions continue to the left forever.

- As a roster: We can try to list all the solution as in $\{6.5, 6, 5.99, 0, -8, -10.3, ...\}$ Since there is an infinite number of solutions trying to write them out like this is not very useful. However, for those inequalities where there is a small number of solutions, the roster method is useful.

b) $3 > y + 8$

•SOLUTION•

$$3 - 8 > y + 8 - 8 \quad \Leftarrow \text{ Subtracted 8 from both sides.}$$

$$-5 > y \quad \Leftarrow \text{ Simplified.}$$

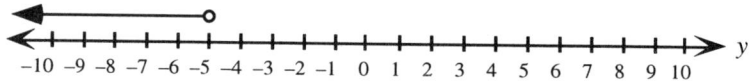

Notice the direction of the arrow is not the same as the direction of the inequality symbol.

c) $3x - 2(x + 1) \leq -6$

•SOLUTION•

$$3x - 2x - 2 \leq -6 \quad \Leftarrow \text{ Distributed the } -2.$$

$$x - 2 \leq -6 \quad \Leftarrow \text{ Combined like terms.}$$

$$x - 2 + 2 \leq -6 + 2 \quad \Leftarrow \text{ Added 2 to both sides.}$$

$$x \leq -4 \quad \Leftarrow \text{ Simplified.}$$

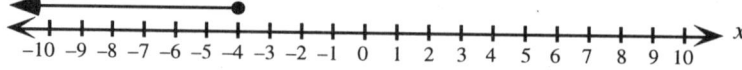

Practice 2.5.1 Solve and graph the solution set.

1. $x + 8 < -2$
2. $2x + 5 - x < -1 + 8$
3. $3 - 15 \leq x - 12$
4. $4y - 3(y + 1) \geq 5$
5. $5x - 4(x - 2) \geq 3$
6. $-6(8 - 1) > -(40 - x)$

Topic 2.5 B — Solving Inequalities Using the Multiplication and Division Properties

The multiplication and division properties of inequality are different from those of equations. Consider the following true inequality: $1 < 2$. If we multiply each side by a positive number, such as 3, we still have a true inequality.

$$1 \ < \ 2 \qquad \Leftarrow \text{ True inequality.}$$

$$3(1) \ < \ 3(2) \qquad \Leftarrow \text{ Multiplied each side by 3.}$$

$$3 \ < \ 6 \qquad \Leftarrow \text{ Inequality is still true.}$$

However, if we multiply each side by a negative number, such as -3, the inequality becomes false.

$$1 \ < \ 2 \qquad \Leftarrow \text{ True inequality.}$$

$$-3(1) \ < \ -3(2) \qquad \Leftarrow \text{ Multiplied each side by } -3.$$

$$-3 \ < \ -6 \qquad \Leftarrow \text{ Inequality is false.}$$

To make the inequality $-3 < -6$ true we must change the direction of the inequality symbol (that is, replace $<$ with $>$). Thus, to keep inequalities equivalent, we must reverse the inequality symbol whenever we multiply by a negative number.

The same situation is true for division. Dividing both sides of an inequality by a positive number produces an equivalent inequality but if we divide by a negative number we must reverse the inequality symbol in order to keep the inequality true.

— Property — ***Multiplication and*** ***division properties of*** ***inequality***	English: Multiplying or dividing both sides of an inequality by a positive number results in an equivalent inequality. Multiplying or dividing both sides of an inequality by a negative number *and reversing the inequality symbol* results in an equivalent inequality. Examples: • $2x > 6$ is equivalent to $\dfrac{2x}{2} > \dfrac{6}{2}$ • $-2x > 6$ is equivalent to $\dfrac{-2x}{-2} < \dfrac{6}{-2}$ Algebra: • $a > b$ is equivalent to $ca > cb$ if $c > 0$ • $a > b$ is equivalent to $ca < cb$ if $c < 0$ • $a > b$ is equivalent to $\dfrac{a}{c} > \dfrac{b}{c}$ if $c > 0$ • $a > b$ is equivalent to $\dfrac{a}{c} < \dfrac{b}{c}$ if $c < 0$

It's important to note that changing the inequality from $3 < x$ to $x > 3$ is **not** an example of changing the direction of the inequality symbol. Instead, these are just two ways of saying the same thing. That is, instead of saying, "three is less than x", we say, "x is more than three."

Example 2.5.2   Solve and graph the solution set.

a) $5x > 15$

•SOLUTION•

$$\frac{5x}{5} > \frac{15}{5} \qquad \Leftarrow \text{ Divided both sides by 5.}$$

$$x > 3 \qquad \Leftarrow \text{ Reduced.}$$

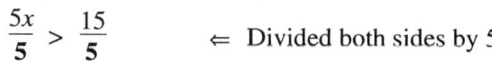

b) $-3t \le 21$

•SOLUTION•

$$\frac{-3t}{-3} \le \frac{21}{-3} \qquad \Leftarrow \text{ Divided both sides by } -3 \textbf{ and reversed the}$$
$$\textbf{inequality symbol.}$$

$$t \ge -7 \qquad \Leftarrow \text{ Reduced.}$$

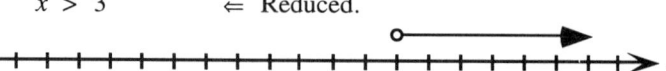

c) $\frac{y}{-3} \ge 1$

•SOLUTION•

$$(-3)\left(\frac{y}{-3}\right) \le (-3)1 \qquad \Leftarrow \text{ Multiplied both sides by } -3 \textbf{ and reversed the}$$
$$\textbf{inequality symbol.}$$

$$y \le -3 \qquad \Leftarrow \text{ Simplified.}$$

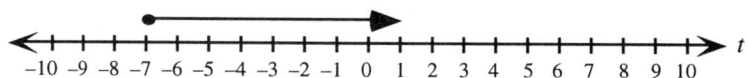

Practice 2.5.2   Solve and graph the solution set.

1. $4y \ge -16$ 2. $8 > -2x$ 3. $\frac{x}{3} < -3$ 4. $\frac{t}{-2} \ge 2$ 5. $\frac{-2x}{3} < 6$ 6. $10 > \frac{5x}{-2}$

Topic 2.5 C — Combining Properties to Solve Inequalities

The procedure we use to solve inequalities is almost exactly the same as the procedure we used to solve equations. The only difference is that we must remember to reverse the inequality symbol if we multiply or divide by a negative number.

— *Procedure* — *Solving inequalities*	***Step 1 LCD*** If fractions are present, multiply each term by the LCD. **If you multiply by a negative you MUST reverse the inequality symbol.**
	Step 2 Simplify Simplify each side of the inequality.
	Step 3 Move Terms Use addition or subtraction to move terms with variables to one side of the inequality and terms without variables to the other side. Then, simplify each side of the inequality again.
	Step 4 Divide Divide both sides of the inequality by the coefficient of the variable. **If you divide by a negative you MUST reverse the inequality symbol.**
	Step 5 Check Check your answer by substituting a solution into the original inequality. If a true statement results, your confidence should increase that your answer is correct. As a further check, substitute a number that is NOT a solution into the original inequality. A FALSE statement should result.

Example 2.5.3 Solve and graph the solution set.

    a) $4 - 2t > 10$

        •SOLUTION•

        Step 1 LCD   No fractions, so skip this step.

        Step 2 Simplify   Both expressions are in simplest form.

        Step 3 Move Terms

          $4 - \mathbf{4} - 2t \; > \; 10 - \mathbf{4}$ $\Leftarrow$ Subtracted 4 from both sides.

               $-2t \; > \; 6$ $\Leftarrow$ Combined like terms.

        Step 4 Divide

            $\dfrac{-2t}{\mathbf{-2}} < \dfrac{6}{\mathbf{-2}}$ $\Leftarrow$ Divided both sides by -2 **and reversed the inequality symbol**.

             $t \; < \; -3$ $\Leftarrow$ Simplified.

        Step 5 Check   The solution is $t < -3$. We leave the check to you.

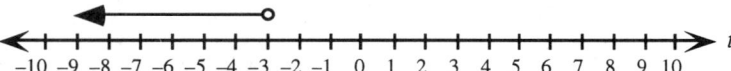

    b) $-x + 3 > 3x + 11$

        •SOLUTION•

        Step 1 LCD   No fractions, so skip this step.

        Step 2 Simplify   Both expressions are in simplest form.

        Step 3 Move Terms

        $-x + x + 3 \; > \; 3x + x + 11$ $\Leftarrow$ Added x to both sides.

               $3 \; > \; \mathbf{4x} + 11$ $\Leftarrow$ Combined like terms.

          $3 - \mathbf{11} \; > \; 4x + 11 - \mathbf{11}$ $\Leftarrow$ Subtracted 11 from both sides.

               $-8 \; > \; 4x$ $\Leftarrow$ Combined like terms.

        Step 4 Divide

            $\dfrac{-8}{\mathbf{4}} > \dfrac{4x}{\mathbf{4}}$ $\Leftarrow$ Divided both sides by 4. (The inequality symbol does *not* change because 4 is positive.)

             $-2 \; > \; x$ $\Leftarrow$ Reduced.

        Step 5 Check   The solution is $-2 > x$. We leave the check to you.

        Be careful graphing this. Since -2 is greater than the solutions, the solutions are values which are less than -2. Therefore, the arrow points to the left.

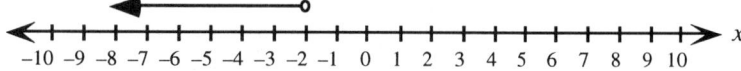

c) $x + 5 > x$

> •SOLUTION•
>
> **Step 1 LCD**   No fractions, so skip this step.
>
> **Step 2 Simplify**   Both expressions are in simplest form.
>
> **Step 3 Move Terms**
>
> $x + 5 - x \; > \; x - x$ ⇐ Subtracted x from both sides.
>
> $5 \; > \; 0$ ⇐ Simplified.
>
> The final inequality is an identity, which means any value of x will make the inequality true. So, the answer is "all real numbers". The graph of the solution set is the entire number line.

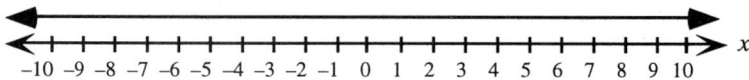

d) $x + 5 < x$

> •SOLUTION•   This is similar to the above problem but it has a very different solution.
>
> **Step 1 LCD**   No fractions, so skip this step.
>
> **Step 2 Simplify**   Both expressions are in simplest form.
>
> **Step 3 Move Terms**
>
> $x + 5 - x \; < \; x - x$ ⇐ Subtracted x from both sides.
>
> $5 \; < \; 0$ ⇐ Simplified.
>
> The final inequality is a contradiction, which means that no value of x will make the inequality true. Therefore, the answer is "no solution". The graph of the solution set is just the number line because no points are included.

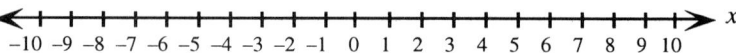

e) $\dfrac{1}{2} - \dfrac{x}{3} \geq x + \dfrac{9}{2}$

> •SOLUTION•
>
> **Step 1 LCD**
>
> $$6\left(\frac{1}{2}\right) - 6\left(\frac{x}{3}\right) \; \geq \; 6(x) + 6\left(\frac{9}{2}\right)$$ ⇐ Multiplied each term by the LCD of all the fractions.
>
> **Step 2 Simplify**
>
> $$3 - 2x \; \geq \; 6x + 27$$ ⇐ Reduced fractions.
>
> **Step 3 Move Terms**
>
> $$3 - 2x + 2x \; \geq \; 6x + 2x + 27$$ ⇐ Added $2x$ to both sides.
>
> $$3 \; \geq \; 8x + 27$$ ⇐ Combined like terms.
>
> $$3 - 27 \; \geq \; 8x + 27 - 27$$ ⇐ Subtracted 27 from both sides.
>
> $$-24 \; \geq \; 8x$$ ⇐ Combined like terms.
>
> **Step 4 Divide**
>
> $$\frac{-24}{8} \; \geq \; \frac{8x}{8}$$ ⇐ Divided both sides by 8.
>
> $$-3 \; \geq \; x$$ ⇐ Reduced fractions.
>
> **Step 5 Check**   The solution is $-3 \geq x$. We leave the check to you.

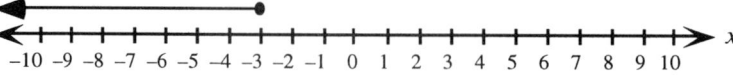

Practice 2.5.3   Solve and graph the solution set.

1. $2x + 5 \leq 7 + x$

2. $-x + 3 > 2x + 5$

3. $m - 6 > m + 5$

4. $2(4 - y) \leq -10$

5. $y + 1 > y - 6$

6. $3(y - 2) \geq 6(-1 - y)$

7. $-4(x - 2) + 12 < 7(1 + 2)$

8. $\dfrac{3}{4} + \dfrac{x}{4} > x - \dfrac{3x}{4}$

9. $\dfrac{1}{5} - \dfrac{x}{3} > \dfrac{1}{2} + \dfrac{2}{3}$

10. $2x - \dfrac{x}{4} \leq \dfrac{x}{2} - \dfrac{1}{3}$

Exercise Set 2.5   The answers to the odd numbered exercises are at the back of the book.

Solve and graph the solution set.

1. $x - 2.5 \geq 4.3$

2. $x - 4.1 \geq -2.1$

3. $y + \dfrac{3}{4} < \dfrac{1}{2}$

4. $y - \dfrac{2}{3} > \dfrac{1}{6}$

5. $\dfrac{y}{2} \geq -3$

6. $\dfrac{x}{4} < 2$

7. $-2.5x > 7.5$

8. $-1.2y < 3.6$

9. $2(2 - x) \geq -2$

10. $3(y - 2) \leq -9$

11. $4(x - 1) + 3 < 3$

12. $2(y + 1) - 7 > 3$

13. $3x + \dfrac{11}{2} \geq \dfrac{75}{10} + x$

14. $5y - \dfrac{5}{2} \leq \dfrac{11}{2} + y$

15. $2(p - 2) + 3(3 - p) > 5$

16. $-2(x - 5) + 3 > 5 - 2(x - 6)$

17. $-(x + 17) - 4x \leq 5(x + 3) - 10x$

18. $3 + 2(x + 1) > 4x - 2(8 + x)$

19. $22 + 16(x + 2) > -4x + 6(x + 9)$

20. $-4(y - 1) + 6(1 - y) \leq 2(y + 5) + 8y$

21. $\dfrac{1}{4}\left(x + \dfrac{1}{2}\right) - \dfrac{1}{3} \leq \dfrac{1}{6}\left(x - \dfrac{1}{4}\right)$

22. $\dfrac{3}{2}\left(y + \dfrac{1}{3}\right) - \dfrac{5}{6} < \dfrac{2}{3}\left(y - \dfrac{1}{2}\right)$

Review Exercises   The answers to all of these exercises are at the back of the book.

23. Write a mathematical model for this sentence: The number of people in the elevator must not exceed 12.

24. Translate to a mathematical model: The sum of 3 and a number.

25. Translate to a mathematical model: Eight less than a number.

26. Graph $x \geq 3$

Simplify.

27. $5(c - 3) - (3 - c)$

28. $\dfrac{2}{5} \div \dfrac{8}{15} \cdot \dfrac{4}{9} + \dfrac{1}{2} \cdot \dfrac{5}{3}$

29. $\dfrac{1}{5}(2p - 5) + \left(\dfrac{p}{2} + 3\right)$

Solve.

30. $0.25(300 - m) + 0.10m = 56.55$

31. $9x = -4x$

32. $3\left(\dfrac{x}{2} - 3\right) - (x - 2) = \dfrac{1}{3}(x - 6) - 5$

33. $5x + 1 = 3 - (3x + 2) + 8x$

34. $2x + 3y = 7$ for y

Chapter 2 Review

Vocabulary to Know

conditional equation — an equation that is true for some values of the variable and not true for other values; $x + 3 = 2$ is a conditional equation.

conditional inequality — an inequality that is true for some values of the variable and not true for other values; $x + 3 > 2$ is a conditional inequality.

contradiction — an equation or inequality that is always false; $x + 1 = x$ is a contradiction..

critical point — the point on a number line that separates the solutions from the non-solutions of an inequality; 3 is the critical point of $x < 3$.

equation — a statement of equality between two expressions.

equivalent equations — equations that have the same solution set.

equivalent inequalities — inequalities that have the same solution set.

formula — a literal equation that represents a rule, a fact, or a way of doing something, such as $p = 2l + 2w$.

identity — an equation or inequality that is always true; $x + 2 = 2 + x$ is an identity..

in terms of — in a literal equation, the isolated variable is solved for in terms of the other variables; in the formula $a = lw$, a is solved for in terms of l and w.

inequality — a statement that one expression is greater than, greater than or equal to, less than, or less than or equal to another.

linear equation in one variable — an equation that contains one variable whose exponent is 1; for example, $2x + 3 = 5$.

literal equation — an equation that involves more than one variable; the formula $a = lw$ is a literal equation.

root of an equation or inequality — a value of a variable that makes an equation or inequality true (same as solution)..

satisfy an equation or inequality — a value satisfies an equation or inequality if it makes the statement true; 2 satisfies $x + 1 = 3$..

solution — a value of a variable that makes an equation or inequality true (same as root)..

solution set — the collection of all solutions to an equation or inequality..

Properties, Definitions, Formulas, and Notation to Understand

Symmetric Property of Equality — Exchanging the expressions on the left and right sides of an equation produces an equivalent equation.	$a = b$ is equivalent to $b = a$ $x = 4$ and $4 = x$ are equivalent.
Addition Property of Equality — Adding the same number (or expression) to both sides of an equation results in an equivalent equation.	$a = b$ is equivalent to $a + c = b + c$ $x - 5 = -12$ is equivalent to $x - 5 + \mathbf{5} = -12 + \mathbf{5}$ $8 - 3x = -2x$ is equivalent to $8 - 3x + \mathbf{3x} = -2x + \mathbf{3x}$
Subtraction Property of Equality — Subtracting the same number (or expression) from both sides of an equation results in an equivalent equation.	$a = b$ is equivalent to $a - c = b - c$ $x + 2 = 6$ is equivalent to $x + 2 - \mathbf{2} = 6 - \mathbf{2}$ $x + 5 = 2x$ is equivalent to $x + 5 - \mathbf{x} = 2x - \mathbf{x}$
Multiplication Property of Equality — Multiplying both sides of an equation by the same *non-zero* number (or expression) results in an equivalent equation.	If $c \neq 0$, then $a = b$ is equivalent to $ca = cb$ $\dfrac{x}{3} = 7$ is equivalent to $\mathbf{3}\left(\dfrac{x}{3}\right) = \mathbf{3}(7)$
Division Property of Equality — Dividing both sides of an equation by the same *non-zero* number (or expression) results in an equivalent equation.	If $c \neq 0$, then $a = b$ is equivalent to $\dfrac{a}{c} = \dfrac{b}{c}$ $3x = 12$ is equivalent to $\dfrac{3x}{\mathbf{3}} = \dfrac{12}{\mathbf{3}}$
Addition Property of Inequality — Adding the same number (or expression) to both sides of an inequality results in an equivalent inequality.	$a > b$ is equivalent to $a + c > b + c$ $x - 5 \geq -12$ is equivalent to $x - 5 + \mathbf{5} \geq -12 + \mathbf{5}$
Subtraction Property Of Inequality — Subtracting the same number (or expression) from both sides of an inequality results in an equivalent inequality.	$a > b$ is equivalent to $a - c > b - c$ $5 + 2x < 3x + 9$ is equivalent to $5 + 2x - \mathbf{2x} < 3x + 9 - \mathbf{2x}$

Multiplication Property of Inequality —

- Multiplying both sides of an inequality by a positive number (or expression) results in an equivalent inequality.

- Multiplying both sides of an inequality by a negative number (or expression) *and reversing the inequality symbol* results in an equivalent inequality.

- If $c > 0$, then $a > b$ is equivalent to $ca > cb$

  $2x > 6$ is equivalent to $\mathbf{5}(2x) > \mathbf{5}(6)$

- If $c < 0$, then $a > b$ is equivalent to $ca < cb$

  $2x > 6$ is equivalent to $\mathbf{-5}(2x) < \mathbf{-5}(6)$

Division Property of Inequality —

- Dividing both sides of an inequality by a positive number (or expression) results in an equivalent inequality.

- Dividing both sides of an inequality by a negative number (or expression) *and reversing the inequality symbol* results in an equivalent inequality.

- If $c > 0$, then $a > b$ is equivalent to $\dfrac{a}{c} > \dfrac{b}{c}$

  $2x > 6$ is equivalent to $\dfrac{2x}{\mathbf{2}} > \dfrac{6}{\mathbf{2}}$

- If $c < 0$, then $a > b$ is equivalent to $\dfrac{a}{c} < \dfrac{b}{c}$

  $-2x > 6$ is equivalent to $\dfrac{-2x}{\mathbf{-2}} < \dfrac{6}{\mathbf{-2}}$

Procedures to Follow

Solving equations.	$2(x - 3) - \dfrac{x}{4} = \dfrac{5x}{6} + 5$
Step 1 LCD If fractions are present, multiply each term by the LCD.	$(\mathbf{24})(\,2(x - 3)\,) - (\mathbf{24})\left(\dfrac{x}{4}\right) = (\mathbf{24})\left(\dfrac{5x}{6}\right) + (\mathbf{24})(5)$
Step 2 Simplify Simplify each side of the equation.	$\mathbf{48}(x - 3) - \mathbf{6}x = 4(5x) + \mathbf{120}$ $\mathbf{48}x - \mathbf{144} - 6x = \mathbf{20}x + 120$ $\mathbf{42}x - \mathbf{144} = 20x + 120$
Step 3 Move Terms Use addition or subtraction to move terms with variables to one side of the equation and terms without variables to the other side. Then, simplify each side again.	$42x - 144 \mathbf{- 20x} = 20x + 120 \mathbf{- 20x}$ $\mathbf{22}x - 144 = 120$ $22x - 144 \mathbf{+ 144} = 120 \mathbf{+ 144}$ $\mathbf{22}x = \mathbf{264}$
Step 4 Divide Divide both sides of the equation by the coefficient of the variable.	$\dfrac{22x}{\mathbf{22}} = \dfrac{264}{\mathbf{22}}$ $x = 12$
Step 5 Check Substitute your answer into the original equation. If a true statement results, your answer is probably correct. If a false statement results, go back over your work to see if you made an error.	$2((12) - 3) - \dfrac{(\mathbf{12})}{4} = \dfrac{5(\mathbf{12})}{6} + 5$ $2(9) - \mathbf{3} = 10 + 5$ $15 = 15$ *Checks*

Graphing an inequality	$x \leq 5$	$-2 > x$
Step 1 Number line Draw a number line and label it with the appropriate variable.		
Step 2 Critical point Determine the critical point and plot it on the number line: • If the inequality symbol is > or <, place an open circle at the critical point. • If the inequality symbol is ≥ or ≤, place a closed circle at the critical point.		
Step 3 Test point Select any other point on the number line to serve as a test point: • If the test point is a solution of the inequality, draw a line from the critical point through the test point and place an arrowhead on that end of the line. • If the test point is NOT a solution of the inequality, draw a line from the critical point away from the test point and place an arrowhead on that end of the line.	Since $1 \leq 5$, we use 1 as a test point 	Since $-2 > -5$, we use -5 as a test point

Solving inequalities	$5 - \dfrac{2}{3}x < 2 - \dfrac{1}{6}(12 - x)$	
Step 1 LCD If fractions are present, multiply each term by the LCD. **If you multiply by a negative you MUST reverse the inequality symbol.**	$(6)5 - (6)\dfrac{2}{3}x < (6)2 - (6)\dfrac{1}{6}(12 - x)$	
Step 2 Simplify Simplify each side of the inequality.	$30 - 4x < 12 - (12 - x)$ $30 - 4x < x$	
Step 3 Move Terms Use addition or subtraction to move terms with variables to one side of the inequality and terms without variables to the other side. Then, simplify each side of the inequality again.	$30 - 4x + 4x < x + 4x$ $30 < 5x$	
Step 4 Divide Divide both sides of the inequality by the coefficient of the variable. **If you divide by a negative you MUST reverse the inequality symbol.**	$\dfrac{30}{5} < \dfrac{5x}{5}$ $6 < x$	
Step 5 Check Check your answer by substituting a solution into the original inequality. If a true statement results, your confidence should increase that your answer is correct. As a further check, substitute a number that is NOT a solution into the original inequality. A FALSE statement should result.	Check $x = 12$ (solution) $5 - \dfrac{2}{3}(12) < 2 - \dfrac{1}{6}(12 - (12))$ $5 - 8 < 2 - 0$ $-3 < 2$ True so 12 is a solution.	Check $x = 0$ (not a solution) $5 - \dfrac{2}{3}(0) < 2 - \dfrac{1}{6}(12 - (0))$ $5 - 0 < 2 - 2$ $5 < 0$ False so 0 is not a solution.

Chapter 2 Review Exercises
The answers to all the exercises are in the back of the book.

1. Is -3 a solution to $5x - (x - 2) = -10$

2. Fill in the box so the first and second equations are equivalent: $x = 3$ and $x - 5 = \square$

3. Fill in the box so the first and second equations are equivalent: $-2 = y$ and $3 - y = \square$

Solve.

4. $6 = -\dfrac{2}{5}x$

5. $\dfrac{x-3}{2} = -7$

6. $-2 = \dfrac{w}{3}$

7. $8x = -x$

8. $-3(2x-5) = -3$

9. $4 - 3(x-7) = 12 - (x-2) + 5$

10. $5 - (3-x) = 6(3+x) - x$

11. $\dfrac{3}{4}x - 5 = -8 - \dfrac{1}{6}(x+4)$

12. $\dfrac{1}{4} - \dfrac{x}{2} = 5\left(x - \dfrac{1}{2}\right)$

13. $x - \dfrac{2}{3}(x+8) = -\dfrac{5}{6}(x+5)$

14. $0.1(x+2) + 5 = 0.2x$

15. $3x - 5(x-2) = 12 - 2x$

16. $8x - 2(5x+1) = 8 - 2(x+3)$

17. $\dfrac{1}{2}x + 3(x-2) = -6$

18. $8 + 3x = 5(1-x) + 3$

19. Use "is solved for" and "in terms of" to discuss the literal equation: $a = 3b + 2c$

20. Show that $-\dfrac{2}{3}x + 2$ is a solution when $2x + 3y = 6$ is solved for y.

Solve for the specified variable.

21. $5x - 2y = 3$ for y

22. $I = \dfrac{e+E}{nr+R}$ for R

23. $19a + 4b = 3(2b+a)$ for b

24. $2m + 3(m-p) = 7(p-m)$ for p

Graph on a number line.

25. $8 > w$

26. $c \le 3$

27. $-2 > m$

28. $w \ge -1$

Write a mathematical model for this sentence

29. The minimum height is 72 inches.

30. At most 8 chairs can be painted at one time.

31. The room must seat more than 20 people.

32. All weights must be under 5 pounds.

33. Translate into English: $8 < x$

34. Translate into English: $x \ge 5$

35. Translate into English:

k

36. Solve $10 - 2(x+3) < 3x - 1$

37. Solve $\dfrac{1}{2}x \le 3(x-2) + \dfrac{1}{3}(x+1)$

38. Solve $8 - (x-2) \le 5x + 4(1-x)$

Chapter 2 Test   The answers to all the exercises are in the back of the book.

1. Is $\dfrac{5}{2}$ a solution of $3 - (x-2) = 5x + 17$

2. Fill in the box so the first and second equations are equivalent: $n = -6$ and $4 + n = \square$

Solve.

3. $3 = \dfrac{n-5}{2}$

4. $5 + 4(x-2) = -11$

5. $1 - (x-3) = 5(x-6) - 20$

6. $\dfrac{1}{2}x - (3-x) = \dfrac{2}{3}x - \dfrac{1}{2}$

7. $8 - 0.1(x-2) = 5 + 0.4(x-3)$

8. $9 - 2(x-3) + 5x = 7x - 4(x-3) - 3$

9. $4x + 7 = 5x - (3-x)$

10. $-3x = 9x$

11. $8 + 2y = 4x$ for y

12. $3x - (5-x) > 8(x-2) + 3$

13. $\dfrac{1}{3}(x-2) \le \dfrac{3}{4}(x-6) - \dfrac{1}{6}$

14. $V = \dfrac{1}{2}w(a-c)$ for a

15. Use "is solved for" and "in terms of" to discuss the literal equation: $w = 5d + 2g$

16. Show that $\dfrac{w-t}{2s}$ is a solution when $w = t + 2sv$ is solved for v.

17. Graph $-4 \le m$

18. Write a mathematical model for this sentence: -5 is less than the temperature.

19. Translate into English:

p

Chapter 3
Using Algebra to Solve Problems

Section 3.1 Modeling with Expressions

Algebra is an important tool that can be used to help solve real–world problems. The basic idea is to take a problem described in English and to translate as much of it as we can into an algebraic equation or inequality, which we can solve. If the translation from English to algebra was a good one, the solution to the equation or inequality should be a big step in finding the solution to the real–world problem. In Chapter 2 we discussed ways of solving equations and inequalities. In this Chapter, we focus on how to do the translations.

Keep in mind that in this section we are asking you to build expressions, not solve equations. This means you will not be "solving for x." Instead, just as the expression techniques in Chapter 1 allowed you to solve equations in Chapter 2 working with expressions in this section will help you succeed with word problems in later sections.

Topic 3.1 A — Expressions Involving Relative Amounts

To describe real–world situations using algebra we must build algebraic expressions that represent English phrases. For example, suppose a classroom has ten more chairs than students. If we let x represent the number of students, then we can write an algebraic expression for the number of chairs by noting that *10 more* suggests we add 10 to the number of students. Therefore, we can use $x + 10$ to represent the number of chairs.

- We refer to *ten more chairs than students* as a **verbal model** because it describes the situation using words.

- We refer to *x + 10* as a **mathematical model** because it describes the situation using mathematics.

When solving real–world problems, we have to be sure that the problem is faithfully described by the verbal model, and that the verbal model is precisely represented by the mathematical model. Then, we can use the mathematical model to help solve the original problem.

To make sure the mathematical model matches the verbal model, it's a good idea to check the translations. Below are two methods that can be used to check the correctness of translations:

- **Relative sizes check** In this method, we use intuition to determine which item should have more and then see if the algebra is consistent with this. In the example where we have ten more chairs than students, intuition tells us there should be more chairs. Comparing the algebraic expressions, we see that $x + 10$ (the expression for the number of chairs) is always larger than x (the expression for the number of students). Since intuition and the algebra are in agreement, we are more confident that the translation is correct.

- **Concrete numbers check** In this method, we make up a number for the given variable and then use intuition and the algebraic expression to calculate the other number. If the numbers agree, our confidence in the translation will increase. In the students/chairs example, let's say we have 30 students. Intuition tells us that there should be 40 chairs (10 more than the number of students). The translation, $x + 10$, also gives 40 when we replace x with 30. Since intuition and the algebra match, we think the translation is correct.

Example 3.1.1　a)　A dance has attracted 8 more men than women. Assume that m represents the number of men. Use the *relative sizes check* and the *concrete numbers check* to see if $m - 8$ may be used to represent the number of women.

•**SOLUTION**• *Relative sizes check*: The problem states that there are more men, m, and the expression for the number of women, $m - 8$, represents a number that is smaller than m. This checks.

Concrete numbers check: If we pretend the number of men, m, is 30 then the number of women, $m - 8$, is 22. This checks because there are more men than women and 30 is more than 22.

b) The number of students at a college is 12 times the number of professors. If we let s represent the number of students, use the *relative sizes check* and the *concrete numbers check* to see if 12s may be used to represent the number of professors.

•**SOLUTION**• *Relative sizes check:* The problem states that there are more students than professors. Since the number of students, s, must be positive, the number of professors, 12s, is *larger* than s. This does *not* check.

Concrete numbers check: If we pretend the number of students, s, is 1200 then the number of professors, 12s, is 14400. This does not check since there must be more students than professors. [†]

Example 3.1.2   a)   A jar contains dimes and nickels. Let n represent the number of nickels. Using n, write an expression for the number of dimes if there are 12 more dimes than nickels.

•**SOLUTION**• You may be able to do this problem by just looking at it, but we want to develop a process that will work for much more complicated problems. We begin by organizing the information in the form of a table which consists of vertical columns and horizontal rows.

The first column is for the names of the items to which the numbers refer; that is, the dimes and nickels.

The second column is for the mathematical models that represent the number of each item:

First column ⟱   *Second column* ⟱

Items	Math Model for number of coins
nickels	
dimes	

We are given that n represents the number of nickels, so n is placed in the *Math Model* column, across from *nickels*.

Items	Math Model for number of coins	
nickels	n	⟸ Given.
dimes		⟸ To be determined.

To fill in the cell for the number of dimes we note that there are 12 more dimes than nickels. If we start with the number of nickels, n, we add 12 to get the number of dimes. That is, the number of dimes may be represented by $n + 12$.

Items	Math Model for number of coins	
nickels	n	⟸ Given.
dimes	$n + 12$	⟸ Translation.

Let's check this using the *Relative sizes* method. Since we are told that there are 12 more dimes than nickels, there are clearly more dimes. If we look at the two expressions, n and $n + 12$, we see that the larger expression, $n + 12$, represents the number of dimes. This makes us comfortable that the translation is correct.

[†]   For your information, the correct expression for the number of professors is $\frac{s}{4}$.

b)   A jar contains dimes and quarters. Let q represent the number of quarters. Using q, write an expression for the number of dimes if there are 3 more quarters than dimes.

•SOLUTION•   Your intuition might lead you to this **incorrect** translation:

Items	Math Model for number of coins	
quarters	q	⇐ Given.
dimes	$q + 3$	⇐ Translation.

Let's check this using a concrete example. If there are 10 quarters then intuition tell us there should be 7 dimes (since there are 3 more quarters than dimes). However, if we let q be 10, then the expression for the dimes, $q + 3$, simplifies to 13, not 7.

Checking with a concrete example may give you a hint as to the proper translation. Since 10 quarters should correspond to 7 dimes maybe the translation should be $q - 3$.

Items	Math Model for number of coins	
quarters	q	⇐ Given.
dimes	$q - 3$	⇐ Translation.

The number of dimes, $q - 3$, is consistent with the check.

c)   Let w represent the number of women at a dance. Using w, write an expression for the number of men if there are 7 less men than women.

•SOLUTION•   The word *less* implies subtraction so we will subtract w and 7. Since subtraction is not commutative we must be careful of the order in which we write the terms. The expression $(7 - w)$ means "w less than 7". But, we want "7 less than w", so the correct expression is $(w - 7)$. [†]

Items	Math Model for number of people	
women	w	⇐ Given.
men	$w - 7$	⇐ Translation.

As a check, intuition tells us that there are more women than men. If you look at the two expressions, the larger expression, w, represents the number of women. This makes us comfortable that the translation is correct.

— Note — *Using a sketch to organize information*	You may find some of the translations in this section frustrating and wonder why we couldn't just, "say it in a straight–forward way." There are two reasons. First, what's straight–forward to one person is not straight–forward to someone else. Consequently, you need to become comfortable with many different English ways of stating what turns out to be the same mathematical sentence. Second, you need to begin to develop techniques for organizing important given information in meaningful and useful ways. Organized information will greatly help you free up attention so that you may clearly think about patterns and connections.

[†]   Note that the phrase "w less than seven" translates to the expression $w - 7$, whereas the sentence "w **is** less than seven" translates as the inequality $w < 7$.

Practice 3.1.2 For each question, be sure to use the *relative sizes check* and the *concrete numbers check* to see if the expressions you create are correct.

1. A cinema sold adult tickets and child tickets. Given that *a* represents the number of adult tickets, use *a* to write an expression that represents the number of child tickets if

    a) there were twelve more child tickets than adult tickets sold.
    b) there were twelve less adult tickets than child tickets sold.
    c) they sold an equal number of adult and child tickets.
    d) the adult tickets sold exceeded the child tickets by 400.
    e) 400 fewer adult tickets than child tickets were sold.
    f) Referring to part e), does the problem say there are more adult or child tickets?
    g) Referring to part e), use your expression to calculate the number of child tickets if there are 900 adult tickets. Does this agree with your answer to part f)?

2. A chemist mixed a strong acid solution with a weak acid solution. If *w* represents the milliliters of weak acid in the final mixture, use *w* to write an expression to represent the milliliters of strong acid in the final mixture if

    a) half the final mixture was weak acid and half was strong acid.
    b) there were 75 more milliliters of weak acid in the final mixture.
    c) the amount of weak acid was 10 milliliters less than the amount of strong acid.
    d) there was 10 less milliliters of strong acid in the final mixture.
    e) Referring to part d), does the problem say there is more strong acid or weak acid?
    f) Referring to part d), use your expression to calculate the amount of strong acid if the final mixture contains 80 ml of weak acid. Does this agree with your answer to part e)?

3. A student decides to invest part of her monthly income in a savings account and part in a mutual fund. If *s* represents the dollars invested in the savings account use *s* to write an expression to represent the dollars invested in the mutual fund if

    a) the mutual fund receives $300 less than the savings account.
    b) the amount in the mutual fund is decreased by $300 you would have the amount in the savings account.
    c) there was $300 more invested in the savings account than in the mutual fund.
    d) she invested 300 fewer dollars in the savings account.
    e) the amount in the mutual fund exceeded the amount in the savings account by $300.
    f) Referring to part e), does the problem say more money is invested in the savings account or the mutual fund?
    g) Referring to part e), use your expression to calculate the amount in the mutual fund if $900 was invested in savings. Does this agree with your answer to part f)?

Topic 3.1 B — A Word About Units

In Section 1.2 we said that a number has two parts: *magnitude* (size), which tells us how much we have and *sign*, which tells us whether the number is greater than or less than 0. In many applied problems, we must consider a third attribute of number, which we call **unit**. Units such as *feet*, *minutes*, or *Mexican pesos* give us a point of reference when we measure things. For example, suppose a man asks you to wash his car and says "I'll give you ten to do the job". When you're done, he thanks you and hands you a dime. After a look of disbelief, you say you were expecting 10 *dollars* not 10 *cents*. You had agreed on the size of the number, and its sign, but you had not discussed the unit of measurement. Obviously, this can cause a lot of misunderstanding. [†]

Paying attention to units will help us check to see if we are combining expressions correctly. For example, we can add 2 + 3 if they represent a *number* of objects.

- we can add 2 students + 3 chairs to get 5 objects.

- we can add 2 dimes + 3 nickels to get 5 coins.

[†]   A *unit* is a multiple of a *standard*. A standard is an agreed upon size for a measurable quantity, such as length or weight. When we make a measurement we make a comparison between some aspect of the object we are measuring and the standard. For example, we often use the foot as a unit of length. In the old days, the standard for the foot was the length of a human foot. If a woman said a tree was 10 feet tall she was saying the tree had a height equivalent in length to 10 of her feet. Since everybody knew how long their own foot was, they could get an idea of how tall the tree was. The problem was that every human foot has a different length and so 10 of her feet was different from 10 of anyone else's feet. As the precision of measurement became more important, the standards changed so that a "foot" was defined the same for everybody. Today, the definitions are quite precise but sometimes difficult to understand. For example, until recently, the standard unit of measure for length was the wavelength of the orange–red line of the spectrum of krypton–86 in a vacuum (one foot was defined as 77,988.8376377953 times this wavelength).

But, if the numbers represent different **measured** quantities, then we cannot add them. For example, if 2 represents the miles you walked and 3 represents the minutes you rested, then it makes no sense to add them. That is, we cannot add 2 miles + 3 minutes.

It's possible to add expressions that have different units, as long as the expressions represent the same fundamental quantity. For example, we can add 2 feet + 3 inches, but we don't get 5. Since feet and inches represent the same fundamental quantity, length, we can convert them to the same unit and then add. Converting 2 feet to inches changes the problem to 24 inches + 3 inches, which simplifies to 27 inches.

Later in this section we will work with geometry problems where we multiply numbers that have units. For example, we will find the area of a square whose side has a length of 8 inches.

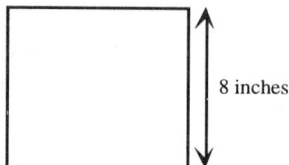

Since the area of a square is the length of one side squared, the area of this square is $(8 \text{ in})^2$ or $(8 \text{ in})(8 \text{ in})$. To simplify this, we multiply $(8) \cdot (8)$ to get 64 and we write $(\text{in}) \cdot (\text{in})$ as in^2. The final product is 64 in^2, which is read *sixty–four square inches* (NOT *sixty–four inches squared*). The unit is square inches; 64 in^2 means we can fill up the big square with 64 smaller squares, each of which is one inch by one inch.

We raise the issue of units now so you can start using them to help you decide if your work in the rest of this section is correct.

Topic 3.1 C — The Whole-Part Concept

In the problems completed in Topic 3.1A, the English phrase was given to you directly. Sometimes the English phrase is not given directly and it's up to you to determine what the unwritten phrase is. For example, let's say a theater sold a total of 156 adult and child tickets. If we let a represent the number of adult tickets, how could we write an expression for the number of child tickets? To do this, we will use the Whole–Part Concept. The **Whole-Part Concept** is based on the fact that if a whole is made up of the sum of its parts then one part may be represented by the difference of the whole and the other parts. For example, suppose we have a jar that contains 20 coins, all dimes and quarters, and we want to know how many quarters are in the jar. If we know there is 1 dime then our intuition tells us there must be 19 quarters; if we know there are 15 dimes, then there must be 5 quarters; and so on. In each case one part may be represented by the difference of the whole and the other part.

Example 3.1.3   a)   A theater sold a total of 156 adult and child tickets. If a represents the number of adult tickets, use a and the Whole–Part Concept to write an expression for the number of child tickets.

•SOLUTION• The *whole* is the total number of tickets; the parts are the number of adult tickets and the number of child tickets.

whole	−	part	⇐ Whole-Part Concept
total number of tickets	−	number of adult tickets	⇐ Replaced the whole and part with specific information. This is the verbal model.
156	−	a	⇐ Translated the verbal model into a mathematical model. Note that 156 and a have the same unit (tickets) and so can be added or subtracted.

— Note — ***Doing work in your head***	If you can do the work in your head and write the expression directly, that's fine. However, it's best to practice with the details in order to prepare for those problems where your intuition is not sufficient.

b) At 9:00 am a fast train leaves Albertville and travels towards Bakersfield, which is 60 miles away. At the same time, a slow train leaves Bakersfield on a parallel track and travels toward Albertville. When the trains meet, the slow train has traveled s miles. Use s and the Whole-Part Concept to write an expression for the distance traveled by the fast train.

•**SOLUTION**• A sketch may help you to visualize the details of this problem. The sketch does not need to be pretty but it should include as much of the given information as possible. For example,

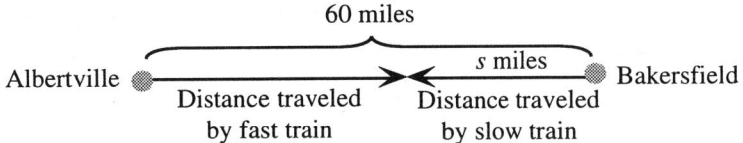

From the sketch, we can see that the *whole* is the 60 miles between the two cities and the *parts* are the distances traveled by each train.

Items	*Distance (miles)*	
slow train	s	⇐ Given.
fast train		⇐ To be determined.

The distance traveled by the fast train can be represented as

whole	−	part	⇐ Whole-Part Concept.
total distance traveled by both trains (miles)	−	distance traveled by slow train (miles)	⇐ Replaced the whole and part with specific information. This is the verbal model.
60	−	s	⇐ Translated the verbal model into a mathematical model. Note that 60 and s have the same unit (miles) and so can be added or subtracted.

The statement of this problem includes some information that is not used to answer the question. The time the trains left their stations, and the fact that one was traveling faster than the other are irrelevant when we write the expression for the distance traveled by the fast train.

— *Note* — *Using a sketch to organize information*	Notice that a sketch helped to organize the given information. A sketch is often useful when geometric ideas, such as distance, are being discussed.

Practice 3.1.3

1. A minor league baseball team sells two types of tickets, choice and bleacher. A total of 385 tickets were sold. If c represents the number of choice tickets, use c to write an expression for the number of bleacher tickets.

2. A florist is making an arrangement that includes r roses and some other flowers. Each rose costs $2.50, and the other flowers each cost $1.25. Use r to write an expression for the number of other flowers in the arrangement if the cost for the arrangement of 22 flowers is $37.50.

3. After fishing for 4 hours a couple notices that the weather is deteriorating and starts traveling the 1.75 miles to shore. At the same time a thunderstorm, moving at 17 mph, reaches the opposite shore of the lake. If d represents the distance the couple has traveled since starting to head back, use d to write an expression for the distance they have left to cover until they reach the shore.

Topic 3.1 D — The Whole-Part Concept and Combining Like Terms

The Whole-Part Concept also applies if there are more than two parts. To find the unknown part in such cases we subtract the known parts from the whole. If we have built the parts correctly they will have like units and so can be combined into a simpler equivalent expression.

Example 3.1.4   a)   A bag of 80 coins contains nickels, dimes, and quarters. There are 5 more nickels than dimes. If d is the number of dimes, use d to write a simplified expression for the number of quarters.

• SOLUTION •   This problem has three items: nickels, dimes, and quarters. We have specific information regarding two of the items (the nickels and dimes), but not the third (the quarters). A table will help organize the information.

Items	*Number of Coins*	
dimes	d	⇐ Given.
nickels	$d + 5$	⇐ 5 more nickels than dimes.
quarters		⇐ To be determined.

We use the Whole–Part Concept to build a verbal model for the number of quarters:

total number of coins	−	number of dimes	−	number of nickels	⇐ Verbal model.
80	−	d	−	$(d + 5)$	⇐ Translation. We need the parentheses around $(d + 5)$ because we are subtracting the entire expression.
80	−	d	−	$d - 5$	⇐ Removed parentheses.
		$75 - 2d$			⇐ Simplified

The number of quarters can be represented by the expression $75 - 2d$.

b)   Three prizes, totaling $2100, are being awarded for a talent show. Write a simplified expression for the value of the first prize if x represents the value of the second prize and the third prize is $400 less than the second prize.

• SOLUTION •   First, we use a table to organize the given information.

Items	*Value (dollars)*	
second prize	x	⇐ Given.
third prize	$x - 400$	⇐ $400 less than the second.
first prize		⇐ To be determined.

Next, we use the Whole–Part Concept to build a verbal model for the first prize:

total value of prizes (dollars)	−	value of second prize (dollars)	−	value of third prize (dollars)	⇐ Verbal model.
2100	−	x	−	$(x - 400)$	⇐ Translation.
2100	−	x	−	$x + 400$	⇐ Removed parentheses.
		$2500 - 2x$			⇐ Combined like terms.

The value of the first prize can be represented by the expression $2500 - 2x$.

Practice 3.1.4

1. A man buys 120 32¢, 28¢ and 20¢ stamps for three different mailings. Write a simplified expression to represent the number of 32¢ stamps if there are 12 more 20¢ stamps than 28¢ stamps and x represents the number of 28¢ stamps.

2. A bus leaves Leafburg and then has two stops before reaching it's destination at Upland which is 235 miles away. The distance from the first to the second stop is half the distance from Leafburg to the first stop. Write a simplified expression to represent the distance from the second stop to Upland if x represents the distance from Leafburg to the first stop.

3. A day care center buys 45 packages of small, medium, and large diapers. Write a simplified expression to represent the number of packages of small diapers they buy if m represents the number of medium packages purchased and 5 less packages of large than packages of medium are purchased.

4. A homeowner combines dish soap, tobacco juice, and water to make a bug spray for her garden. If w is the amount of water in the 2 gallon final mixture, and the amount of dish soap is one–one hundredth the amount of water, write a simplified expression to represent the amount of tobacco juice in the final mixture.

Topic 3.1 E — The Whole-Part Concept and Geometry

The problems that follow use formulas from geometry. Because some of the problems ask you to find an area, we have included some area formulas below.

Figure	Formula	Variables
Square	$a = s^2$	a represents the area. s represents the length of one side.
Triangle	$a = \frac{1}{2}bh$	a represents the area. b represents the length of base. h represents the height.
Circle	$a = \pi r^2$	a represents the area. π represents a constant whose value is approximately 3.14. r represents the radius of the circle.

Example 3.1.5 a) Use x to write a simplified expression for the length of Side B in the figure below.

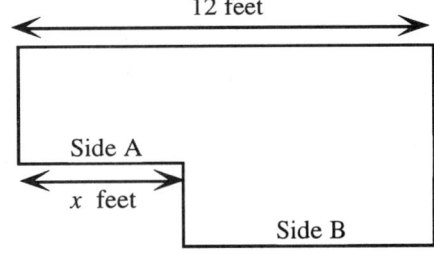

•SOLUTION• It looks like both the top and bottom (sides A and B taken together) have the same length. Therefore, we will think of the whole as the length of the top side (12 feet) and the parts as the lengths of Side A and Side B.

Items	Length (feet)	
Side A	x	⇐ Given.
Side B		⇐ To be determined.

The length of Side B is

whole	–	part	⇐ Whole-Part Concept.
total length (feet)	–	length of Side A (feet)	⇐ Verbal model.
12	–	x	⇐ Translated the verbal model into a mathematical model.

The length of Side B can be represented by the expression $12 - x$.

b) Use x to write a simplified expression for the area of the shaded region between the circle and the square.

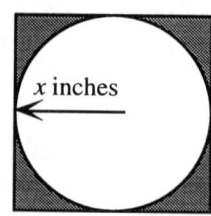

•SOLUTION• To find the shaded area, we can think of the area of the square as the whole and the areas of the circle and shaded region as the parts. The circle has a radius of x inches and the length of a side of the square is $2x$ inches.

2 x inches

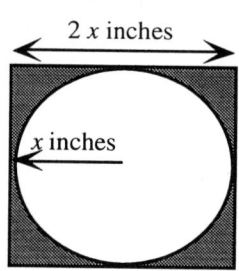

Items	Area (square feet)	
circle	πx^2	⟸ Used area of a circle formula, $a = \pi r^2$
square	$(2x)^2$	⟸ Used area of a square formula, $a = s^2$
shaded region		⟸ To be determined.

The Whole-Part Concept gives us

whole	–	part	⟸ Whole-Part Concept.
area of square (square inches)	–	area of circle (square inches)	⟸ Verbal model.
$(2x)^2$	–	$(3.14)(x^2)$	⟸ Translated the verbal model into a mathematical model.
$4x^2$	–	$3.14x^2$	⟸ Simplified.
$0.86x^2$			⟸ Subtracted.

The area of the shaded part can be represented by the expression $0.86x^2$.

Practice 3.1.5

1. Use x to write a simplified expression for the length of Side A.

2. Use x to write a simplified expression for the length of Side B.

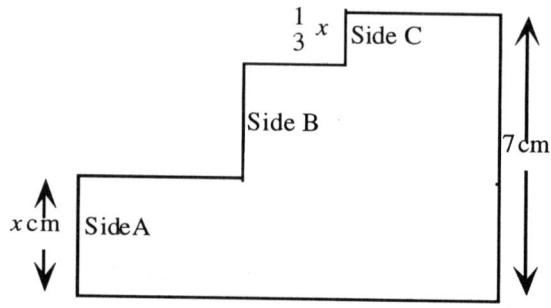

3. Use x to write a simplified expression for the area of the shaded region.

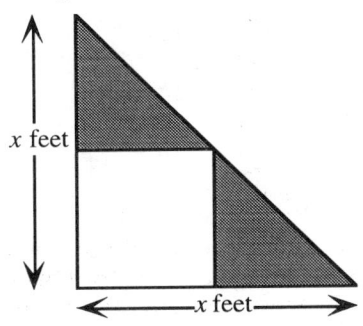

4. Use x to write a simplified expression for the area of the shaded region.

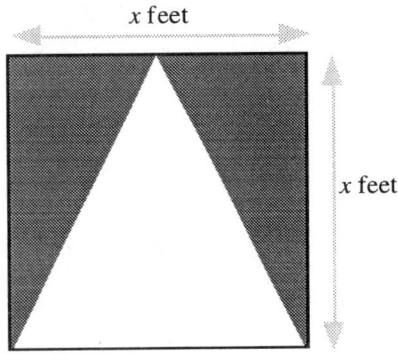

Topic 3.1 F — Representing a Total as the Sum of Expressions

For the last few topics we have been using some strategies to help build expressions. Here is a more formal procedure which includes these strategies.

— Procedure — Building a mathematical model from a verbal model	**Step 1 Organize the information** Decide on the items and build an expression for each item.
	Step 2 Build a verbal model Use English to represent the total as the sum of terms. Remember, only terms with like units may be added.
	Step 3 Build a math model Translate the verbal model into algebra. Because the verbal and math model are so closely related we often do Steps 2 and 3 together.
	Step 4 Simplify Simplify the expression.

If, after reading the problem, you can see your way to the solution, that's fine. If you can't, the procedure will help you focus on all the pieces of information you should consider.

Example 3.1.6   a)   A chemist has two beakers of solution. The first beaker contains 20 ml more solution than the second beaker. If x represents the volume of solution in the first beaker, use x to write a simplified expression for the volume of solution obtained if the two solutions are combined.

•SOLUTION•

Step 1 Organize

Items	Volume (ml)	
first beaker	x	⇐ Given.
second beaker	$x - 20$	⇐ 20 ml less than the first beaker.

Step 2 Verbal model *and* **Step 3 Mathematical Model**   The total volume of solution when the separate solutions are mixed can be represented by the sum of the individual volumes:

Verbal model ⇒   amount in first beaker (ml)   +   amount in second beaker (ml)

Math model ⇒ x + $x - 20$

Step 4 Simplify $x + (x - 20)$ simplifies to $2x - 20$.

The English phrase, "volume of solution obtained if the two solutions are combined" can be represented by the simplified algebraic expression $2x - 20$.

b)   A school has 4 times as many students as teachers. If s represents the number of students, use s to write a simplified expression for the total number of students and teachers.

•SOLUTION•

Step 1 Organize   We have information concerning the students and teachers. Your intuition may lead you to this (INCORRECT) translation:

Items	Number of People	
students	s	⇐ Given.
teachers	$4s$	⇐ 4 times as many students as teachers.

If you check this you will see that it makes no sense. There are more students than teachers but s is less than $4s$ because s is a positive number. If there are 4 times as many students as teachers then there are $\frac{1}{4}$ as many teachers as students. Therefore, the correct translation is this:

Items	Number of People	
students	s	⇐ Given.
teachers	$\frac{1}{4}s$	⇐ $\frac{1}{4}$ as many teachers as students.

Step 2 Verbal model *and* ***Step 3 Mathematical Model***   The total number of students and teachers can be represented by this sum:

Verbal model ⇒   number of students   +   number of teachers

Math model ⇒ s + $\frac{1}{4}s$

Step 4 Simplify $s + \frac{1}{4}s$ simplifies to $\frac{5}{4}s$.

The total number of students and teachers can be represented by $\frac{5}{4}s$.

— Note — **What if the variable represented the number of teachers?**	If the problem had defined the number of teachers as t, then the number of students would be $4t$ and the total number of people would be $t + 4t$, which simplifies to $5t$. This model represents the same situation as $\frac{5}{4}s$ but from a different perspective.

Practice 3.1.6

1.   A nut company combines peanuts and cashews to make a special blend. Given that c represents the pounds of cashews, use c to create a simplified expression which represents the total weight of the mixture if
   a)   the store sold six less pounds of peanuts than cashews.
   b)   one–third as many pounds of peanuts as cashews were sold.
   c)   twice as many pounds of cashews as peanuts were sold.

2.   A bus fare is higher for peak times than for non–peak times. Given that p is the number of daily riders at peak times, use p to write a simplified expression which represents the total number of daily riders if
   a)   there were four peak riders for every non–peak rider.
   b)   there were 75 less peak riders that day.
   c)   there were 75 less non–peak riders that day.

Exercise Set 3.1   The answers to the odd numbered exercises are at the back of the book.

1. Tickets to a haunted house are priced differently for adults and children. Given that a represents the number of tickets sold to adults, create an expression that represents the number of tickets sold to children if

   a) fifty less tickets were sold to children than to adults.
   b) forty more tickets were sold to adults then were sold to children.
   c) there were four times as many child tickets than adult tickets sold.
   d) five child tickets were sold for every adult ticket sold.
   e) there were twice as many adult tickets as child tickets sold.

2. To earn money for a trip to Washington a scout troop decides to sell soft drinks at a football game. Large drinks are sold for $1.25 and small drinks for $.75. Given that s represents the number of small drinks sold create an expression that represents the number of large drinks sold if

   a) four large drinks were sold for every small drink sold.
   b) there were twice as many small drinks as large drinks sold.
   c) there were three times more large drinks than small drinks sold.
   d) two hundred fewer large drinks than small drinks were sold.
   e) eighty more small drinks than large drinks were sold.

3. A student has a chance to transfer part of his credit card balance to a new card with a low introductory rate. He decides to pay as much of his current bill as is convenient and then to transfer the rest to the new card. If t represents the amount he decides to transfer write an expression for the amount he decides to pay off if

   a) the amount he pays off is increased by $1500 you have the amount he transferred.
   b) for each dollar he pays off he transfers four dollars.
   c) he transfers half and pays off half.
   d) he pays off $500 more than he transfers.
   e) he transfers twice as much as he pays off.

4. A student decides to invest part of her monthly income in a savings account and part in a mutual fund. If s represents the amount invested in the savings account then create an expression to represent the amount in the mutual fund if

   a) the mutual fund receives $300 less then the savings account.
   b) she puts the same amount in both accounts.
   c) she invests eight dollars in the savings account for every one dollar she invests in the mutual fund.
   d) she invests $500 less in the savings account.
   e) there is twice as much in the mutual fund as in the savings account.

5. A small engine needs a mixture of gas and oil to run properly. If g is the amount of gas that is added to the final mixture create an expression which represents the amount of oil if

   a) there are 15 parts of gas for each part of oil in the final mixture.
   b) there are 6 more gallons of gas in the final mixture.
   c) one–third as much oil as gas is in the final mixture.
   d) for each milliliter of oil in the final mixture there should be 12 milliliters of gas.
   e) There are 30 less ounces of oil than of gas in the final mixture.

6. An instructor gives a test. If r represents the number of questions a student got right on the test create an expression to represent the number of questions the student got wrong if

   a) the student got 8 questions right for every 1 they got wrong.
   b) the student got 40 more questions right than wrong.
   c) the student got half the questions right and half wrong.
   d) there were five times as many questions answered correctly as incorrectly.
   e) if you increase the number of questions the student got wrong by 18 you have the number of questions the student got right.

7. A student earned $8,500, which she decided to invest in a bond fund that pays 6.75% and a mutual fund that pays 9%. If b represents the amount invested in the bond fund, write an expression for the amount she invested in the mutual fund.

8. A chemistry student mixes w liters of a weak acid solution with a strong acid solution to get 50 liters of solution. Write an expression to represent the liters of strong acid solution.

9. Two cars start from towns 75 miles apart and travel towards each other. If the faster car, which is traveling at 65 mph, has traveled m miles, write an expression for the distance traveled by the slow car when the two cars meet.

10. A student buys a computer and monitor for $1750. After 2 years the computer has depreciated by $550. If originally the computer cost c dollars, write an expression for the original cost of the monitor.

11. The number of hours a project manager works is divided into regular hours, overtime hours and weekend hours. If the project manager worked 75 hours during the week, w hours of which were weekend hours, write a simplified expression to represent the number of overtime hours worked. Assume the manager worked 5 regular hours for every weekend hour worked.

12. A homeowner calculated that her heating costs over a three month period were $408. She noticed that she spent 150 less the third month than the second month. Write a simplified expression to represent the amount spent the first month if she spent c dollars the second month.

13. To help with job interviews a graduating senior spent $285 for four new clothing items. Find the cost for the jacket if the student spent $20 less on the shoes than on the pants and half as much for the shirt as for the pants. Let p represent the cost of the pants.

14. A chemist combines the contents of four beakers into a fifth empty beaker to get 200 milliliters of final solution. Write a simplified expression to represent the milliliters of solution in the fourth beaker if the second beaker held 3 less milliliters than the first beaker and the third beaker held 5 more milliliters then the second. Assume the first beaker held f milliliters of solution.

---

15. Write a simplified expression for the area of the shaded region.

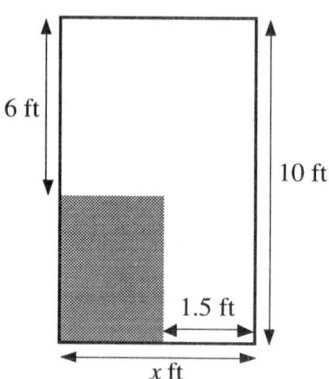

16. Write a simplified expression for the area of the shaded region.

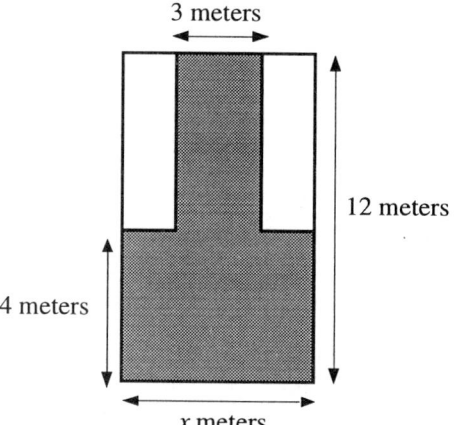

17. Write a simplified expression for the area of the shaded region.

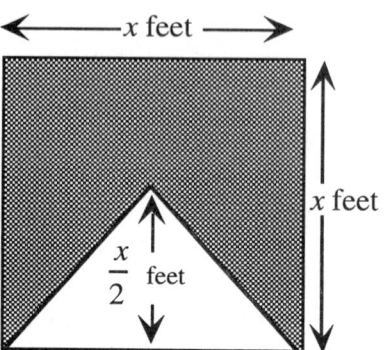

18. Write a simplified expression for the area of the shaded region. Assume both figures are squares.

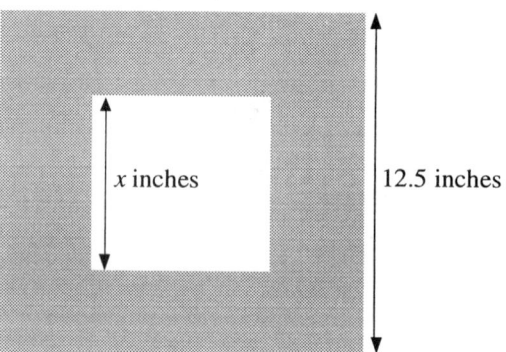

19. Write a simplified expression for the area of the shaded region. Use π for π.

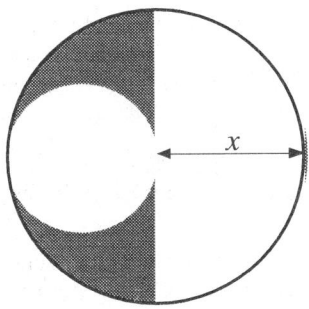

20. Write a simplified expression for the area of the shaded region. Use 3.14 for π.

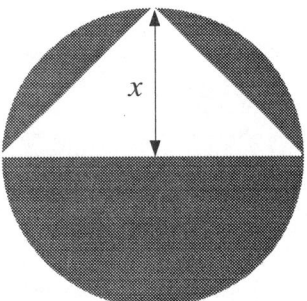

21. Write a simplified expression for the area of the shaded region. Use π for π.

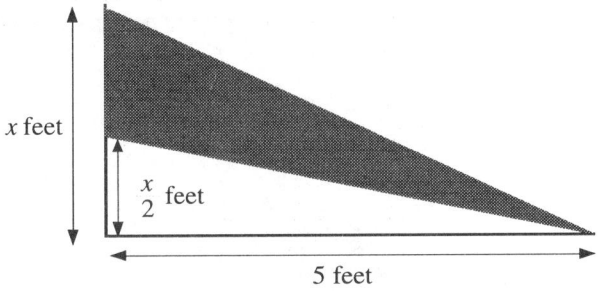

22. Write a simplified expression for the area of the shaded region. Use 3.14 for π.

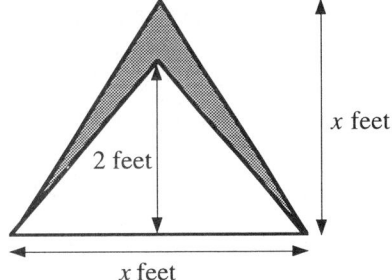

23. A car and a truck leave a town and travel in opposite directions. Given that c represents the distance the car travels, create a simplified expression that represents the total distance driven by the car and truck combined if

   a) the car traveled 200 more miles than the truck.
   b) there were 4 times as many miles traveled by the car as by the truck.
   c) the truck travels two miles more than the car.

24. A student leaves home and walks to school. After class, she returns home. If s represents the distance (miles) she walked to school, create a simplified expression that represents the total distance she walked if

   a) she walked half a mile more to get to school than to return home.
   b) she walked one less mile going to school.
   c) the distance she walked home was three–fourths the distance she walked to school.

25. A dairy decides to make low–fat chocolate milk by mixing whole milk, chocolate syrup, and skim milk. If w represents the gallons of whole milk, create an expression for the total gallons in the mixture if

   a) one–tenth as much chocolate syrup as whole milk and forty more gallons of skim milk than of chocolate syrup was in the final mixture.
   b) three times as much skim milk as whole milk was in the final mixture and 300 less gallons of chocolate syrup as skim milk was in the final mixture.
   c) 48 more gallons of whole milk than skim milk was put into the mixture and one–fourth as much chocolate syrup as skim milk was in the final mixture.

26. A juice company decides to make a fruit punch by mixing orange juice, grape juice and sugar water. If s represents the gallons of sugar water in the mixture, create a simplified expression that represents the total gallons in the mixture if

   a) the mixture has 700 less gallons of orange juice than sugar water and twice as much grape juice as orange juice.
   b) the mixture has 200 more gallons of sugar water than of grape juice and 300 less gallons of orange juice than of grape juice.
   c) the mixture has equal amounts of orange juice and sugar water and half as much grape juice as orange juice.

27. A worker invests m dollars into a money market account. She then splits the remainder of her retirement savings between a bond fund and a mutual fund. Create a simplified expression that represents the total amount she invested if

   a) the amount she placed in the mutual fund was $6000 less than the amount placed in the money market account and she placed 5 times as many dollars in the bond fund as the mutual fund.
   b) the amount invested in the bond fund was half that invested in the mutual fund and the amount invested in the mutual fund was one–third that invested in the money market account.
   c) the amount invested in the bond fund is one–fourth the amount invested in the mutual fund and the amount invested in the bond fund is $15,000 more than the amount invested in the money market account.

28. After graduating and finding a job a student buys a car and begins renting a new apartment. If x represents the amount she is paying back on her student loan create a simplified expression that represents the total of these three monthly payments if

a) the student loan payment is $217 less than the car payment and the monthly rent is three times as much as the car payment.

b) the monthly car payment is one–third the rent payment and the monthly car payment is $60 less than the student loan payment.

c) the rent payment is $422 more than the car payment and the student loan is the same as the car payment..

Review Exercises   The answers to all of these exercises are at the back of the book.

29. Reduce to lowest terms: $\dfrac{27}{90}$ 30. Translate to a mathematical model: One–half of 24.

31. Translate to a mathematical model: Eight more than a number.

32. Translate to a mathematical model: Negative three subtracted from five.

Simplify.

33. $8x - 3(5x - 7)$ 34. $8(3 - n) - (5 - n) - 2(n - 8)$ 35. $\dfrac{2}{3}(m - 3) - \dfrac{1}{2}(m - 2)$

Solve.

36. $8 - (2 - x) = 9x + 14$ 37. $0.15p + 0.25(p + 50) = 8$ 38. $-(x - 3) + 3(x - 2) = 5(x - 1) - 3x$

39. $5x - 3 < x - (2x + 9)$ 40. $-2w < 6$ 41. $8 - 2a \geq 4a - 2(3a + 2)$

Section 3.2 Applying Algebra to Word Problems

Sometimes students are skeptical of word problems because they don't seem "real". For example, a typical introductory algebra word problem might go something like this:

José has a bag that contains dimes and quarters. If he has a total of 784 coins and the value of the coins is $133 how many of each type of coin are there?

In the real world, José would most likely use the automatic coin counting machine at a local bank. So, why do problems like this? We have two reasons:

- First, it takes a lot of time to understand the context of real problems. For example to understand a problem from psychology might take quite a lot of background reading in psychology. Although the problem about José does not involve psychology, the process used to solve the problem uses mathematical tools that are used in psychology. Our hope is that when you get to any course that uses mathematics as a problem solving tool, you will know which mathematical tool to use and how to use it.

- Second, real problems tend to be complex and require higher level mathematics. As you learn more mathematics you will increase your ability to work with more complex, interesting, and realistic problems.

We would like to give you a step by step procedure that you could follow to solve all word problems. Unfortunately, there is none. The best we can do is to give you a general procedure that will help guide you to the solution or many problems.

— Procedure — *General procedure for* *solving word problems*	*Step 1 Organize the information* Read the problem, several times if necessary: • Given: determine what is given. Be sure to include the unit of all numbers. • Find: determine what you are asked to find. • Sketch: draw a sketch if it will help you better understand the problem. • Rates: examine the unit of each given number to determine if a rate is involved. If so, begin construction of an Amount Table. (We will discuss rates and the Amount Table in the next Section.) • Formulas: if standard formulas will be needed, list them. *Step 2 Define unknowns* • Define: assign a definition to a variable; include units, if appropriate. • Construct: use the variable you just defined to build expressions for other unknowns. • Amount Table: if you began an Amount Table in step 1, complete it. *Step 3 Build verbal models* Use English to write relations between the unknowns. *Step 4 Build mathematical models* Translate the verbal models into mathematical models. Because the verbal and mathematical models are so closely related we often do Steps 3 and 4 together. *Step 5 Solve* Solve the mathematical models for the variable. *Step 6 Values of unknowns* Calculate the values of all unknowns. *Step 7 Answer* State the answer to the question. *Step 8 Check* Using the original words of the problem, check to see that the values make sense and are consistent with the original statement of the problem.

Each step of the procedure is designed to help you focus on an important aspect of the problem and its solution. We are trying to give you a general procedure that works for many types of problems, but following our procedure is not the only legitimate way to solve most problems. Sometimes it may be more efficient to rearrange the order of the steps, or to do two steps simultaneously, or to skip some steps when they don't apply. As you become more expert at solving problems you will be better able to judge when to modify the procedure.

Because the procedure works for most of our students and for most problems we will show you all 8 steps in each of our worked examples. However, you may find that for some problems you can skip steps because your intuition enables you to jump ahead. If you choose to do that you must be very careful that your intuition does not lead you astray. For example, you may think you can quickly solve by intuition the problem we presented at the beginning of this book: *A can of juice costs 50 cents. The juice alone costs 40 cents more than the can. How much does the can alone cost?* If you think the answer is 10 cents, your intuition did not lead you to the correct solution. To solve this problem correctly, algebra and a general procedure like the one given above are very helpful.

Example 3.2.1   A can of juice costs 50 cents. The juice alone costs 40 cents more than the can. How much does the can alone cost?

     •SOLUTION• *Step 1 Organize*

Given: • can and juice cost 50 cents.
           • juice alone costs 40 cents more than the can alone.

Find: Cost of can alone.

Sketch: Not helpful.

Rates: Will be discussed in Section 3.3.

Formulas: None needed.

Step 2 Define unknowns

Define: We are asked to find the cost of the can alone so let's say

x represents the cost of the can alone (cents).

— *Note* — ***Using the equal sign in*** ***definitions***	We purposely avoid using = when we assign definitions to variables and expressions. For example, we say x *represents the cost of the can alone* rather than x = *the cost of the can alone* or x = *the can* because we want to be clear about the meaning of the variable x. However, it is not incorrect to write x = *the cost of the can alone* where the equal sign literally means *represents*. This obviously saves time and space as long as you realize that = does not imply an equation which is to be solved. Using symbols to write things faster is fine, but x = *the can* is not a good abbreviation to use. The variable x does not represent the can; it represents some *attribute* of the can that can be represented quantitatively. The attribute could be the cost of the can (as in this problem), or it could be the weight of the can, or the volume of the can, or the density of the can, or the number of cans. We will continue to use the word *represents* in most places, but to save space and time feel free to use = when you are writing definitions by hand.

Construct: We need to construct an expression for the cost of the juice. Since the juice costs 40 cents more than the can we can say

$x + 40$ represents the cost of the juice alone (cents).

(Check: This makes sense since the juice costs more than the can and $x + 40$ is more than x.)

Amount Table: None.

Step 3 Verbal models and *Step 4 Mathematical Models*   There is a relation between the total cost of the can with the juice in it, the cost of the can alone, and the cost of the juice alone

Verbal model $\Rightarrow$	total cost (cents)	=	cost of juice (cents)	+	cost of can (cents)
Math model $\Rightarrow$	50	=	x	+	$(x + 40)$

Step 5 Solve

$$50 = x + (x + 40) \qquad \Leftarrow \text{ Equation to solve.}$$

$$50 = 2x + 40 \qquad \Leftarrow \text{ Combined like terms.}$$

$$10 = 2x \qquad \Leftarrow \text{ Subtracted 40 from both sides.}$$

$$5 = x \qquad \Leftarrow \text{ Divided both sides by 2.}$$

Step 6 Values of unknowns

- x is 5.
- $x + 40$ is $5 + 40$, which simplifies to 45.

Step 7 Answer   The cost of the can, x, is 5 cents.

Step 8 Check

- can and juice cost 50 cents: checks because 5 cents (cost of can) + 45 cents (cost of juice) is 50 cents.
- juice costs 40 cents more than the can: checks because 45 cents (cost of juice) is 40 cents more than 5 cents (cost of can).

Practice 3.2.1

1. A dinner with tip costs $44.00. If the tip was $32 less than the dinner, how much was the dinner alone?

2. There are 5 students for each professor on a college campus. If the total number of students and professors is 1950, how many students and how many professors are there?

3. A family drives to a cabin which is 300 miles away. They drive a while, take a break, and then finish the trip. If the distance driven before the break was 50 miles less than the distance driven after the break, how far did they drive after the break?

Sometimes problems deal with more than two items. To successfully solve problems of this type follow the same procedure.

Example 3.2.2   A student earned $1,200 for college by working three jobs. She made twice as much on the second job as on the first job and $300 less on the third job as on the first job. Find how much she made on each job.

•SOLUTION•

Step 1 Organize

Given: • student earned a total of $1,200.
• student earned twice as much on second job as on the first job.
• student earned $300 less on the third job as on the first job.

Find: • money earned on first job.
• money earned on second job.
• money earned on third job.

Sketch: Not helpful.

Rates: Will be discussed in Section 3.3.

Formulas: None needed.

Step 2 Define unknowns

Define: x represents the money earned on the first job (dollars).

Construct: $2x$ represents the money earned on the second job (dollars).

(Check: This makes sense since the earnings on the second job are more than those on the first job and, since x is positive, $2x$ is more than x.)

$x - 300$ represents the money earned on the third job (dollars).

(Check: This makes sense since earnings on the third job are less than those on the first and $x - 300$ is more than x.)

Amount Table: None.

Step 3 Verbal models and Step 4 Mathematical Models   There is a relation between the money earned on the three jobs:

Verbal model $\Rightarrow$	total money earned (dollars)	=	money earned on first job (dollars)	+	money earned on second job (dollars)	+	money earned on third job (dollars)
Math model $\Rightarrow$	1,200	=	x	+	$2x$	+	$(x - 300)$

Step 5 Solve

$$1200 = x + 2x + (x - 300) \quad \Leftarrow \text{Equation to solve.}$$

$$1200 = 4x - 300 \quad \Leftarrow \text{Combined like terms.}$$

$$1500 = 4x \quad \Leftarrow \text{Added 300 to both sides.}$$

$$375 = x \quad \Leftarrow \text{Divided both sides by 4.}$$

Step 6 Values of unknowns

- x is 375.
- $2x$ is 2(375), which simplifies to 750.
- $x - 300$ is $375 - 300$, which simplifies to 75.

Step 7 Answer

- money earned on the first job, x, is $375.
- money earned on the second job, $2x$, is $750.
- money earned on the third job, $x - 300$, is $75.

Step 8 Check

- student earned a total of $1,200: checks since $375 + $750 + $75 is $1,200.
- student earned twice as much on the second job as on the first job: checks because $750 is twice as much as $375.
- student earned $300 less on the third job as on the first job: checks because $75 is $300 less than $375.

Practice 3.2.2

1. A gas station sells a certain number of gallons of unleaded gas, one third as much super–unleaded and 80 less gallons of premium unleaded as unleaded. Find the gallons of each type of gasoline sold if 410 gallons were sold all together.

2. A community center splits $1200 up into three prizes. The third prize is half as much as the second prize and the second prize is $500 less than the first prize. Find the amount for each prize.

— Study Tip —	
Read the worked examples more than one time	Many students have a difficult time correctly solving word problems. Don't give up; persistence really pays off. As you study our worked out examples don't just read through them one time and expect to understand all that is going on. It typically takes two or three readings for the average student to make sense of all the complexities of many of these worked examples. If you don't understand something, try reading it again to see if you can identify the part that is confusing you and then focus your attention on that part. Don't hesitate to go back to a previous topic and restudy an idea or concept that may be the root of your difficulty in understanding the current problem.
	Rereading confusing parts will help you find information you may have missed in the first read and it will help you brain become more familiar with the details and connections you may have overlooked. It will also help you determine precisely where the difficulty lies. This is especially important if you ask your instructor, a tutor, or a friend for help — the more specific you can be about your confusion the easier it will be for someone else to help you find the answer.

When a problem involves geometric figures, it's useful to create a sketch of the situation and then label the sketch to help keep track of the various parts.

Example 3.2.3   The perimeter of a rectangular section of a park is 560 feet. Two sides that are opposite each other can be seen by the public and so were fenced with expensive wood rails. The other two sides are hidden from view and so were fenced with inexpensive chain link. Find the dimensions of the fenced–in area if the length of a wood fenced side is one–sixth the length of a chain link fenced side.

•SOLUTION•

Step 1 Organize

Given:
- Fence is rectangular in shape.
- Perimeter is 560 feet.
- The length of a wood fenced side is one–sixth the length of a chain link fenced side.

Find: The dimensions of the rectangle (that is, the length and width of the rectangle).

Sketch:

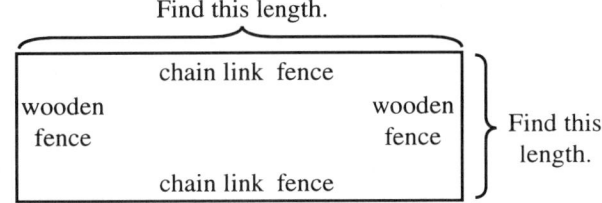

Rates: Will be discussed in Section 3.3.

Formula: Since this problem discusses the perimeter of a rectangle, we will need the formula $p = 2l + 2w$.

Step 2 Define unknowns

Define: We are looking for the length and width of the rectangle. Let's let x represent one of these

x represents the length of a chain link side (feet).

Construct: We need to construct an expression for a wood side. Since a wood side is one–sixth as long as a chain link side, we can say

$\frac{1}{6}x$ represents the length of a wood side (feet). We may write this as $\frac{x}{6}$.

(Check: This makes sense since the wood side is shorter than the chain link side and, since x is positive, $\frac{x}{6}$ is less than x.)

It may help to better visualize the problem if we place these unknowns directly on the sketch.

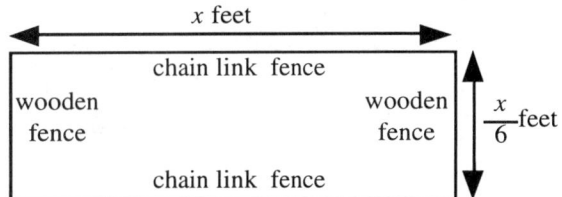

Amount Table: None.

Step 3 Verbal models and Step 4 Mathematical Models   The formula $p = 2l + 2w$ provides us with the relation between the perimeter, the length, and the width.

Verbal model $\Rightarrow$	perimeter of fence (feet)	=	twice length (chain link side) (feet)	+	twice width (wood side) (feet)
Math model $\Rightarrow$	560	=	$2x$	+	$2\left(\frac{x}{6}\right)$

Step 5 Solve

$$560 = 2x + 2\left(\frac{x}{6}\right) \qquad \Leftarrow \text{Equation to solve.}$$

$$560 = 2x + \frac{x}{3} \qquad \Leftarrow \text{Reduced } 2\left(\frac{x}{6}\right) \text{ to } \frac{x}{3}.$$

$$\mathbf{1680 = 6x + x} \qquad \Leftarrow \text{Multiplied each term by the LCD, 3.}$$

$$1680 = \mathbf{7x} \qquad \Leftarrow \text{Combined like terms.}$$

$$240 = x \qquad \Leftarrow \text{Divided both sides by 7.}$$

Step 6 Values of unknowns

- x is 240.

- $\dfrac{x}{6}$ is $\dfrac{240}{6}$, which simplifies to 40.

Step 7 Answer

- The length of one chain link side, x, is 240 feet.

- The length of one wood side, $\dfrac{x}{6}$, is 40 feet.

Step 8 Check

- The wood side is one–sixth the length of the chain link side: checks since 40 feet is one–sixth of 240 feet.
- The perimeter is 560 feet: checks since twice 240 feet plus twice 40 feet is 560 feet.

Practice 3.2.3

1. Find the dimensions of a rectangular parking lot if the length is five more than twice the width and the perimeter is 610 feet.

2. A triathlete runs around a circular lake and then swims side to side through the center. Estimate the diameter of the lake if the total distance she covered running and swimming was 8 miles. Recall that the circumference of a circle is given by $c = \pi d$ (use 3.14 for π and round your answer to the nearest hundredth).

The word *consecutive* means "following in order, without interruption". *Consecutive integers* are integers that follow one another, like 5, 6, 7 or –2, –1, 0. *Consecutive even integers* are even integers (that is, divisible by 2) and one right after the other, like 2, 4, 6. *Consecutive odd integers* are odd integers that are one right after the other, such as –5, –3, –1, 1.

Example 3.2.4   Find two consecutive odd integers whose sum is –36.

•SOLUTION•

Step 1 Organize

Given: • two consecutive odd integers.
• the sum is –36.

Find: • smaller integer.
• larger integer.

Sketch: Not helpful.

Rates: Will be discussed in Section 3.3.

Formulas: None needed.

Step 2 Define unknowns

Define: x represents the smaller odd integer.

Construct: If x represents the smaller odd integer, then $x + 1$ represents the next integer, and $x + 2$ represents the next *odd* integer (every other integer is odd).

$x + 2$ represents the larger *odd* integer.

(Check: This makes sense since $x + 2$ is larger than x.)

Amount Table: None.

Step 3 Verbal models and Step 4 Mathematical Models

Verbal model $\Rightarrow$	sum of integers	=	smaller integer	+	larger integer
Math model $\Rightarrow$	–36	=	x	+	$(x + 2)$

Step 5 Solve

$$-36 = x + (x + 2) \quad \Leftarrow \text{ Equation to solve.}$$

$$-36 = \mathbf{2x} + 2 \quad \Leftarrow \text{ Combined like terms.}$$

$$\mathbf{-38} = 2x \quad \Leftarrow \text{ Subtracted 2 from both sides.}$$

$$-19 = x \quad \Leftarrow \text{ Divided both sides by 2.}$$

Step 6 Values of unknowns

- x is -19.
- $x + 2$ is $-19 + 2$, which simplifies to -17.

Step 7 Answer

- smaller odd integer, x, is -19.
- larger odd integer, $x + 2$, is -17.

Step 8 Check

- consecutive odd integers: checks since -19 and -17 are consecutive odd integers.
- the sum of the integers is -36: checks since $-19 + (-17)$ is -36.

Practice 3.2.4

1. The sum of two consecutive even integers is 106. Find the two integers.

2. Find two consecutive integers whose sum is -67.

3. We are given three consecutive odd integers whose sum is -3. Find all three integers.

Exercise Set 3.2 The answers to the odd numbered exercises are at the back of the book.

Use the General Procedure For Solving Word Problems to help you answer the following questions

1. A washer and dryer set costs $515. The dryer costs $81 less than the washer. Find the cost of each.

2. A stereo set costs $521. The tape deck costs $117 less than the speakers. What is the cost of the tape deck and the cost of the speakers.

3. Lynn has a total of $3422 in her banking accounts. If she has $228 more in her saving account than in her checking account, how much is in each account.

4. A suit costs $168. If the jacket costs $28 more than the pants find the cost of each.

5. A student has a chance to transfer part of his $1835 credit card debt to a new card with a low introductory interest rate. He decides to pay as much of his current bill as is convenient and then to transfer the rest to the new card. Find how much he paid off and how much he transferred if for each dollar he pays off he transfers four dollars.

6. A student decides to invest part of her $219 monthly surplus from her job in a savings account and part in a mutual fund. If she invests two dollars in the savings account for every one dollar she invests in the mutual fund find the amount she deposits in each account every month.

7. A community decides to rent out garden plots in a vacant field. Each garden must have the same width but the length can vary. A resident plans her garden to have a length which is five times the prescribed width. If she uses 108 feet of fencing to enclose her garden what is the prescribed width for each garden?

8. A father wishes to build a sandbox that has a width which is 5 feet shorter than it's length. If he has 32 feet of lumber available for the walls, what should the dimensions be?

9. The sum of two integers is 8. One number is twelve more than the other. Find the numbers.

10. The sum of two integers is 3. One number is five less than the other. Find the numbers.

11. The sum of three integers is 32. The second is three times the first and the third is seven more than the first. Find the integers.

12. The sum of three integers is 42. The second is twice the first and the third is two less than the first. Find the integers.

13. Octavia deposited a total of $3000 in three investment accounts. She has twice as much in her money market account as she has in her savings account and $1000 more in her IRA than in her savings account. How much does she have in each account?

14. Henry sold three times as many VCR's as televisions and 75 more stereos than televisions. If he had a total of 200 sales, how many did he sell of each?

15. A movie sells a total of 422 child, adult, and senior tickets. Find how many of each type was sold if 40 less adult tickets than child tickets were sold and 100 more senior tickets than adults tickets were sold.

16. After figuring out final grades an English teacher noticed he gave 8 less A's then B's and 12 mores C's then A's. If he gave out a total of 23 A's, B's, and C's how many of each where there?

17. A dairy decides to make 700 gallons of chocolate milk by mixing whole milk, skim milk, and chocolate syrup. If 40 more gallons of skim milk than of chocolate syrup was in the final mixture and one–tenth as much chocolate syrup as whole milk was in the final mixture, find how much of each ingredient was included?

18. A juice company decides to make 1200 gallons of fruit punch by mixing orange juice, grape juice and sugar water. If the mixture contained 700 fewer gallons of orange juice than sugar water and twice as much grape juice as orange juice, find how much of each ingredient was included?

19. A worker has $104,000 invested for retirement in three different accounts. If the amount invested in the bond fund is one–fourth the amount invested in the mutual fund and the amount invested in the bond fund is $15,100 more than the amount invested in the money market account find how much is in each account.

20. After graduating and finding a job a student buys a car and begins renting a new apartment. The total of her monthly payments for the car, the apartment and her student loan is $1208. If the student loan payment is $217 less than the car payment and the monthly rent is three times as much as the car payment find the amount the student is paying monthly for each.

21. A nature center is building a new walkway in the shape of a triangle. The second side is a walkway which is 200 feet shorter then the first side while the third side is twice the length of the second side. If the total length of the walkway is 1472 feet, what is the length of the shortest walkway?

22. A park supervisor decides to put a walkway from the lower right corner of rectangular flower garden to the upper left corner. The original rectangular garden had a length that was 75 feet longer then it's width. The new walkway will have a length which is 215 feet longer then the original width. If the original perimeter of the rectangular garden was 718 feet, what is the length of the new walkway?

23. The measures of the angles of any triangle total 180 degrees. If the first angle measures 38 degrees less than the second and the third angle measures 7 degrees more than the first, what is the measure of the second angle?

24. The measures of the angles of any triangle total 180 degrees. If the third angle measures 12 degrees more than the second and the first angle measures 21 degrees more than the third angle, what is the measure of the first angle?

25. Two angles are complementary angles if the sum of their measures is 90 degrees. Assume two angles are complementary and one has a measure which is 26 degrees less than the other. What is the measure of the larger angle?

26. Two angles are supplementary if the sum of their measures is 180 degrees. Assume two angles are supplementary and one angle is three less than twice the other. What is the measure of the smaller angle?

27. An isosceles triangle is a triangle that has two sides of equal length. Assume that an isosceles triangle has a perimeter of 21 inches and that the side which is different from the other two has a length that is one–third the length of the others. Find the length of each side.

28. An equilateral triangle is a triangle where all the sides are of equal length. If the perimeter of an equilateral triangle is 75 inches, what is the length of one side of the triangle?

29. The sum of two consecutive integers is –11. Find the integers.   30. Find two consecutive integers whose sum is 45.

31. Find two consecutive odd integers whose sum is –28.   32. Find two consecutive odd integers whose sum is 168.

33. The sum of three consecutive odd integers is –9. Find the integers.

34. The sum of three consecutive odd integers is 33. Find the integers.

***

Review Exercises   The answers to all of these exercises are at the back of the book.

35. Reduce to lowest terms: $\dfrac{66}{34}$ 36. Factor: $9r - 5r$ 37. Simplify: $\dfrac{2}{3}x + 4 = 5\left(x - \dfrac{1}{2}\right)$

Simplify.

38. $x - (5x - 2x)$

39. $\dfrac{1}{18} + \dfrac{5}{14}$

40. $\dfrac{3}{26}\left[\, 1 + \dfrac{2}{3}\left(1 - \dfrac{1}{3} \right)\right]$

41. $9(2n - 5) - 3(n - 7) - (2 - n)$

42. $\dfrac{5}{6}(8k - 9) - \dfrac{2}{3}\left(\dfrac{k}{2} - 3k \right)$

43. $\left|\, -4\, \right| \cdot (-32) \div (-4)(-3)$

Solve.

44. $0.10(2x - 5) - 2x = -(0.5 + x) - 0.8x$

45. $0.32(500 - m) + 0.29m = 147.97$

46. $3x = 0$

47. Translate to a mathematical model: The quotient of twice a number and 5.

Section 3.3 The Amount Formula: Finding the Amount

In this Section we develop the *Amount Formula*, which is can be an important translation tool for problems that contain rates.

Topic 3.3 A — The Amount Formula

If you drive 55 miles per hour for 3 hours you can calculate the distance you travel by multiplying your speed by the time: $\left(\dfrac{55 \text{ miles}}{1 \text{ hour}} \right) \cdot (3 \text{ hours})$. This simplifies to 165 miles. This is an illustration of the **Amount Formula**, amount = rate • base.

— *Definition* — **Amount Formula**	English: An amount may be found by multiplying a rate by a base.
	Example: A mechanic charges $12 per hour and works for 5 hours. The amount of money he earns is calculated by multiplying his rate, $\left(\dfrac{12 \text{ dollars}}{1 \text{ hour}} \right)$, by the base, (5 hours) to get 60 dollars.
	Algebra: amount = rate • base

Let's look at each part of the Amount Formula:

Rate   A **rate**, such as 55 miles per hour, is the ratio of two quantities that are measured using different units. Many times we use the word "per" to indicate a rate. Some examples of rates are:

- speed of travel is a rate that may be given as *miles per hour (mph)*. A speed of 55 mph literally means you go 55 miles for every 1 hour you travel. This can be written as $\dfrac{55 \text{ miles}}{1 \text{ hour}}$.

- hourly wages is a rate of pay that may be given as *dollars per hour*. A wage of 5 dollars per hour literally means you earn five dollars for every 1 hour you work. This can be written as $\dfrac{5 \text{ dollars}}{1 \text{ hour}}$.

- the cost of an item is a rate that may be given as *dollars per item*. A theater ticket that costs $10 has a rate of 10 dollars per ticket, which literally means 10 dollars for every 1 ticket. This can be written as $\dfrac{10 \text{ dollars}}{1 \text{ ticket}}$.

English descriptions such as *speed*, *hourly wages*, and *cost* do not explicitly use the word "rate" even though they really are referring to rates.

Base   A **base**† is what the rate is based on. The unit of the base is always the same as the unit in the rate's denominator. This gives us a clue as to what the base is. For example, the denominator of the rate $\frac{55 \text{ miles}}{1 \text{ hour}}$ tells us the unit of the base is *hours*, which implies that the base is time. Likewise, the denominator of the rate $\frac{10 \text{ dollars}}{1 \text{ ticket}}$ tells us the unit of the base is *tickets*, which implies that the base is the number of tickets.

Amount   The **amount** is the result of multiplying the rate and the base. The unit for the amount is always the same as the unit in the rate's numerator. This gives a clue as to what the amount is. For example, the numerator of the rate $\frac{55 \text{ miles}}{1 \text{ hour}}$ tells us the unit of the amount is *miles* and this implies that the amount is distance. Similarly, the numerator of the rate $\frac{10 \text{ dollars}}{1 \text{ ticket}}$ tells us the unit of the amount is *dollars* and this implies that the amount is money.

Topic 3.3 B — The Amount Formula and Unit Reduction

When using the Amount Formula it's important to take the units into account. For example, here is how we would calculate the distance traveled by a car going 55 mph for 3 hours:

amount = rate • base   ⇐ Amount Formula.

amount = $\left(\dfrac{55 \text{ miles}}{1 \text{ hour}}\right)$ • (3 hours)   ⇐ Substituted values into formula.

To get the final answer we treat the numbers and units independently:

- we find the *numeric part* of the answer by simplifying $\frac{55}{1} \cdot 3$ to get 165.

- we find the *unit part* of the answer by simplifying $\frac{\text{miles}}{\text{hour}} \cdot \text{hour}$. This simplifies to *miles* because the unit *hour* in the denominator reduces with the unit *hours* in the numerator. Some people would say "the hours cancel" but we prefer to say "the hours reduce" since the word "cancel" can be easily misinterpreted.

To show the unit reduction, we write the entire process like this:

amount = $\left(\dfrac{55 \text{ miles}}{1 \,\text{hour}}\right)$ • 3 hours

amount = 165 miles

Example 3.3.1   Use the Amount Formula, amount = rate • base, to solve. Be sure to properly reduce the units.

a)   A machine can stuff envelopes at a rate of 60 envelopes per minute. Find the number of envelopes stuffed in 15 minutes.

•**SOLUTION**•   The rate is $\frac{60 \text{ envelopes}}{1 \text{ minute}}$ and the base is 15 minutes.

amount	=	rate	•	base	⇐ Amount formula.
work done	=	rate	•	time	⇐ Meaning for this problem.
work done	=	$\left(\dfrac{60 \text{ envelopes}}{1 \,\text{min}}\right)$	•	(15 min)	⇐ Reduced units.
work done	=	900 envelopes			⇐ Multiplied.

The machine can stuff 900 envelopes in 15 minutes.

---

†   Base is used in many different ways in mathematics. For example we used the word base when talking about exponents. Context is used to decide which meaning is needed.

b) Find the value of 17 quarters.

> •SOLUTION• Since one quarter is worth 25 cents, the rate is $\dfrac{25 \text{ cents}}{1 \text{ quarter}}$. Since we have 30 quarters, that is the base. When problems involve money, the amount usually is referred to as the *value*.

amount	=	rate	•	base	⇐ Amount formula.
value	=	value of 1	•	quantity	⇐ Meaning for this problem.
value	=	$\left(\dfrac{25 \text{ cents}}{1 \text{ quarter}}\right)$	•	(30 quarter)	⇐ Reduced units.
value	=	750 cents			⇐ Multiplied.

> The 17 quarters are worth 750 cents. If we were to use dollars instead of cents, the rate would be $\dfrac{0.25 \text{ dollars}}{\text{quarter}}$. In that case, the value of the 17 quarters would be $7.50.

c) A ticket for a play costs $30. Find the value of 15 tickets.

> •SOLUTION• Since one ticket costs 30 dollars, the rate is $\dfrac{30 \text{ dollars}}{1 \text{ ticket}}$. Since we have 15 tickets, that is the base.

amount	=	rate	•	base	⇐ Amount formula.
value	=	value of 1	•	quantity	⇐ Meaning for this problem.
value	=	$\left(\dfrac{30 \text{ dollars}}{1 \text{ ticket}}\right)$	•	(15 ticket)	⇐ Reduced units.
value	=	450 dollars			⇐ Multiplied.

d) Leon earns a 7% commission on every house he sells. How much did he earn on a house he sold for $115,000?

> •SOLUTION• The 7% commission means he earns $7 for every $100 worth of house he sells. Thus, the rate is $\dfrac{7 \text{ dollars earned}}{100 \text{ dollars sold}}$. Since he sold $115,000 worth of house, that is the base.

amount	=	rate	•	base	⇐ Amount formula.
earned	=	commission rate	•	sale price	⇐ Meaning for this problem.
earned	=	$\left(\dfrac{7 \text{ dollars earned}}{100 \text{ dollars sold}}\right)$	•	(115,000 dollars sold)	⇐ Reduced units.
earned	=	8,050 dollars earned			⇐ Simplified.

Practice 3.3.1 Use the Amount Formula to solve. Be sure to properly reduce the units.

1. A wildflower grower bags 125 ounces of seeds that cost 95¢ an ounce. What is the total dollar cost of the seeds?

2. Stalactites are limestone deposits which hang from the ceilings of some caves. The growth rate for a stalactite is approximately 0.08 inches per year. How long will a stalactite have grown after 125 years?

Topic 3.3 C — The Amount Table

The Amount Formula is a powerful tool but it can be confusing when we discuss problems that have more than one item. To help organize the information as problems become more complex, we use the Amount Table.

The **Amount Table** is basically a grid of rows and columns that helps us organize information about the items of importance in a problem (cheap tickets and expensive tickets, dimes and quarters, a fast train and a slow train, etc.). Each individual box in the table is called a **cell** and each cell holds a specific type of information. An empty Amount Table looks like this:

amount =	rate •	base
Descriptions →		
Units →		
↓ Item names ↓		

We write the Amount Formula at the top of the table as a reminder that we can calculate the amount by multiplying the rate and the base.

Suppose a situation involves a car that travels 165 miles by going 55 mph for 3 hours, and a truck that travels 100 miles by going 50 mph for 2 hours. Let's see what the parts of the Amount Table would look like:

Rows of the Amount Table The rows are read across.

- The top row provides descriptions of the amount, rate, and base.

amount =	rate •	base
Descriptions → **distance**	**speed**	**time**

- The second row contains the units of the amount, rate, and base.

amount =	rate •	base
Descriptions → distance	speed	time
Units → **miles**	**miles hour**	**hours**

- The next row holds information about one of the items of importance in the problem. The items are the things or situations the numbers refer to. In this example, the car is considered an item since some of the numbers refer to the car's distance, speed, and time of travel.

amount =	rate •	base
Descriptions → distance	speed	time
Units → miles	miles hour	hours
↓ Item names ↓		
car 165	55	3

- The next row holds information about another item. In this example, the truck is considered another item.

	amount =	rate •	base
Descriptions →	distance	speed	time
Units → ↓ Item names ↓	miles	miles hour	hours
car	165	55	3
truck	**100**	**50**	**2**

If there are additional items there would be additional rows, one for each new item.

Columns of the Amount Table   The columns are read down.

- The first column contains a brief description of the items so we can easily identify them:

	amount =	rate •	base
Descriptions →	distance	speed	time
Units → ↓ Item names ↓	miles	miles hour	hours
car	165	55	3
truck	100	50	2

- The remaining 3 columns contain the labels, units, and values for the amount, rate, and base of each item.

	amount =	rate •	base
Descriptions →	**distance**	**speed**	**time**
Units → ↓ Item names ↓	**miles**	**miles hour**	**hours**
car	**165**	**55**	**3**
truck	**100**	**50**	**2**

Topic 3.3 D — Using the Amount Formula with the General Procedure for Solving Word Problems

We will use the General Procedure for Solving Word Problems introduced in the last section to solve problems that involve the Amount Formula. Although you may be able to do some of the problems without going through the general procedure please practice with the process so that you can use it when the problems get much more difficult.

Example 3.3.2   a)   A cashier collects 276 movie tickets. If 80 of the tickets were $7.50 adult tickets and the rest were $3.50 child tickets, find the total dollars collected.

•SOLUTION•

Step 1 Organize

Given: • 276 total tickets were collected.
- 80 of the tickets were adult tickets.
- each adult ticket cost $7.50.
- each child ticket cost $3.50.

Find: The total dollars collected.

Sketch: Not helpful.

Rates: Because rates are involved, we will construct an Amount Table to help organize the information about the number, value, and price of the two types of tickets.

First, we draw the grid and write in the unit and description of the rate:

	amount	=	rate	•	base
Descriptions →			**price of one ticket**		
Units →			**dollars** **ticket**		
↓ Item names ↓					

The problem discusses two rates, one for adult tickets and one for child tickets. We fill in the information for those cells:

	amount	=	rate	•	base
Descriptions →			price of one ticket		
Units →			dollars ticket		
↓ Item names ↓					
adult tickets			**7.50**		
child tickets			**3.50**		

Next, we fill in the units and descriptions for the base and amount.

• The unit of the denominator of the rate is always the same as the unit of the base. Therefore, the unit of the base must be *tickets* and the base must be number of tickets.

• The unit of the numerator of the rate is always the same as the unit of the amount. Therefore, the unit of the amount must be *dollars* and the amount must be the value of the tickets.

	amount	=	rate	•	base
Descriptions →	**value of tickets**		price of one ticket		**number of tickets**
Units →	**dollars**		dollars ticket		**tickets**
↓ Item names ↓					
adult tickets			7.50		
child tickets			3.50		

Since we are given the number of adult tickets, we can fill in that cell.

	amount	=	rate	•	base
Descriptions →	value of tickets		price of one ticket		number of tickets
Units →	dollars		dollars ticket		tickets
↓ Item names ↓					
adult tickets			7.50		**80**
child tickets			3.50		

We will have to do some calculations to fill in the remaining cells. In the table below we placed ❶, ❷, and ❸ in the blank cells so that we can refer to them later.

	amount =	*rate* •	*base*
Descriptions →	value of tickets	price of one ticket	number of tickets
Units →	dollars	$\dfrac{\text{dollars}}{\text{ticket}}$	tickets
↓ Item names ↓			
adult tickets	❷	7.50	80
child tickets	❸	3.50	❶

Formulas: amount = rate • base.

Step 2 Define unknowns

Define: We are looking for the total amount collected from ticket sales. Let's say

x represents total amount collected from ticket sales.

Construct: There are no other expressions to construct.

Amount Table: We now complete the Amount Table.

❶ is the number of child tickets. We use the Whole-Part Concept to calculate this. The whole is 276 tickets, the known part is 80 adult tickets. Subtracting gives the number of child tickets: 276 − 80, which is 196 tickets.

❷ is the value of the adult tickets. We multiply rate • base to get

$\left(\dfrac{7.50 \text{ dollars}}{1 \text{ ticket}}\right)$ • (80 tickets), which simplifies to 600 dollars.

❸ is the value of the child tickets. We multiply rate • base to get

$\left(\dfrac{3.50 \text{ dollars}}{1 \text{ ticket}}\right)$ • (196 tickets), which simplifies to 686 dollars.

The complete Amount Table looks like this:

	amount =	*rate* •	*base*
Descriptions →	value of tickets	price of one ticket	number of tickets
Units →	dollars	$\dfrac{\text{dollars}}{\text{ticket}}$	tickets
↓ Item names ↓			
adult tickets	**600**	7.50	80
child tickets	**686**	3.50	**196**

Step 3 Verbal models *and* **Step 4 Mathematical Models**   There is a relation between the value of all the tickets, the value of the adult tickets, and the value of the child tickets:

Verbal model ⇒	total value of all tickets (dollars)	=	value of adult tickets (dollars)	+	value of child tickets (dollars)
Math model ⇒	x	=	600	+	686

Step 5 Solve

$$x = 600 + 686 \quad \Leftarrow \text{ Equation to solve.}$$

$$x = 1286 \quad \Leftarrow \text{ Combined like terms.}$$

Step 6 Values of unknowns x is 1286.

Step 7 Answer It's important to answer the question in English. For example, if you were asked at a business meeting, "how much money did we make from tickets sales?" it would not be considered appropriate to answer, "*x* is 1286.". Instead you would answer "A total of $1286 was made from the sale of the tickets."

Step 8 Check

- To check the numerical part of the answer we would go through these same steps again being careful that we did not make a computation error.

- We should check the unit part of the answer. Originally we were looking for the total value of all the tickets. The unit of the answer is dollars, which is what we use to express monetary value. Since the units match we are more comfortable that the answer is correct.

b) A forester fenced in a section of state park with two types of fencing. She used wood to fence the 80 feet that could be seen by the public, at a cost of $24 a foot. She fenced the other 1200 feet, which could not be seen, with chain link at a cost of $6 a foot. Find the total cost of the fence.

•SOLUTION•

Step 1 Organize

Given: • Wood fence is 80 feet long.
• Wood fence costs $24 per foot.
• Chain link fence is 1200 feet long.
• Chain link fence costs $6 per foot.

Find: Cost of the fence.

Sketch:

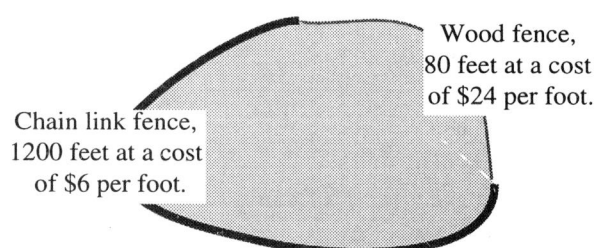

Chain link fence, 1200 feet at a cost of $6 per foot.

Wood fence, 80 feet at a cost of $24 per foot.

Rates: Because rates are involved, we will construct an Amount Table to help organize the information about the length, cost, and price per foot of the two types of fencing.

- The price per foot refers to the cost of one foot of fence. The unit is $\frac{dollars}{foot}$. Since this unit is the ratio of two other units, we can think of the price per foot as a rate.

- The length is measured in feet.

- The cost is measured in *dollars*.

	amount	=	rate	•	base
Descriptions →			**price of one foot**		
Units →			**dollars**		
↓ Item names ↓			**foot**		

The problem discusses two types of fencing, wood and chain link. We fill in the information for those cells:

	amount	=	rate	•	base
Descriptions →			price of one foot		
Units →			dollars / foot		
↓ Item names ↓					
wood			**24**		
chain link			**6**		

Next, we fill in the units and descriptions for the base and amount. The unit of the denominator of the rate is the same as the unit of the base (feet). The unit of the numerator of the rate is the same as the unit of the amount (dollars).

	amount	=	rate	•	base
Descriptions →	**cost of fencing**		costs of one foot		**length of fencing**
Units →	**dollars**		dollars / foot		**feet**
↓ Item names ↓					
wood			24		
chain link			6		

We are given the lengths of each type of fence. We place these in the appropriate cells.

	amount	=	rate	•	base
Descriptions →	cost of fencing		costs of one foot		length of fencing
Units →	dollars		dollars / foot		feet
↓ Item names ↓					
wood			24		**80**
chain link			6		**1200**

We will have to do some calculations to fill in the remaining cells. In the table below we placed ❶ and ❷ in the blank cells so that we can refer to them later.

	amount	=	rate	•	base
Descriptions →	cost of fencing		costs of one foot		length of fencing
Units →	dollars		dollars / foot		feet
↓ Item names ↓					
wood	❶		24		**80**
chain link	❷		6		**1200**

Formulas: amount = rate • base.

Step 2 Define unknowns

Define: x represents the total cost of the fence.

Construct: There are no other expressions to construct.

Amount Table: We use the Amount Formula to complete the Amount Table:

❶ is the cost of the wood fence. We multiply rate • base to get

$$\left(\frac{24 \text{ dollars}}{1 \text{ foot}}\right) \cdot (80 \text{ feet}),$$ which simplifies to 1920 dollars.

❷ is the cost of a chain link fence. We multiply rate • base to get

$$\left(\frac{6 \text{ dollars}}{1 \text{ feet}}\right) \cdot (1200 \text{ feet}),$$ which simplifies to 7200 dollars.

The complete Amount Table looks like this:

	amount =	rate •	base
Descriptions →	cost of fencing	costs of one foot	length of fencing
Units →	dollars	$\dfrac{\text{dollars}}{\text{foot}}$	feet
↓ Item names ↓			
wood	**1920**	24	80
chain link	**7200**	6	1200

Step 3 Verbal models* and *Step 4 Mathematical Models There is a relation between the cost of the entire fence, the cost of the wood fence, and the cost of the chain link fence:

Verbal model ⟹ total cost of fence (dollars) = cost of wood fence (dollars) + cost of chain link fence (dollars)

Math model ⟹ x = 1920 + 7200

Step 5 Solve

$x = 1920 + 7200$ ⟸ Equation to solve.

$x = 9120$ ⟸ Combined like terms.

Step 6 Values of unknowns x is 9120.

Step 7 Answer The total cost of the fence is $9,120.

Step 8 Check We leave the check to you.

— *Note* — ***Practice the process***	Keep in mind that you are trying to learn a process that can be used to solve difficult problems. If you are able to solve the problems without using the general procedure, that's great, but don't do it. The answers are unimportant; you are trying to learn a *process* which will become critical for success in the next section.

Practice 3.3.2 Use the General Procedure For Solving Word Problems to help you answer the following questions.

1. A butcher mixes 20 pounds of inexpensive hamburger which costs $0.89 a pound with some expensive hamburger, which costs $1.10 a pound to make 30 pounds of medium priced hamburger. Find the total cost of the hamburger that went into the mixture.

2. A forest fire was traveling at 0.25 miles a minute for two hours before reaching an area which had been purposely burned to decrease the fuel available. At this point the fire slowed to 0.05 miles per minute and continued to travel for 45 minutes. What was the total distance covered by the fire in the 2 hours 45 minutes under consideration?

Exercise Set 3.3 The answers to the odd numbered exercises are at the back of the book.

Use the Amount Formula to help you answer the following questions. Be sure to properly reduce the units.

1. Find the value of 34 nickels.

2. Find the value of 89 dimes.

3. Movie tickets cost $5.50 each. Find the value of 20 tickets.

4. A blank VCR tape costs $3.50. Find the cost of 50 tapes.

5. Jane receives a 5% commission on every car she sells. How much would she earn if she sold $50,000 worth of cars?

6. A state has a 7% sales tax. Find the tax on $80 worth of merchandise.

Use the General Procedure For Solving Word Problems to help you answer the following questions.

7. A wildflower grower mixes 12 ounces of seeds that cost 95¢ an ounce with a second variety of seeds that cost 49¢ an ounce. If the total mixture contained 2 pounds of seeds, how much did the total mixture cost? (There are 16 ounces in a pound.)

8. A model buys 3 oz of Hydrating Creme at $17.00 an ounce and 2.5 oz of Skin Fortifier, which costs $33.50 more an ounce than the Hydrating Creme. What was the total amount she spent?

9. At the front of his house, John put up a nice fence that cost $15 per linear foot. Along the sides and back he put up a cheaper fence that cost $12 per linear foot. John has a square lot and bought a total of 1500 feet of fencing. Find the total cost of the fence.

10. The long sides of a rectangular room were paneled with expensive walnut, which cost $30 per linear foot, installed. The other two sides, which were each 9 feet long, were covered with cheaper wallpaper at a cost $12 per linear foot. If the perimeter of the room is 42 feet, find the cost of the wall covering.

11. Two trains leave towns 732 miles apart and travel toward each other. The fast train, which travels at 78 mph, leaves at 12 noon. The slow train, which travels at 50 mph, leaves two hours later. How far apart will they be three hours after the fast train started?

12. A train leaves Chicago at 11 am at a speed of 110 mph. Two hours later Joe leaves Chicago going in the opposite direction at a speed of 60 mph. How far apart will they be at 4 pm?

13. A company is buying new computers. An inexpensive system costs $1,099, a midrange system costs $2350 and an expensive system costs $4,900. If they buy four inexpensive systems for each expensive system, and half as many midrange systems as inexpensive systems, find the total amount spent if they purchased 12 expensive systems?

14. If a gas station sells per hour 300 gallons of unleaded gas, one third as much super–unleaded gas, and 80 less gallons of premium unleaded as super–unleaded gas find the number of gallons of gasoline sold over a 16 hour period.

15. A bus leaves Leafburg and passes through two towns before reaching it's destination at Upland. The trip from Leafburg to Upland takes a total of three and a half hours. The bus travels for 45 minutes at 45 mph on the trip between Leafburg and the first town. The bus travels at 70 mph between the first and second town and 55 mph between the second town and Upland. The time from the first to the second town is 45 minutes more than the time from Leafburg to the first town. How far is the bus trip from Leafburg to Upland.

16. Participants in a bike rally must ride from point A to B then from B to C and finally from C back to A along a different trail. One participant rode the course in 3 hours 15 minutes. It took the participant 30 minutes to ride from A to B, and 75 minutes longer to ride from C back to A then it took to ride from A to B. If the participant rode at 14 mph from A to B, 3 mph faster from B to C and 12 mph from C back to A find the length of the course.

17. A dairy decides to make chocolate milk by mixing whole milk, which costs 12¢ a gallon with skim milk which costs 11¢ a gallon and chocolate syrup which costs 5¢ a gallon. If there were 80 more gallons of skim milk than of chocolate syrup in the final mixture and one–twentieth as much chocolate syrup as whole milk in the final mixture, what is the cost of the final mixture if 200 gallons of whole milk was used.

18. A juice company decides to make a fruit punch by mixing 200 gallons of orange juice, which costs them 8¢ a gallon, with grape juice which costs 11¢ a gallon and sugar water which costs 1¢ a gallon. If the mixture contained seven hundred less gallons of orange juice than sugar water and twice as much grape juice as orange juice what was the total cost of the mixture?

Review Exercises   The answers to all of these exercises are at the back of the book.

19. Identify each base and its exponent: $2x^5$

20. Graph $x < 4$

21. Simplify: $12x - 8y + 21x$

22. Simplify: $-(m - 2) - (5 - m) - (m - 1)$

Solve.

23. $\frac{1}{3}(x - 4) = \frac{x}{2} - 1$

24. $0.12(10 - w) + 0.06w = 0.84$

25. $\frac{1}{2}(x - 6) + \frac{2}{3} = 2x - \frac{3}{2}(5 + x)$

26. $\frac{1}{2}(x - 3) = \frac{2}{3}(x + 6) - \frac{11}{2}$

27. $V = \pi r^2 h$ for h

28. $5x - (x - 2) < 2x + 5$

Section 3.4 The Amount Formula: Finding the Rate or Base

In the last section both the rates and the bases were known values. In this section either the rates or the bases will be unknown. The idea will be to use the same procedure you practiced previously, but let a variable represent values you are not explicitly given.

Example 3.4.1   José has a bag that contains dimes and quarters. There are a total of 784 coins in the bag and the value of the coins is $133. Find the number of each type of coin.

•**SOLUTION**•

Step 1 Organize

Given: • total number of coins is 784.
         • total value of coins is $133.

Find: • number of dimes.
         • number of quarters.

Sketch: Not helpful.

Rates: The value of a coin can be written as a rate using the unit $\frac{cents}{coin}$.
         Since rates are involved we will construct an Amount Table. We placed ❶, ❷, ❸, and ❹ in the blank cells so that we can refer to them later when we fill in those quantities.

	amount =	*rate* •	*base*
Descriptions →	value of coins	value of 1 coin	number of coins
Units → ↓ *Item names* ↓	cents	$\frac{cents}{coin}$	coins
dimes	❸	10	❶
quarters	❹	25	❷

Formula: amount = rate • base.

Step 2 Define unknowns

Define: We are looking for the number of dimes and quarters. We can define either of these as the unknown. Let's say

d represents the number of dimes.

Construct: We need to construct an expression for the number of quarters. Since we know the total number of coins (784) and the number of dimes (d), we can use the Whole-Part Concept to construct an expression for the number of quarters.

$784 - d$ represents the number of quarters.

(Check: This makes sense since the number of quarters must be less than the total number of coins and, since d is positive, $784 - d$ is less than 784.)

Amount Table: We now complete the Amount Table.

❶ is the number of dimes, d.

❷ is the number of quarters, $784 - d$.

❸ is the value of the dimes. We multiply rate • base to get

$\left(\frac{10 \text{ cents}}{1 \text{ coin}}\right) \cdot (d \text{ coins})$, which simplifies to $10d$ cents

❹ is the value of the quarters. We multiply rate • base to get

$$\left(\frac{25 \text{ cents}}{1 \text{ coin}}\right) \cdot ((784 - d) \text{ coins}), \text{ which simplifies to } 25(784 - d) \text{ cents}$$

The complete Amount Table looks like this:

	amount	=	rate	•	base
Descriptions →	value of coins		value of 1 coin		number of coins
Units →	cents		$\dfrac{\text{cents}}{\text{coin}}$		coins
↓ Item names ↓					
dimes	**10d**		10		**d**
quarters	**25(784 − d)**		25		**784 − d**

Step 3 Verbal models *and* ***Step 4 Mathematical Models***   We may describe two relations, one between the *number* of coins and the other between the *value* of the coins:

- relation between the *number* of coins:

Verbal model ⇒   *number* of all coins   =   *number* of dimes   +   *number* of quarters

Math model ⇒   784   =   d   +   (784 − d)

When we combine like terms on the right we get 784 = 784. This does not help us solve the problem. This situation comes about because we are trying to use the same piece of information twice: We used the Whole-Part Concept to define the number of quarters as 784 − d; but we also used this same information to say the total number of coins is the sum of the number of dimes and the number of quarters. The fact that d + (784 − d) is 784 can be taken as a confirmation that our expressions for the number of dimes and the number of quarters probably are correct.

- relation between the *value* of the coins:

Verbal model ⇒   *value* of all coins   =   *value* of dimes   +   *value* of quarters
            (cents)   (cents)   (cents)

Math model ⇒   133   =   10d   +   25(784 − d)

We have to be careful about units here. The total value of the coins is given as 133 *dollars* but the rate we used for the coins is *cents* per coin. We cannot add or equate quantities that have different units. Either we change $133 into 13,300 cents or we change 10 cents per coin into $0.10 per coin (and likewise for quarters). We will change $133 to 13,300 cents since most people find whole numbers easier to work with than decimals.

Verbal model ⇒   *value* of all coins   =   *value* of dimes   +   *value* of quarters
            (cents)   (cents)   (cents)

Math model ⇒   13300   =   10d   +   25(784 − d)

We can solve this equation for d.

Step 5 Solve

$$13300 = 10d + 25(784 - d) \quad \Leftarrow \text{ Equation to solve.}$$

$$13300 = 10d + \mathbf{19600 - 25d} \quad \Leftarrow \text{ Distributed the 25.}$$

$$13300 = \mathbf{-15d} + 19600 \quad \Leftarrow \text{ Simplified right side.}$$

$$\mathbf{-6300} = -15d \quad \Leftarrow \text{ Subtracted 19600 from both sides.}$$

$$420 = d \quad \Leftarrow \text{ Divided both sides by } -15.$$

Step 6 Values of unknowns   We fill in the unknown values in the Amount Table.

Descriptions →	value of coins	value of 1 coin	number of coins
Units →	cents	$\dfrac{\text{cents}}{\text{coin}}$	coins
↓ Item names ↓			
dimes	10*d* is 10(**420**) **4200 cents** (**value of dimes**)	10 cents per coin	*d* is **420** (**number of dimes**)
quarters	25(784 − *d*) is 25(784 − **420**) **9100 cents** (**value of quarters**)	25 cents per coin	784 − *d* is 784 − **420** **364** (**number of quarters**)

Step 7 Answer

- number of dimes, *d*, is 420.
- number of quarters, 784 − *d*, is 364.

Step 8 Check

- we have 784 coins: checks because 420 dimes + 364 quarters is 784 coins.
- value of the coins is $133: checks because 4200 cents (value of dimes) + 9100 cents (value of quarters), is 13,300 cents, which is $133.

— *Note* — ***Strategy for equation building***	In problems like the one above there are two models of interest. We are concerned with the *number of* items and with the *value of* those items. There is a very important distinction between *number* and *value*. For example, which is more, 5 nickels or 3 dimes? If *more* means number or weight or volume then the answer is 5 nickels. If *more* means value or buying power then the answer is 3 dimes. It will help you to think about the units of measurement when doing these problems so that you can keep straight number and value.

Practice 3.4.1 Use the General Procedure For Solving Word Problems to help you answer the following questions.

1. Tickets to a baseball game cost $8 each, with a $2 discount for kids under twelve years old. If 620 tickets were sold and $4,400 was collected, how many regular tickets were sold?

2. A church group is sponsoring a recycling drive. On average they are collecting 4 pounds of newspaper per house and $\frac{3}{4}$ pounds of aluminum per house. If at the end of the day they had collected a total of 1710 pounds of material, approximately how many houses did they reach?

3. William can lay 40 rolls of sod an hour when the weather is cool but only 30 when the weather is hot. Find the total amount of sod William laid during cool weather if he worked 10 hours and laid 320 rolls.

In the previous example we found unknown bases. The following example illustrates how to solve problems where the rates are unknown.

Example 3.4.2   Bob and Alice sold tickets for a play. They sold 420 tickets for the play alone and 364 package tickets which included both a dinner and the play. Each package ticket was $15 more than a play ticket alone. If they collected a total of $13,300 how much did each play ticket cost and how much did each package ticket cost?

•SOLUTION•

Step 1 Organize

Given: • 420 play tickets sold.
    • 364 package tickets sold.
    • a package ticket costs $15 more than a play ticket.
    • total value of the tickets is $13,300.

Find: • cost of a play ticket
• cost of a package ticket

Sketch: Not helpful.

Rates: The price of a ticket can be written as a rate using the unit

$\dfrac{dollars}{ticket}$. Since rates are involved we will construct an Amount Table to help organize the information. Since we know the number of tickets sold we fill in the corresponding cells.

	amount	= rate	• base
Descriptions →	value of tickets	value of 1 ticket	number of tickets
Units → ↓ Item names ↓	dollars	$\dfrac{dollars}{ticket}$	tickets
play tickets	❸	❶	420
package tickets	❹	❷	364

Formula: amount = rate • base.

Step 2 Define unknowns

Define: We are looking for the rate of a play ticket and the rate of a package ticket. We can define either of these as the unknown. Let's say

x represents the rate of a play ticket (play alone).

Construct: We need to construct an expression for the rate of a package ticket. Since we know the rate for a package ticket is $15 more than the rate for a play ticket (x), we can construct an expression for the rate of a package ticket.

$(x +15)$ represents the rate of a package ticket (play with dinner).

(Check: This makes sense since the rate of a package ticket is more than the rate of a play ticket and $x + 15$ is more than x.)

Amount Table: We now complete the Amount Table.

❶ is the rate for a play ticket, x.

❷ is the rate for a package ticket $x + 15$.

❸ is the value of all the play tickets. We multiply rate • base to get

$\left(\dfrac{x\,dollars}{1\,ticket}\right)$ • (420 tickets), which simplifies to $420x$ dollars

❹ is the rate of the package tickets. We multiply rate • base to get

$\left(\dfrac{x + 15\,dollars}{1\,ticket}\right)$ • ((364) tickets), which simplifies to $364(x + 15)$ dollars

The complete Amount Table looks like this:

	amount	= rate	• base
Descriptions →	value of tickets	value of 1 ticket	number of tickets
Units → ↓ Item names ↓	dollars	$\dfrac{dollars}{ticket}$	tickets
play tickets	**420x**	**x**	420
package tickets	**364($x + 15$)**	**$x + 15$**	364

Step 3 Verbal models* and *Step 4 Mathematical Models We may describe two relations, one between the *number* of tickets and the other between the *value* of the tickets:

- relation between the *number* of tickets:

Verbal model ⇒	number of all tickets	=	number of play tickets	+	number of package tickets
Math model ⇒	420 + 364	=	420	+	364

When we combine like terms we get 784 = 784. This equation is an identity and so it does not help us solve the problem. Notice also that this equation did not involve the rates so it could not lead to a solution.

- relation between the *value* of the tickets:

Verbal model ⇒	value of all tickets (dollars)	=	value of play tickets (dollars)	+	value of package tickets (dollars)
Math model ⇒	13300	=	420x	+	364(x + 15)

We can solve this equation for x.

— *Note* — **We don't need two models**	We only need one verbal model to solve the problem. However, this problem has two important relations, one dealing with the value of the tickets and the other with the number of tickets. Before we constructed the models it would have been difficult to tell which model is the better one to use.

Step 5 Solve

$$13300 = 420x + 364(x + 15) \quad \Leftarrow \text{Equation to solve.}$$
$$13300 = 420x + \mathbf{364x + 5460} \quad \Leftarrow \text{Distributed 364.}$$
$$13300 = \mathbf{784x} + 5460 \quad \Leftarrow \text{Combined like terms.}$$
$$\mathbf{7840} = 784x \quad \Leftarrow \text{Subtracted 5460 from both sides.}$$
$$10 = x \quad \Leftarrow \text{Divided both sides by 784.}$$

Step 6 Values of unknowns We fill in the unknown values in the Amount Table.

Descriptions →	value of tickets	value of 1 ticket	number of tickets
Units → ↓ *Item names* ↓	dollars	$\dfrac{\text{dollars}}{\text{ticket}}$	tickets
play tickets	420x is 420(10) **4200 dollars** **(value of play tickets)**	x is **10** **10 dollars per ticket** **(rate for a play ticket)**	420 tickets
package tickets	364(x + 15) is 364(**10** + 15) 364(**25**) **9100 dollars** **(value of package tickets)**	x + 15 **10 + 15** **25 dollars per ticket** **(rate for a package ticket)**	364 tickets

Step 7 Answer

- cost of a play ticket, x, is 10 dollars per ticket.

- cost of a package ticket, $x + 15$, is 25 dollars per ticket.

Step 8 Check

- The package rate is $15 more than the play rate. Checks since 25 is 15 more than 10.

- the value of all the tickets is $13,300: checks because $4,200 (value of play tickets) + $9,100 (value of package tickets), is $13,300.

Practice 3.4.2 Use the General Procedure For Solving Word Problems to help you answer the following questions.

1. Tom has a drawer of old stamps. He has 300 of one type of stamp and 460 of a second type of stamp whose value is 14¢ cheaper than the first. If the total value of all the stamps is $156 find the value of each type of stamp.

2. A non–profit agency is monitoring the pledges received through the efforts of two of their telephone solicitors. An experienced solicitor received 30 pledges over a four–hour period while a novice solicitor received 18 pledges over the same time period but at an average of $12 less per pledge. Find the dollars each solicitor averaged per pledge if together they brought in $1080 over the four–hour period.

In the previous problems, we were concerned with two items. Sometimes the number of items exceeds two as the following problem shows.

Example 3.4.3   A school wants to make $10,000 selling seats to a basketball game. It decides to charge different prices for different types of seats. The best seats are court side and there are 100 of those. There are also 500 regular and 200 bleacher seats (farthest from the court). The court side seats should cost twice what the bleachers cost and the regular seats should cost $2 more than the bleachers. What should they charge for each type of seat?

•SOLUTION•

Step 1 Organize

Given: • school wants to make $10,000.
- there are 100 court side seats.
- there are 500 regular seats.
- there are 200 bleacher seats.
- court side seats cost twice what the bleachers cost.
- regular seats cost $2 more than the bleachers.

Find: • cost of a court side seat.
- cost of a regular seat.
- cost of a bleacher seat.

Sketch: Not helpful.

Rates: The price of a seat can be written as a rate using the unit $\frac{\text{dollars}}{\text{seat}}$.
Since rates are involved we will construct an Amount Table.

	amount	=	rate	•	base
Descriptions →	value of seats		value of 1 seat		number of seats
Units →	dollars		$\frac{\text{dollars}}{\text{seat}}$		seats
↓ Item names ↓					
court side	❹		❶		100
regular	❺		❷		500
bleacher	❻		❸		200

Formula: amount = rate • base.

Step 2 Define unknowns

Define: We are looking for the cost of each type of seat. Since the cost of a court side seat and the cost of a regular seat are given in terms of the cost of a bleacher seat, let's say

b represents the cost of a bleacher seat.

Construct: We need to construct expressions for the cost of a court side seat and a regular seat.

- We know a court side seat costs twice what a bleacher seat costs so we can write

    $2b$ represents the cost of a court side seat.

    (Check: This makes sense since the cost of a court side seat is more than the cost of a bleacher seat and, since b is positive, $2b$ is more than b.)

- We know a regular seat costs $2 more than a bleacher seat so we can write

    $b + 2$ represents the cost of a regular seat.

    (Check: This makes sense since the cost of a regular seat is more than the cost of a bleacher seat and $b + 2$ is more than b.)

Amount Table: We now complete the Amount Table.

❶ is the cost of a court side seat, which is $2b$.

❷ is the cost of a regular seat, which is $b + 2$.

❸ is the cost of a bleacher seat, which is b.

❹ is the value of the court side seats. We multiply rate • base to get

$$\left(\frac{2b \text{ dollars}}{1 \text{ seat}}\right) \cdot 100 \text{ seats}, \text{ which simplifies to } 200b \text{ dollars}$$

❺ is the value of the regular seats. We multiply rate • base to get

$$\left(\frac{(b + 2) \text{ dollars}}{1 \text{ seat}}\right) \cdot (500 \text{ seats}), \text{ which simplifies to } 500(b + 2) \text{ dollars}$$

❻ is the value of the bleacher seats. We multiply rate • base to get

$$\left(\frac{b \text{ dollars}}{1 \text{ seat}}\right) \cdot (200 \text{ seats}), \text{ which simplifies to } 200b \text{ dollars}$$

The complete Amount Table looks like this:

	amount	=	rate	•	base
Descriptions →	value of seats		value of 1 seat		number of seats
Units → ↓ Item names ↓	dollars		$\dfrac{\text{dollars}}{\text{seat}}$		seats
court side	**200b**		**2b**		100
regular	**500(b + 2)**		**b + 2**		500
bleacher	**200b**		**b**		200

Step 3 Verbal models *and* **Step 4 Mathematical Models** We may describe two relations, one between the *number* of seats and the other between the *value* of the seats:

- relation between the *number* of seats:

Verbal model ⇒	*number* of all seats	=	*number* of court side seats	+	*number* of regular seats	+	*number* of bleacher seats
Math model ⇒	x	=	100	+	500	+	200

This equation does not involve rates so it will not help us solve the problem. (However, if the question had asked us to find the total number of seats this equation would have given us the solution.)

- relation between the *value* of the seats:

	value of all seats (dollars)	=	*value* of court side seats (dollars)	+	*value* of regular seats (dollars)	+	*value* of bleacher seats (dollars)
Verbal model $\Rightarrow$							
Math model $\Rightarrow$	10,000	=	$200b$	+	$500(b+2)$	+	$200b$

We can solve this equation for b.

Step 5 Solve

$10000 = 200b + 500(b+2) + 200b$	$\Leftarrow$	Equation to solve.
$10000 = 200b + 500b + 1000 + 200b$	$\Leftarrow$	Distributed 500.
$10000 = 900b + 1000$	$\Leftarrow$	Combined like terms.
$9000 = 900b$	$\Leftarrow$	Subtracted 1000 from both sides.
$10 = b$	$\Leftarrow$	Divided both sides by 900.

Step 6 Values of unknowns
We fill in the unknown values in the Amount Table.

	amount	=	rate	$\cdot$	base
Descriptions $\rightarrow$	value of seats		value of 1 seat		number of seats
Units $\rightarrow$	dollars		$\dfrac{\text{dollars}}{\text{seat}}$		seats
$\downarrow$ Item names $\downarrow$					
court side	$200b$ is $200(\mathbf{10})$ **2000 dollars** **(value of court side seats)**		$2b$ is $2(\mathbf{10})$ **20 dollars per seat** **(price of a court side seat)**		100 seats (number of court side seats)
regular	$500(b+2)$ is $500(\mathbf{10}+2)$ **6000 dollars** **(value of regular seats)**		$b+2$ is $\mathbf{10}+2$ **12 dollars per seat** **(price of a regular seat)**		500 seats (number of regular seats)
bleacher	$200b$ is $200(\mathbf{10})$ **2000 dollars** **(value of bleacher seats)**		b is **10 dollars per seat** **(price of a bleacher seat)**		200 seats (number of bleacher seats)

Step 7 Answer

- cost of a court side seat $2b$, is $20.
- cost of a regular seat, $b+2$, is $12.
- cost of a bleacher seat, b, is $10.

Step 8 Check

- school wants to make $10,000: checks because they earn $2,000 on court side seats, $6,000 on regular seats, and $2,000 on bleacher seats.
- court side seats cost twice what the bleachers cost: checks since a court side seat costs $20, which is twice the cost of a $10 bleacher seat.
- regular seats cost $2 more than the bleachers: checks since a regular seat costs $12, which is $2 more than a bleacher seat.

Practice 3.4.3 Use the General Procedure For Solving Word Problems to help you answer the following questions.

1. A theater company makes $3840 on a sold–out performance. In the theater they have a certain number of premium seats, twice as many regular seats as premium seats and 75 more cheap seats as regular seats. If they charge $14 for a premium seat, $12 for a regular seat and $8 for a cheap set how many of each kind of seat were available?

2. A company which manufactures hard drives needs to clear their old inventory of 1500 drives. They bring all of last years units to a computer and peripheral sale. They expect to sell 40 more $215 drives than $450 drives per hour and half as many $360 as $215 drives per hour. If the sale will last 10 hours, how many of each type of drive will they have to sell on average per hour to completely clear their inventory by the end of the sale.

Exercise Set 3.4   The answers to the odd numbered exercises are at the back of the book.

Use the General Procedure For Solving Word Problems to help you answer the following questions.

1. A baseball team sold season tickets for the upcoming year. An adult ticket costs $35, while a child's ticket costs $12. A total of 180 tickets were sold for a total value of $3,425. How many of each kind were sold?

2. At a sport shop inexpensive running shoes sell for $34.99 a pair and expensive running shoes sell for $84.95 a pair. Find how many expensive pairs were sold if the store sold six more pairs of cheap shoes than expensive shoes and the total amount made from sales was $809.64.

3. Bill deposited a total of 68 ten–dollar and five–dollar bills. If the total value is $425, how many of each kind of bill did he deposit?

4. Angela knows she made $642.50 selling two types of magazine subscriptions but she can't remember how many $4.00 magazines and how many $3.75 magazines she sold. If she knows she sold 37 more of the cheap magazines, how many of the expensive magazines did she sell?

5. Every *House and Flower* magazine costs a company a certain amount to print while every *Car and Passenger* magazine costs the company 10¢ less. If the company prints 4000 of each type of magazine, and spends a total of $6400, find how much each magazine costs to print.

6. Beth mixed together 15 pounds of peanuts with twice as many pounds of cashews. The cashews cost $2.25 more a pound than the peanuts. If she spent $157.50, how much did a pound of peanuts cost and how much did a pound of cashews cost?

7. A student has 300 pages of easy reading and 80 pages of difficult reading to do over a three day weekend. If it takes her about 4 times as long to read a page of difficult material as it does for her to read a page of easy material how fast would she have to read each page of difficult material and how fast would she have to read the easy material if she had 31 hours to spend on the reading?

8. A student needs to type two papers over the weekend. The first paper is a technical paper and he allows himself 2 hours to finish it. The second is an opinion paper and he allows himself 45 minutes to finish it. He estimates that his typing speed will drop by about 25 words per minute while working on the technical paper. How many words per minute will he need to type the opinion paper if both papers together are 3765 words?

9. The Department of Recreation built a bike path around a lake. 725 feet of the path was asphalt. The other 525 feet, which went through swampy terrain, was made of wooden planks and cost $20 per linear foot more than the asphalt part. The department has set aside $25,625 to cover the cost of the path. Find the price in dollars per linear foot for each type of path if the project is to come in exactly on budget.

10. A painting and wallpaper company is bidding on a project. They estimate that they will have to cover 1010 square feet with paint and 510 square feet with wallpaper. If the owners want the project to come in around $1400, what should the company charge per square foot for wallpapering to have the price come in at exactly $1400. Assume that wallpapering costs $2.00 more per square foot than does painting.

11. Joe has to make a deposit of $1410 made up of $5 bills, $10 bills, and $20 bills. If he has 4 times as many $5 bills as $20 bills, and three less $10 bills than $5 bills how many of each kind does he have?

12. The manager of a discount store needs to stock her cash registers with $6,500 at the beginning of each day. From experience, she knows that she needs 3 times as many ones as five's and half as many tens as five's. How many of each type of bill should she ask for?

13. Tickets for a basketball game cost $4.50 for the upper seats, $7.00 for the middle seats and $10.50 for the lower seats. If 400 more middle than lower seats are sold, and 100 less upper than lower seats are sold, find the number of each type of seat sold when a total of $6706 was made.

14. Three kinds of stamps are purchased. Sixteen less 28¢ than 20¢ are bought and one–half as many 32¢ as 28¢ are bought. If the total cost of the stamps was $24.96, how many of each type was purchased?

Review Exercises   The answers to all of these exercises are at the back of the book.

15. Prime factor: 1225

16. Graph $x \le -3$

17. Solve: $\dfrac{2}{3}x - \dfrac{9}{2} < x + \dfrac{1}{2}$

Simplify.

18. $8 - \left| 2 - 5 \right|$

19. $-4(n-3) - (-2n)$

20. $y + 6[3y + 9(y + 3) - 12]$

Solve.

21. $5 - (x - 8) - 2x = 18 - (5 - x)$

22. $0.8(p - 1) + 3(p + .2) = 3 - 0.2p$

23. $x - \dfrac{5}{8} = \dfrac{2}{3}(4 - x) + \dfrac{5}{3}x$

---

24. Two small towns have a friendly contest to decide which is larger. Given that a represents the population of Allenville, use a to write an expression that represents the population of Benton if

   a) There are 600 more people living in Benton.
   b) There are 600 less people living in Benton.
   c) The population of Allenville exceeds that of Benton by 200 people.
   d) There are half as many people living in Allenville as in Benton.

Section 3.5 Distance-Rate-Time Word Problems

Many applied problems are based on the distance formula, $d = rt$. This is just a special version of the Amount Formula, amount = rate • base, where the amount is *distance traveled*, the rate is the *speed of travel*, and the base is *time of travel*.

Amount Formula $\Rightarrow$ amount = rate • base

Distance Formula $\Rightarrow$ distance = rate • time

Topic 3.5 A — Word Problems that Involve Adding Distances

Problems that involve adding and subtracting distances have a form very close to the problems you did in the last Section.

Example 3.5.1   Two trains leave cities 320 miles apart and travel towards each other. The slow train, which travels at 40 mph, leaves at 8:00 am. The fast train, which travels at 72 mph, leaves one hour later. At what time do they pass?

   •SOLUTION•

   Step 1 Organize

   Given: • trains begin 320 miles apart.
   • slow train has a rate of 40 mph.
   • slow train leaves at 8:00 am.
   • fast train has a rate of 72 mph.
   • fast train leaves one hour later.

   Find: Time trains pass each other.

   Sketch:

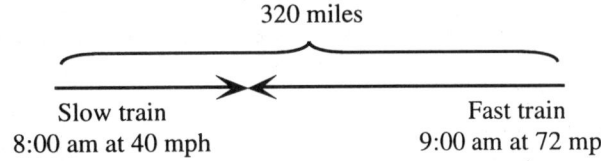

   320 miles

   Slow train   Fast train
   8:00 am at 40 mph   9:00 am at 72 mph

Rates: Rates are involved so we will construct an Amount Table. Since the problem discusses the distance and time of two trains, the trains are the items.

	amount	=	rate	•	base
Descriptions →	distance		speed		time
Units → ↓ Item names ↓	miles		$\dfrac{\text{miles}}{\text{hour}}$		hours
slow train	❸		40		❶
fast train	❹		72		❷

Formula: amount = rate • base

Step 2 Define unknowns

Define: We are asked to find a time of day but the base in the Amount Formula stands for the time the train travels, not a time of day. If we find the number of hours the slow train travels, we can add this to 8:00 am to get the time of day the trains pass. Therefore, let's say

t represents the travel time for the slow train.

Construct: We must construct an expression for the length of time the fast train travels. Since the fast train begins 1 hour later than the slow train, the fast train travels for 1 hour less. Therefore, we can say

$t - 1$ represents the travel time for the fast train.

(Check: This makes sense since the fast train travels for less time than the slow train and $t - 1$ is less than t.)

— Note — **We can define the variable in more than one way**	We could just as correctly define t to be the time of travel for the fast train. In that case, the time for the slow train would be $t + 1$. (Check: This makes sense since the slow train travels for more time than the fast train and $t + 1$ is more than t.)

Amount Table: We now complete the Amount Table.

❶ is the time for the slow train, t.

❷ is the time for the fast train, $t - 1$

❸ is the distance traveled by the slow train. We multiply rate • base to get

$\left(\dfrac{40 \text{ miles}}{1 \text{ hour}}\right) \cdot (t \text{ hours})$, which simplifies to $40t$ miles

❹ is the distance traveled by the fast train. We multiply rate • base to get

$\left(\dfrac{72 \text{ miles}}{1 \text{ hour}}\right) \cdot ((t - 1) \text{ hours})$, which simplifies to $72(t - 1)$ miles

The complete Amount Table looks like this:

	amount	=	rate	•	base
Descriptions →	distance		speed		time
Units → ↓ Item names ↓	miles		$\dfrac{\text{miles}}{\text{hour}}$		hours
slow train	**40t**		40		t
fast train	**72$(t - 1)$**		72		$t - 1$

Step 3 Verbal models and Step 4 Mathematical Models   The sketch below may help you to visualize the relation between the distance the trains started apart, the distance traveled by the slow train, and the distance traveled by the fast train:

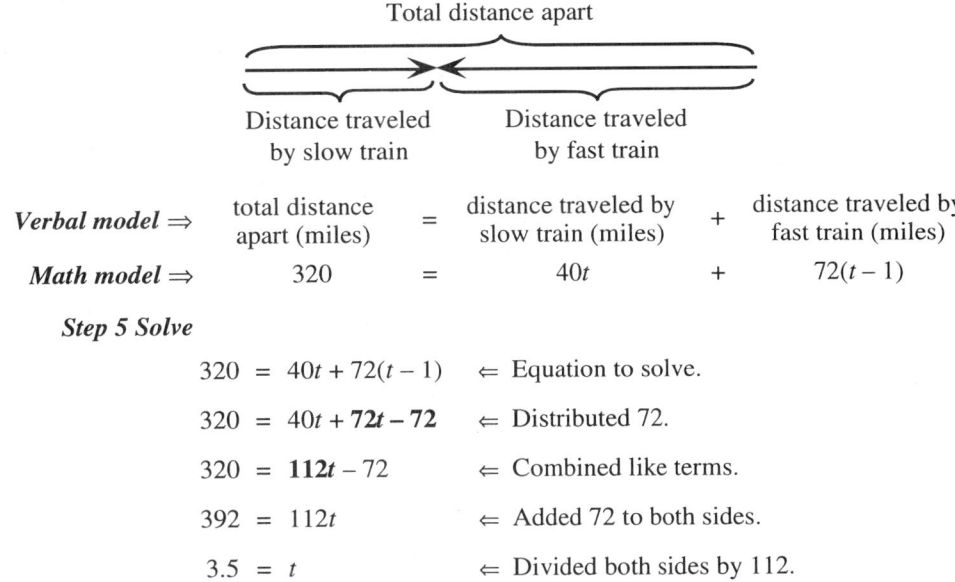

	Verbal model ⇒	total distance apart (miles)	=	distance traveled by slow train (miles)	+	distance traveled by fast train (miles)
	Math model ⇒	320	=	$40t$	+	$72(t-1)$

Step 5 Solve

$$320 = 40t + 72(t-1) \quad \Leftarrow \text{Equation to solve.}$$

$$320 = 40t + \mathbf{72t - 72} \quad \Leftarrow \text{Distributed 72.}$$

$$320 = \mathbf{112t} - 72 \quad \Leftarrow \text{Combined like terms.}$$

$$392 = 112t \quad \Leftarrow \text{Added 72 to both sides.}$$

$$3.5 = t \quad \Leftarrow \text{Divided both sides by 112.}$$

Step 6 Values of unknowns   We fill in the unknown values in the Amount Table.

	amount	=	rate	•	base
Descriptions →	distance		speed		time
Units →	miles		$\dfrac{\text{miles}}{\text{hour}}$		hours
↓ Item names ↓					
slow train	$40t$ is $40(3.5)$ **140 miles** **(distance traveled by slow train)**		40 mph (speed of slow train)		t is **3.5 hours** **(time of travel for slow train)**
fast train	$72(t-1)$ is $72(3.5-1)$ **180 miles** **(distance traveled by fast train)**		72 mph (speed of fast train)		$t-1$ is **3.5 − 1** **2.5 hours** **(time of travel for fast train)**

Step 7 Answer   We found that the time of travel for the slow train, t, is 3.5 hours, which is 3 hours and 30 minutes. If we add 3:30 to 8:00 am we get 11:30 am. This is the time the trains pass each other.

Step 8 Check

- fast train leaves one hour later: checks because the slow train travels for 3.5 hours and the fast train travels for 2.5 hours.
- trains begin 320 miles apart: checks because the slow train travels 140 miles and the fast train travels 180 miles, and 140 + 180 is 320.

Practice 3.5.1

1. Two trains leave towns 732 miles apart and travel toward each other. The fast train, which travels at 78 mph, leaves at 12 noon. The slow train, which travels at 50 mph, leaves two hours later. At what time do they pass each other?

2. A train leaves Chicago at 11 am at a speed of 110 mph. Two hours before Joe had left Chicago going in the opposite direction at a speed of 60 mph. At what time will they be 885 miles apart?

Topic 3.5 B — Word Problems that Involve Equating Distances

This group of problems will use the idea of setting distances equal to each other.

Example 3.5.2 One hour after watching his daughter leave for camp Scott realizes that she forgot her sleeping bag. How long will it will take Scott to catch his daughter's bus if the bus is traveling at 30 miles per hour and Scott can travel in his car at 50 miles per hour?

•SOLUTION•

Step 1 Organize

Given: • car leaves one hour after bus.
        • car travels at a rate of 50 mph.
        • bus travels at a rate of 30 mph.

Find: time it takes car to catch bus.

Sketch:

bus at 30 mph

car at 50 mph, one hour later

Rates: Rates are involved so we will construct an Amount Table. Since the problem discusses the distance and time of the bus and the car, they are the items.

	amount	=	rate	•	base
Descriptions →	distance		speed		time
Units →	miles		$\dfrac{\text{miles}}{\text{hour}}$		hours
↓ Item names ↓					
car	❸		50		❶
bus	❹		30		❷

Formula: amount = rate • base.

Step 2 Define unknowns

Define: We are looking for the time for the car so let's say

        t represents the time for the car.

Construct: We need to construct an expression for the time for the bus. Since the bus begins 1 hour earlier than the car it travels for 1 hour more. We can say

        t + 1 represents the time the bus.

        (Check: This makes sense since the bus travels for more time than the car and *t* + 1 is more than *t*.)

Amount Table: We now complete the Amount Table.

        ❶ is the time for the car, *t*.

        ❷ is the time for the bus, *t* + 1.

        ❸ is the distance traveled by the car. We multiply rate • base to get

        $\left(\dfrac{50 \text{ miles}}{1 \text{ hour}}\right) \cdot (t \text{ hours})$, which simplifies to 50*t* miles

        ❹ is the distance traveled by the bus. We multiply rate • base to get

        $\left(\dfrac{30 \text{ miles}}{1 \text{ hour}}\right) \cdot ((t + 1) \text{ hours})$, which simplifies to 30(*t* + 1) miles

The complete Amount Table looks like this:

	amount	=	*rate*	•	*base*
Descriptions →	distance		speed		time
Units → ↓ Item names ↓	miles		$\frac{miles}{hour}$		hours
car	**50t**		50		**t**
bus	**30(t + 1)**		30		**t + 1**

Step 3 Verbal models* and *Step 4 Mathematical Models Since the car and bus begin and end at the same place, and they follow the same path, they travel the same distance (look at the sketch if you doubt this).

Verbal model ⇒ $\dfrac{\text{distance traveled}}{\text{by bus (miles)}}$ = $\dfrac{\text{distance traveled}}{\text{by car (miles)}}$

Math model ⇒ $30(t + 1)$ = $50t$

Step 5 Solve

$$30(t + 1) = 50t \quad \Leftarrow \text{ Equation to solve.}$$

$$30t + 30 = 50t \quad \Leftarrow \text{ Distributed 30.}$$

$$30 = 20t \quad \Leftarrow \text{ Subtracted } 30t \text{ from both sides.}$$

$$1.5 = t \quad \Leftarrow \text{ Divided both sides by 20.}$$

Step 6 Values of unknowns We fill in the unknown values in the Amount Table.

	amount	=	*rate*	•	*base*
Descriptions →	distance		speed		time
Units → ↓ Item names ↓	miles		$\frac{miles}{hour}$		hours
car	$50t$ 50(**1.5**) **75 miles** **(distance by car)**		50 mph		t **1.5 hours** **(car travel time)**
bus	$30(t + 1)$ 30(**1.5** + 1) **75 miles** **(distance by bus)**		30 mph		$t + 1$ **1.5 + 1** **2.5 hours** **(bus travel time)**

Step 7 Answer Time of travel for the car, t, is 1.5 hours. It will take Scott $1\frac{1}{2}$ hours to catch up with the bus.

Step 8 Check

- The bus travels one hour longer than the car: checks since the bus travels for 2.5 hours and the car travels for 1.5 hours.

- The car and bus travel the same distance: checks since both travel 75 miles.

Practice 3.5.2

1. How long would it take a car going 56 mph to overtake a car going 40 mph that had a half hour head start?

2. It takes a family traveling by car at 50 mph a certain number of hours to travel from their city to a another city where their relatives live. If they fly at 450 mph from their relatives' city back home, it takes 20 hours less time. How far is it between cities?

Exercise Set 3.5   The answers to the odd numbered exercises are at the back of the book.

Use the General Procedure For Solving Word Problems to help you answer the following questions.

1. Since midnight, a forest fire has been traveling through woods at approximately 18 mph. After jumping a road and entering a swampy area the fire slows to approximately 11 mph. Find the time the fire stayed in the swamp if it traveled a total of 89 miles since midnight and the fire remained in the swamp for 90 minutes longer than in the woods.

2. A couple in a rubber raft decides to take a 14.5 mile floating picnic. They begin in a creek, which has a speed of 8 mph, and float for awhile before entering a river where they float at 6 mph until reaching a bridge where they end their trip. If the trip in the creek took 75 minutes less than the trip in the river how long did they travel in the creek?

3. Loretta is driving 232.5 miles to her mother's house for the holidays. Because of road construction she goes 20 mph slower for part of her trip. If she traveled for 3 hours through the road construction and then completed the remainder of her trip in 1.5 hours, find her speed while driving through the construction zone.

4. A plane traveled at a certain speed for 45 minutes before running into turbulence and reducing its speed by 50 mph. Find the plane's rate before the turbulence if it continued at the slower speed for the remaining 1 hour and 15 minutes of the 662.5 mile trip.

5. A truck leaves a city at 9:00 am traveling 40 kph (kilometers per hour). At 11:15 am, another truck leaves the city traveling in the same direction at 90 kph. If radio communication between the trucks is possible only if they are within 65 km (kilometers) of each other, at what time can communication be established?

6. A canoe race team leaves at 9:00 am and travels at 300 feet per minute down river. At 9:15 am a second team traveling at 340 feet per minute begins it's pursuit of the first team. At approximately what time will the second team be within 300 feet of the first?

7. Dana's boat becomes untied and begins floating down a river that has a current of 6 miles an hour. One hour later Dana realizes the boat is missing and begins chasing the boat on a bike path that runs along the river. How fast will she have to ride if she wants to catch the boat in one–half hour?

8. Ray must ride to the post office and return home by 5 pm. He leaves at 4:25 pm and rides at 12 mph arriving at the post office at 4:45 pm. If he leaves again immediately, at what speed should he ride to get home on time?

9. A women needs to needs to meet with a client and return to her office by noon. She leaves her office at 9:00 a.m. and travels at a certain speed for 40 minutes before reaching her client. At 11:30 she leaves her client's office and increases her speed by 10 mph to reach her office exactly at noon. At what speed did she return to her office?

10. To help raise money for a scholarship a college professor decides to run from town A to town B and then back again in ten hours. After running at a constant rate for 3.5 hours he reaches town B; he then decreases his rate by 3 miles per hour and returns to town A just in time. How far is town A from town B?

Review Exercises   The answers to all of these exercises are at the back of the book.

11. Prime factor: 297   12. Write the opposite: $-\dfrac{7}{3}$ 13. Graph $x \geq 1.5$ 14. Simplify: $\dfrac{1}{3}x - 2 \leq 3 - 4\left(x + \dfrac{1}{4}\right)$

15. Name the property that justifies the statement: $6 + (x + 2) = (6 + x) + 2$ 16. Simplify: $0.2x + 8y - 1.4x + 3.5y$

Solve.

17. $0.95 = 0.55(2 - w) + 0.45w$ 18. $0.25(x - 20) + 0.05x = x - 0.7(4 + x)$ 19. $0.05x + 0.2(x + 10) \geq 0.2x - 1$

Section 3.6 Word Problems Involving Percent

In this Section we discuss word problems that involve percent.

Topic 3.6 A — Discount, Tax, Commission, Change, and Totals

To encourage consumers to buy their products, a company may offer a discount. For example, let's say a store advertises 15% off the list (regular) price for a TV. Since the price will be measured in dollars, and percent means *per hundred*, 15% off means the store will reduce the price of the TV by $15 for every $100 of the regular price. You will save $15 on a $100 TV; $30 on a $200 TV; $45 on a $300 TV; and so on. The 15% is called the **discount rate** and has the unit $\dfrac{\text{dollars saved}}{\text{dollars of list price}}$.

Using the same logic, percents in relation to taxes, commissions, gratuities, totals, and many others may be thought of as rates:

- A 5% sales tax is a rate of $\dfrac{\$5 \text{ tax}}{\$100 \text{ of selling price}}$.

- A 7% commission charged by a Realtor for selling a house is a rate of $\dfrac{\$7 \text{ commission}}{\$100 \text{ of selling price}}$.

- A test score of 95% is a rate of $\dfrac{95 \text{ questions answered correctly}}{100 \text{ questions asked}}$.

Notice that in each case, the numerator and denominator are measured using the same unit (for example, dollars) but they refer to different quantities (for example, tax and selling price).

We can use the General Procedure for Solving Word Problems to find the answers to problems involving percents.

Example3.6.1   a)   A stereo went on sale for 12% off the list price. If the sale price was $154, find the list price.

    •SOLUTION•

Step 1 Organize

Given: • 12% discount.
             • sale price was $154.

Find: List price.

Sketch: Not helpful.

Rates: A discount of 12% literally means a rate of $\dfrac{\$12 \text{ saved}}{\$100 \text{ list}}$.

Since rates are involved we will construct an Amount Table.

	amount	=	*rate*	•	*base*
Descriptions →	discount		discount rate		list price
Units →	dollars saved		$\dfrac{\text{dollars saved}}{\text{dollars list}}$		dollars list
↓ *Item names* ↓					
stereo	❷		$\dfrac{12}{100}$		❶

Formula: amount = rate • base.

Step 2 Define unknowns

Define: We are looking for the list price so let's say

      x represents the list price (dollars).

Construct: There are no other expressions to construct.

Amount Table: We now complete the Amount Table.

❶ is the list price, x.

❷ is the discount. We multiply rate • base to get

$$\left(\frac{12 \text{ dollars saved}}{\$100 \text{ dollars list}}\right) \bullet (x \text{ dollars list}), \text{ or } \frac{12}{100} x \text{ dollars saved}$$

Since most people write percents as decimals, we will write 0.12 instead of $\frac{12}{100}$. Written as a decimal, a 12% discount literally means $\frac{0.12 \text{ dollars saved}}{1 \text{ dollar list}}$. That is, you save 12 cents for every dollar of list price.

	amount	=	rate	•	base
Descriptions →	discount		discount rate		list price
Units → ↓ Item names ↓	dollars saved		$\dfrac{\text{dollars saved}}{\text{dollars list}}$		dollars list
stereo	**0.12x**		0.12		**x**

Step 3 Verbal models and ***Step 4 Mathematical Models*** The sale price is calculated by subtracting the discount from the list price

Verbal model ⇒	sale price (dollars)	=	list price (dollars)	−	discount (dollars)
Math model ⇒	154	=	x	−	$0.12x$

Step 5 Solve

$$154 = x - 0.12x \quad \Leftarrow \text{ Equation to solve.}$$

$$154 = 0.88x \quad \Leftarrow \text{ Subtracted } 1.00x - 0.12x.$$

$$175 = x \quad \Leftarrow \text{ Divided both sides by 0.88.}$$

Step 6 Values of unknowns We fill in the unknown values in the Amount Table.

	amount	=	rate	•	base
Descriptions →	discount		discount rate		list price
Units → ↓ Item names ↓	dollars saved		$\dfrac{\text{dollars saved}}{\text{dollars list}}$		dollars list
stereo	0.12x **0.12(175)** **21 dollars**		0.12		x **175 dollars**

Step 7 Answer The list price, x, is $175.

Step 8 Check

- the discount is 12%: checks because saving $21 on an item that regularly costs $175 corresponds to a discount rate of $\frac{\$21 \text{ saved}}{\$175 \text{ list}}$, which simplifies to 0.12, or 12%.

- the sale price is $154: checks because the list price, $175, minus the discount, $21, is $154.

b) Gina scored 90% on a test. If she got 4 problems wrong, how many problems were on the test?

•SOLUTION•

Step 1 Organize

Given: • score is 90%.
• 4 wrong answers.

Find: Number of problems on the test.

Sketch: Not helpful.

Rates: A score of 90% literally means a rate of $\dfrac{90 \text{ questions correct}}{100 \text{ questions asked}}$.

	amount	=	rate	•	base
Descriptions →	questions correct		score		questions asked
Units →	questions correct		$\dfrac{\text{questions correct}}{\text{questions asked}}$		questions asked
↓ Item names ↓					
test	❷		$\dfrac{90}{100}$		❶

Formula: amount = rate • base.

Step 2 Define unknowns

Define: We are looking for the number of questions on the test so let's say x represents the number of questions on the test.

Construct: There are no other expressions to construct.

Amount Table: We now complete the Amount Table.

❶ is the number of questions, x.

❷ is the number of questions answered correctly. We multiply rate • base to get

$$\left(\frac{90 \text{ questions correct}}{100 \text{ questions asked}}\right) \cdot (x \text{ questions asked}), \text{ or } \frac{90}{100}x \text{ questions correct}$$

Again, we will write the fraction $\dfrac{90}{100}$ as the decimal 0.9.

	amount	=	rate	•	base
Descriptions →	questions correct		score		questions asked
Units →	questions correct		$\dfrac{\text{questions correct}}{\text{questions asked}}$		questions asked
↓ Item names ↓					
test	**0.9x**		0.9		**x**

Step 3 Verbal models and Step 4 Mathematical Models
The total number of questions is the sum of the number of correctly and incorrectly answered questions.

Verbal model ⇒	total number of questions asked	=	number answered correctly	+	number answered incorrectly
Math model ⇒	x	=	$0.9x$	+	4

Step 5 Solve

$$x = 0.9x + 4 \quad \Leftarrow \text{ Equation to solve.}$$

$$0.1x = 4 \quad \Leftarrow \text{ Subtracted } 0.9x \text{ from both sides.}$$

$$x = 40 \quad \Leftarrow \text{ Divided both sides by 0.1.}$$

Step 6 Values of unknowns   We fill in the unknown values in the Amount Table.

	amount	=	rate	•	base
Descriptions →	questions correct		score		questions asked
Units → ↓ *Item names* ↓	questions correct		$\dfrac{\text{questions correct}}{\text{questions asked}}$		questions asked
stereo	$0.9x$ **0.9(40)** **36 correct**		0.9		x **40 asked**

Step 7 Answer   The number of questions asked, *x*, is 40.

Step 8 Check

- the percent correct is 90%: checks because 36 of the 40 questions asked were correctly answered, and $\dfrac{36}{40}$ simplifies to 0.9, which is 90%.

- 4 questions were incorrectly answered: checks because 36 were answered correctly, and adding the 4 wrong answers gives a total of 40 questions.

Practice 3.6.1

1. Jeff left a 15% tip on his lunch. If his tip was $1.35, find the price of his lunch.

2. Twenty percent of the members of the Sumitville school board voted for a new gymnasium. If 3 members voted for the gym, find the number of members on the board.

3. Liz bought a dress marked 25% off. If her discount was $19, find the original price of the dress.

4. In a test of a voice recognition computer program, 598 words were correctly identified but 52 words were misinterpreted. What is the percent accuracy of the program?

Topic 3.6 B — Simple Interest Formula

When you borrow money for a student loan or to buy a house the amount you borrow is called the **principal**. Eventually, you must pay back the principal, plus **interest**. Interest is money that is paid for using someone else's money (as if the money were being rented). If the interest is calculated using the original principal, it's called **simple interest**. When the interest is based on both the original principal and any unpaid interest, it's called **compound interest**. In this book, we will discuss only simple interest.

We calculate simple interest using the **Simple Interest Formula**

$$i = prt$$

where

i   represents the interest, which is the amount of money you pay for the use of the money you borrow.

p   represents the principal, which is the amount of money you borrow.

r   represents the rate of interest, which is the part of the principal you pay to use the money for a certain period of time. If the unit of time is *years*, the rate is called the **annual percentage rate** or **APR**. The rate is usually expressed as a percent. For example, if you take out a loan with an APR of 8% the rate literally means $\dfrac{8 \text{ dollars owed}}{100 \text{ dollars borrowed per 1 year}}$. Note that there are two units in the denominator, *dollars borrowed* and *year*.

t   represents the length of time you borrow the money. The unit of time must be the same as the unit of time used in the denominator of the rate. For example, if the rate were given as 1.5% per *month* then the time must be measured in *months*; if the rate were given as 18% per year then the time must be measured in years.

If you are the lender instead of the borrower, rather than *paying* interest you *earn* interest. In either case, the mathematics is the same.

Let's use this formula to calculate the simple interest you would owe after 2 years if you borrowed $1,200 at a rate of 16% per year:

interest = principle • rate • time

$$\text{interest} = (1200 \text{ dollars borrowed}) \cdot \frac{16 \text{ dollars owed}}{100 \text{ dollars borrowed per } 1 \text{ year}} \cdot (2 \text{ years})$$

interest = 384 dollars

You will pay $384 to use the $1,200 for 2 years. At the end of 2 years you will owe $1,200 + $384 or $1,584.

— Note — *Simple interest and the* *real world*	Banks, credit card companies, loan companies, etc., typically use *compound interest* instead of simple interest. The mathematics of compound interest is an interesting topic you will cover in future algebra, economics, and business courses.

Example 3.6.2 a) Michelle wants to borrow $5,000 to repair wind damage to the roof of her house. She agrees to pay simple interest on the money at a rate of 8% per year and she agrees to repay the money in 9 months. How much will she have to repay?

•**SOLUTION**• The principal is 5,000 dollars borrowed; the rate is

$$\frac{8 \text{ dollars owed}}{100 \text{ dollars borrowed per year}},$$

which we will write as the decimal 0.08; the time is 9 months. Since the unit of time used in the rate is *years*, before we can use the Simple Interest Formula we must convert 9 months into years by multiplying by $\frac{1}{12}$:

$$(9 \text{ months}) \cdot \left(\frac{1 \text{ year}}{12 \text{ months}} \right), \text{ which simplifies to } 0.75 \text{ year}$$

Now, we can use the Simple Interest Formula.

$i = prt$	⇐ Simple Interest Formula.
$i = 5,000(0.08)(0.75)$	⇐ Substituted values.
$i = 300$	⇐ Simplified.

Michelle would pay $300 interest on the $5,000 loan. At the end of 9 months, she would have to pay back $5,000 + $300 or $5,300.

b) Decide how much you would have to invest at 4% simple interest to earn $240 in 3 years.

•**SOLUTION**• We are given the amount of interest we wish to earn, the rate, and the time, and we are asked to find the principal.

$i = prt$	⇐ Simple Interest Formula.
$240 = p(0.04)(3)$	⇐ Equation to solve.
$240 = p(0.12)$	⇐ Multiplied 0.04 and 3.
$2000 = p$	⇐ Divided both sides by 0.12.

You would need to invest $2000 to earn $240 simple interest in 3 years at an annual rate of 4%.

c) Find the interest rate if you took out a simple interest loan for $800 and had to pay $400 in interest at the end of 4 years,

• SOLUTION •

$$i = prt \quad \Leftarrow \text{ Simple Interest Formula.}$$

$$400 = (800)r(4) \quad \Leftarrow \text{ Equation to solve.}$$

$$240 = 3200r \quad \Leftarrow \text{ Multiplied 800 and 4.}$$

$$0.075 = r \quad \Leftarrow \text{ Divided both sides by 3200.}$$

Finally, we convert 0.075 to a percent by multiplying by 100% to get 7.5%.

Practice 3.6.2

1. To buy a used car Elaine borrowed $2500 at a simple interest rate of 9% to be paid in 4 years. Find the total amount she will have to repay in 4 years.

2. To buy a computer Kevin borrowed $2200 at a simple interest rate of 15% for 8 months. Find the total amount he will have to repay in 8 months.

3. How much would you have to invest at 9.5% simple interest to earn $285 in 2 years.

4. How much would you have to invest at 4.9% simple interest to earn $122.50 in 5 months.

5. If you took out a simple interest loan for $1500 and had to pay $750 in simple interest in 5 years, what was the annual rate of interest?

Topic 3.6 C — Solving Algebra Word Problems Dealing with Simple Interest

We may use the General Procedure for Solving Word Problems to develop a solution for algebra word problems that deal with simple interest. Since these problems involve more than just evaluating the Simple Interest Formula we will use the following modified version of the Amount Table to help organize the information:

	interest =	*principal* •	*rate* •	*time*
Description →	money owed in interest	money borrowed	interest rate	time money borrowed
Units → ↓ *Item names* ↓	dollars interest	dollars borrowed	dollars interest / dollars borrowed per year	years

This Amount Table has an extra column because the amount (money owed in interest) is based on *two* quantities: the money borrowed (the principal) and the time. That is, the interest you owe is higher if you borrow more money, and it's higher the longer you keep the money. This is reflected in the simple interest rate, which has two units in its denominator. For example, a simple interest rate such as 8% per year literally means

$$\frac{8 \text{ dollars interest}}{100 \text{ dollars borrowed per year}}$$

The denominator tells us that the rate is based on not one but two bases: the principal (dollars borrowed) and the time (year). Therefore, we can think of the Simple Interest Formula as a version of the Amount Formula that has *two* bases.

Example 3.6.3   Maria wants to make a one year investment of her $784 tax refund in a combination of bonds, which pay 10% per year, and her brother's new company, which he says will pay 25% per year. How much should she invest in bonds and how much in her brother's business so that she can earn $133 in interest income at the end of one year? [†]

•SOLUTION•

Step 1 Organize

Given: • total of $784 to invest.
   • total earnings is $133.
   • bonds pay 10% per year.
   • company pays 25% per year.

Find: • amount to invest in bonds.
   • amount to invest in the company.

Sketch: Not helpful.

Rates:

	interest	=	principal	•	rate	•	time
Description →	money earned in interest		money invested		interest rate		time money invested
Units → ↓ *Item names* ↓	dollars interest		dollars invested		$\dfrac{\text{dollars interest}}{\text{dollars invested per year}}$		years
bonds	❸		❶		$\dfrac{10}{100}$ or 0.10		1
company	❹		❷		$\dfrac{25}{100}$ or 0.25		1

Formula: interest = principal • rate • time

Step 2 Define unknowns

Define: We want to calculate the money invested in bonds and the money invested in the company. Let's say

   b represents the money invested in bonds (dollars)

Construct: We need to construct an expression for the money invested in the company. Since we know the total money invested (784) and the money invested in bonds (*b*), we can use the Whole-Part Concept to construct an expression for the money invested in the company.

   784 – *b* represents the money invested in the company (dollars)

   (Check: This makes sense since the money invested in bonds must be less than the total money invested and, since *b* is positive, 784 – *b* is less than 784.)

Amount Table: We now complete the Amount Table.

   ❶ is the money invested in bonds, *b*.

   ❷ is the money invested in the company, 784 – *b*.

   ❸ is the interest earned on the bonds. We multiply principal • rate • time to get 0.10*b*

   ❹ is the interest earned on the company. We multiply principal • rate • time to get 0.25(784 – *b*)

[†]   Why doesn't she put all the money into the company, which pays 25% per year? Because it's usually the case that the higher the rate of return the greater the risk involved. Remember, these are *predicted* yields, not guaranteed yields. If her brother's business fails she will earn no interest and probably will lose her original investment as well.

The complete Amount Table looks like this:

interest =	principal •	rate •	time
Description → money earned in interest	money invested	interest rate	time money invested
Units → dollars interest _↓ Item names ↓_	dollars invested	$\dfrac{\text{dollars interest}}{\text{dollars invested per year}}$	years
bonds **0.10b**	**b**	0.10	1
company **0.25(784 − b)**	**784 − b**	0.25	1

Step 3 Verbal models _and_ **Step 4 Mathematical Models**   We may describe two relations, one between the money _invested_ and the other between the money _earned_:

• relation between money _invested_:

Verbal model ⇒   total amount _invested_ (dollars) = amount _invested_ in bonds (dollars) + amount _invested_ in the company (dollars)

Math model ⇒   784   = b + $784 - b$

This equation simplifies to 784 = 784, which is an identity. It will not help us solve the problem. But, it does tell us that our expressions for the amount invested in bonds and the amount invested in the company probably are correct since their sum is the amount invested.

• relation between money _earned_:

Verbal model ⇒   total amount _earned_ (dollars) = amount _earned_ from bonds (dollars) + amount _earned_ from the company (dollars)

Math model ⇒   133   = $0.10b$ + $0.25(784 - b)$

Step 5 Solve

$$133 = 0.10b + 0.25(784 - b) \quad \Leftarrow \quad \text{Equation to solve.}$$

$$133 = 0.10b + \mathbf{196 - 0.25b} \quad \Leftarrow \quad \text{Distributed 0.25.}$$

$$133 = \mathbf{-0.15b} + 196 \quad \Leftarrow \quad \text{Combined like terms.}$$

$$\mathbf{-63} = -0.15b \quad \Leftarrow \quad \text{Subtracted 196 from both sides.}$$

$$420 = b \quad \Leftarrow \quad \text{Divided both sides by } -0.15.$$

Step 6 Values of unknowns   We fill in the unknown values in the Amount Table.

	interest =	principal •	rate •	time
Description →	money earned in interest	money invested	interest rate	time money invested
Units → _↓ Item names ↓_	dollars interest	dollars invested	$\dfrac{\text{dollars interest}}{\text{dollars invested per year}}$	years
bonds	$0.10b$ 0.10(**420**) **42 dollars** **(money earned from bonds)**	b **420 dollars** **(money invested in bonds)**	0.10	1
company	$0.25(784 - b)$ 0.25(784 − **420**) **91 dollars** **(money earned from company)**	$784 - b$ 784 − **420** **364 dollars** **(money invested in company)**	0.25	1

Step 7 Answer

- money invested in bonds is $420.

- money invested in the company $364.

Step 8 Check

- interest earned is $133: checks because Maria earned $42 from the bonds and $91 from the company and 42 + 91 is 133.

- money invested is $784: checks because Maria invested $420 in bonds and $364 in the company and 420 + 364 is 784.

— Note — *Deja vu*	The equations used in this problem may look familiar to you. They are basically the same as those for the coins problem discussed in Section 3.4. Below are the two problems with the commonalties highlighted: *José has a bag that contains **dimes** and **quarters**. There are a total of **784** coins in the bag and the value of the coins is **$133**. Find the number of each type of coin.* $$13300 = 10d + 25(784 - d)$$ *Maria wants to make a one year investment of her **$784** tax refund in a combination of bonds, which pay **10%** per year, and her brother's new company, which he says will pay **25%** per year. How much should she invest in bonds and how much in her brother's business so that she can earn **$133** in interest income at the end of one year?* $$133 = 0.10b + 0.25(784 - b)$$ It would be a good idea for you to study these two problems to see how they are similar. When you begin to see the connections you will be on the way to a deeper understanding of the mathematics we use to describe them. Many problems that look different on the surface turn out to have the same structure when we describe them mathematically.

Practice 3.6.3

1. Sam plans to split $12,000 between two accounts. The first pays 8% APR while the second pays 6% APR. If the combined interest earned after 1 year from both accounts is $930, how much was put into each account?

2. To help with retirement a nurse put some money into an account paying 7% and $8500 less into a second account which pays 18%. If the combined interest earned after one year from both accounts was $1220, how much was put into each account?

In the problem above we found the interest earned by mixing the principal in a certain way. In the next problem we see the effect of choosing a certain mix for the interest rates.

Example 3.6.4 After a successful year, a printer decides to put some of his profits into two retirement accounts. He puts $7,500 in a bond mutual fund and $2,500 in a stock mutual fund. The stock mutual fund has an expected simple interest rate of return that is 8 percentage points higher than the expected rate for the bond mutual fund; taken together, the two accounts earn $18,000 in simple interest after 30 years. What is the rate of interest of each account?

•SOLUTION•

Step 1 Organize

Given
- rate of stock is 8 percentage points more than rate of bonds.
- interest earned is $18,000.
- amount invested in bonds is $7,500.
- amount invested in stocks is $2,500.
- time is 30 years.

Find
- rate of bonds.
- rate of stocks.

Sketch: Not helpful.

Rates:

	interest	=	*principal*	•	*rate*	•	*time*
Description →	money earned in interest		money invested		interest rate		time money invested
Units → ↓ Item names ↓	dollars interest		dollars invested		$\dfrac{\text{dollars interest}}{\text{dollars invested per year}}$		years
bonds	❸		7500		❶		30
stocks	❹		2500		❷		30

Formula: interest = principal • rate • time.

Step 2 Define unknowns

Define: We are looking for the rate of interest of each account. Let's say

b represents the interest rate of the bonds.

Construct: The stock has a rate that is 8 percentage points more than that of the bonds. To make the calculations easier, we will write 8% as 0.08. Therefore, we can write

$b + 0.08$ represents the interest rate of the stock.

(Check: This makes sense since the stock interest rate is more than the bond interest rate and $b + 0.08$ is more than b.)

Amount Table: We now complete the Amount Table.

❶ is the interest rate of the bonds, b.

❷ is the interest rate of the stocks, $b + 0.08$.

❸ is the interest earned on the bonds. We multiply principal • rate • time to get 225,000b.

❹ is the interest earned on the stocks. We multiply principal • rate • time to get 75,000($b + 0.08$).

The complete Amount Table looks like this:

	interest	=	*principal*	•	*rate*	•	*time*
Description →	money earned in interest		money invested		interest rate		time money invested
Units → ↓ Item names ↓	dollars interest		dollars invested		$\dfrac{\text{dollars interest}}{\text{dollars invested per year}}$		years
bonds	**225,000b**		7500		**b**		30
stocks	**75,000($b + 0.08$)**		2500		**$b + 0.08$**		30

Step 3 Verbal models *and* Step 4 Mathematical Models
We may describe two relations, one between the rates and the other between the interest earned:

- relation between rates: The problem tells us the stocks have a simple interest rate that is 8 percentage points higher than the bonds. This can be written as

Verbal model ⇒	rate of stocks (percent)	=	rate of bonds (percent)	+	8%
Math model ⇒	$b + 0.08$	=	b	+	0.08

This is an identity and will not help solve the problem.

- relation between interest earned: The total interest earned must be the sum of the interest earned from both accounts. This can be written as

Verbal model $\Rightarrow$ total amount *earned* $=$ amount *earned* from $+$ amount *earned* from
(dollars)   bonds (dollars)   stocks (dollars)

Math model $\Rightarrow$ $18,000$ $=$ $225,000b$ $+$ $75,000(b + 0.08)$

We can solve this equation.

Step 5 Solve

$18,000 = 225,000b + 75,000(b + 0.08)$ $\Leftarrow$ Equation to solve.

$18,000 = 225,000b + 75,000b + 6,000$ $\Leftarrow$ Distributed the 75,000.

$18,000 = 300,000b + 6,000$ $\Leftarrow$ Combined like terms.

$12,000 = 300,000b$ $\Leftarrow$ Subtracted 6,000 from both sides.

$0.04 = b$ $\Leftarrow$ Divided both sides by 300,000.

Step 6 Values of unknowns — We fill in the unknown values in the Amount Table.

	interest	=	principal	•	rate	•	time
Description →	money earned in interest		money invested		interest rate		time money invested
Units → ↓ Item names ↓	dollars interest		dollars invested		$\dfrac{\text{dollars interest}}{\text{dollars invested per year}}$		years
bonds	$225,000b$ is $225,000(\mathbf{0.04})$ **9,000 dollars** **(money earned in bonds)**		7500 dollars (money invested in bonds)		b is **0.04** **(interest rate on bonds)**		30
stocks	$75,000(b + 0.08)$ is $75,000(\mathbf{0.12})$ **9,000 dollars** **(money earned in stocks)**		2500 dollars (money invested in stocks)		$b + 0.08$ **0.04 + 0.08** **0.12** **(interest rate on stock mutual fund)**		30

Step 7 Answer

- the rate of the bonds is 4%.

- the rate of the stocks is 12%.

Step 8 Check

- rate of stocks is 8 percentage points more than rate of bonds: checks because the bonds have a 4% rate and 8 percentage points more than this is 12%, the rate of the stocks.

- interest earned is $18,000: checks because the interest earned by the bonds is $9,000 and the interest earned by the stocks is $9,000. The sum is $18,000.

Practice 3.6.4

1. Calvin invested $6000 in a safe account and $5000 in a riskier account. The riskier account pays 10 percentage points more in simple interest per year than the safe account. If the two accounts will earn $25,400 in interest at the end of 20 years, what is the rate of each account?

2. Larry invested $3000 in bonds and $8000 in stocks. The return on the stocks was equivalent to a simple interest rate that was 9 percentage points more than the bonds. If the investments earned $16,560 in 12 years, what is the equivalent simple rate of each investment?

Topic 3.6 D — The Amount Formula and Chemical Concentrations

We can use the Amount Formula to help solve problems that deal with chemical concentrations. For example, would you rather have a 70 pound bar that is 12% gold or a 50 pound bar that is 17% gold? The 70 pound bar is clearly more ore, but does it contain more gold? Let's use an Amount Table to find the quantity of gold in each bar.

- The rate for the big bar is 12%, which means $\dfrac{12 \text{ pounds of gold}}{100 \text{ pounds of ore}}$.

- The base for the big bar is the 70 pounds of ore that we have.

- The amount is the quantity of gold in the big bar.

If we use similar logic for the little bar we can construct the following Amount Table:

	amount =	rate •	base
Descriptions →	quantity of gold	concentration of gold	weight of bar
Units → ↓ Item names ↓	pounds gold	$\dfrac{\text{pounds gold}}{\text{pounds bar}}$	pounds bar
big bar	❶	$\dfrac{12}{100}$ or 0.12	70
little bar	❷	$\dfrac{17}{100}$ or 0.17	50

We use the Amount Formula to calculate the pounds of gold in each bar:

❶ $\left(\dfrac{12 \text{ pounds gold}}{100 \text{ pounds bar}}\right) \cdot (70 \text{ pounds bar})$, which simplifies to 8.4 pounds gold

❷ $\left(\dfrac{17 \text{ pounds gold}}{100 \text{ lb bar}}\right) \cdot (50 \text{ pounds bar})$, which simplifies to 8.5 pounds gold

The big bar has more "stuff" but the little bar has more gold, and it's the gold that we're interested in. We would rather have the little bar.

Other types of chemical concentrations will follow the same pattern. For example,

- Whole milk is about 4% butterfat by volume. Literally, this means there are 4 gallons of butterfat in every 100 gallons of whole milk. This can be written as the rate $\dfrac{4 \text{ gallons butterfat}}{100 \text{ gallons milk}}$.

- A 15% acid solution means there are 15 gallons of acid in every 100 gallons of solution. We write the rate as $\dfrac{15 \text{ gallons acid}}{100 \text{ gallons solution}}$.

Example 3.6.5 a) Use an Amount Table to help you decide which has more acid, 20 gallons of a 15% acid solution or 15 gallons of a 20% acid solution.

•SOLUTION• We will write the rates as decimals to make the calculations easier.

	amount =	rate •	base
Descriptions →	quantity of acid	concentration of acid	volume of solution
Units → ↓ Item names ↓	gallons acid	$\dfrac{\text{gallons acid}}{\text{gallons solution}}$	gallons solution
weak solution	❶	$\dfrac{15}{100}$ or 0.15	20
strong solution	❷	$\dfrac{20}{100}$ or 0.20	15

We use the Amount Formula to calculate the gallons of acid in each solution:

❶ $\left(\dfrac{0.15 \text{ gal acid}}{1 \text{ gal solution}}\right) \cdot (20 \text{ gal solution})$, which simplifies to 3 gal acid

❷ $\left(\dfrac{0.20 \text{ gal acid}}{1 \text{ gal solution}}\right) \cdot (15 \text{ gal solution})$, which simplifies to 3 gal acid

Both contain the same quantity of acid even though there are different quantities of solution.

b) Use an Amount Table to determine which has more water: 40 gallons of a 25% orange juice solution or 100 gallons of an 80% orange juice solution?

•**SOLUTION**• We are asked for the quantities of water rather than orange juice. If the solution is 25% juice it is 100% – 25% or 75% water. Therefore,

- for the 25% solution we use 75% as the rate for *water*.
- for the 80% solution we use 20% as the rate for *water*.

	amount =	*rate* •	*base*
Descriptions →	quantity of water	concentration of water	volume of solution
Units → ↓ *Item names* ↓	gallons water	$\dfrac{\text{gallons water}}{\text{gallons solution}}$	gallons solution
weak solution	❶	$\dfrac{75}{100}$ or 0.75	40
strong solution	❷	$\dfrac{20}{100}$ or 0.20	100

We use the Amount Formula to calculate the gallons of water in each solution:

❶ $\left(\dfrac{0.75 \text{ gal water}}{1 \text{ gal solution}}\right) \cdot (40 \text{ gal solution})$, which simplifies to 30 gal water

❷ $\left(\dfrac{0.20 \text{ gal water}}{1 \text{ gal solution}}\right) \cdot (100 \text{ gal solution})$, which simplifies to 20 gal water

The 40 gallon solution has more water.

Practice 3.6.5 Use an Amount Table to help you solve each problem:

1. Which has more butterfat, 35 gallons of whole milk (4% butterfat by volume) or 68 gallons of 2% milk (2% butterfat).

2. Which has more antifreeze, 6 gallons of a 35% solution or 2 gallons of pure antifreeze.

Topic 3.6 E — Mixing Chemicals

When chemicals of different concentrations are mixed together the result is a combination of the original solutions. The key to solving problems involving mixing chemicals is to focus your attention on the individual substances within the mixtures. This is illustrated in the next example.

Example 3.6.6 A manufacturer requires 784 gallons of a solvent containing 17% alcohol. In their warehouse they have two chemical mixtures. Chemical A is 10% alcohol and chemical B is 25% alcohol. How many gallons of A and B should be combined to obtain the desired solvent?

•**SOLUTION**•

Step 1 Organize

Given: • final mixture contains 784 gallons.
- final mixture is 17% alcohol.
- Chemical A is 10% alcohol.
- Chemical B is 25% alcohol.

Find: • number of gallons of Chemical A needed.
      • number of gallons of Chemical B needed.

Sketch:

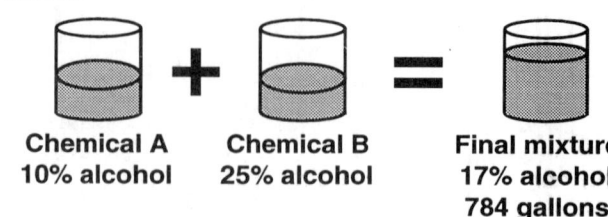

Chemical A	**Chemical B**	**Final mixture**
10% alcohol	**25% alcohol**	**17% alcohol**
		784 gallons

Rates: Concentrations given as percents are rates so we will construct an Amount Table. We have 3 items for this problem Chemical A, Chemical B, and the Final Mixture.

	amount =	*rate* •	*base*
Descriptions →	volume of alcohol	concentration of alcohol	volume of mixture
Units → ↓ *Item names* ↓	gallons alcohol	$\dfrac{\text{gallons alcohol}}{\text{gallons mixture}}$	gallons mixture
Chemical A	❸	$\dfrac{10}{100}$ or 0.10	❶
Chemical B	❹	$\dfrac{25}{100}$ or 0.25	❷
final mixture	❺	$\dfrac{17}{100}$ or 0.17	784

Formula: amount = rate • base.

Step 2 Define unknowns

Define: We are looking for the volume of each chemical. Let's say

a represents the volume of Chemical A (gallons).

Construct: We need to construct an expression for the volume of Chemical B so we use the Whole–Part Concept to say

$784 - a$ represents the volume of Chemical B (gallons).

(Check: This makes sense since the volume of Chemical B must be less than the total volume of the mix and, since a is positive, $784 - a$ is less than 784.)

Amount Table: We now complete the Amount Table.

❶ is the volume of Chemical A, a.

❷ is the volume of Chemical B, $784 - a$.

❸ is the volume of alcohol in Chemical A. We multiply rate • base to get $0.10a$ gallons of alcohol.

❹ is the volume of alcohol in Chemical B. We multiply rate • base to get $0.25(784 - a)$ gallons of alcohol.

❺ is the volume of alcohol in the final mixture. We multiply rate • base to get $(0.17) \cdot (784)$, which simplifies to 133 gallons of alcohol (rounded).

The complete Amount Table looks like this:

	amount	=	rate	•	base
Descriptions →	volume of alcohol		concentration of alcohol		volume of mixture
Units →	gallons alcohol		$\dfrac{\text{gallons alcohol}}{\text{gallons mixture}}$		gallons mixture
↓ Item names ↓					
Chemical A	**0.10a**		0.10		**a**
Chemical B	**0.25(784 − a)**		0.25		**784 − a**
final mixture	**133**		0.17		784

Step 3 Verbal models** and **Step 4 Mathematical Models   There are 3 totals to help us build a model. We can define relations dealing with the total gallons of the *chemicals* or with the total gallons of *alcohol* in the chemicals or with the total gallons of *water* in the chemicals:

- Total gallons of **chemicals**: The final mixture is made by pouring Chemical A and Chemical B into a single container. Therefore, the total number of gallons in the final mixture must be the sum of the number of gallons of Chemical A and the number of gallons of Chemical B.

	total volume of		volume of		volume of
Verbal model ⇒	final mixture (gallons)	=	Chemical A (gallons)	+	Chemical B (gallons)
Math model ⇒	784	=	a	+	$784 - a$

This is an identity which will not help us solve the problem.

- Gallons of **alcohol**: Again, because we form the final mixture by combining Chemical A and Chemical B, we can say that the number of gallons of **alcohol** in the final mixture is the sum of the number of gallons of **alcohol** in Chemical A and the number of gallons of **alcohol** in Chemical B.

	total volume of		volume of **alcohol** in		volume of **alcohol** in
Verbal model ⇒	**alcohol** in final mixture (gallons)	=	Chemical A (gallons)	+	Chemical B (gallons)
Math model ⇒	133	=	$0.10a$	+	$0.25(784 - a)$

We can solve this equation for a.

- Gallons of **water**: Since we can solve the above equation for a we really don't need to consider the water relation. However, we could solve the problem using the water relation. Since the final mixture is 17% alcohol it must be 83% water; therefore, there are 0.83(784) gallons of water (rounded) in the final mixture. Since Chemical A is 10% alcohol it must be 90% water; since Chemical B is 25% alcohol it must be 75% water.

	total volume of		volume of **water** in		volume of **water** in
Verbal model ⇒	**water** in final mixture (gallons)	=	Chemical A (gallons)	+	Chemical B (gallons)
Math model ⇒	0.83(784)	=	$0.90a$	+	$0.75(784 - a)$

Step 5 Solve   Let's solve the alcohol equation.

$$133 = 0.10a + 0.25(784 - a) \quad \Leftarrow \text{Equation to solve.}$$

$$133 = 0.10a + \mathbf{196 - 0.25a} \quad \Leftarrow \text{Distributed the 0.25.}$$

$$133 = \mathbf{-0.15a} + 196 \quad \Leftarrow \text{Combined like terms.}$$

$$\mathbf{-63} = -0.15a \quad \Leftarrow \text{Subtracted 196 from both sides.}$$

$$420 = a \quad \Leftarrow \text{Divided both sides by } -0.15.$$

Step 6 Values of unknowns   We fill in the unknown values in the Amount Table.

	amount	=	rate	•	base
Descriptions →	volume of alcohol		concentration of alcohol		volume of mixture
Units → ↓ Item names ↓	gallons alcohol		$\dfrac{\text{gallons alcohol}}{\text{gallons mixture}}$		gallons mixture
Chemical A	$0.10a$ is $0.10(\mathbf{420})$ **42 gallons** (alcohol in Mix A)		0.10 gallons alcohol per 1 gallon mixture		a is **420 gallons** (volume of Mix A)
Chemical B	$0.25(784 - a)$ is $0.25(784 - \mathbf{420})$ **91 gallons** (alcohol in Mix B)		0.25 gallons alcohol per 1 gallon mixture		$784 - a$ is $784 - \mathbf{420}$ **364 gallons** (volume of Mix B)
final mixture	133 gallons (alcohol in final mix)		0.17 gallons alcohol per 1 gallon mixture		784 gallons (volume of final mix)

Step 7 Answer

- number of gallons of Chemical A needed is 420.
- number of gallons of Chemical B needed is 364.

Step 8 Check

- we have to make 784 gallons of mixture: checks because 420 gallons of Chemical A plus 364 gallons of Chemical B is 784 gallons.

- A and B form a mixture that is 17% alcohol: checks because the final mixture has (42 gallons of alcohol from Chemical A) + (91 gallons of alcohol from Chemical B). This gives 133 gallons of alcohol in the final mixture. Dividing this by the total gallons of final mixture, 784, gives $\dfrac{133}{784}$. If we do the division and round to two decimal places we get 0.17, which is 17%.

— Note — *Deja vu all over again*	The equation we solved in this problem may look familiar to you. It's basically the same as for the José coins problem and the Maria investments problem discussed in Section 3.4. José coins: $13300 = 10d + 25(784 - d)$, where d is number of dimes. Maria investments: $133 = 0.10b + 0.25(784 - b)$, where b is money invested in bonds. Chemicals: $133 = 0.10a + 0.25(784 - a)$, where a is volume of Chemical A. Even though the problems involve different situations, mathematically they are almost identical.

Practice 3.6.6

1. A chemist requires 100 milliliters of 28% alcohol solution. On hand are two mixtures, one consisting of 25% alcohol and the other 30% alcohol. How many milliliters of each kind should be combined to obtain the desired mixture?

2. A steel company needs 60 tons of 8% iron ore. They have ore consisting of 7% iron and 10% iron. How much of each kind should be combined to obtain the desired mixture?

— Caution — *Adding percents can be tricky*	You have to be careful when working with chemical concentrations because your intuition might lead you to the wrong conclusions. For example, if we add a gallon of 4% milk to a gallon of 2% milk we don't get 2 gallons of 6% milk; we get two gallons of 3% milk

Exercise Set 3.6   The answers to the odd numbered exercises are at the back of the book.

Use the General Procedure For Solving Word Problems to help you answer the following questions.

1. Elise bought a coat marked '40% off'. If the discount price was $90, what was the original price?

2. Bob bought a car that was discounted 30%. If he paid $10,500 for the car, what was the original price?

3. Jeff paid a 15% tip on his lunch. If his total bill (with the tip) was $13.80, what was the original price of his lunch?

4. Millie paid 6% tax on her computer. If the total bill (including the tax) was $1325, what was the price of the computer?

5. In a recent survey, 520 people said they would go away on a vacation if they could afford it, but 280 people said they would not. What percent of the people said they would not go away on a vacation?

6. Four hundred forty eight people said that they were satisfied with their long distance carrier while 672 said they were not. What percent said they were satisfied?

7. The property tax rate in Culver City is 4% of the assessed value of a house. If the property tax on a house is $2000, find the assessed value of the house.

8. The sales clerk of a discount store told John that he could buy a TV for 70% of the list price. If the discounted price was $350, what was the list price?

9. How much simple interest would you earn after one year on an investment of $3,000 at an APR of 5%?

10. Bill invested $4,000 for 8 months. If the annual percentage rate of interest was 6%, how much did he earn in simple interest?

11. Lance borrowed $2,400 for 7 months. If the APR charged by the bank was 8%, how much interest will he owe when the loan is due?

12. How much simple interest would you earn after one year on an investment of $4,500 at an APR of $8\frac{3}{4}$%?

13. How much you would have to invest at 5.5% simple interest to earn $341 in 1 year.

14. How much you would have to invest at 8.5% simple interest to earn $76.50 in 1 year.

15. You invest $5,500 in a simple interest account at an APR of 4%. At the end of one year, what is the value of the account?

16. You invest $3,000 in a simple interest account at an APR of 5%. At the end of one year, what is the value of the account?

17. If you took out a simple interest loan for $1500 and had to pay $750 in simple interest in 1 year, what was the annual rate of interest?

18. Ben borrowed $850 and had to pay $34 in simple interest after 1 year, What was the annual rate of interest?

19. Owen has $25,000 to invest. She plans to invest part of it in stocks that are paying 18% and the rest in a savings account that pays 7%. How much must she invest in each account if she wants to earn $2850 at the end of the year?

20. Andy has $12,000 to invest. He plans to invest part of it in bonds that are paying 20% and the rest in a certificate of deposit that pays 8%. How much should he invest in each account if he wants to earn $1440 at the end of the year?

21. Jack borrowed $8000 for his school bills. Part was borrowed from the bank at 12% and the rest from his uncle at 6%. How much did he borrow from each source if the interest he owed at the end of the year was $600?

22. Jenny borrowed $6500 for her car. Part was borrowed from the bank at 7% and the rest from her dad at 5%. How much did she borrow from each source if the interest she owed at the end of the year was $415?

23. Brandon has $36,000 to invest. He plans to invest part in a safe account at 6.5% and the rest in a riskier account paying 18%. How much should he invest in each account if he wants to earn $4640 simple interest at the end of the year?

24. Bridgett borrowed $2500 for her school bill. Part was borrowed from her friend at 6% and the rest from the credit union at 10%. How much did she borrow from each source if the interest she owed at the end of the year was $230?

25. Bill invested $12,000 in a safe account and $18,000 in a riskier account. The riskier account simple interest rate is twice the simple interest rate of the safe account. If the two accounts will earn $2,880 in interest in 1 year, what is the rate of each account?

26. Joyce invested $15,000 in a risky account and $20,000 in a safer account. The safe account simple interest rate is one–third the risky simple interest rate. If the two accounts will earn $4,500 in 1 year, what is the rate of each account?

27. Elaine invested $6,000 at 8% in a money market account, $15,000 in her credit union, and $12,000 in her pension fund. The simple interest rate paid by the pension fund is twice the rate paid by the credit union. If the interest earned by all three accounts at the end of one year is $2430, what is the rate paid by the pension fund and the credit union?

28. Sam invested $8,000 at 12% in a retirement account, $4,000 in his savings account and $10,000 in a high risk account. The simple interest rate paid by the high risk account is three times the rate paid by the savings account. If the interest earned by all three accounts is $2,660, what is the rate paid by the savings account and the high risk account?

Use an Amount Table to help you answer the following questions.

29. Which has the most hydrochloric acid, 15 liters of a 25% acid solution or 25 liters of a 10% acid solution.

30. Which has the most pure alcohol, 22 liters of a 12% alcohol solution or 30 liters of a 9% alcohol solution.

31. Which has the most water, 60 milliliters of a 30% acid solution or 75 milliliters of a 25% acid solution.

32. Which has the most water, 120 milliliters of a 55% iodine solution or 100 milliliters of a 46% iodine solution.

33. Which has the most acid, 40 milliliters of a 30% acid solution, 50 milliliters of a 24% acid solution or 60 milliliters of a 20% acid solution.

34. Which has the least iodine, 80 milliliters of a 45% iodine solution, 90 milliliters of a 40% iodine solution or 70 milliliters of a 50% iodine solution.

35. How much of a 40% acid solution must be mixed with a 25% solution to produce 30 gallons of a 32% solution?

36. How many gallons of a 6% boric acid solution should be mixed with 25 gallons of a 12% boric acid solution to produce a solution that is 10% acid?

37. A chemist needs 30 liters of 60% acid solution. On hand he has a 40% solution and pure acid (100% solution). How much of each kind should be combined to obtain the desired mixture?

38. Bill has 25 ounces of a 20% hydrochloric acid solution. How much water should he add to dilute it to a 16% solution?

39. A mining company needs 60 tons of 6% iron ore. They have ore consisting of 12% iron and 4% iron. How much of each kind should be combined to obtain the desired mixture?

40. Vicki has on hand a 20% alcohol solution and some pure alcohol. How much of each should she mix to obtain 40 liters of a 70% alcohol solution?

Review Exercises   The answers to all of these exercises are at the back of the book.

41. Simplify: $x - 2[5x + 3(x - 8) + 7]$

42. Graph $x > -10$

43. Solve: $5 + \dfrac{x}{4} + (x + 3) = \dfrac{x}{6} - \dfrac{2}{3}$

44. Solve: $0.01(5) + 0.05(a + 10) = 0.03a$

45. Solve: $0.4(7.5) - 0.2x = -(x - 3) + 0.8x$

46. Solve: $-0.5x < 2$

---

47. A machine can sort stones by size at the rate of 300 pounds per minute. How many pounds can be sorted in $2\dfrac{1}{2}$ hours?

Chapter 3 Review

Vocabulary to Know

amount — in the Amount Formula, the product of the rate and the base.

Amount Formula — amount = rate • base.

Amount Table — a grid to help organize information to be used in an Amount Formula.

annual percentage rate (APR) — interest rate for a year.

APR (annual percentage rate) — interest rate for a year.

base — in the Amount Formula, the quantity that the rate is based on.

cell — the intersection of a row and a column in an Amount Table.

compound interest — interest that is calculated using both the original principal and any unpaid interest.

Distance Formula d = rt — version of the Amount Formula where d represents distance, r represents rate or speed, and t represents time.

interest — money earned for the use of your money or money owed by you for the use of someone else's money.

mathematical model — a representation that uses mathematics to describe an original object, operation, procedure, relation, etc..

principal — money borrowed for a loan or given for an investment.

rate — the ratio of two quantities that are measured using different units.

Simple Interest Formula i = prt — where *i* represents interest, *p* represents principal, *r* represents rate, and *t* represents time.

simple interest — interest that is calculated using the original principal.

unit — in the measurement of physical quantities, a multiple of a standard; feet, pounds, and hours are units.

verbal model — a representation that uses written or oral language to describe an original object, operation, procedure, relation, etc..

Whole-Part Concept — if a whole is formed from the sum of two parts then the difference of the whole and one part is the other part.

Properties, Definitions, Formulas, and Notation to Understand

Amount Formula — an amount may be found by multiplying a rate by a base.	amount = rate • base A mechanic charges $12 per hour and works for 5 hours. The amount of money he earns is calculated by multiplying his rate, $\left(\frac{12 \text{ dollars}}{1 \text{ hour}}\right)$, by the base, (5 hours) : $a = \left(\frac{12 \text{ dollars}}{1 \text{ hour}}\right) \cdot (5 \text{ hours})$ $a = 60$ dollars
Distance Formula — version of the Amount Formula used in problems dealing with distances, speeds, and times.	$d = rt$ A car travels at a speed of 50 mph for 4 hours. The distance covered is $d = \left(\frac{50 \text{ miles}}{1 \text{ hour}}\right) \cdot (4 \text{ hours})$ $d = 200$ miles
Simple Interest Formula — version of the Amount Formula used in problems dealing with borrowing or investing money.	$i = prt$ The simple interest on a loan of $500 at an APR of 8% for 9 months is $i = (500 \text{ dollars}) \cdot \frac{8 \text{ dollars}}{100 \text{ dollars per year}}) \cdot \left(\frac{9}{12} \text{ year}\right)$ $i = 30$ dollars

Procedures to Follow

Procedure for building a mathematical model from a verbal model.	A company has 25 fewer executives than factory workers. If *x* represents the number of executives, use *x* to write a simplified expression for the total number of factory workers and executives.
Step 1 Organize the information Decide on the items and build an expression for each item.	Items | Number of people executives | *x* factory workers | *x* + 25
Step 2 Build a verbal model Use English to represent the total as the sum of terms.	number of factory workers + number of executives
Step 3 Build a math model Translate the verbal model into algebra. Because the verbal and math model are so closely related we often do Steps 2 and 3 together.	(*x* + 25) + *x*
Step 4 Simplify Simplify the expression.	2*x* + 25

Procedure for general procedure for solving word problems.	Dean drove for 3 hours on the highway and then turned on to a smaller road and decreased his speed by 20 mph and drove for 2 more hours. If the total distance of his trip was 270 miles, what was his speed on the highway?

Step 1 Organize the information Read the problem, several times if necessary:

- Given: determine what is given. Be sure to include the unit of all numbers.

- Find: determine what you are asked to find.

- Sketch: draw a sketch if it will help you better understand the problem.

- Rates: examine the unit of each given number to determine if a rate is involved. If so, begin construction of an Amount Table.

Given:
- drove 3 hours on highway.
- drove 2 hours at 20 mph less on small road.
- total trip was 270 miles.

Find: Speed while on the highway.

Sketch:

270 miles

Highway for 3 hours → Small road 20 mph slower for 2 hours

Rates:

	amount	=	rate	•	base
Descriptions →	distance		speed		time
Units → ↓Item names↓	miles		$\dfrac{\text{miles}}{\text{hour}}$		hours
highway	❸		❶		3
small road	❹		❷		2

- Formulas: if standard formulas will be needed, list them.

Formulas: amount = rate • base

Step 2 Define unknowns

- Define: assign a definition to a variable; include units, if appropriate.

- Construct: use the variable you just defined to build expressions for other unknowns.

- Amount Table: if you began an Amount Table in Step 1, complete it.

Define: x represents speed on highway.

Construct: $x - 20$ represents speed on small road.

Amount Table:
❶ x
❷ $x - 20$
❸ $3x$
❹ $2(x - 20)$

Step 3 Build verbal models Use English to write relations between the unknowns.

$$\text{Total distance} = \text{distance highway} + \text{distance small road}$$

Step 4 Build mathematical models Translate the verbal models into mathematical models. Because the verbal and mathematical models are so closely related we often do Steps 3 and 4 together.

$$270 = 3x + 2(x - 20)$$

Step 5 Solve Solve the mathematical models for the variable.

$270 = 3x + 2x - 40$

$270 = 5x - 40$

$310 = 5x$

$62 = x$

Step 6 Values of unknowns Calculate the values of all unknowns.

❶ x is 62 mph, speed on highway

❷ $x - 20$ is $62 - 20$, or 42 mph, speed on small road

❸ $3x$ is 3(62), or 186 miles, distance on highway

❹ $2(x - 20)$ is 2(62 − 20), or 84 miles, distance on small road

Step 7 Answer State the answer to the question.

Speed on the highway, x, is 62 mph.

Step 8 Check   Using the original words of the problem, check to see that the values make sense and are consistent with the original statement of the problem.

- drove at 20 mph less on small road: checks because speed on highway is 62 mph and speed on small road is 42 mph.

- total trip was 270 miles: checks because distance on highway was 186 miles and distance on small road was 84 miles and 186 + 84 is 270.

Chapter 3 Review Exercises   The answers to all the exercises are in the back of the book.

1. Two telephone solicitors are selling magazine subscriptions. If e represents the number of subscriptions the more experienced solicitor sells, use e to write an expression for the number of subscriptions the less experienced solicitor sells if

   a) The more experienced solicitor sells 18 more subscriptions.
   b) The less experienced solicitor sells half as many subscriptions.
   c) They sell the same number of subscriptions.
   d) The less experienced solicitor sells 20 fewer subscriptions.

2. A computer reseller is recording how many monitors she sells. If f represents the number of 15 inch monitors she sells use f to represent the number of 17 inch monitors she sells if

   a) The number of 17 inch monitors she sells is 6 more than the number of 15 inch monitors.
   b) The number of 15 inch monitors exceeded the number of 17 inch monitors by 8.
   c) She sold three 15 inch monitors for every 17 inch monitor.
   d) She sold three 17 inch monitors for every 15 inch monitor.

3. Beth has a 40 pound mixture of peanuts, almond, and cashews. If she has 8 more pounds of almonds than peanuts write a simplified expression to represent the pounds of cashews she has. Use p to represent the pounds of peanuts she has.

4. A student earned $1200 for college by working three jobs. She made twice as much on the second job as on the first job. If s represents the amount made on the second job write an expression for the amount she made on the third job.

5. A theater company has a certain number of premium seats, twice as many regular seats as premium seats and 75 more cheap seats than regular seats. If p represents the number of premium seats write a simplified expression to represent the total number of seats in the theater.

6. Write an expression for the circumference of the inner circle.

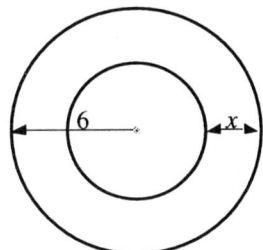

7. A movie sells child, adult, and senior tickets. If c represents the number of child tickets sold, use c to write a simplified expression which represents the total number of people attending the movie if

   a) twice as many child tickets as adult tickets were sold and 200 fewer senior tickets than child tickets were sold.
   b) 40 fewer adult tickets than child tickets were sold and 100 more senior than adults tickets were sold.
   c) one–third as many senior tickets as adult tickets were sold and 24 more adult than child tickets were sold.

8. A family decides to surround a swing set with 4 inch by 6 inch boards. They will then fill the area inside the boards with sand. If they want the width to be one–third the length, and they have 68 linear feet of boards, what should be the dimensions of the structure?

9. After spending $41 for school supplies a student notices she spent $16 more for a backpack than for a calculator and half as much for a mechanical pencil as for the calculator. Find the amount she spent on each item.

10. Find the dimensions of a rectangle which has a length that is 13 inches less than twice the width and the perimeter is 67 inches longer than the length.

11. The sum of three consecutive even integers is 60, find the larger integer.

12. In most areas the amount of water which percolates through the ground to recharge the ground water reservoir is about 4 inches per year. How much water will have been added to the ground water reservoir after 18 years?

13. Find the total cost of a peanut/cashew mixture if the 15 pounds of peanuts that went into the mixture cost $2.00 a pound and the 8 pounds of cashews that went into the mixture cost $2.25 more a pound than the peanuts.

14. A coin changing machine will return coins when you insert paper money. If you insert a $5 bill and get back 32 coins, how many quarters and how many dimes do you receive?

15. For selling $6190.40 worth of tires a salesperson earns a $475 commission. If expensive tires cost $291.96 for a set of four and cheap tires cost $125.80 for a set of four, and the salesperson sold 16 more cheap sets of tires, how many expensive sets of tires did the salesperson sell?

16. A father decides to help at a school carnival. He is given the task of selling 180 bags of doughnuts over the 5 hours the carnival is open. After selling the same number of bags each hour for the first 3 hours he realizes he needs to increase the number he is selling by 15 bags per hour to sell all the doughnuts by the end of the carnival. If he succeeds in selling all 180 bags, and he sells the same number of bags per hour each of the last two hours, how many bags per hour was he selling for the first 3 hours?

17. I have $15.25 worth of nickels, dimes and quarters. There are 17 more quarters than dimes and 20 less nickels than dimes. How many of each type of coin do I have?

18. Due to snow Wayne is having a hard time getting to school to take a test. He traveled at 12 mph on the freeway for a certain period of time before leaving the freeway and trying some back roads where he was able to average 20 mph. If he spent half an hour more on the freeway than on the back roads how long did he travel on the back roads during his 14 mile trip to school?

19. Lauren and Steve need to travel from their home to the home of a friend. Lauren leaves at 8 am but Steven sleeps in and so leaves one and a half hours later but drives 30 mph faster. If they both arrive at the friend's house at 12 noon, how fast was each traveling?

20. Seth bought a suit coat which was marked off 45%. If he paid $385, what was the original price of the coat?

21. How much would you have to invest at 6% simple interest to earn $1000 in interest in 5 years?

22. A couple earns $955 in interest by putting $2000 in an interest earning checking account and $9500 in a stock account whose rate of return is 4 percentage points higher. What is the rate for each account?

23. Mary has $10,000 to invest. She invests $8,000 in a coffee stand and the rest in an Internet stock. At the end of the year her accounts had increased by a total of $2000. What was the rate of return for each investment if the Internet stock had a rate of return 120% more than the coffee stand?

24. Which has more alcohol, 23 gallons of an 18% solution or 25 gallons of a 16% solution?

25. A company wishes to market a drink that is 10% juice. To make the new drink they mix 210 gallons of a solution that is 4% juice with another type of drink that is 30% juice. How many gallons will be in the final mixture?

Chapter 3 Test   The answers to all the exercises are in the back of the book.

1. A student needs to type 2 papers over the weekend. If f stands for the number of pages the first paper will be, use f to represent the number of pages the second paper will be if

   a) The first paper is 12 pages more than the second.
   b) The second paper is 4 pages less than the first.
   c) The second paper contains three–fourths as many pages as the first.
   d) The first paper contains half as many pages as the second.

2. The rectangular perimeter of a nature park is 560 feet. Two sides that are opposite each other were fenced with wood rails while the other two sides were fenced with chain link. If w represents the length of one of the sides bounded with wood rails create a simplified expression to represent the length of one side for which the chain link was used.

3. The manager of a discount store needs to stock her cash registers at the beginning of each day. From experience, she knows that she needs 3 times as many ones as five's and half as many tens as five's. If f represents the number of five dollar bills create a simplified expression using f to represent the total number of bills.

4. In the figure below, the shaded regions are the same size. Write a simplified expression for the area of one of the shaded regions.

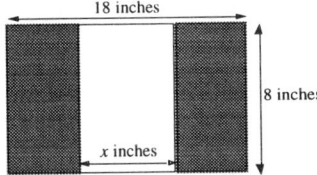

5. A contractor bought three colors of landscaping stones for a new house. If g represents the pounds of gray stones, use g to write a simplified expression that represents the total pounds of stones purchased if

   a) there were 400 pounds more white stones than gray stones and there were twice as many pounds of pink stones as white stones.
   b) there were half as many pounds of pink stones as gray stones and 500 more pounds of gray stones than white stones.
   c) there were equal weights of pink and gray stones and the pounds of pink stones was half that of the white stones.

6. To clear a residential driveway of snow a company charges $25. To clear the driveway and shovel all steps and sidewalks they charge an additional amount. For clearing 38 driveways alone and 27 driveways, steps, and sidewalks the company makes a total $2030. How much is the additional cost to do the steps and sidewalks?

7. A bus driver is recording the number of passengers she has on each of three different runs. The number of passengers on the first run is three times the number on the second run and the number on the second run is 18 less than the number of passengers on the third run. If she carried a total of 23 passengers, how many passengers did she have on the third run?

8. The perimeter of a triangle is two more than twice the length of its longest side. Find the perimeter if the length of the shortest side is half the length of the longest side and the length of the medium side is one less than the average of the lengths of the other two sides.

9. The sum of three consecutive even integers is 2 more than the smallest integer. Find the integers.

10. Latonya is driving to her mother's house for the holidays. Because of road construction she can only travel at a speed of 45 mph for $2\frac{1}{4}$ hours she drives through the construction area. What was the length of the road construction?

11. One hundred eighty hours of overtime is to be split between two groups of workers. One group has 4 more workers than the other. If the group with less workers is to get 3 hours per worker and the other group will get 5 hours per worker how many workers are in each group?

12. A clerk is filling shelves with 210 cans of oil. She fills 9 shelves with the larger cans and three less shelves with the smaller. If she can fit 10 less cans per shelf with the larger size, how many larger cans fit on a shelf.

13. A restaurant sells 3 different types of lunch specials. Special A costs $3 more than special B and special B is twice as expensive as special C. If the total income from 30 Special C's, 21 Special B's and 18 Special A's was $972, what is the price of each special?

14. Two trains are 296 miles apart and traveling towards each other on parallel tracks. The first train travels at 27 mph for one–third the time of the second train. If the second train travels at 65 mph find the distance traveled by the second train before they begin to pass.

15. Would it be cheaper to borrow $500 for 2 years at 20% simple interest or $200 for 2 years at 50% simple interest?

16. Sue invests $5,000 equally in two accounts, one paying 3 percentage points more than the other. If she earns $175 in simple interest after one year, what was the interest rate of each account?

17. Light cream is 20% butter fat while heavy cream is 36% milk fat. How much whole milk, which is 4% milk fat, must be mixed with 1 gallon of heavy cream to make a container of light cream?

Chapter 4
Mathematical Models

Section 4.1 Introduction to Four Mathematical Models

In this chapter we expand on the idea of "model" by more fully developing alternative ways of representing real world situations, including through the use of numbers, algebraic symbols, and pictures.

Topic 4.1 A — Mathematical Models

Suppose we wish to examine the relation between distance and time for a car going 50 mph. Intuition tells us that the longer the car travels the more distance it will cover. Creating a model for this situation will help us quantify the relation so that we can make precise predictions about the distance traveled for a given time, and vice versa. We might model this situation in several ways:

- Verbal model: Every hour, the total distance traveled by the car increases by 50 miles.

    This representation uses English to describe the relation between the distance the car has traveled and the time it has been traveling.

- Math model in the form of an *equation* or *formula*: $y = 50x$.

    This representation uses algebraic symbols to show a general relation between distance, represented by y, and time, represented by x. We can use the formula to predict the value for y from a given value for x, and vice versa.

- Math model in the form of a *data table*:

x time (hours)	y distance (miles)
0	0
1	50
2	100
4	200

    This representation shows that specific times, listed in the left column, are related to specific distances, listed in the right column. For example, in the highlighted row, the time 1 hour corresponds to the distance 50 miles. We can use the data table to predict the values for y from the given values for x, and vice versa.

- Math model in the form of a *graph* (these will be discussed in great detail in this Chapter)

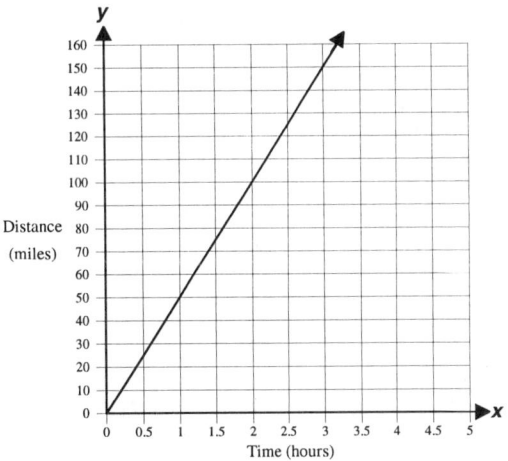

    This representation is a picture of the relation between distance, represented by the vertical number line, and time, represented by the horizontal number line. We can use the graph to predict the value for y from a given value for x, and vice versa.

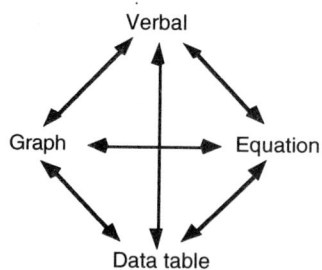

Verbal

Graph ⟷ Equation

Data table

These four types of models are different ways of describing the same situation and are linked to each other, as the diagram on the left shows. Each type of model has advantages and disadvantages depending on the situation and what we are asked to find. In fact, different courses, or different instructors teaching the same course, typically prefer different ways of portraying the same relation. For example, one instructor may make a point using an equation while another instructor may use a graph to make the same point.

Topic 4.1 B — Input and Output Variables

In Chapter 3 we discussed building mathematical models based on the Amount Formula, *amount = rate • base*. In this chapter we will broaden the usefulness of this formula by modifying it in the following ways:

- Rather than using a single variable to represent one element of the formula, say the base, we will use two variables, one to represent the amount and the other to represent the base.

- Rather than having the variable represent a single but unknown value, the two variables will represent two sets of numbers that are related to each other in a specific way.

- Rather than defining the rate as the ratio of two fixed quantities, we will define the rate in such a way that it describes how the two variables are changing relative to each other.

By tradition we use y to represent the amount, m to represent the rate, and x to represent the base. Using these variables, the amount formula looks like this:

$$\textbf{amount} \quad = \quad \textbf{rate} \quad • \quad \textbf{base}$$
$$y \quad = \quad m \quad • \quad x$$

where

- y is called the **output variable** or the **dependent variable** and represents the **output**. The output is a set of values we wish to predict from our knowledge of other values. In the example of a car traveling at 50 mph, the output is the set of distances the car may travel (100 miles, $87\frac{1}{2}$ miles, 158.75 miles, etc.). In an equation, the output variable is usually isolated on the left. For example, in $y = 50x$, the output is represented by y.

- x is called the **input variable** or the **independent variable** and represents the **input**. The input is the set of values we will use to predict the output. In the car example, the input is the set of times the car may travel (2 hours, $1\frac{3}{4}$ hours, 3.175 hours, etc.). In an equation, the input variable usually is involved in mathematical operations and typically appears on the right. For example, in $y = 50x$, the input is represented by x.

- m is called the **rate of change**; it tells us how the output changes in relation to changes in the input. For example, in the car equation $y = 50x$, the rate of change is 50 mph. This rate of change tells us that the output changes by +50 miles when the input changes by +1 hour. We will discuss rate of change in much more detail in Section 4.3.

Whether a quantity is considered input or output depends on the situation. For example, the formula $c = \pi d$ shows the relation between the circumference c of a circle and its diameter, d. If a farmer wanted to know how much fencing would be needed to enclose a circular pasture he might measure the diameter of the pasture and then use the formula to predict the circumference, since measuring the diameter would be easier than measuring the circumference. In that case, he wants to predict the value of c so c is the output variable; he will be predicting c from his knowledge of d, so d is the input variable. However, if the farmer wanted to measure the distance across a circular lake, he might find it easier to measure the lake's circumference and then use $d = \frac{c}{\pi}$ to predict the diameter. In this case, circumference is the input and diameter is the output. The formulas below model the same relation but from two different perspectives.

output   input   output   input

$$c = \pi d \qquad\qquad d = \frac{c}{\pi}$$

Topic 4.1 C — Translating a Verbal Model into an Equation

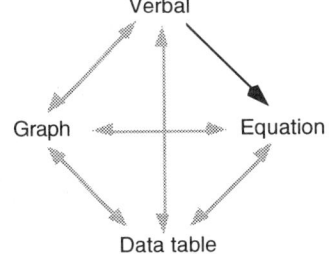

Suppose we need to rent a moving truck and notice the following advertisement:

> **Opening month special.**
>
> **Rent our trucks for $1.25 per mile.**
>
> **No daily charge.**

We could consider this description a verbal model because it uses English to describe a relation between the miles the truck is driven and the cost to drive those miles. This model is easy to understand but to facilitate the calculation of the cost for different distances driven we need an equivalent mathematical model such as a graph, a data table, or an equation. Let's build an equation.

The process for building an equation is similar to the General Procedure we used in Chapter 3 to solve word problems.

— Procedure — *Building an equation,* $y = mx$, *from a verbal* *description*	***Step 1 Organize the information*** Read the problem, several times if necessary: • Given: determine what is given. Be sure to include the unit of all numbers. • Find: determine what you are asked to find. • Sketch: draw a sketch if it will help you better understand the problem. • Rate of change: With these problems it may not be initially clear what the amount (the output) and the base (the input) should be. Therefore, we usually delay constructing an Amount Table until after the variables and the rate of change have been precisely defined or identified. • Formula: $y = mx$ ***Step 2 Define unknowns*** • Identify the output (what you are trying to predict) and its unit. This will be represented by y. • Identify the input (what you are using to predict the output) and its unit. This will be represented by x. • Identify the rate of change, m, and its unit. The rate of change is the ratio $\dfrac{\text{change in output}}{\text{change in input}}$. The numerator must have the same unit as the output; the denominator must have the same unit as the input. • Amount Table: At this point it may be useful to organize the information in an Amount Table. ***Step 3 Verbal model*** This is given in the statement of the problem. ***Step 4 Mathematical model*** Substitute the value of m into the equation, $y = mx$.

Let's apply this procedure to the truck problem:

Step 1 Organize the information

Given: The charge for the truck is $1.25 per mile.

Find: An equation that models this situation.

Step 2 Define unknowns   As a customer, we would most likely want to predict the cost of the truck. Since the quantity we wish to predict is usually the output, we will define

- output: Let y represent the cost (dollars). Keep in mind that the variable y does not represent a single cost but a set of possible costs that are associated with the distances we might drive.

- input: We want to predict the cost using the distance we drive the truck. Therefore, we let x represent the distance driven (miles). Like the output variable, the input variable x does not represent a single distance but a set of possible distances we might drive.

- rate of change: The rate of change is defined as $\dfrac{\text{change in output}}{\text{change in input}}$. The output, cost, changes by \$1.25 when the input, distance, changes by 1 mile. Therefore, the rate of change is $\dfrac{1.25 \text{ dollars}}{1 \text{ mile}}$.

We can organize this information into an Amount Table as follows:

| | *amount* | = | *rate* | • | *base* |
| | *output* | = | *rate of change* | • | *input* |

Descriptions →	*cost*	*charge*	*distance driven*
Units →	*dollars*	$\dfrac{dollars}{mile}$	*miles*
↓ Item names ↓			
truck	y	1.25	x

Step 3 Verbal model   Given.

Step 4 Mathematical model   The math model (in the form of an equation) is $y = 1.25x$. Remember, x does not represent a single unknown quantity, as it did in Chapter 3. Instead, it represents an infinite number of possible distances such as 5 miles, 8.2 miles, 103.45 miles, etc. Likewise, the variable y represents an infinite number of possible costs such as \$6.25, \$10.25, \$129.3125, etc. [†]

Example 4.1.1   a)   The manager of a theater needs to keep track of the money she takes in as tickets to a show are sold throughout the week. Given that the price of a single ticket is \$7.50, build a model for this situation in the form of an equation.

•SOLUTION• ***Step 1 Organize the information***

Given: Each ticket costs \$7.50.

Find: An equation that models this situation.

Step 2 Define unknowns

- output: We want to predict the money collected so we let y represent the money collected (dollars).

- input: We will use the number of tickets sold to predict the cost so we let x represent the number of tickets sold.

- rate of change: We are given that each ticket costs \$7.50, which means the output, money collected, changes by \$7.50 every time the input, number of tickets sold, changes by 1. Therefore, the rate of change is $\dfrac{7.50 \text{ dollars}}{1 \text{ ticket}}$.

[†]   Even though the model $y = 1.25x$ allows x to represent any value, in reality the values for x cannot be negative because we cannot have a negative distance. There is also a practical maximum value that x can take on but it is not clear what that might be. One million miles is obviously too many; 2,000 miles is large but within reason. Likewise, the values for y cannot be negative and there is a practical limit on the cost. We will discuss this in detail later.

An Amount Table may be used to organize the information:

	amount	=	rate	•	base
	output	=	rate of change	•	input
Descriptions →	**money collected**		**price**		**tickets sold**
Units →	**dollars**		$\dfrac{dollars}{ticket}$		**tickets**
↓ Item names ↓					
tickets	y		7.5		x

Step 3 Verbal model   Given.

Step 4 Mathematical model $y = 7.5x$

b)   A machine can form 5 plastic cups in 2 minutes. Build an equation that will determine the time it takes the machine to form any number of cups.

•SOLUTION• *Step 1 Organize the information*

Given: 5 cups can be formed in 2 minutes.

Find: An equation that determines the time it takes to form any number of cups.

Step 2 Define unknowns

- output: We want to predict the time so we let y represent the time (minutes) it takes to form the cups.

- input: We will use the number of cups made to predict the time so we let x represent the number of cups made.

- rate of change: The problem tells us that it takes 2 minutes to form 5 cups. Therefore, the output, time, changes by 2 minutes, when the input, number of cups, changes by 5 cups. The rate of change is $\dfrac{2 \text{ minutes}}{5 \text{ cups}}$.

An Amount Table may be used to organize the information:

	amount	=	rate	•	base
	output	=	rate of change	•	input
Descriptions →	**time**		**rate**		**cups made**
Units →	**minutes**		$\dfrac{minutes}{cup}$		**cups**
↓ Item names ↓					
cups	y		$\dfrac{2}{5}$		x

Step 3 Verbal model   Given.

Step 4 Mathematical model $y = \dfrac{2}{5}x$

c)   A machine can form 5 plastic cups in 2 minutes. Build an equation that will determine the number of cups produced for any given amount of time.

•SOLUTION• *Step 1 Organize the information*

Given: 5 cups can be formed in 2 minutes.

Find: An equation that determines the number of cups made in any given amount of time.

Step 2 Define unknowns This situation is similar to the one described in the above example except we are to make a prediction about the number of cups formed rather than the time.

- output: We want to predict the number of cups so we let y represent the number of cups.

- input: We will use the time to predict the number of cups so we let x represent the time (minutes).

- rate of change: The problem tells us that it takes 2 minutes to form 5 cups. Therefore, the output, cups, changes by 5 cups, every time the input, time, changes by 2 minutes. The rate of change is $\dfrac{5 \text{ cups}}{2 \text{ minutes}}$.

An Amount Table may be used to organize the information:

	amount	=	rate	•	base
	output	=	rate of change	•	input

Descriptions →	**cups made**	**rate**	**time**
Units →	**cups**	$\dfrac{cups}{minute}$	**minutes**
↓ Item names ↓			
cups	y	$\dfrac{5}{2}$	x

Step 3 Verbal model Given.

Step 4 Mathematical model $y = \dfrac{5}{2}x$

Practice 4.1.1 Be sure to define your variables for each of the following problems.

1. A train left a station and traveled at 78 mph. Build an equation that will predict the distance the train has traveled given the time since it left the station.

2. A painter can paint 3 rooms every 10 hours. Build an equation that will predict the time it will take her to paint any number of rooms.

3. A painter can paint 3 rooms every 10 hours. Build an equation that will predict the number of rooms he can paint in a given amount of time.

4. The electric company charges $0.06 per kilowatt hour of electricity used. Build an equation that will determine the cost for different amounts of electricity used.

Topic 4.1 D — Translating an Equation into a Data Table

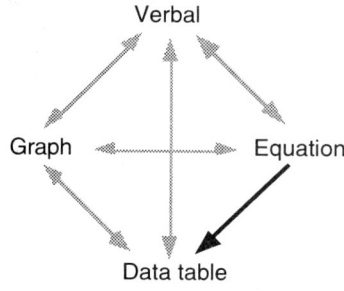

An equation like $y = mx$ is a nice general model, but sometimes we would like some specific input and output values. For example let's return to the truck rental situation. Say instead of wanting to rent the truck we owned the truck. If we rented it to a number of different people we would like a way to keep track of the different miles driven and the related costs. One common way to keep track of related sets of numbers is to use a data table.

A **data table** is a grid with rows and columns that helps organize input and output values. Input values are written in the left column and the corresponding output values are written in the right column. For example, the outline of a data table for the truck rental problem might look like the following:

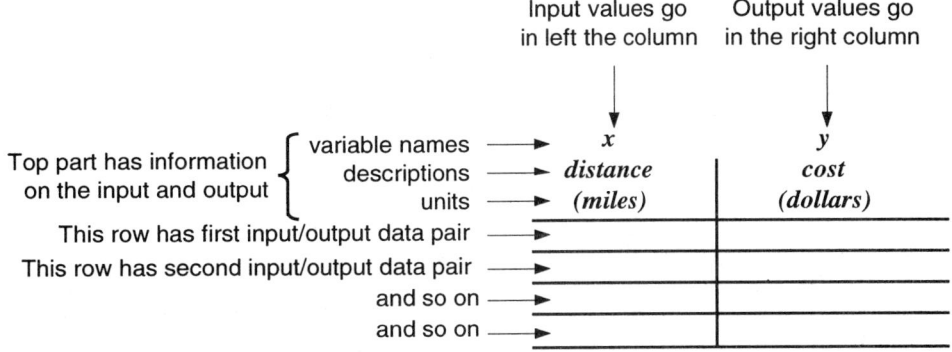

x distance (miles)	y cost (dollars)
10	
20	
35	
50	

Now, let's enter the actual data. Let's say the first customer drove 10 miles, the second 20 miles, the third 35 miles, and the fourth 50 miles. We place these values in the left (distance) column.

We can use the equation $y = 1.25x$ to calculate the output value for each input value. Below is the calculation for the first distance, 10 miles:

$$y = 1.25x \quad \Leftarrow \text{ Equation.}$$

$$y = 1.25(\textbf{10}) \quad \Leftarrow \text{ Substituted 10 for the input.}$$

$$y = \textbf{12.5} \quad \Leftarrow \text{ Simplified.}$$

We write the value 12.5 to the right of its input value and in the right (cost) column. We follow the same procedure for the other input values.

x distance (miles)	y cost (dollars)	Calculation using $y = 1.25\,x$
10	12.50	$\leftarrow y = 1.25(10)$
20	25.00	$\leftarrow y = 1.25(20)$
35	43.75	$\leftarrow y = 1.25(35)$
50	62.50	$\leftarrow y = 1.25(50)$

To save space, we sometimes construct data tables using a horizontal format:

	First pair of input/output	Second pair	Third pair	Fourth pair
Input goes in top row ⟶ *x (distance in miles)*	10	20	35	50
Output goes in bottom row ⟶ *y (cost in dollars)*	12.50	25.00	43.75	62.50

If we want to write a single input/output pair we use **ordered pair notation**. This notation consists of the input and output written inside parentheses and separated by a comma. For example, the first data pair in the above table would be written as (10,12.50); the second pair would be written as (20,25.00); and so on. Data written in this way are called *ordered* pairs because they always are written with the input first and then the output, as in (input,output).

Example 4.1.2   A nature center is having a fund raiser where each participant is charged $2.50. Build an equation that will predict the money earned for various numbers of participants. Then construct a data table that has data for a family of 4, a group of 20, and a bus carrying 32 participants.

•SOLUTION•   *Step 1 Organize the information*

Given: Each participant is charged $2.50.

Find: An equation that predicts the money earned for various numbers of participants.

Step 2 Define unknowns

- output: We want to predict the money earned so we let y represent the money earned (dollars).

- input: We will use the number of participants to predict the money earned so we let x represent the number of participants.

- rate of change: The output, money earned, changes by $2.50 when the input, number of participants, changes by 1. Therefore, the rate of change is $\dfrac{2.50 \text{ dollars}}{1 \text{ participant}}$.

An Amount Table may be used to organize the information:

| | amount | = | rate | • | base |
| | output | = | rate of change | • | input |

Descriptions →	**money earned**	**charge**	**people participating**
Units → ↓ Item names ↓	**dollars**	$\dfrac{\textbf{dollars}}{\textbf{participant}}$	**participants**
fund raiser	y	2.5	x

Step 3 Verbal model   Given.

Step 4 Mathematical model $y = 2.5x$

Given inputs of 4, 20, and 32 the data table should look like the following:

x **number of participants**	y **money earned (dollars)**
4	10
20	50
32	80

Practice 4.1.2   Be sure to define your variables for each of the following problems.

1. After jumping a road, a forest fire is traveling away from the road at approximately 28 feet per minute. Build an equation which predicts the distance the fire has traveled from the road given the time since the fire jumped the road. Build a data table and record the input/output pairs for inputs of 20 minutes, 50 minutes, 1.5 hours, and 3 hours.

2. A women decides to place part of an inheritance into a bond which pays 8.5% per year. Create a equation which predicts the interest the women receives after one year for any amount of money invested. Then, in a data table record the interest she will earn if she invests $1,000, $5,000, and $15,000.

3. The cost of electricity to run 2 strings of holiday mini-lights is about $2.50 a month. Create an equation which predicts the cost to run a string of lights given the number of strings. Construct a data table that shows the monthly cost for 8, 15, 50, and 100 strings.

Topic 4.1 E — Translating a Data Table into a Graph

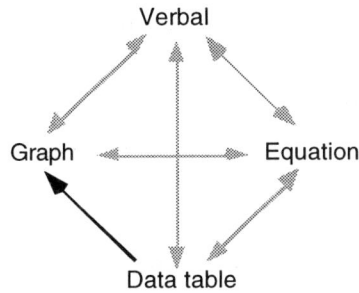

We know the data tables discussed above model relations between input and output because we were given verbal models of the relations and we created the data tables from those. However, sometimes we want to see if a quantitative relation between two things exists and so we construct the data table by collecting numerical data. The data table helps us organize the information but it may not help us see if a relation exists. Representing the data in a graph may help us to decide if there is a relation, and, if so, what the relation might be.

Recall that a graph is a visual representation of a set of data. In Section 2.5 we used the number line to graph sets of data that corresponded to the solutions of inequalities like $x > 3$.

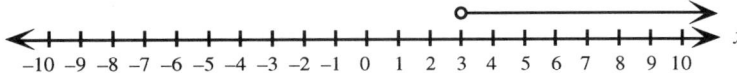

To graph a set of data that consists of input and output pairs we need two number lines, one for input and another for output.

Let's use the truck rental situation to illustrate how to construct a graph from a data table. Here is the data table we constructed:

x distance (miles)	y cost (dollars)
10	12.50
20	25.00
35	43.75
50	62.50

Since there are two sets of numbers we need to plot the values from each set on its own number line.

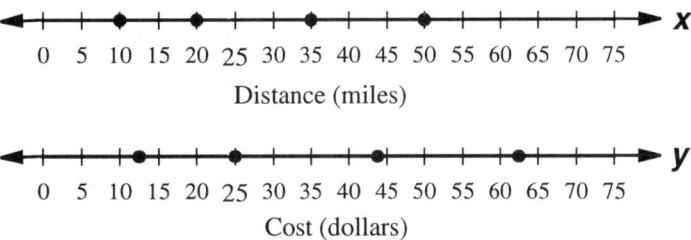

These two graphs represent the distances and the costs but they do not show us which distance corresponds to which cost; that is, the separate number lines do not show us the relation between the distance (input) and the cost (output). To show the correspondence, we rotate the output number line 90 degrees and place both zero points at the same location.

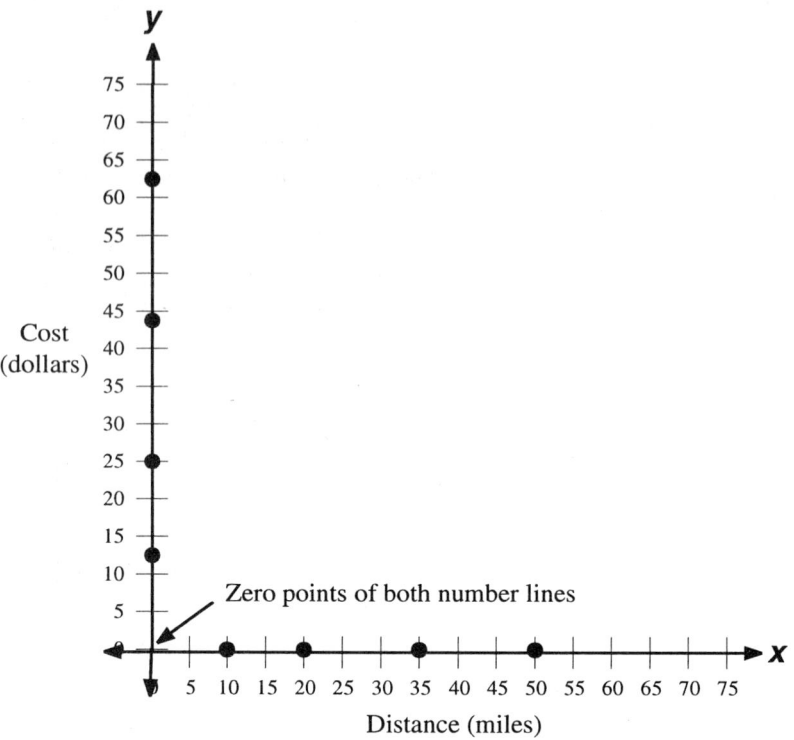

This picture displays all the data but it still doesn't tell us much about the relation between the input and output since it's not obvious which distance corresponds to which cost. To relate the information from both variables we create a new point for each *pair* of data. For example, the first row in the data table says a distance of 10 miles corresponds to a cost of $12.50. To show this correspondence on the graph we mentally draw a vertical line from the input,10, and a horizontal line from the output, 12.5. The point where these two lines meet shows us that an input of 10 corresponds to an output of 12.5.

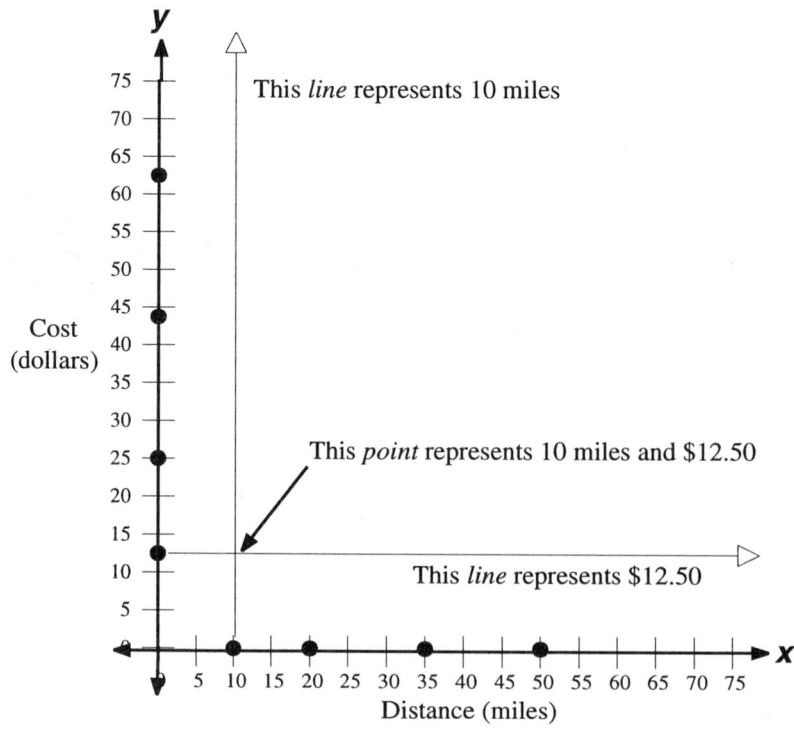

Locating points like this is called **plotting** the points. The location of a point on the graph is called the **coordinates** of the point. We would say that 10 is the ***x* coordinate** of the ordered pair (10,12.50) and 12.50 is the ***y* coordinate**.

The horizontal number line is called the **horizontal axis** or **x axis**, and the vertical number line is called the **vertical axis** or **y axis.** The **origin** is the point where the zeros of the number lines coincide; the coordinates of the origin are (0,0). The number lines taken together are referred to as an **x-y coordinate system**[†]. The two-dimensional flat surface the x-y coordinate system rests upon is called the **coordinate plane**. Graphs plotted on x-y coordinate systems are called **x-y graphs**.

To make it easier to plot points we usually overlay a rectangular grid on the coordinate system. To make it easier to read the locations of points we often write their coordinates next to the dots on the grid. If we do this, and then plot the remaining three points in the data table we get the following graph:

x distance (miles)	y cost (dollars)
10	12.50
20	25.00
35	43.75
50	62.50

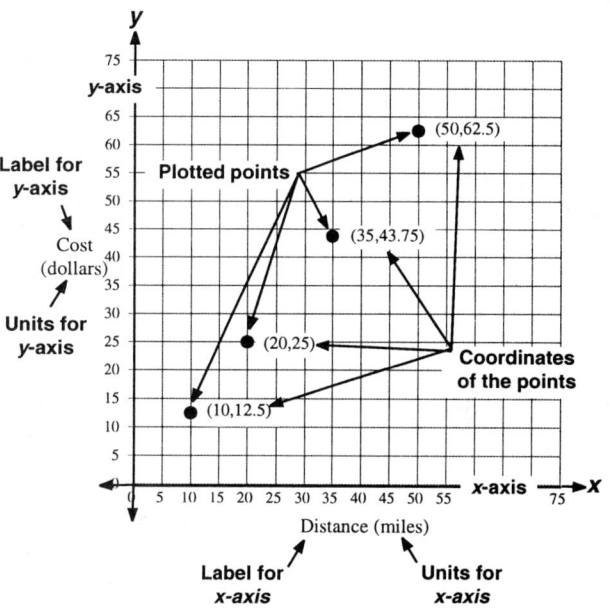

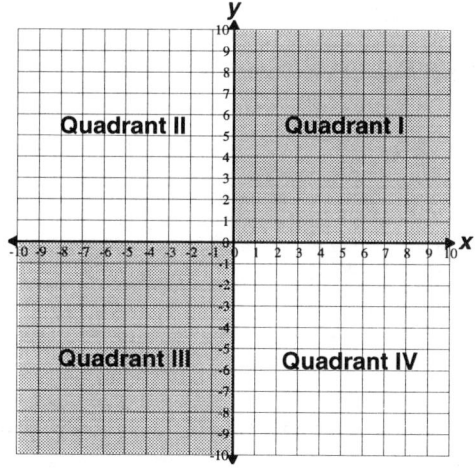

The above graph only shows the positive portions of the x– and y-axes because all the numbers in the data table are positive. If negative data are present we simply extend each axis in its negative direction. When we do that, the coordinate plane is broken up into four parts called Quadrants. We use the Roman numerals I, II, III, and IV as labels for the Quadrants, as shown on the left. Except for the points that fall exactly on one of the axes, all the points on the plane fall into one of these four Quadrants.

[†] The x-y coordinate system is also called the Cartesian coordinate system, in honor of the French mathematician and philosopher Rene Descartes. x-y graphs are also called Cartesian graphs.

Example 4.1.3 On the coordinate system provided, graph the data given in the following data table which describes the relation between the number of dimes we might have and their value.

x number of dimes	y value (cents)
0	0
5	50
7	70
12	120

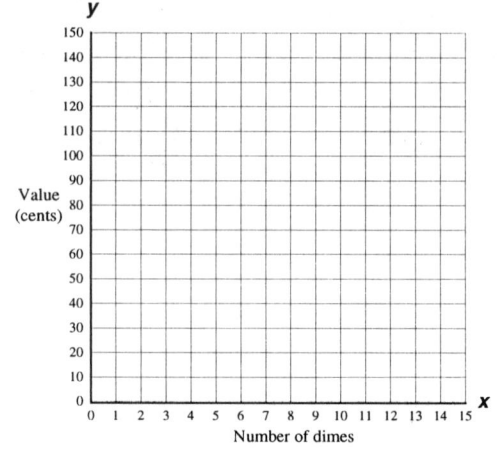

•SOLUTION•

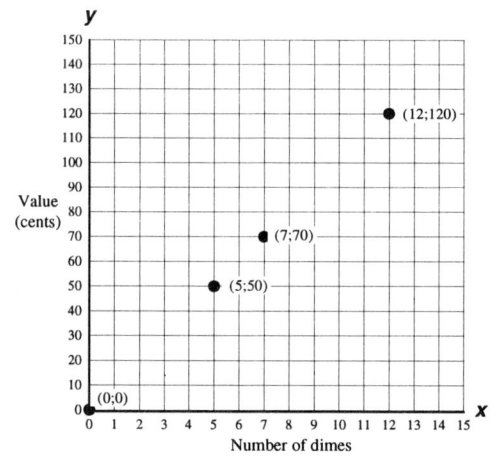

Practice 4.1.3 Graph the data.

1.

x time (minutes)	y distance (feet)
20	560
50	1400
90	2,520
180	5,040

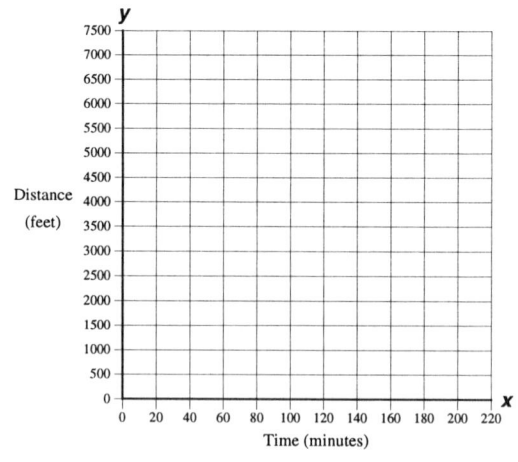

2.

x amount invested (dollars)	y interest earned (dollars)
1,000	85
5,000	425
15,000	1,275

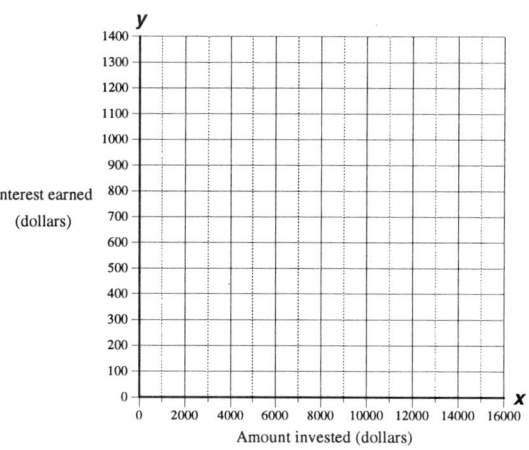

Topic 4.1 F — General Translations Between Types of Models

The verbal descriptions, equations, data tables, and graphs we have just discussed are all different ways of representing a relation between input and output. The chart below summarizes the notation used when working with input and output.

— Note — *Summary of input and* *output ideas*	*Input*	*Output*
	• in a formula, we usually supply the values for the input variables.	• in a formula, we usually calculate the values of the output variable using the values for the input variable.
	• in a formula, the input variable usually is involved in mathematical operations and typically appears on the right side of the equation	• in a formula, the output variable is usually isolated on the left side of the equation.
	INPUT VARIABLE *X* IS ON RIGHT AND NOT ISOLATED. $y = 55x$ **OUTPUT VARIABLE *Y* IS ISOLATED ON LEFT.**	
	• the input variable is generally represented by x.	• the output variable is generally represented by y.
	• in a data table, input values appear on the left. | *Input*	*Output*
---	---
first input value	first output value
second input value	second output value		• in a data table, output values appear on the right.
	• in an ordered pair, the input value appears on the left. (**input value**,output value)	• in an ordered pair, the output value appears on the right. (input value,**output value**)
	• on a graph, input values appear on the horizontal axis.	• on a graph, output values appear on the vertical axis.

Example 4.1.4 A shopping mall estimates that on average 18 people enter the mall every minute.

a) Translate the verbal description into an equation where the input is the time (minutes) since the mall opened and the output is the number of people who have entered the mall.

•SOLUTION•

| | amount | = | rate | • | base |
| | output | = | rate of change | • | input |

	people in the mall	*speed at which people enter mall*	*time since opening*
Units →	*people*	*people / minute*	*minutes*
↓ *Item names* ↓			
mall	*y*	18	*x*

The equation is $y = 18x$.

b) Use the equation to complete the data table below.

x time (minutes)	*y* people
10	
90	
180	
	5,526
	9,180

•SOLUTION• We use the equation $y = 18x$ to calculate the unknown values. For the first three rows we are given the input (x) and we must calculate the output. The calculation for the first row would look like the following:

$y = 18x$ ⇐ Equation to evaluate.

$y = 18(\mathbf{10})$ ⇐ Substituted 10 for the input.

$y = 180$ ⇐ Simplified.

Keep in mind that even though we don't show the units they are always there. For example the product 18(10) really means $\left(\dfrac{18\ \text{people}}{1\ \text{minute}}\right) \cdot 10\ \text{minutes}$, which simplifies to 180 people.

The last two rows give us the output and we must calculate the corresponding input.

$y = 18x$ ⇐ Equation to evaluate.

$\mathbf{5{,}526} = 18x$ ⇐ Substituted 5,526 for the output.

$307 = x$ ⇐ Divided both sides by 18.

The completed data table is given below:

x time (minutes)	*y* people
10	**180**
90	**1,620**
180	**3,240**
307	5,526
510	9,180

c) Plot the points given in the data table on the following grid:

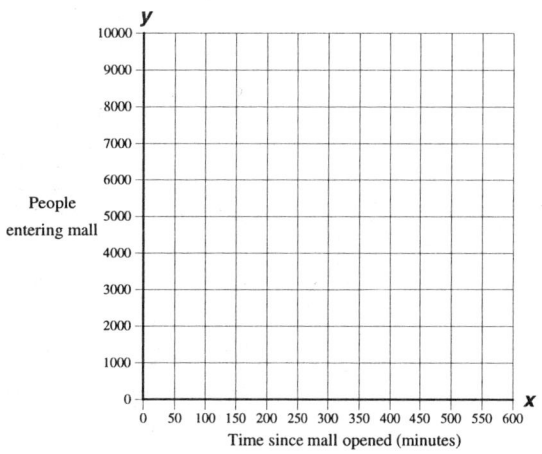

•SOLUTION•

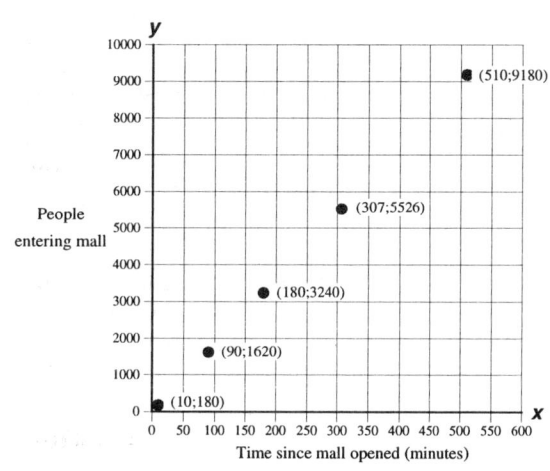

Practice 4.1.4

1. For the first six hours after opening, 13 people on average leave the mall per minute.

a) Create an equation from the verbal description. Let the input be the minutes since the mall opened and the output be the number of people who have left the mall.

b) Complete the following data table:

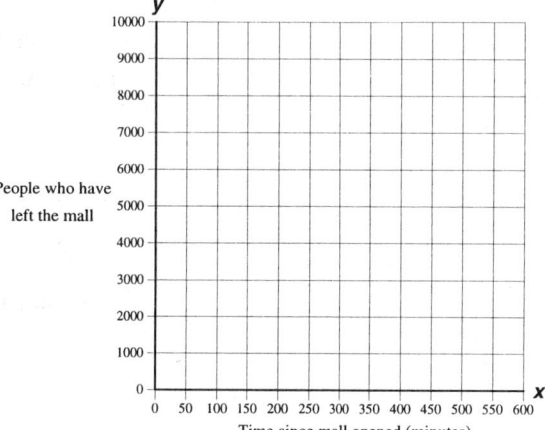

x time since mall opened (minutes)	y people who have left the mall
60	
120	
180	
	2,990
	4,004

c) Plot the points on the given grid.

2. A carpet cleaner can finish a job every 1.5 hours.

a) Create an equation from the verbal description where hours worked is the input and jobs completed is the output.

b) Complete the following data table:

x time worked (hours)	y jobs completed
8	
40	
120	
	50
	100

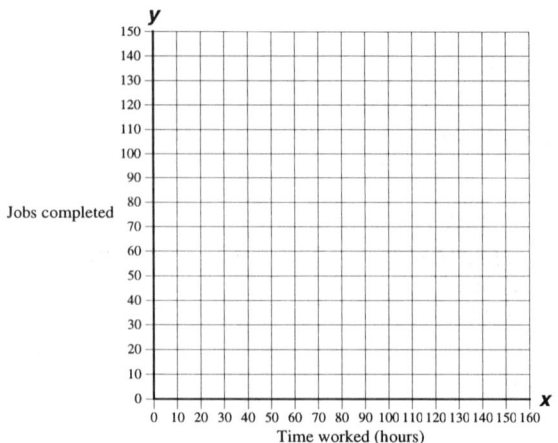

c) Plot the points on the given grid.

3. Ben begins a weight loss program with the goal of losing 3.5 pounds per month.

a) Create an equation from the verbal description using months into the program as the input and total weight loss in pounds as the output. Assume he reaches his goal each month and maintains all previous weight loss.

b) Complete the following data table:

x time into program (months)	y weight loss (pounds)
5	
12	
15	
	24.5
	49.0

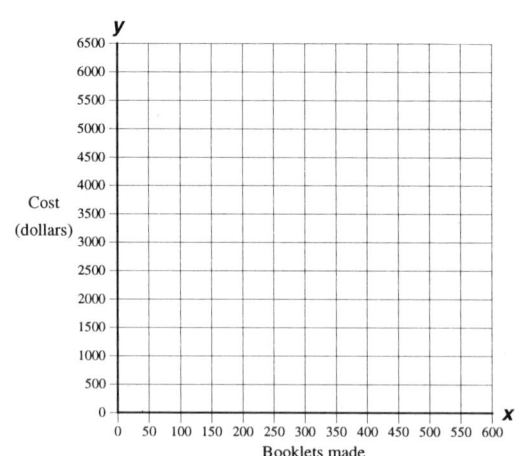

c) Plot the points on the given grid.

4. A writer puts together a 75 page booklet for sale at sports shows. For each book, the cover and binding costs $2.50 and the copying cost is 12¢ per page.

a) Create an equation from the verbal description where the output is the cost for producing the booklets and the input is number of booklets made.

b) Complete the following data table:

x booklets made	y cost (dollars)
100	
200	
500	
	575
	4,991

c) Plot the points on the given grid.

Exercise Set 4.1   The answers to the odd numbered exercises are at the back of the book.

Be sure to define your variables when you answer the following questions.

1. An oil slick spreads at a rate of 0.25 square miles per hour. Build an equation that will predict the size of the oil slick given the time since the spill.

2. Dana's boat becomes untied and begins drifting down a river that has a current of 6 miles per hour. Build an equation that will predict how long it has been since the boat became untied given the distance the boat has traveled.

3. Robert is taking part in a bicycle ride to raise money for AIDS research. After breakfast he begins riding at 12 mph. Build an equation that will predict the distance Robert has ridden given the time since he left. Then construct a data table that includes the following times: 2.5 hours, 3.75 hours, and 5.5 hours.

4. A train leaves a station at 8 am and travels 89 mph. Build an equation that will predict the train's distance from the station given the time (minutes) since it left the station. Then construct a data table that includes the following times: 21 minutes, 36 minutes, and 2 hours.

5. A roofer can tear off and shingle 1 roof in 4 days. Create a equation which outputs the number of roofs completed given the time the roofer has worked. In a data table, record the number of roofs completed if he worked 68, 92, and 284 days.

6. It takes a roofer 12 days to tear off and shingle 3 roofs. Create a equation which outputs the time worked given the number of roofs completed. In a data table, record the number of roofs completed if he worked 5, 7, and 13 days.

7. Graph the data.

x time worked (days)	y roofs completed
68	17
92	23
284	71

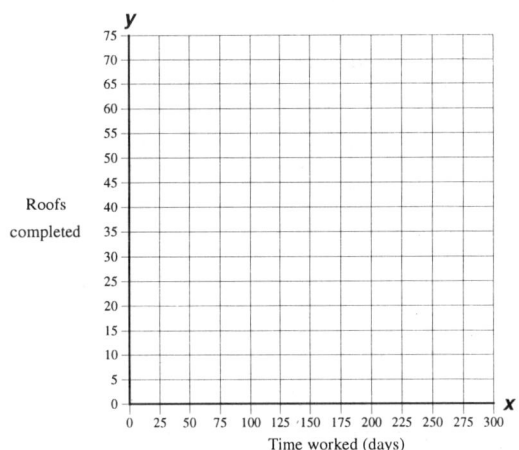

8. Graph the data.

x packages of lights	y cost (dollars)
8	10.00
15	18.75
50	62.50
100	125.00

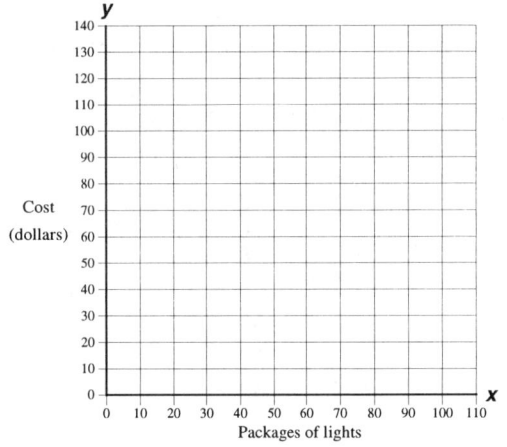

9. A model buys skin fortifying creme at $50.60 an ounce. If the input is the ounces of cream purchased and the output is the cost, complete the following models.

   a) Create an equation from the verbal description.
   b) Complete the following data table:

x creme purchased (ounces)	y cost (dollars)
4.0	
12.0	
16.0	
	298.54
	500.94

   c) Plot the points on the given grid.

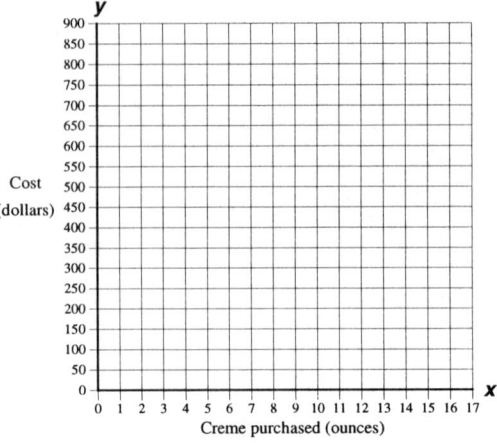

10. A women buys hydrating creme at $17.10 an ounce. If the input is the ounces of cream purchased and the output is the cost, complete the following models.

    a) Create an equation from the verbal description.
    b) Complete the following data table:

x creme purchased (ounces)	y cost (dollars)
4.0	
12.0	
16.0	
	299.25
	495.90

    c) Plot the points on the given grid.

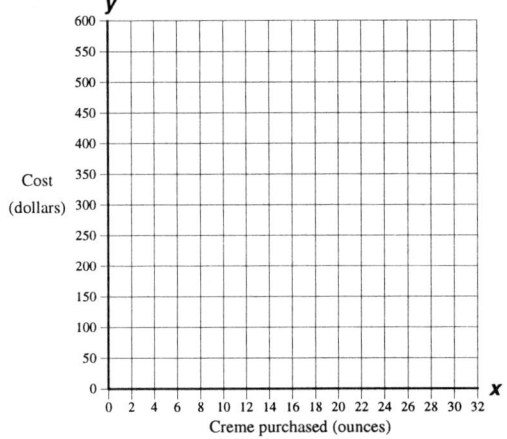

11. A family starts their 1–year old, 5–year old and 7–year old in child care. The prices are $130 a week for children under 5 and $110 a week for children 5 and over. If time since starting child care is the input and total cost for child care is the output, complete the following models.

    a) Create an equation from the verbal description.
    b) Complete the following data table:

x time since starting child care (weeks)	y cost (dollars)
4	
26	
52	
	4,900
	15,050

    c) Plot the points on the given grid.

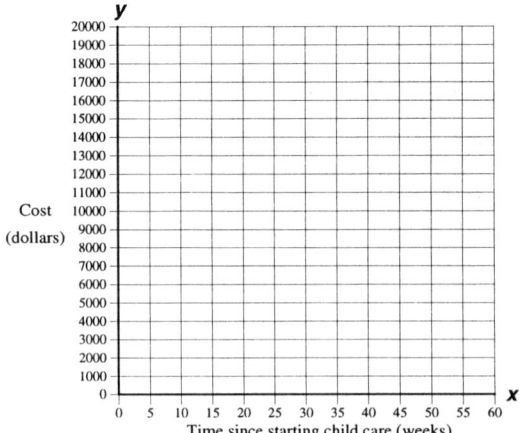

12. A family buys two used vehicles. It costs about $11.40 a week for gas for the first vehicle and about $18.75 a week for gas for the second. If the input is weeks since purchase and the output is total cost for fuel, complete the following models.

   a) Create an equation from the verbal description.
   b) Complete the following data table:

x time since vehicle purchased (weeks)	y cost of gas (dollars)
4	
12	
52	
	512.55
	994.95

   c) Plot the points on the given grid.

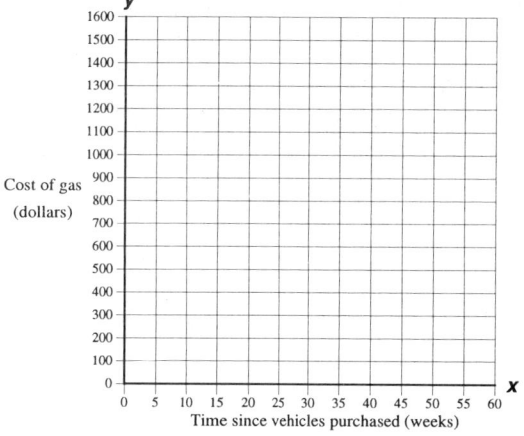

13. A chemist is pouring a 35% alcohol solution into a larger container. If the input is the milliliters of solution in the larger container and the output is the milliliters of alcohol in the larger container complete the following models.

   a) Create an equation from the verbal description.
   b) Complete the following data table:

x solution in container (milliliters)	y alcohol in container (milliliters)
10	
50	
200	
	28.0
	38.5

   c) Plot the points on the given grid.

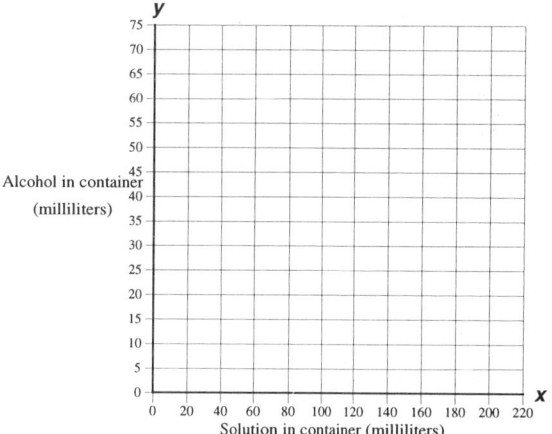

14. A worker is pouring whole milk into a large vat. Whole milk is 4% butterfat. If the input is the gallons of whole milk poured into the vat and the output is the gallons of butterfat in the vat, complete the following models.

   a) Create an equation from the verbal description.
   b) Complete the following data table:

x milk in vat (gallons)	y butterfat in vat (gallons)
50	
100	
250	
	15
	25

   c) Plot the points on the given grid.

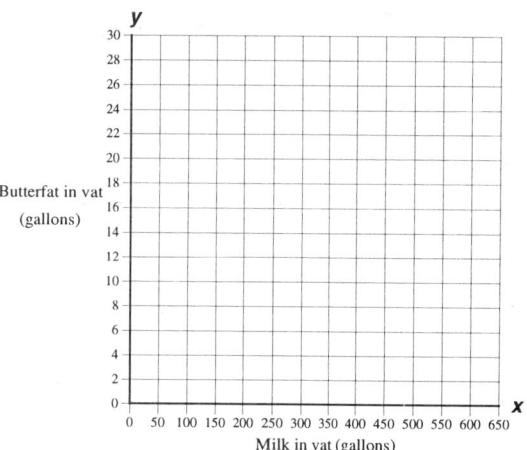

15. To count the vehicles which travel on a road the traffic control department places a rubber hose across the road. Every time a vehicle's tires depress the hose, air is forced from the hose and causes a counter to increase by 1. If the input is the number showing on the counter inside the box and the output is the number of vehicles to have used the road, build the following models.

   a) Create an equation from the verbal description.
   b) Complete the following data table:

x counter reading	y vehicles
1,400	
2,116	
1,714	
	400
	1,000

   c) Plot the points on the given grid.

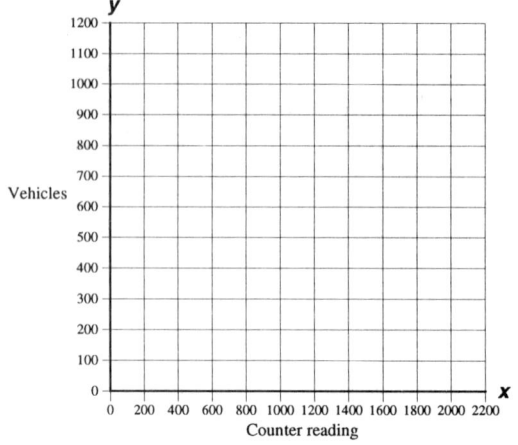

16. A stoplight is located on the entrance ramp of a freeway. Every 7 seconds the light flashes green and one car is allowed through. If the input is the number of cars ahead of you, and the output is seconds you will have to wait to enter the freeway, create the following models. (Assume when you enter the line a car has just left.)

   a) Create an equation from the verbal description.
   b) Complete the following data table:

x cars ahead of you	y time until you can enter freeway (seconds)
7	
24	
11	
	119
	238

   c) Plot the points on the given grid.

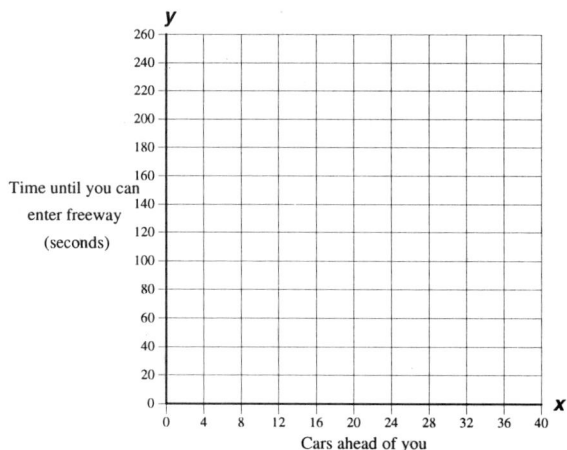

Review Exercises   The answers to all of these exercises are at the back of the book.

17. Simplify: $2[3 + 4(2 + 5) - 3^2]$

18. Prime factor: 196

19. Simplify: $\dfrac{9}{49} + \dfrac{3}{21}$

20. Simplify: $30 - \left| 3 - 45 \right| \div 7 + 5 \cdot \left| -2 \right|$

21. Graph: $7 \geq t$

22. Factor: $5t - t$

23. Solve: $3\left(\dfrac{x}{2} - 3\right) - (x - 2) = \dfrac{1}{3}(x - 6) - 5$

24. Solve: $8x - (3 - x) < 5x - 11$

25. Graph 325 on a number line. Use an interval from –300 to 500 in steps of 100.

26. Ben knows that the gas mileage for his van is 7 miles per gallon less when he is pulling a trailer. On a recent 1168 mile trip, the van used 56 gallons of gas while pulling the trailer and 22 gallons while not pulling the trailer. Find the gas mileage (miles per gallon) for the van when it is not pulling the trailer.

27. Jeremy paid 7% tax on a purchase. If the tax was $21.35, how much was the purchase?

Section 4.2 The Linear Equation

In the previous section we took a quick look at representing a verbal description using an equation, a data table, and a graph. For the next few sections we will expand on each of these. We begin by discussing the link between the verbal description and a specific type of equation called the linear equation.

Topic 4.2 A — Constructing the Linear Equation

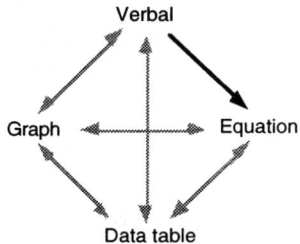

An equation that can be written in the form $y = b + mx$ [†] is called a **linear equation in two variables**. In this equation,

- y represents the output.
- x represents the input.
- m represents a constant that describes how the input and output change relative to each other; it is a rate of change, like miles per hour.
- b represents a constant that tells us the value of the output when the input is 0; it usually represents the "initial" value of the output.

Notice that if the value of b is 0, the linear equation simplifies to $y = mx$, which is the two variable amount formula we discussed in the last section.

The linear equation is useful in situations where we start at some initial value and then change at a constant rate. As an example, let's put a small twist on the truck rental situation of the previous section. Let's say that in addition to the mileage charge the rental company charges a $10 processing fee when you rent the truck. The advertisement might then look like this:

> **Opening month special.**
> **Rent our trucks for**
> **$10 down**
> **plus $1.25 per mile**

We want to build a model that predicts the cost given the distance the truck is driven.

- Since we want to predict cost we will let y represent the cost (dollars).
- Since we want to predict cost from the distance we will let x represent the distance driven (miles).
- The advertisement says that we start with the initial cost of $10.00.
- The ad also says that the cost changes by $1.25 for every mile we drive. This is the rate of change $\dfrac{\$1.25}{1 \text{ mile}}$.

We can model this situation as follows:

$$\begin{aligned} \text{total cost} &= \text{initial cost} + (\text{cost per mile}) \cdot (\text{miles}) \\ y &= \$10.00 + \left(\frac{\$1.25}{1 \text{ mile}}\right) \cdot (x \text{ miles}) \\ y &= 10 + 1.25x \end{aligned}$$

In terms of the linear equation, $y = b + mx$, the $10.00 is the initial value, b, and the $1.25 per mile is the rate of change, m.

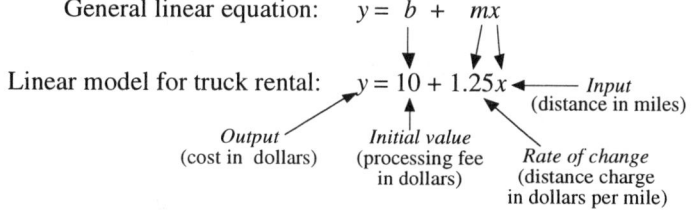

General linear equation: $y = b + mx$

Linear model for truck rental: $y = 10 + 1.25x$ ← Input (distance in miles)

Output (cost in dollars)   Initial value (processing fee in dollars)   Rate of change (distance charge in dollars per mile)

[†]   This equation is also commonly written as $y = mx + b$. Notice these are equivalent since only the order of the terms has been changed.

As another example, suppose an electric company charges you $0.06 for every kilowatt hour (KWH) of electricity you use, plus a flat fee of $5.00 per month for connecting your house to the electrical grid. We want to build a model that predicts the cost from the amount of electricity used.

- Since we want to predict cost, we will let y represent the cost (dollars).

- Since we want to base the prediction on the electricity used we will let x represent the amount of electricity used (KWH).

- The verbal statement tells us we start with an initial value (the flat fee) of $5.00; therefore, b is 5.

- The verbal statement also tells us that the cost changes by $0.06 for every change of 1 KWH; therefore, the rate of change, m, is $\dfrac{0.06 \text{ dollars}}{1 \text{ KWH}}$, or 0.06.

Putting this all together gives us the following:

General linear equation: $y = b + mx$

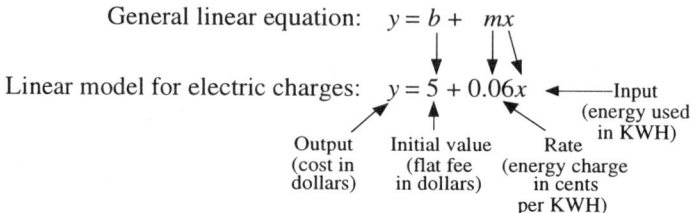

Linear model for electric charges: $y = 5 + 0.06x$ ⟵——Input (energy used in KWH)

Output (cost in dollars)   Initial value (flat fee in dollars)   Rate (energy charge in cents per KWH)

For some situations you may be able to easily identify the output, the input, the initial value, and the rate of change and then simply write down the linear equation that models the situation. If you have difficulty doing this, use the procedure for building an equation that we discussed in the previous section. The only new piece will be identifying the initial value, b. Here is the modified procedure:

| — Procedure —

Building an equation,

$y = b + mx$ *, from a verbal description* | **Step 1 Organize the information**   Read the problem, several times if necessary:

• Given: determine what is given. Be sure to include the unit of all numbers.
• Find: determine what you are asked to find.
• Sketch: draw a sketch if it will help you better understand the problem.
• Rate of change: With these problems it may not be initially clear what the amount (the output) and the base (the input) should be. Therefore, we usually delay constructing an Amount Table until after the variables and the rate of change have been precisely defined or identified.
• Formula: $y = b + mx$.

Step 2 Define unknowns

• Identify the output (what you are trying to predict) and its unit. This will be represented by y.
• Identify the input (what you are using to predict the output) and its unit. This will be represented by x.
• **Identify the initial value, b, and its unit. This constant must have the same unit as the output.**
• Identify the rate of change, m, and its unit. The rate of change is the ratio $\dfrac{\text{change in output}}{\text{change in input}}$. The numerator must have the same unit as the output; the denominator must have the same unit as the input.
• Amount Table: At this point it may be useful to organize the information in an Amount Table.

Step 3 Verbal model   This is given in the statement of the problem.

Step 4 Mathematical model   Substitute the **values of b and m into the linear equation**, $y = b + mx$. |

Example 4.2.1   a)   A scout troop decides to raise money by selling candy at a sporting event. The candy costs $600 and the troop plans to charge $0.75 for each candy bar. Build a linear equation that models this situation.

•**SOLUTION**• *Step 1 Organize the information*

Given: • $600 spent for candy.
         • Each bar will be sold for $0.75.

Find: A linear equation that models this situation.

Step 2 Define unknowns

• output: The troop most likely wants to predict the money they take in so we will let *y* represent the money taken in (dollars).

• input: The money taken in will be predicted from the number of bars they sell. Therefore, we will let *x* represent the number of bars.

• initial value: Before any bars are sold, the troop must spend $600. Since the output is the money taken in, and the $600 is money that they spent, we will write the initial value as –600, meaning that they start off with a debt of $600.

• rate of change: The money taken in changes by $0.75 when the bars sold changes by 1. Therefore, the rate of change is $\dfrac{\$0.75}{1 \text{ bar}}$, which we will write as 0.75.

We can use a modified Amount Table to organize the information.

	output	=	initial value	+	rate of change	•	input
	y	=	*b*	+	*m*	•	*x*
Descriptions →	money taken in		cost of bars		cost per bar		number of bars
Units →	dollars		dollars		$\dfrac{\text{dollars}}{\text{bar}}$		bars
↓ Item names ↓							
bars	*y*		–600		0.75		*x*

Step 3 Verbal model   Given.

Step 4 Mathematical model $y = -600 + 0.75x$. Note that the values for the input and output are restricted at both ends.

• The troop cannot sell less than 0 candy bars, which means they cannot lose more than $600.

• Common sense tells us that the money taken in cannot be extremely large, say $100,000. However, we cannot determine the maximum value for the output because we do not know how many bars they bought (the maximum value of the input). This also means that we cannot determine their profit because we do not know their cost per bar.

b)   Since 1994 the number of lawyers employed by a firm has remained stable at fourteen. Construct a linear equation that will predict the number of lawyers given the number of years since 1994.

•**SOLUTION**• *Step 1 Organize the information*

Given: • 14 lawyers employed in 1994.
         • Number of lawyers has remained constant.

Find: A linear equation that models this situation.

Step 2 Define unknowns

- output: Since we want to predict the number of lawyers we will let y represent the number of lawyers.

- input: Since the output will be predicted by the year we will let x represent the number of years since 1994.

- initial value: We begin at 1994 with 14 lawyers. Therefore the initial value is 14.

- rate of change: The number of lawyers does not change from year to year. Mathematically, we can say that the number of lawyers changes by 0 when the year changes by 1. Therefore, the rate of change is $\dfrac{0 \text{ lawyers}}{1 \text{ year}}$, which we will write as 0. Notice this does not say the firm has no lawyers; it says the number of lawyers is not *changing* from year to year.

We can use a modified Amount Table to organize the information.

	output	=	initial value	+	rate of change	•	input
	y	=	b	+	m	•	x
Descriptions →	number of lawyers		number of lawyers in 1994		change in number of lawyers each year		number of years since 1994
Units → ↓ Item names ↓	lawyers		lawyers		$\dfrac{\text{lawyers}}{\text{year}}$		years
bars	y		14		0		x

Step 3 Verbal model   Given.

Step 4 Mathematical model $y = 14 + 0x$, which we would write as $y = 14$.

Practice 4.2.1   Write a linear equation for each of the following situations.

1. The cost to lease a car is $2,095 down and $325 a month. Write a linear equation that will return the total cost to lease the car if you input the number of months the car is leased.

2. In 1994 a family had their second child. The size of their family has not increased since then. If the number of years since 1994 is the input, and the number of children in the family is the output, write a linear equation that will output the number of children in the family given the number of years since 1994.

3. From 1970 to 1986 lead emissions in the United States dropped by approximately 13 million tons per year. If 219 million tons of lead were emitted in 1970 write a linear equation that will output the millions of tons of lead emitted if the input is the number of years since 1970.

4. Since 1970 Americans have been reducing their average whole milk consumption by $\dfrac{7}{10}$ of a gallon each year. Make a linear equation that will predict the gallons of whole milk the average American will consume given the number of years since 1970 if the average American drank 25 gallons of whole milk in 1970.

Topic 4.2 B — Using Linear Equations to Solve Problems

We can use the techniques for constructing linear equations from verbal descriptions to help solve problems.

Example 4.2.2   To rent a big screen TV you must pay $400 down and $175 a month.

a) Construct a linear equation that predicts cost from the rental time.

•SOLUTION• We let y represent the cost (dollars) and x represent the rental time (months). The initial cost is $400 and the total cost changes by $175 each month. Therefore, the linear equation is $y = 400 + 175x$.

b) Use the linear equation to find how much will you spend if you keep the TV for a year and a half.

> •**SOLUTION**• We are given an input (time) and we are asked to find it's corresponding output (cost). Before using the formula we must convert $1\frac{1}{2}$ years into 18 months so the unit of the input matches the unit of the denominator of the rate of change.
>
> $$y = 400 + 175x \qquad \Leftarrow \text{ Equation.}$$
>
> $$y = 400 + 175(\mathbf{18}) \qquad \Leftarrow \text{ Replaced the input variable with 18.}$$
>
> $$y = 3550 \qquad \Leftarrow \text{ Simplified.}$$
>
> It will cost \$3,550 to rent the TV for a year and a half.

c) If the purchase price for the TV is \$6000, use the linear equation to find how many months it will take before the rental price will be the same as the purchase price.

> •**SOLUTION**• We are given the output and asked to find the corresponding input.
>
> $$y = 400 + 175x \qquad \Leftarrow \text{ Equation.}$$
>
> $$\mathbf{6000} = 400 + 175x \qquad \Leftarrow \text{ Replaced the output variable with 6000.}$$
>
> $$32 = x \qquad \Leftarrow \text{ Solved for } x$$
>
> If you rent the TV for 32 months, which is 2 years and 8 months, you will have spent the same amount as if you had purchased the TV.

Example 4.2.3   In 1970, 527 women earned law degrees in the U. S. From then until 1985 the number of women earning law degrees increased by approximately 941 every year.

a) Build a linear equation that will predict the number of women law school graduates if you input the number of years since 1970.

> •**SOLUTION**• We let y represent the number of woman law graduates and x represent the number of years since 1970. The initial value in 1970 is 527 graduates and each year since then the change in the number of graduates has been 941. Therefore, the linear equation is $y = 527 + 941x$.

b) Use the linear equation to estimate the number of graduates in 1977.

> •**SOLUTION**• We calculate the output using the given input. Because x is the number of years since 1970, the x value for 1977 is 7.
>
> $$y = 527 + 941x \qquad \Leftarrow \text{ Linear equation.}$$
>
> $$y = 527 + 941(\mathbf{7}) \qquad \Leftarrow \text{ Replaced the input variable with 7.}$$
>
> $$y = 7114 \qquad \Leftarrow \text{ Simplified.}$$
>
> The model says there were 7,114 women law graduates in 1977.

c) Use the linear equation to find the first year there were 5,000 women graduates.

> •**SOLUTION**• We replace y with 5000 and solve for the year.
>
> $$y = 527 + 941x \qquad \Leftarrow \text{ Linear equation.}$$
>
> $$\mathbf{5000} = 527 + 941x \qquad \Leftarrow \text{ Replaced the output variable with 5000.}$$
>
> $$4.8 = x \qquad \Leftarrow \text{ Solved for } x \text{ and rounded to the nearest tenth.}$$
>
> Because x is the number of years since 1970, the model predicts that 5000 women would graduate with law degrees in 1974.8, which we would round off to 1975.

Practice 4.2.3

1. A computer originally costs $2700. Every month after it's purchase it's value decreases by $45.

   a) Build a linear equation that will predict the current value of the computer given the time since its purchase.
   b) Use your linear equation to find how many months it will take before the computer is worth $0.
   c) Use your linear equation to find the value of the computer 1.5 years after its purchase.

2. In 1993 the population of the United States was approximately 259 million. For the next 50 years it's projected the population will increase by 2.4 million each year.

   a) Build a linear equation that will predict the population of the United States in millions given the number of years since 1993.
   b) Use your linear equation to estimate the year the population will reach 300 million.
   c) Use your linear equation to find the anticipated population in 1999.

3. Since 1985 the cost to purchase a license and registration for your automobile has increased approximately $8.68 a year. The cost in 1985 was $118.

   a) Build a linear equation that will predict the cost for a license and registration given the number of years since 1985.
   b) Use your linear equation to predict the cost for license and registration in the year 2000.
   c) Use your linear equation to predict the year if will cost about $300 for license and registration expenses.

4. Since 1970 the United States has been producing about 3 million tons less carbon monoxide each year. To answer the following questions, use the fact that in 1970 the United States produced 127 million tons of carbon monoxide.

   a) Build a linear equation that will predict the millions of tons of carbon monoxide produced in the United States if you input the number of years since 1970.
   b) Although it probably will not happen, use your linear equation to decide what year the carbon monoxide emissions will drop to 0 million tons a year.
   c) Use your linear equation to predict the million tons of carbon monoxide produced in 2010.

Topic 4.2 C — From the Equation to English

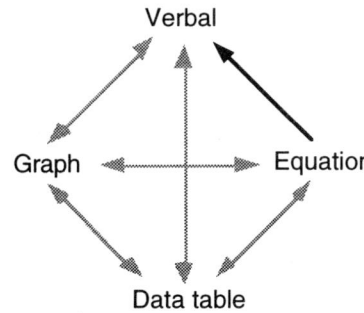

Sometimes we are given a linear equation and we must state its meaning verbally. When doing this, it's important to keep the following two points in mind.

- When you translate from algebra, which traditionally does not show the units, to English, where units are expected, you will have to decide what units are implied by the situation and which numbers they should go with.

- Think carefully about the meaning of the rate of change. The rate of change is defined as the ratio $\frac{\text{change in output}}{\text{change in input}}$. This tells us how the output changes as the input changes.

As an example, suppose we are given the equation $y = 80 - 2x$, and told that this models the relation between the air left in a SCUBA tank and the time since the dive began. Specifically,

- y represents the cubic feet of air left in the tank.

- x represents the number of minutes since the dive began.

Given this information we can determine the meaning of the constants in the equation.

- 80 has the unit cubic feet of air because it must have the same unit as the output. It's the amount of air in the tank when the dive began because it's the value of the output when the input is 0.

- -2, which we can think of as $\frac{-2}{1}$, has the unit $\frac{\text{cubic feet of air}}{\text{minute}}$ because the numerator of the rate of change has the same unit as the output while the denominator has the same unit as the input. Therefore, -2 must mean $\frac{-2 \text{ cubic feet of air}}{1 \text{ minute}}$. The negative sign means the air is leaving the tank. In English, we would say the diver is using 2 cubic feet of air every minute.

Therefore, the equation tells us that a diver started a dive with 80 cubic feet of air and used the air at a rate of 2 cubic feet per minute.

If the English meaning of an equation is not immediately obvious, try using the following procedure.

— Procedure — **Building a verbal model** **from the linear equation** $y = b + mx$	**Step 1 Input/output** Identify the input and the output, including their units. **Step 2 Initial value** Identity the initial value and its unit. (Remember, the initial value has the same unit as the output.) **Step 3 Rate of change** Identify the rate of change and its unit. The numerator has the same unit as the output and the denominator has the same unit as the input. **Step 4 English** Write an English description using the information from steps 1 through 3.

Example 4.2.4   a)   The equation $y = 477 + 34.70x$ predicts the dollar cost (y) to insure an automobile given the number of years since 1985 (x). Use English to express a meaning for this equation.

•**SOLUTION**• ***Step 1 Input/output***   The input is the number of years since 1985 and has the unit years. The output is cost and has the unit dollars.

Step 2 Initial value   The initial value is the value of the output when the input is 0, and it has the same unit as the output. Therefore, it's $477.

Step 3 Rate of change   The rate of change is the number multiplied by the input variable. We can write this number, 34.70, as $\dfrac{\$34.70}{1 \text{ year}}$.

Step 4 English   This equation says insurance cost $477 in 1985 and has been rising by $34.70 each year since then.

b)   The linear equation $y = 15 + \dfrac{5}{11}x$ predicts the pressure under the ocean in pounds per square inch (y) given the depth below the surface in feet (x). Use English to express a meaning for this equation.

•**SOLUTION**• ***Step 1 Input/output***   The input is the depth below the surface and has the unit feet. The output is the underwater pressure and has the unit pounds per square inch.

Step 2 Initial value   The initial value is the value of the output when the input is 0 and it has the same unit as the output. Therefore, it's 15 pounds per square inch.

Step 3 Rate of change   The rate of change is $\dfrac{5 \text{ pounds per square inch}}{11 \text{ feet}}$.

Step 4 English description   This linear equation says the pressure at the surface of the ocean is 15 pounds per square inch and the pressure increases by 5 pounds per square inch for every 11 feet you go below the surface.

Practice 4.2.4

1. The model $y = 75 + 35x$ predicts the cost to rent a dining room set (dollars) from the months the set is rented. Express in English a meaning for this equation.

2. The equation $y = 7277 - 400x$ models the thousands of tons of lead emitted (y) each year (x) in the U. S. from 1986 to 1992, with year 1986 being year 0. Express in English a meaning for this equation.

3. The model $y = 16 - \dfrac{1}{32}x$ predicts the gas left in a car's gas tank (gallons) from the miles the car has driven since the last fill up. Express in English a meaning for this equation.

Exercise Set 4.2   The answers to the odd numbered exercises are at the back of the book.

1. Bean Publishing is trying to determine how much it will cost to print a new book. The charge to set up the press for the first printing is $9,000. In addition, they estimate that paper, ink, and labor will cost $3 for each book printed. Build a linear equation that models this situation. Define the variables, including their units.

2. In 1970 the average U. S. manufacturing worker was making about $3.52 an hour. Since then their unadjusted hourly earnings have increased about 37¢ each year. Write a linear equation that will return the average hourly earnings for manufacturing workers if you input the number of years since 1970. Define the variables, including their units.

3. To help the environment, a homeowner composts his leaves, grass, and sticks. Over time microorganisms break down the organic material into humus that can be used in his garden. He begins with a pile that is 6 feet in height and the pile decreases in height by 4 inches every week. Build a linear equation that will output the height of the pile given the number of weeks since the pile was built. Define the variables, including their units.

4. A SCUBA diver's tank holds 80 cubic feet of air. The diver uses 2 cubic feet of air per minute. Make a linear equation that will output the cubic feet of air left in the tank given the number of minutes the diver has been underwater. Assume the diver begins with a full tank. Define the variables, including their units.

5. The cost to lease a car is $1500 down and $275 a month.

    a) Write a linear equation that will predict the total cost to lease the car if you input the number of months the car is leased.
    b) If you plan to spend $8,000 on the lease, use your linear equation to decide how many months can you rent the car.
    c) If the purchase price for the car is $18,999 use your linear equation to decide when the cost to lease and the purchase price be the same.

6. The cost to rent a TV is $50 down and $20 a month.

    a) Build a linear equation that will predict the total cost to rent the TV if you supply the number of months the TV is rented.
    b) If you plan to spend only $200 in rental costs, use your linear equation to decide how long can you rent the TV.
    c) If the purchase price for the TV is $310 use your linear equation to decide when the rental cost will be the same as the purchase price.

7. Unadjusted for inflation, the average U. S. worker was making about $3.09 an hour in 1970. Since then the hourly earnings have increased about 35¢ each year.

    a) Build a linear equation that will predict the average unadjusted hourly earnings given the number of years since 1970.
    b) Use your linear equation to decide when the average U. S. worker will earn about $12.00 an hour.
    c) Use your linear equation to decide what the average U. S. worker will earn as an hourly wage in 2013.

8. If we adjust for inflation with 1982 used as the base year [†] then in 1970 the average U. S. worker was making about $8.42 an hour . Since then their hourly earnings, have decreased about 4¢ each year.

    a) Build a linear equation that will output the average adjusted hourly earnings if you input the number of years since 1970.
    b) Use your linear equation to estimate when the average U. S. worker will be earning an adjusted income of $6.00 an hour.
    c) Use your linear equation to estimate what the average U. S. worker will earn as an adjusted hourly wage in 2013.

9. In 1970 approximately 58% of incoming freshman at U. S. colleges had an average grade in high school of B− to B+. Since 1970 the percentage has remained relatively constant.

    a) Build a linear equation that will output the percent of incoming freshman with an average grade of B− to B+ if you input the number of years since 1970.
    b) Use the linear equation to predict the percent of incoming freshman with an average grade of B− to B+ in 1994.
    c) Use your linear equation to decide when the percent of incoming freshman at U. S. colleges with an average grade in high school of B− to B+ be around 58%.

[†]   The idea here is that if your income goes up $1 but the price of everything you buy goes up $2 then your increased income is actually worth less in terms of it's purchasing power.

10. The percent of incoming freshman at U. S. colleges with an average high school grade of D has remained relatively constant since 1970.

    a) Create a linear equation that will output the percent of incoming freshman with an average high school grade of D given the number of years since 1970 if in 1970 one percent of incoming freshman had an average grade of D.
    b) Use your linear equation to predict the percent of incoming freshman with an average high school grade of D in 1992.
    c) Use your linear equation to predict the year that about one percent of incoming freshman will have an average high school grade of D.

11. Every year since 1985 the cost of driving a car has risen $2.50 per mile (taking into account all expenses such as gas, maintenance, depreciation etc.); in 1985, it cost an average of $27.40 for every mile driven.

    a) Write a linear equation that will predict the cost per mile driven given the number of years since 1985.
    b) Use your linear equation to predict the average cost per mile driven in the year 2005.
    c) Use your linear equation to predict when it will cost the average automobile owner $100 for every mile driven.

12. Since 1985 the cost of insuring an automobile has risen $34.70 per year while in 1985 it cost about $490 to insure an automobile for one year.

    a) Write a linear equation that will predict the cost of insurance given the number of years since 1985.
    b) Use your linear equation to predict the cost of auto insurance in the year 2007.
    c) Use your linear equation to predict when it will cost the average automobile owner $1000 for auto insurance for the year.

13. Since 1970 the percent of male freshman attending U. S. colleges has been decreasing by 0.39% per year.

    a) Create a linear equation that will predict the percent of freshman who are male given the number of years since 1970 if, in 1970, 54% of incoming freshman were male.
    b) Use your linear equation to predict the year the percent of incoming freshman who are male will be 40%.
    c) Use your linear equation to predict what percent of freshman will be male in the year 2000.

14. Since 1970 the percent of female freshman at U. S. colleges has been increasing by 0.39% per year.

    a) Create a linear equation that will output the percent of freshman who are female given the number of years since 1970 if 46% of incoming freshman were female in 1970.
    b) Use your linear equation to predict the year the percent of incoming freshman who are female will be 60%.
    c) Use your linear equation to predict what percent of freshman will be female in the year 2000.

15. The distance in miles you are from a flash of lightening can be found by counting the seconds between the lightning flash and the thunder and dividing the seconds by 5.

    a) Create a linear equation that will output the distance you are from the lightening flash given the number of seconds between the flash and the thunder.
    b) If you count 18 seconds between the flash and the thunder use your linear equation to find your distance from the lightening.
    c) Use your linear equation to estimate how long it would take to hear the thunder if you were 1.6 miles from the lightening.

16. One bucket of water can make up to thirty buckets of dry snow (also known as "powder" snow).

    a) Create a linear equation that will predict the buckets of water given the buckets of powder snow you have.
    b) Use your linear equation to predict how many buckets of water you would have if you melted 100 buckets of powder snow.
    c) Use your linear equation to predict the buckets of powder snow you could make from 5 buckets of water.

17. A student earns a base salary of $125 a week plus a commission of $1.50 for each item she sells.

    a) Build a linear equation that will output her total weekly earnings given the number of items she sells in one week.
    b) Use your linear equation to find her weekly salary if she sells 115 items.
    c) Use your linear equation to find the number of items she must sell in order to earn $500 in one week.

18. A novice diver is more anxious and uses air faster then usual. If a diver started with 80 cubic feet of air and used air at the rate 3.25 cubic feet a minute answer the following questions.

    a) Build a linear equation that will output the air left in the tank given the minutes into the dive.
    b) Use your linear equation to estimate the number of minutes before the novice diver runs out of air?
    c) If the novice diver has been underwater for 12 minutes use your linear equation to estimate how much air is left in the tank?

Write an English explanation of each of the following linear equations.

19. The model $y = 2100 - 35x$ predicts the current value of a new computer in dollars (y) from the months since the computer was purchased (x). Express in English a meaning for this equation.

20. The model $y = 19200 - 200x$ relates the current value of a new automobile in dollars (y) to the months since the car was purchased (x). Express in English a meaning for this equation.

21. The equation $y = 25 - \frac{7}{10}x$ models the gallons of whole milk consumed as a beverage (y) each year (x) in the U. S. from 1970 to 1991, with year 1970 being year 0. Express in English a meaning for this equation.

22. The equation $y = 5 + \frac{4}{10}x$ models the gallons of low-fat milk consumed as a beverage (y) each year (x) in the U. S. from 1970 to 1991, with year 1970 being year 0. Express in English a meaning for this equation.

23. The model $y = 91879 + 1087x$ relates the thousand of tons of carbon monoxide released in the U. S. (y) to the year between 1940 and 1970 (x), with 1940 being year 0. Express in English a meaning for this equation.

24. The equation $y = 128474 - 1297x$ models the thousand of tons of carbon monoxide emitted (y) each year (x) in the U. S. from 1970 to 1994 with year 1970 being year 0. Express in English a meaning for this equation.

25. The formula $y = \frac{1}{2}x$ shows the relation between the radius of a circle (y) and the diameter (x) for the same circle. Express in English a meaning for this equation.

26. The formula $y = 2\pi x$ shows the relation between the circumference of a circle (y) in inches and the radius (x) for the same circle in inches. Express in English a meaning for this equation.

27. The equation $y = 6.59 - .09x$ models the federal minimum hourly wage since 1969 in 1996 dollars. Express in English a meaning for this equation.

28. The equation $y = 1.50 + 0.11x$ models the unadjusted federal minimum hourly wage since 1969. Express in English a meaning for this equation.

29. Stalactites are formations found hanging from the ceiling of some caves. Although the growth rate of stalactites varies, the equation $y = 12.5x$ is sometimes used to approximate the age of the stalactite. In this equation y represents the age (in years) of the stalactite and x represents its length (in inches). Express in English a meaning for this equation.

30. The equation $y = 0.8x$ models the growth of coral (y) in inches given the number of years since the coral began to grow (x). Express in English a meaning for this equation.

Review Exercises   The answers to all of these exercises are at the back of the book.

31. Simplify: $\frac{33}{50} \cdot \frac{90}{22}$ 32. Solve: $5 - x = 3(x - 2) + 5 - 4x$ 33. Graph: $1.5 > x$

34. Find the coordinate for each point shown on the number line:

$$-8 \ -7 \ -6 \ -5 \ -4 \ -3 \ -2 \ -1 \ 0 \ 1 \ 2 \ 3 \ 4 \ 5 \ 6 \ 7 \ 8$$

35. Graph 850 on a number line. Use an interval from 0 to 1050 in steps of 150.

36. Identify the base and its exponent and then simplify: -2^4 37. Find five consecutive integers whose sum is 165.

Section 4.3 The Graph of a Linear Equation

Topic 4.3 A — The Relation Between Data Tables and Graphs

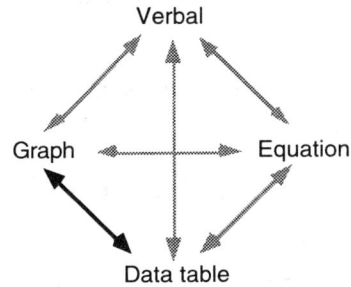

Verbal

Graph

Equation

Data table

Suppose we are working at an archeological dig where a nine-inch long femur (thigh bone) of a prehistoric woman is uncovered. We would like to estimate the woman's height but all we have is the bone. We think there is a relation between femur length and height but we do not know what the relation is. To get an approximate idea, we decide to measure the femur length and height of four women working at the dig and record the data. Since we will supply the length of a femur, we will define the input as *femur length in inches*; since we want to predict height, we will define *height in inches* as the output. The data are organized in the following data table:

	x *femur length* *(inches)*	*y* *height* *(inches)*
first woman ⇒	13.0	54
second woman ⇒	15.5	60
third woman ⇒	20.5	72
fourth woman ⇒	18.0	66

To determine whether these data show a pattern we plot the data on an *x-y* coordinate system.

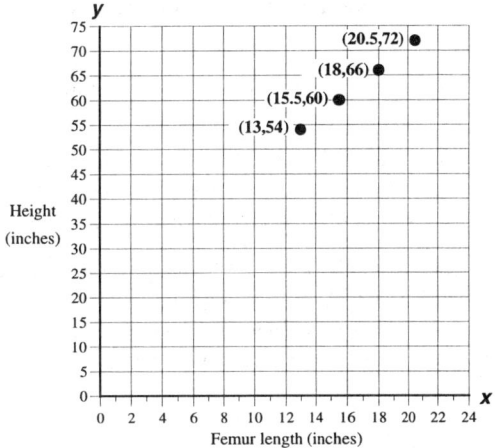

As you can see, there does seem to be a pattern in the data. Unfortunately, none of the women measured have a femur length of 9 inches so there is no direct way of using the measurements to predict the woman's height. To remedy this, we collect data from another 30 women and plot these data on the same coordinate system.

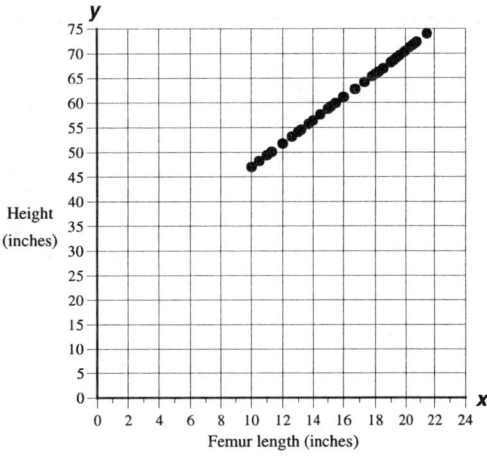

With this much data the individual points seem to run together to form a straight line. When all the data fall on a straight line we say the relation between the input and output is a **linear relation**.

In practice, we may not be able to plot this many points because it may be too difficult or time consuming to collect that much data. For most situations, we plot enough points to give us a good idea of what the graph probably looks like and then we sketch the graph by connecting the plotted points with a smooth line. We are assuming that the points on the line that fall between the data points we actually plotted are correct because we have reason to believe there is a pattern here.

For this example, we probably would have guessed that the graph is a straight line after plotting the first 4 points. Therefore, we connect the points with a straight line with arrowheads on the ends to indicate that we think the graph will continue in both directions beyond the points we measured. †

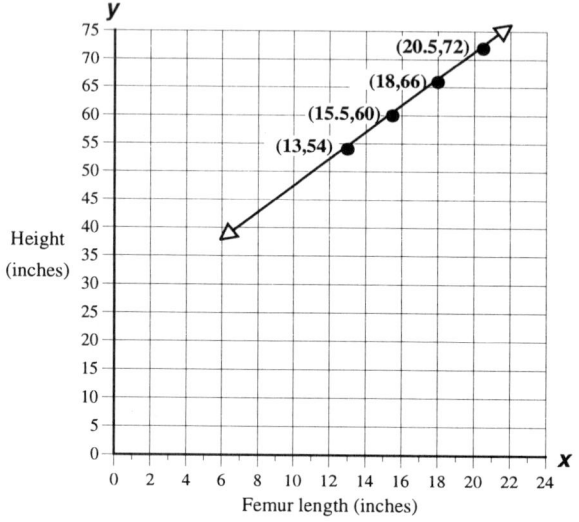

— *Note* —

Connecting points with a straight line may not be correct

We choose to connect the data points with a straight line because from our knowledge of biology we have reason to believe that the relation between height and femur length is linear. However, there are many other ways we can connect the data points as the following two graphs illustrate:

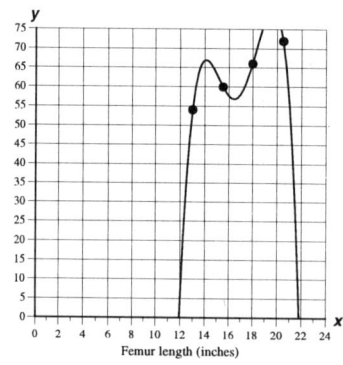

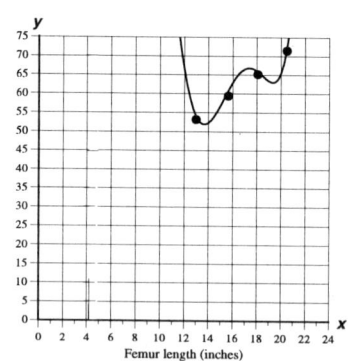

Now we can get a reasonable estimate for the height of the prehistoric woman. Since she had a femur length of 9 inches we find 9 inches on the input axis, move up to the line and then left to the output axis and read the value 45 inches. We estimate the height of the prehistoric woman was about 45 inches.

---

† In real applications, this may or may not be a good assumption. For example, a negative femur length makes no sense, nor does a femur length of 100 inches.

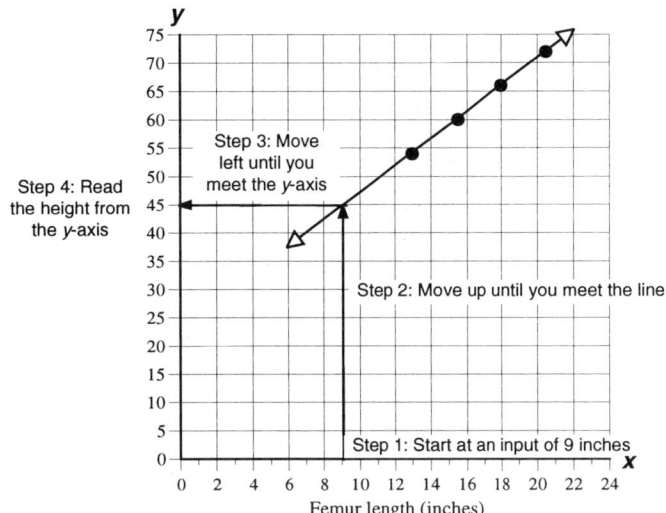

Notice that we are predicting a value that falls outside the set of data that we used to construct the graph. Doing this is called **extrapolation**. Predicting a value that falls between the measured data points is called **interpolation**. A conjecture made through interpolation is often considered "safer" than one made through extrapolation since extrapolation assumes the relation will continue beyond the information we have been given.

Now suppose we had uncovered a mummified body of a woman encased in cloth. In that case it would be easy to estimate the women's height but difficult to measure her femur length. Say we estimated the women's height to be 63 inches. We could estimate her femur length by starting at 63 inches on the y-axis, moving right to the line, and then down to the x-axis. You can see from the graph we would expect a woman 63 inches tall to have a femur length of somewhere between 16 and 17 inches.

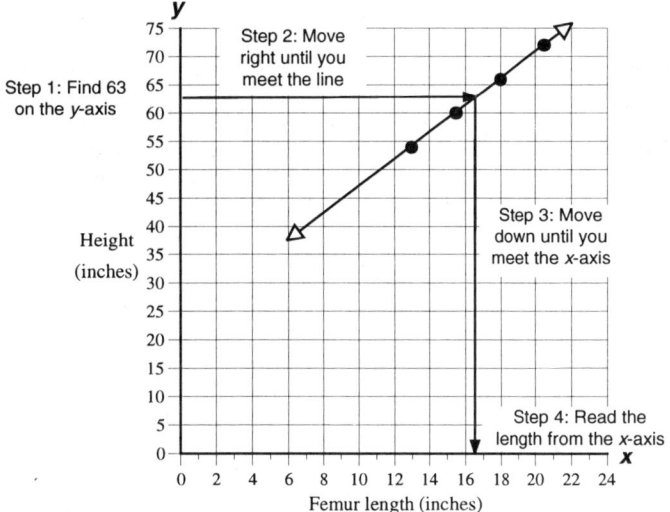

Example 4.3.1   On the coordinate system provided, plot and label each ordered pair given in the data table. If you feel the graph can be approximated by a straight line, connect the points and use the graph to fill in the missing input or output values.

a)

x	y
−4	9
−2	5
1	−1
5	
	−3
0	

•SOLUTION• First, we plot each given point. Since they appear to fall on the same line we connect them.

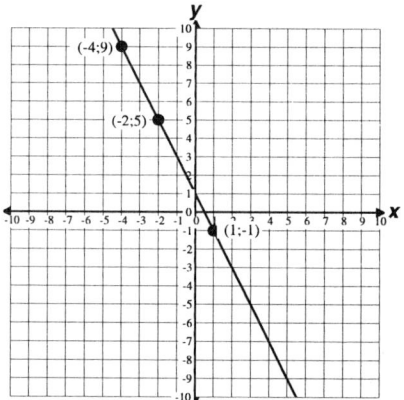

To fill in the missing values in the data table, we begin with each given input or output value and use the graph to find the second coordinate in the ordered pair. For example, to find the output that corresponds to the input 5 we start at 5 on the *x*-axis, move down to the line, and then left to the *y*-axis to find the output value, −9.

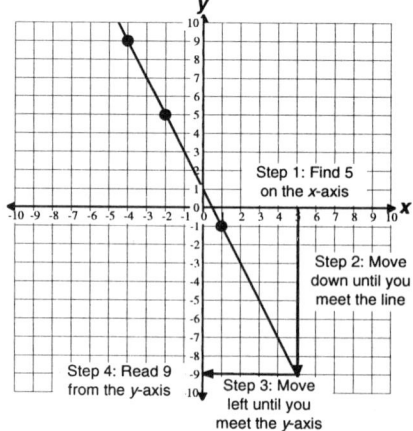

We fill in the rest of the values in a similar way.

x	y
−4	9
−2	5
1	−1
5	**−9**
2	−3
0	**1**

b)

x	y
3	7
2	−5
−4	−1
−7	
	0
6	

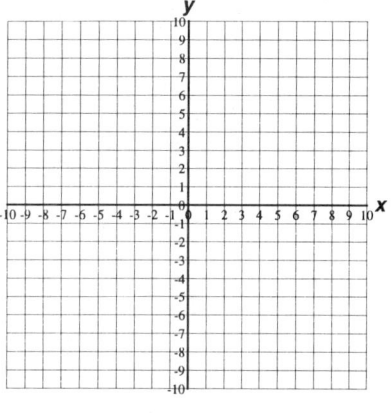

•**SOLUTION**• We plot each given point.

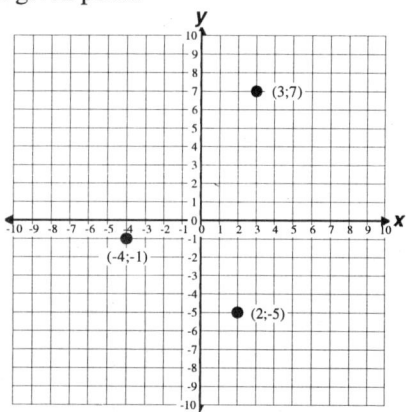

These points do not appear lie on a straight line so we will not connect them. Given these data it would be difficult to predict the output given the input, and vice versa. We would need more points to see if a relation between input and output really does exist.

Practice 4.3.1 Plot the following on the given *x-y coordinate* system. Then, if possible, use a linear model to fill in the missing input or output values.

1.

x	y
6	−2
4	−1
0	1
	4
	2
1	

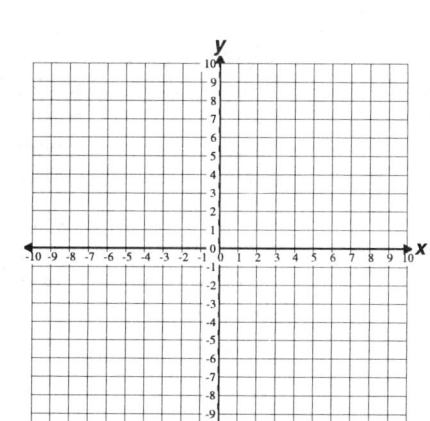

2.

x	y
4	−1
−2	−3
10	−5
	6
−8	
0	

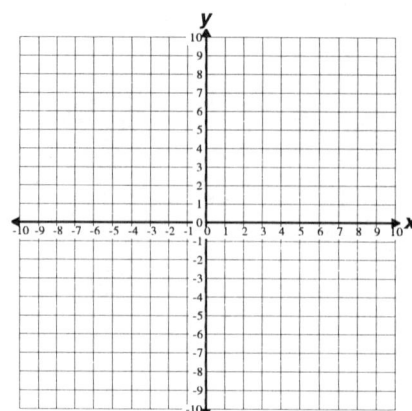

3.

x	y
2	5
−2	−7
1	2
3	
−1	
	−10

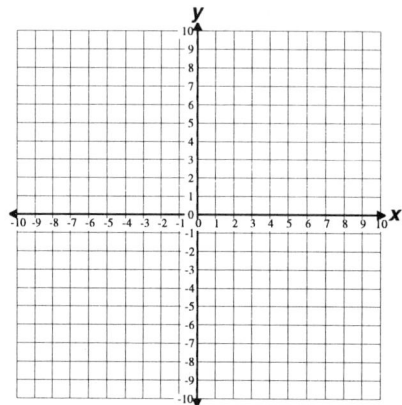

4.

x	y
1	5
−6	5
−2	5
0	
	−3
8	

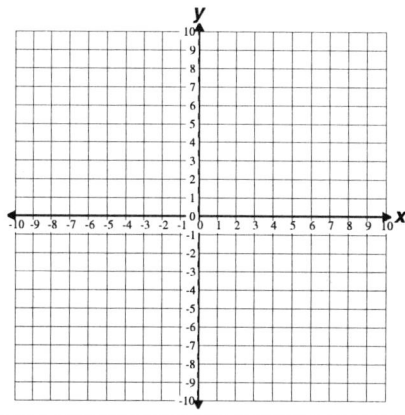

Topic 4.3 B — Graphing Real Data

The **scale** of an *x-y* graph refers to the size and placement of the numbers on the axes. With some data sets it's easy to use a standard scale where we show values from −10 to 10 on the *x-* and *y*-axes and use a grid spacing of 1 unit. However, in many applied problems, the data sets involve large or small numbers which we could not locate on a standard grid. In those cases we may have to adjust the scale of one or both axes.

We may adjust the scale in several ways, such as disregarding unimportant information, adjusting the units, or transforming inputs or outputs. For example, below is a graph that approximates unadjusted average earnings for U. S. workers for the years 1970 to 1991.

x time (calendar year)	y average hourly wage (dollars)
1970	3.25
1980	6.50
1985	8.50
1990	10.00
1991	10.25

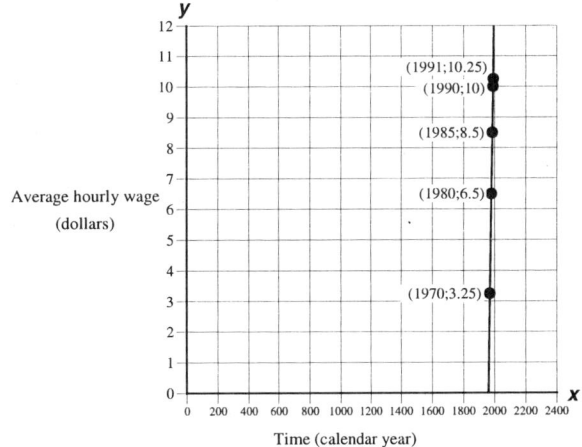

This graph is correct but not very useful because the data are all bunched up. To make the graph more useful, we will zoom in on the years between 1970 and 1991 by adjusting the scale of the horizontal axis so that each division represents 5 years. Also, because we are not interested in years before 1970, we will omit the part of the grid that falls between 0 and 1965. In the graph below, we placed break marks ⟫ just to the right of 0 to show that part of the axis is missing.

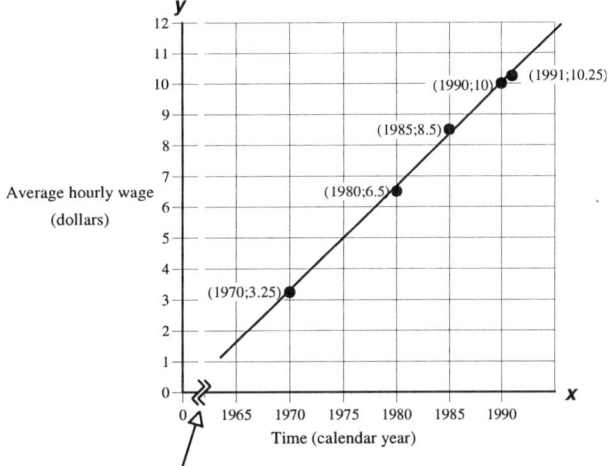

Break marks indicate that data between 0 and 1965 are omitted.

Note that because we are using real data the straight line we drew to represent the relation does not pass through all the points. The relation is not exactly linear but it is really close. We drew the best straight line we could using the given data.

Another way of making the numbers easier to work with is to change the definition of the input. For example, let's define the input as *number of years since 1970* instead of *calendar year*. Using this definition, 1970 becomes year 0, 1975 becomes year 5, and so on. This allows us to work with numbers like 0 and 5 instead of 1970 and 1975. Below is the transformed data table:

x time (number of years since 1970)	y average hourly wage (dollars)
0	3.25
10	6.50
15	8.50
20	10.00
21	10.25

We have plotted the new graph below. It still has all the information of the original graph but it's displayed in a more meaningful way. [†]

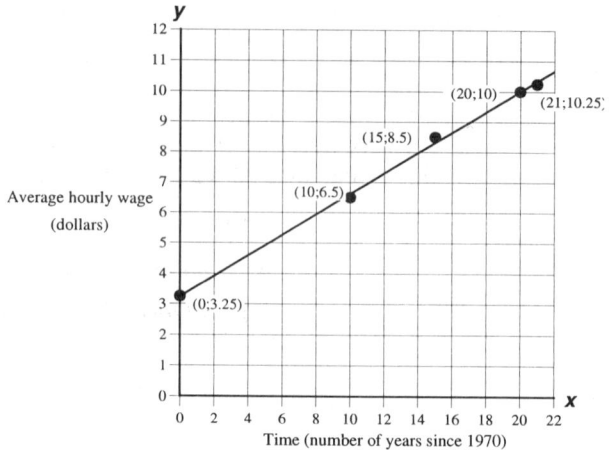

Note that we changed the label on the input axis to show that the numbers now represent time in number of years since 1970. Also, we no longer needed to break the axis between 0 and 1965.

Another problem in working with real data occurs when numbers are very large or small. In those cases we may change the unit of one or both axes to count by hundreds or thousands instead of ones. For example consider the following set of data:

x	y
0	3000
20	2000
60	0

Graphing these data using a grid size of 1 would make the graph unreadable. A more reasonable graph can be made by adjusting the grid sizes used on both axes. For the input axis we might want to go up to a number that is close to 60 and easy to divide up; 80 is a good choice since we can break up 80 into 8 units of 10. Therefore, we will draw an input axis with 0 at its center and which goes to –80 on the left and +80 on the right and has grid lines every 10 units. [†] The input axis now will look like this:

$$\text{-80 -70 -60 -50 -40 -30 -20 -10 \quad 0 \quad 10 \quad 20 \quad 30 \quad 40 \quad 50 \quad 60 \quad 70 \quad 80} \quad X$$

For the y-axis, we will use a grid line spacing of 1000. But, instead of writing the labels 1000, 2000, 3000, etc., we will use labels of 1, 2, 3, etc., and make a note that we changed the units from ones to one-thousands.

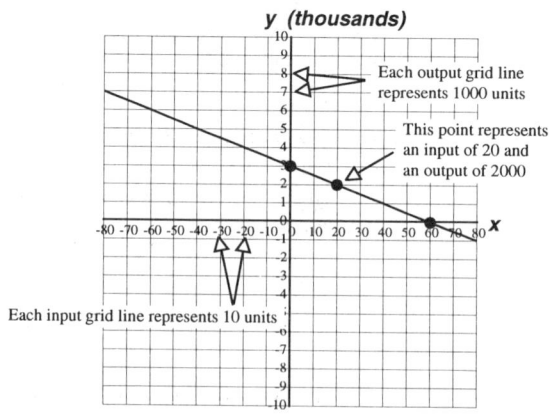

[†] This is a common practice in modeling. You may feel uncomfortable with us arbitrarily assigning 0 to 1970, but the calendar we use in the U. S. had its 0 defined not by "the beginning of time" or because of some mathematical principle but by a historical event, the birth of Jesus of Nazareth. Many people do not use this birthday as the 0 point for their calendar. For example, 1970 in the Moslem calendar is 1348 and in the Hebrew calendar 1970 is 5730.

[†] Our choice is somewhat arbitrary; we could just as correctly draw the input axis from 0 to 60 in steps of 5 or -20 to 70 in steps of 4.

Example 4.3.2   Graph the data shown below.

x (calendar year)	1977	1981	1985	1989
y (number of women earning bachelor's degrees)	427,000	463,000	499,000	535,000

The scale of the input axis should be *number of years since 1977* and the scale for the output axis should be *number of women earning bachelor's degrees* measured in thousands. Since all the values are positive use only Quadrant I.

•SOLUTION•   First, we write the data table using the transformed data:

x (number of years since 1977)	0	4	8	12
y (number of women earning bachelor's degrees in thousands)	427	463	499	535

Next, we draw the axes and grid and plot the transformed data:

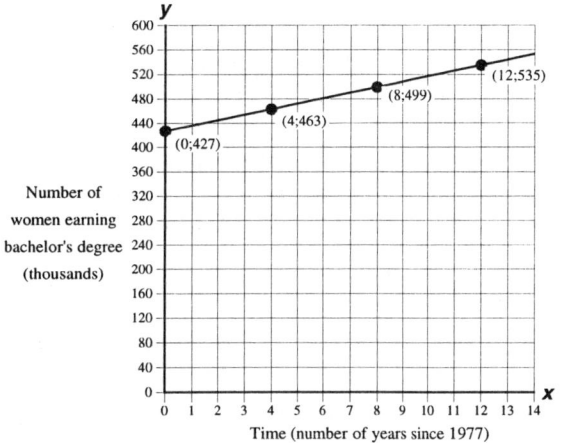

Practice 4.3.2   Graph the following linear relations. Remember that your graphs may look different from those in the answer key if you used a scale that was different from the one we chose to use.

1.   Graph the following data; make 1975 year 0 and use the label "hundreds" on the y-axis.

x time (calendar year)	y number of women earning degrees in dentistry
1975	170
1978	461
1981	752
1984	1043

2.   Graph the following data; use the label "hundreds" on the x-axis and use the label "thousands" on the y-axis.

x number of damaged items	y profit (dollars)
1200	68,000
2800	52,000
600	74,000
1900	61,000

Topic 4.3 C — Answering Questions Using the Graph of a Data Table

Sometimes we can use the idea of estimating from a graph to answer real-world questions.

Example 4.3.3   A diver notices that 5 minutes into a dive she has 45 cubic feet of air left in her tank. Five minutes later she notices that she has 30 cubic feet of air left.

a)   Create a data table with *time into the dive* as the input and *air left in the tank* as the output.

•SOLUTION• The verbal description provides us with two data points.

- She has 45 cubic feet of air 5 minutes into the dive; this is the point (5,45).

- Five minutes later is 10 minutes into the dive. Now, she has 30 cubic feet of air. This is the point (10,30).

x *time into dive* *(minutes)*	*y* *air in tank* *(cubic feet)*
5	45
10	30

b)   Graph the data on the given coordinate system (assume a linear relation).

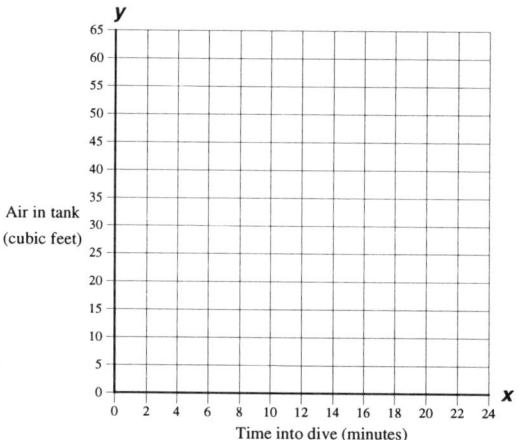

•SOLUTION• We plot the two points on the coordinate system. Since we are to assume that the relation is linear we can connect the points with a straight line. Notice that only Quadrant I is used because negative time or negative air is not meaningful.

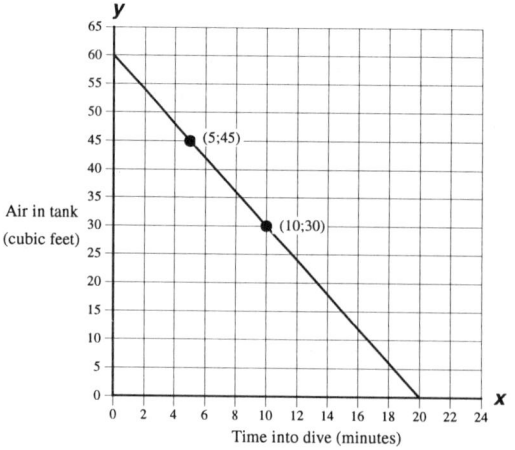

c)   When will the diver run out of air?

•**SOLUTION•** She will run out of air when she has 0 cubic feet of air in the tank. From the graph we see that an output of 0 corresponds to an input of 20. Therefore, she will run out of air 20 minutes into the dive.

d)   How much air was in the tank at the beginning of the dive?

•**SOLUTION•** The beginning of the dive is when the time into the dive is 0. From the graph we see that an input of 0 corresponds to an output of 60. Therefore, she has 60 cubic feet of air at the beginning of the dive.

e)   How much air is in the tank 15 minutes into the dive?

•**SOLUTION•** From the graph we can see that an input of 15 minutes corresponds to an output of 15 cubic feet of air.

f)   After how many minutes will she have 10 cubic feet of air left?

•**SOLUTION•** From the graph we can see that an output of 10 cubic feet of air corresponds to an input of about 17 minutes.

Practice 4.3.3

1.   A school buys candy bars in bulk and then sells them individually at a sporting event. When they sold 100 bars they made a profit of –$525. When they sold 400 bars they made a profit of –$300. Assume that bars sold, in hundreds, is the input and profit is the output.

a)   Create a data table for this situation.
b)   Assume a linear relation and graph the relation using the coordinate system on the right.
c)   How much did they spend to buy the bars?
d)   Approximately how many bars do they need to sell before they will break even (that is, have a profit of $0)?
e)   To make a profit of $100 how many bars must they sell?
f)   If they sell 1000 bars what will be their profit?

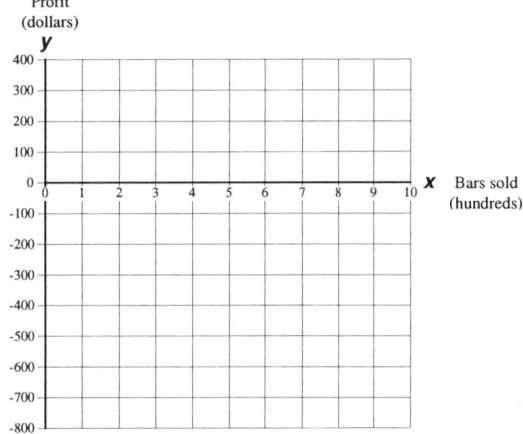

2.   Byron begins to record information about his weight loss a few weeks after he begins his weight loss program. After 27 days he weighs 209 pounds. After 42 days he weighs 204 pounds. If days into the program is the input and current weight is the output use this information to answer the following questions.

a)   Create a data table for this situation.
b)   Assume a linear relation and graph the relation.
c)   After how many days will Byron reach his target weight of 190 pounds?
d)   What will be Byron's weight 60 days into the program?
e)   How much did Byron weigh when he began the program?

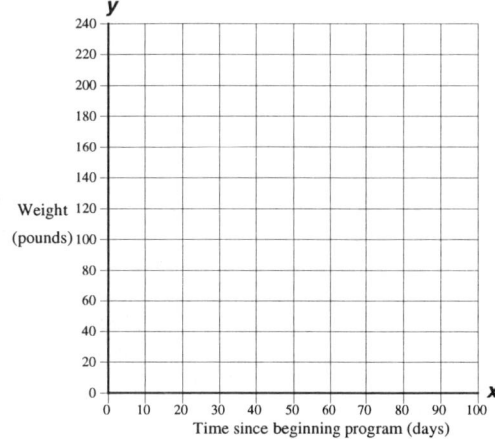

Topic 4.3 D — Using a Data Table to Graph a Linear Equation

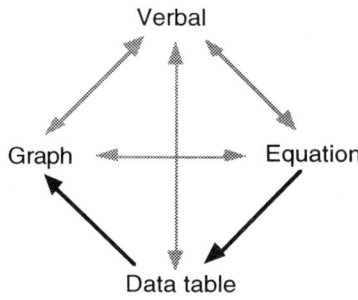

You may have noticed that newspapers and magazines often present mathematical information in the form of graphs instead of equations. This is because most readers are more comfortable with pictures than with algebraic symbols. In the future you may be called upon to represent a linear equation as a graph.

It's important to realize that a linear equation can be written in several different, but equivalent forms. For example, the linear equation $y = 1 + 2x$ can be written as follows:

$$y = 2x + 1 \qquad \Leftarrow \text{ Used Commutative Property of Addition.}$$
$$\text{or} \qquad y - 2x = 1 \qquad \Leftarrow \text{ Subtracted } 2x \text{ from both sides.}$$
$$\text{or} \qquad 0 = 2x - y + 1 \qquad \Leftarrow \text{ Moved all terms to the right side.}$$

Even though y is not isolated in some of the above equations, we usually work under the assumption that x is the input and y is the output.

To graph a linear equation we begin by constructing some specific input/output pairs and organizing these in a data table. An easy way to find specific pairs is to choose three small, integer, values as input and then to calculate the corresponding output values. We enter these input/output pairs into the data table and then draw the graph by plotting the points and connecting them with a straight line.

For example suppose we are given the linear equation $y = 2x + 1$. We select values for x, substitute them into the equation $y = 2x + 1$, and then simplify to obtain the corresponding values for y. If we select the values 0, 2, and −1 for x, the data table would look like the following:

x	y	*Calculated output , y, by evaluating 2x + 1*
0	1	$\Leftarrow 2(\mathbf{0}) + 1$
2	5	$\Leftarrow 2(\mathbf{2}) + 1$
−1	−1	$\Leftarrow 2(\mathbf{-1}) + 1$

Notice that these ordered pairs are solutions to the equation $y = 2x + 1$ because if we substitute an input value for x and the corresponding output value for y the result is a true statement.

If we plot the above solutions on an x-y coordinate system we get the following:

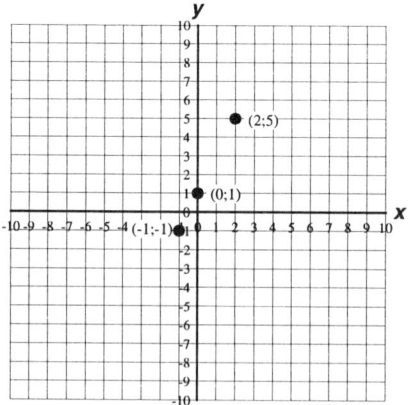

Of course, we are not forced into selecting 0, 2, and −1 for input; we could have chosen any value for x. This means there are an infinite number of choices for x and consequently an infinite number of corresponding y's. Let's plot 50 more solutions:

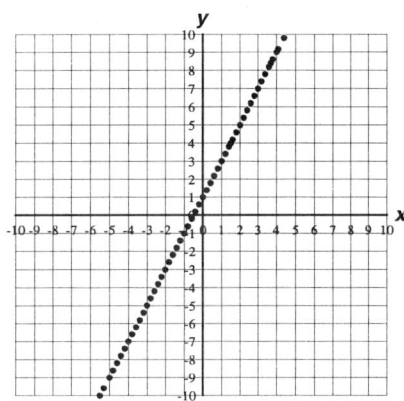

We can see a clear pattern emerging. There are so many points that they seem to blur into each other and they all appear to lie on a straight line. In fact, every solution to the equation $y = 2x + 1$ lies on this straight line. Rather than plotting many points, we usually plot a few and then connect the points with a straight line to indicate all the solutions to the equation.

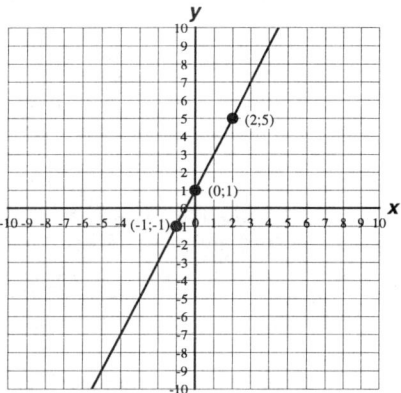

— *Property*— *Graph of a linear equation*	English: The graph of a linear equation, $y = b + mx$ is a straight line.
	Example: The graph of $y = 2x + 1$ is a straight line.
	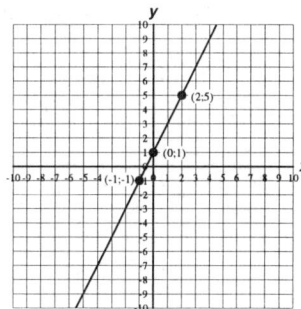
	Algebra: If m and b are constants, then the graph of any equation that can be written in the form $y = b + mx$ is a straight line.

Example 4.3.4   a)   Graph $y = x - 3$

• SOLUTION • We choose three values for input, calculate the output, and write the data in a data table. Let's use 0, 4, and –4 for input values.

x	y	Calculated output , y, by evaluating x – 3
0	–3	⇐ **(0)** – 3
4	1	⇐ **(4)** – 3
–4	–7	⇐ **(–4)** – 3

We plot the points and connect them with a straight line.

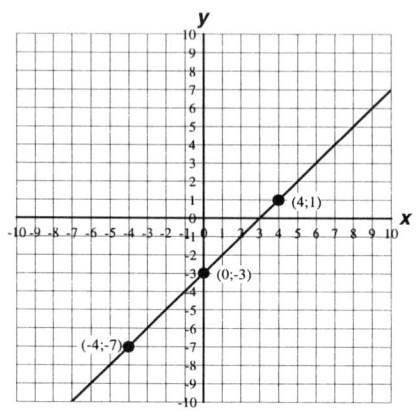

— Note— **The third point is a check**	Note that we only need two points to graph a straight line. We calculate and plot a third point as a check. If all three points don't line up we know we have made an error.

b)   Graph $x + 2y = 4$

• SOLUTION • Let's choose input values of 0, –2, and 6 and find the corresponding output values:

x	y	Calculated output , y, by solving x + 2y = 4

0	2	⇐ **(0)** + 2y = 4
		          2y = 4
		           y = 2

–2	3	⇐ **(–2)** + 2y = 4
		          2y = 6
		           y = 3

6	–1	⇐ **(6)** + 2y = 4
		          2y = –2
		           y = –1

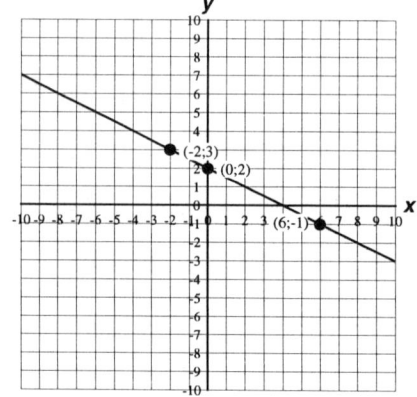

c)   Graph $y = 6$

•**SOLUTION**•   The equation says that the output is always 6, regardless of the input. Therefore, we can choose any value we want for x and its corresponding y value will be 6. We will choose values for x of –4, 1, and 5.

x	y
–4	6
1	6
5	6

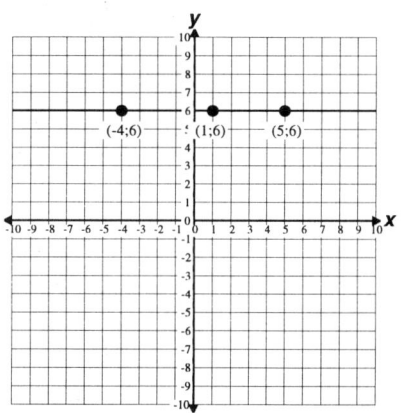

d)   Graph $y = -300 + 5x$

•**SOLUTION**•   First, we will generate three ordered pairs by choosing input values, substituting them into the equation, and solving for y. We almost always use an input value of 0 because it usually makes the calculations easier. The other input values we select depend on the situation and our own preferences. In this case, we will choose values for the input that will generate a wide range of values for the output. We will use 20 and 60 for the input.

x	y
0	–300
20	–200
60	0

Since the input values go from 0 to 60 we will use an increment of 10 on the input axis. Since the y-values are so large we will use an increment of 40 on the output axis. This is a judgment call—we could have used 10 or 60 or 100.

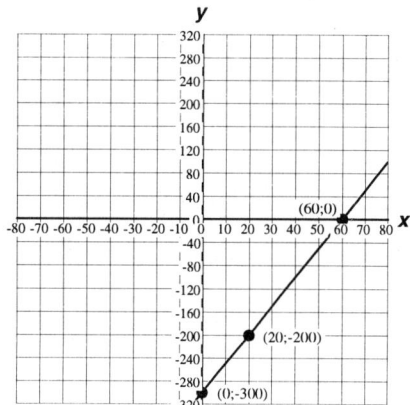

If we had chosen different scales for the axes or different input values the graphs would look a bit different but they would contain similar information. Below is an alternative data table and graph for $y = -300 + 5x$:

x	y
−40	−500
0	−300
70	50

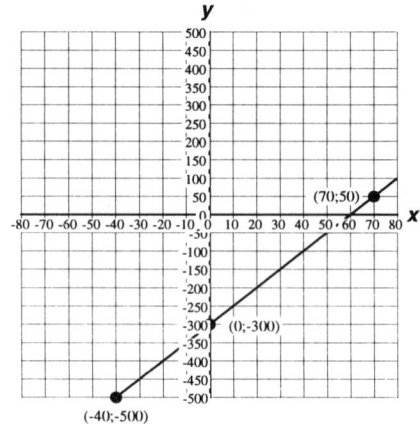

Practice 4.3.4 Graph the following linear relations. Remember that answers on your graphs may look different from those in the answer key if you used a scale that was different from the one we chose to use.

1. $y = x + 2$
2. $y = 3000 + 50x$. Use one-thousand as the unit on the y-axis
3. $y = -2$
4. Graph $y = 1000x - 10,000$. Use one-thousand as the unit on the y-axis.
5. $-x + y + 2 = 0$

Topic 4.3 E — The Intercepts of a Linear Equation

Two special points on a linear graph are where the line crosses the x and y axes. These points are called the intercepts of the line.

- The **x-intercept** is the point where the line meets the x-axis. We find this intercept by substituting 0 for y and then solving for x. This point always has coordinates $(a,0)$, where a is a constant.

- The **y-intercept** is the point where the line meets the y-axis. We find this intercept by substituting 0 for x and then solving for y. This point always has coordinates $(0,b)$, where b is a constant.

Since we know that one of the components of an intercept is always 0, we often use abbreviated terminology. For example, if the y-intercept of a graph is the point $(0,3)$ many people would refer to the y-intercept simply as 3.

Example 4.3.5   Find the x-intercept and the y-intercept:

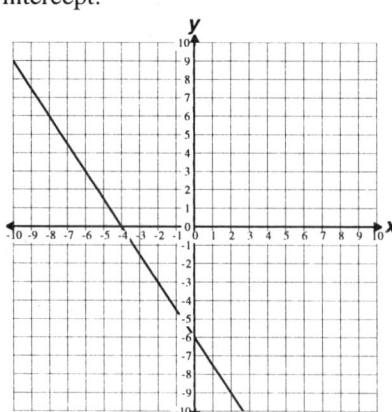

•SOLUTION• The x-intercept is $(-4,0)$. The y-intercept is $(0,-6)$.

Practice 4.3.5 Find the *x*-intercept and the *y*-intercept of the following graphs.

1.

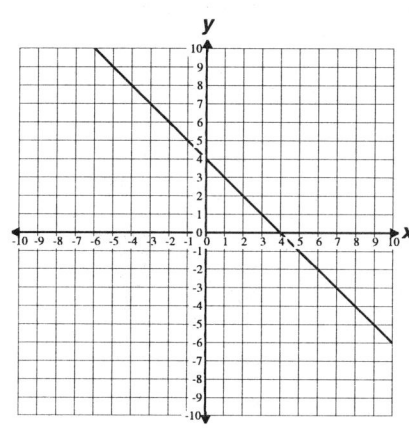

2.

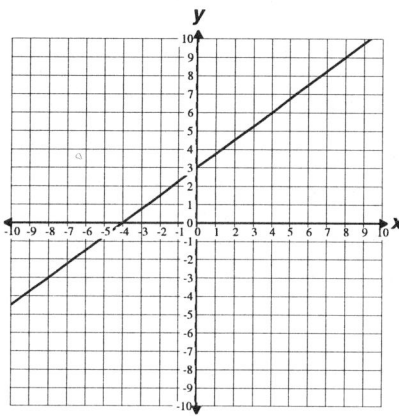

3.

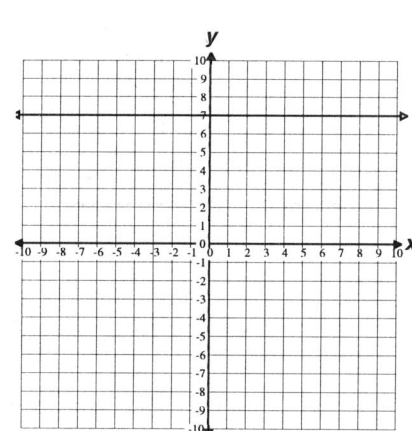

4.

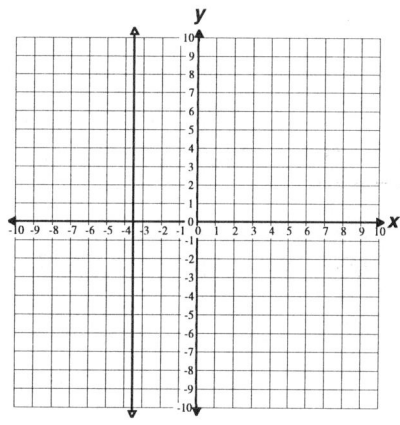

5.

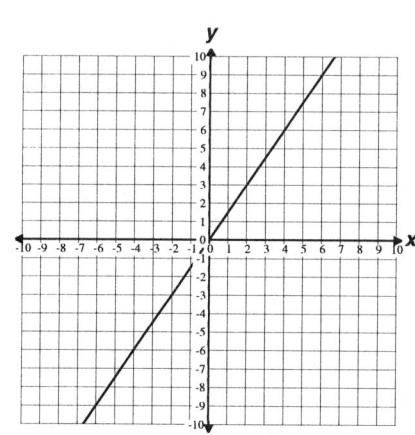

6.

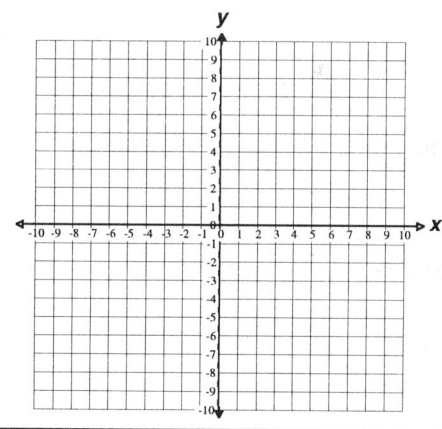

Using the *x*– and *y*-intercepts to graph a line is sometimes called the **intercept method of graphing a linear equation**.

— Procedure — *Using the intercept method to graph a linear equation*	**Step 1 *y-intercept*** In the equation, substitute 0 for *x* and solve for *y*. This ordered pair $(0,b)$ is the *y*-intercept.
	Step 2 *x-intercept* In the equation, substitute 0 for *y* and solve for *x*. This ordered pair $(a,0)$ is the *x*-intercept.
	Step 3 *Plot* Plot the two points on an *x-y* coordinate system and connect the points with a straight line.

Example 4.3.6 Use the intercept method to graph $x + \frac{1}{2}y = 4$.

• **SOLUTION** • ***Step 1 y-intercept*** Substitute 0 for x and solve for y to get $y = 8$. This means the y-intercept is (0,8).

Step 2 x-intercept Substitute 0 for y and solve for x to get $x = 4$. This means the x-intercept is (4,0).

Step 3 Plot Plot the two points and connect them with a straight line.

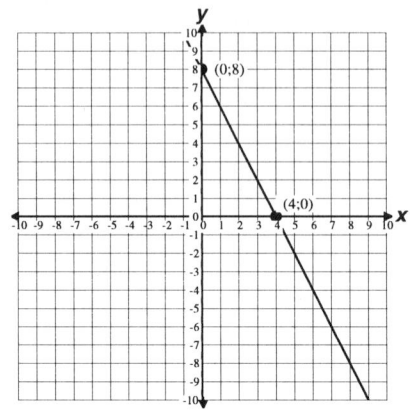

As a check, it would be a good idea to determine a third point and plot it.

Practice 4.3.6 Construct an x-y coordinate system where both axes go from –10 to 10 in steps of 1. Then, use the intercept method to graph the following linear equations.

1. $4x + 2y = 12$ 2. $6 + 2y = 2x$ 3. $-4x + 3y - 12 = 0$ 4. $y - 2x = 0$

Exercise Set 4.3   The answers to the odd numbered exercises are at the back of the book.

Plot the following on the given *x-y coordinate* system. Then, if possible, use a linear model to fill in the missing input or output values.

1.

x	y
–8	–2
–6	4
2	6
	3
	1
	0

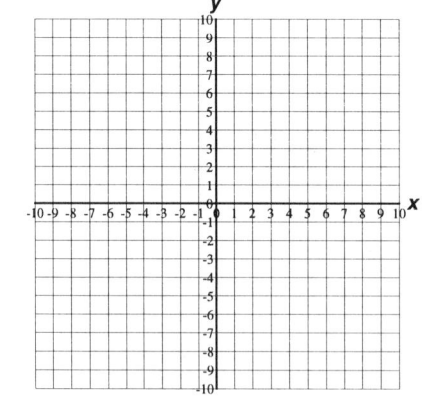

2.

x	y
0	−1
1	−4
−9	−2
	5
	1
	−2

3.

x	y
5	−8
1	−4
−3	0
	4
	1
	−1

4.

x	y
3	−2
0	1
−3	4
	8
	−6
	0

5.

x	y
1	−1
4	8
−1	−7
	5
0	
	−10

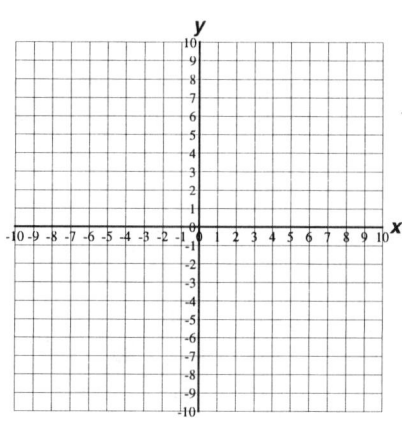

6.

x	y
−4	−7
0	1
3	7
	−3
	5
−1	

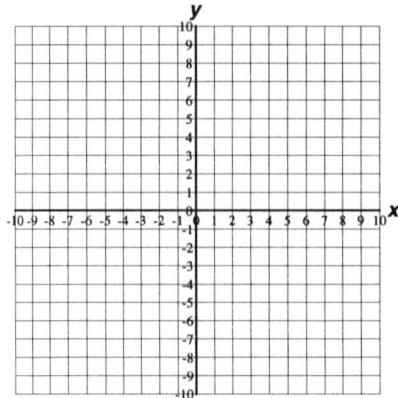

7.

x	y
5	8
0	3
−5	−2
	0.5
	2.5
−8.5	

8.

x	y
−1	−3
0	−2
5	3
−4.5	
	−2.5
	4

9.

x	y
−3	−2
−3	5
−3	0
	1
4	
	2

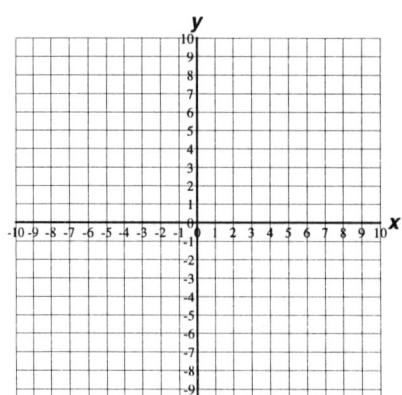

10.

x	y
1	7
1	6
1	−7
	4
−3	
	−6

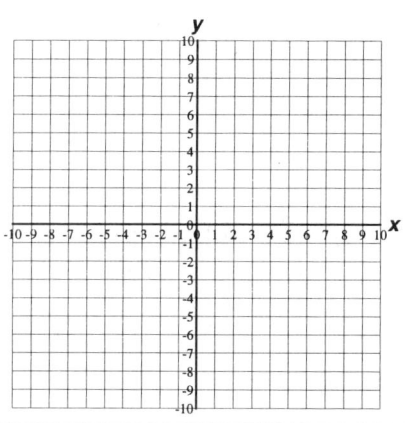

Graph the following linear relations.

11. $y = -500 + 25x$. Use a scale of hundreds for the y-axis.

12. $y = 1900 - 100x$. Use a scale of hundreds for the y-axis.

13. Make the year 1960 year 0 and use unit thousands on the y-axis. Use Quadrant I only.

x (calendar year)	1960	1961	1968	1974
y (number of men earning master's degrees)	47,000	55,000	111,000	159,000

14. Make the year 1979 year 0 and use unit thousands on the y-axis. Use Quadrant I only.

x (calendar year)	1979	1982	1985	1987
y (number of men earning master's degrees)	151,000	147,100	143,200	140,600

15. $y = -20x + 2400$

16. $y = -15x - 450$

17.

x temperature (degrees Fahrenheit)	y oil used to heat house (gallons)
45	25
20	40
15	43
10	46

18.

x temperature (degrees Fahrenheit)	y cost to heat house (dollars)
45	87.50
20	140.00
15	150.50
10	161.00

19. Make 1940 year zero and use a scale of millions on the y-axis.

x time (calendar year)	y carbon monoxide emitted (tons)
1940	92,000,000
1950	103,000,000
1960	114,000,000
1970	125,000,000

20. Make 1975 year zero and use a scale of millions on the *y*-axis.

x time (calendar year)	*y* carbon monoxide emitted (tons)
1975	129,000,000
1980	122,500,000
1985	116,000,000
1990	109,500,000

To answer the following questions, create a data table for the situation and then graph the data on the given coordinate system.

21. A sign at a local bank switches between displaying the current temperature in degrees Centigrade and degrees Fahrenheit. In the morning, a driver notices the sign shows 20° Centigrade and then 68° Fahrenheit. On his way home from work he notices the same sign blink 35° Centigrade and then 95° Fahrenheit. Assume degrees Centigrade is the input and degrees Fahrenheit is the output.

a) About what Fahrenheit temperature is –29° Centigrade?
b) Normal temperature for a human is 98.6° Fahrenheit. About what Centigrade temperature is this?
c) Would you consider a 40° Centigrade a hot day?
d) About what Centigrade temperature is 0° Fahrenheit?

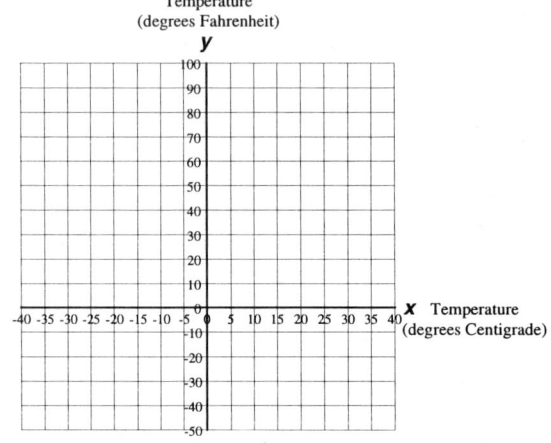

22. While planning a trip to Minnesota, a French businesswoman reads that the temperature in Minneapolis ranges between –31° C (–35° F) in the coldest winters and 40°C (104° F) in the hottest summers. Assume she wants to predict degrees Centigrade from degrees Fahrenheit.

a) Water freezes at 32° Fahrenheit. At what temperature in degrees Centigrade will water freeze?
b) If you prefer the temperature in the building where you live to be around 70° Fahrenheit what Centigrade temperature would you set a thermostat at?
c) About what Fahrenheit temperature is –40° Centigrade?
d) Would you feel comfortable swimming in a pool that was 30 degrees Centigrade?

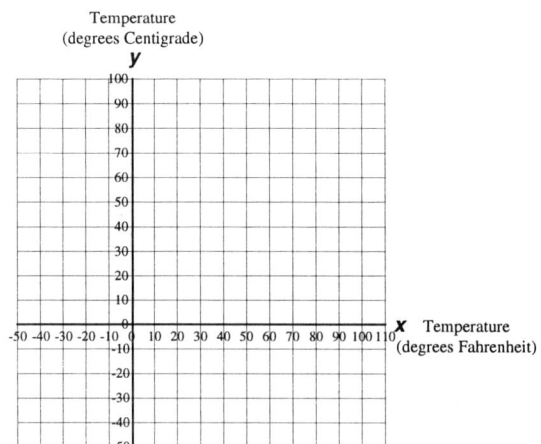

23. Two hours after a snowfall begins a girl runs outside and pushes a ruler through the snow until it touches the ground. She measures the snow depth to be 8 inches. Four hours after the snowfall began she again pushes the ruler to the ground and finds the depth to be 9 inches. Assume time since the snowfall began is input, depth of the snow is the output and that the snow falls at a constant rate throughout the snowfall.

a) How much snow was on the ground before the snowfall began?

b) If the snowfall continues for 8 hours what would be the snow depth on the ground?

c) If the snowfall continues for 12 hours how much **new snow** would have fallen?

d) How long would the snow have to fall before there would be 14 inches of snow on the ground?

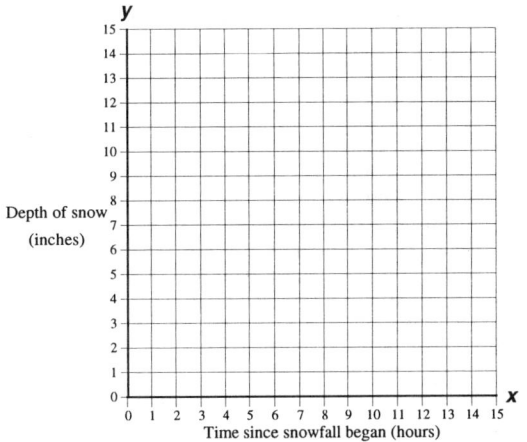

24. After hearing a thunderstorm is coming a meteorology student goes outside to make sure her rain gauge is dry. There is about 2 hours of light rain before the thunderstorm begins . After the thunderstorm has been going for 30 minutes she runs outside and notices the gauge now reads 2.75 inches. After 45 minutes she again runs outside again and sees the gauge now holds 3.75 inches. Assume minutes since the thunderstorm began is the input and inches of rain collected is the output.

a) If the thunderstorm goes for an hour what will the rain gauge read?

b) How much light rain fell before the thunderstorm began?

c) How many minutes will the thunderstorm have gone on before there is 5 inches of rain in the rain gauge?

d) After how many minutes will 1 inch of rain fall during the thunderstorm?

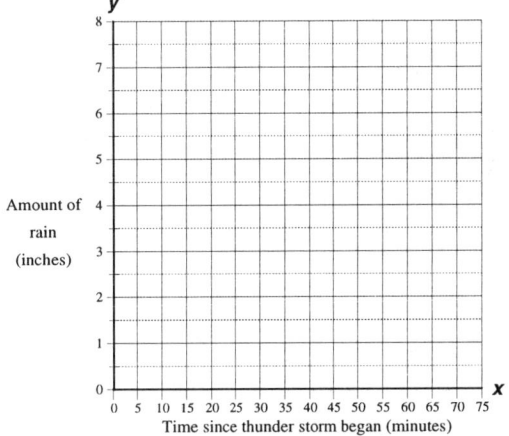

25. Even though the minimum hourly wage has been steady or rising since 1968, in terms of its purchasing power it actually fell from 1968 to 1989 [†]. For example, in 1974 the minimum hourly wage had a purchasing power (in 1996 dollars) of about $6.26 while in 1982 the purchasing power had fallen to about $5.27. Assume number of years since 1968 is the input and the minimum hourly wage in terms of purchasing power is the output.

a) What was the minimum hourly wage (in terms of 1996 dollars) in 1968?

b) What year did the minimum hourly wage (in terms of 1996 dollars) drop to $5.00?

c) What was the minimum hourly wage (in terms of 1996 dollars) in 1989?

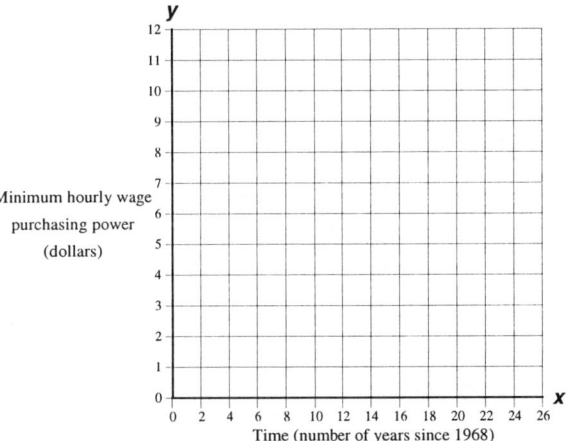

[†] The idea here is that if the minimum wage stays the same but, due to inflation, the cost of what you buy goes up (or if the cost goes up faster than the minimum wage goes up) you have less purchasing power.

26. The most consistent decrease in the purchasing power of the minimum hourly wage was between 1979 and 1989. For example 4 years after 1979 it was about \$5.36 while 8 years after 1979 it was \$4.60. Assume number of years since 1979 is the input and the minimum hourly wage in terms of 1996 dollars is the output.

a) What year did the minimum hourly wage (in terms of 1996 dollars) drop to \$5.00?
b) What was the minimum hourly wage (in terms of 1996 dollars) in 1989, the last year of decline?
c) What was the minimum hourly wage (in terms of 1996 dollars) in 1979?

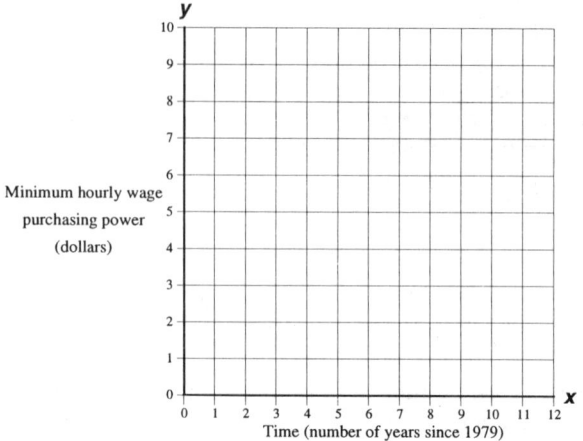

Construct an x-y coordinate system where both axes go from -10 to 10 in steps of 1. Then, calculate three ordered pairs and draw the graphs of the linear equations.

27. $x - 3y = 6$ 28. $4x - y = 8$ 29. $x = 7$ 30. $x = -4$ 31. $y = -6x + 6$

32. $y = 4x - 4$ 33. $y = -3$ 34. $y = 8$ 35. $x + y - 2 = 0$ 36. $x - y + 7 = 0$

Find the x-intercept and the y-intercept of the following graphs.

37.

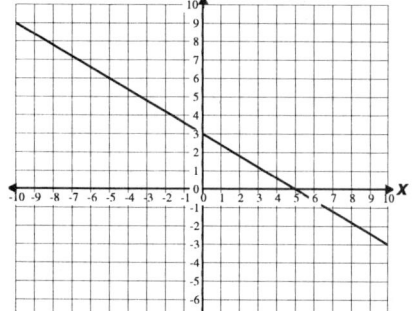

38.

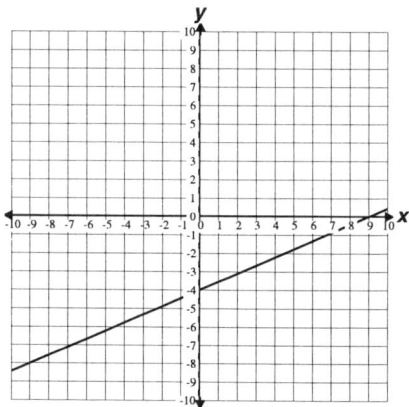

39.

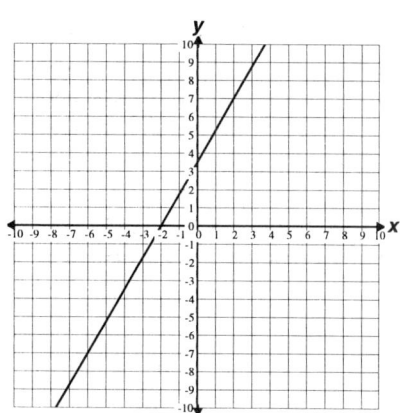

40.

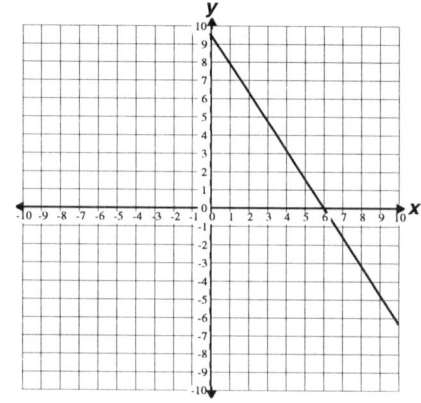

41.

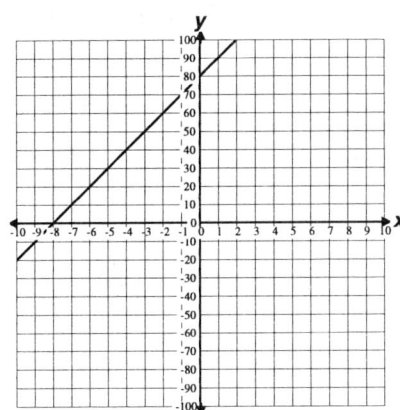

42.

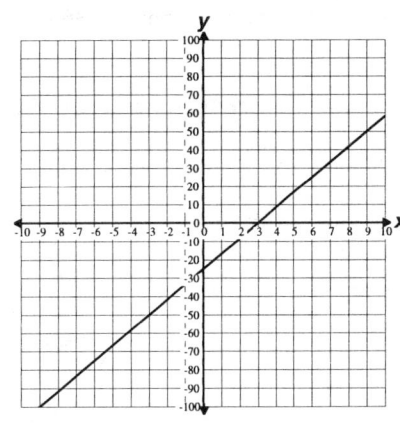

43.

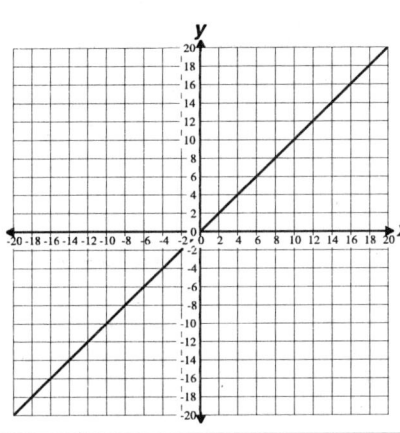

44.

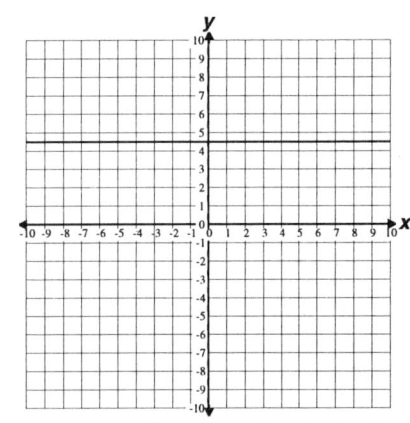

Construct an *x-y* coordinate system where both axes go from −10 to 10 in steps of 1. Then, use the intercept method to graph the following linear equations.

45. $y = \frac{1}{2}x + 3$ 46. $y = \frac{1}{3}x - 3$ 47. $2x - 8y = 4$ 48. $4x - 2y = 6$

49. $y = -\frac{1}{4}x$ 50. $y = \frac{2}{3}x$ 51. $y = -8 - 8x$ 52. $y = 10 + 5x$

Review Exercises   The answers to all of these exercises are at the back of the book.

53. Evaluate $5 - (x + 2y)$ when x is −2 and y is −3.   54. Solve $i = prt$ for p 55. Reduce to lowest terms: $\frac{124}{440}$

56. Simplify: $y - 4[2y + 5(y - 6) + 15]$ 57. Solve: $\frac{3x}{2} - \frac{1}{2} = \frac{1}{4}(x + 3)$ 58. Find the reciprocal: $\frac{3x}{2}$

59. A couple decides to split up the work while painting three rooms of their house. If the bedroom has 52 less square feet than the family room and the kitchen has 18 more square feet then the bedroom find the square footage of the bedroom. Assume the total square footage of the rooms discussed is 220 square feet.

60. A landscaper has to bring to a job two types of bricks, which weigh a total of 446 pounds. One brick weighs 2 pounds more than the other. If he loads 41 light bricks and 55 heavier bricks how much does a light brick weigh?

Section 4.4 Slope and Some of Its Uses

Topic 4.4 A — Using a Graph to Create a Linear Equation

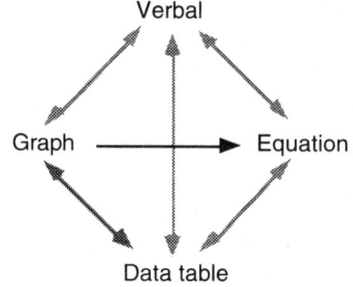

In Section 4.2 we defined the parts of the linear equation, $y = b + mx$, as follows:

- y represents the output.
- x represents the input.
- m represents a constant that describes how the input and output change relative to each other; it is a rate of change, like miles per hour.
- b represents a constant that tells us the value of the output when the input is 0; it usually represents the "initial" value of the output.

To build a linear equation from a graph we need to determine values for b and m. Using the vocabulary of graphs, b is known as the y-intercept since it's the y-value of the point where the line crosses the y-axis. You can see this by realizing that if x is 0, as it must be if the point is on the y-axis, then the equation of the line becomes

$$\begin{array}{lll} y = b + mx & \Leftarrow & \text{Linear equation.} \\ y = b + m(0) & \Leftarrow & \text{Substituted 0 for } x. \\ y = b & \Leftarrow & \text{Simplified to find the output.} \end{array}$$

This means that as long as we can determine where the line meets the y-axis, we can fill in the value for b.

Using the vocabulary of graphs, m is known as the **slope** of the line. It is defined as the ratio of the change in the output to the corresponding change in the input when going from one point on the line to another. For example suppose we want to calculate the slope of the following line:

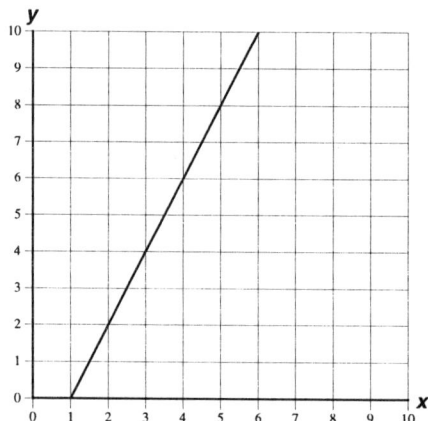

Let's say that we start at some point on the line, say (2,2) and move to another point on the line, say (5,8). To do this we must increase the output by 6 units while increasing the input by 3 units.

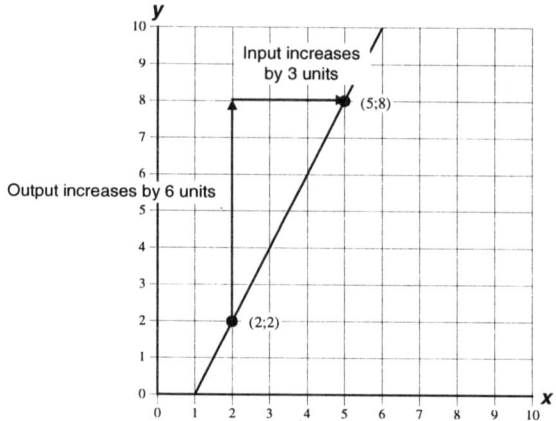

Since the slope is the ratio of the change in output to the change in input, the slope for this line is $\frac{6}{3}$ or 2. Some people like to remember this as **rise** over **run** since we are comparing the change in the vertical distance (the rise) to the change in the horizontal distance (the run).

Once we have determined the y-intercept, b, and the slope, m, we substitute the values in the equation $y = b + mx$. Here is a procedure you can use to build a linear equation from a graph.

— Procedure — *Writing the Linear* *Equation of a Graph.*	**Step 1 Find b** Find the value of y where the line crosses the y-axis; this is b. **Step 2 Find m** Locate a point on the line where you can easily determine the coordinates (usually, this is the intersection of two grid lines). Starting at that point, move vertically and then horizontally until you reach another easily identifiable point on the line. Write the ratio of the vertical distance you moved to the horizontal distance; that is, write $\dfrac{\text{change in } y \text{ (rise)}}{\text{change in } x \text{ (run)}}$. This is m. **Step 3 Equation** Substitute the values found in steps 1 and 2 in the equation $y = b + mx$.

Example 4.4.1   a)   Construct the equation of the line shown in the following graph:

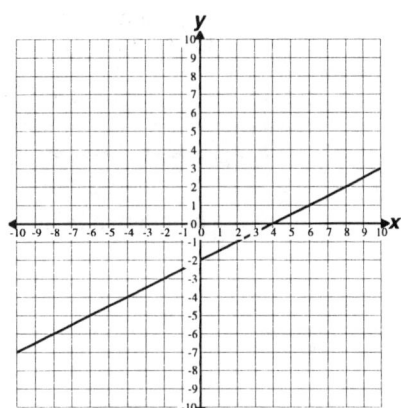

•SOLUTION• **Step 1 Find b**   The y-intercept is $(0,-2)$, so b is -2.

Step 2 Find m   The points $(0,-2)$ and $(2,-1)$ are easy to locate on the line. We start at $(0,-2)$ and as we move to $(2, -1)$ the y value changes by 1 while the x value changes by 2. Therefore, the slope is $\frac{1}{2}$.

Step 3 Equation   The equation is $y = -2 + \frac{1}{2}x$

b)   Construct the equation of the line shown in the following graph:

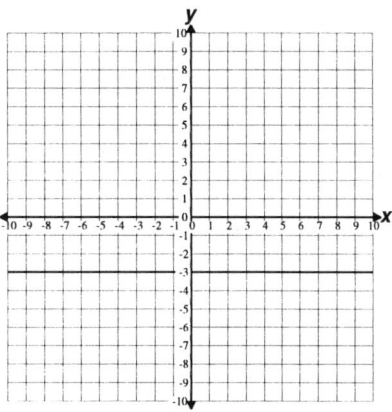

•SOLUTION• **Step 1 Find b**   The y-intercept is $(0,-3)$, so b is -3.

Step 2 Find m   We will use the points $(1,-3)$ and $(2,-3)$. We start at $(1,-3)$ and as we move to $(2,-3)$ the vertical change is from -3 to -3, which is 0. The horizontal change is from 1 to 2, or 1. Therefore, the slope is $\frac{0}{1}$, which is 0.

Step 3 Equation   The equation is $y = -3 + 0x$, which simplifies to $y = -3$.

c)   Construct the equation of the line shown in the following graph:

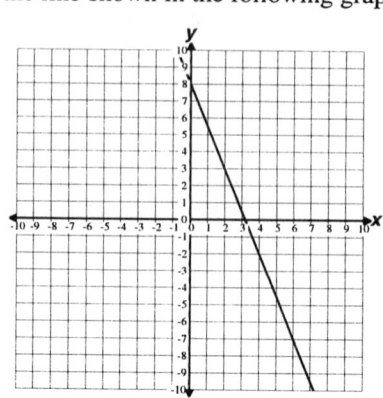

•SOLUTION• **Step 1 Find b**   The y-intercept is $(0,8)$, so b is 8.

Step 2 Find m   Let's start at the point $(2,3)$ and move to the point $(6,-7)$. As we move, the output goes from 3 to -7, a change of -10, while the input goes from 2 to 6, a change of 4. Therefore, the slope is $\frac{-10}{4}$, which simplifies to $-\frac{5}{2}$.

Step 3 Equation   The equation is $y = 8 - \frac{5}{2}x$

Practice 4.4.1   Construct the equation of the line shown in the graph:

1.

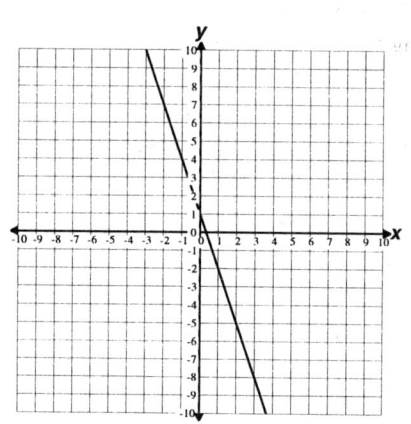

2.

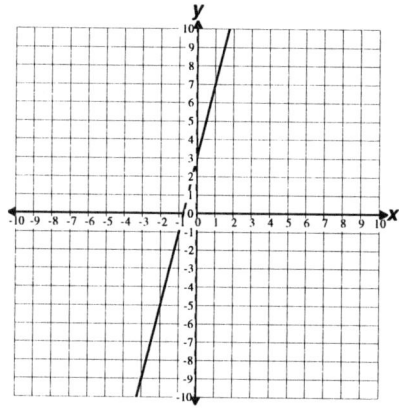

3.

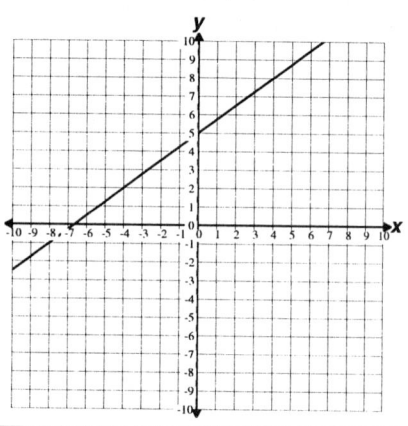

4.

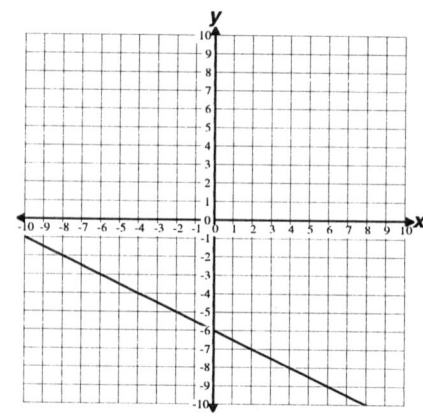

Topic 4.4 B — Using a Graph to Create a Verbal Statement

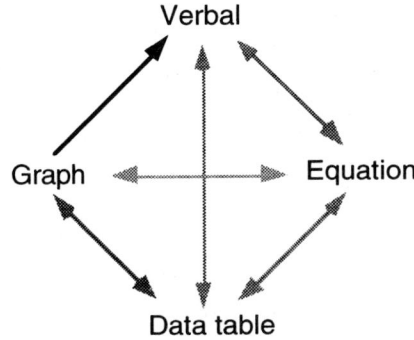

When a graph represents a real situation we should be able to explain the situation to others by gathering information directly from the graph.

Example 4.4.2   a)   Discuss the meaning of the graph.

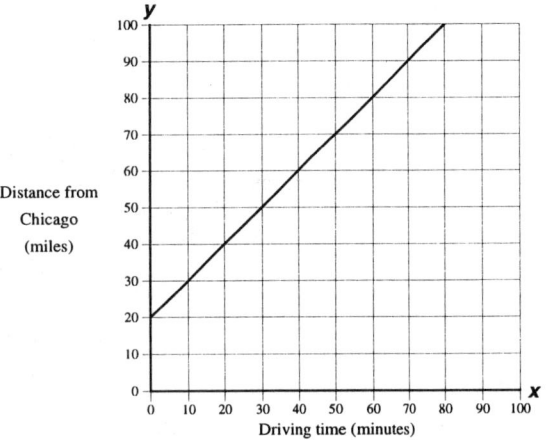

•**SOLUTION**• The y-intercept is (0,20). Since the output is distance from Chicago and the input is driving time, (0,20) means the person was 20 miles from Chicago when he or she started driving.

To see how distance from Chicago changes as driving time changes we begin at one point, say (0,20) and move to another point, say (10,30). The distance changes by 10 miles while the time changes by 10 minutes, which is $\dfrac{10 \text{ miles}}{10 \text{ minutes}}$ or 1 mile per minute. This is the speed the person was traveling. We know the person was traveling away from Chicago because every minute he or she is 1 mile farther away.

In summary, the graph says the person started 20 miles from Chicago and traveled away from Chicago at a speed of 1 mile per minute (which is 60 mph).

Note that we have found the y-intercept and slope of the graph so we could easily write the equation as $y = x + 20$.

b)   Discuss the meaning of the graph.

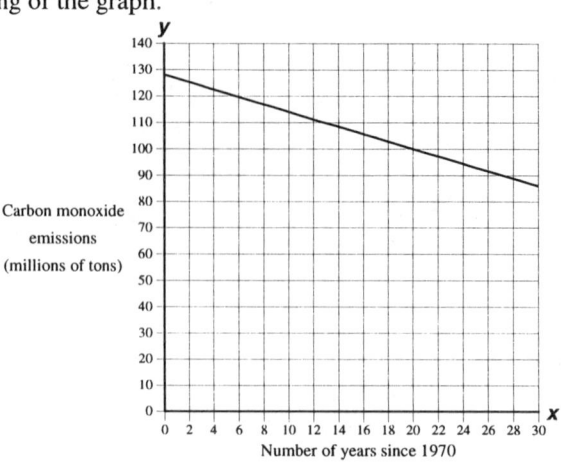

Carbon monoxide emissions (millions of tons)

Number of years since 1970

•SOLUTION• We estimate the y-intercept to be about $(0,128)$. This means that in 1970 there were about 128 million tons of carbon monoxide emitted.

To see how the emissions have changed, we begin at one point, say the intercept $(0,128)$ and move to another easily identifiable point, say $(20,100)$. The emissions have gone down by 28 million tons while 20 years have passed, which we may write as $\dfrac{-28 \text{ million tons}}{20 \text{ years}}$. This reduces to $-\dfrac{7}{5}$ million tons of emissions per year. This means that carbon monoxide emissions have been decreasing by $\dfrac{7}{5}$ million tons per year since 1970. (Of course, this is only an approximation. If we had started at say $(5,120)$ and moved to $(20,100)$ the change would be $\dfrac{-20}{15}$ which reduces to $-\dfrac{4}{3}$ million tons of emissions per year. Reading data from a graph involves judgments and approximations that may lead to different answers.)

In summary, the graph says that in 1970 128 million tons of carbon monoxide were emitted. Since then, emissions have dropped by $\dfrac{7}{5}$ million tons per year.

Note that we have found the y-intercept and slope of the graph so we could easily write the equation as $y = 128 - \dfrac{7}{5}x$.

Practice 4.4.2 Discuss the meaning of each graph.

1.

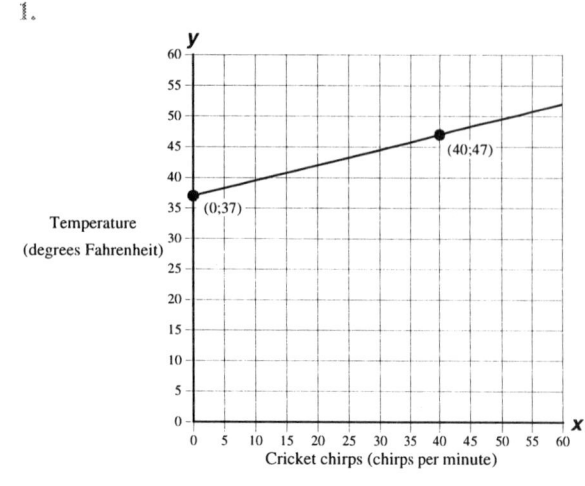

Temperature (degrees Fahrenheit)

Cricket chirps (chirps per minute)

2.

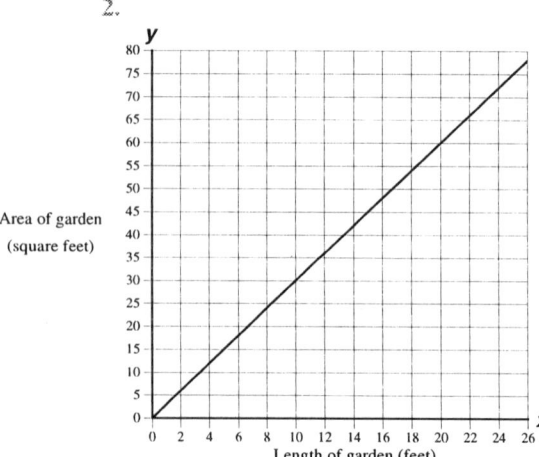

Area of garden (square feet)

Length of garden (feet)

3.

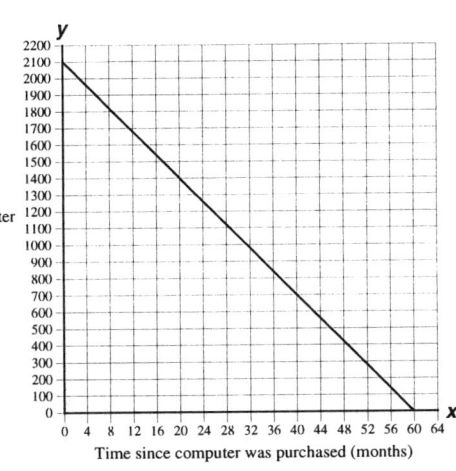

4.

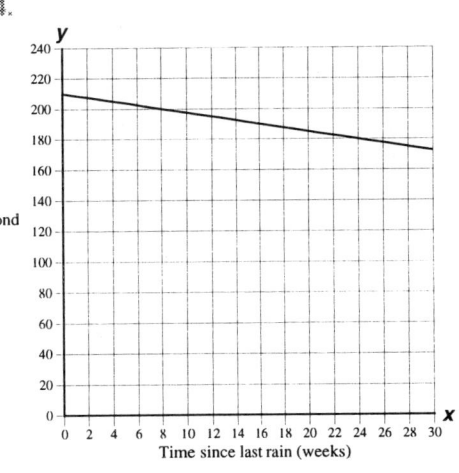

Topic 4.4 C — Constructing Graphs Given a Point and the Slope

We can graph a straight line given one point and the slope of the line by using the following procedure:

— *Procedure* —	**Step 1**	Plot the given point.
Graphing a line given a point and the slope	**Step 2**	Write the slope in the form $\dfrac{\text{change in } y}{\text{change in } x}$.
	Step 3	Starting at the given point,
		• move vertically the amount indicated in the numerator of the slope.
		• move horizontally the amount indicated in the denominator of the slope.
		• draw a dot.
	Step 4	Draw a straight line through the points.

Example 4.4.3   a)   Graph the straight line that passes through (2,–5) with a slope of 4.

•SOLUTION• **Step 1**   Plot (2,–5)

Step 2   Write the slope as $\dfrac{4}{1}$.

Step 3   Starting at (2,–5), move up 4 and right 1 and then draw a dot. This should be at coordinates (3,–1).

Step 4   Draw a straight line through the points.

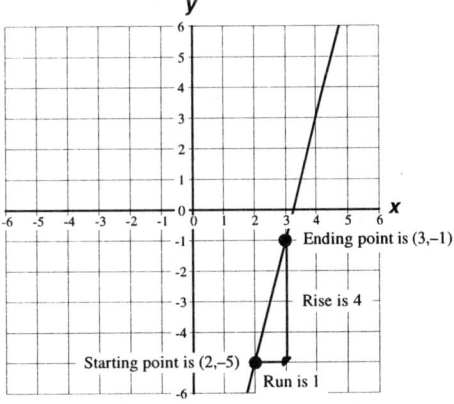

b)   Graph the straight line that passes through (–4,4) with a slope of $-\frac{7}{3}$.

•SOLUTION• *Step 1*   Plot (–4,4)

Step 2   We can write the slope as $\frac{-7}{3}$ or as $\frac{7}{-3}$, both of which are equivalent to $-\frac{7}{3}$. We will use $\frac{-7}{3}$.

Step 3   Starting at (–4,4), move down 7 and right 3 and then draw a dot. This should be at coordinates (–1,–3).

Step 4   Draw a straight line through the points.

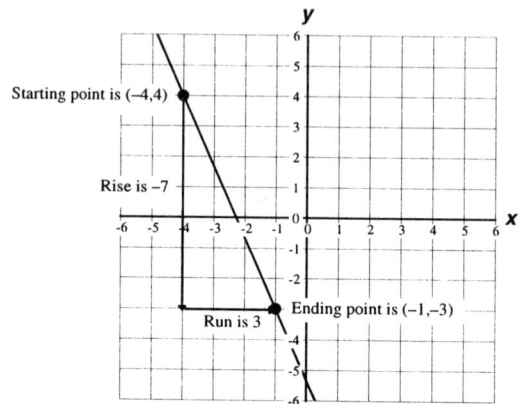

c)   Graph the straight line described by the following situation. "A laundry charges $3 per pound for washing clothes, plus a handling charge of $2." Use *Weight of clothes (pounds)* for the input and *Charge (dollars)* for the output.

•SOLUTION• First, we must determine two points from the verbal description.

   •   The laundry charges a handling charge of $2 regardless of the weight of the clothes; this gives the point (0,2), which is the *y*-intercept.

   •   The laundry charges $3 per pound of clothes; this gives a rate of change (slope) of $\frac{3 \text{ dollars}}{1 \text{ pound}}$.

Step 1   Plot (0,2).

Step 2   Write the slope as $\frac{3}{1}$.

Step 3   Starting at (0,2) move up 3 and right 1 and then draw a dot. This should be at coordinates (1,5).

Step 4   Draw a straight line through the points.

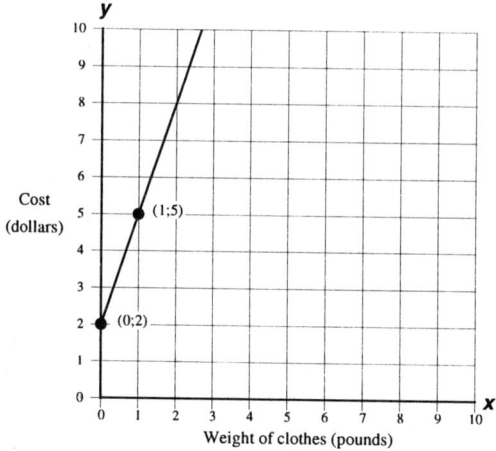

Practice 4.4.3 Graph the straight line that passes through the given point and which has the given slope.

1. (0,7) with a slope of $\frac{2}{3}$

2. (3,3) with a slope of –2

3. (–4,2) with a slope of 0

4. Graph the straight line described by the following situation. Assume year is the input and 1990 is year 0. Assume boats constructed annually is the output. "In 1990 we built 3 boats. Since then we have been able to increase our production by 2 boats a year."

5. Graph the straight line described by the following situation. Assume the day since the convention started is the input and the first day of sales is day 0. Assume daily sales in dollars is the output. "I know we had sold $50 worth of note pads the first day and that sales increased by $25 a day for the next 8 days."

Topic 4.4 D — The Slope Formula

The slope formula gives us an algebraic way of calculating the slope of a line when we are given two specific points that the line passes through. The slope formula looks like this:

$$m = \frac{y_2 - y_1}{x_2 - x_1} \quad \begin{array}{l} \Leftarrow \text{ rise} \\ \Leftarrow \text{ run} \end{array}$$

where m is the slope, and (x_1, y_1) and (x_2, y_2) are the coordinates of the two points. In the formula, we use the letter x to indicate the input values and the letter y for the output values. We use the subscripts 1 and 2 to differentiate between the points we are using to calculate the slope. The subscripts 1 and 2 have no *numerical* meaning — they are just used as labels to indicate two different pieces of data. In the formula, the numerator, $y_2 - y_1$, is the rise and the denominator, $x_2 - x_1$, is the run. In a data table or on a graph, the points (x_1, y_1) and (x_2, y_2) would be indicated as follows:

	x	y	
input 1 ⇒	x_1	y_1	⇐ output 1
input 2 ⇒	x_2	y_2	⇐ output 2

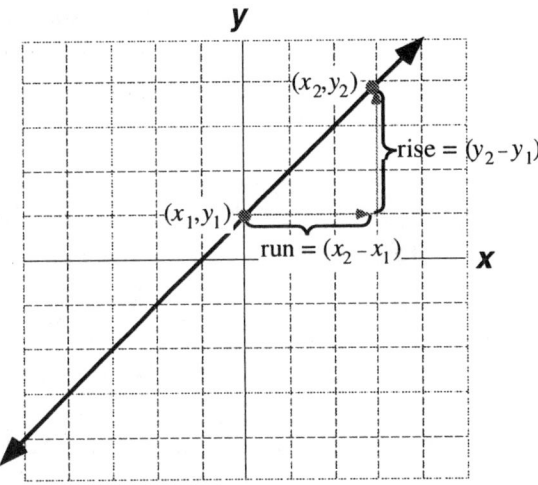

Example 4.4.4   a)   Calculate the slope of the straight line that passes through the points (2,0) and (4,–1).

•**SOLUTION**• In the formula, substitute (2,0) for (x_1, y_1) and (4,–1) for (x_2, y_2).

$$m = \frac{y_2 - y_1}{x_2 - x_1} \quad \Leftarrow \text{ Slope formula.}$$

$$m = \frac{-1 - 0}{4 - 2} \quad \Leftarrow \text{ Substituted values into the formula.}$$

$$m = \frac{-1}{2} \quad \Leftarrow \text{ Simplified.}$$

We can write the slope as $\frac{-1}{2}$ or $-\frac{1}{2}$, or –0.5.

b) Calculate the slope of the straight line that passes through the following points.

x	y
−4	−5
2	7
6	15

•SOLUTION• We can use any two points to calculate the slope. If we choose (−4,−5) as the first pair then x_1 is −4 and y_1 is −5. If we choose (2,7) as the second pair then x_2 is 2 and y_2 is 7.

$$m = \frac{y_2 - y_1}{x_2 - x_1} \quad \Leftarrow \text{ Slope formula.}$$

$$m = \frac{7 - (-5)}{2 - (-4)} \quad \Leftarrow \text{ Substituted values.}$$

$$m = \frac{12}{6} \quad \Leftarrow \text{ Simplified.}$$

$$m = \frac{2}{1} \quad \Leftarrow \text{ Wrote with denominator of 1.}$$

The slope is $\frac{2}{1}$ or 2.

Note that it makes no difference which points we use or in which order we use them. Below is the slope calculation done in three different ways:

Using (6,15) for (x_1,y_1) and (2,7) for (x_2,y_2)	Using (2,7) for (x_1,y_1) and (−4,−5) for (x_2,y_2)	Using (−4,−5) for (x_1,y_1) and (6,15) for (x_2,y_2)
$m = \dfrac{7 - 15}{2 - 6}$	$m = \dfrac{-5 - 7}{-4 - 2}$	$m = \dfrac{15 - (-5)}{6 - (-4)}$
$m = \dfrac{-8}{-4}$	$m = \dfrac{-12}{-6}$	$m = \dfrac{20}{10}$
$m = 2$	$m = 2$	$m = 2$

Practice 4.4.4 Calculate the slope of the line that passes through the given points.

1. (6,3) and (5,12)   2. (6,0) and (9,5)

3.

x	y
−4	−7
0	−5
2	−4

4.

x	y
−2	−50
1	−5
3	25

Topic 4.4 E — Using Slope With Units

Consider the following data table and graph, which illustrates how people in the U. S. changed their consumption of low-fat milk between 1970 and 1990. †

x time (number of years since 1970	y lowfat milk (gallons per person)
0	5.0
5	7.0
10	9.0
15	11.0
20	13.0

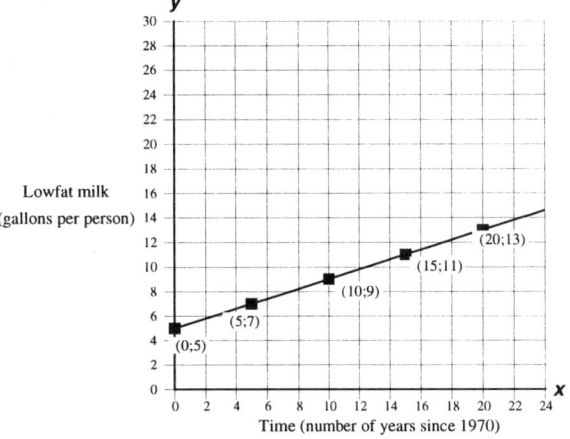

To find the slope of the line we choose any two points, say (0,5) and (10,9) and substitute their coordinates into the formula. (There is nothing special about these two points; any two points, in any order, will give the same slope.) We will use (0,5) for (x_1,y_1) and (10,9) for (x_2,y_2). We substitute these values into the slope formula and simplify. Notice how we keep the units in applied problems.

$$m = \frac{y_2 - y_1}{x_2 - x_1} \qquad \Leftarrow \text{Slope formula.}$$

$$m = \frac{9 \text{ gallons} - 5 \text{ gallons}}{10 \text{ years} - 0 \text{ years}} \qquad \Leftarrow \text{Substituted values into the formula.}$$

$$m = \frac{4 \text{ gallons}}{10 \text{ years}} \qquad \Leftarrow \text{Simplified.}$$

In real world problems its common to write the slope with a denominator of 1. Therefore, we will write the slope as $\frac{0.4 \text{ gallons}}{1 \text{ year}}$, or 0.4 gallons per year. This says that low fat milk consumption has increased 0.4 gallons per year. That is, the average person drank four tenths more gallons of low fat milk per year, every year between 1970 and 1990.

Example 4.4.5   a)   The following data show how whole milk consumption changed between 1970 and 1990. Calculate the slope of the graph and discuss its meaning.

x time (number of years since 1970	y whole milk (gallons per person)
0	25.0
5	21.5
10	18.0
15	14.5
20	11.0

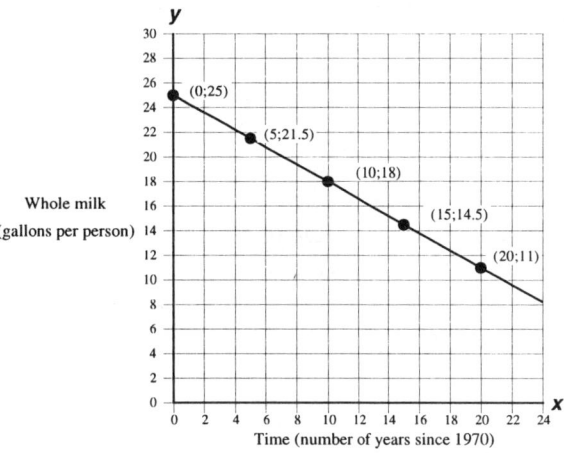

---

†   Source: <u>Statistical Abstract of the United States</u>, 1995, Table No. 220, Per Capita Consumption of Selected Beverages, by Type.

•SOLUTION• We will arbitrarily pick (0,25) for (x_1,y_1) and (20,11) for (x_2,y_2). Substituting into the formula and simplifying gives the following:

$$m = \frac{y_2 - y_1}{x_2 - x_1} \qquad \Leftarrow \text{ Slope formula.}$$

$$m = \frac{11 \text{ gal} - 25 \text{ gal}}{20 \text{ years} - 0 \text{ years}} \qquad \Leftarrow \text{ Substituted values ands units into the formula.}$$

$$m = \frac{-14 \text{ gal}}{20 \text{ years}} \qquad \Leftarrow \text{ Simplified.}$$

$$m = \frac{-0.7 \text{ gallons}}{1 \text{ year}} \qquad \Leftarrow \text{ Wrote with denominator of 1.}$$

The negative sign in the slope, $\frac{-0.7 \text{ gallons}}{1 \text{ year}}$, says that whole milk consumption *decreased* as time passed from 1970 to 1990. Each year, people drank 0.7 gallons *less* of whole milk than the previous year.

b) Gable Electronics found that when $4,000 was spent on advertising profits were $27,000. They found that profits climbed to $34,500 when $7,000 was spent on advertising but fell to $20,750 when the advertising budget was $1,500. Assume dollars spent on advertising is the input and profit is the output. Calculate the slope of the line that would connect the points and discuss the meaning of the slope.

•SOLUTION• We will organize information in a data table to make it easier to understand.

x (money spent on advertising in dollars)	4,000	7,000	1,500
y (profit earned in dollars)	27,000	34,500	20,750

We will arbitrarily pick (4000, 27000) for (x_1,y_1) and (7000,34500) for (x_2,y_2). Substituting into the formula and simplifying gives the following:

$$m = \frac{y_2 - y_1}{x_2 - x_1} \qquad \Leftarrow \text{ Slope formula.}$$

$$m = \frac{\$34500 \text{ profit} - \$27000 \text{ profit}}{\$7000 \text{ advertising} - \$4000 \text{ advertising}} \qquad \Leftarrow \text{ Substituted values and units into the formula.}$$

$$m = \frac{\$7500 \text{ profit}}{\$3000 \text{ advertising}} \qquad \Leftarrow \text{ Simplified.}$$

$$m = \frac{\$2.5 \text{ profit}}{\$1 \text{ advertising}} \qquad \Leftarrow \text{ Wrote with denominator 1.}$$

The implication here is that every dollar spent on advertising corresponds to an increase of $2.50 in profits.

Practice 4.4.5   Calculate the slope of the line that would connect the given points and discuss the meaning of the slope.

1.

x time (number of years since 1985)	y beef consumed per person (pounds)
0	74
3	68
6	62

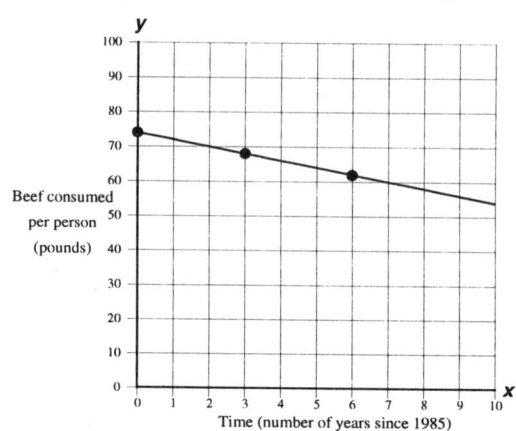

2. In 1980 there were about 1,044,000 structural fires in the U. S.. Six years later the number had fallen to about 848,400 and twelve years later to about 652,800. Assume years is the input and structural fires is the output.

3.

x (number of tickets sold)	20	50	80	115
y (profit earned in dollars)	–60	150	360	605

4. Eighty–four dollars was earned in annual interest when $700 was invested. When $12,800 was invested $1536 in annual interest was earned. The annual interest earnings dropped to $600 when $5000 was invested. Assume dollars invested is the input and interest earned is the output.

Topic 4.4 F — The Size and the Sign of the Slope

The slope of a line gives us an idea of both the "steepness" and the "uphill or downhill nature" of the line. For example, let's look again at the low fat milk and whole milk graphs and compare the slopes of the lines.

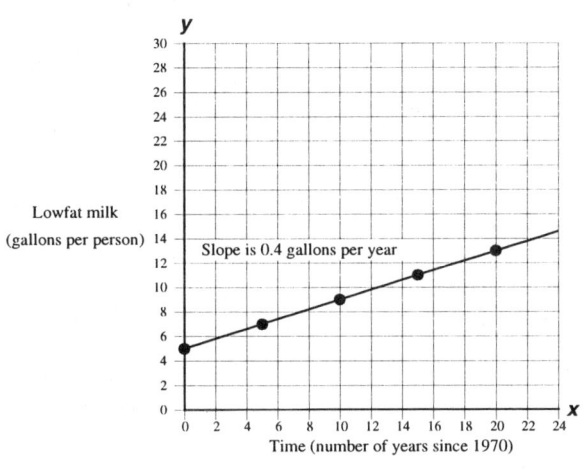

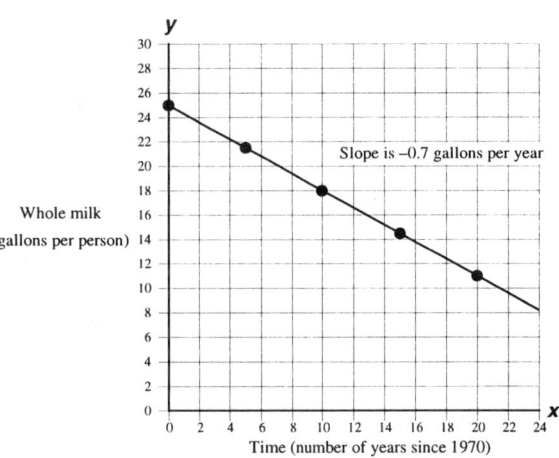

For each line, note the relation between the tilt and the calculated slope:

* The slope of the low fat milk line is 0.4 gallons per year and its graph runs *up* hill from left to right.

* The slope of the whole milk line is –0.7 gallons per year and its graph runs *down* hill from left to right.

— Note — *Working left to right,* *uphill is positive and* *downhill is negative*	• A straight line that runs uphill from left to right has positive slope. A positive slope means that output increases as input increases. • A straight line that runs downhill from left to right has negative slope. A negative slope means that output decreases as input increases. 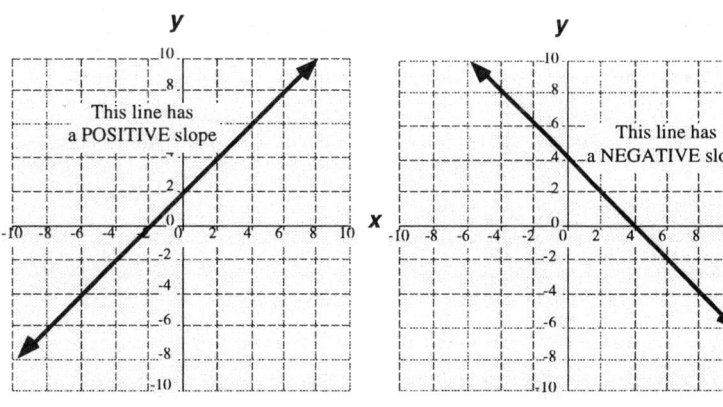

There is a also a relation between the steepness of a straight line and the absolute value of the slope. Look at the values of the slopes of the following lines:

Negative Slopes

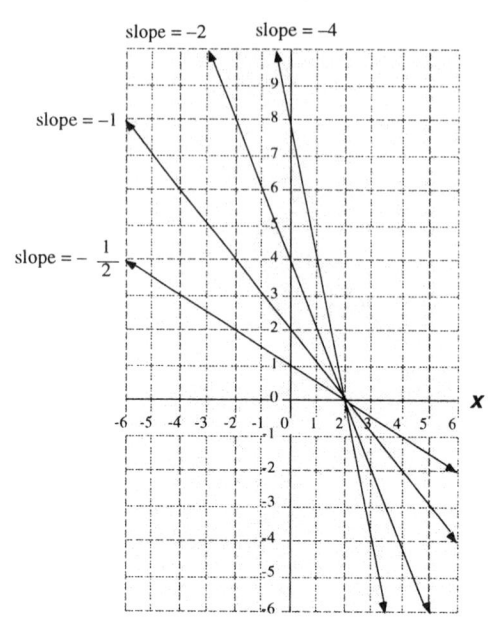

Positive Slopes

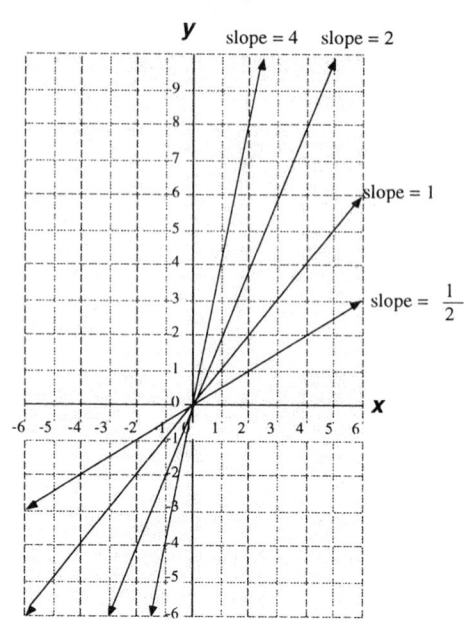

— Note —	The larger the absolute value of a slope, the steeper the straight line. That is, steeper straight lines show a greater change in output (y) for a given change in input (x).
Steepness of slopes	

Topic 4.4 G — Two Special Cases Involving Slope

Whenever you look at a general mathematical idea it's important to consider the special cases. With slope the case of an unchanging output (the y-values remain the same) and the case of an unchanging input (the x-values remain the same) are the special cases. Let's see what happens to m when the output or input does not change.

Below are data which show the relation between the number of U. S. Senators and the number of years since 1980.

x time (number of years since 1980)	y number of U. S. Senators
0	100
5	100
10	100
15	100

Number of U. S. Senators

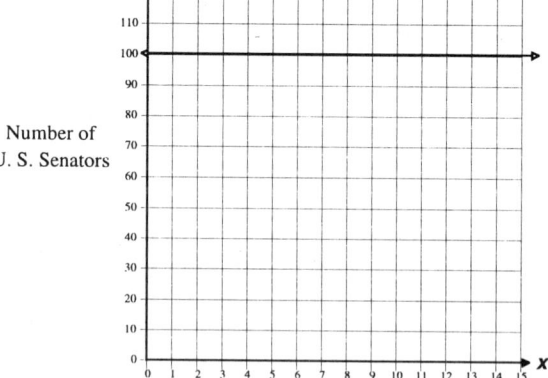

Time (number of years since 1980)

We will use (0,100) and (5,100) to calculate the slope:

$$m = \frac{100 \text{ senators} - 100 \text{ senators}}{5 \text{ years} - 0 \text{ years}} \qquad \Leftarrow \text{Substituted values.}$$

$$m = \frac{0 \text{ senators}}{5 \text{ years}} \qquad \Leftarrow \text{Simplified.}$$

$$m = \frac{0 \text{ senators}}{1 \text{ year}} \qquad \Leftarrow \text{Wrote with denominator of 1.}$$

When the input changes but the output does not, we say the line has zero slope (it's neither positive nor negative). This does not mean the output is zero; it means the *change* in the output is zero when the input changes. In other words, the number of U. S. Senators remained constant between 1980 and 1995.

Note that the graph is a horizontal line; again, this tells us that the output is constant while the input is changing.

The second situation occurs when the input doesn't change but the output does. Below are data that show the relation between the cost of a can of juice and the age of the purchaser at a convenience store.

x cost of juice (dollars)	y age of buyer (years)
1.05	19
1.05	16
1.05	54
1.05	31

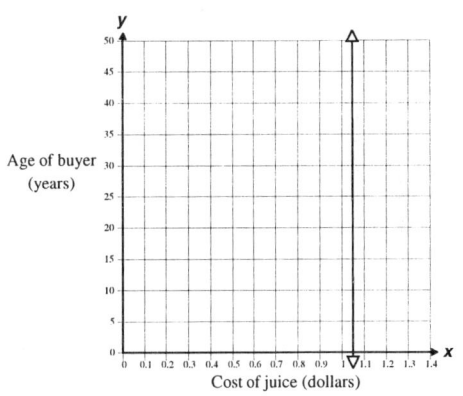

We will use (1.05,19) and (1.05,16) to calculate the slope:

$$m = \frac{19 \text{ years} - 16 \text{ years}}{1.05 \text{ dollars} - 1.05 \text{ dollars}} \qquad \Leftarrow \text{Substituted values.}$$

$$m = \frac{3 \text{ years}}{0 \text{ dollars}} \qquad \Leftarrow \text{Simplified.}$$

Since we cannot divide by 0 this slope is undefined. Notice that with an undefined slope we lose the ability to predict output from input and vice versa. For example if we were told that the juice cost $1.05 we would have no way of determining the age of the purchaser because the data does not show a relation between age and cost of juice.

| — Note —

Horizontal lines have 0
slope and vertical lines
have undefined slope | • The slope of a horizontal line is 0 because the y component of every point on the line is the same (that is, in going from one point on the line to another the change in y is 0).

• The slope of a vertical line is undefined because the x component of every point on the line is the same (that is, in going from one point on the line to another the change in x is 0).

Be careful with the vocabulary! Using everyday language, we might say that a horizontal line has "no slope" because it's completely flat. However, *no slope* and *0 slope* are two very different things. *No slope* means the slope is undefined (a vertical line) whereas *0 slope* means there is a 0 rise no matter what the run (a horizontal line). The point is that *0* and *no* do not always mean the same thing. [†] |

[†] No and zero don't always mean the same thing in English either. For example, if the number of dogs in the room is 0 we can correctly say that "there are no dogs in the room." However, if the temperature outside were 0 degrees, we would not say that there is "no temperature"!

Example 4.4.6   a)   Decide if the slope of a line defined by the following data table is 0 or undefined.

x	y
−1	−3
0	−3
8	−3

•SOLUTION• Since the output is always the same, the slope is 0.

b)   Decide if the slope of a line defined by the following data is 0 or undefined. Use units to describe the meaning of the slope

x *(height of a car buyer in inches)*	67	67	67
y *(cost of a new car in dollars)*	10,219	15,201	12,922

•SOLUTION• Since the input is always the same, the slope is undefined. In this case the undefined slope tells us that knowing that the height of a car buyer is 67 inches does not allow us to predict the amount he or she will pay for a new car.

Practice 4.4.6   Decide if the slopes of the lines described by the following data are 0 or undefined. If appropriate, use units and describe the meaning of the slope.

1.

x	10	20	30
y	157	157	157

2.

x *(number of stock shares purchased)*	100	100	100
y *(price per share (dollars)*	14.5	12	11.75

3.

x	y
40	143
40	263
40	383
40	508

4.

x calories consumed	y weight (pounds)
1200	163
1350	163
1175	163
1500	163

Topic 4.4 H — Slopes of Parallel and Perpendicular Lines

Parallel straight lines are lines that never intersect. For example,

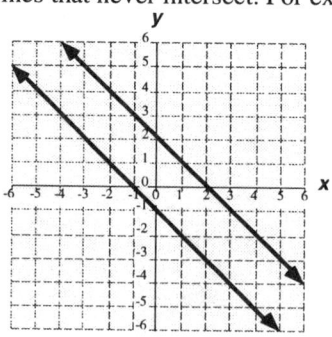

Using the above picture we can see that both lines have a slope of −1 since they increase 1 in the x direction while decreasing 1 in the y direction. Because both lines both have a slope of −1 they are parallel.

| — Definition —

Slopes of parallel lines | English: Parallel lines have the same slope.

Example: The lines shown in the following graph have the same slope and therefore they are parallel.

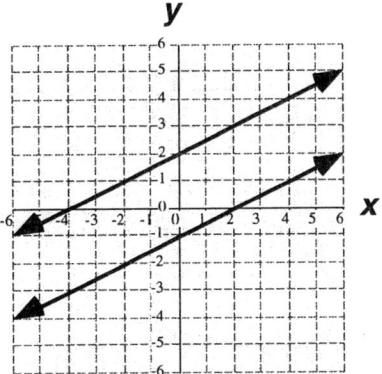

Algebra: Two straight lines with slopes m_1 and m_2 are parallel if $m_1 = m_2$. |

Perpendicular lines are lines that cross at right angles (90 degrees). To find the relation between the slopes of perpendicular lines, let's begin with the line shown below.

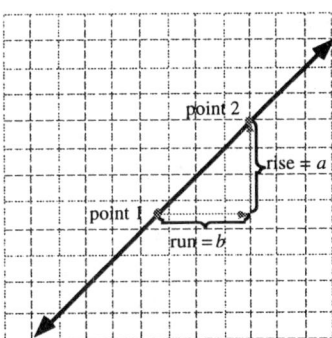

From the figure, we can see that the slope of this line is $\dfrac{a}{b}$ because to go from point 1 to point 2 we run b units and rise a units. If we rotate the line 90 degrees we will create a new line that is perpendicular to the original line.

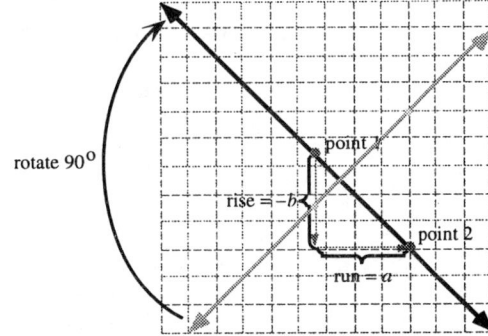

The slope of the new line is $\dfrac{-b}{a}$ because to go from point 1 to point 2 we rise (actually drop) $-b$ units and run a units. The rise of the original line has become the run of the new line; the run of the original line has become the rise of the new line, but with the opposite sign. The slopes of the two perpendicular lines are $\dfrac{a}{b}$ and $\dfrac{-b}{a}$, which are negative reciprocals of each other.

— Definition —	English: The slopes of perpendicular lines are negative reciprocals. As a result, the product of the slopes of perpendicular lines is always −1.
Slopes of perpendicular *lines*	Examples:

- If one line has a slope of $\frac{2}{3}$ then the slope of a perpendicular line is $-\frac{3}{2}$.

- If one line has a slope of −2 then the slope of a perpendicular line is $\frac{1}{2}$.

Algebra: We can state the condition on the slopes of two perpendicular line in two different ways:

- If one non-vertical straight line has a slope of m, then the slope of a straight line perpendicular to the given line is $-\frac{1}{m}$.

OR

- Two straight lines with slopes m_1 and m_2 are perpendicular if $m_1 m_2 = -1$.

Example 4.4.7　a)　Given that *Line 1* passes through the points (2,4) and (1,5) and *Line 2* passes through (−3,2) and (−7,−2). Are these lines parallel, perpendicular, or neither?

•SOLUTION•　We calculate the slope of each line:

$$m_1 = \frac{5 - 4}{1 - 2}, \text{ which reduces to } -1$$

$$m_2 = \frac{-2 - 2}{-7 - (-3)}, \text{ which reduces to } 1$$

Since the slopes of the lines are negative reciprocals, the lines are perpendicular. OR, since the product of the slopes is −1, the lines are perpendicular.

b)　Given the following data for two lines:

Line 1

x	y
5	−2
−5	−7

Line 2

x	y
1	−3
5	−7

Are these lines parallel, perpendicular, or neither?

•SOLUTION•　We calculate the slope of each line:

$$m_1 = \frac{-7 - (-2)}{-5 - 5}, \text{ which reduces to } \frac{1}{2}$$

$$m_2 = \frac{-7 - (-3)}{5 - 1}, \text{ which reduces to } -1$$

The slopes are not the same, so the lines are not parallel. The slopes are not negative reciprocals, so the lines are not perpendicular. These lines are neither parallel nor perpendicular.

Practice 4.4.7　Decide whether the two given straight lines are parallel, perpendicular or neither.

1. *Line 1* passes through the points (0,7) and (−1,4) and *Line 2* passes through the points (0,1) and (−1,−2).

2. *Line 1* passes through the points (4,2) and (0,3) and *Line 2* passes through the points (−1,−2) and (0,2).

3. *Line 1* comes from the data table　　　　　*Line 2* comes from the data table

x	y
2	4
4	8

x	y
2	−3
4	1

4. *Line 1* comes from the data table

x	0	1
y	1	0

Line 2 comes from the data table

x	1	0
y	0	1

Exercise Set 4.4   The answers to the odd numbered exercises are at the back of the book.

Construct the equation of the line shown in the graph.

1.

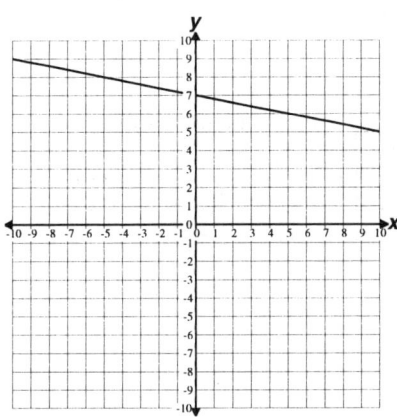

2.

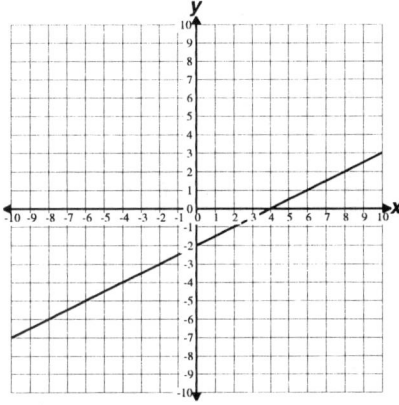

3.

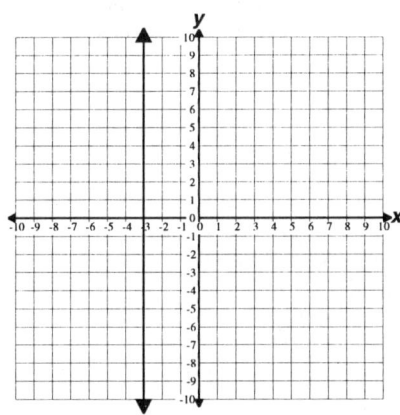

4.

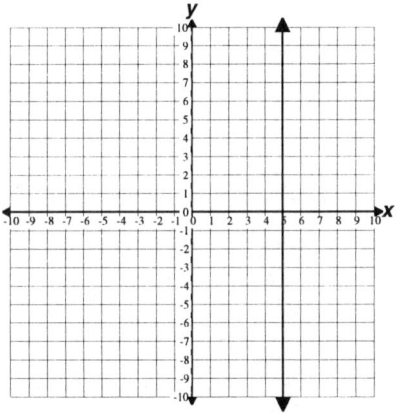

5.

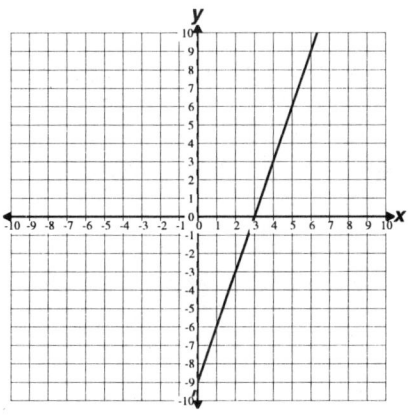

6.

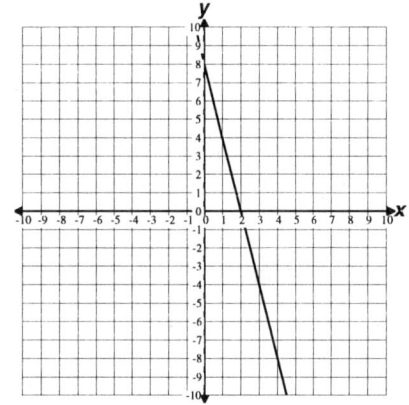

7.

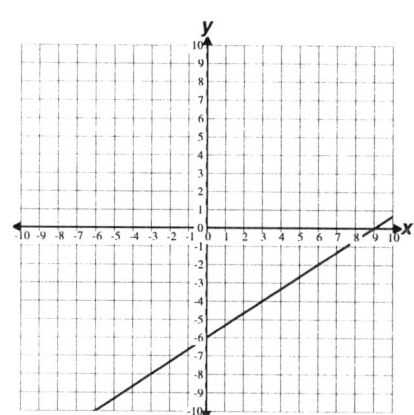

8.

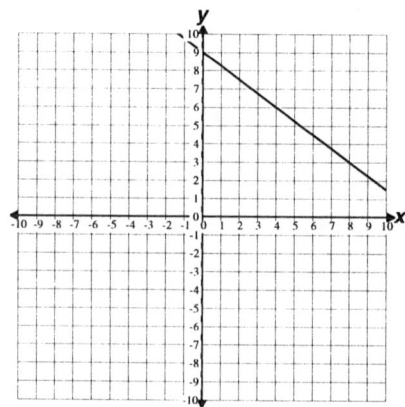

9.

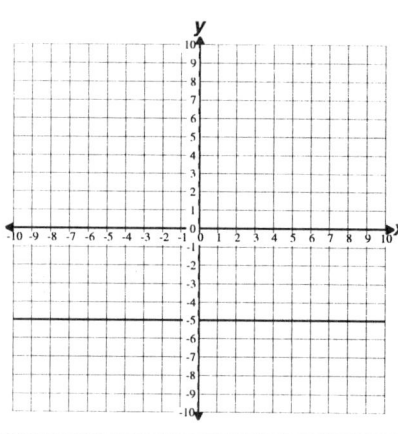

10.

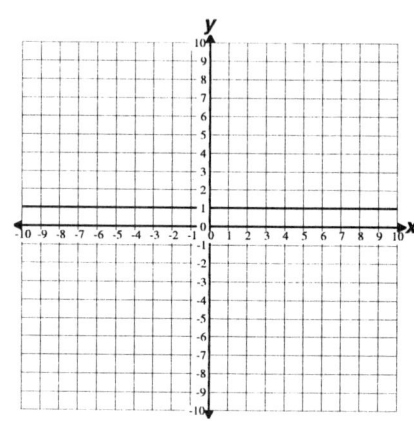

Discuss the meaning of each graph.

11.

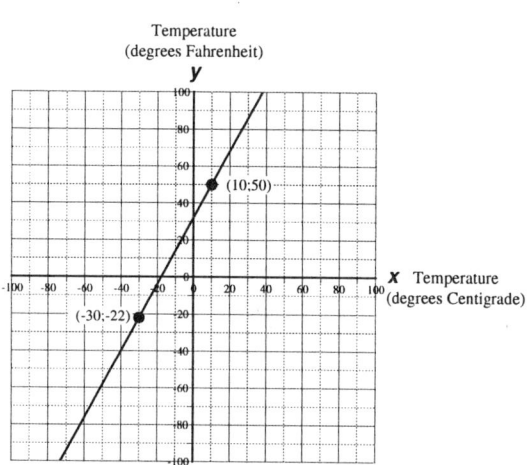

12.

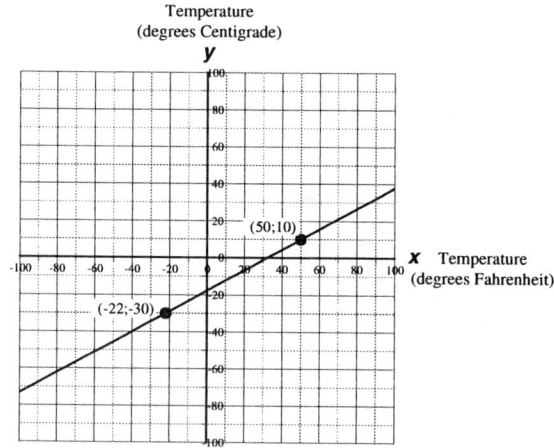

13.
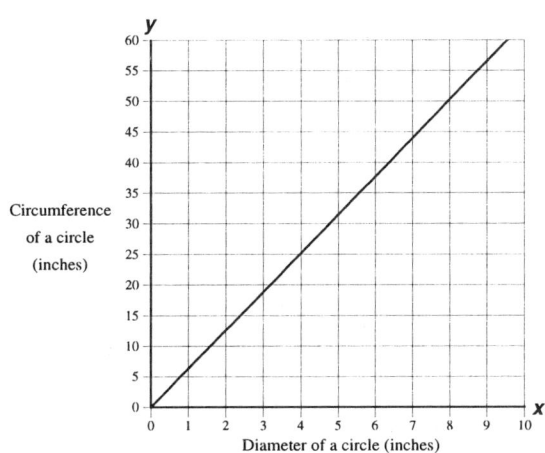

14.
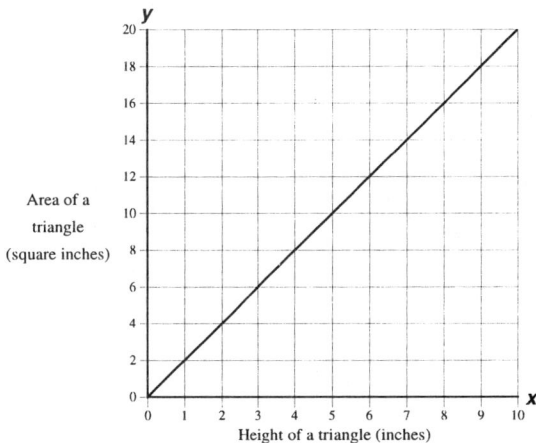

15.
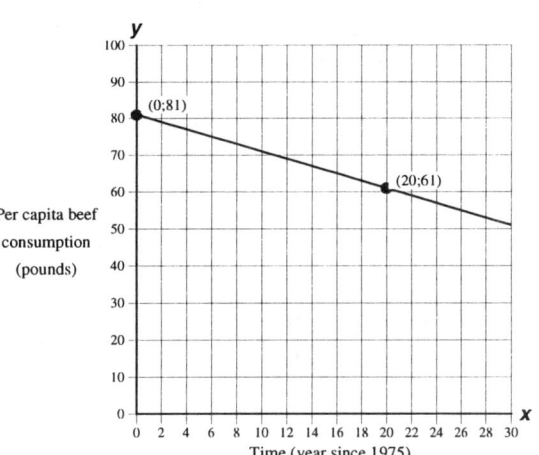

16.
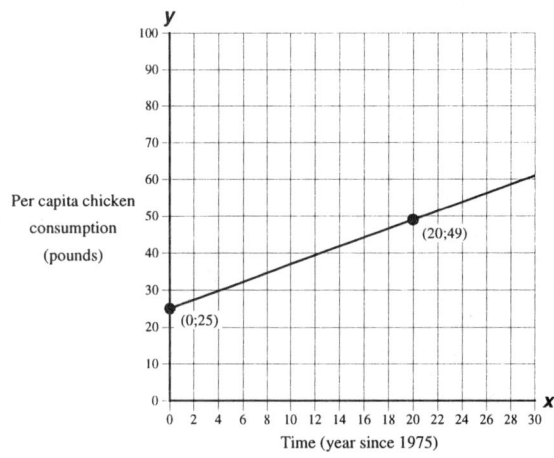

17.
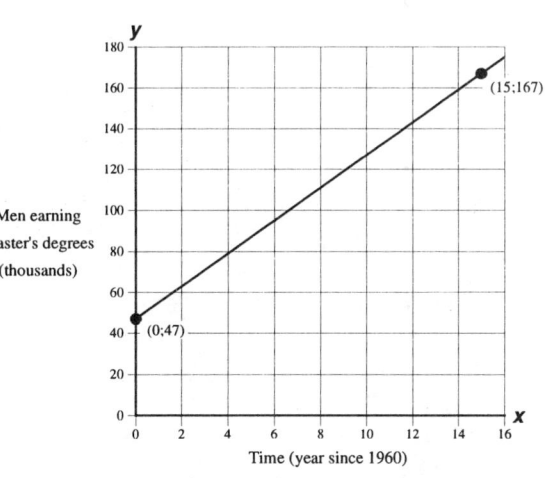

18.
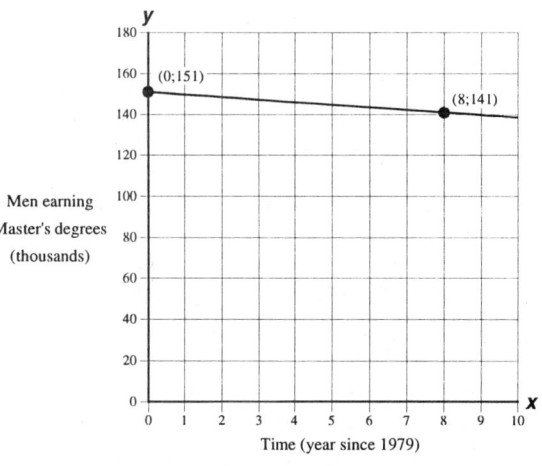

Graph the straight line that passes through the given point and which has the given slope.

19. (6,5) with a slope of $\frac{1}{4}$

20. (1,7) with a slope of –1

21. (–4,–2) with a slope of $\frac{-1}{2}$

22. (–5,1) with a slope of $-\frac{2}{5}$

23. (1,3) with a slope of 0

24. (0,0) with a slope of 3

---

25. Graph the straight line described by the following situation. Assume month since starting the program is the input and pounds lost is the output. "She began the weight loss program at 150 pounds. Since beginning she has lost two pounds a month for the last 10 months."

26. Graph the straight line described by the following situation. Assume month since starting the weight training is the input and pounds lifted is the output. "She began the weight training program bench pressing 45 pounds. Since beginning she has increased the weight she can press by 8 pounds a month."

27. Graph the straight line described by the following situation. Assume miles from camp is the output and minutes since the dust storm was spotted is the input. "The dust storm is 25 miles from camp and is moving towards camp at half a mile a minute."

28. Graph the straight line described by the following situation. Assume miles from the road is the output and minutes since the fire jumped the firebreak is the input. "The forest fire is 18 miles from the road but has just jumped the break and is heading towards the road at a quarter mile every minute."

Calculate the slope.

29.

x	−1	0	1
y	−6	2	10

30.

x	5	7	9
y	−21	−29	−37

31. (8,7) and (10,3).   32. (−3,5) and (−1,9).   33. (0,−2) and (3,5).   34. (4,8) and (7,13).

35.

x	y
4	1
0	2
−4	3

36.

x	y
−5	−5
0	−7
10	−11

Calculate the slope and discuss it's meaning. Make sure to include the units.

37.

x (gallons of solution)	28	37.5	90	98
y (gallons of acid)	6.16	8.25	19.8	21.56

38.

x (gallons of solution)	1	5	10	15
y (gallons of acid)	0.8	4.0	8.0	12.0

39.

x time since investment (years)	y value of investment (dollars)
1	7400
2	6800
3	6200
4	5600
5	5000

40.

x time since purchased (years)	y value of condominium (dollars)
2	44200
4	43400
7	42200
12	40200
15	39000

41.

x dry treatment creme (ounces)	y price (dollars)
0.75	10.35
1.20	16.56
8.00	110.40

42.

x wrinkle reducer (ounces)	y price (dollars)
0.75	22.05
1.20	35.28
8.00	235.20

43. In 1970 the average hourly earnings for retail workers was $2.60. Hourly earnings increased to $3.65 in 1975 and were $5.75 by 1985. Assume the number of years since 1970 is the input and average hourly earnings is the output.

44. In 1970 the average hourly earnings for construction workers was $5.60. Hourly earnings increased to $7.65 in 1975 and were $11.75 by 1985. Assume the number of years since 1970 is the input and average hourly earnings is the output.

45. In 1964 the winning Olympic time for the men's 100 meter freestyle was about 53.46 seconds. In 1984 it was about 50.06 seconds and in 1992 it was about 48.70 seconds. Assume the year is the input and the winning time in seconds is the output.

46. In 1960 the winning Olympic time for the women's 100 meter freestyle was about 61.8 seconds. In 1968 it was about 59.8 seconds and in 1984 it was about 55.8 seconds. Assume the year is the input and the winning time in seconds is the output.

47.

x (time before road construction (hours))	2	3	5
y (distance before road construction (miles))	126	189	315

48.

x (time during road construction (hours)	0.75	1.25	1.50
y (distance during road construction (miles))	20.25	33.75	55.5

Without using the formula, decide if the slopes of the following data are 0 or undefined. If appropriate, use units and describe the meaning of the slope.

49.

x day of the week (number since Sunday)	y pay earned (dollars)
1	128
3	128
5	128

50.

x time since work began (hours)	y pay per hour (dollars)
0	6.50
3	6.50
8	6.50

51.

x	2	8	12
y	5	5	5

52.

x	–5	–2	6
y	–4	–4	–4

53.

x (cost to park a car in dollars)	25	25	25
y (grade point average)	3.1	2.6	3.85

54.

x (cost of vehicle in dollars)	15,599	15,599	15,599
y (size of gas tank in gallons)	12	16	13

55.

x	y
–2	–1
–2	0
–2	1

56.

x	y
15	–5
15	5
15	15

57.

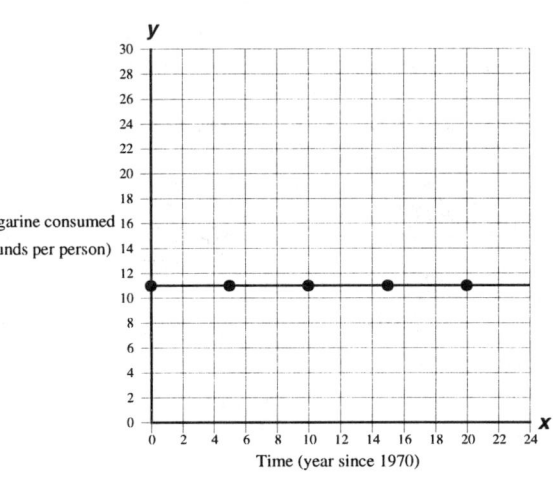

58.

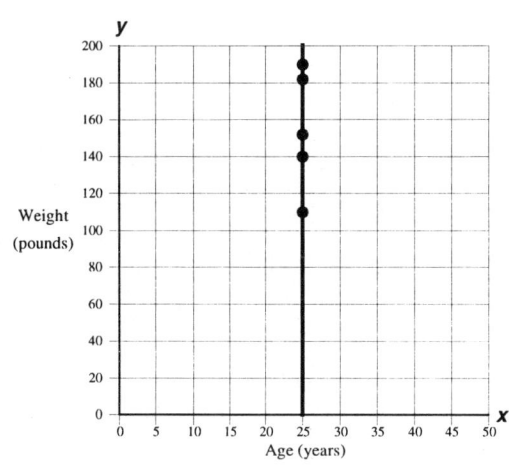

Decide whether the two given straight lines are parallel, perpendicular or neither.

59. *Line 1* passes through the points (–2,1) and (2,13) and *Line 2* passes through the points (–2,–5) and (1,4).

60. *Line 1* passes through the points (1,10) and (–1,4) and *Line 2* passes through the points (0,2) and (3,1).

61. *Line 1* comes from the data table

x	y
2	7
-3	-8

Line 2 comes from the data table

x	y
-3	3
6	0

62. *Line 1* comes from the data table

x	2	-3	1
y	7	-8	4

Line 2 comes from the data table

x	-3	6	0
y	3	0	2

63.

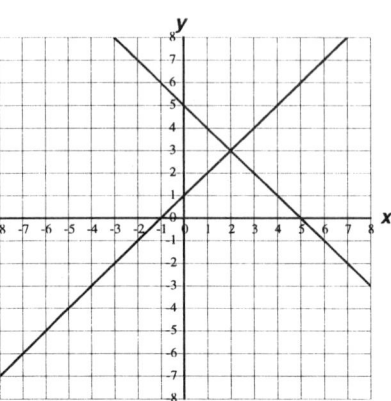

64.

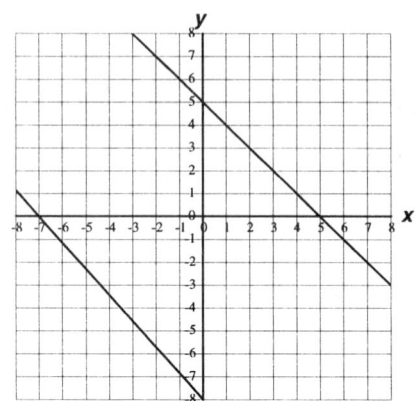

Review Exercises   The answers to all of these exercises are at the back of the book.

65. Simplify: $\dfrac{9}{15} \div \dfrac{12}{25}$

66. Simplify: $\dfrac{1}{18} + \dfrac{5}{12}$

67. Simplify: $8x^2 - 3x + 5x^2 - x$

68. Solve: $-\dfrac{x}{6} = \dfrac{2}{3}x$

69. Graph: $-2 \le x$

70. Solve: $-x \le 5$

71. A child care center enrolls infants and toddlers. If t represents the number of toddlers write a simplified expression for the total number of children if:

   a) there are 3 more infants than toddlers.
   b) the number of toddlers is 10 less than the number of infants.
   c) there are half as many infants as toddlers.

72. After making a down payment of one–third the purchase price for a new flat screen TV a women still owes $760. How much did the TV originally cost?

73. A math class decides that if 80% or more of the students would like to include the average of their four regular tests as their final test grade then the instructor will agree. If 28 students decide to accept the idea and this exactly meets the 80% cut off, how many students are in the class?

Section 4.5 Other Forms of the Linear Equation

Topic 4.5 A — Slope-Intercept Form of a Straight Line

We have seen that the graph of the linear equation $y = b + mx$ is a straight line where m is the slope and b is the y-intercept. If the linear equation is written with the term containing slope first and the y-intercept second, as in $y = mx + b$, then we say the equation is in **slope-intercept form** .

Example 4.5.1   a)   Write the equation of a line with slope –5 and *y*-intercept (0,3). Use slope-intercept form.

•SOLUTION• $y = -5x + 3$.

b)   Find the slope and *y*-intercept of the line $y = 3x - 2$.

•SOLUTION• We write the equation in slope-intercept for to get $y = 3x + (-2)$. The slope is 3 and the *y*-intercept is (0,–2).

e)   Find the slope and *y*-intercept of the line $x + y = 3$.

•SOLUTION• We put the equation into $y = mx + b$ form by solving for *y* to get $y = -x + 3$. Since $-x$ means $-1x$, the slope is –1; the *y*-intercept is (0,3).

Practice 4.5.1

1.   Write the equation of a line with slope –6 and *y*-intercept (0,2). Use slope-intercept form.

2.   Find the slope and *y*-intercept of the line $y = x - 7$.   3.   Find the slope and *y*-intercept of the line $y = 8 - x$.

4.   Find the slope and *y*-intercept of the line $2x + y = 5$.   5.   Find the slope and *y*-intercept of the line $8 - 6x + 2y = 0$.

Topic 4.5 B — Building the Slope-Intercept Form of a Line Given Two Ordered Pairs

Hanna believes that there is a relation between the number of snow blowers she sells and the amount of snow that falls during the Winter. She found that she sold 29 snow blowers when 48 inches of snow fell last year, but only 21 show blowers when 36 inches fell the year before. She would like to develop an equation that she can use to predict her sales given the predicted snow fall from the National Weather Service. Let's organize the information in a data table and graph it:

x total snow fall (inches)	*y* snow blowers sold
48	29
36	21

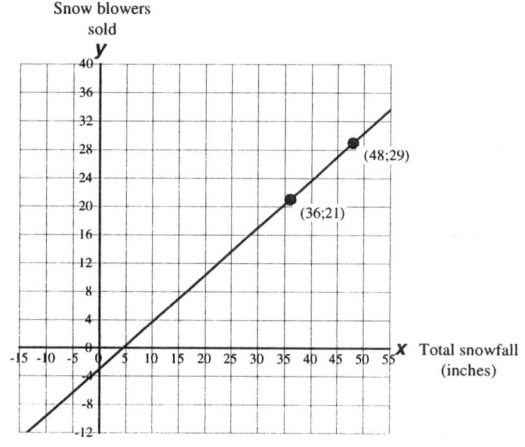

To construct the linear equation $y = mx + b$ we must find the slope of the line and the *y*-intercept. To find the slope, we use the slope formula as follows:

$$m = \frac{29 - 21}{48 - 36}, \text{ which simplifies to } \frac{2}{3}.$$

To find the *y*-intercept we can supply three of the unknowns in the slope-intercept form of the linear equation and then solve for b. We know that *m* is $\frac{2}{3}$, and we can use one of the data points, say (36,21) for *x* and *y*.

$$y = mx + b \qquad \Leftarrow \text{ Slope-intercept form.}$$

$$21 = \frac{2}{3}(36) + b \qquad \Leftarrow \text{ Substituted values for } y, m, \text{ and } x.$$

$$-3 = b \qquad \Leftarrow \text{ Solved for } b.$$

The equation is $y = \frac{2}{3}x - 3$. Hanna can use this equation to predict the number of snow blowers she will sell given the predicted inches of snow for next Winter.

In the above example, we used two data points to construct the equation of a straight line that passed through the points. We can generalize this procedure as follows:

— *Procedure* — **Building a linear equation from two ordered pairs**	***Step 1 Slope*** Use the slope formula to calculate the slope. ***Step 2 y-intercept*** In the slope-intercept form of the linear equation, $y = mx + b$, • substitute the slope for m. • substitute for x and y the coordinates of one point. • solve for b. ***Step 3 Equation*** Write the equation using the values of m and b found above.

Example 4.5.2   Find the equation of the line that goes through the given points. Write the equation in slope-intercept form.

a)

x	−4	−2
y	3	−4

•SOLUTION• The two points are (−4,3) and (−2,−4)

Step 1 Slope

$$m = \frac{-4 - 3}{-2 - (-4)}, \text{ which simplifies to } \frac{-7}{2}$$

Step 2 y-intercept   We use the point (−4,3) in the slope-intercept formula.

$y = mx + b$ ⟸ Slope-intercept form.

$3 = \frac{-7}{2}(-4) + b$ ⟸ Substituted values for y, m, and x.

$-11 = b$ ⟸ Solved for b.

Step 3 Equation

$y = \frac{-7}{2}x + (-11)$ ⟸ Substituted values for m and b.

$y = \frac{-7}{2}x - 11$ ⟸ Simplified.

Practice 4.5.2   Find the equation of the line that goes through the given points. Write the equation in slope-intercept form.

1. (−3,13) and (1,5)

2.

x	−1	0	1
y	−6	2	10

3.

x	y
−6	40
−2	32
4	20

Topic 4.5 C — Equations of Horizontal and Vertical Lines

In Section 4.4 you learned that horizontal lines have a slope of 0. Since every non-vertical straight line has the form $y = mx + b$ the equation of a horizontal line is $y = (0)x + b$, which simplifies to $y = b$. Notice this says the output is always the same as the y-intercept no matter what the input.

— Definition — ***Equation of a horizontal line***	English: The equation of a horizontal straight line contains only the output variable and the y-intercept. It does not contain the input variable. Example: $y = 3$ Algebra: If y is the input variable and b represents the y-intercept, then the equation of a horizontal straight line is $y = b$

In Section 4.4 you learned that a vertical straight line has undefined slope. In addition, unless the line coincides with the y-axis, a vertical straight line has no y-intercept. If we use these ideas in $y = mx + b$ we get

$$y = (\text{undefined})x + (\text{does not exist}).$$

Even though the above equation contains an undefined quantity and a number that does not exist, vertical lines do exist. A characteristic of vertical straight lines is that they always have the same input no matter what the output. This suggests that the input variable, x, is always a constant.

— Definition — ***Equation of a vertical line***	English: The equation of a vertical straight line contains only the input variable and the x-intercept. It does not contain the output variable. Example: The equation $x = 3$ models the line in the following graph: 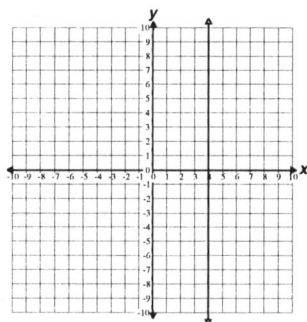 Algebra: If x is the input variable and k represents the x-intercept, then the equation of a vertical straight line is $x = k$

Topic 4.5 D — Point-Slope Form of a Straight Line

Another common form for the linear equation is called the **point-slope form** . We can develop this form from the formula for the slope of a line between (x_1, y_1), a *specific* point, and (x, y), a *general* point on the line.

$$m = \frac{y - y_1}{x - x_1} \qquad \Leftarrow \text{ Slope formula.}$$

$$(x - x_1)m = (x - x_1)\frac{y - y_1}{x - x_1} \qquad \Leftarrow \text{ Multiplied both sides by the LCD, } (x - x_1).$$

$$(x - x_1)m = y - y_1 \qquad \Leftarrow \text{ Reduced expression on the right by the factor } (x - x_1).$$

By tradition, we write the terms with y on the left so the formula is $y - y_1 = m(x - x_1)$.

— *Definition* —
Point-slope form of a line
	English: In a linear model, the difference of two outputs is equal to the product of the slope and the difference of the two corresponding inputs.
	
	Example: The line that passes through (2,5) with slope 3 can be represented by
	
	$$y - 5 = 3(x - 2).$$
	
	Algebra: The equation of a straight line can be written as $y - y_1 = m(x - x_1)$
	
	where
	    m represents the slope of the line.
	    (x,y) represents a general point on the line.
	    (x_1,y_1) represents a specific point on the line.

Example 4.5.3

a) Use the point-slope form of a line to find the equation of the line that goes through (3,5) and has slope –2.

•**SOLUTION**•

$$y - y_1 = m(x - x_1) \quad \Leftarrow \text{ Point-slope form.}$$

$$y - 5 = -2(x - 3) \quad \Leftarrow \text{ Substituted 5 for } y_1, 3 \text{ for } x_1, \text{ and } -2 \text{ for } m.$$

$$y - 5 = -2x + 6 \quad \Leftarrow \text{ Removed parentheses.}$$

$$y = -2x + 11 \quad \Leftarrow \text{ Added 5 to both sides.}$$

b) Use the point-slope form to find the equation of the line that goes through (1,4) and (–3,–6).

•**SOLUTION**• First, we calculate the slope.

$$m = \frac{-6 - 4}{-3 - 1}, \text{ which simplifies to } \frac{5}{2}$$

Next, we substitute into the formula the slope and either given point.

$$y - y_1 = m(x - x_1) \quad \Leftarrow \text{ Point-slope form.}$$

$$y - 4 = \frac{5}{2}(x - 1) \quad \Leftarrow \text{ Chose (1,4) for the point and substituted 4 for } y_1,$$
$$\qquad\qquad\qquad\qquad\qquad 1 \text{ for } x_1, \text{ and } \frac{5}{2} \text{ for } m.$$

$$y - 4 = \frac{5}{2}x - \frac{5}{2} \quad \Leftarrow \text{ Removed parentheses.}$$

$$y = \frac{5}{2}x + \frac{3}{2} \quad \Leftarrow \text{ Added 4 (which is } \frac{8}{2}) \text{ to both sides.}$$

c) Find the equation of a line perpendicular to $y = -\frac{1}{2}x - 3$ and which passes through (4,0).

•**SOLUTION**• Perpendicular lines have slopes that are negative reciprocals. The slope of the given line is $-\frac{1}{2}$ so the slope of the line we want is 2.

$$y - y_1 = m(x - x_1) \quad \Leftarrow \text{ Point-slope form.}$$

$$y - 0 = 2(x - 4) \quad \Leftarrow \text{ Substituted 0 for } y_1, 4 \text{ for } x_1, \text{ and 2 for } m.$$

$$y = 2x - 8 \quad \Leftarrow \text{ Simplified.}$$

Practice 4.5.3

1. Use the point-slope form to find the equation of the line that goes through (3,–5) and which has slope –3.

2. Find the equation of the line perpendicular to the line $y = 3x + 2$ and which goes through the point (–6,1).

3. Find the equation of a line parallel to $x + y = 3$ and which passes through (2,5)

4. Use the point-slope form to find the equation of the line that passes through the given points.

x	–6	8	–1
y	–10	32	5

5. Use the point-slope form to find the equation of the line that goes through (3,6) and (–3,2).

Topic 4.5 E — Standard Form of a Straight Line

All of the forms of a straight line we have discussed may be written in a general way, which we call the **standard form of a linear equation in two variables**.

— *Definition* — *Standard form of a linear equation in two variables*	English: A linear equation in two variables is in standard form when it is written as the product of a constant and the input added to the product of a constant and the output and this sum is set equal to a constant.
	Example: The equation $3x + 7y = 2$ is written in standard form
	Algebra: The equation of a straight line can be written in standard form as
	$Ax + By = C$
	where
	A, B, and C are real numbers and A and B are not both 0.
	x is a variable which represents the input.
	y is a variable which represents the output.

Note the following about $Ax + By = C$:

- If we solve for y we have $y = -\dfrac{A}{B}x + \dfrac{C}{B}$. If we think of $-\dfrac{A}{B}$ as m and $\dfrac{C}{B}$ as b then this has the form $y = mx + b$, which is the slope-intercept form of a non-vertical straight line.

- If A is 0 we have $By = C$. Solving for y we have $y = \dfrac{C}{B}$. This has the form $y = b$, which is the equation of a horizontal line.

- If B is 0 we have $Ax = C$. Solving for x we have $x = \dfrac{C}{A}$. This has the form $x = k$, which is the equation of a vertical line.

We will use standard form in the next chapter when we discuss systems of equations.

Topic 4.5 F — Solving Problems Using the Linear Equation

During the last few sections we have used the linear equation to model many different situations. Here is some practice using the linear model to solve problems.

Example 4.5.4 A parking ramp opens for business at 7:00 a.m. Below is the record of the number of cars that entered the low and the time since the lot opened.

x time since lot opened (minutes)	y number of cars in the lot
15	143
30	263
45	383
60	508

a) Build the linear equation for this situation.

•SOLUTION• Using the data in the table we find that the slope is 8 and the y-intercept is (0,23). Therefore, the linear equation is $y = 8x + 23$.

b) Discuss the meaning of the equation.

•SOLUTION• The equation implies that 23 cars were already in the lot when it opened and that after opening 8 cars per minute entered the lot.

c) How many cars will be in the lot at 8:10 a.m.

•SOLUTION• In this question we are given a time (input) and asked to find the number of cars (output). We will substitute 70 for x and solve for y to get 583. We expect that at 8:10 a.m. there will be 583 cars in the lot.

d) If the parking lot holds a total of 780 cars, after how many minutes will the lot be full? Assume no one leaves the lot.

•SOLUTION• This question gives us an output (number of cars) and asks for an input (time). We substitute 780 for y and solve for x to get 94.625. For this situation, it probably does not make sense to report the answer to the nearest thousandth of a minute. Rounding to the nearest minute gives 95 minutes. But, if we plug 95 back into the equation we get 783 cars in the lot, which is not possible because the lot only holds 780 cars. If we just drop the decimal part of the answer and use 94 minutes the equation predicts 775 cars will be in the lot. This is possible, but it means we still have 5 spots open. If we report the answer as 94.5 minutes (94 minutes and 30 seconds) the model predicts there will be 779 cars in the lot, which is very close to full. As in other rounding situations there is no "mathematical" way to decide what is the best answer. So, here are some *reasonable* answers:

- The lot will fill 94 minutes 37.5 seconds after it opens.
- The lot will fill about 94 and a half minutes after opening.
- The lot will be full 95 minutes after opening.

Practice 4.5.4 For the following problems build the linear equation, discuss it's meaning in terms of m and b, and then use the linear equation to answer the questions given.

1. The following data concern the hourly earnings for a typical beginning worker in the retail industry.

   a) Build the linear equation for this situation.
   b) Discuss the meaning of the equation.
   c) Predict the hourly earnings in 2010.
   d) When will service workers make $20.00 an hour?

x number of years since 1970	y retail worker wages (dollars per hour)
5	3.57
10	4.67
20	6.87

2. The following graph shows how average hourly wages for construction workers has changed over the past 30 years.

   a) Build the linear equation for this situation.
   b) Discuss the meaning of the equation.
   c) Predict the hourly earnings in 2010.
   d) When will construction workers make $20.00 an hour?

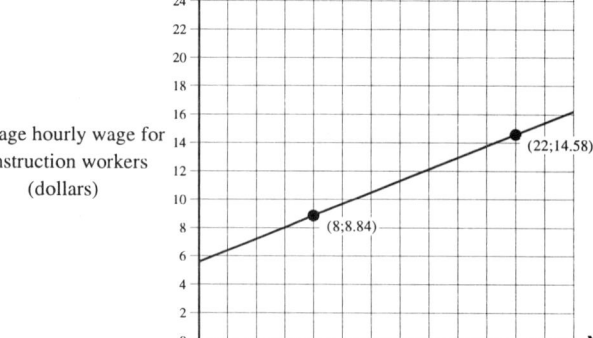

3. In 1970 the average hourly earnings, in current dollars, for service workers was $2.38. In 1985 hourly earnings had risen to $7.78 and by 1990 to $9.58. Assume number of years since 1970 is the input and average hourly earnings is the output.

   a) Build the linear equation for this situation.
   b) Discuss the meaning of the equation.
   c) Predict the hourly earnings in 2010.
   d) When will service workers make $20.00 an hour?

Exercise Set 4.5   The answers to the odd numbered exercises are at the back of the book.

Write the equation of a straight line with the given characteristics. Use slope-intercept form.

1. slope -1 and $y-$ intercept $(0,-4)$
2. slope 1 and y-intercept $(0,8)$

3. slope 0 and y-intercept $(0,0)$
4. slope is 0 and y-intercept $(0,1)$

5. slope $\dfrac{3}{7}$ and $y-$ intercept $(0,-1)$
6. slope $\dfrac{-3}{2}$ and y-intercept $(0,11)$

7. slope undefined and line passes through $(-2,-3)$
8. slope 0 and line passes through the point $(-5,-7)$

9. slope 0 and line passes through the point $(6,-3)$
10. slope undefined and line passes through $(4,-1)$

11. slope -1 and line passes through the point $(0,-1)$
12. slope is -2 and line passes through the point $(0,0)$

13. slope $\dfrac{-4}{9}$ and line passes through the point $(0,0)$
14. slope is $\dfrac{5}{2}$ and line passes through the point $(0,6)$

Find the slope and y-intercept of the line.

15. $y = 9$
16. $x = 2$
17. $y = 5 - x$
18. $y = -\dfrac{1}{2}x - 2$
19. $y = \dfrac{1}{3}x - 10$

20. $y = 2x$
21. $x = 0$
22. $y = -x$
23. $1 - 3y = 12x$
24. $-2 - y = 2x$

25. $-2x = 4y - 1$
26. $-x = -2y + 5$
27. $2x + 2y - 4 = 0$
28. $3x - 3y + 6 = 0$

Find the equation of the line that goes through the given points. Write the equation in slope-intercept form.

29.

x	y
-3	-24
1	-12
5	0

30.

x	y
-2	-7
0	3
2	13

31.

x	6	8	10
y	9	10	11

32.

x	-3	0	6
y	-7	-6	-4

33. $(-10,24), (-5,14), (-2,8), (2,0)$

34. $(-1,3), (3,-13), (6,-25), (12,-50)$

35.

x	-1	4	5
y	1	1	1

36.

x	-5	-3	1
y	-2	-2	-2

37.

x	y
-4	-32
-3	-24
1	8
4	32

38.

x	y
-2	2
-1	1
0	0
1	-1

39.

x	7	7	7
y	1	2	3

40.

x	-1	-1	-1
y	-4	2	6

41. (5,6.25), (10,6.35), (15,6.45)

42. (5,3.15), (10,3.40), (15,3.65)

43.

x	y
–2	1109.2
5	881
9	750.6

44.

x	y
–2	89705
5	97314
9	101662

45.

x	–15	–10	–5
y	8.57	8.02	7.47

46.

x	5	6	11
y	5.17	4.98	4.03

Use the point-slope form to find the equation of the line with the given characteristics.

47. Line passes through (–1,–2) and has slope 4.

48. Line passes through (–3,3) and has slope –2.

49.

x	–2	0	2
y	–1	–5	–9

50.

x	–1	1	2
y	14	2	–4

51.

x	y
6	5
3	3
–9	–5

52.

x	y
–8	–5
4	–2
8	–1

53.

x	y
–6	159.1
–3	155.2
–1	152.6

54.

x	y
–4	118.8
–2	104
6	44.8

55.

x	21	16	22
y	1.25	1.25	1.25

56.

x	–8	–1	15
y	6.1	6.1	6.1

57. Line passes through (–2,2) and is parallel to $y = 5x$

58. Line passes through (8,1) and is parallel to $x + y = \dfrac{3}{2}$

59. Line passes through (0,0) and is perpendicular to $x – 2y = 4$

60. Line passes through (0,5) and is perpendicular to $y = –x + 1$

61. The following graph shows the relation between the cost of leasing a car and the time of the lease.

a) Build the linear equation for this situation.
b) Discuss the meaning of the equation in terms of the down payment and the cost per month
c) The car costs $16,750 if purchased. How long would it take for the leasing costs to equal the purchase price?
d) How much would it cost to lease the car for 3 years?

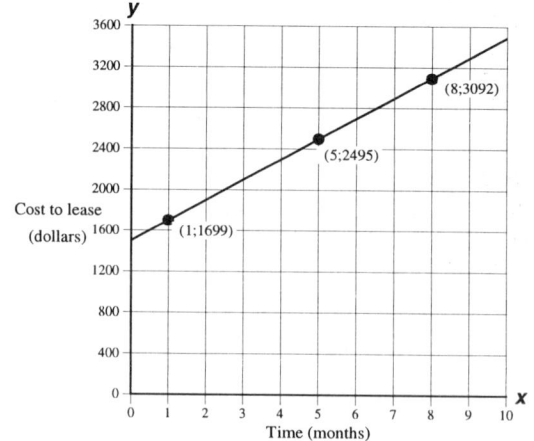

62. This graph shows the relation between the current value of a computer and the time elapsed since it was purchased.

   a) Build the linear equation for this situation.
   b) Discuss the linear equation in terms of the purchase price and the depreciation † per month.
   c) When will the value of the computer be $0?
   d) What will be the value of the computer after 3 years?

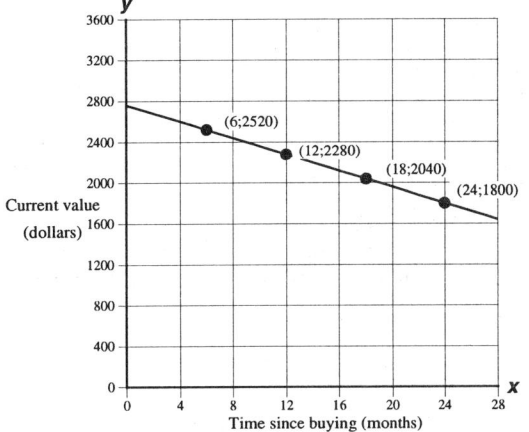

63. The data below show the relation between carbon monoxide emissions in the U. S. and the years between 1945 and 1970.

x (number of years since 1940)	5	15	25	30
y (millions of tons of carbon monoxide)	97.5	108.5	119.5	125.0

   a) Build the linear equation for this situation.
   b) Discuss the meaning of the linear equation.
   c) Before 1940 no records were kept by the U. S. government. Predict the millions of tons released in 1935.
   d) If the trend were to continue when would emissions have exceeded 150 million tons?
   e) If the trend were to continue predict when emissions would have doubled from their 1940 level?

64. The following data show the relation between carbon monoxide emissions in the U. S. and the years between 1975 and 1995.

x (number of years since 1970)	5	10	15	25
y (millions of tons of carbon monoxide)	121.5	115.0	108.5	95.5

   a) Build the linear equation for this situation.
   b) Discuss the meaning of the linear equation.
   c) The U. S. government began gathering data on carbon monoxide in 1940. In 1940 about 92 million tons of carbon monoxide was released. What year does the model for this problem predict emission levels returned to the 1940 level?
   d) When does the model predict carbon monoxide emissions will drop to 0?
   e) How many tons of carbon monoxide do you predict will be released in 2005.

65. Three students who attend a college are discussing their fees after registering for their next semester. They know that part of the cost is for the student service fee and that this fee is the same for every student who registers. The rest of the individual cost pays for classes and is based solely on the number of credits they registered for. The first student registered for 3 credits and was charged $296. The second student registered for 15 credits and was charged $920 dollars. The third student registered for 12 credits and paid $764. Assume credits taken is the input and dollars charged is the output.

   a) Build the linear equation for this situation.
   b) Discuss the meaning of the linear equation.
   c) What is the cost of registering for 8 credits?
   d) If a student had $1000 to spend, how many credits could they register for?

---

† Depreciation represents the idea that as most products age they lose economic value. For example, a computer becomes less valuable over time due to wear and tear and because better computers come into the market.

66. A couple is comparing their electric bills for the last three months. They know that part of the bill is a charge for being attached to the electrical grid and that this fee does not change from month to month. The rest of the bill is the cost for the actual electricity they use. Three months ago they used 1015 kilowatt hours of electricity and paid $103.33. Two months ago they used 646 kilowatt hours and paid $70.12. Last month they used 1840 kilowatt hours and paid $177.58. Assume kilowatt hours of electricity used is the input and monthly cost is the output.

   a) Build the linear equation for this situation.
   b) Discuss the meaning of the linear equation.
   c) If the bill for the current month is $184.42 how many kilowatt hours of electricity did they use?
   d) The month the homeowner used 646 kilowatt hours was September. Assume that a year later everything stayed the same during September except that the homeowners had switched from an electric hot water heater to a gas hot water heater. If they now used 563 kilowatt hours of electricity, how much was the electric hot water costing per month?

67. From 1970 to 1993 the percent of incoming college freshman who were female was approximately linear. Use the data below to answer the following questions.

x *time* *(number of years since 1970)*	*y* *female college students* *(percent of incoming students)*
0	46
5	48
10	50

   a) Build the linear equation for this situation.
   b) Discuss the meaning of the linear equation.
   c) In what year will 60% of the incoming college freshman be female?
   d) What percent of incoming college freshman were female in 1990?
   e) Although it is hazardous to predict so far into the past, when does the model say no incoming freshman were female?
   f) Although it is hazardous to predict so far into the future, when does the model predict all incoming freshman will be female?

68. From 1970 to 1993 the percent of incoming college freshman who were male was approximately linear. Use the data below to answer the following questions.

x *time* *(number of years since 1970)*	*y* *male college students* *(percent of incoming students)*
0	54
5	52
10	50

   a) Build the linear equation for this situation.
   b) Discuss the meaning of the linear equation.
   c) In what year were 60% of the incoming freshman male?
   d) What percent of incoming freshman were male in 1990?
   e) Although it is hazardous to predict so far into the future when does the model say no incoming freshman will be male?
   f) Although it is hazardous to predict so far into the past when does the model predict all incoming freshman were male?

69. A company found that when they spent $4000 on advertising profits were $27,000. The profits climbed to $34,500 when $7000 was spent but fell to $20,750 when the advertising budget was $1500.

   a) Determine if these data can be reasonably modeled using a linear equation.
   b) Build the linear equation for this situation.
   c) Predict the profit if the company spends $5,500 on advertising.
   d) Predict the amount the company should spend on advertising if it wants profit to be $40,000.

70. A grocery store found that when they allocated 2.5 linear feet of shelf space to a cereal monthly sales of that cereal were $8750. When 3.25 linear feet was allocated monthly sales were $11,375. The figures for two other months are sales of $9625 and $5,250 for space of 2.75 and 1.5 linear feet respectively.

    a) Determine if these data can be reasonably modeled using a linear equation.
    b) Build the linear equation for this situation.
    c) How much shelf space should be allocated to produce sales of $14,000?
    d) How many dollars in sales would be produced by 5 linear feet of shelf space?

71. Research has determined that as you age it may be better to moderate your average heart rate when exercising. The graph below shows the relation between target heart rate and age.

    a) Build the linear equation for this graph.
    b) Estimate the target heart rate for someone who is 28 years old.
    c) Estimate the target heart rate for someone who is 80 years old.
    d) If a doctor tells a patient to have an average target heart rate of 130 beats per minute while exercising, what do you estimate the patient's age to be?

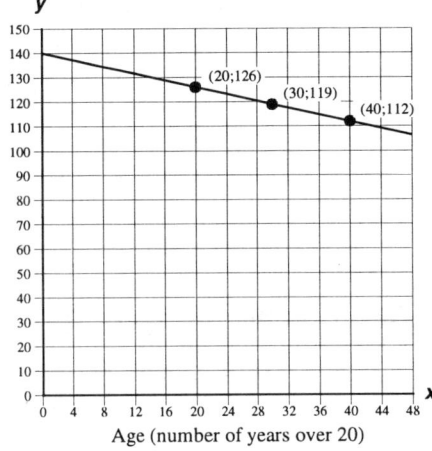

Heart rate (beats per minute)

Age (number of years over 20)

72. The dietary guidelines endorsed by the U. S. Department of Agriculture suggest that only a certain percentage of your daily calories should come from fat. The graph below shows the suggested relation between calories from fat and total calorie intake.

    a) Build the linear equation for this graph.
    b) Estimate the suggested calories from fat for a person whose total daily calorie intake is 2,000 calories
    c) If a person ingests 500 calories from fat how many total calories from all sources should they ingest?

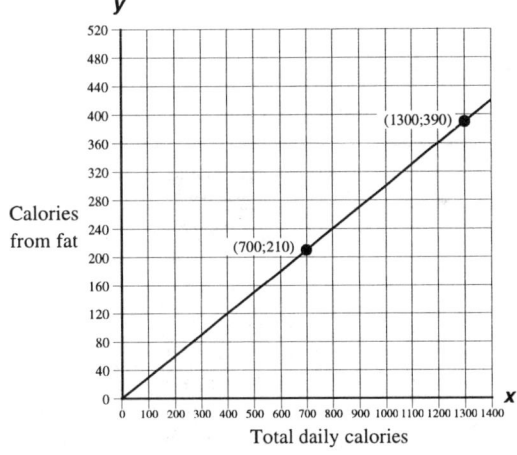

Calories from fat

Total daily calories

Review Exercises The answers to all of these exercises are at the back of the book.

73. Simplify: $8 - \left| -24 \right| \div 6 + 5 \bullet \left| -2 \right|$

74. Simplify: $12x - (8x - x)$

75. Simplify: 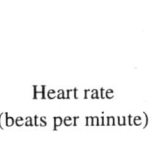 $18 - \dfrac{66 \bullet \frac{1}{3}}{\frac{3}{2} + 4}$

76. Solve: $8 - 3(x + 4) = 2x - 5(x - 1) - 9$

77. Graph: $-2 \le x$

78. Solve: $10 - 3(x - 2) < 5x$

79. Write a mathematical model for this sentence: 4 is the minimum value of the variable.

80. A man begins the 254.5 mile trip home after promising to make it there by 5 pm. After traveling at 40 mph for 1 hour he looks at the time, increases his speed by 26 mph, and makes it home exactly at 5 pm. What time did he begin his trip?

Section 4.6 Linear Regression

This Chapter is about creating mathematical models that can be used to represent real world relations. Typically, models are created by collecting data and then building an equation or a graph that can be used to help us better understand the data and to make predictions.

Unfortunately, when we graph data collected from real world observations we rarely get a perfectly straight line. There are two major reasons for this:

- First, there may be no relation between the input and the output, or, if there is, it may not be linear.

- Second, measurements in the real world are never exact. So, even if the relation between input and output exists and really is perfectly linear, the data we collect will not fall exactly on a straight line because we cannot measure or plot points perfectly.

However, even though the graph of a relation may not be exactly a straight line it may be close enough so that we are comfortable treating it as if it were perfectly straight. People have developed the idea of linear regression to help deal with such situations.

Topic 4.6 A — Linear Regression

Linear regression is a mathematical procedure that helps us determine how close a relation is to being linear and it gives us methods for calculating the values of b and m in the linear equation $y = b + mx$. In this book we will not discuss the details of how linear regression works, but we will use it to help us solve problems. [†]

For example, consider the following data table and graph which show the number of structural fires in the U. S. between 1980 and 1994.

x time (number of years since 1980)	y structural fires (thousands)
0	1065
1	1028
2	945
3	869
4	848
5	860
6	800
7	758
8	747
9	688
10	624
11	640
12	638
13	621
14	614

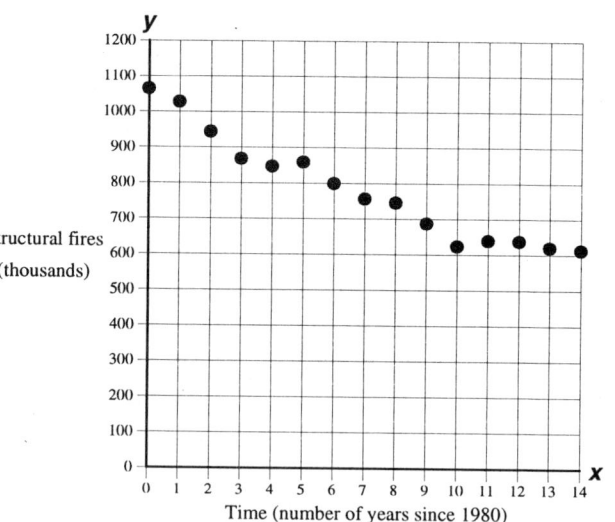

We can see that the plotted points do not fall exactly on a straight line but there does seem to be a general downward trend in the number of fires as time goes on. If we decide to model the data using a straight line then the question becomes, which of the many possible straight lines we could draw through the points would be the "best" line?

[†] The details of linear regression are typically covered in introductory statistics courses.

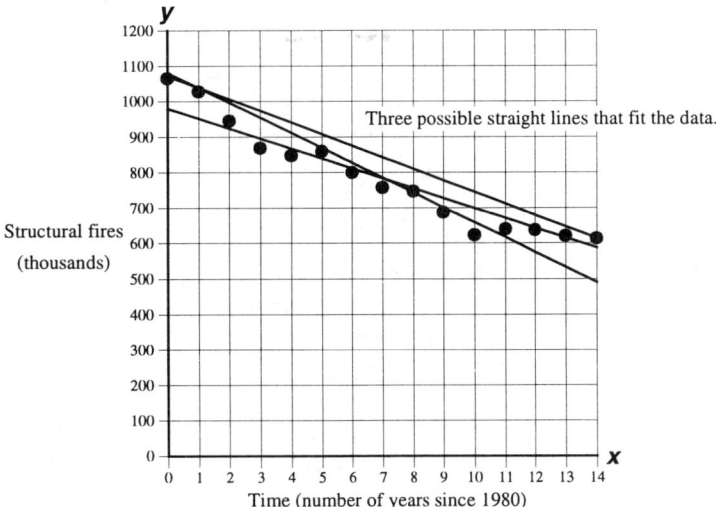

Three possible straight lines that fit the data.

Linear regression gives us a procedure for determining the straight line that best fits the data. This line is called the **regression line**.

Because the calculations involved with linear regression are tedious, a computer or graphing calculator program is used to determine the values of m and b in the linear equation $y = b + mx$. When we run the program using the structural fires data the computer generates the following values: $m = -32$ and $b = 1011$. Therefore, the "best" linear model for the data given is $y = 1011 - 32x$. We can graph this line on the same coordinate system as the points to get the following:

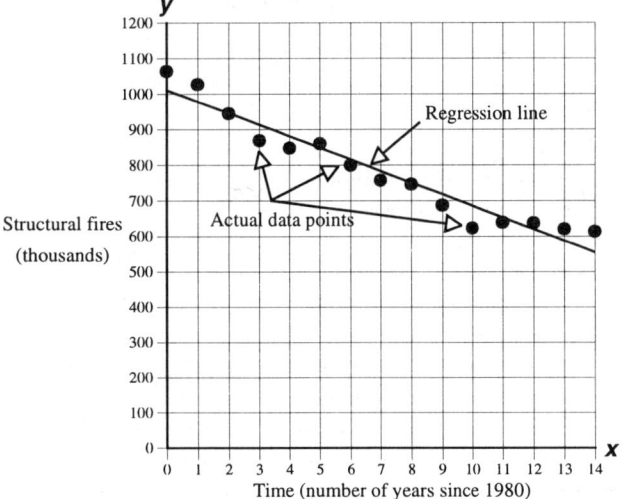

Regression line

Actual data points

Notice that most of the actual data points do not fall on the line, but the line does seem to be a good approximation to the trend shown by the data. There are other mathematical models that could be used to describe the data, but we are choosing a linear model because it looks good and it's easy to work with. [†]

[†] We get a better "fit" if we use higher powers of x. For example, the model $y = 1.7x^2 + 5.7x + 1063$ more accurately describes the data. But, this model is more complicated to work with. If we are willing to use a model that has x^{15} as its highest power we can get an exact fit to the data. However, such a model is very difficult to work with and it still might not exactly represent the real relationship because of measurement error or the fact that an exact relationship may not really exist.

Let's do some examples that generate linear equations from real data sets.

Example 4.6.1   The data table and graph below show the millions of tons of waste generated annually by everyone in the United States between 1960 and 1994. The linear regression equation for these data is $y = 82 + 3.6x$.

x time (number of years since 1960)	y waste (millions of tons)
0	87.8
5	103.4
10	121.9
15	128.0
20	151.5
25	164.4
26	170.7
27	178.1
28	184.2
29	191.4
30	198.0
31	196.8
32	203.0
33	206.5
34	209.0

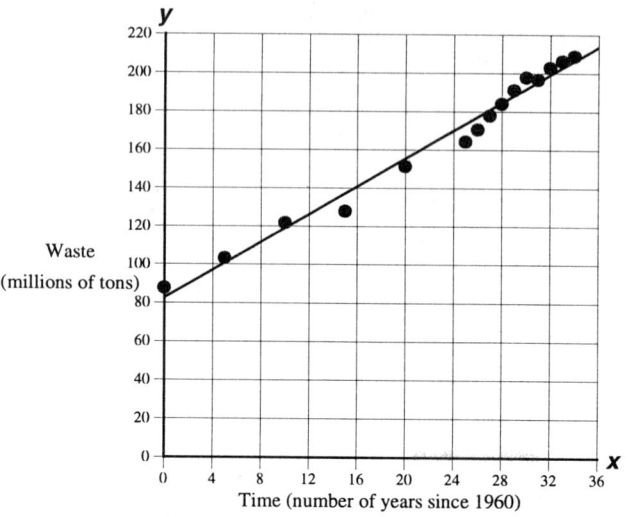

a)   Discuss the meaning of the linear equation.

•SOLUTION• The regression line crosses the y-axis at 82, which means that in 1960, 82 million tons of waste was generated in the United States. The slope of the regression line is 3.6, which means that the amount of waste generated increased by about 3.6 million tons each year between 1960 and 1996.

b)   Estimate the waste generated in 1950.

•SOLUTION• Because 1960 is year 0, 1950 is year –10. After substituting –10 for x and simplifying we get 46 million tons of waste.

c)   Estimate the waste that will be generated in 2010.

•SOLUTION• 2010 is year 50. After substituting 50 for x and simplifying we get 262 million tons of waste.

c)   When will the amount of waste exceed 250 million tons a year?

•SOLUTION• After substituting 250 for y and solving for x we get 47. Since 1960 is year 0, year 47 is 1960 + 47, which is 2007.

This next example brings up an important point about real models. Sometimes the model does not make sense for certain values of input and/or output.

Example 4.6.2 Due to city road construction, the access by customers to a convenience store will become more difficult. Before the construction begins, the store owner decides to see if there is a relation between the number of vehicles that pass the store during business hours and the number of people who make a purchase in the store. The owner collects the data in the following table and uses his computer to generate the linear regression model: $y = -18 + 0.02x$

x vehicles passing store	y paying customers
10615	190
9015	160
12031	225
11001	200
14140	260

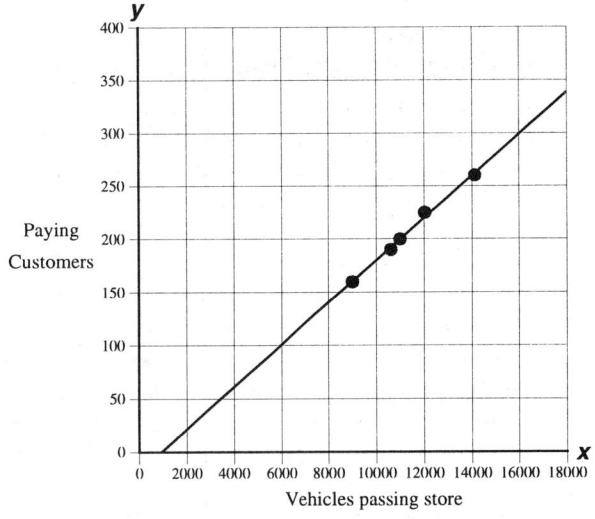

a) Interpret the meaning of the linear equation.

• SOLUTION • The constant, −18, suggests that when no cars pass the store the owner has negative 18 paying customers, which makes no sense because we cannot have a negative number of customers. It's apparent that this model is not meaningful if the number of vehicles is small. The rate of change, 0.02 or $\frac{2 \text{ customers}}{100 \text{ vehicles}}$, says there are 2 paying customers for every 100 vehicles that pass the store.

b) If 13,000 cars pass the store how many paying customers would you expect?

• SOLUTION • When we substitute 13,000 for x we get 242 for y.

c) The day after construction begins and access is limited, 9,778 cars pass the store and 169 paying customers use the store. Do you think the owner is losing business due to the restriction in access?

• SOLUTION • If we substitute 9,778 for x we obtain 178 (rounded) for y. This is the number of customers we would expect under normal circumstances. The difference between what we expect (178) and the actual number (169) is 9 customers. There is a drop, but it does not seem to be large enough to definitely say that the limited access is causing a loss of business. [†]

d) A week after access is limited, 148 paying customers use the store. If the number of vehicles to pass the store is 8,605, do you think the owner is losing business due to the change in access?

• SOLUTION • Before construction we would have expected 8300 cars to pass the store if 148 paying customers used the store (we find this by substituting 148 for x and then solving for y). The difference between 8,605 cars and 8300 cars is 305 cars. At first glance, this may seem significant. On the other hand the model predicts that if 8,605 cars had passed the store before construction 154 paying customers should have used the store. In fact 148 did. This is a difference of 6 customers, even smaller than before. We probably would not feel the business has been affected.

[†] There are mathematical procedures that can be used to tell us if this drop is a real drop or just due to chance. A statistics course would describe this in detail.

Practice 4.6.2

1.

The data table shows the enrollment in United States public elementary and secondary schools over a 124 year period. The regression equation is $y = 0.3x + 6.7$.

a) Interpret the meaning of the equation.
b) Estimate when enrollment will exceed 50 million.
c) Estimate enrollment in the year 2000.
d) What does the model say the change in enrollment was between 1870 and 1900?

x Number of years since 1869	y Enrollment in secondary and elementary schools (millions)
0	7.562
20	12.723
40	17.814
60	25.678
80	25.112
100	45.550
120	40.543
124	43.465

2.

The data table shows the average number of public school days from 1869 to 1959. The regression equation is $y = x + 72$

a) Interpret the meaning of the equation.
b) In Japan the average number of school days is 192. If the trend in the United States had continued beyond 1959 when would the United States have reached 192 days?
c) When does the model predict that the number of school days will be double the amount in 1872? Compare this value to the value from the table.
d) Compare the predicted rise from 1919 to 1949 to the actual rise given in the table.

x Number of years since 1869	y Average number of days attended per pupil
0	78.4
10	81.1
20	86.3
30	99.0
40	113.0
50	121.2
60	143.0
70	151.7
80	157.9
90	160.2

3. The phrase *comparable worth* describes the idea that jobs can be ranked according to factors such as work conditions, supervisor roles, responsibility, skills, etc. and that salaries can be set comparably across job descriptions. The table shows the evaluation points and salary for a number of jobs. The regression line for these data is $y = 12431.71 + 3.50x$.

a) Interpret the meaning of the equation.
b) The job description, "Institutional Attendant" earned 600 points. What would be an appropriate salary?
c) A clerk typist is making $14,900. Do you consider this in line with other jobs that earned 340 points?
d) As the result of a re-evaluation, a job is awarded 95 more points. What is a fair increase in salary per year?

x Evaluation (points)	y Yearly salary (dollars)	Job title
970	15704	Electrician
500	13984	Semiskilled laborer
370	14196	Motor equipment operator
220	13380	Janitor
250	13153	Laborer
1350	18472	Senior engineering technician
470	14193	Senior janitor
2040	20642	Revenue agent
370	13614	Engineering aide
1200	16869	Electrician supervisor
1865	17341	Registered nurse
1065	15194	Licensed practical nurse
490	12958	Senior clerk typist
220	13844	Cook's helper
805	15559	Eligibility technician
940	13894	Principal clerk stenographer

Topic 4.6 B — Using the Regression Line to Build a Linear Equation

Sometimes you may be called upon to interpret and use a graph that shows the line of best fit, but you may not be given the equation. In cases like this the best approach is to construct the equation using two points that fall on the line and which are far apart from each other. We usually do not use actual measured points; rather, we use points that fall exactly on the best fit line.

Let's look again at the graph of structural fires since 1980 and select a pair of points whose coordinates we can easily identify. We will choose (0.5,1000) and (12.5,600). [†]

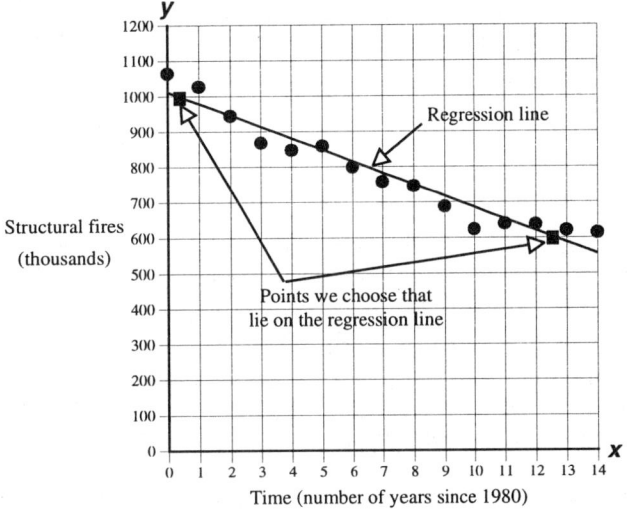

We can use these points to build the linear equation. To find the value of m we use the slope formula as follows:

$$m = \frac{600 - 1000}{12.5 - 0.5}, \text{ which simplifies to } -33$$

To find b (the y-intercept) we substitute -33 for m and one of the points for x and y. We will use the point (0.5,1000):

$$y = b - 33x \qquad \Leftarrow \qquad \text{Linear equation.}$$
$$\mathbf{1000} = b - 33(\mathbf{0.5}) \Leftarrow \qquad \text{Substituted values for } x \text{ and } y.$$
$$b = 1017 \qquad \Leftarrow \qquad \text{Simplified and rounded.}$$

The equation we obtain is $y = 1017 - 33x$. This corresponds well to the equation for the line of best fit generated by the computer, which was $y = 1011 - 33x$. The equations are not exactly the same because we had to estimate the coordinates of the points we used to do the calculations.

After finding the linear equation, it's a good idea to select a third point and see if the model predicts correctly. For example, from the graph we see that an input of 14 corresponds to an output of a little over 600. Using the equation $y = 1017 - 33x$ we get 609 with an input of 14. This suggests that the equation is a good model.

Of course, once we have this linear equation we would be able to use all the tools we developed earlier to make predictions and solve problems.

[†]   Remember, since we are reading points from a graph we are only getting approximate values. If you select different points or if your estimates of the coordinates of the points are different from ours you will get slightly different values for m and b.

Example 4.6.3   The EPA began keeping records on lead emissions in the United States in 1970. The actual data points and the regression line are displayed in the following graph.

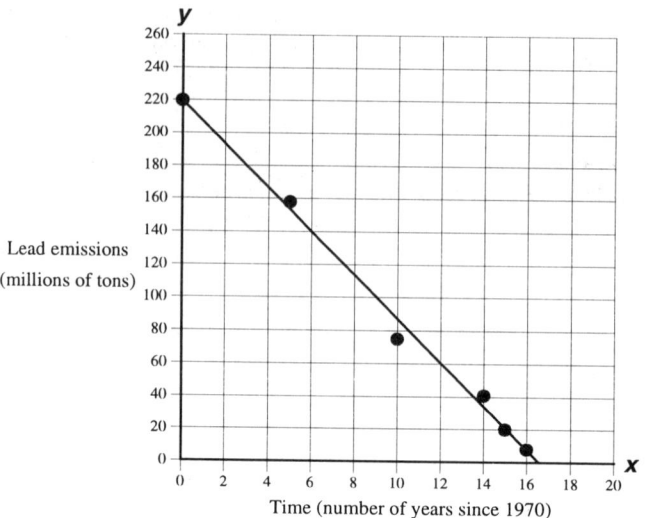

a)   Does the given line reasonably model the given data points?

   •SOLUTION• Yes, the data points seem to lie very close to the regression line.

b)   Use the regression line to construct the linear equation.

   •SOLUTION• We chose the points (0,220) and (15,20) to work with. We use the slope formula to calculate m:

$$m = \frac{20 - 220}{15 - 0} \text{, which simplifies to } -13.3$$

   One of the points we chose, (0,220), is the *y*-intercept. Therefore, the equation of the line is $y = 220 - 13.3x$.

   To check the equation, we will use an input of 12. From the graph, we see that the corresponding output is 60. Using the equation, we substitute 12 for *x* and simplify to get 60.4 for *y*. Since the value from the graph and the predicted value from the equation are very close we are comfortable that the equation is correct.

   The model says that in 1970 lead emissions were about 220 million tons and between 1970 and 1986 emissions dropped by about 13.3 million tons per year.

c)   Predict lead emissions in 1960.

   •SOLUTION• Since 1960 is 10 years before 1970, we use an input value of −10 in the equation. After substituting and simplifying we get 353 for *y*. Therefore, the model predicts that there were 353 million tons of lead emissions in 1960.

d)   Predict the drop in emissions from 1968 to 1972

   •SOLUTION• The model predicts that 1968 emissions were about 246.6 million tons while 1972 emissions were 193.4 million tons. The drop in emissions is 53.2 million tons. We could also answer this question by noting that the slope of the line says emissions are dropping 13.3 million tons per year and 13.3 • 4 is 53.2.

e) Predict what the emissions were in the year 1900.

•**SOLUTION**• The year 1900 corresponds to an x value of –70. Substituting this for x and simplifying gives us 1,151 million tons of lead emissions. However, there were very few automobiles in 1900 and autos are the largest source of lead emissions. This is an example of the danger in extrapolating outside the set of values of the given data – the model may not be reliable there.

Example 4.6.4   Automatic teller machines (ATM's) account for a growing number of banking transactions in the United States. The graph below shows the number of ATM's in relation to the number of years since 1980 and the regression line for those data.

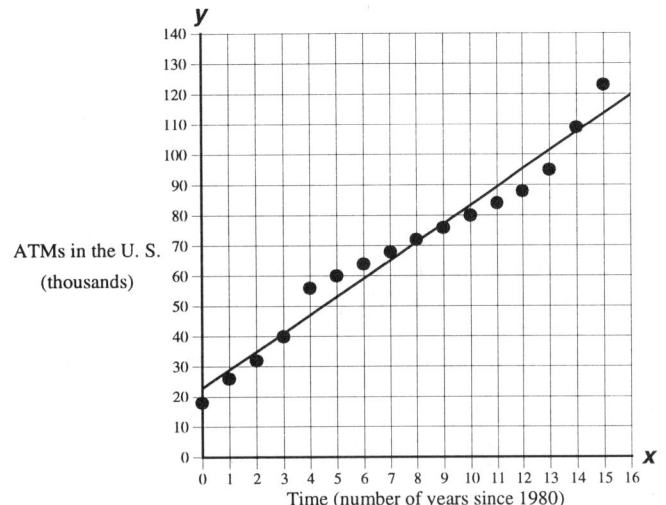

ATMs in the U. S. (thousands)

Time (number of years since 1980)

a) Build a linear equation for the regression line.

•**SOLUTION**• If we use the points (1.2,30) and (14.5,110) we get the equation $y = 23 + 6x$. The equation implies that in 1980 there were about 23,000 ATM machines and since then the number of ATM's has increased by about 6,000 machines each year.

b) Predict when the number of ATM machines will be double the number in 1990.

•**SOLUTION**• We see from the regression line on the graph that in 1990 there were approximately 83,000 machines. Double this is 166,000. If we substitute 166 for y in the regression equation and solve for x we get 24 (rounded). Because x represents the number of years since 1980, 24 corresponds to the year 2004.

c) The predicted U. S. population in 2010 is 297,716,000. Will there be one ATM for every 1,000 people if the current trend continues?

•**SOLUTION**• We would need 297,716 machines to have one ATM for every 1,000 people. Since 2010 is 30 years after 1980, we substitute 30 for x in the equation. Simplifying, we get an output of 203,000 machines. There will not be 1 machine for every 1,000 people in 2010.

d) Predict the year the first ATM machined were available.

•**SOLUTION**• We substitute 0 for y and solve for x to get –4 (rounded). Since x represents the number of years since 1980, –4 corresponds to 1976.

Practice 4.6.4 If it seems that the data points are well modeled by the regression line, use the line to answer the questions.

1. The following graph shows the cost of gas and oil for the average driver in the United States.

   a) If a linear model is appropriate, use two points on the regression line to build the regression equation.
   b) Interpret the meaning of the equation.
   c) Predict the cost of gas and oil in the year 2000.
   d) What do you expect the change in cost to be between 2005 and 2006?
   e) Before 1975 when did the cost exceed $5.00?

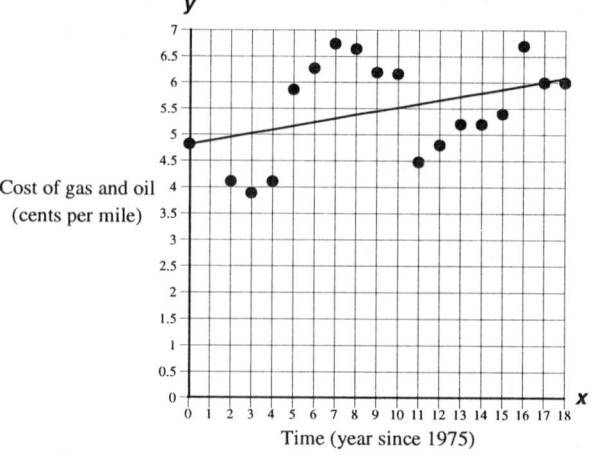

Cost of gas and oil (cents per mile)

Time (year since 1975)

2. The following graph shows enrollment in public elementary and secondary schools since 1869. Round to the tenths place.

   a) If a linear model is appropriate, use two points on the regression line to build the regression equation.
   b) Interpret the meaning of the equation.
   c) Use the graph to estimate the change in enrollment from 1889 to 1929. Compare this to the change given by the regression model.
   d) Use the graph to estimate the change in enrollment from 1949 to 1969. Compare this to the change given by the model.
   e) Estimate when the number of students will again reach the level shown on the graph for 1969.

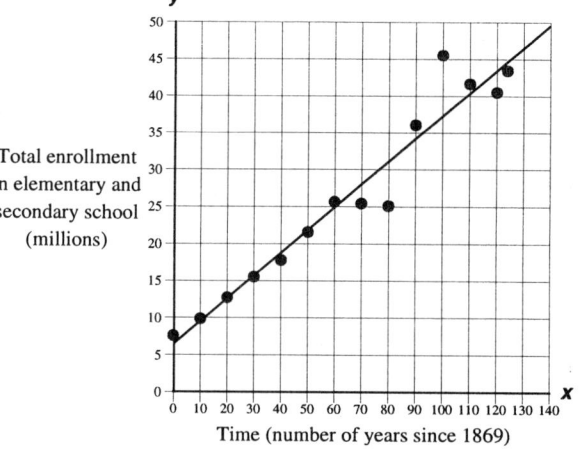

Total enrollment in elementary and secondary school (millions)

Time (number of years since 1869)

3. A city decides to put on a two week "Summer in the Park" series of shows. The data below show the time needed for workers to clean the park in relation to the number of people who attended other similar events.

   a) If a linear model is appropriate, use two points on the regression line to build the regression equation.
   b) Interpret the meaning of the equation.
   c) How long will the cleanup take if 400 people attend?
   d) After one show it took 2 hours and five minutes for the workers to clean up. Estimate the number of people attending.
   e) The last show will feature the Fourth of July fireworks display and the city expects over 600 people to attend. How many extra people will need to help to get the park clean in three hours the following morning?

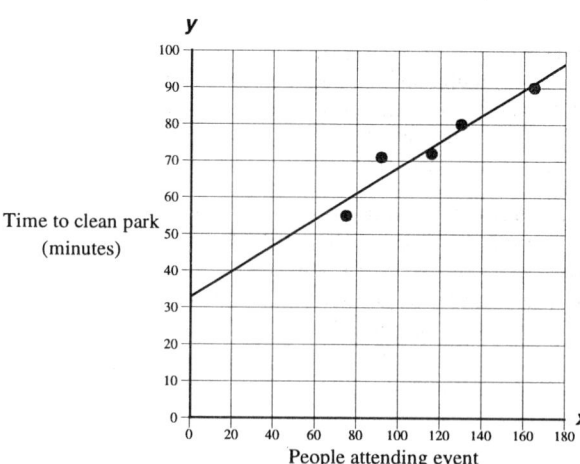

Time to clean park (minutes)

People attending event

Exercise Set 4.6   The answers to the odd numbered exercises are at the back of the book.

1. A teacher wonders if she can predict final grades by looking at students' verbal ability scores before entering the class. The data table to the left gives the data for 10 incoming freshmen. The regression equation is
   $y = 49 + 0.5x$

   a) Interpret the meaning of the equation.
   b) Estimate the verbal score that corresponds to a final grade of 100%.
   c) If the cutoff for passing the class is 70% would you predict a student entering with a verbal test score of 43 will pass the class?
   d) A student who enters with a verbal test score of 25% ends up receiving a final grade of 98%. What does this suggest?

x Verbal test score (percent)	y Final math test grade (percent)
41	65
20	55
31	68
68	81
90	95
82	84
48	73
55	79
72	83
72	76

2. A teacher wonders if he can predict final exam scores by using the average of three midterm exam scores. He collected the data shown on the left and his computer generated the following regression equation:
   $y = -17 + 1.5x$.

   a) Interpret the meaning of the equation.
   b) What average midterm exam score would correspond to a score of 90% on the final exam?
   c) A student goes into the final needing at least a 76% to pass the class. If his average test score before the final is 65% do you predict he will pass the class?
   d) Say a student had an average test grade of 100% what is her predicted final test grade?

x Average of 3 tests (out of 100)	y Final exam score (out of 125)
95	123.0
91	113.5
72	92.5
86	100.5
81	107.5
83	115.0
52	65.0
74	103.5
53	58.5
80	106.5
83	112.0
85	111.5
83	114.0
95	118.5
62	67.5
29	24.5
76	112.0

3. The data to the left shows the relation between how much a car dealer will pay you when you trade-in a certain make of car and how much that same dealer will charge you to buy an identical car. The regression equation is $y = 1564.00 + 1.07x$.

   a) Interpret the meaning of the equation.
   b) If the retail cost of the car was $17,775 how much did the dealer pay for the trade-in?
   c) Use the model to predict the cost if the trade-in was $13,800 and compare this to the table.

x Trade–in price (dollars)	y Retail price (dollars)
11200	13550
11300	13650
14350	16900
13700	16200
13800	16300
14150	16675
14250	16775
17300	20075

4. The data to the left shows the relation between how much a car dealer will pay you when you trade-in a certain make of truck and how much that same dealer will charge you to buy the truck that you traded-in. The regression equation is $y = 1255.00 + 1.08x$.

a) Interpret the meaning of the equation.
b) If the trade-in price was $20,025 what would be the retail price?
c) A student, who has worked out the linear model for his make of truck, is offered $15,600 as a trade-in price. If he believes the dealership will sell the truck for $18,200 is this a fair trade-in price?

y Trade–in price (dollars)	Retail price (dollars)
14650	17025
15650	18075
16000	18450
17600	20225
17900	20525
19350	22050

5. The data to the left shows the Olympic gold medal times for the men's 400–meter freestyle swim between 1924 and 1996. The regression equation is $y = 301.88 - 1.17x$.

a) Interpret the meaning of the equation.
b) Estimate the year that the Olympic gold medal time will be three and half minutes.
c) Use the equation to find the drop between 1924 and 1936. Then compare this estimated drop to the actual drop as given by the data table.
d) Use the equation to predict the drop in times between 1984 and 1996. Then compare this estimated drop to the actual drop as given by the data table. What does this suggest about winning times?

x Number of years since 1924	y Times for men's 400 meter freestyle swim (seconds)
0	304.2
4	301.6
8	288.4
12	284.5
24	281.0
28	270.7
32	267.3
36	258.3
40	252.2
44	249.0
48	240.27
52	231.93
56	231.31
60	231.23
64	226.95
68	225.00
72	227.97

6. The data to the left shows the Olympic gold medal times for the women's 400–meter freestyle swim. The regression equation is $y = 349.78 - 1.66x$.

a) Interpret the meaning of the equation.
b) Estimate the year that the Olympic gold medal time will be four minutes.
c) Use the equation to find the drop between 1924 and 1936. Then compare this estimated drop to the actual drop as given by the data table.
d) Use the equation to predict the drop in times between 1984 and 1996. Then compare this estimated drop to the actual drop as given by the data table. What does this suggest about winning times?

x Number of years since 1924	y Times for women's 400 meter freestyle swim (seconds)
0	362.2
4	342.8
8	328.5
12	326.4
24	317.8
28	312.1
32	294.6
36	290.6
40	283.3
44	271.8
48	259.04
52	249.89
56	248.76
60	247.10
64	243.85
68	247.18
72	247.25

7. From 1960 to 1994 the number of U. S. women earning a bachelors degree has increased approximately linearly. The regression equation for the data in the table is $y = 165 + 14x$.

    a) Interpret the meaning of the equation.
    b) Estimate the year the number of women graduating with bachelor degrees was double the number graduating in 1960.
    c) In 1995 535,000 men graduated with bachelor degrees. What year did this happen for women?

x Number of years since 1960	y Number of women earning bachelors degrees (thousands)
0	138
5	212
10	341
15	418
20	456
25	497
30	560
35	657

8. From 1960 to 1994 the number of women earning a doctorate degree has increased approximately linearly. The regression equation is $y = 0.25 + 0.45x$.

    a) Interpret the meaning of the equation.
    b) Predict the year the number of women earning bachelor degrees will exceed 27,000 the number of degrees earned by men in 1995.
    c) Will the number of women earning doctorate degrees be at least 20,000 by the year 2000?

x Number of years since 1960	y Number of women earning doctorate degrees (thousands)
0	1
5	2
10	4
15	7
20	10
25	11
30	14
35	16

9. The following graph shows the woman's Olympic high jump gold medal performance between 1928 and 1996.

    a) If a linear model is appropriate, use two points on the regression line to build the regression equation.
    b) Interpret the meaning of the equation.
    c) Predict when the gold medal performance will be 2.2 meters.
    d) In 1928 the men's gold medal height was 1.9 meters. What year does the linear equation predict women exceeded this height? Compare your answer with the graph to see if this may have occurred.
    e) Due to the second World War no Olympics were held in 1940 and 1944. Predict what the winning heights would have been had the Olympics been held during those years.

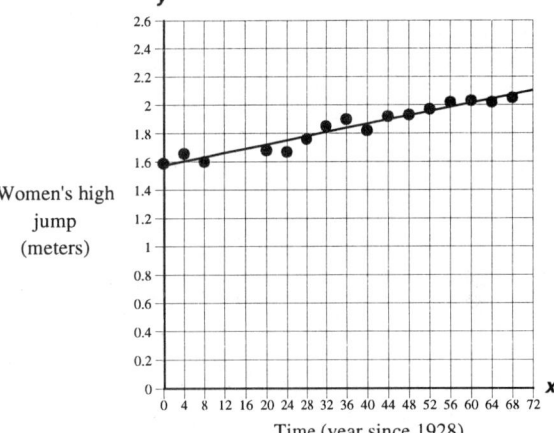

10. The following graph shows the men's Olympic high jump gold medal performance between 1928 and 1996.

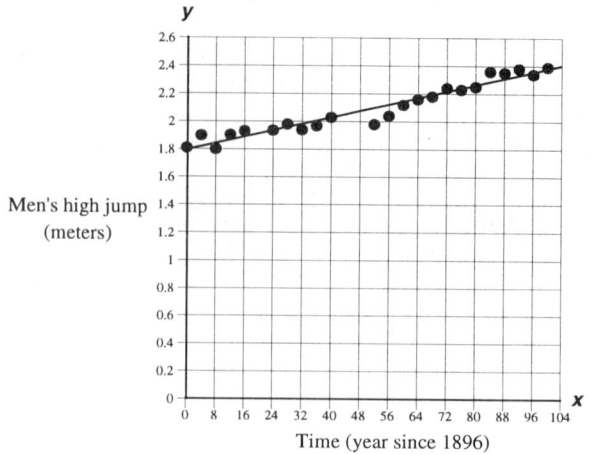

   a) If a linear model is appropriate, use two points on the regression line to build the regression equation.
   b) Interpret the meaning of the equation.
   c) Predict the year the men's gold medal performance will be 2.5 meters.
   d) In 1928, the first year of the women's high jump, the difference between the gold medal height for men and women was 0.35 meters. In 1996 the winning women's height was 2.05 meters. Has the difference between the men's and women's performance increased or decreased over the last 68 years? Use the predicted value to find the men's height.
   e) Due to the second World War there is no gold medal height in 1940 and 1944. Predict what the winning heights would have been had the Olympics been held during those years.

11. The following graph shows the minimum hourly wage in the United States between 1954 and 1997.

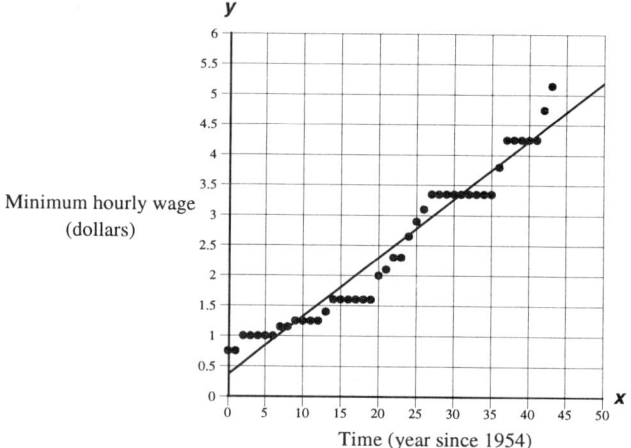

   a) If a linear model is appropriate, use two points on the regression line to build the regression equation.
   b) Interpret the meaning of the equation.
   c) Predict when the minimum hourly wage will reach $6.00.
   d) Predict when the minimum hourly wage will be three times it's 1958 value.
   e) Between 1969 and 1977 the minimum hourly wage went through an uncommon rise. Use the regression equation to find the dollar increase between 1969 and 1977 and compare this to the actual rise shown on the graph.

12. Since 1968, even though the minimum hourly wage has been increasing, its buying value has been going down because of inflation. The graph below shows the minimum hourly wage adjusted for inflation for 1968 through 1996.

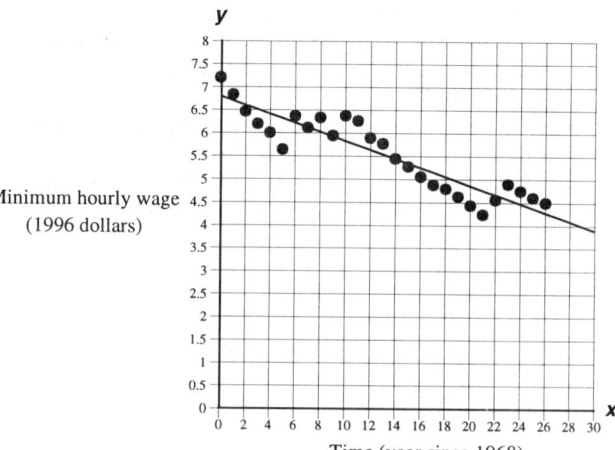

   a) If a linear model is appropriate, use two points on the regression line to build the regression equation.
   b) Interpret the meaning of the equation.
   c) When does the regression equation predict the value of the minimum hourly wage would have lost $2 in relation to it's 1968 predicted value?
   d) Estimate from the data points the dollar drop between 1968 and 1973. Compare this to the drop predicted by the model.
   e) If the trend shown by the regression line where to continue beyond 1996, when would the value of the minimum hourly wage be $3.50?.

13. The graph below compares the price an auto dealer will pay you if you trade-in an older used car and the price they will charge you if you buy this same car from them.

a) If a linear model is appropriate, use two points on the regression line to build the regression equation.
b) Interpret the meaning of the equation.
c) What does the model predict the retail price would be for a car traded in at $6,000.
d) A car traded in for $3600 is being offered for $5560. Does the model predict there is room to negotiate the dealership to a lower price?
e) The model predicts that a car traded-in for $0 can be sold by the dealer for $813; could this be a correct assumption?

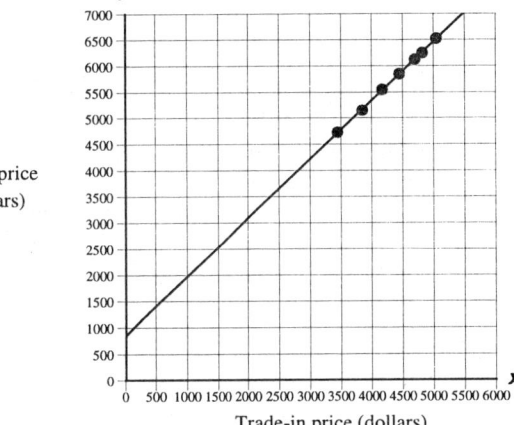

Retail price (dollars)

Trade-in price (dollars)

14. The following graph shows the relation between the trade-in price of a particular make car and its age relative to 1997.

a) If a linear model is appropriate, use two points on the regression line to build the regression equation.
b) Interpret the meaning of the equation.
c) Predict the 1997 trade-in price for a 1989 car of this make.
d) Assuming the relation remains the same in the future, find the trade-in price of a 1997 car in 1999.
e) What is the model year for a car that had 0 trade in value in 1997?

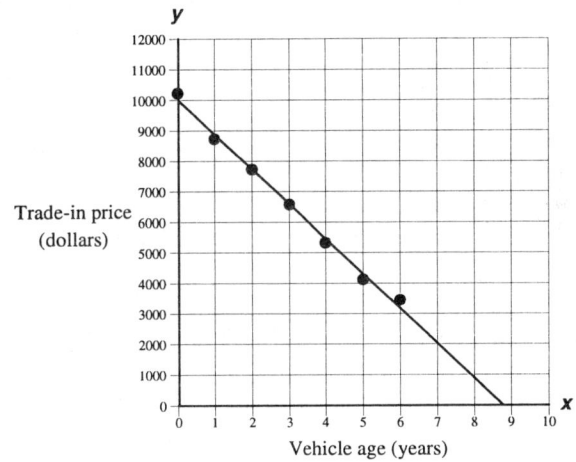

Trade-in price (dollars)

Vehicle age (years)

15. The following graph shows the cases of hepatitis B in the United States between 1980 and 1994.

a) If a linear model is appropriate, use two points on the regression line to build the regression equation.
b) Interpret the meaning of the equation.
c) Predict the year the number of cases of hepatitis B will reach about 25,000. (The number of cases around 1983).
d) When does the model suggest hepatitis B will be eradicated in the United States?
e) What does the model suggest is the drop in cases of hepatitis B between 1975 and 1980.

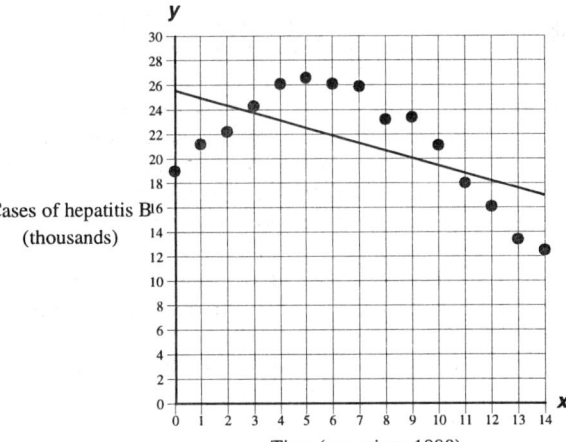

Cases of hepatitis B (thousands)

Time (year since 1980)

16. The following graph shows the number of cases of measles in the United States between 1980 and 1994.

    a) If a linear model is appropriate, use two points on the regression line to build the regression equation.
    b) Interpret the meaning of the equation.
    c) What year will the number of cases of measles exceed 10,000.
    d) Is the drop in cases of measles between 1988 and 1992 the same as the drop between 1978 and 1981?
    e) Predict when the number of cases of measles was under 1000.

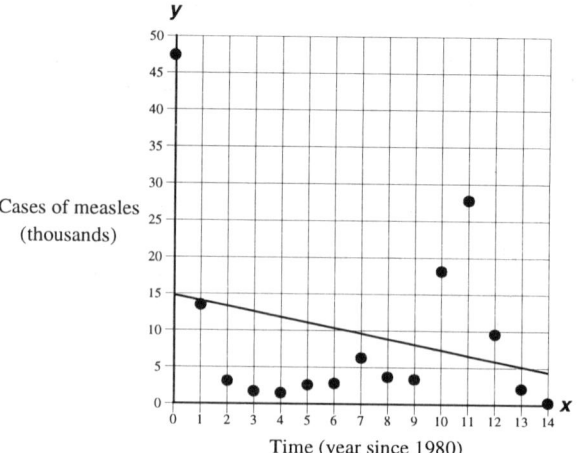

Time (year since 1980)

Review Exercises   The answers to all of these exercises are at the back of the book.

17. Name the property that justifies the statement: $13 \cdot 25 \cdot 47 \cdot 0 = 0$

18. Solve: $0.16a = 0.2(10) + 0.12(a + 5)$

19. Identify each base and its exponent, and then simplify if possible: $(2x)^2$

20. Solve: $\dfrac{4}{5}(x - 15) = x - \left(5 + \dfrac{x}{5}\right)$

21. Graph the straight line that passes through (6,2) with slope 1.

22. Solve: $3 < -\dfrac{x}{2}$

23. A child decides to split some candy between herself and her brother. For every piece she gives her brother she gives herself 2 pieces. Also, she begins by giving herself an extra 8 pieces as compensation for doing the work of splitting the candy up. If she ended up with 32 pieces, how many pieces were originally available?

Section 4.7 Building Inequality Models

Topic 4.7 A — Mathematical Models That Are Not Based On Input/Output

We have found that the linear equation, $y = b + mx$, models a relation between input, represented by x, and output, represented by y. In the usual situation, we were given or could measure x (say, the femur length of a woman) and then we used the model (equation, data table, or graph) to predict the value of the output (the height of the woman).

In this section we look at linear models from a different perspective. We still use the idea that there is a relation between two quantities but our goal will not be to predict the value of one when we are given a value for the other. That is, we will not consider one quantity *input* and the other *output*. In this section we will think of the relation between the quantities as a **constraint** or a condition on the quantities.

For example, suppose that after paying for food, clothes, rent, utilities, taxes, and miscellaneous necessities, Kathy has $200 left over from her pay at the end of each month. She decides that she will put part in a savings account and spend the remainder on fun things. If we let

  s represent the dollars she saves each month

  f represent the dollars she spends on fun things each month

then we can use the Equality Property to write the mathematical model for this situation as

$$\left(\begin{array}{c}\text{money}\\\text{for fun}\end{array}\right) + \left(\begin{array}{c}\text{money}\\\text{for savings}\end{array}\right) = \left(\begin{array}{c}\text{total money}\\\text{available}\end{array}\right)$$
$$f \qquad + \qquad s \qquad = \qquad 200$$

We can think of the $200 as a constraint on the money she saves and the money she uses for fun (they have to total exactly $200).

Before we graph this model we need to decide which variable to put on which axis. With input/output models we always use the x-axis for the input. But here we have nothing designated as input so it does not matter which variable is represented by which axis. We will arbitrarily put f on the horizontal axis and s on the vertical axis.

— *Note* — *Using different letters for variables*	Since we no longer have input and output we will generally not use x and y for the variables. Rather, we will use a letter that helps us remember what each variable stands for (f for fun and s for savings).

We can use the intercept method to quickly graph this model.

f *money spent on fun (dollars)*	s *money saved (dollars)*
0	200
200	0

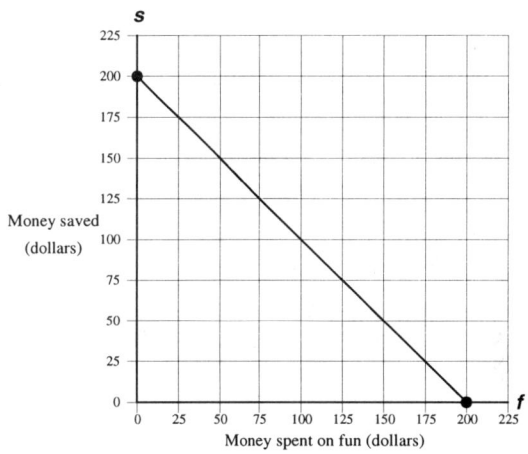

As was the case with input/output graphs, all the points on the line represent solutions to the mathematical model, $f + s = 200$. For example

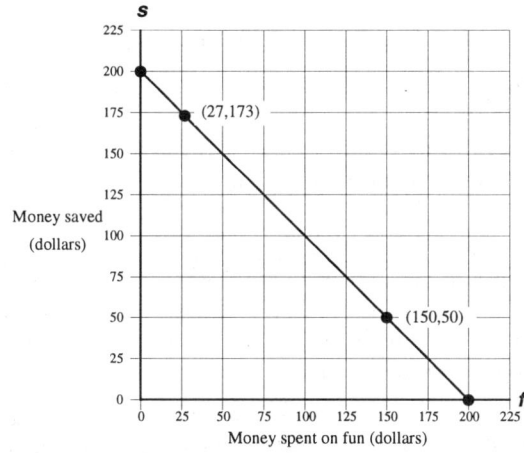

- (27,173) is on the line because $27 + 173 = 200$. This represents spending $27 on fun and saving $173.

- (150,50) is on the line because $150 + 50 = 200$. This represents spending $150 on fun and saving $50.

What do the points that are NOT on the line represent? We did not consider these points with input/output graphs because they had no meaning. But with constraint graphs these points do have meaning. These are points where the dollars spent on fun plus the dollars saved do not add up exactly to 200. For example,

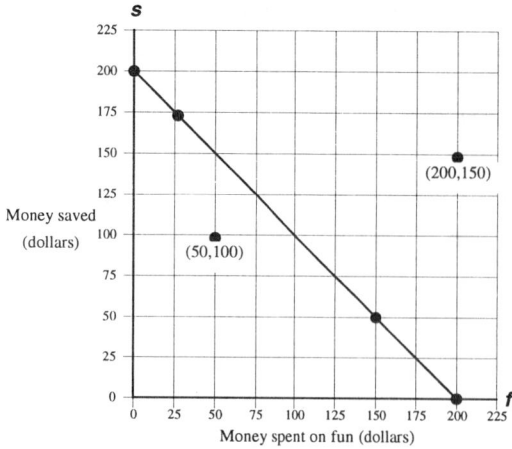

- (50,100) is NOT on the line because $50 + 100 \neq 200$. This represents spending $50 on fun and saving $100. What happens to the rest of the money? It just sits in Kathy's purse collecting dust.

- (200,150) is NOT on the line because $200 + 150 \neq 200$. This represents spending $200 on fun and saving $150. That would be great, but it's not possible since Kathy only has $200 available.

The line $f + s = 200$ breaks up the graph into two **regions**- or portions of the coordinate system:

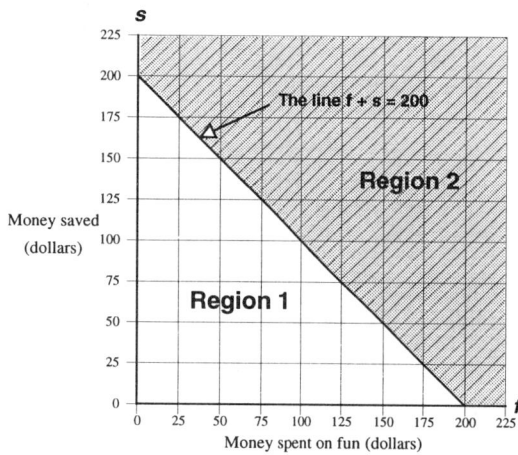

- **Region 1** contains the points that fall below the line. This region represents values for f and s that are possible in Kathy's situation. That is, she can choose to spend less than 200 on fun and savings.

- **Region 2** contains the points that fall above the line. This region represents values for f and s that are not possible in the given situation. That is, since Kathy has only $200 left to spend she may not spend more than $200 on fun and savings combined.

Topic 4.7 B — Constructing Graphs of Inequalities

To mathematically represent the constraint that Kathy can use *up to* $200, we use inequalities. That is, instead of graphing the *equation*

$$f + s = 200$$

which says that she has *exactly* $200 to use for savings and fun we will graph the *inequality*

$$f + s \leq 200$$

which says that she has *at most* $200 to use for savings and fun.

We have seen that the graph of a linear equation, such as $f + s = 200$, is a straight line. Each point on the line represents a possible solution to the equation. Likewise, the graph of a linear inequality shows all the points that are solutions to the inequality. To see what the graph of $f + s \leq 200$ looks like we have plotted 100 points on the following graph:

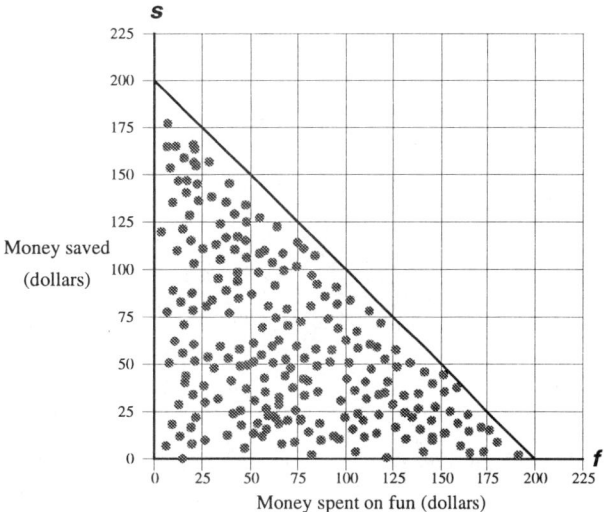

Notice that all the points that are solutions to the inequality $f + s \leq 200$ are either on the line or below it. It turns out that every solution to the inequality will fall on or below the line and every point that falls above the line is NOT a solution. To show that every point in the region below the line is a solution we shade the region:

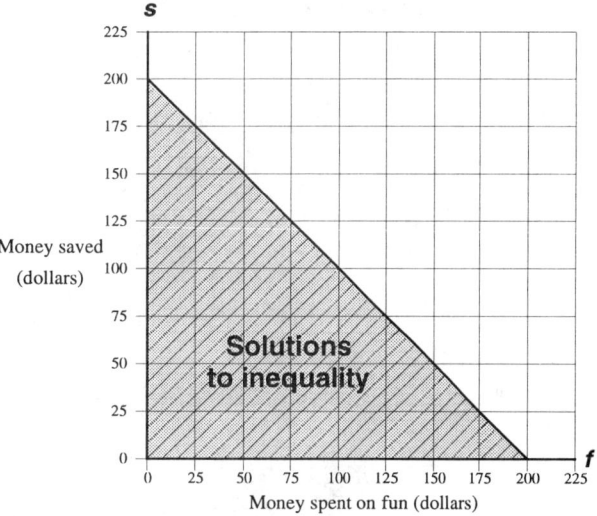

A similar result is obtained for all linear inequalities. All the points that are solutions fall into one of the regions of the graph. That is, all the solutions are either above the line or below the line, and sometimes on the line. Compare these mathematical models and their graphs:

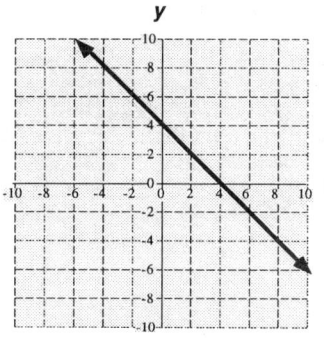

- $x + y = 4$ is a linear equation whose graph is a straight line. Points on the line are solutions, points not on the line are not solutions.

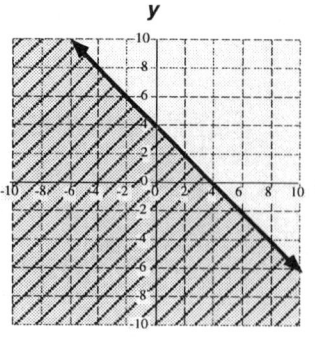

- $x + y \leq 4$ is a linear inequality whose graph is the straight line $x + y = 4$ and the region below it. Points on the line and points below the line are solutions. Points above the line are NOT solutions.

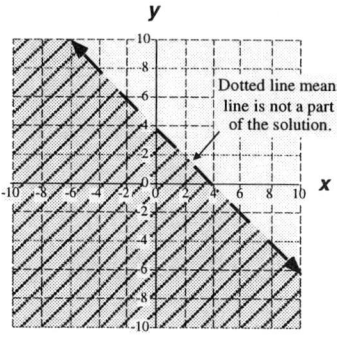

- $x + y < 4$ is a linear inequality whose graph is the region below the line, but not including the line itself. We know the line is not included because the inequality symbol is *less than*, <, rather than *less than or equal to*, ≤. We show that the line is not included by making it dotted. Points below the line are solutions. Points on or above the line are NOT solutions.

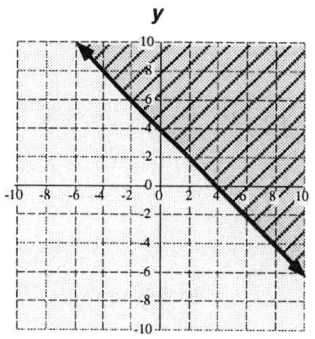

- $x + y \geq 4$ is a linear inequality whose graph is the straight line $x + y = 4$ and the region above it:

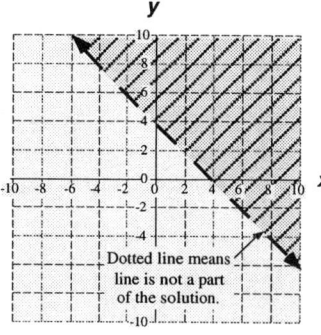

- $x + y > 4$ is a linear inequality whose graph is the region above the straight line, but not including the line itself. We know the line is not included because the inequality symbol is > rather than ≥.

Since the graphs of linear inequalities always consist of a line (dotted or solid) and a shaded region we can quickly construct them using the following procedure:

— *Procedure* — *Graphing a linear* *inequality*	***Step 1 Graph the boundary*** To find the boundary that separates the solutions from the non-solutions, graph the linear *equality*. If the inequality symbol is ≥ or ≤, the points on the line are part of the solution; in that case, draw a solid line. Otherwise, draw a dotted line to show that the points on the line are not part of the solution. ***Step 2 Shade the Region*** To determine which region contains the solutions, choose a **test point** that is NOT on the line. The point (0,0) usually is a good choice since it's easy to locate. If the test point makes the inequality true, shade the side of the line that contains the point. Otherwise, shade the other side of the line.

Let's see how this procedure works.

Example 4.7.1   a)   Graph $x + 2y \le 4$.

•**SOLUTION**•

Step 1 Graph the boundary   The boundary is the line given by the equation $x + 2y = 4$. We will use the intercept method to construct the graph.

x	y
0	2
4	0

Since the original inequality says *less than or equal to* (that is, $\le$ rather than $<$), points on the line are solutions and so the line on the graph will be solid.

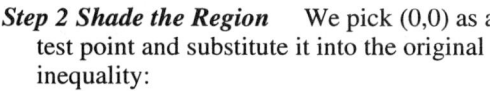

Step 2 Shade the Region   We pick (0,0) as a test point and substitute it into the original inequality:

$x + 2y \le 4$ ⟸ Original inequality.

$(0) + 2(0) \le 4$ ⟸ Substituted 0 for x and y.

$0 \le 4$ ⟸ Simplified.

Since $0 \le 4$ is true, the test point (0,0) is a solution and therefore ALL points in the region that contains (0,0) are solutions. We indicate this by shading below the line.

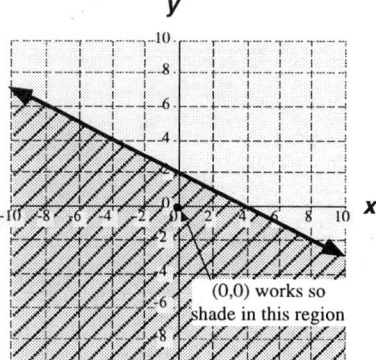

(0,0) works so shade in this region

b)   Graph $2x - 3y < 6$

•**SOLUTION**•

Step 1 Graph the boundary   The boundary is the line given by the equation $2x - 3y = 6$.

x	y
0	−2
3	0

Since the original inequality says *less than* (that is, $<$ rather than $\le$), points on the line are *not* solutions and so the line on the graph will be *dotted*.

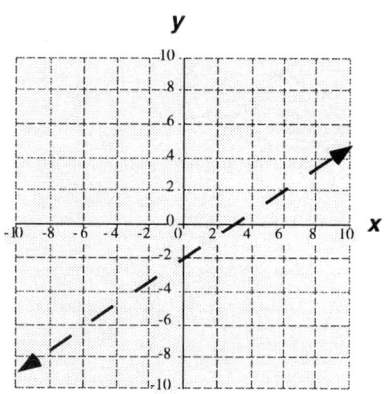

Step 2 Shade the Region We pick (0,0) as a test point and substitute it into the original inequality:

$$2x - 3y \leq 6 \impliedby \text{Original inequality.}$$

$$2(0) - 3(0) \leq 6 \impliedby \text{Substituted 0 for } x \text{ and } y.$$

$$0 \leq 6 \impliedby \text{Simplified.}$$

Since $0 \leq 6$ is true, the test point (0,0) is a solution and therefore ALL points in that region are solutions. We indicate this by shading above the line.

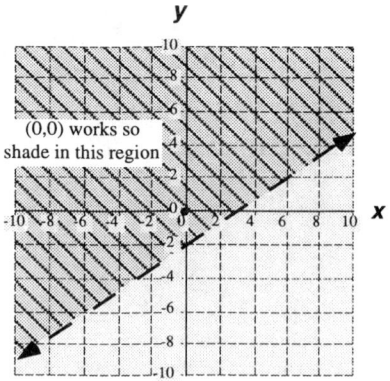

c) Graph $x > 6$

•SOLUTION•

Step 1 Graph the boundary The boundary is the line given by the equation $x = 6$. This is a vertical line which intersects the x-axis at 6. The line will be dotted since the original problem uses > rather than ≥.

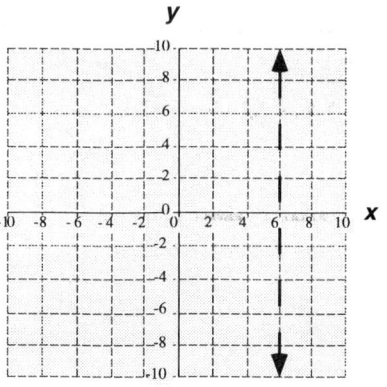

Step 2 Shade the Region We pick (0,0) as a test point and substitute it into the original inequality:

$$x > 6 \impliedby \text{Original inequality.}$$

$$(0) > 6 \impliedby \text{Substituted 0 for } x.$$

Since $0 > 6$ is false, the test point (0,0) is *not* a solution and therefore ALL points in the *other* region are solutions. We indicate this by shading to the right of the line.

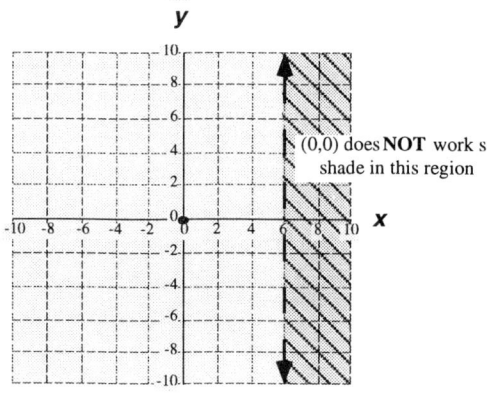

d) Graph $y \geq x$

•SOLUTION•

Step 1 Graph the boundary The boundary is the line given by the equation $y = x$. Since the original inequality uses ≥ (rather than >) points on the line are solutions and so the line is solid.

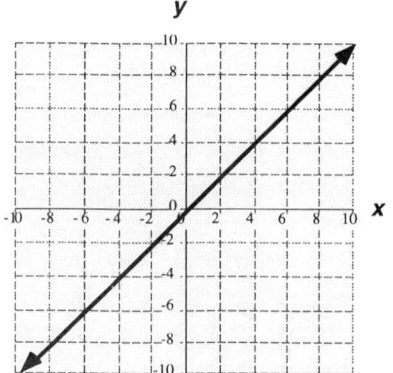

Step 2 Shade the Region We can't use (0,0) as a test point this time because it's on the line. Let's use (0,1) instead.

$$y > x \quad \Leftarrow \text{ Original inequality.}$$

$$(1) > (0) \quad \Leftarrow \text{ Substituted 0 for } x \text{ and } y.$$

Since $1 \geq 0$ is true, the test point (0,1) is a solution and therefore ALL points in that region are solutions. We indicate this by shading above the line.

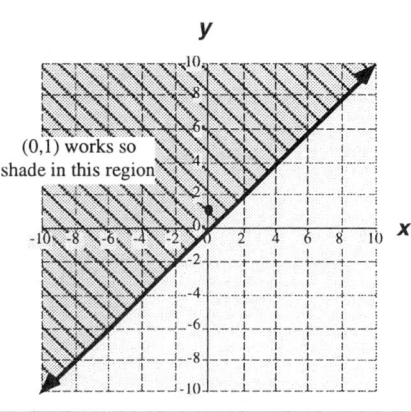

(0,1) works so shade in this region

Practice 4.7.1 Graph.

1. $y > x + 1$ 2. $y < 2x - 3$ 3. $y \geq 5x + 4$ 4. $y \leq x - 1$ 5. $y > \dfrac{x}{2}$ 6. $y \leq \dfrac{2x}{3}$

Topic 4.7 C — Applications of Two-Variable Inequalities

Two-variable inequalities appear in many real-world situations where the variables are related and constraints are involved.

Example 4.7.2 To comply with University academic environment standards, the perimeter of a rectangular classroom in new construction must be greater than 800 feet.

a) If w represents the width of a room and l the length, construct a mathematical model for the constraint on the perimeter.

•**SOLUTION**• The perimeter of a rectangle is $p = 2l + 2w$. The problem states that the perimeter must be greater than 800 feet. So, we have

$$\begin{aligned} \text{perimeter} &> 800 \\ 2l + 2w &> 800 \end{aligned}$$

b) Construct a graph of the model.

•**SOLUTION**•

Step 1 Graph the boundary We graph $2l + 2w = 800$ using the intercept method.

l length (feet)	w width (feet)
0	400
400	0

Since the inequality uses $>$ rather than $\geq$ the boundary line is dotted.

Step 2 Shade the Region We pick (0,0) as a test point and substitute it into the original inequality:

$$2l + 2w > 800 \quad \Leftarrow \text{ Original inequality.}$$

$$2(0) + 2(0) > 800 \quad \Leftarrow \text{ Substituted 0 for } x \text{ and } y.$$

$$0 > 800 \quad \Leftarrow \text{ Simplified.}$$

Since $0 \geq 800$ is false, the test point (0,0) is not a solution and so all points in the other region are solutions. We indicate this by shading above the line.

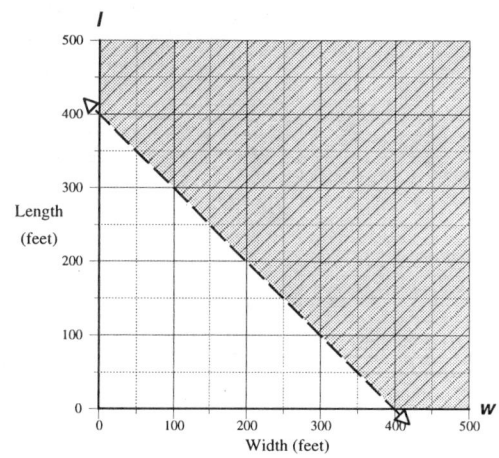

c) Discuss the meaning of points located in different regions of the graph, including exactly on the boundary between the regions.

•SOLUTION•

- The region below the boundary shows rooms that do not meet the standard because they have perimeters that are under 800 feet.
- The boundary (dotted line) shows rooms that do not meet the standard because they have perimeters of *exactly* 800 feet and the standard says the perimeter must be *greater than* 800 feet.
- The region above the boundary shows rooms with perimeters that meet the standard. Note that even though all these lengths and widths fit the model $2l + 2w > 800$ some are not *physically possible* because they have 0 or negative values (for example we cannot have a room that is –2 feet wide). Also, some of the lengths and widths are not *practical* (for example, we would not want a room that is 1 foot wide and 800 feet long, even though it meets the standard).

Practice 4.7.2

1. To keep the cost within his budget, Bill is building a house with a perimeter that is no more than 175 feet.

    a) If w represents the width and l represents the length, construct a mathematical model for the constraint on the perimeter.
    b) Construct a graph of the model.
    c) Discuss the meaning of points located in the region above the boundary.
    d) Discuss the meaning of the points on the boundary.

2. Joe has no more than 100 feet of fencing to make a rectangular dog pen.

    a) If w represents the width and l represents the length, construct a mathematical model for the constraint on the perimeter.
    b) Construct a graph of the model.
    c) Discuss the meaning of points located in the region below the boundary.
    d) Discuss the meaning of the points on the boundary.

Example 4.7.3 It takes a machine 1 minute to form 5 ceramic cups and 1 minute to form a plate. The machine can operate for a maximum of 400 minutes per shift.

a) If c represents the number of cups and p the number of plates, construct a mathematical model for the constraint imposed on the machine.

•SOLUTION• The machine can work for a maximum of 400 minutes so it must be true that

$$\left(\begin{array}{c}\text{time spent} \\ \text{on cups}\end{array}\right) + \left(\begin{array}{c}\text{time spent} \\ \text{on plates}\end{array}\right) \leq \left(\begin{array}{c}\text{maximum} \\ \text{time allowed}\end{array}\right)$$

To find the time spent on cups and plates we will construct an Amount Table. The unit for the rate of change could be either $\dfrac{\text{items}}{\text{minute}}$ or $\dfrac{\text{minutes}}{\text{item}}$. However, since the constraint is on the number of minutes the machines can work we want the amount to represent time. Therefore, we will use the rate with time in the numerator, $\dfrac{\text{minutes}}{\text{item}}$.

	amount	=	rate	•	base
	output	=	rate of change	•	input
Descriptions →	time needed to make items		time to make 1		number to be made
Units →	minutes		$\dfrac{\text{minutes}}{\text{item}}$		items
↓ Item names ↓					
cups	❶		$\dfrac{1 \text{ min}}{5 \text{ cup}}$		c
plates	❷		$\dfrac{1 \text{ min}}{1 \text{ plate}}$		p

We use the Amount Formula to find the time for each item, ❶ and ❷:

❶ is $\frac{1}{5}c$, or $\frac{c}{5}$

❷ is $\frac{1}{1}p$, or just p.

Using these values the model becomes

$$\left(\begin{array}{c} \text{time spent} \\ \text{on cups} \end{array}\right) + \left(\begin{array}{c} \text{time spent} \\ \text{on plates} \end{array}\right) \leq \left(\begin{array}{c} \text{maximum} \\ \text{time allowed} \end{array}\right)$$

$$\frac{c}{5} \qquad + \qquad p \qquad \leq \qquad 400$$

b)   Construct a graph of the model.

•**SOLUTION**•

Step 1 Graph the boundary   The boundary is the line given by the equation

$\frac{c}{5} + p = 400$. We use the intercept method to graph this.

c (number of cups)	p (number of plates)
0	400
2000	0

Step 2 Shade the Region   We pick (0,0) as a test point and substitute it into the original inequality:

$\frac{c}{5} + p \leq 400 \quad \Leftarrow$ Original inequality.

$\frac{(0)}{5} + (0) \leq 400 \quad \Leftarrow$ Substituted 0 for x and y.

$0 \leq 400 \quad \Leftarrow$ Simplified.

Since $0 \leq 400$ is true, the test point (0,0) is a solution and therefore all points in its region are solutions. We indicate this by shading below the line.

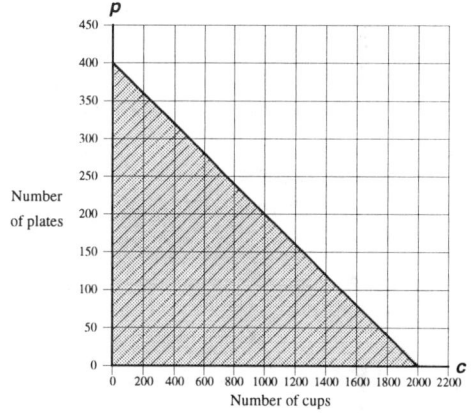

c)   How many plates and cups can the machine produce per shift?

•**SOLUTION**• All points in the shaded region are solutions to the inequality $\frac{c}{5} + p \leq 400$.

However, only non-negative numbers can be used since we cannot produce a negative number of plates or cups. Possible solutions include

*   (400,0), which is 400 cups and 0 plates.

*   (0,2000), which is 0 cups and 2000 plates.

*   (200,1000), which is 200 cups and 1000 plates.

*   (20,10), which is 20 cups and 10 plates. With this solution the machine stands idle for some of the time.

*   (83.5,100), which is $83\frac{1}{2}$ cups and 100 plates. We assume here that the machine can stop half way through the process of making a cup and pick up where it left off during the next shift. This may or may not be a reasonable assumption

Practice 4.7.3

1. It takes 8 hours to make a deluxe model VCR and only 5 hours to make a standard model. Let *d* represent the number of deluxe models and *s* represent the number of standard models. The maximum number of hours worked is 40. Construct a mathematical model of this situation and then graph it.

2. It takes 30 minutes to bind a novel and 4 minutes to bind a magazine. If *n* represents the number of novels bound and *m* the number of magazines bound, construct a mathematical model if the maximum time worked will be 480 minutes . Graph the model.

3. Sally has a collection of dimes and quarters worth at least $6.50. If *d* represents the number of dimes and *q* represents the number of quarters, construct a mathematical model for the constraint on the value of the collection. Graph the model.

4. Jim has a collection of $10 CD's and $8 cassettes that is worth at least $200. If *d* represents the number of CD's and *c* represents the number of cassettes, construct a mathematical model for the constraint on the value of the collection. Graph the model.

Exercise Set 4.7   The answers to the odd numbered exercises are at the back of the book.

Graph.

1. $y \geq -5x + 10$ 2. $y \leq -3x + 6$ 3. $x \leq -3$ 4. $y \geq 4$

5. $y > \dfrac{2}{3}x + 4$ 6. $y > \dfrac{4}{5}x + 4$ 7. $y > -x$ 8. $y \leq x$

9. To prepare for a distance run Amy decides to run 10 miles or less per day. She runs part of the way at 10 mph and walks the rest of the way at 5 mph.

    a) If *r* represents the time she runs and *w* represents the time she walks, construct a mathematical model for the constraint on the distance.
    b) Construct a graph of the model with *r* on the horizontal axis.
    c) Discuss the meaning of points located in the region below the boundary.
    d) Discuss the meaning of points on the boundary.

10. Two graduate students decide to test their new communication device by driving for two hours in opposite directions and then attempting to talk to each other. Assume they do not wish to be more than 200 miles apart at the end of the two hours.

    a) If *x* represents the speed of the first student and *y* represents the speed of the second student construct a mathematical model for the constraint on the distance.
    b) Construct a graph of the model with *x* on the horizontal axis.
    c) Discuss the meaning of points located in the region below the boundary.
    d) Discuss the meaning of points on the boundary.

11. William has a savings jar into which he puts any $1 and $5 bills he has when he gets home. He knows the value of the bills in the jar is at least $100.

    a) If *o* represents the number of $1 bills and *f* represents the number of $5 bills, construct a mathematical model for the constraint on the value of the bills in the jar.
    b) Construct a graph of the model with *o* on the horizontal axis.
    c) Discuss the meaning of points located in the region above the boundary.
    d) Discuss the meaning of points on the boundary.

12. Students are selling tickets to a play. Regular tickets cost $4.00 each and senior citizen tickets cost $3.75 each. To break-even the students will need to make more than $600.

    a) If *r* represents the number of regular tickets sold and *s* represents the number of senior tickets sold, construct a mathematical model for the constraint on the total value of the tickets sold which exceeds the break-even amount.
    b) Construct a graph of the model with *a* on the horizontal axis.
    c) Discuss the meaning of points located in the region above the boundary.
    d) Discuss the meaning of points on the boundary.

13. Kendra wants to pay at most $400 in interest at the end of a year by borrowing from two sources. She will pay an uncle 5% per year in simple interest and a bank 8% per year in simple interest.

    a) If f represents the amount she borrows from the uncle and e represents the amount borrowed from the bank, construct a mathematical model for the constraint on the interest she must pay. Assume that even though she will not borrow all the money from the uncle she still wishes to consider all possible mixes.
    b) Construct a graph of the model with f on the horizontal axis.
    c) Discuss the meaning of points located in the region above the boundary.
    d) Discuss the meaning of points on the boundary.

14. Due to tax concerns Yolanda wants to keep her interest earnings to under $600 for the year. Assume Yolanda will invest some of her money in an account that is expected to pay 12% in simple interest this year and another account which is expected to pay 10% in simple interest this year.

    a) If x represents the amount she invests in the 12% account and y represents the amount she invests in the 10% account, construct a mathematical model for the constraint on the interest she will earn.
    b) Construct a graph of the model with x on the horizontal axis.
    c) Discuss the meaning of points located in the region below the boundary.
    d) Discuss the meaning of points on the boundary.

15. To stay within their budget a couple wants to keep their weekly daycare expenses at or under $240. Expenses for the toddler are $3 per hour and for the baby are $4 per hour. Let t be the number of hours the toddler is in daycare per week and b be the number of hours the baby is in daycare per week. Construct a mathematical model of the constraint on weekly dollars spent, graph the model and then use the graph to describe in English the meaning of the points (30,30), (40,30), (40,40).

16. To keep the appearance they are not interested in profit a public television station has decided to limit the profit they make from a mug and tee shirt sale to under $2000. They make $8 on each tee shirt and $10 on each set of 6 mugs they sell. Use t to represent the number of tee shirts sold and m to represent the number of mug sets sold. Construct a mathematical model for the constraint on the total profit they make, graph the model, and then use the graph to describe in English the meaning of the points (300,350), (125,100), and (100,100).

17. Justin wishes to deposit $800 in one account and $1,000 in a second account so as to earn a minimum of $80 in simple interest in one year. Let e represent the rate for the $800 deposit and t represent the rate for the $1,000 deposit. Construct a mathematical model for the constraint on the amount of interest earned at different rates of interest, graph the model, and then use the graph to describe in English the meaning of the points (6,8), (4,4.5), and (5,4).

18. Claudia wishes to make a minimum of $40 in simple interest by depositing some money in a mutual fund that pays 8% simple interest and depositing additional funds in a savings account that pays 5% simple interest. Let m represent the amount deposited in the mutual fund and s represent the amount deposited in the savings account. Construct a mathematical model for the constraint on the amount of interest earned and then graph the model. Use the graph to describe in English the meaning of the points (150,400), (400,300), and (400,200).

19. On her trip out west, Tara wants to cover more than 350 miles today. She will make part of the trip at 50 mph and the rest at 70 mph. Let f represent the time she travels at 50 mph and s represent the time she travels at 70 mph. Construct a mathematical model for the constraint on the total distance she travels, graph the model, and then use the graph to describe in English the meaning of the points (3.5, 2.5), (4,2), (0,7).

20. Due to bad weather a plumber wishes to limit his travel distance to under 36 miles. He allows himself half an hour to drive to one job and three-quarters of an hour to drive to a second job. Let x represent the speed he drives to the first job and y represent the speed he drives to the second job. Construct and graph a mathematical model to represent the possible distances he could travel. Use the graph to describe in English the meaning of the points (35,25), (30,40), (20,30).

Review Exercises The answers to all of these exercises are at the back of the book.

21. Simplify: $3x - 2[4x - 3(x - 5) + 6]$

22. Solve: $\frac{1}{2}(x - 3) = \frac{2}{3}(x + 6) - \frac{11}{2}$

23. Find the slope and y–intercept of the line described by $y = 2x - 1$

24. Graph: $0 < x$

25. Find the equation of the line that goes through (2,5) and (5,–1).

26. Solve: $-0.5 > 3 - \frac{x}{2}$

27. A coffee retailer is having a hard time selling enough regular coffee. He decides to mix 45 pounds of the regular coffee, which sells for $6 a pound, with some special blend coffee, which sells for $45 a pound. How many pounds of special blend must he use if he wants to sell the mix for $12.50 a pound?

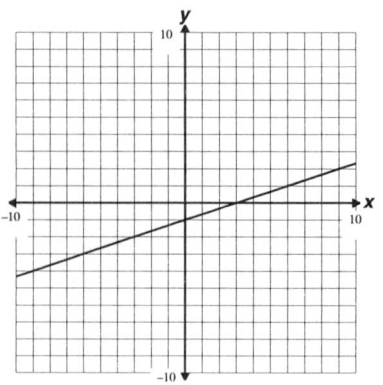

28. Find the *x*-intercept and the *y*-intercept of the graph.

Section 4.8 Relations and Functions

Topic 4.8 A — Relations

In everyday usage, we know that a *relation* is a connection or correspondence between things. For example, the blood relation between a parent and a child. In mathematics we define a **relation** as a correspondence between two sets of numbers, such as input and output. We can describe relations using several different notations:

- Describing a relation using *set notation*: The set {(2,1), (–3,0), (6,–4)} is a relation. The ordered pair (2,1) shows that an input of 2 corresponds to an output of 1. Likewise, the input –3 corresponds to an output of 0, and the input 6 corresponds to the output –4.

- Describing a relation using a *data table*: The data table

Input	Output
2	–5
3	0
–1	2

is a relation because it shows the correspondence between input and output. For example, an input of 2 corresponds to an output of –5.

- Describing a relation using an *equation*: The equation $y = 2x + 3$ is a relation because it shows the correspondence between the input represented by x and the output represented by y. For example, if x is 4, then we can use the equation to calculate the corresponding output, $2(4) + 3$, which simplifies to 11.

- Describing a relation using as an *inequality*: The inequality $y > 2x + 3$ is a relation because it shows the correspondence between the input represented by x and the output represented by y.

  - Describing a relation using a *graph*: The graph is a relation because each point on the line shows the correspondence between the input, indicated by the x-axis, and the output, indicated by the y-axis. For example, if the input is –4 we can see from the graph that the output is –3.

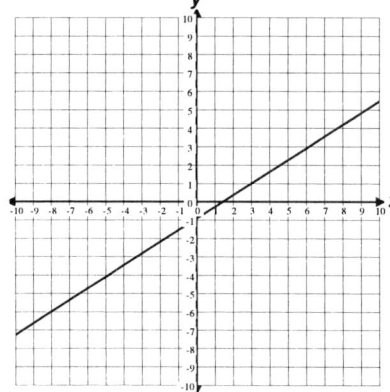

The **domain** of a relation is defined as the set of values for the *input*. The **range** is the set of values for the *output*. For example,

- the domain of the relation {(2,1), (–3,0), (6,–4)} is the set {2, –3, 6} because these are the *x* values of the ordered pairs. The range is the set of *y* values {1,0,–4}.

- the domain of the relation given by the data table

Input	Output
5	–5
3	0
–1	2
0	7

  is the set {5,3,–1,0} because those are the input values. The range is the set {–5,0,2,7}.

- the domain of the relation $y = 2x + 3$ is the set of real numbers because the input variable, *x*, can take on any value. The range is also the set of real numbers because we can substitute any real number for *y* and solve for *x*.

- the domain of

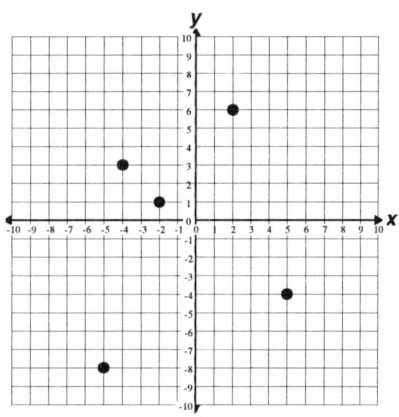

  is {–5,–4,–2,2,5} and the range is {–8,–4,1,6}.

Example 4.8.1   a)   Determine the domain and range of {(5,0), (3,–2), (–1,4), (5,7)}

- **SOLUTION** • The domain is the set of *x* values, which is {5,3,–1}. The range is the set of *y* values, which is {0,–2,4,7}.

b)   Determine the domain and range of $y = x$.

- **SOLUTION** • Since we can replace *x* with any number, the domain is the set of real numbers. Since *y* can be any value, the range is also the set of real numbers.

c)   Determine the domain and range of the relation described by the following data table:

Input	Output
–1	8
–3	2
–3	5

- **SOLUTION** • The domain is the set of input values, which is {–1,–3}. The range is the set of output values, which is {8,2,5}.

d) Determine the domain and range of the relation described by the following graph:

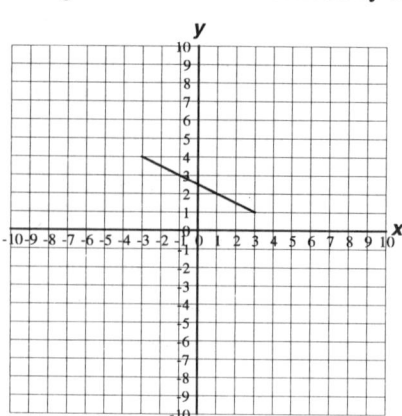

• SOLUTION • The input starts at –3, ends at 3, and includes all values in between. Symbolically, we can write this as $-3 \le x \le 3$, which literally means "negative three is less than or equal to x and x is less than or equal to 3. Using set builder notation, we can write the domain as $\{x \mid -3 \le x \le 3\}$.

The output goes from 1 to 4. Again, using set builder notation, we can write the range as $\{y \mid 1 \le y \le 4\}$.

Practice 4.8.1 Determine the domain and range.

1. (2,3) (5,–20) (6,4) (–1,–3)

2. (0,0) (4,–5) (–8,2)

3. $y = 3x$

4.

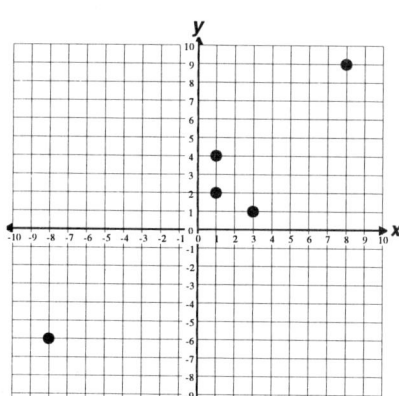

5.

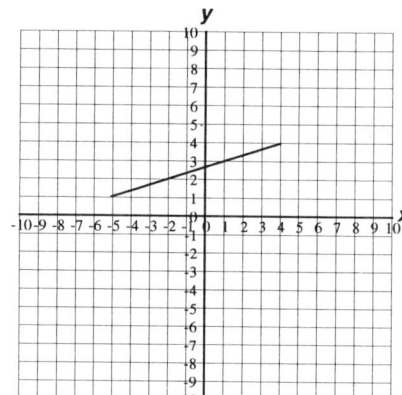

6.

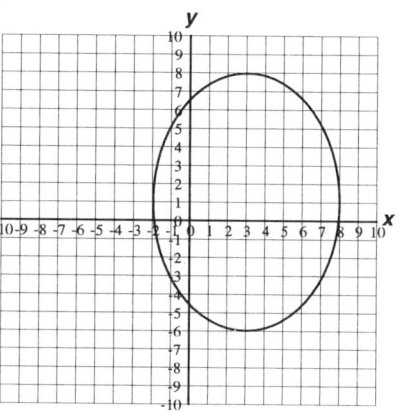

Topic 4.8 B — Functions

A **function** is a *relation* where every input corresponds to **only one** output. For example, the relation between time of day (input) and temperature (output) is a *function* because at any given time the temperature cannot be two different values. When people discuss a function they often say, "the output is a function of the input", so we might say "temperature is a function of time of day."

— .*Caution* — ***Function does not imply causality***	When we say "temperature is a function of time of day" we are implying that the two are related in some way, but we are not necessarily saying that the input *causes* the output. That is, we are not saying that time of day *causes* temperature changes (the sun and winds do that).

Many of the input/output applications we have studied so far are functions. For example:

- the distance, d, traveled by a car going 50 mph is a function of the time of travel, t. We may model this as $d = 50t$. We say that d is a function of t and that the equation $d = 50t$ represents a function.

- the interest, i, earned on an investment is a function of the principal, the interest rate, and the time. We may model this as $i = (500)(0.05)t$ for a principal of $500 and a rate of 5% per year. We say that i is a function of t and that the equation $i = (500)(0.05)t$ represents a function.

The mathematical models for these functions give us a rule to follow so we can find the output when we are given the input. For example, $d = 50t$ tells us to multiply the input, t, by 50 to get the output, d.

The fact that each input generates only one output makes functions important in applied problems because this enables us to make unambiguous predictions. For example, if $d = 50t$, then a car that has been driving 3 hours (t) will have gone 150 miles (d). Notice we have only one prediction that makes sense.

In future mathematics classes you will study many different types of functions including linear, quadratic, logarithmic, exponential, and trigonometric functions. All of the linear two-variable graphs we have discussed so far, except for the vertical straight lines, describe linear functions. A relation whose graph is a vertical straight line is *not* a function because a single value of x corresponds to many different values of y.

Example 4.8.2　　a)　Determine whether this relation is a function: $\{(2,1), (-1,5), (1,1)\}$

　　　　　•**SOLUTION**• This is a function since each input corresponds to only one output.

　　　b)　Determine whether this relation is a function: $y = x$

　　　　　•**SOLUTION**• This is a function since each value we select for x (input) will generate only one possible value for y (output).

　　　c)　Determine whether the relation described by the following data table is a function:

Input	Output
−1	8
2	3
6	−1
2	5

　　　　　•**SOLUTION**• This is NOT a function since an input of 2 corresponds to an output of 3 and to an output of 5.

Practice 4.8.2　Determine whether the relation is a function.

1. $\{(4,5), (2,2), (4,3)\}$　　　2. $\{(1,1), (3,-2), (1,4)\}$　　　3. $y = x^2$　　　4. $y = x^3$

5.

x	y
−1	3
0	3
1	3

6.

x	y
−2	3
−2	5
−2	9

Graphs that are curves may be functions, or not. If every input has only one corresponding output then we have a function. For example, the following is a graph of a function:

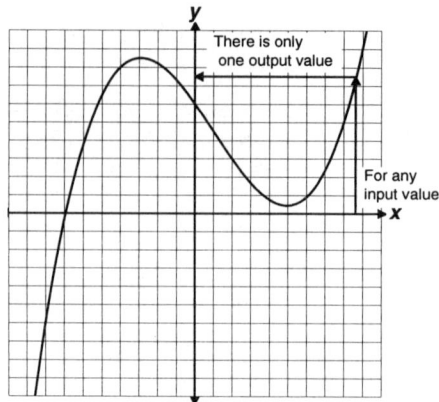

But, the following is NOT the graph of a function because some values of x correspond to more than one value of y,

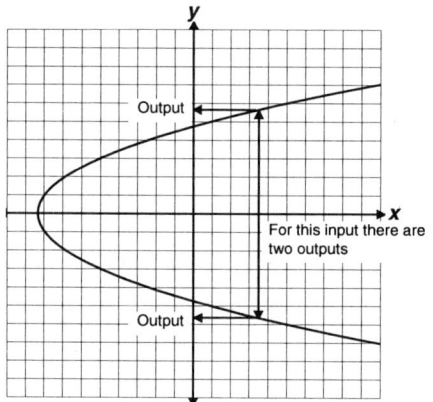

An easy way to see if a graph represents a function is to use the *vertical line test*.

— *Procedure* — *Vertical line test for a* *function*	If *any* vertical straight line can cross a graph more than once then the graph is NOT a function.

Example 4.8.3   a)   Is this the graph of a function?

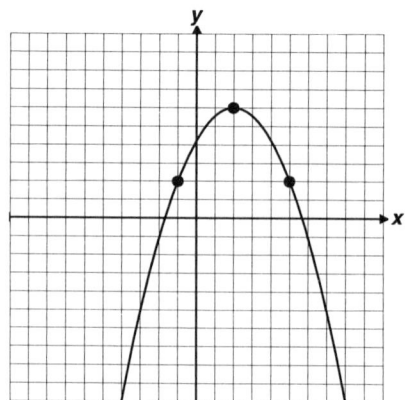

•SOLUTION• We can draw a vertical straight line anywhere and it will not hit the graph more than one time. Therefore, this is the graph of a function. Each value of x corresponds to only one value of y.

b) Is this the graph of a function?

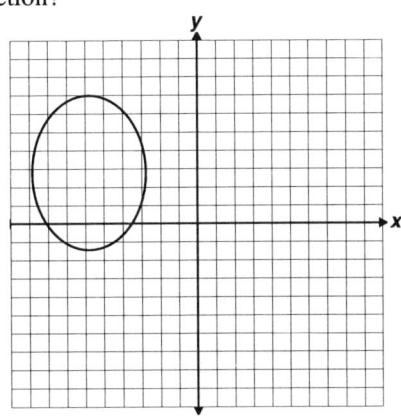

• **SOLUTION** • This is not the graph of a function because some values of x correspond to more than one value of y. For example, if we draw the vertical line $x = 4$ it will hit the graph twice, once at $(4,7)$ and again at $(4,-2)$.

c) Is this the graph of a function?

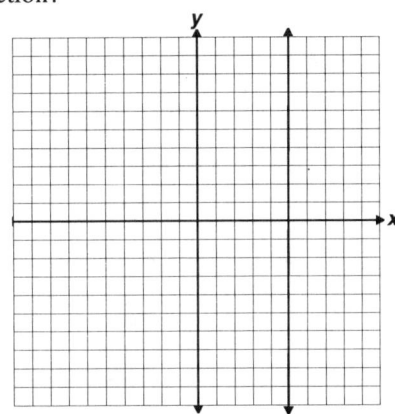

• **SOLUTION** • This is not the graph of a function because it's a vertical straight line. When x is 5, y can be any value.

Practice 4.8.3 Determine whether the given graph represents a function.

1.

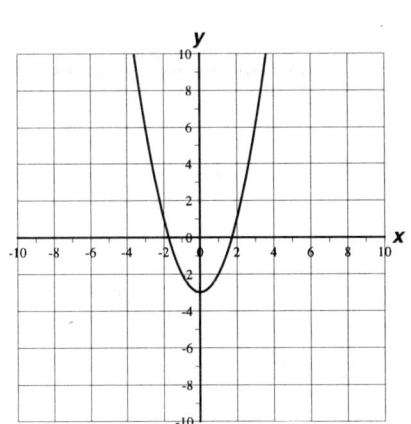

2.

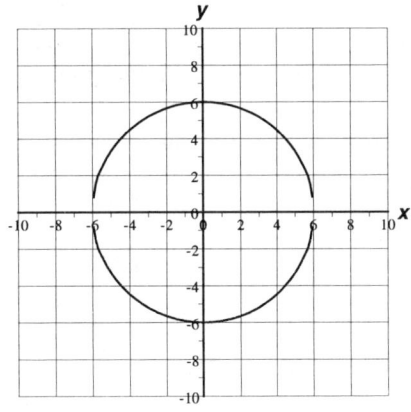

3.

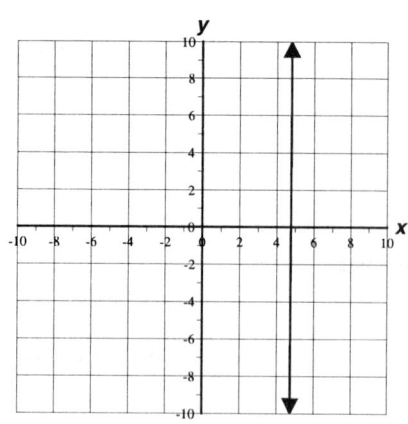

4.

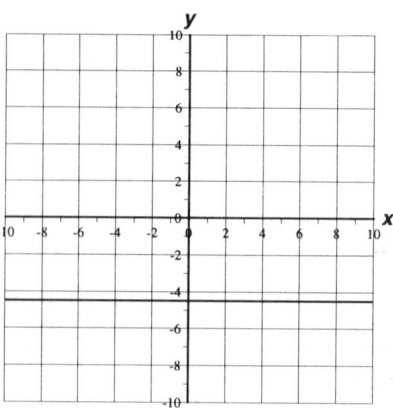

Topic 4.8 C — Function Notation

The model $w = -0.7t + 25$ shows the relation between whole milk consumption, w, and the number of years since 1970, t. This relation is a function because every value of t corresponds to only one value of w.

Mathematicians many times write such formulas using **function notation**. In this notation, we write the output variable with the input variable attached on the right and enclosed in parentheses. Using this notation,

$$w = -0.7t + 25 \text{ becomes } w(t) = -0.7t + 25$$

The symbol $w(t)$ is read "w of t" and literally means "the value of the function when the input has a value of t". The output is now represented by $w(t)$ rather than just w. Notice that the parentheses do **not** imply multiplication. Instead, the parentheses are used to quickly tell us what the input variable is. Thus, $w(t)$ tells us we have an output that depends on the value of the input variable, t.

In the whole milk example we used the variables w and t. Of course, with other situations we could use different letters or other symbols that make sense. For example,

*   to show that height (h) is a function of femur length (l) we might use $h(f) = 2.4f + 22.8$.

*   to show that cost (c) is a function of the number of items bought (n) we might write $c(n) = 2n + 1$.

To talk about functions in general mathematicians use the notation $f(x)$. Here, the f means we have a function, x is the input for that function, and $f(x)$ is the output.

— .Caution — *Different uses of* *parentheses*	We have used parentheses in several different ways. You have to look at the context in which parentheses are used to determine their meaning.
	Example — **Explanation**
	$w - (-t)$ — Parentheses make the meaning of the subtraction clearer (writing $w - -t$ might be confusing).
	$w - (t + 2t)$ — Parentheses indicate which operation is to be done first (the addition inside the parentheses).
	$(w)(t)$ — Parentheses indicate multiplication (they tell us to multiply the values of w and t).
	$2(w + t)$ — Parentheses indicate a factor so we can use the Distributive Property.
	$w(t) = 2t + 1$ — Parentheses indicate a function.

Topic 4.8 D — Evaluating Functions

When a function is represented as an equation, the equation tells us the procedure to follow to generate the output from a given input. For example $f(x) = 2x + 1$ tells us we can find the output, $f(x)$, by multiplying the input by 2 and then adding 1. Thus,

- $f(3) = 2(3) + 1$, which simplifies to 7.

- $f(a) = 2(a) + 1$, which simplifies to $2a + 1$.

— Procedure —	*Step 1*	Replace the variable with the given value.
Evaluating a function	*Step 2*	Simplify.

Example 4.8.4   a)   Given $f(x) = x^2 - 1$, find $f(-5)$.

•SOLUTION• The function tells us to square the input and then subtract 1. The notation $f(-5)$ tells us to do this for an input of –5.

$$f(x) = x^2 - 1 \qquad \Leftarrow \text{ Function.}$$
$$f(-5) = (-5)^2 - 1 \qquad \Leftarrow \text{ Replaced } x \text{ with } -5.$$
$$f(-5) = \mathbf{25} - 1 \qquad \Leftarrow \text{ Simplified the expression on the right.}$$
$$f(-5) = 24 \qquad \Leftarrow \text{ Subtracted.}$$

b)   Given $g(x) = 3(x - 5)$, find $g(8)$.

•SOLUTION•

$$g(x) = 3(x - 5) \qquad \Leftarrow \text{ Function.}$$
$$g(\mathbf{8}) = 3((\mathbf{8}) - 5) \qquad \Leftarrow \text{ Replaced } x \text{ with } 8.$$
$$g(8) = 3(3) \qquad \Leftarrow \text{ Subtracted inside parentheses.}$$
$$g(8) = 9 \qquad \Leftarrow \text{ Multiplied.}$$

c)   Given $h(x) = 3x^2 + 2x - 3$, find $h(0)$.

•SOLUTION•

$$h(x) = 3x^2 + 2x - 3 \qquad \Leftarrow \text{ Function.}$$
$$h(\mathbf{0}) = 3(\mathbf{0})^2 + 2(\mathbf{0}) - 3 \qquad \Leftarrow \text{ Replaced } x \text{ with } 0.$$
$$h(0) = \mathbf{0 + 0} - 3 \qquad \Leftarrow \text{ Simplified the expression on the right.}$$
$$h(0) = -3 \qquad \Leftarrow \text{ Added and subtracted.}$$

Practice 4.8.4   Find the value of the function for the input value indicated.

1.   Given $f(x) = 3x - 4$, find $f(-3)$ 2.   Given $g(x) = -(x + 7)$, find $g(4)$ 3.   Given $h(x) = 3x^2 + 2x - 5$, find $h(3)$

4.   Given $h(x) = 5x^3 + 2$, find $h(-2)$ 5.   Given $g(x) = 4x + 3$, find $g\left(\dfrac{1}{2}\right)$

It's important to keep in mind the meaning of $f(x)$ notation. For example, $f(3)$ represents the value of the function (the output) when the input is 3.

Example 4.8.5   a)   What is the meaning of $f(-2)$?

•SOLUTION• $f(-2)$ is the value of the function when the input is –2.

b)   What is the meaning of $f(3) = 7$?

•SOLUTION• This says that 7 is the value of the function when the input is 3.

c)   What is the meaning of $\dfrac{f(4) - f(2)}{4 - 2}$?

> •SOLUTION• $f(4)$ is the output when the input 4.
>
> $f(2)$ is the output when the input is 2.
>
> The expression represents the quotient of the difference in the outputs and the difference of the inputs. It's an example of the slope formula, $m = \dfrac{y_2 - y_1}{x_2 - x_1}$.

Practice 4.8.5   Write in English the meaning of the given equation.

1. $h(5) = 1$ 2. $f(-3) = 4$ 3. $g(-1.6) = 2.4$ 4. $g(3.7) = 5.5$

Since functions are relations, they can be represented in several different ways including sets of ordered pairs, data tables, graphs, and equations.

Example 4.8.6   a)   Use the graph to complete the data table.

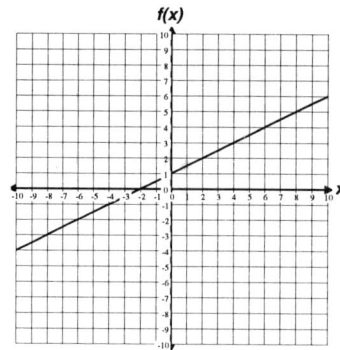

x	$f(x)$
-4	
	0
	2
0	

> •SOLUTION• From the graph we can see that
>
> - an input of -4 has an output of -1.
> - an output of 0 corresponds to an input of -2.
> - an output of 2 came from an input of 2.
> - an input of 0 leads to an output of 1.
>
> The completed data table looks like this:

x input	$f(x)$ output
-4	-4
-2	0
2	2
0	1

b)   Write the following as a set of ordered pairs: $h(5) = 0$, $h(0) = -5$, $h(-2) = 7$ and $h(6) = 7$.

> •SOLUTION• $h(5) = 0$ tells us that the output is 0 when the input is 5. As an ordered pair this would be written $(5,0)$. Likewise, the other ordered pairs would be $(0,-5)$ $(-2,7)$, and $(6,7)$.

c)   Plot on an x–y coordinate system: $f(2) = 0$, $f(-3) = 1$, $f(0) = 2$ and $f(5) = -3$

> •SOLUTION• The input is the value in the parentheses and the output is the value that $f(x)$ equals.

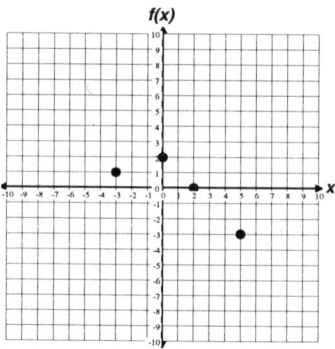

Practice 4.8.6

1. Use the graph to complete the table.

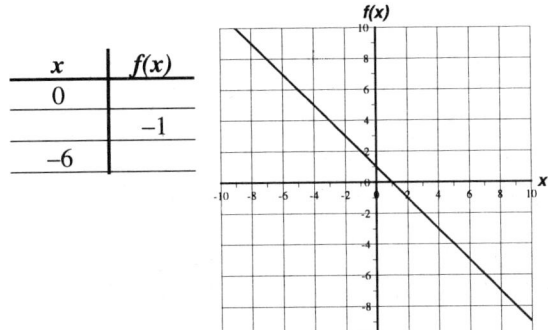

x	$f(x)$
0	
	-1
-6	

2. Use the graph to complete the table.

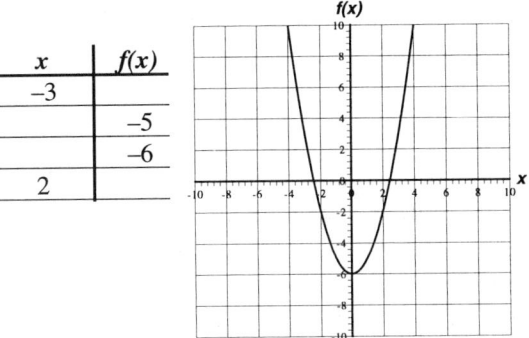

x	$f(x)$
-3	
	-5
	-6
2	

3. Write as a set of ordered pairs: $h(2) = -1$, $h(5) = 2$, $h(6) = 3$ and $h(0) = -3$

4. Plot on a rectangular coordinate system: $f(0) = 2$, $f(5) = -2$, $f(3) = 6$ and $f(-6) = -4$

— Note — *f(x) notation may seem* *awkward*	The $f(x)$ notation may seem awkward to you at first. However, you will encounter it so often in your future mathematics courses it's important for you to understand it and practice working with it now.

Topic 4.8 E — Function Notation and Applications

Function notation allows us to communicate ideas and questions in a condensed form. For example, in the whole milk model, $w(t) = -0.7t + 25$, the input is *number of years since 1970* and the output is *gallons of whole milk consumed per person*. We can interpret $w(t)$ as asking, "how much whole milk is consumed t years after 1970?". Using this interpretation, $w(1)$ asks "how much whole milk is consumed 1 year after 1970?" To answer the question we replace t with 1 and simplify.

$$w(t) = -0.7t + 25 \qquad \Leftarrow \text{Function.}$$

$$w(1) = -0.7(1) + 25 \qquad \Leftarrow \text{Replaced } t \text{ with 1.}$$

$$w(1) = 24.3 \qquad \Leftarrow \text{Simplified.}$$

The answer is usually written in the form $w(1) = 24.3$. In English this says "in 1971 the average person drank 24.3 gallons of whole milk."

Example 4.8.7   Given that $w(t) = -0.7t + 25$, answer the following:

a) Find the amount of whole milk consumed in 1976. That is, find $w(6)$.

•SOLUTION•

$$w(t) = -0.7t + 25 \qquad \Leftarrow \text{Function.}$$

$$w(6) = -0.7(6) + 25 \qquad \Leftarrow \text{Replaced } t \text{ with 6.}$$

$$w(6) = -0.7 + 25 \qquad \Leftarrow \text{Multiplied.}$$

$$w(6) = 20.8 \qquad \Leftarrow \text{Added.}$$

So $w(6) = 20.8$, which says 20.8 gallons of whole milk were consumed per person in 1976.

b) What is the meaning of $w(4)$?

•SOLUTION• Since 1970 is year 0, $w(4)$ is the amount of whole milk consumed per person in 1974.

c) What is the meaning of $w(15) - w(16)$?

> •SOLUTION• $w(15) - w(16)$ is the difference in whole milk consumption between 1985 (15 years after 1970) and 1986 (16 years after 1970).

d) What is the meaning of $w(9) = 18.7$?.

> •SOLUTION• Since 1979 is year 9, this says that the average person drank 18.7 gallons of whole milk in 1979.

e) What is the meaning of $w(35) = 0.5$?

> •SOLUTION• 35 years after 1970 is 2005. If the model is true, then in 2005 the average person will be drinking half a gallon of whole milk a year. This implies that the average person will stop drinking whole milk within the year since the slope says people will drink 0.7 gallons less per year.

Practice 4.8.7 The taxes taken out of a salesman's check can be represented by $t(w) = 0.3w$ where w represents his wages.

1. Find the amount of taxes he pays on wages of $250.

2. Express in English the meaning of $t(750)$.

3. Express in English the meaning of $t(450) - t(200)$.

4. Express in English the meaning of $t(1000) = 300$.

We conclude this section by using functions to model a real situation.

Example 4.8.8 The Boy Scouts are running an aluminum can recycling drive. Each pound of aluminum they collect can be sold for $0.05. The cost of renting a truck to collect the cans and deliver them to a recycling plant is $25.

a) Let x represent the weight of the aluminum collected (pounds) and y the profit earned (dollars). Build a function that shows profit as a function of the weight of aluminum.

> •SOLUTION• The rate of change is $0.05 per pound and the base is x pounds. The Scouts will earn $0.05x$ dollars for the x pounds of cans they collect. But, they will have to pay out $25 for the truck rental. Combining these we have
>
> $$\begin{array}{ccccc} \text{profit} & = & \text{money earned} & - & \text{money expended} \\ y(x) & = & 0.05x & - & 25 \end{array}$$

b) Calculate $y(100)$ and discuss it's meaning.

> •SOLUTION•
>
> $$y(x) = 0.05x - 25 \qquad \Leftarrow \text{Function.}$$
>
> $$y(100) = 0.05(100) - 25 \qquad \Leftarrow \text{Replaced } x \text{ with 100.}$$
>
> $$y(100) = 5 - 25 \qquad \Leftarrow \text{Multiplied.}$$
>
> $$y(100) = -20 \qquad \Leftarrow \text{Subtracted.}$$
>
> $y(100)$ is –20, which means that when 100 pounds of aluminum are collected, the profit is –20 dollars. That is, they have spent $20 more than they have earned.

c) Calculate $y(x) = 0$ and discuss it's meaning.

> •SOLUTION• Here we were asked to substitute 0 for $y(x)$.
>
> $$y(x) = 0.05x - 25 \qquad \Leftarrow \text{Function.}$$
>
> $$0 = 0.05x - 25 \qquad \Leftarrow \text{Replaced } y(x) \text{ with 0.}$$
>
> $$25 = 0.05x \qquad \Leftarrow \text{Added 25 to both sides.}$$
>
> $$500 = x \qquad \Leftarrow \text{Divided both sides by 0.05.}$$
>
> Since $y(x)$ represents the profit, $y(500) = 0$ means the profit is 0 when 500 pounds of aluminum have been collected. At this point they have spent all the money collected on the truck. Any additional money earned is profit.

d)   Calculate the amount of aluminum the group has to collect to make a profit of $30.

•**SOLUTION**• We substitute 30 for $y(x)$ and solve for x.

$$y(x) = 0.05x - 25 \quad \Leftarrow \text{ Function.}$$

$$\mathbf{30} = 0.05x - 25 \quad \Leftarrow \text{ Replaced } y(x) \text{ with 30.}$$

$$55 = 0.05x \qquad\quad \Leftarrow \text{ Added 25 to both sides.}$$

$$1100 = x \qquad\qquad \Leftarrow \text{ Divided both sides by 0.05.}$$

When 1100 pounds of aluminum have been collected the group will have made $30.

Practice 4.8.8

1. The cost of producing VCR's is $225 for each unit plus $1500 for operating expenses.

a)   If n represents the number of units made and c the cost, build a function that shows cost as a function of the number of units made.
b)   How units must be made if the total cost is $7125?
c)   Find $c(100)$.
d)   Find $c(150)$

2. The VCR company has to pay property tax of $15 on each unit in the warehouse at the end of the year, as well as $7500 property tax on the warehouse itself.

a)   If n represents the number of units in the warehouse at the end of the year and t represents the tax, build a function showing tax as a function of the number of units in the warehouse.
b)   Find $t(12)$ and state what it represents.
c)   What does $t(250) - t(125)$ represent?
d)   If the taxes are $7755, how many units are in the warehouse?

Exercise Set 4.8   The answers to the odd numbered exercises are at the back of the book.

Determine the domain and range.

1. $(-1,4)$ $(-2,8)$ $(2,5)$ $(8,32)$ 2. $(3,4)$ $(4,4)$ $(5,4)$ 3. $y + x + 2 = 0$ 4. $2x = y - 4$

5.

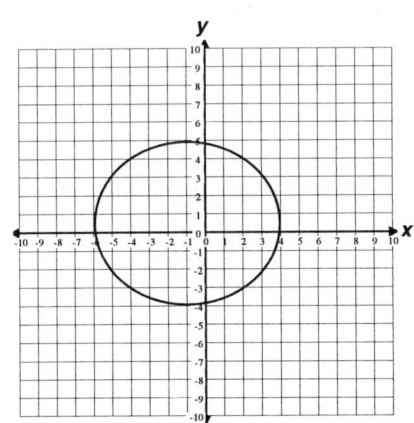

6.

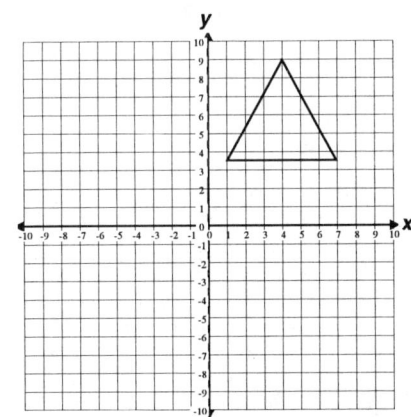

7.

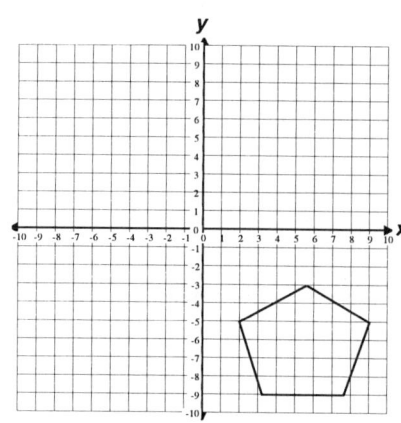

8.

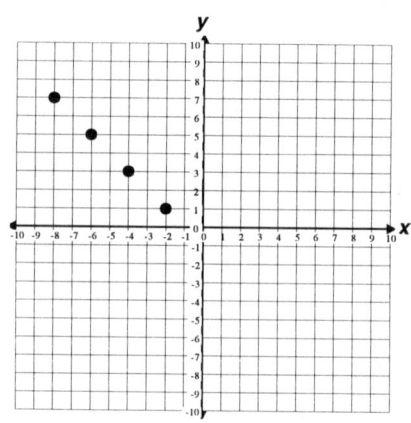

Determine whether the relation is a function.

9. {(1,–6), (4,–6), (8,12)}   **10.** {(2,9), (3,7), (2,–5)}   **11.** {(0,1), (1,1), (1,0)}   **12.** {(9,9), (8,8), (7,7)}

13.

x	y
12	6
21	–3
32	3

14.

x	y
8	1
6	1
4	1

Determine whether the given graph represents a function.

15.

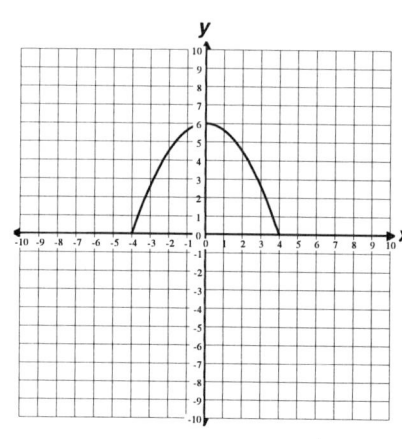

16.

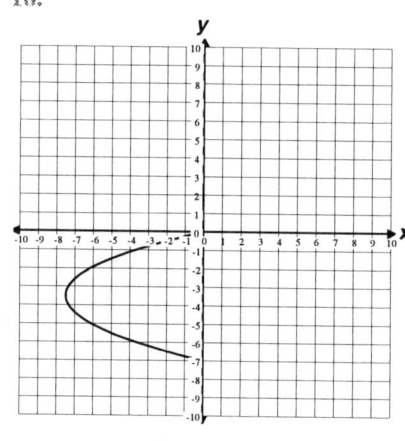

17.

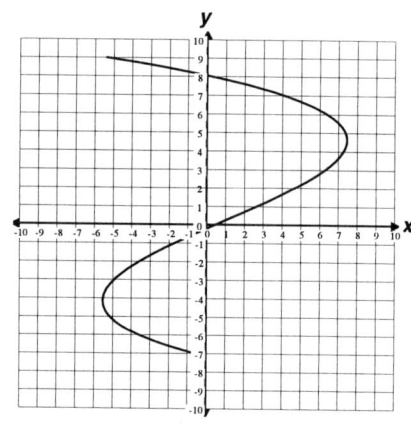

18.

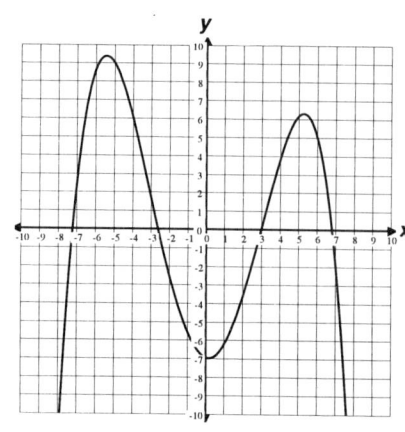

19.

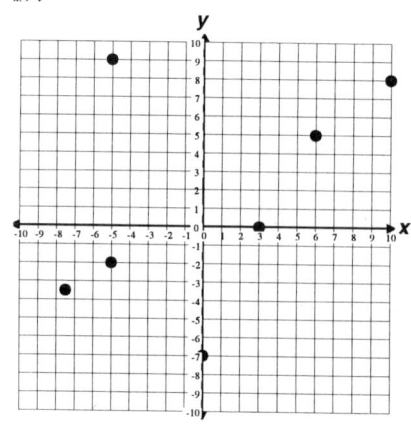

20.

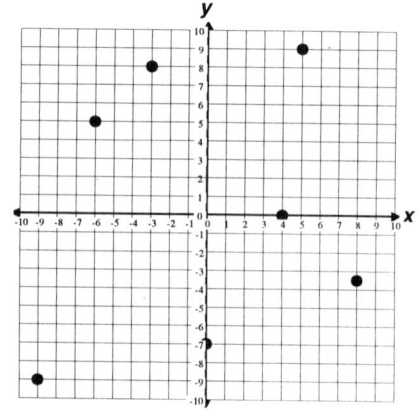

Find the value of the function for the input value indicated.

21. Given $f(x) = x^3 - x$, find $f(-1)$.

22. Given $g(x) = \dfrac{x-2}{2-x}$, find $g(0)$.

23. Given $a(x) = (x+5)(x-5)$, find $a(3)$.

24. Given $a(x) = |-x+7|$, find $a(20)$.

25. Given $h(y) = y^2 - y + 6$, find $h(-1)$.

26. Given $h(y) = \dfrac{y - \frac{1}{y}}{1-y}$, find $h(2)$.

27. Given $b(t) = t^3 + \sqrt{t}$, find $b(4)$.

28. Given $m(p) = p^3 - \sqrt{2+p}$, find $m(2)$.

29. Given $l(h) = \dfrac{h-2}{1-h}$ find $l\left(\dfrac{1}{2}\right)$.

30. Given $g(v) = \dfrac{v + \frac{1}{4}}{2-v}$, find $g\left(\dfrac{1}{2}\right)$.

Use the graph to complete the table.

31.

x	f(x)
	2
	5
	-4

32.

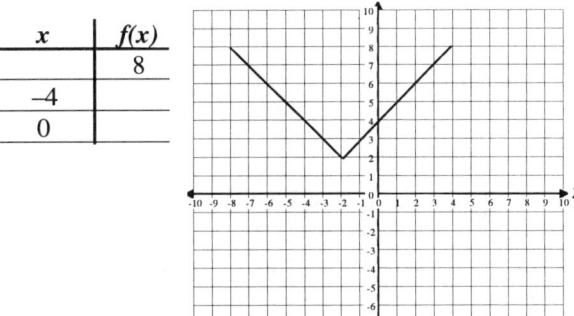

x	f(x)
	8
-4	
0	

33.

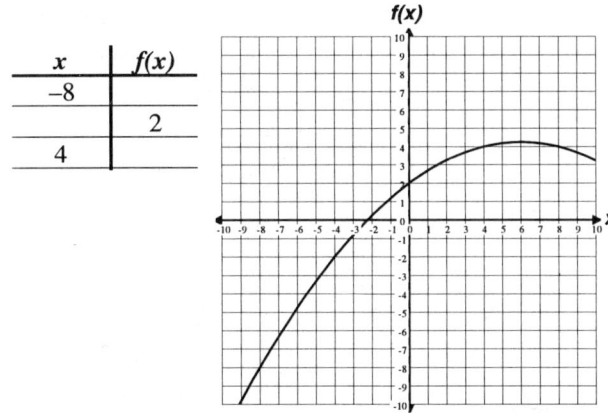

x	f(x)
-8	
	2
4	

34.

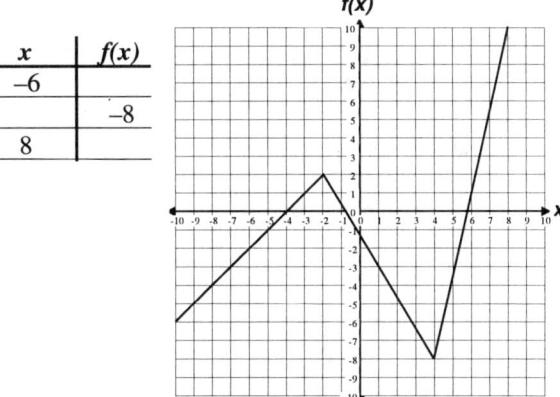

x	f(x)
-6	
	-8
8	

35. Write as a set of ordered pairs: $m(-4) = 3$, $m(0) = 1$, $m(-1) = 2$ and $m(6) = -4$

36. Write as a set of ordered pairs: $t(2) = 4$, $t(3) = -2$, $t(1) = 0$ and $t(-1) = -5$

37. Plot on an x-y coordinate system: $f(-4) = 0$, $f(-3) = 1$, $f(0) = 2$ and $f(5) = 3$

38. Plot on an x-y coordinate system: $f(-3) = 10$, $f(0) = 4$, $f(3) = 2$, and $f(7) = 10$

39. The distance a canoe floats in a river can be represented by $d(t) = 7t$. where t is the time the canoe has been floating, in hours, and d is the distance the canoe has traveled in miles.
    a) Find how far the canoe has floated after 135 minutes.
    b) Express in English the meaning of $d(1.5)$.
    c) Express in English the meaning of $d(1) - d(0.75)$.
    d) Express in English the meaning of $d(\frac{4}{3}) = 9\frac{1}{3}$.

40. The dollar cost c of s seats to a basketball game can be represented by $c(s) = 35s$.

    a) Find the cost of 6 seats.
    b) Express in English the meaning of $c(2)$.
    c) Express in English the meaning of $c(7) - c(3)$.
    d) Express in English the meaning of $c(21) = 735$.

41. The number of structural fires in the United States (in thousands) since 1980 can be represented by $f(t) = -32.6t + 1044$ where t is the year since 1980.

    a) Express in English the meaning of $f(12)$.
    b) Express in English the meaning of $f(8) - f(4)$.
    c) Express in English the meaning of $f(t) = 700$.
    d) Express in English the meaning of $f(15) = 555$.

42. The estimated population of the United States from 1996 to 2050 in millions can be represented by $p(t) = 2.4t + 265$ where t is the year since 1996.
    a) Express in English the meaning of $p(24)$.
    b) Express in English the meaning of $p(16) - p(11)$.
    c) Express in English the meaning of $p(t) = 400$.
    d) Express in English the meaning of $p(30) = 337$.

Review Exercises   The answers to all of these exercises are at the back of the book.

43. Evaluate the expression when x is -2 and y is -3: $-y + \dfrac{2x - 1}{x - y}$ 44. Simplify: $9x - 2(6x + 5)$

45. Solve: $\dfrac{x}{2} - \dfrac{x}{3} + 2 = \dfrac{5x}{6} - 10$ 46. Graph: $y = 3x + 1$ 47. Use the intercept method to graph $3x + 2y = 6$

48. Find the equation of a line parallel to $y = 2x - 3$ and which passes through $(3,1)$.   49. Graph: $y \geq 3x - 2$

50. The sum of two consecutive even integers is 122. Find the integers.

51. Construct the equation of the line shown in the graph.

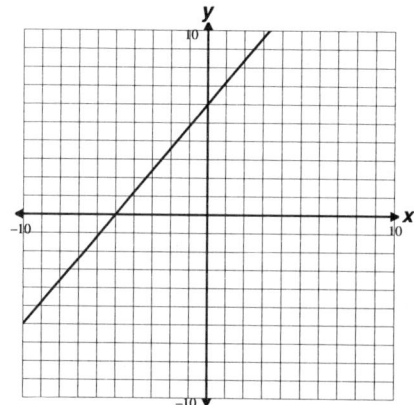

Chapter 4 Review

Vocabulary to Know

constraint — a limiting value in a problem.

coordinates — the location of a point on an *x-y* coordinate system, written in the form (input,output).

coordinate plane — the two-dimensional flat surface the *x-y* coordinate system rests upon.

data table — a grid used to organize data.

domain — the set of values allowed for the input variable.

extrapolation — estimating from outside the given data set.

function notation — notation used to indicate a function where the output variable is written in the form *f(x)* rather than *y*.

function — a relation where every input corresponds to only one output.

horizontal axis — the horizontal number line of an *x-y* coordinate system.

intercept method of graphing a linear equation — graphing a straight line by calculating and plotting its intercepts.

interpolation — estimating from within the given data set.

linear equation in two variables — an equation that contains two variables, each of which has an exponent of 1; the *x-y* graph of such an equation is a straight line; $y = b + mx$ is a linear equation in two variables.

linear regression — a mathematical procedure that determines the best linear equation which will fit a given set of input/output data.

linear relation — a relation between input and output where the *x-y* graph is a straight line.

ordered pair notation — an input value and its corresponding output value written in the form (input,output).

origin — the point of intersection of the zeros of the horizontal and vertical axes of an *x-y* coordinate system; the coordinates of the origin are (0,0).

plotting — locating points on a grid, such as an *x-y* coordinate system.

point-slope form — the equation of a line written in the form $y - y_1 = m(x - x_1)$.

range — the set of values allowed for the output variable.

region — a portion of the *x-y* plane.

regression line — the line which best fits the data of an input/output relation.

relation — a correspondence between two sets of numbers, such as input and output.

rise — the change in output when going from one point to another on a line.

run — the change in input when going from one point to another on a line.

scale — the size and placement of the numbers on the axes of an *x-y* graph.

slope — the ratio of the change in output to the change in input; slope describes the "steepness" of a line; the slope of a line that passes through (x_1,y_1) and (x_2,y_2) is

$$m = \frac{y_2 - y_1}{x_2 - x_1} \quad \begin{array}{l} \Leftarrow \text{ rise} \\ \Leftarrow \text{ run} \end{array}.$$

slope-intercept form — the form of a linear relation written as $y = mx + b$.

standard form of a linear equation in two variables — the equation of a straight line written in the form $Ax + By = C$.

test point — when graphing a linear inequality, a point used to see if a region contains solutions to the inequality.

vertical axis — the vertical number line of an *x-y* coordinate system.

x-intercept — an ordered pair whose output value is 0; the point where a graph crosses the *x*-axis.

x-axis — the horizontal number line of a *x-y* coordinate system.

x-coordinate — the input value of an ordered pair; the *x*-coordinate of (2,3) is 2.

x-y coordinate system — number lines drawn at right angles to each other and intersecting at their 0 points.

x-y graph — a graph drawn on an *x-y* coordinate system.

y-intercept — an ordered pair whose input value is 0; the point where a graph crosses the *y*-axis.

y-axis — the vertical number line of an *x-y* coordinate system.

y-coordinate — the output value of an ordered pair; the *y*-coordinate of (2,3) is 3.

Properties, Definitions, Formulas, and Notation to Understand

parts of an x-y graph

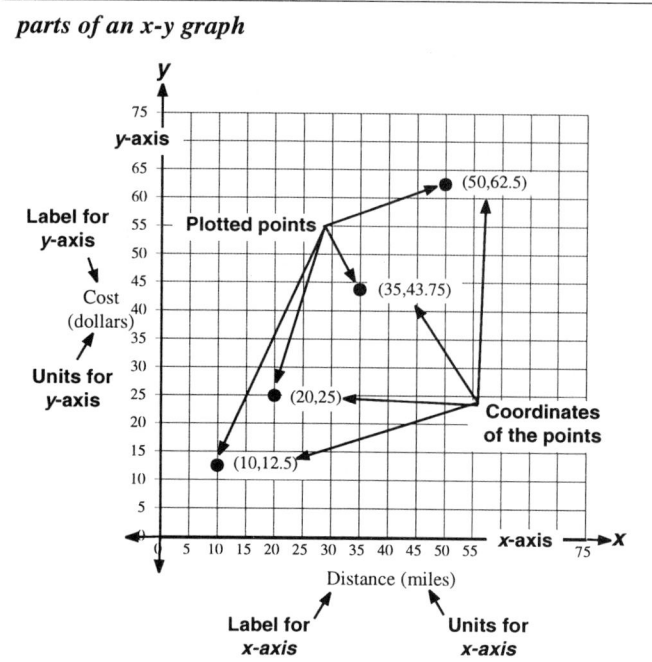

quadrants of an x-y graph — except for the points that fall exactly on one of the axes, all the points on the plane fall into one of these four Quadrants.

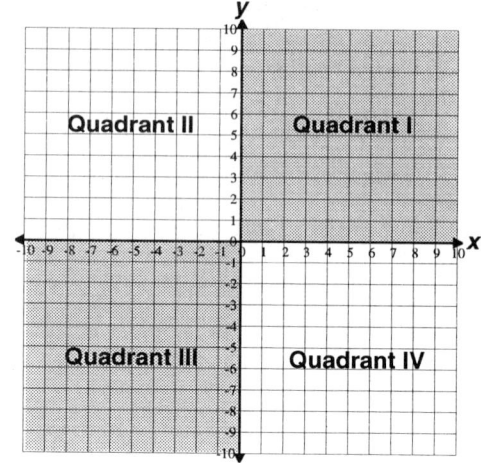

input and output	**Input**	**Output**
	• in a formula, we usually supply the values for the input variables.	• in a formula, we usually calculate the values of the output variable using the values for the input variable.
	• in a formula, the input variable usually is involved in mathematical operations and typically appears on the right side of the equation	• in a formula, the output variable usually is isolated on the left side of the equation.

INPUT VARIABLE *X* IS ON RIGHT AND NOT ISOLATED ⎤

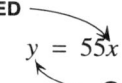

⎿ OUTPUT VARIABLE *Y* IS ISOLATED ON LEFT

	• the input variable generally is represented by *x*.	• the output variable generally is represented by *y*.
	• in a data table, input values appear on the left.	• in a data table, output values appear on the right.

Input	Output
first input value	first output value
second input value	second output value

	• in an ordered pair, the input value appears on the left as in (**input value**,output value)	• in an ordered pair, the output value appears on the right as in (input value,**output value**)
	• on a graph, input values appear on the horizontal axis.	• on a graph, output values appear on the vertical axis.

linear equation — an equation whose x-y graph is a straight line.	An equation written in the form $y = b + mx$ is a linear equation
	
	where
	    y represents the output.
	    x represents the input.
	    b represents a constant that tells us the value of the output when the input is 0; it usually represents the "initial" value of the output.
	    m represents a constant that describes how the input and output change relative to each other; it is a rate of change, like miles per hour.

slope of a line — the ratio of the change in the output to the corresponding change in the input when going from one point on the line to another. The formula is
 	
$$m = \frac{y_2 - y_1}{x_2 - x_1} \quad \begin{array}{l} \Leftarrow \text{ rise} \\ \Leftarrow \text{ run} \end{array}$$
 	
where
 	
    (x_1, y_1) is one point on the line.
 	
    (x_2, y_2) is another point on the line.
 	
m is 0 for a horizontal line
 	
m is undefined for a vertical line
	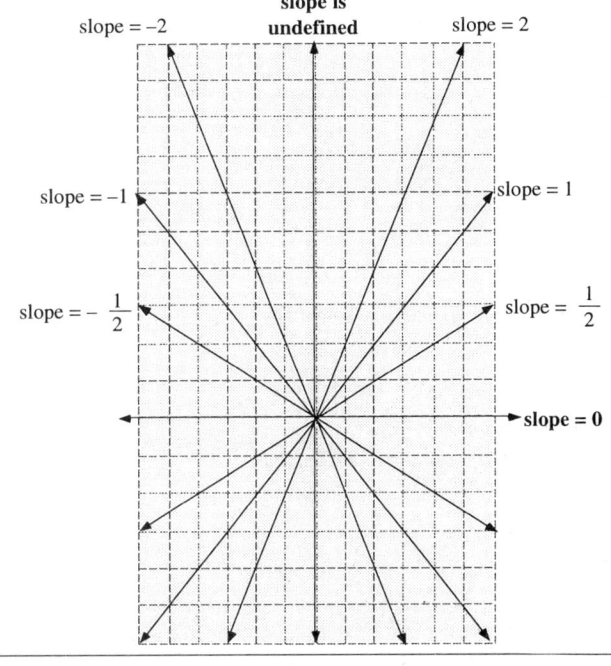

slopes of parallel lines — Parallel lines have the same slope.	Two straight lines with slopes m_1 and m_2 are parallel if $m_1 = m_2$.
slopes of perpendicular lines — The slopes of perpendicular lines are negative reciprocals. As a result, the product of the slopes of perpendicular lines is always –1.	We can state the condition on the slopes of two perpendicular line in two ways: • If one non-vertical straight line has a slope of m, then the slope of a straight line perpendicular to the given line is $-\dfrac{1}{m}$. OR • Two straight lines with slopes m_1 and m_2 are perpendicular if $m_1 m_2 = -1$.
equation of a horizontal line — The equation of a horizontal straight line contains only the output variable and the y-intercept. It does not contain the input variable.	If y is the input variable and b represents the y-intercept, then the equation of a horizontal straight line is $y = b$. $y = 3$ is a horizontal line 3 units above the x– axis.
equation of a vertical line — The equation of a vertical straight line contains only the input variable and the x-intercept. It does not contain the output variable.	If x is the input variable and k represents the x-intercept, then the equation of a vertical straight line is $x = k$ $x = 3$ is a vertical line 3 units to the right of the y– axis.

slope-intercept form	The equation of a straight line can be written as $$y = mx + b$$ where x is a variable that represents the input y is a variable that represents the output m is a constant that represents the slope of the line. b is a constant that represents the y-component of the y-intercept. The line with slope 3 and y-intercept (0,–2) can be represented by $y = 3x - 2$.
point-slope form of a line	The equation of a straight line can be written as $$y - y_1 = m(x - x_1)$$ where m represents the slope of the line. (x,y) represents a general point on the line. (x_1,y_1) represents a specific point on the line. The line with slope 3 and which goes through (2,5) can be represented by $y - 5 = 3(x - 2)$.
standard form of a linear equation in two variables	The equation of a straight line can be written in standard form as $Ax + By = C$ where A, B, and C are real numbers and A and B are not both 0. x is a variable that represents the input. y is a variable that represents the output. $3x + 4y = -2$ is written in standard form.

Procedures to Follow

Procedure for building an equation, y = b + mx, from a verbal description.	A dry cleaning store charges customers $1.50 per pound of clothes plus a $3 handling charge for each order. Build a linear equation that will predict the cost given the pounds of clothes left for cleaning.

Step 1 Organize the information   Read the problem, several times if necessary:
 	
• Given: determine what is given. Be sure to include the unit of all numbers.
 	
• Find: determine what you are asked to find.
 	
• Sketch: draw a sketch if it will help you better understand the problem.
 	
• Rate of change: With these problems it may not be initially clear what the amount (the output) and the base (the input) should be. Therefore, we usually delay constructing an Amount Table until after the variables and the rate of change have been precisely defined or identified.
 	
• Formula: $y = b + mx$.
	Given: • $3 service charge.
	       • $1.50 per pound charge.
	
	  Find: A linear equation that will predict the cost given the pounds of clothes.

Step 2 Define unknowns

- Identify the output (what you are trying to predict) and its unit. This will be represented by y.

- Identify the input (what you are using to predict the output) and its unit. This will be represented by x.

- Identify the initial value, b, and its unit. This constant must have the same unit as the output.

- Identify the rate of change, m, and its unit. The rate of change is the ratio $\dfrac{\text{change in output}}{\text{change in input}}$. The numerator must have the same unit as the output; the denominator must have the same unit as the input.

- Amount Table: At this point it may be useful to organize the information in an Amount Table.

- output: We want to predict cost so y will represent the cost (dollars).

- input: We will predict the cost from the weight of the clothes so x will represent the weight of the clothes (pounds).

- initial value: The service charge is $3.

- rate of change: Each pound of clothes costs $1.50.

Step 3 Verbal model This is given in the statement of the problem.	Given.
Step 4 Mathematical model Substitute the values of b and m into the linear equation, $y = b + mx$.	$y = 1.5x + 3$

Procedure for building a verbal model from the linear equation $y = b + mx$.

The equation $y = -\dfrac{1}{3}x + 2$ represents the value of a copier (in thousands of dollars) x years after it was purchased. Use English to express a meaning for this equation.

Step 1 Input/output Identify the input and the output, including their units.	- input is the number of years since the copier was bought. - output is the value of the copier in thousands of dollars.
Step 2 Initial value Identify the initial value and its unit. The initial value has the same unit as the output.	The initial value is the value of the output when the input is 0. That is 2 thousand dollars.
Step 3 Rate of change Identify the rate of change and its unit. The numerator has the same unit as the output and the denominator has the same unit as the input.	The rate of change is $\dfrac{-1 \text{ thousand dollars}}{3 \text{ years}}$
Step 4 English Write an English description using the information from Steps 1 through 3.	The copier originally was purchased for $2,000 and loses $1,000 in value every 3 years (or about $333 per year).

Procedure for using the intercept method to graph a linear equation.

Graph: $y = 2x - 4$

Step 1 y-intercept In the equation, substitute 0 for x and solve for y. This ordered pair $(0,b)$ is the y-intercept.	$y = 2(\mathbf{0}) - 4$ $y = -4$ The y-intercept is $(0,-4)$
Step 2 x-intercept In the equation, substitute 0 for y and solve for x. This ordered pair $(a,0)$ is the x-intercept.	$\mathbf{0} = 2x - 4$ $2 = x$ The x-intercept is $(2,0)$
Step 3 Plot Plot the two points on an x-y coordinate system and connect the points with a straight line.	

Procedure for writing the linear equation of a graph.	Construct the equation of the line shown below
	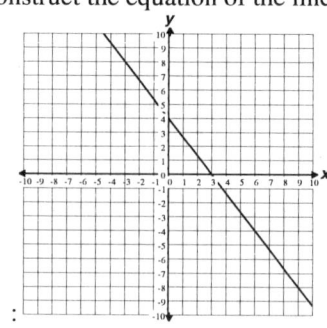
Step 1 Find b Find the value of y where the line crosses the y-axis; this is b.	The y-intercept is $(0,4)$; therefore, b is 4.
Step 2 Find m Locate a point on the line where you can easily determine the coordinates (usually, this is the intersection of two grid lines). Starting at that point, move vertically and then horizontally until you reach another easily identifiable point on the line. Write the ratio of the vertical distance you moved to the horizontal distance; that is, write $\dfrac{\text{change in } y \text{ (rise)}}{\text{change in } x \text{ (run)}}$. This is m.	Starting at $(0,4)$ and moving to $(3,0)$ we move down 4 units (the rise is -4) and over 3 units (the run is 3). Therefore, the slope is $\dfrac{-4}{3}$.
Step 3 Equation Substitute the values found in Steps 1 and 2 in the equation $y = b + mx$.	$y = 4 + \dfrac{-4}{3} x$ or $y = 4 - \dfrac{4}{3}x$

Procedure for graphing a line given a point and the slope.	Graph the line with slope 2 and which passes through $(6,1)$
Step 1 Plot the given point.	See below.
Step 2 Write the slope in the form $\dfrac{\text{change in } y}{\text{change in } x}$.	$m = \dfrac{2}{1}$
Step 3 Starting at the given point, • move vertically the amount indicated in the numerator of the slope. • move horizontally the amount indicated in the denominator of the slope. • draw a dot.	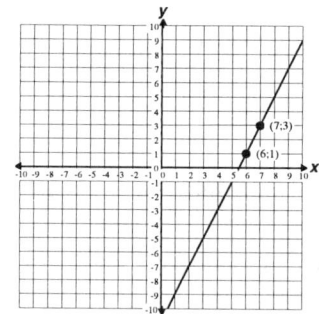
Step 4 Draw a straight line through the points.	

Procedure for building a linear equation from two ordered pairs.	Find the equation of the line that passes through the points $(2,3)$ and $(-1,5)$.
Step 1 Slope Use the slope formula to calculate the slope.	$m = \dfrac{3-5}{2-(-1)}$, which simplifies to $\dfrac{-2}{3}$
Step 2 y-intercept In the slope-intercept form of the linear equation, $y = mx + b$, • substitute the slope for m. • substitute for x and y the coordinates of one point. • solve for b.	Substituting 3 for y, $\dfrac{-2}{3}$ for m, and 2 for x gives $3 = \dfrac{-2}{3}(2) + b$ and solving gives $\dfrac{13}{3} = b$
Step 3 Equation Write the equation using the values of m and b found above.	$y = \dfrac{-2}{3} x + \dfrac{13}{3}$

Procedure for graphing a linear inequality.	Graph: $y \leq x - 2$
Step 1 Graph the boundary To find the boundary that separates the solutions from the non-solutions, graph the linear *equality*. If the inequality symbol is $\geq$ or $\leq$, the points on the line are part of the solution; in that case, draw a solid line. Otherwise, draw a dotted line to show that the points on the line are not part of the solution.	Graphing $y = x - 2$ gives 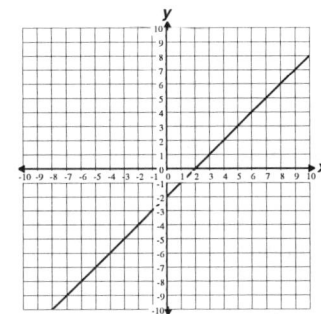 The boundary line is solid because the inequality is less than or **equal to**.
Step 2 Shade the Region To determine which region contains the solutions, choose a test point that is NOT on the line. The point (0,0) usually is a good choice since it's easy to locate. If the test point makes the inequality true, shade the side of the line that contains the point. Otherwise, shade the other side of the line.	Substituting (0,0) into $y \leq x - 2$ gives $0 \leq 0 - 2$, which is false. Therefore, we shade the region that does not contain (0,0). 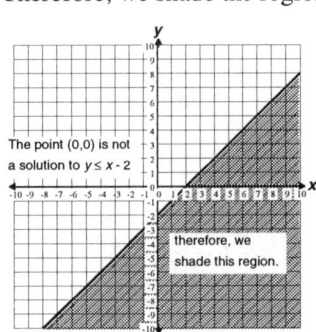

Vertical line test for a function.	Is this the graph of a function? 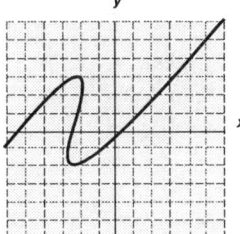
If *any* vertical straight line can cross a graph more than once then the graph is NOT a function.	This is *not* a function because a vertical line can cross the graph at more than one point: 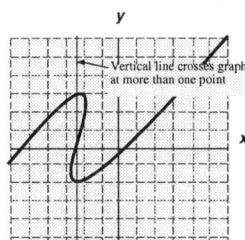

Procedure for evaluating a function.	Given $f(x) = 2x^2 + 5x - 1$, find $f(-3)$
Step 1 Replace the variable with the given value.	$f(-3) = 2(-3)^2 + 5(-3) - 1$
Step 2 Simplify.	$f(-3) = 18 - 15 - 1$, which simplifies to $f(-3) = 2$

Chapter 4 Review Exercises   The answers to all the exercises are in the back of the book.

1. A machine can stuff envelopes at a rate of 80 envelopes per minute. Build an equation that will predict the number of envelopes stuffed after the machine as worked for *x* minutes.

2. A machine can stuff envelopes at a rate of 80 envelopes per minute. Build an equation that will predict the time it will take the machine to stuff any number of envelopes.

3. An airplane starts on the ground and ascends at a rate of 500 feet per minute. Build an equation that will predict the height of the plane (feet) given the number of minutes after takeoff. Then construct a data table that has data for the plane for times of 2 minutes, 5 minutes, and 10 minutes.

4. An airplane starts on the ground and ascends at a rate of 500 feet per minute. Build an equation that will predict the time since the plane took off (minutes) given height of the plane (feet). Then construct a data table that has data for the plane for heights of 100 feet, 800 feet, and 1000 feet.

---

5. For the last 2 hours a group has been selling soft drinks at a fair at the rate of 30 drinks every 15 minutes.

   a) Create an equation from the verbal description. Let the input be the minutes since the group started selling drinks and the output be the number of drinks sold.

   b) Complete the following data table:

x time since started (minutes)	y number of drinks sold
120	
240	
600	
	540
	870

   c) Plot the points on the following grid:

   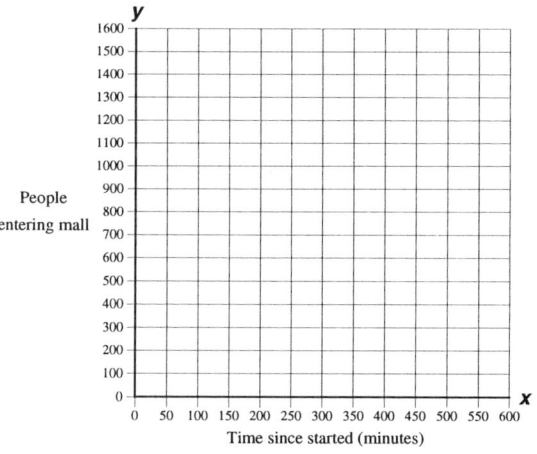

---

6. A student finds that she spends on average $3.80 whenever she eats lunch at school.

   a) Create an equation from the verbal description. Let the input be the number of lunches she buys and the output be the total dollars she has spent.

   b) Complete the following data table:

x number of lunches bought	y amount spent (dollars)
5	
20	
120	
	494
	760

   c) Plot the points on the following grid:

   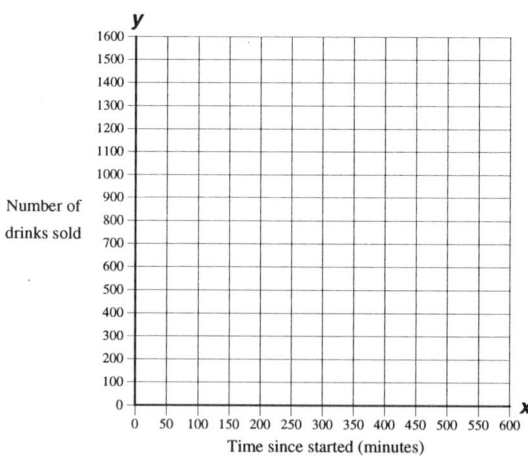

---

7. In 1932 the men's Olympic gold metal distance for the discus was 49.5 meters. Since then the distance has been increasing about 0.33 meters per year. Write a linear equation which will output the current gold metal distance given the year since 1932.

8. Since 1985 the amount of paper and paperboard recovered for recycling has been increasing by 1.64 million tons per year. Build a linear model using the year since 1985 as the input and the millions of tons of recovered paper and paperboard as the output. The amount recovered in 1985 was 13 million tons.

9. In 1982 the motion picture industry had annual receipts of 12.76 billion dollars. Since then the annual receipts have increased by 3.33 billion dollars per year.

    a) Build a linear equation that will predict the annual receipts for the motion picture industry given the year since 1982.
    b) Use your linear equation to predict when annual receipts will top 50 billion.
    c) Use your linear equation to predict the annual receipts in 2010.

10. Bert lives 420 miles from his parents. After visiting them he returns to his home driving at a constant speed of 65 mph.

    a) Build a linear equation that predicts the miles he has left to drive given the number of hours since leaving his parents.
    b) Use your linear equation to estimate the distance left to his apartment 3.25 hours after leaving.
    c) If Bert wants to take a break after driving for 250 miles, how long after leaving his parents will this be?

11. The model $y = 72 + x$ shows the relation between y, the average number of days for the school term for primary and middle school until 1959, and x, the number of years since 1869. Write an English explanation for this equation.

---

12. Plot on an x-y coordinate system and use a linear model to fill in the missing input or output values.

x	y
-2	-6
1	3
2	6
-1	
-3	

13. Plot on an x-y coordinate system and use a linear model to fill in the missing input or output values.

x	y
-3	9
0	3
4	-5
-2	
1	

14. The following data compare the fuel efficiency (miles per gallon) of the most fuel inefficient cars and the annual fuel cost (dollars). Graph the data using a scale of thousands of dollars on the y–axis.

x mileage (miles per gallon)	y annual fuel cost (thousand of dollars)
9	1841
11	1557
12	1446
13	1341

15. Discuss the meaning of the graph.

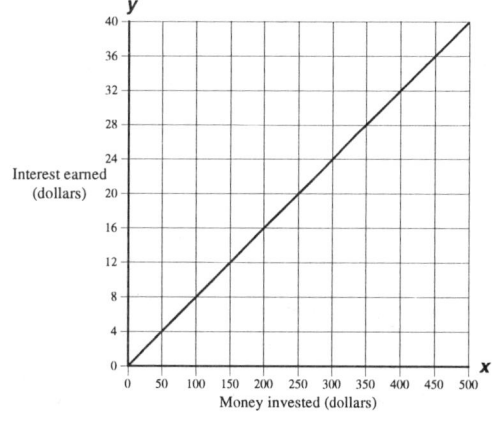

---

16. The Environmental Protection Agency (EPA) estimates how much it will cost each year for fuel given the miles per gallon (mpg) your car gets in the city. Due to differences in the cars one model does not work well for all cars. The best car in 1996 got 44 mpg while driving in the city. The annual fuel cost was $392. The fifth best car out of the top five got 33 mpg (city) and the annual fuel cost was $513. If mpg is the input and annual fuel cost is the output use this information to answer the following questions.

    a) Create a data table for this situation.
    b) Assume a linear relation and graph the relation using the coordinate system on the right.
    c) It's common for a small car to get around 28 mpg while driving in the city. What would the annual fuel cost be?
    d) If you did all city driving and you wanted your annual fuel costs to be under $650 a year, what mpg should you look for in a car?
    e) The worst car in the survey got 9 mpg (city) what do you estimate the annual costs to be?
    f) The worst car actually had an annual cost of $1841. Discuss some reasons the prediction is not close to the actual value.

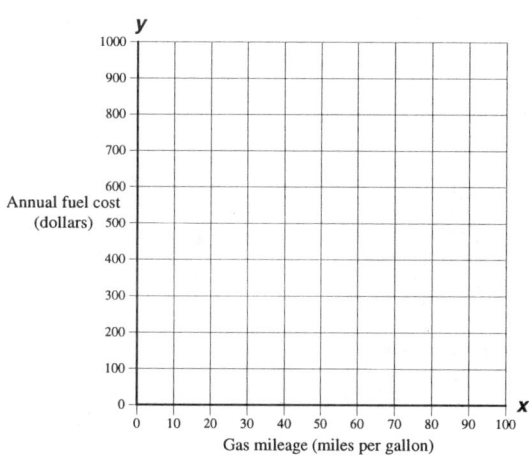

17. Graph: $y = -x + 2$

18. Graph: $6x + 2y = 8$

19. Find the x-intercept and the y-intercept:

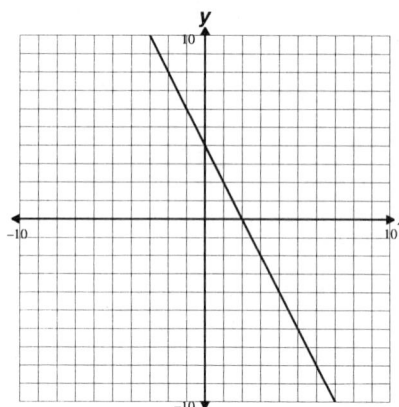

20. Find the x-intercept and the y-intercept:

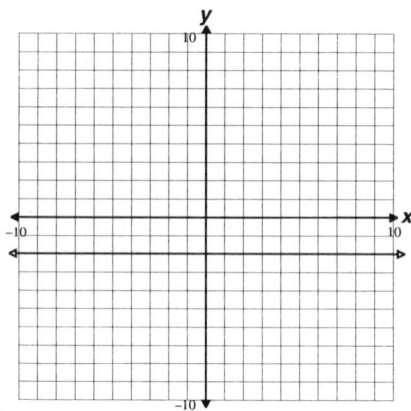

21. Use the intercept method to graph $2x - y = 4$

22. Use the intercept method to graph $y = -3x + 6$

23. Construct the equation of the line shown in the graph.

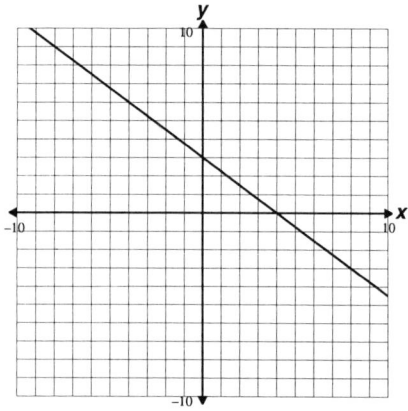

24. Construct the equation of the line shown in the graph.

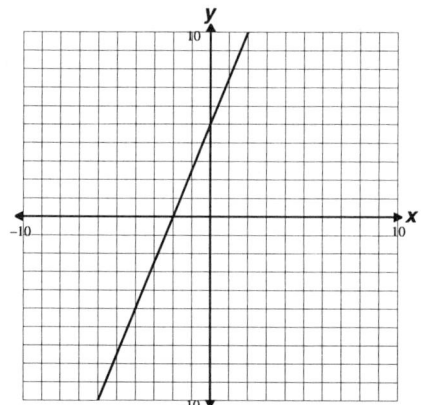

25. Graph the straight line that passes through (3,6) with slope -1

26. Graph the straight line that passes through (4,4) with slope 0.

27. Calculate the slope of the straight line that passes through (1,–1) and (–3,3).

28. Calculate the slope of the straight line described by the following data table:

x	y
7	2
1	5

29. Calculate the slope of the line that would connect the points given in the data table and discuss the meaning of the slope.

x number of years since 1970	y domestic fuel consumption (billions of gallons)
0	95
11	115
20	131
25	140

30. The cost for $100,000 of term life insurance for males and females of the same age is as follows: At 30 years of age the price for a male is $12.95 per month while for a female it's $12.12 per month. At 40 years of age a male policy is $15.65 per month and a female policy is $13.74. At 50 years of age it's $30.10 for a male and $40.47 for a female. At 60 years of age the price is $64.80 for a male and $43.23 for a female. Assume the price for a male policy is the input and the price for a female policy is the output. Calculate the slope of the line that would connect the given points and discuss the meaning of the slope.

31. Decide whether these two straight lines are parallel, perpendicular or neither: *Line 1* passes through the points (1,3) and (–4,4) and *Line 2* passes through the points (4,–2) and (5,3).

32. Decide whether these two straight lines are parallel, perpendicular or neither:

Line 1 comes from the data table

x	y
6	0
–2	5

Line 2 comes from the data table

x	y
9	1
1	6

33. Find the slope and *y*-intercept of the line described by $y = x - 4$

34. Find the slope and *y*-intercept of the line described by $x - 3y = 6$

35. Write the equation of a line with slope –2 and *y*-intercept (0,3)

36. Find the equation of the line that goes through (0,4) and (1,1).

37. Find the equation of the line that goes through (–2,3) and (5,3).

38. Find the equation of a line parallel to $y = 3x - 9$ and which passes through (2,1).

39. Find the equation of a line perpendicular to $y = -2x - 5$ and which passes through (2,7).

40. The data table shows the projected life expectancy for women from 1960 to 2010.

    a) Build a linear equation for this situation.
    b) Discuss the meaning of the equation.
    c) Predict the life expectancy for women in 2050.
    d) When will the life expectancy for women be 85 years?

x number of years since 1960	y life expectancy for women (years)
0	74.5
30	78.7
50	81.5

41. The data table shows the public elementary and secondary estimated finances for the period 1983 to 1995. The regression equation is $y = 14.5x + 118.7$.

    a) Interpret the meaning of the equation.
    b) Compare the values for 1985 and 1990 to the table values.
    c) Using only the model, when should finances be double those of 1985?

x number of years since 1983	y school finances (billions of dollars)
0	125.2
1	133.4
2	147.0
3	161.1
4	170.7
5	184.4
7	218.2
10	268.8
12	286.0

42. The following graph shows the median income for U. S. families from 1947 to 1994.

    a) If a linear model is appropriate, use two points on the regression line to build the regression equation.
    b) Interpret the meaning of the equation.
    c) Predict the median income in the year 2000.
    d) The change between 1950 and 1970 was actually $16,633. What does the model predict the difference to be?
    e) When will the median income exceed $50,000?

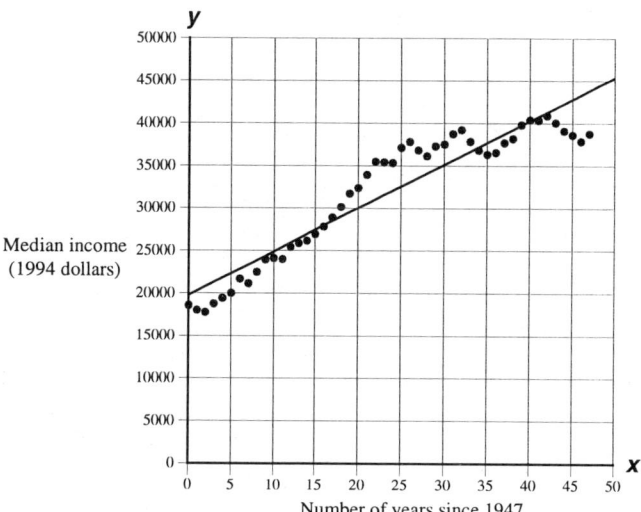

Median income (1994 dollars)

Number of years since 1947

43. The graph shows the relationship between tuna caught
    by the U. S. fishing industry between 1985 and 1993.
    a) If a linear model is appropriate, use two points on
       the regression line to build the regression equation.
    b) Interpret the meaning of the equation.
    c) Estimate the catch in 1980.
    e) When does the model predict the pounds caught
       will return to the 1988 level.

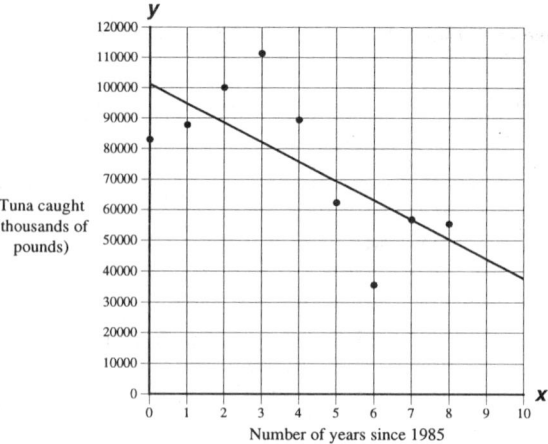

44. Is the relation shown in the data table a function?

x	y
3	0
0	3
−1	5
−7	−2
−1	6

45. Is this relation a function? {99,3),(–5,4),(7,–2),(2,–1),(10,–3)

46. Graph: $y > 3x - 2$

47. Determine the domain and range of {(6,2),(3,5),(6,0),8,5),(1,2)}

48. Graph: $x < 3$

49. Given $f(x) = 3x - 9$, find $f(-5)$.

50. Given $f(x) = 2x^2 + x - 4$, find $f(-3)$.

51. Determine the domain and range of

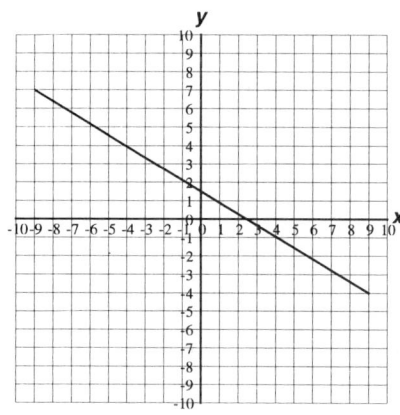

52. Determine the domain and range of

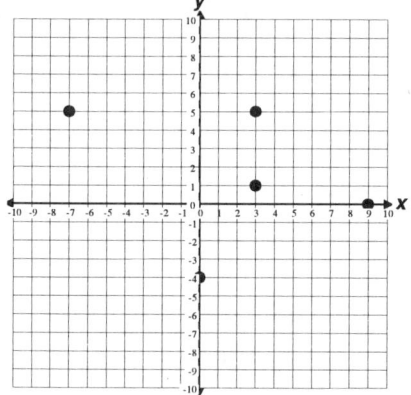

53. Is the following relation a function?

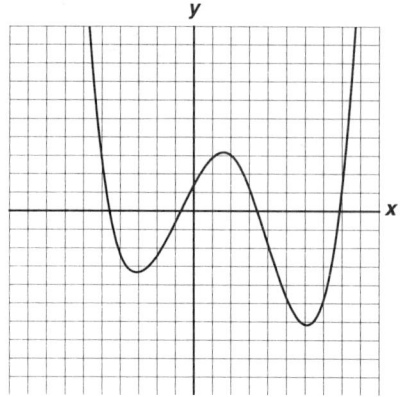

54. Is the following relation a function?

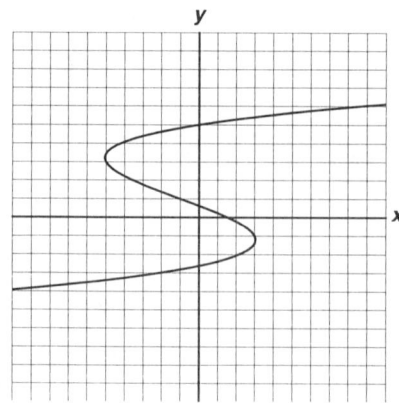

55. Use the graph to complete the data table:

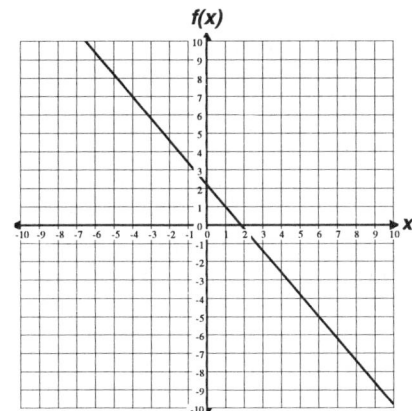

f(x)

x	$f(x)$
−4	
1	
6	

56. Write as a set of ordered pairs: $g(1) = 2$, $g(4) = 5$, $g(-1) = 0$ and $g(0) = 1$

57. The taxes (dollars) taken out of a clerk's check can be represented by $t(w) = 0.3w$ where w represents his wages (dollars).

    a) Find the taxes he pays on wages of $500.
    b) Express in English the meaning of $t(2000)$.
    c) Express in English the meaning of $t(750) - t(225)$.
    d) Express in English the meaning of $t(500) = 150$.

58. The equation $d(t) = 400 - 65t$ represents the distance (miles) you are from home after driving for t hours.

    a) Find the distance left to travel after 2 hours and 45 minutes.
    b) Express in English the meaning of $d(1) - d(2)$.
    c) Express in English the meaning of $d(t) = 100$.
    d) Express in English the meaning of $d(1.25) = 318.75$.

59. In 1970 54% of incoming college freshman were male. From then until 1993 the percent of incoming freshman who are male dropped by $\frac{2}{5}$ of a percentage point each year. For the following questions, let t represent the number of years since 1970 and m the percent of incoming freshman who are male:

    a) Build a function that shows the percent of incoming male freshman as a function of the year.
    b) Find the year in which $m(t) = 50$. Interpret the result in English.
    c) Calculate $m(-5)$. Interpret the result in English.
    d) Calculate $m(10) - m(15)$. Interpret the result in English.

Chapter 4 Test   The answers to all the exercises are in the back of the book.

1. A hot air balloon is ascending at a rate of 200 feet per minute. Build a model for this situation in the form of an equation.

2. In 1982 annual receipts from businesses engaged in laundry and other cleaning services was 10.2 billion dollars. Since then the annual receipts have increased by 0.83 billion per year. Write a linear equation which will output the annual receipts in billions of dollars given the year since 1982.

3. For the same car, an automobile dealer will pay one price if he buys it and a different price if he sells it. Assume that for a certain make of car the dealer charges you $1564 plus $1.07 for every $1 he paid for the trade–in.

    a) Build a linear equation that will predict the cost you pay to buy a car given the price the dealer paid for the same car.
    b) Use your linear equation to predict your cost if the dealer paid $3200 for a car.
    c) If the dealer will pay you $1800 for your car, how much will he sell it for?

4. By the end of the term an instructor will have assigned 400 points on the basis of 6 tests. To receive an A a student must have received a minimum of 360 test points. A student uses the equation $y = 400 - 7x$, where x is the number of tests, as a model for her particular situation in class. Write an English explanation for this equation.

5. A rectangle is to have a length which is twice it's width.

   a) Create an equation from the verbal description. Let the input be the width and the output be the length.

   b) Complete the following data table:

x width (inches)	y length (inches)
2	
8	
5	
	12
	15

   c) Plot the points on the following grid:

   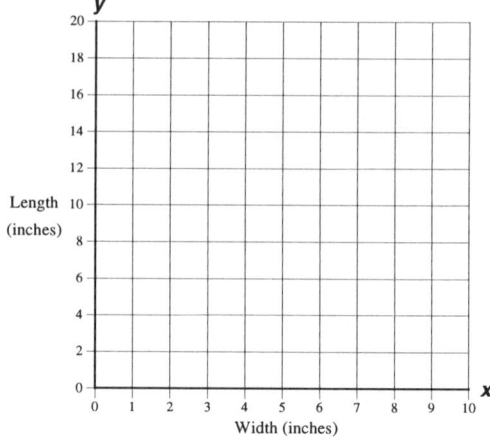

6. It costs the U. S. Department of the Treasury as much to make a $50 bill as it does to make a $1 bill. During a recent year they produced 3,960,000,000 one dollar bills at a total cost of $14,652,000. During the same year they produced 360,000,000 fifty dollar bills at a total cost of $1,332,000. Use this information to answer the following questions.

   a) Create a data table for this situation using number of bills produced in hundreds of millions as the input and total cost to produce in millions of dollars as the output.

   b) Assume a linear relation and graph the relation.

   c) Use the graph to estimate the cost of producing 1,080,000,000 bills.

   d) Use the graph to estimate the number of bills that could be produced for a cost of $9 million.

   e) What are the advantages and disadvantages of using a graph versus a linear equation for information like this?

   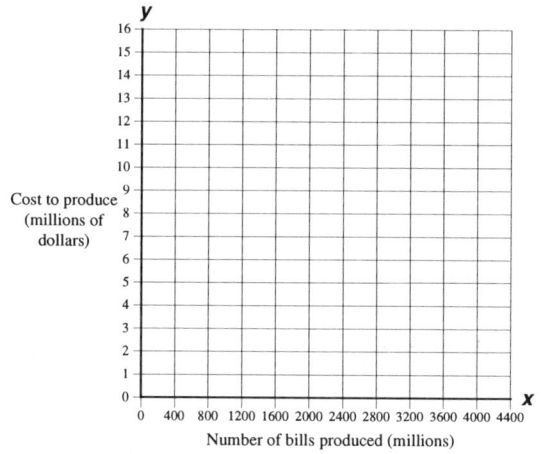

7. Plot on an x-y coordinate system and use a linear model to fill in the missing input or output values.

x	y
−3	8
3	2
8	−3
−2	
5	

8. The following data show the relation between the face value of currency printed by the U. S. Government and the total cost to produce the one, five, twenty, and fifty dollar bills. Graph the data using a scale of billions of dollars for the x–axis and millions of dollars on the y axis.

x face value (dollars)	y total cost to produce (dollars)
3,960,000,000	14,652,000
2,160,000,000	7,992,000
1,080,000,000	3,996,000
360,000,000	1,332,000

9. Find the *x*-intercept and the *y*-intercept:

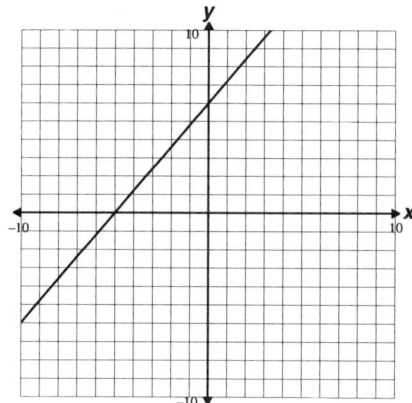

10. Construct the equation of the line shown in the graph.

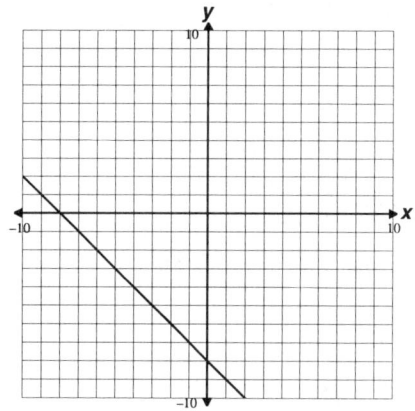

11. Discuss the meaning of each graph.

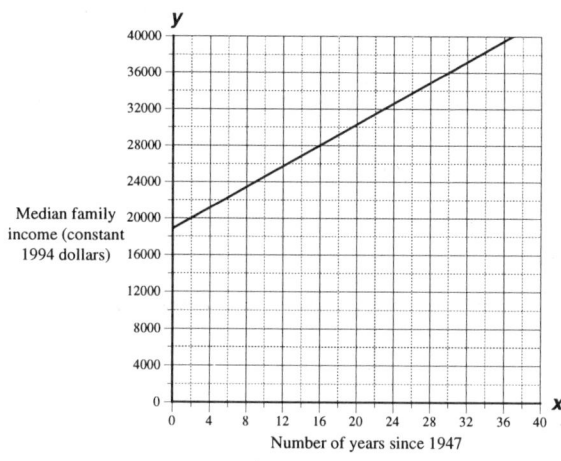

Number of years since 1947

12. The data table below shows the annual fuel cost for the worst "gas guzzlers" in 1996, according to the EPA. Calculate the slope of the line that would connect the given points and discuss the meaning of the slope.

x mileage (miles per gallon)	*y* annual fuel cost (dollars)
9	1825
11	1577
12	1453
13	1329

13. Graph the straight line that passes through (–3,1) with slope –2

14. Calculate the slope of the straight line that passes through (7,1) and (–2,–4).

15. Decide whether these two straight lines are parallel, perpendicular or neither: *Line 1* passes through the points (2,–4) and (5,5) and *Line 2* passes through the points (8,0) and (–1,3).

16. Find the slope and *y*-intercept of the line described by $5 - y = 2x$

17. Find the equation of the line that goes through (2,–5) and (–3,5).

18. Find the equation of a line parallel to $y = 3x - 9$ and which passes through (0,4).

19. Determine the domain and range of {(3,1),(5,3),(3,5),(0,5),(1,3)}

20. Determine whether this relation is a function: {2,–1),(3,2),(–1,0),(2,9)}

21. The data table shows the amount Americans spent on dining out between 1975 and 1994. The regression equation is $y = 10.5x + 53.4$

    a) Interpret the meaning of the equation.
    b) Use the regression equation to predict the amount spent in 1985 and compare this with the actual amount spent as listed in the table.
    c) If the trend continues predict the amount spent dining out in the year 2000.

x number of years since 1975	*y* expenditures dining out (billions of dollars)
0	60.4
1	68.0
2	75.0
5	105.9
10	150.0
15	218.0
19	258.4

22. Graph: $y \leq 2x - 5$

23. Graph $y = 2x - 5$ on an x-y coordinate system.

24. Determine the domain and range of

25. Determine whether this graph represents a function:

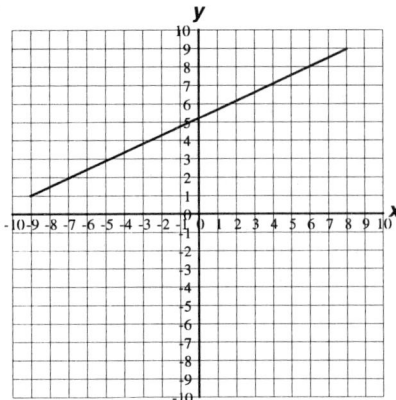

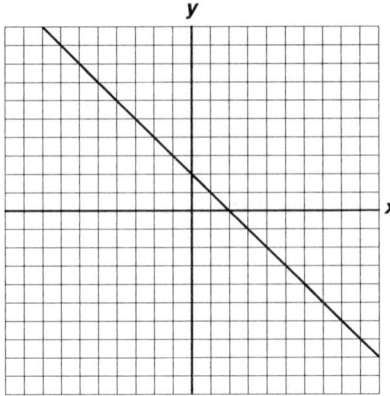

26. Given $f(x) = x^2 - 3x + 2$, find $f(-2)$.

27. Plot on an x-y coordinate system: $f(5) = -2, f(3) = 8, f(-8) = 3$

28. Since 1980 the number of automatic teller machines (ATM's) in the U. S. has been growing by about 6,000 per year. Assume there were 23,000 machines in 1980. Let t represent the year since 1980 and A represent the number of ATM's in thousands.

    a) Build a function that shows the number of ATM's as a function of the year since 1980.
    b) Calculate $A(8)$. Interpret the result in English.
    c) Find the year in which $A(t) = 80$. Interpret the result in English.
    d) Find the year in which $A(t) = 2A(5)$. Interpret the result in English.

Chapter 5
Systems of Equations

Section 5.1 Solving Systems by Graphing

A **system** of equations is a set of two or more equations that are considered together. Typically, each equation represents a different situation but all the situations are related to each other in some way. For example,

A freight train and a passenger train must share the same track. The freight train leaves a station at 9 am and travels 50 mph. One hour later, the passenger train leaves the station and travels 60 mph in the same direction. To avoid a crash, at what time and where must the railroad dispatcher move the freight train to a side track?

We can model this situation using two equations, one for each train. To answer the questions, we must solve the equations simultaneously.

Topic 5.1 A — The Solution of a System

Let's discuss the mathematics of systems by solving a simple problem. Suppose we are looking for two numbers whose difference is 6 and whose sum is 1. If we let x represent the larger number and y represent the smaller we can build a mathematical model for each condition on the numbers:

Condition 1 — The difference is 6   We can model this condition using $x - y = 6$. This is a linear equation that has infinitely many solutions. Let's create a data table that lists some solutions and then draw the graph.

x	y
−1	−7
0	−6
1	−5
1.5	−4.5
2	−4
2.5	−3.5
3	−3
4	−2
4.5	−1.5
5	−1
6	0
7	1

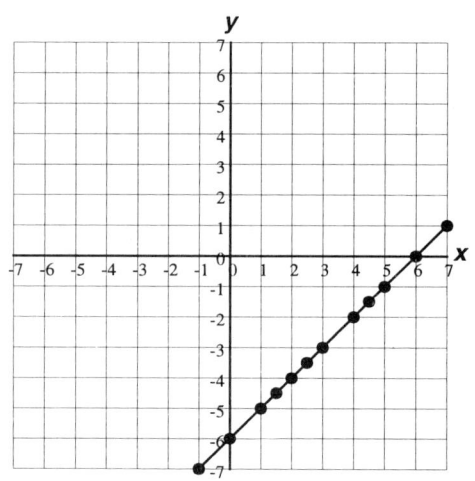

Note that every point on the line, not just the ordered pairs we plotted, is a solution to $x - y = 6$.

Condition 2— The sum is 1   We can use $x + y = 1$ to model this condition:

x	y
–1	2
–4	5
–5	6
–5.5	6.5
0	1
1	0
1.5	–0.5
2	–1
2.5	–1.5
3	–2
4	–3
4.5	–3.5
5	–4
6	–5
7	–6

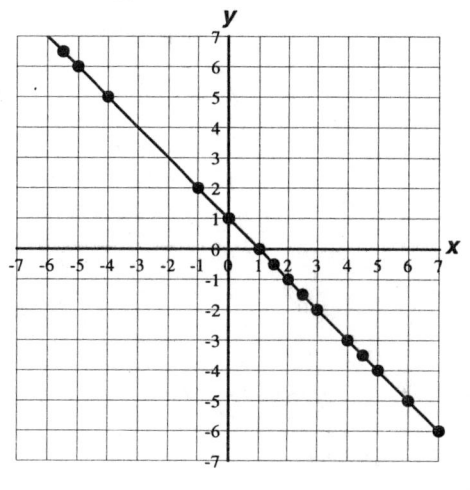

Again, note that every point on the line, not just the ordered pairs we plotted, is a solution to $x + y = 1$.

If we want to know which numbers satisfy *both* conditions, we must consider the equations together, as a system. We write the equations of the system as follows:

$$\left\{ \begin{array}{rcl} x - y &=& 6 \\ x + y &=& 1 \end{array} \right\} \begin{array}{l} \mathbf{❶} \Leftarrow \text{this says the difference is 6} \\ \mathbf{❷} \Leftarrow \text{this says the sum is 1} \end{array}$$

We use the braces { } to indicate that the equations are to be considered together and we use the labels ❶ and ❷ to help refer to each equation more easily.

The graph of $x - y = 6$ displays all the pairs of numbers whose difference is 6; the graph of $x + y = 1$ displays all the pairs of numbers whose sum is 1. If we plot both graphs on the same coordinate system we can see all the solutions for each equation at once. (Remember, every point on each line is a solution to at least one of the equations.)

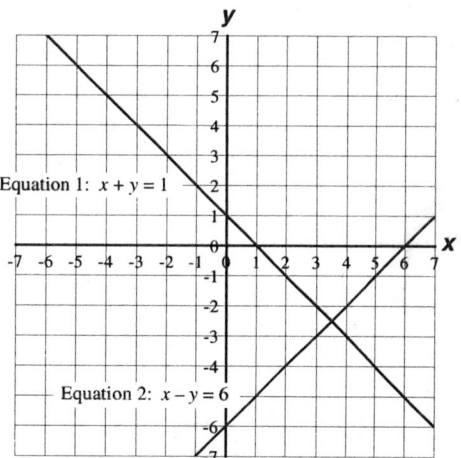

The numbers we seek have to be a solution to *both* equations. This is the point where the lines intersect: (3.5,–2.5) We say that this point is the **solution of the system** since it satisfies both equations. We may check the solution to the system by checking the ordered pair in each equation.

Check solution in equation ❶:

Left expression   ?=?   Right expression

$$x - y = 6 \qquad \Leftarrow \text{Equation to check.}$$

$$(3.5) - (-2.5) \quad | \quad 6 \qquad \Leftarrow \text{Substituted 3.5 for } x \text{ and } -2.5 \text{ for } y.$$

$$6 \quad | \quad 6 \qquad \Leftarrow \text{Checks.}$$

Check solution in equation ❷:

Left expression   ?=?   Right expression

$$x + y = 1 \qquad \Leftarrow \text{Equation to check.}$$

$$(3.5) + (-2.5) \quad | \quad 1 \qquad \Leftarrow \text{Substituted 3.5 for } x \text{ and } -2.5 \text{ for } y.$$

$$1 \quad | \quad 1 \qquad \Leftarrow \text{Checks.}$$

Example 5.1.1   a)   Does (2,1) solve $\begin{cases} x + y = 3 & ❶ \\ \frac{1}{2}x - 2y = -1 & ❷ \end{cases}$

> •SOLUTION• If (2,1) is a solution to the system it will make both equations true. We substitute to see if this is so.

Equation ❶	Equation ❷	Comment
$x + y = 3$	$\frac{1}{2}x - 2y = -1$	⇐ Equations to check.
$(2) + (1) \quad \| \quad 3$	$\frac{1}{2}(2) - 2(1) \quad \| \quad -1$	⇐ Substituted 2 for x and 1 for y in both equations.
$3 \quad \| \quad 3$	$-1 \quad \| \quad -1$	⇐ Simplified. Both equations check.

> Since (2,1) solves both equations, it's a solution of the system.

b)   Does (1, –3) solve $\begin{cases} y = x - 4 & ❶ \\ 2x + 6y = -14 & ❷ \end{cases}$

> •SOLUTION• Substitute (1,–3) into each equation and see if a true statement results.

Equation ❶	Equation ❷	Comment
$y = x - 4$	$2x + 6y = -14$	⇐ Equations to check.
$(-3) \quad \| \quad (1) - 4$	$2(1) + 6(-3) \quad \| \quad -14$	⇐ Substituted 1 for x and –3 for y in both equations.
$-3 \quad \| \quad -3$	$-16 \quad \| \quad -14$	⇐ Simplified. Only one equation checks.

> The solution checks in equation ❶ but not in equation ❷. The ordered pair does not solve the system because it does not make *both* equations true.

Practice 5.1.1

1. Does (1,4) solve $\begin{cases} x - y = -3 & ❶ \\ x - 2y = -7 & ❷ \end{cases}$

2. Does (3,3) solve $\begin{cases} 3x + y = 10 & ❶ \\ 2x - 3y = -3 & ❷ \end{cases}$

3. Does (4,3) solve $\begin{cases} x - y = 1 & ❶ \\ 2x - 3y = 1 & ❷ \end{cases}$

4. Does (2,–1) solve $\begin{cases} 2x + y = 3 & ❶ \\ 3x - 2y = 8 & ❷ \end{cases}$

A system that can be represented by the graphs of two straight lines is called a **two by two linear system** or a **2 by 2 linear system**. The *2 by 2* comes from the fact that 2 equations are used to describe the system and 2 variables are involved. The word *linear* tells us the graphs are straight lines. If the system consisted of 3 equations and 3 variables it would be called a 3 by 3 system.

The graphical procedure we use to solve a system can be summarized as follows:

— *Procedure* — *Solving a 2 by 2 system by graphing*	**Step 1** Graph each equation. **Step 2** Identify the coordinates of the points where the graphs intersect. These points are the solutions to the system.

Example 5.1.2   a)   Find the solution to the system:

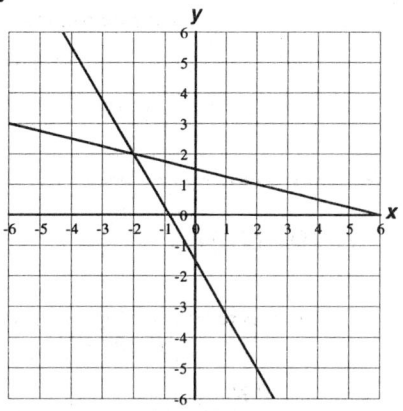

•**SOLUTION**• The lines cross at (–2,2). Since that is the only point they have in common, it's the solution to the system.

b)   Use graphing to find the solution to the system:

Line one		Line two	
x	*y*	*x*	*y*
–6	–10	0	–8
0	4	–3	–3

•**SOLUTION**• We graph the lines and locate the point of intersection.

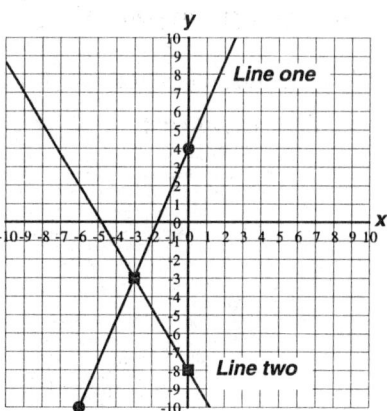

The solution is (–3,–3)

c)   Use graphing to find the solution to the system:

$$\left\{ \begin{array}{rcrcl} 2x & + & y & = & 7 \\ x & - & y & = & 8 \end{array} \right\} \begin{array}{l} \text{❶} \\ \text{❷} \end{array}$$

•SOLUTION• We create a data table for each equation and then plot the points.

Equation ❶ $2x + y = 7$ Equation ❷ $x - y = 8$

x	y
0	7
3.5	0

x	y
0	−8
8	0

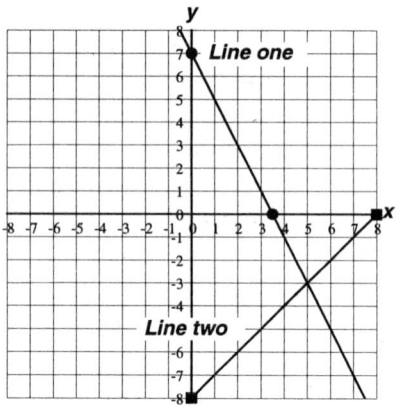

The solution is (5,−3).

___

Practice 5.1.2   Find the solution to the system.

1.

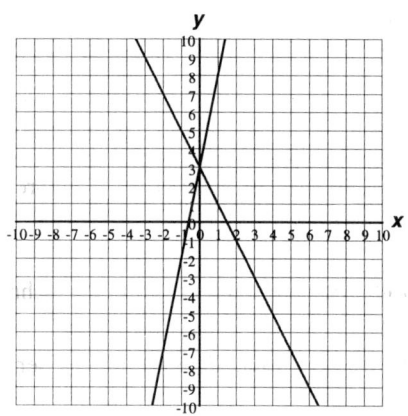

2.

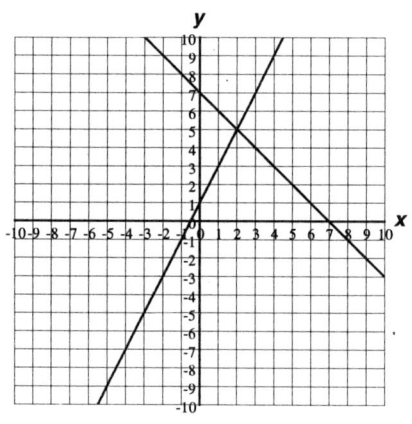

3.   **Line one**

x	y
1	3
0	4

Line two

x	y
0	−2
5	3

4. $$\left\{ \begin{array}{rcrcl} x & + & y & = & -1 \\ x & - & 2y & = & 8 \end{array} \right\} \begin{array}{l} \text{❶} \\ \text{❷} \end{array}$$

Like single equations, some systems have no solution or multiple solutions.

Example 5.1.3   Use graphing to solve.

a)   Find two numbers whose sum is 1 and whose sum is 5.

    •**SOLUTION**• Your intuition may tell you that this does not seem possible. Let's see what the mathematics tells us. We let x represent the larger number and y represent the smaller number and write the equations that describe this situation:

$$\begin{cases} x + y = 1 \\ x + y = 5 \end{cases} \begin{array}{l} ❶ \Leftarrow \text{this says the sum is 1} \\ ❷ \Leftarrow \text{this says the sum is 5} \end{array}$$

Next, we use the intercept method to graph each equation:

Equation ❶, $x + y = 1$

x	y
0	1
1	0

Equation ❷, $x + y = 5$

x	y
0	5
5	0

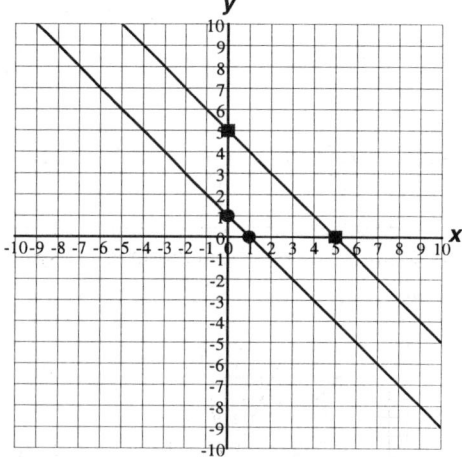

It looks like these lines will never intersect (they are parallel). Therefore, the system has no solution. This is reasonable because we can't have two numbers with a sum of 1 *and* a sum of 5.

b)   The difference of one number and twice another number is 6. Find the numbers if three less than half the first number is the second number.

    •**SOLUTION**• We let x represent the larger number and y represent the smaller number and write the equations that describe this situation.

$$\begin{cases} x - 2y = 6 \\ \frac{1}{2}x - 3 = y \end{cases} \begin{array}{l} ❶ \Leftarrow \text{condition from first sentence} \\ ❷ \Leftarrow \text{condition from second sentence} \end{array}$$

Next, we calculate some points and plot the graphs:

Equation ❶, $x - 2y = 6$

x	y
4	−1
6	0

Equation ❷,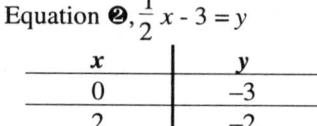

x	y
0	−3
2	−2

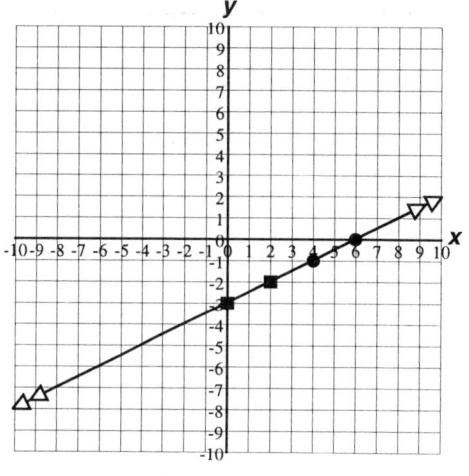

These lines intersect everywhere (they are the same line). Therefore, the system has an infinite number of solutions. Since we can't list all the ordered pairs that make this system true we will use variable expressions for the input and output coordinates.

- We will represent the input as x, to show that the input can be any value.

- For the output, we will use an expression that is equivalent to y. Since equation ❷ tells us that $y = \frac{1}{2}x - 3$ for this system, we will write the output as $\frac{1}{2}x - 3$.

Using this notation, we write the solution as the ordered pair $(x, \frac{1}{2}x - 3)$. This implies that the input may be any number, but the output must always be 3 less than half the input, which is the constraint imposed by equation ❷. Note that if we transform both equations into slope–intercept form we get $y = \frac{1}{2}x - 3$.

Practice 5.1.3 Find the numbers that meet the given conditions by building a system of equations and then solving the system graphically.

1. The difference of two numbers is 1 and the larger number is 1 more than the smaller number. Let y represent the larger number.

2. The sum of two numbers is 3 and their difference is 7. Let x represent the larger number.

3. The sum of the two numbers is 4 and twice the smaller plus twice the larger is –5. Let x represent the larger number.

— Note — ***Slopes of parallel lines***	The slope of a line is an measure of how it's oriented on the x–y plane. Therefore, if the lines are parallel, they must have the same slope. Systems consisting of lines with the same slope have either no solution (the lines are parallel but do not overlap) or an infinite number of solutions (the lines overlap).

Topic 5.1 B — Types of Systems and Equations

Systems of equations may be classified as consistent or inconsistent:

- a **consistent system** has at least one solution. For a 2 by 2 linear system, this means the graphs intersect at one point or the graphs lie on top of each other (that is, they intersect at all points).

- an **inconsistent system** has no solutions; in this case, conditions have been placed on the variables that are impossible to meet. For a 2 by 2 linear system, this means the graphs are parallel lines that are not on top of each other. The lines never intersect.

We use a different vocabulary when we talk about the *equations* of a system.

- if the equations are equivalent they are called **dependent**. When written in slope–intercept form, dependent equations are identical and, therefore, their graphs are identical.

- if the equations are not equivalent they are called **independent**. When written in slope–intercept form, either the slopes or the y–intercepts (or both) of independent equations are different and, therefore, their graphs are different.

Since the solutions of 2 by 2 linear systems are the points where the graphs intersect, we have three possibilities:

Possibility 1   The lines intersect at one point.

- The *system* is consistent (it has a solution).

- The *equations* are independent (the graphs are different).

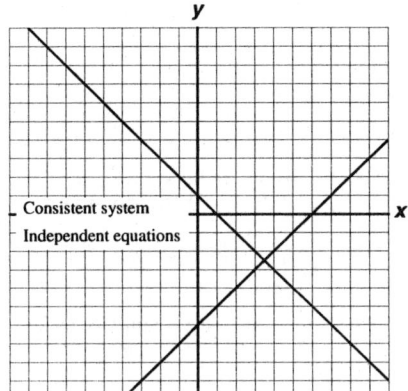

Possibility 2   The lines do not intersect.

- The *system* is inconsistent (it has no solution).

- The *equations* are independent (the graphs are different).

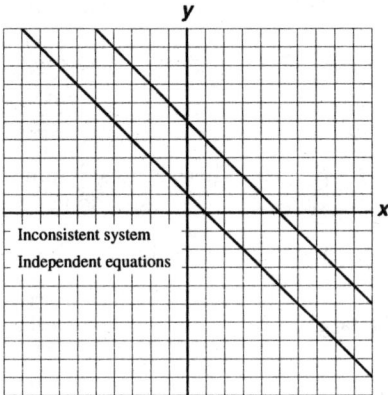

Possibility 3   The lines lie on top of each other.

- The *system* is consistent (it has a solution).

- The *equations* are dependent (the graphs are the same).

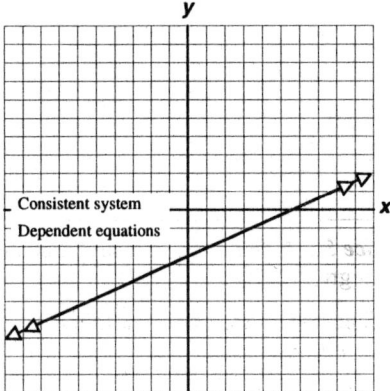

— Note —	Try to keep the vocabulary straight:
Systems are consistent or inconsistent while equations are dependent or independent	• *Systems* are either consistent (there is at least one solution) or inconsistent (there are no solutions). • *Equations* are either dependent (they are equivalent) or independent (they are not equivalent).

Example 5.1.4   • Classify each system as consistent or inconsistent.
   • Classify the equations of each system as independent or dependent.

a)

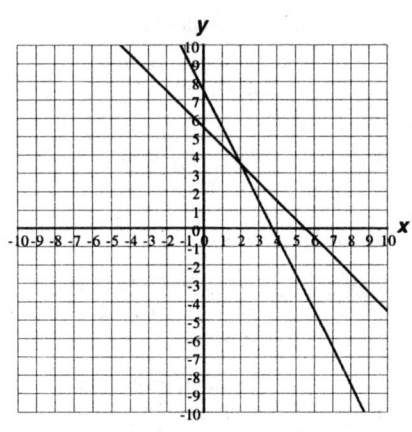

•**SOLUTION**•   • System: There is a solution so the system is consistent.
   • Equations: The lines do not coincide so the equations are independent.

b)

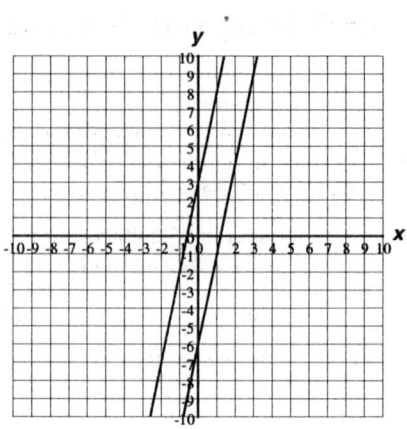

•SOLUTION•
- System: There is no solution so the system is inconsistent.
- Equations: The lines do not coincide so the equations are independent.

c)

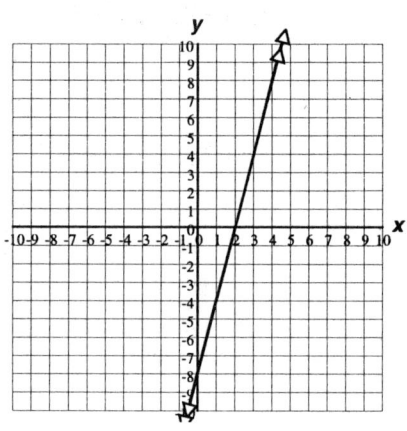

•SOLUTION•
- System: There is a solution (actually, there are an infinite number of solutions) so the system is consistent.
- Equations: The lines coincide so the equations are dependent.

Practice 5.1.4 Classify each system as consistent or inconsistent and the equations of each system as dependent or independent.

1.

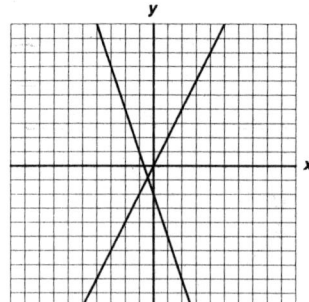

2.

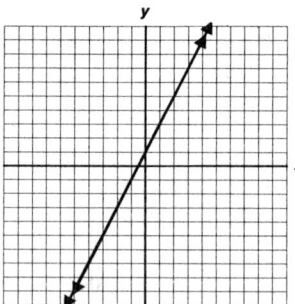

3.

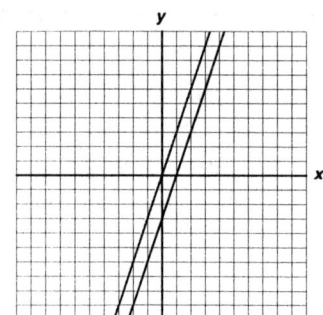

4.

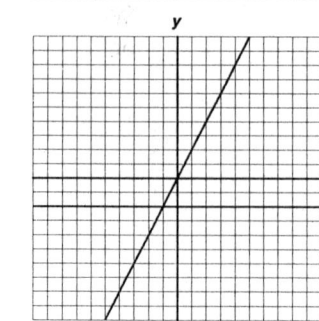

Topic 5.1 C — Problem Solving With Systems

In real–world situations, two by two systems can be used to help us choose between two alternatives.

Example 5.1.5   A student in a biology class decides to hire a tutor. One tutor, Matt, asks $100 for the semester and then charges $5 per hour. A second tutor, Keisha, charges a flat rate of $8 per hour.

a)   Model this situation with a 2 by 2 system of linear equations. Define the input as *tutoring time (hours)* and the output as *cost (dollars).*

•SOLUTION•   We can use the techniques we developed in Chapter 4 to build the equations .

Step 1 Organize the information

Given: • Matt charges a flat fee of $100.
   • Matt charges $5 per hour.
   • Keisha charges $8 per hour.

Find: A system of equations that models the data.

Step 2 Define unknowns   The problem gives us the variable definitions.

x represents the tutoring time (hours).

y represents the cost (dollars)

Step 3 Find the initial values and the rates

Recall that a linear equation has the form $y = b + mx$ where b is the initial value and m is the rate. The initial value for Matt is $100 (his flat fee) and his rate is $5 per hour. The initial value for Keisha is $0 (she has no flat fee) and her rate is $8 per hour.

We can organize this information in a table that is very similar to the Amount Table we used to help solve word problems in Chapter 4. The only difference is that we need an extra column for the constant value.

	y	=	b	+	m	•	x
Descriptions →	cost		flat fee		rate		time
Units → ↓ Item names ↓	dollars		dollars		$\dfrac{\text{dollars}}{\text{hour}}$		hours
Matt	y		100		5		x
Keisha	y		0		8		x

Step 4 Mathematical models

$$\left\{ \begin{array}{lcl} y & = & 100 + 5x \\ y & = & 0 + 8x \end{array} \right\} \begin{array}{l} ❶ \\ ❷ \end{array}$$

b)   Solve the system by graphing and explain what the solution means.

•SOLUTION•   We graph both equations on the same coordinate system by generating some ordered pairs and then plotting them.

Equation ❶, $y = 100 + 5x$

x time (hours)	y cost (dollars)
0	100
20	200

Equation ❷, $y = 0 + 8x$

x time (hours)	y cost (dollars)
0	0
20	160

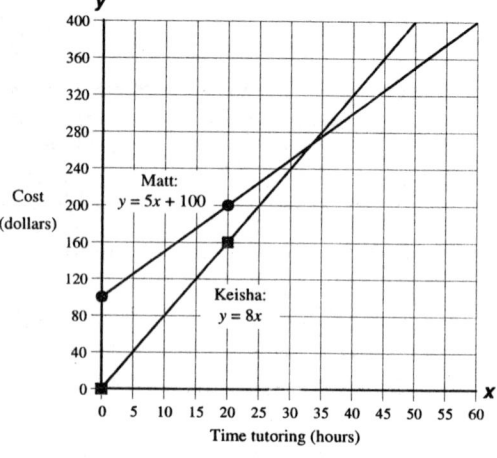

From the graph, we estimate that the lines cross at about (33,270). That is, if they tutor you for 33 hours, the cost for both tutors is the same, $270.

c)   Which tutor is the better buy?

•**SOLUTION**• It depends. The vertical axis indicates cost so the lower line represents the lower-cost tutor. For a few hours of tutoring, Keisha is cheaper because her line is below Matt's. But, since Keisha's rate (the slope of her line) is greater, her cost rises more quickly. Eventually, Keisha's line crosses Matt's and she becomes more expensive than he.

Therefore, for input values before the intersection of the two lines, Matt's line is above Keisha's, which means Matt costs more. After the intersection, Keisha's line is above Matt's and so she costs more for input values above that point.

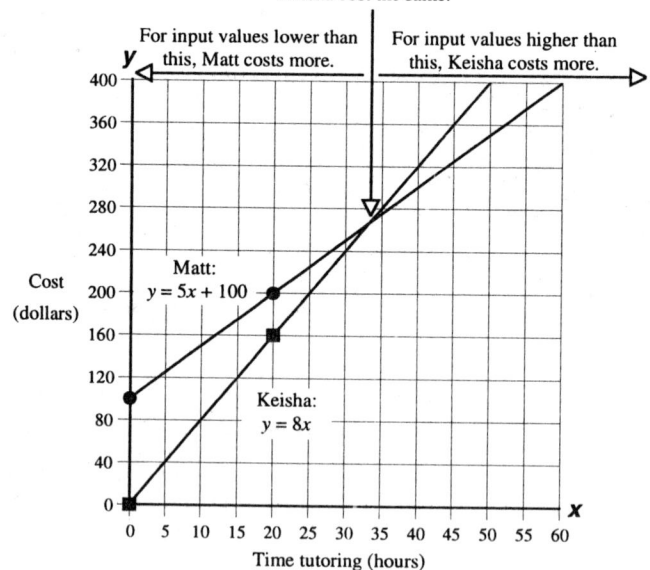

If you plan to use fewer than 33 hours of tutoring Keisha is cheaper; if you plan to use more than 33 hours of tutoring, Matt is cheaper.

Practice 5.1.5

1. Antoine needs to do an annual inventory check at his store. He can pay his employee $20 an hour or he can pay a worker from a temporary employment agency a flat fee of $70 plus $10 an hour.

   a) Model this situation with a 2 by 2 system of linear equations using time as the input and cost as the output.
   b) Graph the system. The input axis should go from 0 to 10 hours in steps of 1 hour; the output axis should go from $0 to $200 in steps of $20.
   c) Find the point of intersection and explain what it means.
   d) Which worker would cost less money?

2. Company A charges $200 plus $50 a day to rent a back hoe, while Company B charges $350 plus $35 a day.

   a) Model this situation with a 2 by 2 system of linear equations using time as the input and cost as the output.
   b) Graph the system. The input axis should go from 0 to 12 days in steps of 1 day; the output axis should go from $0 to $1000 in steps of $100.
   c) Find the point of intersection and explain what it means.
   d) Which company would cost less money?

3. Below is a graph of Monthly Finance Charge versus Money Borrowed for two credit cards.

   a) How much money would you have to borrow for the cost of the cards to be the same?
   b) Explain the meaning of the points before and after the intersection of the lines.
   c) If you borrow $200, how much more expensive is Card 2?

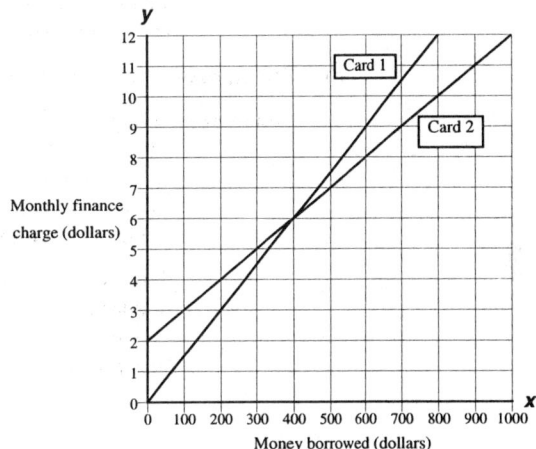

Exercise Set 5.1   The answers to the odd numbered exercises are at the back of the book.

1. Does (4,0) solve $\begin{cases} x - y = 4 & \text{❶} \\ 2x - 2y = 8 & \text{❷} \end{cases}$

2. Does (0,−1) solve $\begin{cases} x + 2y = 5 & \text{❶} \\ 2x + 3y = -3 & \text{❷} \end{cases}$

3. Does (2,−3) solve $\begin{cases} x - y = 5 & \text{❶} \\ 2x - y = -1 & \text{❷} \end{cases}$

4. Does (−1,−1) solve $\begin{cases} x + y = -2 & \text{❶} \\ 2x - y = -1 & \text{❷} \end{cases}$

Find the solution to the system.

5.

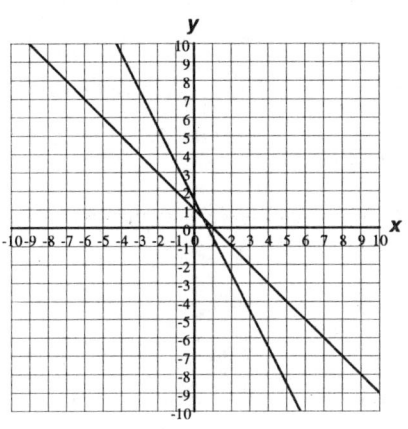

6.

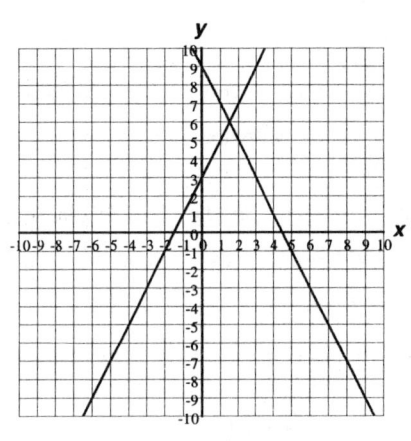

7.

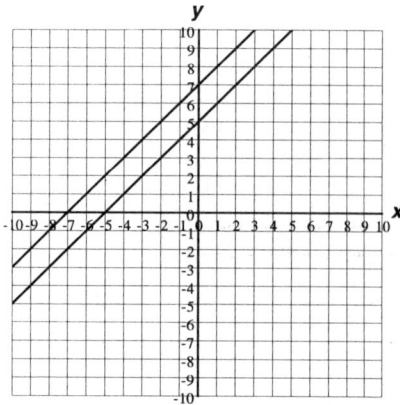

8.

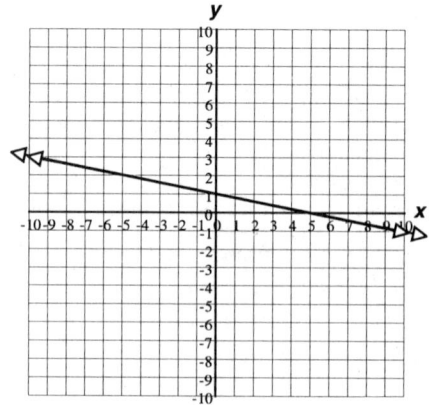

9.

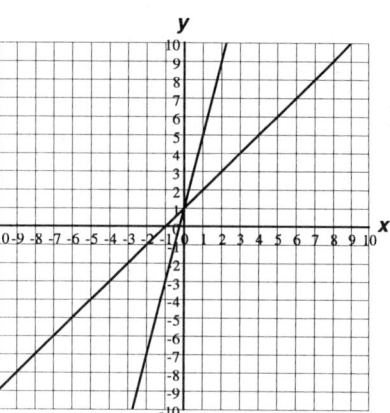

10.

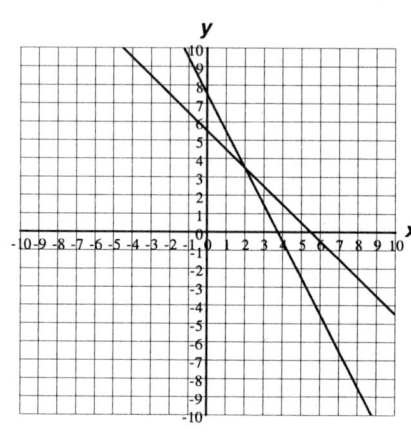

11.

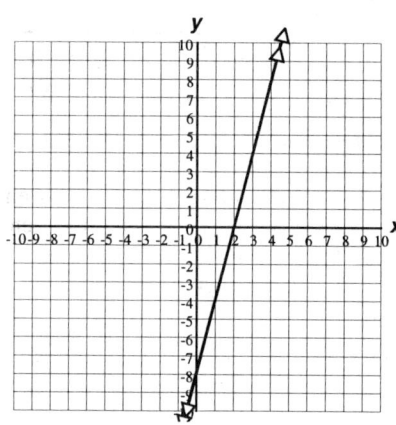

12.

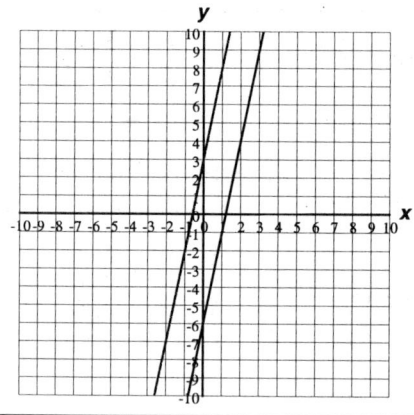

Find the numbers that meet the given conditions by building a system of equations and then solving the system graphically.

13. The sum of two numbers is 1 and twice the larger plus three times the smaller is –3. Let x represent the larger number.

14. The difference of two numbers is –5 and twice the smaller plus twice the larger is 2. Let y represent the larger number.

15. One number is 5 more than twice another and three times the larger minus six times the smaller is 15. Let y represent the larger number.

16. One number is 5 more than another and the difference of the numbers is 1. Let y represent the larger number.

17. The sum of the two numbers is –6 and the larger is nine more than twice the smaller. Let y represent the larger number.

18. The sum of the two numbers is 7 and their difference is 3. Let x represent the larger number.

19. The sum of two numbers is –1 and 8 minus the first number equals the second number. Let x represent the first number.

20. The difference of two numbers is 9 and three times the larger number is 27 more than three times the smaller number. Let x represent the larger number.

Classify each system as consistent or inconsistent and classify the equations of each system as dependent or independent.

21.

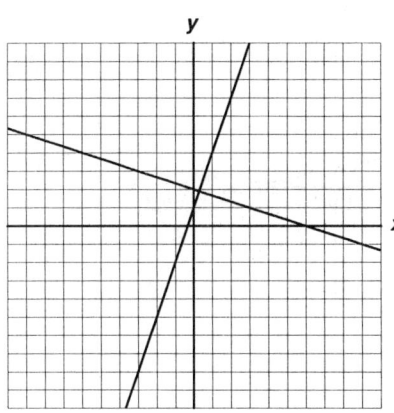

22.

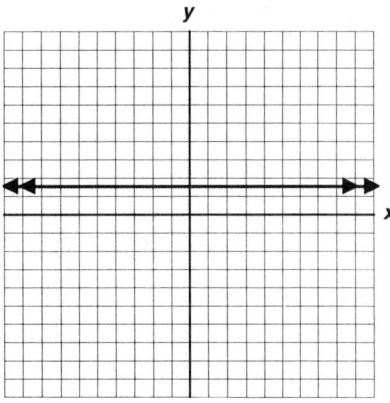

23.

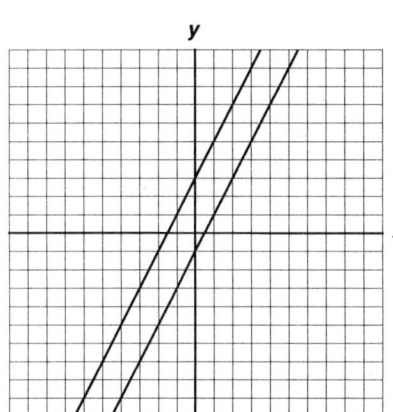

24.

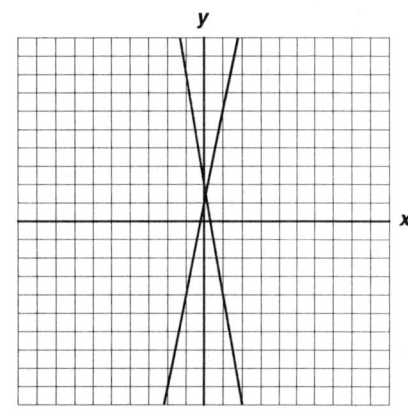

25. Given the system $\left\{ \begin{array}{l} y = 2x + 1 \\ y = 2x - 2 \end{array} \right\}$) **❶** **❷**

    a) Graph the system.
    b) Solve the system.
    c) Classify the system as consistent or inconsistent.
    d) Classify the equations as independent or dependent.

26. Given the system $\left\{ \begin{array}{l} y = 4x + 1 \\ 8x - 2y = -2 \end{array} \right\}$ **❶** **❷**

    a) Graph the system.
    b) Solve the system.
    c) Classify the system as consistent or inconsistent.
    d) Classify the equations as independent or dependent.

27. Given the system $\left\{ \begin{array}{l} y = \dfrac{3}{4}x - 6 \\ y = -\dfrac{3}{2}x + 3 \end{array} \right\}$ **❶** **❷**

    a) Graph the system.
    b) Solve the system.
    c) Classify the system as consistent or inconsistent.
    d) Classify the equations as independent or dependent.

28. Given the system $\left\{ \begin{array}{l} y = 2 - \dfrac{3}{4}x \\ y = \dfrac{1}{2}x - 8 \end{array} \right\}$ **❶** **❷**

    a) Graph the system.
    b) Solve the system.
    c) Classify the system as consistent or inconsistent.
    d) Classify the equations as independent or dependent.

29. A car can be leased from Company A for $250 a month and $4500 down. The same car can be leased from Company B for $500 down and $500 a month.

    a) Model this situation with a 2 by 2 system of linear equations.
    b) Graph the system. The input axis should go from 0 to 20 months in steps of 2 months; the output axis should go from $0 to $10,000 in steps of $1000.
    c) Find the point of intersection and explain what it means.
    d) Which plan would cost less money?

30. Sara needs to rent a car for a day. Econo Cars charges $15 a day and $0.30 for every mile driven. Luxo Trans charges $75 a day with unlimited miles.

    a) Model this situation with a 2 by 2 system of linear equations.
    b) Graph the system. The input axis should go from 0 to 240 miles in steps of 20 miles; the output axis should go from $0 to $100 in steps of $10.
    c) Find the point of intersection and explain what it means.
    d) Which plan would cost less money?

31. The following is a graph of Cost and Revenue versus Number of units Sold for a small business.

    a) What is the break even point (that is, how many units must be sold for the cost to equal to the revenue)?
    b) Explain the meaning of the points before and after the break even point.
    c) If the business sells 30 units, how much profit do they make?

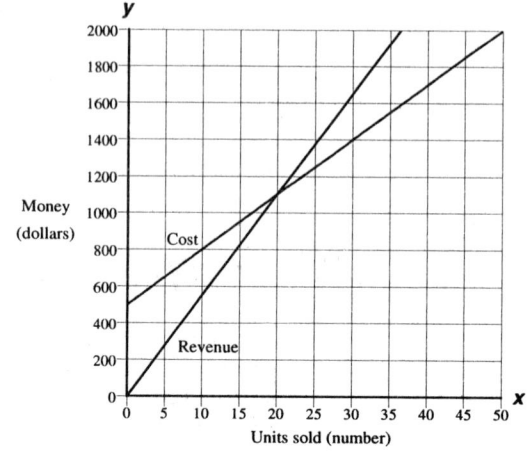

32. The following graph shows the money that can be earned by selling cars for two different dealerships.

    a) For the income from both jobs to be the same, how many cars must be sold?
    b) Explain the meaning of the points before and after the intersection of the lines.
    c) On the sale of 4 cars, how much more would a salesperson at Penny's Plymouth make than a salesperson at Phil's Ford?

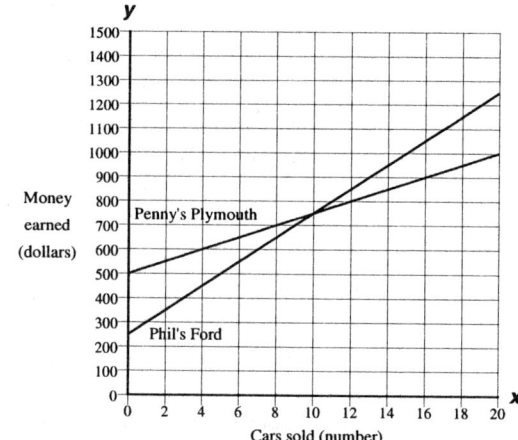

Review Exercises The answers to all of these exercises are at the back of the book.

33. Find the coordinate for each point shown on the number line:

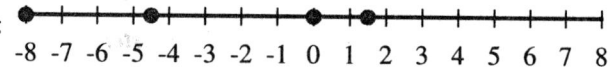

-8 -7 -6 -5 -4 -3 -2 -1 0 1 2 3 4 5 6 7 8

34. Identify the base and its exponent: $-x^6$

35. Graph the straight line that passes through $(-1, 9)$ with slope $-\dfrac{5}{3}$.

36. Solve: $\dfrac{2}{5}(x+5) - \dfrac{1}{3}(x+2) = \dfrac{x}{10} + 1$

37. Solve: $\dfrac{2}{3}(x-4) + 1 = \dfrac{1}{2}x + \dfrac{1}{6}(x-3)$

38. You wish to combine the money you have in 3 accounts into a single passbook account. Write an expression to represent the amount in the passbook account if the amount in the savings account exceeds the amount in the checking account by $280 and the amount in the Christmas account is one–fourth the amount in the savings account. Use c to represent the amount in the checking account.

39. A couple has $15,000 to invest and wants to earn $1200 in interest in one year. If they invest in one account at a 13% rate of return and a second account at a 5% rate of return, how much should they put in each account?

40. Calculate the slope of the straight line that passes through $(2, 3)$ and $(5, -1)$.

41. Use the graph to complete the data table:

x	f(x)
0	
	–2
8	

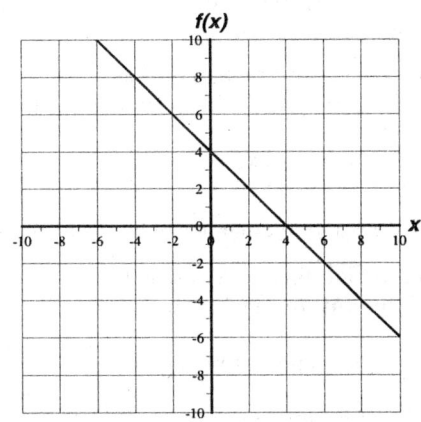

Section 5.2 Solving Systems Using Substitution

As we saw in Section 5.1, graphs are useful for visualizing relations and for finding approximate solutions to systems. Unfortunately, graphs are not a good way to get exact answers. In this section and the next we discuss two algebraic methods of finding the exact solutions to system.

Topic 5.2 A — Using the Substitution Method to Solve Consistent Systems

Suppose we seek two numbers whose sum is 3 and where one of the numbers is 5 less than the other. To find the numbers, we will construct a 2 by 2 system with x representing the larger number and y the smaller. The equations for the system are constructed so that each represents a constraint imposed on the numbers:

Constraint 1   English description: The sum of the numbers is 3.

Mathematical model: $x + y = 3$

Constraint 2   English description: One number is 5 less than the other.

Mathematical model: $y = x - 5$

Taken together as a system we have $\left\{ \begin{array}{rcl} x + y &=& 3 \\ y &=& x - 5 \end{array} \right\}$ ❶ ❷

Let's see how we can solve the system algebraically.

Equation ❷, $y = x - 5$, says that for this system y and $x - 5$ represent the same number. Therefore, in this system any place we see a y we can replace it with its equivalent, $x - 5$. Watch what happens to equation ❶ when we replace y with its equivalent, $x - 5$.

Original system

$$\left\{ \begin{array}{rcl} x + y &=& 3 \\ y &=& x - 5 \end{array} \right\} \begin{array}{l} ❶ \\ ❷ \end{array}$$

Replacing y in equation ❶ with x – 5 from equation ❷

$$x + (x - 5) = 3$$

We now have an equation with a single variable, x, which we can solve.

$2x - 5 = 3$	⇐ Combined like terms.
$2x = 8$	⇐ Added 5 to both sides.
$x = 4$	⇐ Divided both sides by 2.

We found that 4 is the value of x which satisfies the *system*.

To find the value of y that satisfies the system we substitute 4 for x into either original equation and solve for y. Let's use equation ❷:

$$y = x - 5 \qquad \Leftarrow \text{Equation ❷, which we will solve for } y.$$

$$y = (4) - 5 \qquad \Leftarrow \text{Substituted 4 for } x.$$

$$y = -1 \qquad \Leftarrow \text{Simplified.}$$

The system will be satisfied when x is 4 and y is -1. That is, if we substitute these values into either equation we will wind up with a true statement.

Check solution in equation ❶:

$$\textit{Left expression} \quad ?=? \quad \textit{Right expression}$$

$$\begin{array}{c|c} x + y = & 3 \qquad \Leftarrow \text{Equation to check.} \\ (4) + (-1) & 3 \qquad \Leftarrow \text{Substituted 4 for } x -1 \text{ for } y. \\ 3 & 3 \qquad \Leftarrow \text{Checks.} \end{array}$$

Check solution in equation ❷:

$$\textit{Left expression} \quad ?=? \quad \textit{Right expression}$$

$$\begin{array}{c|c} y = & x - 5 \qquad \Leftarrow \text{Equation to check.} \\ (-1) & (4) - 5 \qquad \Leftarrow \text{Substituted } -1 \text{ for } y \text{ and 4 for } x. \\ -1 & -1 \qquad \Leftarrow \text{Checks.} \end{array}$$

We found that 4 and -1 are the two numbers whose sum is 3 and where one number is 5 more than the other. If we were to graph the two equations we would find that they intersect at the point $(4, -1)$.

We call this the **substitution method of solving a system** because we substitute an expression for one of the variables. We can summarize the process of solving a system by substitution as follows:

— Procedure —	
Solving a 2 by 2 linear system by substitution	**Step 1 Solve** If necessary, solve one of the equations for one of the variables.
	Step 2 Substitute Substitute the expression found in step 1 for all occurrences of the variable in the other equation and solve.
	Step 3 Back Substitute Substitute the value found in step 2 into either *original* equation and solve.
	Step 4 Check The solution is the ordered pair found in steps 2 and 3. Check your ordered pair by substituting into *both* equations.

Example 5.2.1   Use substitution to solve $\left\{ \begin{array}{rcrcr} x & - & y & = & 4 \\ 2x & + & 6y & = & -16 \end{array} \right\} \begin{array}{l} ❶ \\ ❷ \end{array}$

•SOLUTION•   We can solve for either variable in either equation. We choose to solve for x in equation ❶ because it has a coefficient of 1. We could just as correctly solve equation ❷ for x but that would involve more steps.

Step 1 Solve   We solve equation ❶ for x.

$$x - y = 4 \qquad \Leftarrow \text{Equation ❶.}$$

$$x = 4 + y \qquad \Leftarrow \text{Added } y \text{ to both sides.}$$

Step 2 Substitute   We substitute $4 + y$ for x in equation ❷ and solve for y.

$$2x + 6y = -16 \qquad \Leftarrow \text{Equation ❷.}$$

$$2(\mathbf{4 + y}) + 6y = -16 \qquad \Leftarrow \begin{array}{l}\text{Substituted in equation ❷ the} \\ \text{expression we found for } x \text{ from} \\ \text{equation ❶}\end{array}$$

$$\mathbf{8 + 2y} + 6y = -16 \qquad \Leftarrow \text{Distributed the 2.}$$

$$8 + \mathbf{8y} = -16 \qquad \Leftarrow \text{Combined like terms.}$$

$$8y = \mathbf{-24} \qquad \Leftarrow \text{Subtracted 8 from both sides.}$$

$$y = -3 \qquad \Leftarrow \text{Divided both sides by 8.}$$

Step 3 Back Substitute   We substitute -3 for y in equation ❶ and solve for x.

$$x - y = 4 \qquad \Leftarrow \text{Equation ❶.}$$

$$x - (\mathbf{-3}) = 4 \qquad \Leftarrow \text{Substituted } -3 \text{ for } y.$$

$$x + \mathbf{3} = 4 \qquad \Leftarrow \text{Changed subtraction to addition.}$$

$$x = 1 \qquad \Leftarrow \text{Subtracted 3 from both sides.}$$

The solution is the ordered pair $(1,-3)$. We leave the check to you.

Note that it makes no difference which variable we substitute for. For example, we could have solved this problem by first solving equation ❶ for y and then substituting into equation ❷.

Practice 5.2.1   Use substitution to solve.

1. $\left\{ \begin{array}{rrrr} x & + & y & = & -5 \\ & & y & = & 2x + 1 \end{array} \right\} \begin{array}{l}❶ \\ ❷\end{array}$ 2. $\left\{ \begin{array}{rrrr} 2x & + & y & = & 9 \\ & & x & = & y + 3 \end{array} \right\} \begin{array}{l}❶ \\ ❷\end{array}$ 3. $\left\{ \begin{array}{rrrr} 2x & + & 3y & = & 7 \\ x & + & y & = & 3 \end{array} \right\} \begin{array}{l}❶ \\ ❷\end{array}$

Sometimes solving for one of the variables in one of the equations involves more than one step.

Example 5.2.2   Use substitution to solve $\left\{ \begin{array}{rrrr} -3x & - & 4y & = & -8 \\ 2x & + & 4y & = & 6 \end{array} \right\} \begin{array}{l}❶ \\ ❷\end{array}$

•**SOLUTION**• **Step 1 Solve**   We solve equation ❷ for x.

$$2x + 4y = 6 \qquad \Leftarrow \text{Equation ❷.}$$

$$2x = 6 - \mathbf{4y} \qquad \Leftarrow \text{Subtracted } 4y \text{ from both sides.}$$

$$x = \mathbf{3 - 2y} \qquad \Leftarrow \text{Divided both sides by 2.}$$

Step 2 Substitute   We substitute $3 - 2y$ for x in equation ❶ and solve for y.

$$-3x - 4y = -8 \qquad \Leftarrow \text{Equation ❶.}$$

$$-3(\mathbf{3 - 2y}) - 4y = -8 \qquad \Leftarrow \text{Substituted } 3 - 2y \text{ for } x.$$

$$\mathbf{-9 + 6y} - 4y = -8 \qquad \Leftarrow \text{Distributed the } -3.$$

$$-9 + \mathbf{2y} = -8 \qquad \Leftarrow \text{Combined like terms.}$$

$$2y = 1 \qquad \Leftarrow \text{Added 9 to both sides.}$$

$$y = \frac{1}{2} \qquad \Leftarrow \text{Divided both sides by 2.}$$

Step 3 Back Substitute We substitute $\frac{1}{2}$ for y in equation ❷ and solve for x.

$$2x + 4y = 6 \qquad \Leftarrow \text{ Equation ❷.}$$

$$2x + 4\left(\frac{1}{2}\right) = 6 \qquad \Leftarrow \text{ Substituted } \frac{1}{2} \text{ for } y.$$

$$2x + 2 = 6 \qquad \Leftarrow \text{ Reduced.}$$

$$2x = 4 \qquad \Leftarrow \text{ Subtracted 2 from both sides.}$$

$$x = 2 \qquad \Leftarrow \text{ Divided both sides by 2.}$$

You should verify that $\left(2, \frac{1}{2}\right)$ is a solution for both equations.

Practice 5.2.2 Use substitution to solve each system.

1. $\begin{cases} 3x - 6y = -6 \\ -2x + 5y = 6 \end{cases} \begin{matrix} ❶ \\ ❷ \end{matrix}$ 2. $\begin{cases} 4x + 2y = 6 \\ 3x - 5y = -2 \end{cases} \begin{matrix} ❶ \\ ❷ \end{matrix}$ 3. $\begin{cases} 2x + 4y = -16 \\ 3x + 2y = -8 \end{cases} \begin{matrix} ❶ \\ ❷ \end{matrix}$

The equations of a system may be written in many different forms. However, the basic ideas for solving the system remains the same.

Example 5.2.3 Use substitution to solve $\begin{cases} d = 30t \\ d = 50t - 50 \end{cases} \begin{matrix} ❶ \\ ❷ \end{matrix}$

•**SOLUTION**• ***Step 1 Solve*** The equations are already solved for d.

Step 2 Substitute We substitute $30t$ for d in equation ❷.

$$d = 50t - 50 \qquad \Leftarrow \text{ Equation ❷.}$$

$$\mathbf{30t} = 50t - 50 \qquad \Leftarrow \text{ Substituted } 30t \text{ for } d.$$

$$\mathbf{-20t} = -50 \qquad \Leftarrow \text{ Subtracted } 50t \text{ from both sides.}$$

$$t = 2.5 \qquad \Leftarrow \text{ Divided both sides by } -20.$$

Step 3 Back Substitute We substitute 2.5 for t in equation ❶.

$$d = 30t \qquad \Leftarrow \text{ Equation ❶.}$$

$$d = 30(\mathbf{2.5}) \qquad \Leftarrow \text{ Substituted 2.5 for } t.$$

$$d = 75 \qquad \Leftarrow \text{ Multiplied.}$$

The solution is $(2.5, 75)$. We leave the check to you.

Practice 5.2.3 Use substitution to solve each system.

1. $\begin{cases} t = p - 250 \\ 2t + 3p = 1125 \end{cases} \begin{matrix} ❶ \\ ❷ \end{matrix}$ 2. $\begin{cases} w + t = 10 \\ 0.05w + 0.25t = 1.50 \end{cases} \begin{matrix} ❶ \\ ❷ \end{matrix}$

Topic 5.2 B — Systems With Infinitely Many Solutions or No Solution

Not all systems have solutions, and not all solutions consist of a single point.

Example 5.2.4   a)   Solve the system $\left\{\begin{array}{rcl} 6x & - & 3y & = & 12 \\ 2x & - & y & = & 4 \end{array}\right\}$ ❶ ❷

•**SOLUTION**• ***Step 1 Solve***   We solve equation ❷ for y.

$$2x - y = 4 \qquad \Leftarrow \text{Equation ❷ , which we will solve for } y.$$

$$-y = -2x + 4 \qquad \Leftarrow \text{Subtracted } 2x \text{ from both sides.}$$

$$y = 2x - 4 \qquad \Leftarrow \text{Divided both sides by } -1.$$

Step 2 Substitute   We substitute $2x - 4$ for y in equation ❶.

$$6x - 3y = 12 \qquad \Leftarrow \text{Equation ❶.}$$

$$6x - 3(2x - 4) = 12 \qquad \Leftarrow \text{Substituted } 2x - 4 \text{ for } y.$$

$$6x - 6x + 12 = 12 \qquad \Leftarrow \text{Distributed } -3.$$

$$12 = 12 \qquad \Leftarrow \text{Combined like terms.}$$

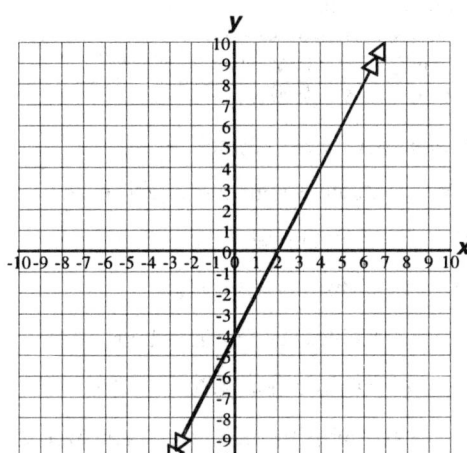

We attempted to find an ordered pair that satisfies both equations by combining the information from the equations into a single equation. We ended up with the identity 12 = 12. This is a system's way of telling us the equations are the same and so their graphs intersect everywhere.

To check that the equations are the same, we can write them both in $y = mx + b$ form:

Equation ❶		***Equation ❷***	
$6x - 3y = 12$	$\Leftarrow$ Equation to solve.	$2x - y = 4$	$\Leftarrow$ Equation to solve.
$-3y = -6x + 12$	$\Leftarrow$ Subtracted $6x$.	$-y = -2x + 4$	$\Leftarrow$ Subtracted $2x$.
$y = 2x - 4$	$\Leftarrow$ Divided by -3.	$y = 2x - 4$	$\Leftarrow$ Divided by -1.

The equations are identical, which means they are dependent and their graphs coincide.

Since we can't list all the ordered pairs that make this system true we will use variable expressions for the input and output coordinates.

- We will represent the input as x, to show that the input can be any value.

- For the output, we will use an expression that is equivalent to y. Since $y = 2x - 4$ for this system, we will write the output as $2x - 4$.

Therefore, we write the solution as the ordered pair $(x, 2x - 4)$. This implies that the input may be any number, but the output must always be 4 less than twice the input.

b) Solve the system $\begin{cases} -2x + y = -3 \\ -4x + 2y = 8 \end{cases}$ ❶ ❷

•SOLUTION• Since y in equation ❶ has a coefficient of 1, it will be easiest to solve for that variable.

Step 1 Solve

$$-2x + y = -3 \quad \Leftarrow \text{ Equation ❶ , which we will solve for } y.$$

$$y = 2x - 3 \quad \Leftarrow \text{ Added } 2x \text{ to both sides.}$$

Step 2 Substitute We substitute $2x - 3$ for y in equation ❷.

$$-4x + 2y = 8 \quad \Leftarrow \text{ Equation ❷.}$$

$$-4x + 2(\mathbf{2x - 3}) = 8 \quad \Leftarrow \text{ Substituted } 2x - 3 \text{ for } y.$$

$$-4x + \mathbf{4x - 6} = 8 \quad \Leftarrow \text{ Distributed the 2.}$$

$$-6 = 8 \quad \Leftarrow \text{ Added like terms.}$$

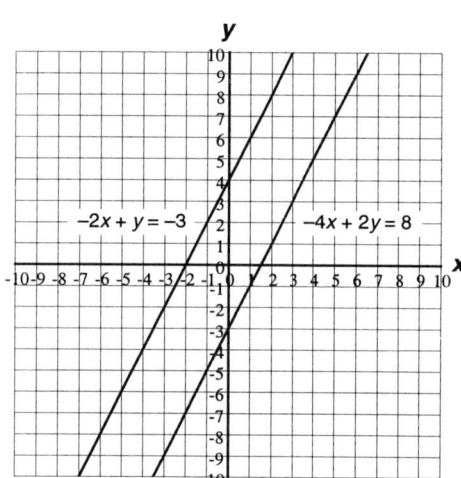

We used algebra to attempt to answer the question, "What solutions do the equations have in common?". We ended up with the contradiction $-6 = 8$. This is a system's way of telling us that the equations have no solutions in common. Therefore, the system has no solution. We can see this easily from the graph.

As a check, we can write equations ❶ and ❷ in $y = mx + b$ form:

Equation ❶	**Equation ❷**
$-2x + y = -3$ $\Leftarrow$ Equation to solve.	$-4x + 2y = 8$ $\Leftarrow$ Equation to solve.
$y = 2x - 3$ $\Leftarrow$ Added $2x$.	$2y = 4x + 8$ $\Leftarrow$ Added $4x$.
	$y = 2x + 4$ $\Leftarrow$ Divided by 2.

The equations show the same slope, 2, but different y–intercepts. This means the graphs are parallel and do not intersect. The system is inconsistent and the equations are independent.

Practice 5.2.4 Use substitution to solve each system.

1. $\begin{cases} 3x + y = 7 \\ -6x - 2y = -14 \end{cases}$ ❶ ❷

2. $\begin{cases} 4x + 2y = 5 \\ 2x + y = 1 \end{cases}$ ❶ ❷

3. $\begin{cases} x = 2y + 4 \\ -2x + 4y = -1 \end{cases}$ ❶ ❷

4. $\begin{cases} -3y = 7 - x \\ -3x + 9y = -21 \end{cases}$ ❶ ❷

Let's summarize our findings about the solutions to systems.

| — *Note* —

Possible solutions to
linear systems | After the first substitution,

• if you end up with a conditional equation, such as $x = 2$, the equations have one solution in common and their graphs intersect at one point (whose x component is 2). The system is consistent and the equations are independent.

• if you end up with an identity, such as $12 = 12$, the equations are equivalent and their graphs lie exactly on top of each other. There are an infinite number of solutions to the system. The general solution is written as an ordered pair with x as the input and an expression containing x as the output. The system is consistent and the equations are dependent.

• if you end up with a contradiction, such as $-6 = 8$, the equations have no solutions in common and their graphs are parallel lines. There is no solution to the system and therefore the system is inconsistent. The equations are independent. |

Topic 5.2 C — Problem Solving With Systems

Example 5.2.5   Below is displayed the Monthly Finance Charge versus Money Borrowed for two credit cards, TruValue and SavesOLot:

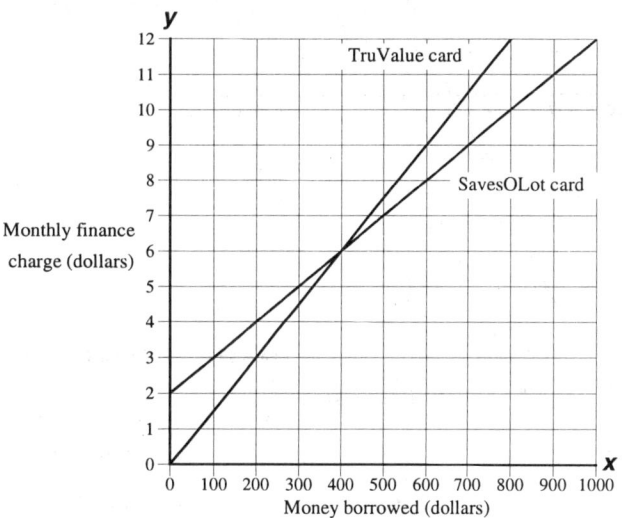

Given that two points on the graph are

SavesOLot card

x *amount borrowed* *(dollars)*	y *cost per month* *(dollars)*
0	2
100	3

TruValue card

x *amount borrowed* *(dollars)*	y *cost per month* *(dollars)*
0	0
100	1.50

use substitution to find the amount you would have to borrow so that SavesOLot and TruValue cost the same.

> •**SOLUTION**• To solve the system algebraically we first construct a linear model, $y = mx + b$, for each credit card. From the data tables we can see that the constant (vertical intercept) for SavesOLot is (0,2) and the constant for TrueValue is (0,0).

We calculate the rate (slope) as follows:

SavesOLot	What we did	TruValue
$m = \dfrac{3-2}{100-0}$ $m = \dfrac{1}{100}$ $m = 0.01$	⇐ Substituted values ⇒ into slope formula and then simplified.	$m = \dfrac{1.50-0}{100-0}$ $m = \dfrac{1.50}{100}$ $m = 0.015$

The equations for the two cards are as follows:

SavesOLot: $y = 0.01x + 2$

TrueValue: $y = 0.015x$

Now, we must solve the system. We will use substitution.

$$\left\{ \begin{array}{rcl} y &=& 0.01x + 2 \\ y &=& 0.015x \end{array} \right\} \begin{array}{l} \text{❶ SavesOLot} \\ \text{❷ TruValue} \end{array}$$

$\mathbf{0.015x} = 0.01x + 2$ ⇐ Substituted $0.015x$ for y in equation ❶.

$\mathbf{0.005x} = 2$ ⇐ Subtracted $0.01x$ from both sides.

$\qquad x = 400$ ⇐ Divided both sides by 0.005.

Now we will substitute 400 for x in either original equation. We will use equation ❷.

$\qquad y = 0.015x$ ⇐ Original equation ❷.

$\qquad y = 0.015(\mathbf{400})$ ⇐ Substituted 400 for x.

$\qquad y = 6$ ⇐ Multiplied.

The solution to the system is (400,6). This means if you borrow $400 then both cards cost $6 per month.

Practice 5.2.5

1. Given the following data, find the number of units that must be sold for cost to equal revenue and the revenue generated by those units.

Cost

x number of units sold	y money (dollars)
0	2520
10	2870

Revenue

x number of units sold	y money (dollars)
0	0
10	750

2. A bricklayer can pay $11.95 a month for cell phone service if he pays $99 for the phone. A second company will charge only $2.36 for the phone but they also charge $17.99 a month for the service. After how many months are the costs for both services equal?

3. Given the following data, find when the divers will have the same amount of air in their tanks and how much air that is.

Joni

x time diving (minutes)	y air in tank (cubic feet)
0	52
40	0

Diane

x time diving (minutes)	y air in tank (cubic feet)
0	80
20	0

Exercise Set 5.2   The answers to the odd numbered exercises are at the back of the book.

Use substitution to solve.

1. $\begin{cases} x + y = 9 \\ 2x + 3y = 22 \end{cases}$ ❶ ❷
2. $\begin{cases} x - y = 1 \\ 2x - 3y = -1 \end{cases}$ ❶ ❷
3. $\begin{cases} 6x - 5y = -5 \\ x - y = -2 \end{cases}$ ❶ ❷

4. $\begin{cases} x + y = 5 \\ 3x + 2y = 6 \end{cases}$ ❶ ❷
5. $\begin{cases} x + y = 1 \\ 2x + 4y = 10 \end{cases}$ ❶ ❷
6. $\begin{cases} 4x + 3y = 12 \\ x + y = 1 \end{cases}$ ❶ ❷

7. $\begin{cases} 4x - 2y = 2 \\ 3x + 3y = 15 \end{cases}$ ❶ ❷
8. $\begin{cases} 6x + 3y = 0 \\ 2x + 3y = 8 \end{cases}$ ❶ ❷
9. $\begin{cases} 7x + 3y = 11 \\ 2x - 4y = 8 \end{cases}$ ❶ ❷

10. $\begin{cases} 3x + 6y = 9 \\ 5x + 4y = 3 \end{cases}$ ❶ ❷
11. $\begin{cases} n + d = 25 \\ 0.05n + 0.10d = 2.25 \end{cases}$ ❶ ❷

12. $\begin{cases} g = w + 225 \\ 3w + 3g = 975 \end{cases}$ ❶ ❷
13. $\begin{cases} 4x - y = 7 \\ -4x + y = -7 \end{cases}$ ❶ ❷
14. $\begin{cases} x + 5y = 2 \\ 2x + 10y = 1 \end{cases}$ ❶ ❷

15. $\begin{cases} 3x + y = 7 \\ -6x - 2y = 10 \end{cases}$ ❶ ❷
16. $\begin{cases} 3x + 2y = 6 \\ 6x + 4y = 12 \end{cases}$ ❶ ❷
17. $\begin{cases} x + 4y = 3 \\ -2x - 8y = -6 \end{cases}$ ❶ ❷

18. $\begin{cases} 3y = 7 - x \\ -2x - 6y = 4 \end{cases}$ ❶ ❷
19. $\begin{cases} 4x + 2y = 2 \\ 2x + y = 1 \end{cases}$ ❶ ❷
20. $\begin{cases} x = 2y - 4 \\ -2x + 4y = -8 \end{cases}$ ❶ ❷

21. Given the following data, find when the two jobs will pay the same salary and what that salary is.

Job 1

x number of cars sold	y money earned (dollars)
0	250
1	300

Job 2

x number of cars sold	y money earned (dollars)
0	500
10	750

22. Given the following data, find when the cost of buying and renting are the same and what that cost is.

Buying

x time (months)	y cost (dollars)
0	2500
1	2950

Renting

x time (months)	y cost (dollars)
0	1000
1	1600

23. Given the following data about two cars returning home from a trip, find when the two cars are the same distance from home and how long they drove to get to that point.

Car 1

x driving time (hours)	y distance from home (miles)
0	358
1	313

Car 2

x driving time (hours)	y distance from home (miles)
0	382
1	327

24. Given the following data about a car and a small plane traveling to Dallas, find when they are the same distance from Dallas and what that distance is.

Car

x travel time (hours)	y distance from Dallas (miles)
0	560
1	490

Plane

x travel time (hours)	y distance from Dallas (miles)
0	980
1	840

Review Exercises   The answers to all of these exercises are at the back of the book.

25. Evaluate the expression when x is -2 and y is -3: $xy^2 - xy$

26. Given $f(x) = x - 1$, find $f(3)$.

27. Simplify: $5w - (3w - 3)$

28. Simplify: $5y - 3[2y - 4(y + 8) -2]$

29. Graph: $-5 \leq w$

30. A nut company combines peanuts, which cost $1.12 a pound, and 40 pounds of cashews, which cost $2.65 more a pound than the peanuts, to make a special blend. Find the total cost of the mixture if the store used 6 less pounds of cashews than peanuts.

31. Sue paid $42.60, including 6.5% sales tax, for a sweater. What was the price of the sweater alone?

32. Graph the straight line that passes through $(0,3)$ with slope -2.

33. Determine whether this relation is a function: $\{(4,3),(8,2),(3,1),(4,9)\}$

Section 5.3 Using the Addition Method to Solve Systems

In this section we develop a second algebraic method of solving systems. Depending on the system we need to solve, this new method may be more efficient than the Substitution Method.

Topic 5.3 A — The Addition Method of Solving a System

Let's develop this third method of solving a system by trying to solve

$$\left\{ \begin{array}{rcl} x + y & = & 3 \\ x - y & = & 5 \end{array} \right\} \begin{array}{l} ❶ \\ ❷ \end{array}$$

We know from the Addition Property of Equality that if we add the *same* real number to both sides of an equation the resulting equation will be equivalent to the original. For example, if we add 5 to both sides of equation ❶ we get $x + y + 5 = 3 + 5$, which simplifies to $x + y + 5 = 8$. Unfortunately, this does not seem to get us anywhere since there are still two unknowns.

Now, let's be a bit clever. We know that *for this system $x - y$* and 5 represent the same real number because equation ❷ says $x - y = 5$. So, when we add "5" to both sides of equation ❶ we will add $x - y$ to the left side and 5 to the right side. Since they both represent the same real number the resulting equation will be equivalent to the original.

$$\begin{array}{rcll} x + y & = & 3 & ❶ \\ + \quad x - y & = & 5 & ❷ \\ \hline 2x + 0y & = & 8 & ❶ + ❷ \end{array}$$

We have combined information from both equations into a single equation. An important characteristic of the new equation is that the coefficient of the y term is 0. This means that the new equation can be written with only one variable, x. This allows us to solve the equation to find a unique value for x.

$$\begin{array}{ll} 2x + 0y = 8 & \Leftarrow \text{ Equation to solve. This is equation ❶ + ❷.} \\ 2x = 8 & \Leftarrow \text{ Simplified.} \\ x = 4 & \Leftarrow \text{ Divided both sides by 2.} \end{array}$$

We found that 4 is the value of x that satisfies the *system*.

To find the value of y that satisfies the system we substitute 4 for x into either *original* equation and solve for y.

$$x + y = 3 \qquad \Leftarrow \text{ One of the } original \text{ equations. This is equation } \mathbf{❶}.$$

$$(\mathbf{4}) + y = 3 \qquad \Leftarrow \text{ Substituted 4 for } x.$$

$$y = -1 \qquad \Leftarrow \text{ Subtracted 4 from both sides.}$$

We found that $(4,-1)$ is the solution for this system. That is, if we substitute these values into either equation we will wind up with a true statement. If we were to graph the equations they would intersect at the point $(4,-1)$.

The method we used to solve the system is called the **Addition Method** because we added the two given equations to form a single equation. Some people refer to this method as the **Elimination Method** because when we add the equations one of the variables is eliminated.

Let's try this method on another system.

Example 5.3.1   Use the Addition Method to solve the system $\left\{ \begin{array}{rcrcr} x & + & 2y & = & -1 \\ 2x & - & 2y & = & 10 \end{array} \right\} \begin{array}{l} \mathbf{❶} \\ \mathbf{❷} \end{array}$

•**SOLUTION**• We add the equations and then solve the resulting equation.

$$\begin{array}{rcrcrl} x & + & 2y & = & -1 & \mathbf{❶} \\ + \quad 2x & - & 2y & = & 10 & \mathbf{❷} \\ \hline 3x & + & 0y & = & 9 & \mathbf{❶ + ❷} \end{array} \qquad \Leftarrow \text{ Added the two equations.}$$

$$3x + 0y = 9 \qquad \Leftarrow \text{ Equation to solve. This is equation } \mathbf{❶ + ❷}.$$

$$3x = 9 \qquad \Leftarrow \text{ Simplified.}$$

$$x = 3 \qquad \Leftarrow \text{ Divided both sides by 3.}$$

Now we know the value of x which solves the system. To find the value of y we substitute 3 for x in either original equation and solve for y

$$x + 2y = -1 \qquad \Leftarrow \text{ Equation } \mathbf{❶}.$$

$$(\mathbf{3}) + 2y = -1 \qquad \Leftarrow \text{ Substituted 3 for } x.$$

$$2y = -4 \qquad \Leftarrow \text{ Subtracted 3 from both sides.}$$

$$y = -2 \qquad \Leftarrow \text{ Divided both sides by 2.}$$

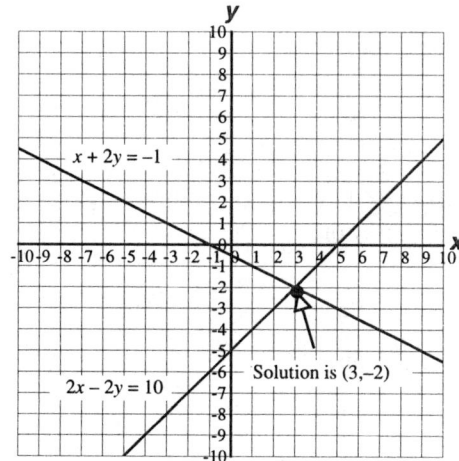

The solution to the system is the ordered pair $(3,-2)$. If we graph the system we can see that this is where the lines intersect.

We can algebraically check the solution by substituting $(3,-2)$ into each equation and simplifying. In both cases we should end up with an identity. We leave the check to you.

Practice 5.3.1 Use the Addition Method to solve the system.

1. $\left\{ \begin{array}{rcrcr} 2x & - & y & = & -8 \\ x & + & y & = & -1 \end{array} \right\} \begin{array}{l} \mathbf{❶} \\ \mathbf{❷} \end{array}$
2. $\left\{ \begin{array}{rcrcr} 3x & + & y & = & 3 \\ -3x & + & 2y & = & 3 \end{array} \right\} \begin{array}{l} \mathbf{❶} \\ \mathbf{❷} \end{array}$
3. $\left\{ \begin{array}{rcrcr} 4x & + & 3y & = & 1 \\ -4x & + & 2y & = & -26 \end{array} \right\} \begin{array}{l} \mathbf{❶} \\ \mathbf{❷} \end{array}$

Topic 5.3 B — Using the Addition Method When One Equation Is Transformed

In the previous examples we were fortunate that the coefficients of one of the variables of the original equations were opposites because when we added the two equations one of the variables was eliminated. This left us with a single variable equation which we could solve.

What do we do if this doesn't happen? For example, let's add the equations in the following system where the coefficients of neither variable are opposites:

$$
\begin{array}{rrrrl}
2x & - & y & = & 8 \quad ❶ \\
+ \quad x & + & 3y & = & -10 \quad ❷ \\
\hline
3x & + & 2y & = & -2 \quad ❶ + ❷
\end{array}
$$

Adding the equations didn't get us anywhere because neither new coefficient is 0 and so the new equation still has two variables.

The y coefficients would add to 0 if the coefficient of y in equation ❶ were -3 rather than -1. We can transform the coefficient of y in equation ❶ to -3 by multiplying equation ❶ by 3. This is permitted because the Multiplication Property of Equality says we may multiply an equation by any non–zero number and the result will be an equivalent equation. Let's see what happens when we do this:

$$2x - y = 8 \qquad \Leftarrow \text{Equation ❶, which we will transform.}$$

$$(3)2x - (3)y = (3)8 \qquad \Leftarrow \text{Multiplied each term by 3.}$$

$$6x - 3y = 24 \qquad \Leftarrow \text{Simplified.}$$

Now, let's add the equations of the new, equivalent, system:

$$
\begin{array}{rrrrl}
6x & - & 3y & = & 24 \quad ❶ \text{ (transformed)} \\
+ \quad x & + & 3y & = & -10 \quad ❷ \\
\hline
7x & + & 0y & = & 14 \quad ❶ + ❷
\end{array}
$$

By making the y coefficients opposites, the coefficient of the new y term becomes 0 and so we are left with a single–variable equation, which we can solve.

$$7x + 0y = 14 \qquad \Leftarrow \text{Equation to solve, (transformed ❶) + ❷.}$$

$$7x = 14 \qquad \Leftarrow \text{Simplified.}$$

$$x = 2 \qquad \Leftarrow \text{Divided both sides by 7.}$$

Now we know the value for x which solves the system. To find the value of y we substitute 2 for x in either original equation and solve for y.

$$x + 3y = -10 \qquad \Leftarrow \text{Equation ❷.}$$

$$(2) + 3y = -10 \qquad \Leftarrow \text{Substituted 2 for } x.$$

$$3y = -12 \qquad \Leftarrow \text{Subtracted 2 from both sides.}$$

$$y = -4 \qquad \Leftarrow \text{Divided both sides by 3.}$$

The solution of the system is $(2, -4)$.

Deciding which coefficients to make opposites does not affect the solution. Let's try the same problem but this time let's make the x coefficients opposites.

Example 5.3.2   a)   Make the x coefficients opposites and then use the Addition Method to solve

$$\left\{ \begin{array}{rcrcr} 2x & - & y & = & 8 \\ x & + & 3y & = & -10 \end{array} \right\} \begin{array}{l} ❶ \\ ❷ \end{array}$$

•**SOLUTION**• To make the x coefficients opposites we could multiply equation ❶ by $-\frac{1}{2}$ or we could multiply equation ❷ by -2. It will be easier to do the latter.

$$\left\{ \begin{array}{rcrcr} 2x & - & y & = & 8 \\ x & + & 3y & = & -10 \end{array} \right\} \begin{array}{l} ❶ \\ ❷ \end{array} \quad \Leftarrow \text{ System to solve.}$$

$$\left\{ \begin{array}{rcrcr} 2x & - & y & = & 8 \\ (-2)x & + & (-2)3y & = & (-2)(-10) \end{array} \right\} \begin{array}{l} ❶ \\ ❷ \end{array} \quad \Leftarrow \text{ Multiplied each term of equation ❷ by } -2.$$

$$\left\{ \begin{array}{rcrcr} 2x & - & y & = & 8 \\ -2x & - & 6y & = & 20 \end{array} \right\} \begin{array}{l} ❶ \\ ❷ \end{array} \quad \begin{array}{l} \Leftarrow \text{ Simplified. Look at how the sign of each} \\ \text{term in equation ❷ changed.} \end{array}$$

$$\begin{array}{r} 2x \ - \ y \ = \ 8 \quad ❶ \\ + \ \underline{-2x \ - \ 6y \ = \ 20} \quad ❷ \\ 0x \ - \ 7y \ = \ 28 \quad ❶ + ❷ \end{array} \quad \Leftarrow \text{ Added equations vertically.}$$

We can solve this new equation for y.

$$-7y = 28 \qquad \Leftarrow \text{ Equation to solve, ❶ + (transformed ❷).}$$

$$y = -4 \qquad \Leftarrow \text{ Divided both sides by } -7.$$

Now we know the value for y which solves the system. To find the value of x we substitute -4 for y in either original equation and solve for x. We will use the original equation ❷ because the x in that equation has a coefficient of 1. But, we could have just as correctly used equation ❶.

$$x + 3y = -10 \qquad \Leftarrow \text{ Original equation ❷.}$$

$$x + (-4)3 = -10 \qquad \Leftarrow \text{ Substituted } -4 \text{ for } y.$$

$$x - 12 = -10 \qquad \Leftarrow \text{ Multiplied.}$$

$$x = 2 \qquad \Leftarrow \text{ Added 12 to both sides.}$$

The solution is the same as before, $(2,-4)$.

b)   Use the Addition Method to solve $\left\{ \begin{array}{rcrcr} 3x & + & 6y & = & 9 \\ x & + & 2y & = & 3 \end{array} \right\} \begin{array}{l} ❶ \\ ❷ \end{array}$

•**SOLUTION**• Since the x in equation ❷ has a coefficient of 1, we will multiply each term of equation ❷ by -3 to make the coefficients of x opposites. We could multiply equation ❶ by $-\frac{1}{3}$ but this would introduce fractions, which would make the work more difficult.

$$\left\{ \begin{array}{rcrcr} 3x & + & 6y & = & 9 \\ (-3)x & + & (-3)2y & = & (-3)3 \end{array} \right\} \begin{array}{l} ❶ \\ ❷ \end{array} \quad \Leftarrow \text{ Multiplied each term of equation ❷ by } -3.$$

$$\left\{ \begin{array}{rcrcr} 3x & + & 6y & = & 9 \\ -3x & - & 6y & = & -9 \end{array} \right\} \begin{array}{l} ❶ \\ ❷ \end{array} \quad \Leftarrow \text{ Simplified.}$$

$$\begin{array}{r} 3x \ + \ 6y \ = \ 9 \quad ❶ \\ + \ \underline{-3x \ - \ 6y \ = \ -9} \quad ❷ \\ 0x \ + \ 0y \ = \ 0 \quad ❶ + ❷ \end{array} \quad \Leftarrow \text{ Added equations vertically.}$$

$$0 = 0 \qquad \Leftarrow \text{ Simplified.}$$

The equation we end up with is an identity. This means that all ordered pairs which make equation ❶ true also make equation ❷ true. Since we can't list all the ordered pairs that make this system true we will use variable expressions for the input and output coordinates.

- We will represent the input as x, to show that the input can be any value.

- For the output, we will use an expression that is equivalent to y. We find this expression by solving one of the original equations for y:

$$x + 2y = 3 \qquad \Leftarrow \text{ Original equation ❷.}$$

$$2y = -x + 3 \qquad \Leftarrow \text{ Subtracted } x \text{ from both sides.}$$

$$y = -\frac{1}{2}x + \frac{3}{2} \qquad \Leftarrow \text{ Divided both sides by 2.}$$

We may write the general output as $-\frac{1}{2}x + \frac{3}{2}$.

The solution to the system is written as the general ordered pair $\left(x, -\frac{1}{2}x + \frac{3}{2}\right)$. This implies that the input may be any number, but the output must always be $-\frac{1}{2}x + \frac{3}{2}$.

Practice 5.3.2 Use the Addition Method to solve the system.

1. $\begin{cases} x - 4y = 3 & ❶ \\ -4x + 2y = -12 & ❷ \end{cases}$

2. $\begin{cases} x + 3y = 18 & ❶ \\ -5x + y = 6 & ❷ \end{cases}$

3. $\begin{cases} x + 4y = 2 & ❶ \\ -2x - 8y = 5 & ❷ \end{cases}$

4. $\begin{cases} -6x + 2y = 5 & ❶ \\ 12x - 4y = -10 & ❷ \end{cases}$

In Section 4.5 we said that an equation written in the form $Ax + By = C$ is said to be in standard form. We may also define standard form for a system.

— Definition — *Standard form of a 2 by 2 linear system*	English: Terms containing variables are written on the left, with the variables in alphabetical order. The constant terms are written on the right. Example: The following system is written in standard form: $2x + 9y = 4$ $8x - y = 1$ Algebra: A 2 by 2 linear system is in standard form when it's written as $Ax + By = C$ $Dx + Ey = F$ where x and y are variables and A, B, C, D, E, and F are constants.

If we are given a system that is not in standard form, we should put it in standard form before using the Addition Method.

Example 5.3.3   Use the Addition Method to solve $\begin{Bmatrix} y + 2 = 4x \\ x + y = 3 \end{Bmatrix}$ ❶
❷

•**SOLUTION**• We will put the system into standard form and then use the Addition Method.

$$\begin{Bmatrix} y + 2 = 4x \\ x + y = 3 \end{Bmatrix} \begin{matrix} ❶ \\ ❷ \end{matrix}$$ ⇐ System to solve.

$$\begin{Bmatrix} -4x + y = -2 \\ x + y = 3 \end{Bmatrix} \begin{matrix} ❶ \\ ❷ \end{matrix}$$ ⇐ In equation ❶, subtracted $4x$ and 2 to put system into standard form.

$$\begin{Bmatrix} -4x + y = -2 \\ (-1)x + (-1)y = (-1)3 \end{Bmatrix} \begin{matrix} ❶ \\ ❷ \end{matrix}$$ ⇐ Multiplied each term of equation ❷ by -1 so the coefficients of y become opposites. (We could have multiplied equation ❷ by 4 if we had wanted to eliminate the x terms.)

$$\begin{Bmatrix} -4x + y = -2 \\ -x - y = -3 \end{Bmatrix} \begin{matrix} ❶ \\ ❷ \end{matrix}$$ ⇐ Simplified.

$$\begin{aligned} -4x + y &= -2 \quad ❶ \\ + \quad \underline{-x - y} &= \underline{-3} \quad ❷ \\ -5x + 0y &= -5 \quad ❶ + ❷ \end{aligned}$$ ⇐ Added equations vertically.

We can solve this equation for x.

$-5x = -5$ ⇐ Equation to solve, ❶ + (transformed ❷).

$x = 1$ ⇐ Divided both sides by -5.

Now we know the value of x which solves the system. To find the value of y we substitute 1 for x in either original equation and solve for y. We will use the original equation ❶.

$y + 2 = 4x$ ⇐ Original equation ❶.

$y + 2 = 4(\mathbf{1})$ ⇐ Substituted 1 for x.

$y = 2$ ⇐ Simplified.

The solution is (1,2). You should check it by substituting 1 for x and 2 for y in both original equations.

Practice 5.3.3 Use the Addition Method to solve the system.

1. $\begin{Bmatrix} y - 1 = 2x \\ x + y = 7 \end{Bmatrix} \begin{matrix} ❶ \\ ❷ \end{matrix}$ 2. $\begin{Bmatrix} x + 2 = -y \\ 2x - y = -1 \end{Bmatrix} \begin{matrix} ❶ \\ ❷ \end{matrix}$ 3. $\begin{Bmatrix} -5 = y - x \\ 2x - y = -1 \end{Bmatrix} \begin{matrix} ❶ \\ ❷ \end{matrix}$

Topic 5.3 C — Using the Addition Method When Both Equations Are Transformed

Let's try the Addition Method to solve the following system:

$$\begin{Bmatrix} -3x + 2y = -1 \\ 2x + 5y = 45 \end{Bmatrix} \begin{matrix} ❶ \\ ❷ \end{matrix}$$

To make the x–coefficients opposites we could multiply equation ❶ by $\frac{2}{3}$, or we could multiply equation ❷ by $\frac{3}{2}$. In either case, we would cause fractions to be introduced into the system. This isn't wrong, but it makes finding the solution more cumbersome and more prone to error. Instead of multiplying just one equation by a fraction, we will multiply each equation by some integer that will make one of the variables have opposite coefficients. For example, we could multiply each term in equation ❶ by 2 and each term in equation ❷ by 3. The x coefficients will then be opposites.

$$\begin{matrix} ❶ \\ ❷ \end{matrix} \left\{ \begin{matrix} -3x & + & 2y & = & -1 \\ 2x & + & 5y & = & 45 \end{matrix} \right\} \begin{matrix} \cdot 2 \\ \cdot 3 \end{matrix} \Rightarrow \left\{ \begin{matrix} (2)(-3x) & + & (2)2y & = & (2)(-1) \\ (3)2x & + & (3)5y & = & (3)45 \end{matrix} \right\}$$

This simplifies to

$$\begin{matrix} ❶ \\ ❷ \end{matrix} \left\{ \begin{matrix} -6x & + & 4y & = & -2 \\ 6x & + & 15y & = & 135 \end{matrix} \right\}$$

Alternatively, we could make the y coefficients opposites by multiplying each term in equation ❶ by 5 and each term in equation ❷ by –2.

$$\begin{matrix} ❶ \\ ❷ \end{matrix} \left\{ \begin{matrix} -3x & + & 2y & = & -1 \\ 2x & + & 5y & = & 45 \end{matrix} \right\} \begin{matrix} \cdot 5 \\ \cdot -2 \end{matrix} \Rightarrow \left\{ \begin{matrix} (5)(-3x) & + & (5)2y & = & (5)(-1) \\ (-2)2x & + & (-2)5y & = & (-2)45 \end{matrix} \right\}$$

This simplifies to

$$\begin{matrix} ❶ \\ ❷ \end{matrix} \left\{ \begin{matrix} -15x & + & 10y & = & -5 \\ -4x & + & -10y & = & -90 \end{matrix} \right\}$$

The choice of which coefficients to make opposites is yours. An easy way to make coefficients of one of the variables opposites is to do the following:

— *Procedure* — *Deciding what to multiply* *by when transforming* *both equations in a system*	*Step 1* Decide which variable you wish to eliminate. Let's say it's x. *Step 2* Multiply equation ❶ by the coefficient of x in equation ❷ and multiply equation ❷ by the *opposite* of the original coefficient of x in equation ❶. We use the opposite here in order to make the signs of the x coefficients different.

For example, to make the x coefficients opposites in the system

$$\left\{ \begin{matrix} -3x & + & 2y & = & -1 \\ 2x & + & 5y & = & 45 \end{matrix} \right\} \begin{matrix} ❶ \\ ❷ \end{matrix}$$

we would multiply equation ❶ by 2 (the coefficient of x in equation ❷) and we would multiply equation ❷ by 3 (the opposite of the coefficient of x in equation ❶). Alternatively, to make the y coefficients opposites we would multiply equation ❶ by 5 (the coefficient of y in equation ❷) and equation ❷ by –2 (the opposite of the coefficient of y in equation ❶).

Before solving some systems, let's practice making coefficients opposites.

Example 5.3.4   a)   Transform the equations in $\left\{ \begin{matrix} 4x & - & 3y & = & 7 \\ 3x & + & 2y & = & 1 \end{matrix} \right\} \begin{matrix} ❶ \\ ❷ \end{matrix}$ so that the x coefficients are opposites.

• SOLUTION • We multiply equation ❶ by 3 (the coefficient of x in equation ❷). We multiply equation ❷ by –4 (the *opposite* of the coefficient of x in equation ❶).

$$\begin{matrix} ❶ \\ ❷ \end{matrix} \left\{ \begin{matrix} 4x & - & 3y & = & 7 \\ 3x & + & 2y & = & 1 \end{matrix} \right\} \begin{matrix} \cdot 3 \\ \cdot -4 \end{matrix} \Rightarrow \left\{ \begin{matrix} (3)(4x) & - & (3)3y & = & (3)(7) \\ (-4)3x & + & (-4)2y & = & (-4)1 \end{matrix} \right\}$$

This simplifies to

$$\begin{matrix} ❶ \\ ❷ \end{matrix} \left\{ \begin{matrix} 12x & - & 9y & = & 21 \\ -12x & - & 8y & = & -4 \end{matrix} \right\}$$

b) Transform the equations in $\left\{\begin{array}{rrr} 4x & - & 3y & = & 7 \\ 3x & + & 2y & = & 1 \end{array}\right\} \begin{array}{c} \mathbf{0} \\ \mathbf{2} \end{array}$ so that the y coefficients are opposites.

•**SOLUTION**• We multiply equation **①** by 2, the coefficient of y in equation **②**.
We multiply equation **②** by 3, the *opposite* of the coefficient of x in equation **①**.

$$\begin{array}{c} \mathbf{0} \\ \mathbf{2} \end{array}\left\{\begin{array}{rrrrr} 4x & - & 3y & = & 7 \\ 3x & + & 2y & = & 1 \end{array}\right\}\begin{array}{c} \bullet 2 \Rightarrow \\ \bullet 3 \Rightarrow \end{array}\left\{\begin{array}{rrrrr} (2)(4x) & - & (2)3y & = & (2)(7) \\ (3)3x & + & (3)2y & = & (3)1 \end{array}\right\}$$

This simplifies to

$$\begin{array}{c} \mathbf{0} \\ \mathbf{2} \end{array}\left\{\begin{array}{rrrrr} 8x & - & 6y & = & 14 \\ 9x & + & 6y & = & 3 \end{array}\right\}$$

Practice 5.3.4 For each system,

a) Transform the equations so that the x coefficients are opposites.
b) Transform the equations so that the y coefficients are opposites.

1. $\left\{\begin{array}{rrrrr} 4x & - & 5y & = & -2 \\ 3x & - & 2y & = & -5 \end{array}\right\}\begin{array}{c} \mathbf{0} \\ \mathbf{2} \end{array}$

2. $\left\{\begin{array}{rrrrr} 6x & + & 5y & = & 8 \\ 2x & + & 8y & = & 1 \end{array}\right\}\begin{array}{c} \mathbf{0} \\ \mathbf{2} \end{array}$

Once we have opposite coefficients we can solve the system using addition.

Example 5.3.5 Solve the system $\left\{\begin{array}{rrrrr} 2x & + & 3y & = & -2 \\ 5x & - & 2y & = & -24 \end{array}\right\}\begin{array}{c} \mathbf{0} \\ \mathbf{2} \end{array}$

•**SOLUTION**• We choose to eliminate the x variables so we will multiply equation **①** by 5 and equation **②** by –2 and then add.

$$\left\{\begin{array}{rrrrr} (5)2x & + & (5)3y & = & (5)(-2) \\ (-2)5x & - & (-2)2y & = & (-2)(-24) \end{array}\right\}\begin{array}{c} \mathbf{0} \\ \mathbf{2} \end{array} \quad \Leftarrow \text{Multiplied equation } \mathbf{0} \text{ by 5 and}$$
equation **②** by –2.

$$\left\{\begin{array}{rrrrr} 10x & + & 15y & = & -10 \\ -10x & + & 4y & = & 48 \end{array}\right\}\begin{array}{c} \mathbf{0} \\ \mathbf{2} \end{array} \quad \Leftarrow \text{Simplified. Look carefully at the signs}$$
of each term.

$$\begin{array}{rrrrrl} & 10x & + & 15y & = & -10 & \mathbf{0} \\ + & -10x & + & 4y & = & 48 & \mathbf{2} \\ \hline & 0x & + & 19y & = & 38 & \mathbf{0} + \mathbf{2} \end{array} \quad \Leftarrow \text{Added equations vertically.}$$

We can solve this equation for y.

$19y = 38$ ⟸ Equation to solve, (transformed **①**) + (transformed **②**).

$y = 2$ ⟸ Divided both sides by 19.

Now we know the value for y which solves the system. To find the value of x we substitute 2 for y in either original equation and solve for x. It looks like neither equation will be easier, so we will use the original equation **①**.

$2x + 3y = -2$ ⟸ Original equation **①**.

$2x + 3(2) = -2$ ⟸ Substituted 2 for y.

$2x + 6 = -2$ ⟸ Multiplied.

$2x = -8$ ⟸ Subtracted 6 from both sides.

$x = -4$ ⟸ Divided both sides by 2.

The solution is $(-4, 2)$. We leave the check to you.

Practice 5.3.5 Solve the system.

1. $\begin{cases} 4x - 5y = -2 \text{ ❶} \\ 3x - 2y = -5 \text{ ❷} \end{cases}$ 2. $\begin{cases} 6x + 5y = 8 \text{ ❶} \\ 5x + 6y = 14 \text{ ❷} \end{cases}$ 3. $\begin{cases} 6x - 3y = 9 \text{ ❶} \\ -8x + 4y = -12 \text{ ❷} \end{cases}$

Don't let fractions distract you. If the original system contains fractions, multiply each equation by its LCD and reduce. This will convert all the fractions into integer values, which are easier to work with.

Example 5.3.6   Solve the system $\begin{cases} \dfrac{2}{3}x + \dfrac{y}{2} = -3 \text{ ❶} \\ \dfrac{x}{4} + \dfrac{3y}{2} = \dfrac{-15}{4} \text{ ❷} \end{cases}$

• SOLUTION • We want to clear the fractions first by multiplying each equation by its own LCD. The LCD of equation ❶ is 6 and the LCD of equation ❷ is 4.

$\begin{cases} (6)\dfrac{2}{3}x + (6)\dfrac{y}{2} = (6)(-3) \text{ ❶} \\ (4)\dfrac{x}{4} + (4)\dfrac{3y}{2} = (4)\dfrac{-15}{4} \text{ ❷} \end{cases}$ ⇐ Multiplied each equation by its LCD.

$\begin{cases} 4x + 3y = -18 \text{ ❶} \\ x + 6y = -15 \text{ ❷} \end{cases}$ ⇐ Simplified. Look carefully at how the fractions reduced.

$\begin{cases} 4x + 3y = -18 \text{ ❶} \\ (-4)x + (-4)6y = (-4)(-15) \text{ ❷} \end{cases}$ ⇐ Multiplied every term in equation ❷ by –4. This made the coefficients of x opposites.

$\begin{cases} 4x + 3y = -18 \\ -4x + -24y = 60 \end{cases}$ ❶ ❷   ⇐ Simplified..

$\begin{array}{r} 4x + 3y = -18 \text{ ❶} \\ + \underline{-4x - 24y = 60} \text{ ❷} \\ 0x + -21y = 42 \text{ ❶ + ❷} \end{array}$ ⇐ Added equations vertically.

We can solve this equation for y.

$-21y = 42$ ⇐ Equation to solve, (transformed ❶) + (transformed ❷).

$y = -2$ ⇐ Divided both sides by –21.

Now we know the value for y which solves the system. To find the value of x we substitute –2 for y in either original equation and solve for x. It looks like neither equation will be easier, so we will use the original equation ❷.

$\dfrac{x}{4} + \dfrac{3y}{2} = \dfrac{-15}{4}$ ⇐ Original equation ❷.

$\dfrac{x}{4} + \dfrac{3(-2)}{2} = \dfrac{-15}{4}$ ⇐ Substituted –2 for y.

$\dfrac{x}{4} - 3 = \dfrac{-15}{4}$ ⇐ Simplified.

$(4)\left(\dfrac{x}{4}\right) - (4)(3) = (4)\left(\dfrac{-15}{4}\right)$ ⇐ Multiplied each term by the LCD, 4.

$x - 12 = -15$ ⇐ Reduced.

$x = -3$ ⇐ Added 12 to both sides.

The solution is (–3,–2). We leave the check to you.

Practice 5.3.6 Solve the system.

1. $\begin{cases} \dfrac{3}{4}x - \dfrac{y}{4} = 2 \\ \dfrac{3}{5}x + \dfrac{y}{5} = 2 \end{cases}$ ❶ ❷

2. $\begin{cases} \dfrac{7}{6}x - \dfrac{5}{6}y = 1 \\ \dfrac{2}{9}x + \dfrac{4}{9}y = 2 \end{cases}$ ❶ ❷

3. $\begin{cases} \dfrac{x}{6} + \dfrac{y}{3} = 3 \\ \dfrac{x}{4} + \dfrac{y}{2} = 1 \end{cases}$ ❶ ❷

We can summarize the Addition Method for solving 2 by 2 systems as follows:

— *Procedure* — ***Addition method for*** ***solving systems***	***Step 1 Standard Form*** Put both equations into the form $ax + by = c$. ***Step 2 Clear Fractions*** Use the LCD to clear fractions from each equation. ***Step 3 Transform Equations*** Transform into opposites the coefficients of one of the variables by multiplying each term in one or both equations by a constant. ***Step 4 Add*** Add the equations. ***Step 5 Solve*** Solve the resulting equation. ***Step 6 Substitute*** Substitute the solution from step 5 into either *original* equation and solve for the variable. ***Step 7 Check*** Check the solution by substituting into *both* equations.
— *Note* — ***Which method is best?***	Students sometimes wonder which method, Addition or Substitution, is best. Both will always work, but as a general rule, if the coefficient of one of the variables is 1 or –1 then substitution is usually easier.

Exercise Set 5.3   The answers to the odd numbered exercises are at the back of the book.

1. $\begin{cases} x + 2y = 11 \\ 3x - 2y = 1 \end{cases}$ ❶ ❷

2. $\begin{cases} 4w - 6r = -2 \\ 5w + 6r = 11 \end{cases}$ ❶ ❷

3. $\begin{cases} 4x - 5y = 3 \\ -4x + y = -7 \end{cases}$ ❶ ❷

4. $\begin{cases} x + 4y = 9 \\ 3x - 4y = -5 \end{cases}$ ❶ ❷

5. $\begin{cases} 3x - y = 3 \\ -5x + y = -7 \end{cases}$ ❶ ❷

6. $\begin{cases} x + 2y = -1 \\ -x - 3y = 1 \end{cases}$ ❶ ❷

7. $\begin{cases} 4x + y = 7 \\ x - 4y = -11 \end{cases}$ ❶ ❷

8. $\begin{cases} x - 5y = 7 \\ 3x + y = 5 \end{cases}$ ❶ ❷

9. $\begin{cases} -2x + y = -3 \\ x - 3y = 14 \end{cases}$ ❶ ❷

10. $\begin{cases} 6x - 8y = 4 \\ -3x + 4y = 5 \end{cases}$ ❶ ❷

11. $\begin{cases} 3x - y = 7 \\ -9x + 3y = -21 \end{cases}$ ❶ ❷

12. $\begin{cases} x - 8y = 20 \\ -4x + y = -18 \end{cases}$ ❶ ❷

13. $\begin{cases} x + 5y = 3 \\ 2x + 10y = 6 \end{cases}$ ❶ ❷

14. $\begin{cases} 2x + 3y = 0 \\ -4x - 6y = 5 \end{cases}$ ❶ ❷

15. $\begin{cases} 2x - 8y = 18 \\ -4x + y = -21 \end{cases}$ ❶ ❷

16. $\begin{cases} x - 2y = 4 \\ -3x + 6y = -12 \end{cases}$ ❶ ❷

17. $\begin{cases} 2y = 14 - 3x \\ 3x - 4y = 8 \end{cases}$ ❶ ❷

18. $\begin{cases} 4x - 8 = 3y \\ -4x + 3y = -8 \end{cases}$ ❶ ❷

19. $\begin{cases} x = 3 - 2y \\ -5x - 10y = -15 \end{cases}$ ❶ ❷

20. $\begin{cases} 3y = 5x - 3 \\ x + 6y = 27 \end{cases}$ ❶ ❷

21. $\begin{cases} 6x - 5y = -3 \\ 9x + 2y = 5 \end{cases}$ ❶ ❷

22. $\begin{cases} 4x + 3y = 2 \\ 5x + 2y = 6 \end{cases}$ ❶ ❷

23. $\begin{cases} 3x - 9y = 6 \\ 5x - 15y = 10 \end{cases}$ ❶ ❷

24. $\begin{cases} 4x + 3y = 6 \\ 3x - 5y = 19 \end{cases}$ ❶ ❷

25. $\begin{cases} 6x = 3y + 5 \\ 4y = 7 + 8x \end{cases}$ ❶ ❷

26. $\begin{cases} 5x - 14 = -2y \\ 2x - 5y - 23 = 0 \end{cases}$ ❶ ❷

27. $\begin{cases} 2x - 3y = 5 \\ 3x + 5y = 17 \end{cases}$ ❶ ❷

28. $\begin{cases} 3x - 9y = 7 \\ 5x - 15y = 2 \end{cases}$ ❶ ❷

29. $\begin{cases} \dfrac{x}{2} + \dfrac{2}{5}y = \dfrac{7}{5} \\ \dfrac{x}{2} - \dfrac{2}{3}y = \dfrac{1}{3} \end{cases}$ ❶ ❷

30. $\begin{cases} \dfrac{2}{3}x - \dfrac{y}{4} = \dfrac{1}{12} \\ \dfrac{x}{2} - \dfrac{3}{10}y = \dfrac{-1}{2} \end{cases}$ ❶ ❷

31. $\begin{cases} \dfrac{x}{4} + \dfrac{y}{2} = 1 \\ \dfrac{2}{3}x + \dfrac{4}{3}y = 1 \end{cases}$ ❶ ❷

32. $\begin{cases} \dfrac{x}{2} + \dfrac{2}{3}y = \dfrac{1}{3} \\ \dfrac{x}{2} - \dfrac{2}{5}y = \dfrac{7}{5} \end{cases}$ ❶ ❷

33. $\begin{cases} \dfrac{5}{4}x - \dfrac{3}{4}y = 3 \\ \dfrac{3}{7}x + \dfrac{5}{7}y = 2 \end{cases}$ ❶ ❷

34. $\begin{cases} \dfrac{x}{3} + \dfrac{2}{3}y = 2 \\ \dfrac{x}{6} + \dfrac{y}{3} = 1 \end{cases}$ ❶ ❷

35. $\begin{cases} \dfrac{x}{4} + \dfrac{y}{2} = \dfrac{9}{2} \\ \dfrac{x}{6} + \dfrac{y}{3} = 3 \end{cases}$ ❶ ❷

36. $\begin{cases} \dfrac{4}{7}x - \dfrac{3}{2}y = \dfrac{-17}{2} \\ \dfrac{1}{6}x + \dfrac{y}{3} = -\dfrac{1}{6} \end{cases}$ ❶ ❷

37. $\begin{cases} \dfrac{x}{4} - \dfrac{y}{2} = -1 \\ \dfrac{3}{4}x + \dfrac{y}{2} = 1 \end{cases}$ ❶ ❷

38. $\begin{cases} \dfrac{x}{3} - \dfrac{2}{3}y = 1 \\ \dfrac{-2}{3}x + \dfrac{4}{3}y = -2 \end{cases}$ ❶ ❷

Review Exercises   The answers to all of these exercises are at the back of the book.

39. Graph −800 on a number line. Use an interval from −900 to 0 in steps of 100.

40. Is −3 a solution of $\dfrac{2}{3}x - 8(x+2) = 3 - x$?

41. Solve: $4(x-3) \geq 8 - x$

42. A bookshelf has a total of 31 books about topics A, B, and C. There are 4 less books about C than about B. If x represents the number of books about C, create a simplified expression to represent the number of books about A.

43. Find the x-intercept and the y-intercept:

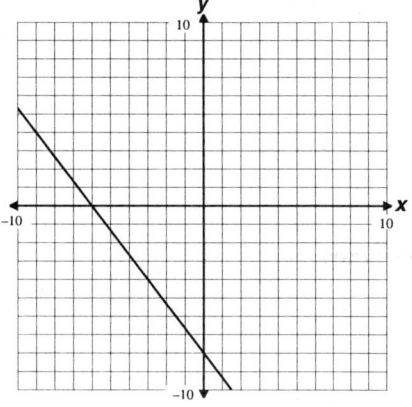

44. Construct the equation of the line shown in the graph.

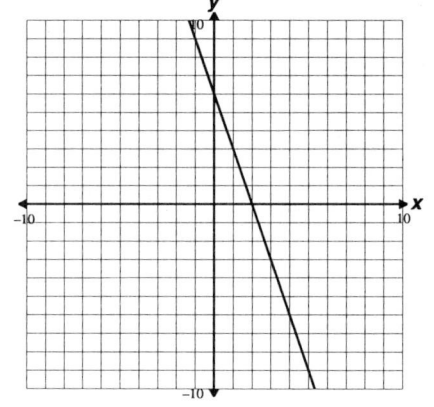

45. Find the equation of the line that goes through (0,−2) and (3,1).

46. Find the equation of a line parallel to $y = x$ and which passes through (4,2).

47. The sum of two consecutive even integers is −14, find the two integers.

48. Two campaign workers are soliciting votes by going door to door. On average the first worker reaches 15 houses per hour and the second reaches 19 houses per hour. If the first worker began two hours before the second, how long did each work if they reached a total of 183 houses.

Section 5.4 Solving Word Problems Using Systems

In Chapter 4, we used a one variable/one equation approach to solve traditional algebra word problems. For example, we solved the following problem:

Find two consecutive odd integers whose sum is –36.

We defined x as one number and used $x + 2$ as the other. Since the sum of the two numbers is –36, we wrote the equation $-36 = x + (x + 2)$ and then solved for x.

We can also use a 2 by 2 system to solve a problem like this. If we define x as the smaller number and y as the larger then the constraints on the numbers (their sum and the fact that they are consecutive odd integers) may be modeled using the following system:

$$\left\{ \begin{array}{rcl} x + y &=& -36 \\ y &=& x + 2 \end{array} \right\} \quad \begin{array}{l} \text{❶} \Leftarrow \text{first constraint says the sum of the two numbers is } -36 \\ \text{❷} \Leftarrow \text{second constraint says the numbers are consecutive odd integers} \end{array}$$

This system is easily solved using the Substitution Method.

So, why didn't we use systems to solve word problems right from the start? Two reasons:

- Systems are more difficult to solve than single equations. We had to go through a lot of algebra before you acquired the tools needed to solve systems.

- Some problems must be solved using one variable. For example, the following problem discusses a single number and cannot logically be done using two variables.

    Twice a number is 6 less than five
    times the number. Find the number.

In this section we will practice solving word problems using a two variable approach. Some of the problems we discuss can be solved using either one or two variables, but some can be solved only by using a 2 by 2 system.

Topic 5.4 A — Using Systems with the General Procedure for Solving Word Problems

The General Procedure for Solving Word Problem is reproduced below for your reference. We changed the word *variable* to *variables* because the models we now develop will involve more than one variable.

— Procedure — *General procedure for* *solving word problems*	*Step 1 Organize the information* Read the problem, several times if necessary:
	• Given: determine what is given. Be sure to include the unit of all numbers. • Find: determine what you are asked to find. • Sketch: draw a sketch if it will help you better understand the problem. • Rates: examine the unit of each given number to determine if a rate is involved. If so, begin construction of an Amount Table. • Formulas: if standard formulas will be needed, list them.
	Step 2 Define unknowns
	• Define: assign definitions to *variables*; include units, if appropriate. • Construct: use the *variables* you just defined to build expressions for other unknowns. • Amount Table: if you began an Amount Table in step 1, complete it.
	Step 3 Build verbal models Use English to write relations between the unknowns.
	Step 4 Build mathematical models Translate the verbal models into mathematical models. Because the verbal and mathematical models are so closely related we often do Steps 3 and 4 together.
	Step 5 Solve Solve the mathematical models for the *variables*.
	Step 6 Values of unknowns Calculate the values of all unknowns.
	Step 7 Answer State the answer to the question.
	Step 8 Check Using the original words of the problem, check to see that the values make sense and are consistent with the original statement of the problem.

Let's solve a problem using the two variable approach. The example given below is the same one we solved using one variable in Section 3.4.

Example 5.4.1   José has a bag that contains dimes and quarters. There are a total of 784 coins in the bag and the value of the coins is $133. Find the number of each type of coin.

• SOLUTION •

Step 1 Organize

Given: • total number of coins is 784.
        • total value of coins is $133.

Find: • number of dimes.
        • number of quarters.

Sketch: Not helpful.

Rates: The value of a coin can be written as a rate using the unit $\frac{\text{cents}}{\text{coin}}$.
        Since rates are involved we will construct an Amount Table.

	amount =	*rate* •	*base*
Descriptions →	value of coins	value of 1 coin	number of coins
Units →	cents	$\frac{\text{cents}}{\text{coin}}$	coins
↓ *Item names* ↓			
dimes	❸	10	❶
quarters	❹	25	❷

Formula: amount = rate • base.

Step 2 Define unknowns

Define: We are looking for the number of dimes and quarters. We will use a variable to represent each of these numbers:

d represents the number of dimes.

q represents the number of quarters.

Construct: Not needed.

Amount Table: We now complete the Amount Table.

❶ is the number of dimes, d.

❷ is the number of quarters, q.

❸ is the value of the dimes. We multiply rate • base to get

$\left(\dfrac{10 \text{ cents}}{1 \text{ coin}}\right) \cdot (d \text{ coins})$, which simplifies to $10d$ cents

❹ is the value of the quarters. We multiply rate • base to get

$\left(\dfrac{25 \text{ cents}}{1 \text{ coin}}\right) \cdot (q \text{ coins})$, which simplifies to $25q$ cents

The complete Amount Table looks like this:

	amount	=	rate	•	base
Descriptions →	value of coins		value of 1 coin		number of coins
Units →	cents		$\dfrac{\text{cents}}{\text{coin}}$		coins
↓ Item names ↓					
dimes	**10*d***		10		***d***
quarters	**25*q***		25		***q***

Step 3 Verbal models and Step 4 Mathematical Models

We may describe two relations, one between the *number* of coins and the other between the *value* of the coins:

- relation between the *number* of coins:

Verbal model ⇒   *number* of all coins   =   *number* of dimes   +   *number* of quarters

Math model ⇒   784   = d + q

- relation between the *value* of the coins: Note that since we have chosen to use cents per coin as the rate we must write the value of all the coins in cents (that is, we write the value as 13,300 cents rather than 133 dollars).

Verbal model ⇒   *value* of all coins (cents)   =   *value* of dimes (cents)   +   *value* of quarters (cents)

Math model ⇒   13300   = $10d$ + $25q$

We must solve the system

$$\begin{cases} 784 = d + q \\ 13300 = 10d + 25q \end{cases}$$ ❶ ⇐ constraint on number of coins
 ❷ ⇐ constraint on value of coins

Step 5 Solve   We will use substitution since it's easy to solve ❶ for d to get $d = 784 - q$. Then, we substitute for d in equation ❷ and solve for q as follows:

$$13300 = 10(\mathbf{784 - q}) + 25q \quad \Leftarrow \text{ Substituted } 784 - q \text{ for } d \text{ in equation } ❷.$$

$$13300 = \mathbf{7840 - 10q} + 25q \quad \Leftarrow \text{ Distributed 10.}$$

$$13300 = 7840 + \mathbf{15q} \quad \Leftarrow \text{ Combined like terms.}$$

$$\mathbf{5460} = 15q \quad \Leftarrow \text{ Subtracted 7840 from both sides.}$$

$$364 = q \quad \Leftarrow \text{ Divided both sides by 15.}$$

To find d, we substitute 364 for q in equation ❶ and then solve for d to get 420.

Step 6 Values of unknowns   We fill in the unknown values in the Amount Table.

Descriptions → *Units* → ↓ *Item names* ↓	value of coins cents	value of 1 coin $\dfrac{\text{cents}}{\text{coin}}$	number of coins coins
dimes	10d is 10(**420**) **4200 cents** **(value of dimes)**	10 cents per coin	d is **420** **(number of dimes)**
quarters	25q is 25(**364**) **9100 cents** **(value of quarters)**	25 cents per coin	q is **364** **(number of quarters)**

Step 7 Answer

- number of dimes, d, is 420.
- number of quarters, q, is 364.

Step 8 Check

- we have 784 coins: checks because 420 dimes + 364 quarters is 784 coins.
- value of the coins is \$133: checks because 4200 cents (value of dimes) + 9100 cents (value of quarters), is 13,300 cents, which is \$133.

Before doing some practice exercises, let's compare the one variable and two variable approaches for the coins problem.

One Variable (see Section 3.4)	*Two Variables*
Step 2 Define unknowns	**Step 2 Define unknowns**
Define: We defined d as the number of dimes.	Define: We defined d as the number of dimes and q as the number of quarters.
Construct: We used the Whole-Part Concept to construct an expression for the number of quarters: $784 - d$.	Construct: Not needed.
Amount Table: We used the above definitions and the Amount Formula to construct the following expressions: • $10d$ represents the value of the dimes. • $25(784 - d)$ represents the value of the quarters.	Amount Table: We used the above definitions and the Amount Formula to construct the following expressions: • $10d$ represents the value of the dimes. • $25q$ represents the value of the quarters.
Step 3 Verbal models	**Step 3 Verbal models**
We wrote a single relation involving the value of the dimes and the quarters.	We wrote a relation involving the *number* of coins and another relation involving the *value* of the coins.
$$\begin{array}{ccc} \text{Total value} & \text{value of} & \text{value of} \\ \text{of coins} = & \text{dimes} + & \text{quarters} \end{array}$$	$$\begin{array}{ccc} \text{Total } number & number \text{ of} & number \text{ of} \\ \text{of coins} = & \text{dimes} + & \text{quarters} \end{array}$$ $$\begin{array}{ccc} \text{Total } value & value \text{ of} & value \text{ of} \\ \text{of coins} = & \text{dimes} + & \text{quarters} \end{array}$$
Step 4 Mathematical models We translated the verbal model into a mathematical model that was a single equation. $$13300 = 10d + 25(784 - d)$$	**Step 4 Mathematical models** We translated the verbal model into a mathematical model that was a system. $$\left\{ \begin{array}{rcl} 784 &=& d + q \\ 13300 &=& 10d + 25q \end{array} \right\}$$

In most situations you may choose the method that you find easiest. For now, however, practice using the two variable approach so that you will have a choice in the future.

Practice 5.4.1 Use the General Procedure for Solving Word Problems and two variables to help you answer the following questions.

1. Sandy has a ten–pound box that contains two types of candy. The chocolate candy is worth $3.50 a pound and the butterscotch candy is worth $2.75 a pound. If the total value of the box is $30.50, how many pounds of each type of candy does she have?

2. Lynne has a bag of 32 coins, all nickels and quarters. If the total value of the coins is $5.00, how many of each type of coin does she have?

Topic 5.4 B — More Word Problems

Below are three more examples of using systems to solve word problems.

Example 5.4.2 Lynn flies from Roseville to Tucson at 550 mph. The next day, she drives back to Roseville at a rate of 50 mph. If the total traveling time by plane and car is 36 hours, find the distance from Roseville to Tucson. Use a two–variable approach.

• **SOLUTION** • *Step 1 Organize*

Given: • rate by plane is 550 mph.
   • rate by car is 50 mph.
   • total traveling time is 36 hours.

Find: distance from Roseville to Tucson

Sketch:

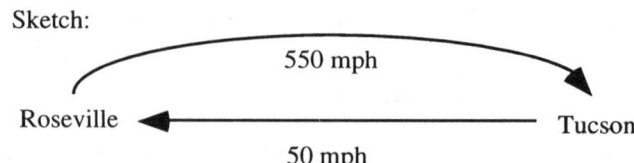

Roseville ⟵—————————————— Tucson

50 mph

Rates: Miles per hour is a rate so we will construct an Amount Table.

	amount	=	rate	•	base
Descriptions →	distance		rate		time
Units → ↓ Item names ↓	miles		$\dfrac{\text{miles}}{\text{hour}}$		hours
car	❶		50		❸
plane	❷		550		❹

Formulas: amount = rate • base.

Step 2 Define unknowns

Define: Since the problem asks for a distance we will define

c represents distance traveled by car (miles)
p represents distance traveled by plane (miles)

Construct: Not needed.

Amount Table: We now complete the Amount Table.

❶ is the distance for the car, c.
❷ is the distance for the plane, p.

We need to calculate the time (the base) for each item. Since

distance = rate • time

if we divide both sides by the rate the time will be isolated:

$$\frac{\text{distance}}{\text{rate}} = \frac{\cancel{\text{rate}} \cdot \text{time}}{\cancel{\text{rate}}}$$

$$\frac{\text{distance}}{\text{rate}} = \text{time}$$

We calculate the time for the car and for the plane and enter the expressions into the table under the base column:

	amount	=	rate	•	base
Descriptions →	distance		rate		time
Units → ↓ Item names ↓	miles		$\dfrac{\text{miles}}{\text{hour}}$		hours
car	c		50		$\dfrac{c}{50}$
plane	p		550		$\dfrac{p}{550}$

Step 3 Verbal models *and* **Step 4 Mathematical Models**   We have defined two variables so we need two equations. Each equation will correspond to a constraint imposed by the problem.

- *Relation between distance traveled by car and distance traveled by plane:* The problem does not state this explicitly but common sense says that the distance from Roseville to Tucson (by plane) is the same as the distance from Tucson to Roseville (by car, assuming a straight road). That is,

Verbal model $\Rightarrow$ $\quad\dfrac{\text{distance traveled}}{\text{by car (miles)}}\quad=\quad\dfrac{\text{distance traveled}}{\text{by plane (miles)}}$

Math model $\Rightarrow$ $\qquad\qquad c \qquad\qquad = \qquad\qquad p$

- *Relation between rate of car and rate of plane*: The rates involve no variables so this relation will not help us solve the problem.

- *Relation between time of car and time of plane*: The constraint on the time is that the total travel time by car and plane is 36 hours. This implies

Verbal model $\Rightarrow$ $\quad\dfrac{\text{total travel time}}{\text{(hours)}}\quad = \quad\dfrac{\text{time of travel}}{\text{for car (hours)}}\quad + \quad\dfrac{\text{time of travel}}{\text{for plane (hours)}}$

Math model $\Rightarrow$ $\qquad 36 \qquad = \qquad \dfrac{c}{50} \qquad + \qquad \dfrac{p}{550}$

Step 5 Solve Substitution is the easiest method to use.

$$\left\{ \begin{array}{l} c \;=\; p \\ 36 \;=\; \dfrac{c}{50} \;+\; \dfrac{p}{550} \end{array} \right\} \begin{array}{l} ❶ \\ ❷ \end{array} \qquad \Leftarrow \text{System to solve.}$$

$$36 = \dfrac{c}{50} + \dfrac{c}{550} \qquad \Leftarrow \text{Substituted } c \text{ for } p \text{ in equation ❷.}$$

$$(550)(36) = (550)\left(\dfrac{c}{50}\right) + (550)\left(\dfrac{c}{550}\right) \qquad \Leftarrow \text{Multiplied each term by 550, the LCD}$$

$$19800 = 11c + c \qquad \Leftarrow \text{Simplified.}$$

$$19800 = 12c \qquad \Leftarrow \text{Combined like terms.}$$

$$1650 = c \qquad \Leftarrow \text{Divided both sides by 12}$$

Now we know the value for c which solves the system. To find the value of p we substitute 1650 for c in either original equation and solve for p. Equation ❶ is the obvious choice.

$$c = p \qquad \Leftarrow \text{Original equation ❶.}$$

$$1650 = p \qquad \Leftarrow \text{Substituted 1650 for } c.$$

Step 6 Values of unknowns

	amount	=	rate	$\bullet$	base
Descriptions →	distance		rate		time
Units →	miles		$\dfrac{\text{miles}}{\text{hour}}$		hours
↓ Item names ↓					
car	c **1650 miles** **(distance car travels)**		50		$\dfrac{c}{50}$ $\dfrac{\mathbf{1650}}{\mathbf{50}}$ **33 hours** **(time car travels)**
plane	p **1650 miles** **(distance plane travels)**		550		$\dfrac{p}{550}$ $\dfrac{\mathbf{1650}}{\mathbf{550}}$ **3 hours** **(time plane travels)**

Step 7 Answer The distance traveled by both the car and the plane is 1650. This must be the distance between Roseville and Tucson.

Step 8 Check

- total travel time is 36 hours: checks because the time for the car (33 hours) plus the time for the plane (3 hours) is 36 hours.

- distance going must be the same as the distance returning: checks because the car and the plane both traveled 1650 miles. You could also say that since it took Lynn 3 hours to fly, and the plane was traveling 550 mph, then Lynn traveled 550 • 3 or 1650 miles. In the same way 33 hours of driving at 50 mph is 50 • 33, which is also 1650 miles.

Practice 5.4.2 Use the General Procedure for Solving Word Problems and two variables to help you answer the following questions.

1. Latonya is driving to her mother's house for the holidays. Because of road repairs she averaged only 50 mph for part of her trip and 62 mph for the rest of the trip. If she traveled for 2 hours more at 62 mph than she did at 50 mph and the total trip was 740 miles, how far did she travel at 62 mph?

2. A plane traveled at 240 mph before running into turbulence and reducing its speed to 200 mph. If the plane traveled at the faster speed for an hour and a half longer than it did at the slower speed and the total trip was 1460 miles, how long was it flying through the turbulence?

3. A freight train and a passenger train must share the same track. The freight train leaves a station at 9 am and travels 50 mph. One hour later, the passenger train leaves the station and travels 60 mph in the same direction. To avoid a crash, at what time and where must the railroad dispatcher move the freight train to a side track?

The previous problem could be solved using either one or two variables. The next two problems cannot be solved with a single variable.

Example 5.4.3 A plane can fly 1400 miles in 2 hours when it's flying with a constant tail wind but it takes the plane 2.5 hours to fly the same distance with the same constant speed head wind. Find the speed of the wind and the air speed of the plane.

•SOLUTION•

Step 1 Organize First, be sure you understand the words of the problem. A *tail wind* means the wind is blowing from the direction of the plane's tail and so it pushes the plane and makes it go faster relative to the ground (like going downstream on a river). A *head wind* means the wind is blowing against the forward motion of the plane and so it makes the plane go slower. The *air speed* of a plane is its speed relative to the air, not the ground. This is how fast the engines can move the plane when there is no wind.

Given: • plane flies 2 hours with wind.
• plane flies 2.5 hours against wind.
• distance traveled in either case is 1400 miles.
Find: Speed of air and air speed of plane.
Sketch: None.
Rates: Miles per hour implies a rate so we will construct an Amount Table.

Descriptions →	*amount*	=	*rate*	•	*base*
	distance		rate		time
Units →	miles		$\frac{\text{miles}}{\text{hour}}$		hours
↓ Item names ↓					
flying with wind	❸		❶		2
flying against wind	❹		❷		2.5

Formulas: amount = rate • base.

Step 2 Define unknowns

Define: Since the problem asks for speeds we will define

w represents speed of wind.
p represents air speed of the plane.

Construct: When the plane is going with the wind its ground speed is the result of the force generated by its engines and the force of the wind. That is, its speed is $p + w$. When the plane is moving against the wind the ground speed of the plane is reduced by the speed of the wind. That is, its speed is $p - w$.

❶ $p + w$ is the ground speed flying with the wind.

❷ $p - w$ is the ground speed flying against the wind.

Amount Table: We use the Amount Formula to calculate ❸ and ❹.

❸ $(p + w)2$ is the distance traveled when flying with the wind.

❹ $(p - w)2.5$ is the distance traveled when flying against the wind.

	amount	=	rate	•	base
Descriptions →	distance		rate		time
Units → ↓ *Item names* ↓	miles		$\dfrac{\text{miles}}{\text{hour}}$		hours
flying with wind	$(p + w)2$		$p + w$		2
flying against wind	$(p - w)2.5$		$p - w$		2.5

Step 3 Verbal models *and* ***Step 4 Mathematical Models***   We have defined two variables so we need two equations. Each equation will correspond to a constraint imposed by the problem.

• Flying with the wind:

Verbal model $\Rightarrow$	total distance (miles)	=	distance flying with wind (miles)
Math model $\Rightarrow$	1400	=	$(p + w)2$

• Flying against the wind:

Verbal model $\Rightarrow$	total distance (miles)	=	distance flying against wind (miles)
Math model $\Rightarrow$	1400	=	$(p - w)2.5$

Step 5 Solve   We must solve the system

$$\begin{cases} 1400 = (p + w)2 & ❶ \\ 1400 = (p - w)2.5 & ❷ \end{cases}$$

We will solve this using the Addition Method.

$$\begin{cases} 1400 = 2p + 2w \\ 1400 = 2.5p - 2.5w \end{cases} \quad \Leftarrow \text{Removed parentheses in each equation.}$$

$$\begin{cases} 3500 = 5p + 5w \\ 2800 = 5p - 5w \end{cases} \quad \Leftarrow \begin{array}{l} \text{Multiplied equation ❶ by 2.5 and equation ❷ by 2} \\ \text{to make the coefficients of } w \text{ opposites.} \end{array}$$

$$\begin{array}{r} 3500 = 5p + 5w \\ + \underline{2800 = 5p - 5w} \\ 6300 = 10p + 0w \end{array} \quad \Leftarrow \text{Added equations vertically.}$$

We can solve this equation for p.

$$6300 = 10p \qquad \Leftarrow \text{ Equation to solve, (transformed ❶) + (transformed ❷).}$$

$$630 = p \qquad \Leftarrow \text{ Divided both sides by 10.}$$

To find w, we substitute 630 for p in either original equation. We will use equation ❶.

$$1400 = ((\mathbf{630}) + w)2 \qquad \Leftarrow \text{ Substituted 630 for } p \text{ in equation ❶.}$$

$$1400 = 1260 + 2w \qquad \Leftarrow \text{ Distributed the 2.}$$

$$140 = 2w \qquad \Leftarrow \text{ Subtracted 1260 from both sides.}$$

$$70 = w \qquad \Leftarrow \text{ Divided both sides by 2.}$$

Step 6 Values of unknowns

	amount	=	rate	•	base
Descriptions →	distance		rate		time
Units →	miles		$\dfrac{\text{miles}}{\text{hour}}$		hours
↓ Item names ↓					
flying with wind	$(p + w)2$ **(630 + 70)2** **1400 miles** **(distance flying with wind)**		$p + w$ **630 + 70** **700 mph** **(speed with wind)**		2
flying against wind	$(p - w)2.5$ **(630 − 70)2.5** **1400 miles** **(distance flying against wind)**		$p - w$ **630 − 70** **560 mph** **(speed against wind)**		2.5

Step 7 Answer   The air speed of the plane is 630 mph and the wind speed is 70 mph.

Step 8 Check

- the plane travels 1400 miles in 2 hours going with the wind: checks because speed with the wind is (700 mph) times (2 hours) is 1400 miles.

- the plane travels 1400 miles in 2.5 hours going against the wind: checks because speed against the wind is (560 mph) times (2.5 hours) is 1400 miles.

Practice 5.4.3   Use the General Procedure for Solving Word Problems and two variables to help you answer the following questions.

1. A plane leaves Miami and flies 3575 miles to Vancouver in 5.5 hours. During the flight it experiences a tail wind. On the return trip, there is a head wind of equal strength and it takes 6.5 hours to fly back to Miami. Find the speed of the wind and the speed of the plane.

2. A boat can travel 21 miles downstream (with the current) in 3 hours but it takes the boat 7 hours to go the same distance upstream (against the current). Find the speed of the current and the speed of the boat.

Example 5.4.4   Two machines are needed to produce plastic cups and plates. Machine A stamps out the form and machine B paints on a design. It takes 5 minutes for machine A to form a cup and 1 minute for a plate. It takes 3 minutes for machine B to paint a cup and 1 minute for a plate. If machine A can operate for 400 minutes per shift and machine B can operate for 300 minutes per shift how many plates and cups can be produced per shift?

•SOLUTION• *Step 1 Organize*

Given: • machine A forms a cup in 5 minutes.
         • machine A forms a plate in 1 minute.
         • machine A operates for 400 minutes.
         • machine B paints a cup in 3 minutes.
         • machine B paints a plate in 1 minute.
         • machine B operates for 300 minutes.

Find: Number of plates and number of cups produced.

Rates: Minutes to form a cup implies a rate so we will construct an Amount Table. We can choose the rate to be

$$\frac{\text{minutes}}{\text{item}}, \text{ such as } \frac{5 \text{ minutes}}{\text{cup}}$$

or

$$\frac{\text{items}}{\text{minute}}, \text{ such as } \frac{1 \text{ cup}}{5 \text{ minutes}}.$$

We will use $\frac{\text{minutes}}{\text{item}}$ to avoid the introduction of fractions.

	amount =	*rate* •	*base*
Descriptions →	time to produce	rate	number produced
Units →	minutes	$\frac{\text{minutes}}{\text{item}}$	items
↓ Item names ↓			
machine A cups	❺	5	❶
machine A plates	❻	1	❷
machine B cups	❼	3	❸
machine B plates	❽	1	❹

Formulas: amount = rate • base.

Step 2 Define unknowns

Define: We are looking for the number of cups and plates produced so we will define

          c represents number of cups produced
          p represents number of plates produced

Construct: Not needed.

Amount Table: We now complete the Amount Table.

          ❶ and ❸ are the number of cups, *c*.
          ❷ and ❹ are the number of plates, *p*.
          ❺, ❻, ❼, and ❽ are calculated using the Amount Formula.

	amount	= rate	• base
Descriptions →	time to produce	rate	number produced
Units →	minutes	$\dfrac{\text{minutes}}{\text{item}}$	items
↓ Item names ↓			
machine A cups	$5c$	5	c
machine A plates	$1p$	1	p
machine B cups	$3c$	3	c
machine B plates	$1p$	1	p

Step 3 Verbal models** and **Step 4 Mathematical Models   We have defined two variables so we need two equations. Each equation will correspond to a constraint imposed by the problem.

- *Constraint on time machine A can work*: The time machine A spends working is the time it spends on cups plus the time it spends on plates:

Verbal model $\Rightarrow$ $\begin{array}{ccc}\text{total time for machine A} \\ \text{(minutes)}\end{array} = \begin{array}{c}\text{time for machine A} \\ \text{to do cups (minutes)}\end{array} + \begin{array}{c}\text{time for machine A} \\ \text{to do plates (minutes)}\end{array}$

Math model $\Rightarrow$ $400 = 5c + 1p$

- *Constraint on time machine B can work*: This is the same idea as for machine A:

Verbal model $\Rightarrow$ $\begin{array}{ccc}\text{total time for machine B} \\ \text{(minutes)}\end{array} = \begin{array}{c}\text{time for machine B} \\ \text{to do cups (minutes)}\end{array} + \begin{array}{c}\text{time for machine B} \\ \text{to do plates (minutes)}\end{array}$

Math model $\Rightarrow$ $300 = 3c + 1p$

Step 5 Solve   We will solve the system using the Addition Method.

$\begin{cases} 400 = 5c + 1p & \text{❶} \\ 300 = 3c + 1p & \text{❷} \end{cases}$ $\Leftarrow$ System to solve.

$\begin{cases} 400 = 5c + 1p & \text{❶} \\ (-1)300 = (-1)3c + (-1)1p & \text{❷} \end{cases}$ $\Leftarrow$ Multiplied each term in equation ❷ by -1
to make the coefficients of p opposites.

$\begin{cases} 400 = 5c + p & \text{❶} \\ \mathbf{-300} = -3c - p & \text{❷} \end{cases}$ $\Leftarrow$ Simplified.

$\begin{array}{rrcl} & 400 = & 5c + & p \quad \text{❶} \\ + & \underline{-300 = } & \underline{-3c - } & \underline{p} \quad \text{❷} \\ & 100 = & 2c + & 0p \quad \text{❶ + ❷} \end{array}$ $\Leftarrow$ Added equations vertically.

We can solve this equation for c.

$100 = 2c$ $\Leftarrow$ Equation to solve, ❶ + (transformed ❷).

$50 = c$ $\Leftarrow$ Divided both sides by 2.

To find p, we substitute 50 for p in either original equation. We will use equation ❶.

$400 = 5(\mathbf{50}) + p$ $\Leftarrow$ Substituted 50 for c in equation ❶.

$400 = \mathbf{250} + p$ $\Leftarrow$ Multiplied.

$150 = p$ $\Leftarrow$ Subtracted 250 from both sides.

Step 6 Values of unknowns

	amount	=	rate	•	base
Descriptions →	time to produce		rate		number produced
Units →	minutes		$\dfrac{\text{minutes}}{\text{item}}$		items
↓ Item names ↓					
machine A cups	$5c$ $5(50)$ 250 (number of cups)		5		c 50 (number of cups)
machine A plates	$1p$ $1(150)$ 150 (number of plates)		1		p 150 (number of plates)
machine B cups	$3c$ $3(50)$ 150 (number of cups)		3		c 50 (number of cups)
machine B plates	$1p$ $1(150)$ 150 (number of plates)		1		p 150 (number of plates)

Step 7 Answer   The machines produce 50 cups and 150 plates.

Step 8 Check

- total time for machine A is 400 minutes: checks because it spends 250 minutes on cups and 150 minutes on plates, which is a total of 400 minutes.

- total time for machine B is 300 minutes: checks because it spends 150 minutes on cups and 150 minutes on plates, which is a total of 300 minutes.

Practice 5.4.4 Use the General Procedure for Solving Word Problems and two variables to help you answer the following questions.

1. Machine A takes 30 minutes to paint the body of a sedan and 20 minutes to paint the body of an economy car. Machine B can paint the fenders on a sedan in 10 minutes and the fenders on an economy car in 5 minutes. If machine A can work for 350 minutes before it needs to be cleaned and machine B for 100 minutes, how many of each type of car can be painted?

2. It takes 8 hours to build the electronics for a deluxe sound system but only 3 hours for a standard model. The cabinet for the deluxe model takes 3 hours to build, while the cabinet for the standard model takes only 1 hour. If the electronic plant works 54 hours a week and the cabinet makers only 19 hours per week, how many of each type will be made?

Exercise Set 5.4   The answers to the odd numbered exercises are at the back of the book.

Use the General Procedure for Solving Word Problems and two variables to help you answer the following questions.

1. Machine A bound 45 more books than machine B. If together they bound 145 books, how many did each bind?

2. Maria and Salina sell shoes. Maria's sales were $150 greater than Salina's sales. Together they sold $600 worth of shoes. How many dollars worth of shoes did Maria sell?

3. Lynn has a total of $3325 in two banking accounts. If she has $325 more in her savings account than in her checking account, how much is in her checking account?

4. Together Sue and Joe sell 38 tickets to a raffle. If Sue sells ten more tickets than Joe, how many tickets did each sell?

5. The perimeter of a rectangle is 292 feet. If the length is 2 feet more than three times the width, what are the dimensions of the rectangle?

6. The area of a trapezoid is 135 cm^2 and its height is 10 cm. If the length of one base is 5 less than 3 times the length of the other, find the length of each base.

7. Robert has 25 bills, some $1 bills and some $20. If the total value is $215, how many of each kind does he have?

8. Jamey has a deposit made up of $10 bills and $20 bills. He has a total of 72 bills. If he deposits a total of $1120, how many $10 bills did he deposit?

9. Joe bought 25 pounds of gourmet coffee for his restaurant, some at $6.60 a pound and the rest at $9.50 a pound. If the total value of the coffee is $199.80, how many pounds of the $6.60 coffee did he buy?

10. Pilar's Pizza Shop served 125 customers during the lunch hour. Some of the customers paid $4.50 for the pizza bar only, while the rest paid $5.75 for the salad/pizza bar combination. If total sales was $681.25, how many customers bought the combination?

11. A train leaves Mason City at 11 am at an average speed of 110 mph. Two hours later a car leaves Mason City going in the opposite direction at an average speed of 60 mph. At what time will they be 985 miles apart?

12. Two cyclists start from the same point and ride in opposite directions. One cyclist rides three times as fast as the other. In five hours they are 140 miles apart. Find the rate of each cyclist.

13. A bicyclist leaves Vernon at 8 am and travels 10 mph. Thirty minutes later, a second bicyclist leaves Vernon and travels along the same road and in the same direction at 7 mph. At what time will they be 17 miles apart?

14. A plane leaves Huntington at 3 pm and travels due east at 240 mph. Fifteen minutes later another plane leaves Huntington flying the same route but at 200 mph. At what time will they be 100 miles apart?

15. A plane can fly 1050 miles in 3 hours when it is flying with a tail wind but it takes the plane 3.5 hours to fly the same distance with a head wind. Find the speed of the wind and the air speed of the plane.

16. A boat can travel 42 miles downstream in 4 hours but it can only travel 22 miles upstream in the same amount of time. Find the speed of the current and the speed of the boat.

17. Machine A can print a novel in 30 minutes and a magazine in 4 minutes. Machine B can bind a novel in 5 minutes, but takes only 1 minute to bind a magazine. If machine A works for 480 minutes and machine B for 95 minutes, how many novels and how many magazines will be printed and bound?

18. One week Joe sold 5 standard refrigerators and 6 deluxe models for a total sales of $9425. The next week business was slow and he sold only 3 standard models and 2 deluxe models for a total sales of $3975. What was the price for each model?

Review Exercises   The answers to all of these exercises are at the back of the book.

19. Name the property that justifies the statement: $x5 = 5x$ 20. Graph: $x \geq -7$ 21. Solve: $\dfrac{2}{3}x - 5(x+1) > \dfrac{3x}{2}$

22. Find the width of a rectangle if its width is one–quarter of its length and its perimeter is 130 inches.

23. Ben must ride to the post office and return home by 5 pm if he wants to catch his bus for his night class. He leaves at 4:25 pm and rides at 12 mph arriving at the post office at 4:45 pm. If he leaves again immediately, and rides at a speed of 17 mph will he get home on time?

24. Decide whether these two straight lines are parallel, perpendicular or neither: *Line 1* passes through the points (1,3) and (0,4) and *Line 2* passes through the points (0,–2) and (5,3).

25. Determine the domain and range of {(3,2), (5,–1), (3,0), (5,–2)}

Chapter 5 Review

Vocabulary to Know

2 by 2 linear system — a system of two unknowns and two equations where the graphs are straight lines.

Addition Method of solving a system— a method of solving a system of equations where the equations are added in order to eliminate one of the variables (same as Elimination Method).

consistent system — a system that has at least one solution; for a 2 by 2 linear system, this means the graphs intersect at one point or the graphs are on top of each other.

dependent equations — the equations of a system are dependent if their graphs are the same; for a 2 by 2 linear system this means there are infinitely many solutions (the lines are on top of each other).

Elimination Method of solving a system— a method of solving a system of equations where the equations are added in order to eliminate one of the variables (same as Addition Method).

inconsistent system — a system that has no solutions; for a 2 by 2 linear system, this means the graphs are parallel lines that are not on top of each other.

independent equations — the equations of a system are independent if their graphs are different; for a 2 by 2 linear system this means there is one solution (the lines cross at one point), or no solutions (the lines are parallel.

solution of a system — a set of ordered pairs that satisfies all equations of the system.

Substitution Method of solving a system — a method of solving a system of equations where one variable is replaced by an equivalent expression.

system of equations — two or more equations that are considered together.

two by two linear system — a system of two unknowns and two equations where the graphs are straight lines.

Properties, Definitions, Formulas, and Notation to Understand

Possible solutions to linear systems

Possibility 1   The lines intersect at one point.

- The *system* is consistent (it has a solution).

- The *equations* are independent (the graphs are different).

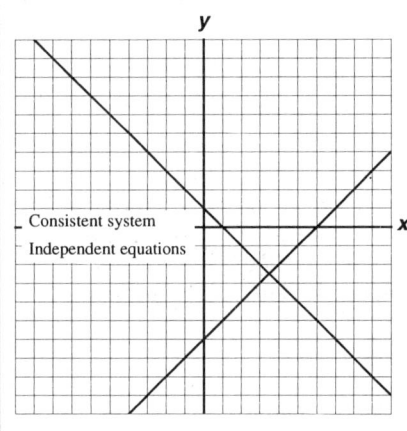

Consistent system
Independent equations

Possibility 2   The lines do not intersect.

- The *system* is inconsistent (it has no solution).

- The *equations* are independent (the graphs are different).

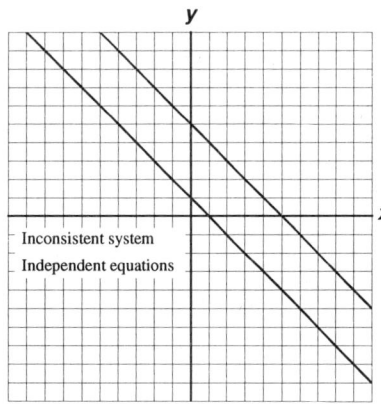

Inconsistent system
Independent equations

Possibility 3 The lines lie on top of each other.

- The *system* is consistent (it has a solution).

- The *equations* are dependent (the graphs are the same).

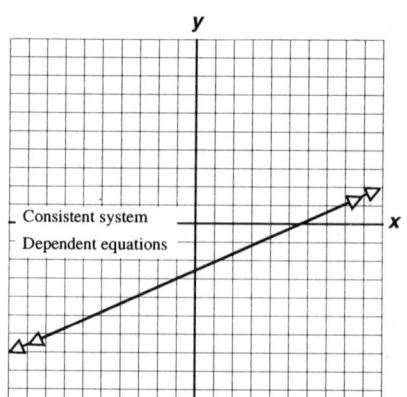

Consistent system
Dependent equations

Systems are either consistent (there is at least one solution) or inconsistent (there are no solutions)

Equations are either dependent (they are equivalent) or independent (they are not equivalent).

Possible situations when solving a 2 by 2 linear system by substitution

- After the first substitution, if you end up with a conditional equation, such as $x = 2$, the system has one solution (whose x component is 2). The graphs intersect at one point. The system is consistent and the equations are independent.

$$\left\{ \begin{array}{rcl} x + y & = & 3 \\ y & = & 2x \end{array} \right\} \begin{array}{l} \text{❶} \\ \text{❷} \end{array}$$

Substituting $2x$ for y in ❶ gives $x + 2x = 3$, or $x = 1$.

Substituting 1 for x in ❷ gives $y = 2(\mathbf{1})$, or $y = 2$.

The solution is (1,2).

- After the first substitution, if you end up with an identity, such as $12 = 12$, the system has infinitely many solutions. The general solution is written as an ordered pair with x as the input and an expression containing x as the output. The graphs are on top of each other. The system is consistent and the equations are dependent.

$$\left\{ \begin{array}{rcl} x + y & = & 3 \\ y & = & 3 - x \end{array} \right\} \begin{array}{l} \text{❶} \\ \text{❷} \end{array}$$

Substituting $3 - x$ for y in ❶ gives

$x + 3 - x = 3$, which simplifies to $3 = 3$.

The equations are dependent and the solution is $(x, 3 - x)$.

- After the first substitution, if you end up with a contradiction, such as $-6 = 8$, the system has no solution. The graphs are parallel lines and therefore do not intersect. The system is inconsistent and the equations are independent.

$$\left\{ \begin{array}{rcl} x + y & = & 3 \\ y & = & 4 - x \end{array} \right\} \begin{array}{l} \text{❶} \\ \text{❷} \end{array}$$

Substituting $4 - x$ for y in ❶ gives

$x + 4 - x = 3$, which simplifies to $4 = 3$.

The equations are independent and there is no solution.

Standard form of a 2 by 2 linear system — terms containing variables are written on the left, with the variables in alphabetical order. The constant terms are written on the right.

A 2 by 2 linear system is in standard form when it's written as

$$ax + by = c$$
$$dx + ey = f$$

where x and y are variables and a, b, c, d, e, and f are constants.

The following system is written in standard form:

$$2x + 9y = 4$$
$$8x - y = 1$$

Procedures to Follow

Procedure for solving a 2 by2 system by graphing.	Solve: $\left\{ \begin{array}{rcl} x + y & = & 3 \\ x - y & = & 1 \end{array} \right\}$ ❶ ❷
Step 1 Graph each equation.	
Step 2 Identify the coordinates of the points where the graphs intersect. These points are the solutions to the system.	The lines intersect at (2,1).

Procedure for solving a 2 by 2 linear system by substitution.	Solve: $\left\{ \begin{array}{rcl} x + y & = & -2 \\ 2x - y & = & -7 \end{array} \right\}$ ❶ ❷
Step 1 Solve If necessary, solve one of the equations for one of the variables.	Solving ❶ for y gives $y = -x - 2$
Step 2 Substitute Substitute the expression found in Step 1 for all occurrences of the variable in the other equation and solve.	Substituting $-x - 2$ for y in ❷ gives $2x - (-x - 2) = -7$ $3x = -9$ $x = -3$
Step 3 Back Substitute Substitute the value found in Step 2 into either *original* equation and solve.	Substituting -3 for x in ❶ gives $(-3) + y = -2$ $y = 1$
Step 4 Check The solution is the ordered pair found in Steps 2 and 3. Check your ordered pair by substituting into *both* equations.	The solution is $(-3,1)$, which satisfies both equations: $(-3) + (1) = -2$ ❶ $\mid$ $2(-3) - (1) = -7$ ❷ $-2 = -2$ $\mid$ $-7 = -7$

Procedure for deciding what to multiply by when transforming both equations in a system.	Solve: $\left\{ \begin{array}{rcl} -2x + 3y & = & 8 \\ 5x + 2y & = & -1 \end{array} \right\}$ ❶ ❷	
Step 1 Decide which variable you wish to eliminate. Let's say it's x.	To eliminate x	To eliminate y
Step 2 Multiply equation ❶ by the coefficient of x in equation ❷ and multiply equation ❷ by the *opposite* of the original coefficient of x in equation ❶.	Multiply ❶ by 5 and ❷ by 2 $\left\{ \begin{array}{rcl} -10x + 15y & = & 40 \\ 10x + 4y & = & -2 \end{array} \right\}$	Multiply ❶ by 2 and ❷ by -3 $\left\{ \begin{array}{rcl} -4x + 6y & = & 16 \\ -15x - 6y & = & 3 \end{array} \right\}$

Procedure for addition method for solving systems.	Solve: $\left\{ \begin{array}{rcl} y & = & \frac{2}{3}x - 1 \\ x + y & = & 4 \end{array} \right\}$ ❶ ❷
Step 1 Standard Form Put both equations into the form $ax + by = c$.	$\left\{ \begin{array}{rcl} \frac{-2}{3}x + y & = & -1 \\ x + y & = & 4 \end{array} \right\}$ ❶ ❷

Step 2 Clear Fractions Use the LCD to clear fractions from each equation.	Multiply ❶ by 3 to get $$\begin{cases} -2x + 3y = -3 & \text{❶} \\ x + y = 4 & \text{❷} \end{cases}$$
Step 3 Transform Equations Transform into opposites the coefficients of one of the variables by multiplying each term in one or both equations by a constant.	Multiply ❷ by 2 to get $$\begin{cases} -2x + 3y = -3 & \text{❶} \\ 2x + 2y = 8 & \text{❷} \end{cases}$$
Step 4 Add Add the equations.	$5y = 5$
Step 5 Solve Solve the resulting equation.	$y = 1$
Step 6 Substitute Substitute the solution from Step 5 into either *original* equation and solve for the variable.	Substitute 1 for y in ❷ to get $x + 1 = 4$ $x = 3$ The solution is $(3,1)$
Step 7 Check Check the solution by substituting into *both* equations.	Check for ❶ $\quad\quad$ Check for ❷ $1 = \frac{2}{3}(3) - 1 \quad\quad 3 + 1 = 4$ $1 = 1$ Checks $\quad\quad 4 = 4$ Checks

Procedure for general procedure for solving word problems.	In the Campus Cafe, 2 tacos and a drink cost \$3.90 while 5 tacos and 2 drinks cost \$9.30. What is the price of a taco and what is the price of a drink?
Step 1 Organize the information Read the problem, several times if necessary: • Given: determine what is given. Be sure to include the unit of all numbers. • Find: determine what you are asked to find. • Sketch: draw a sketch if it will help you better understand the problem. • Rates: examine the unit of each given number to determine if a rate is involved. If so, begin construction of an Amount Table. • Formulas: if standard formulas will be needed, list them.	Given: • 2 tacos and 1 drink cost \$3.90 $\quad\quad\quad$ • 5 tacos and 2 drinks cost \$9.30 Find: Price of a taco. Sketch: Not helpful Rates: We are looking for price per taco, which is a rate that has units $\frac{\text{cents}}{\text{item}}$. Since rates are involved we construct an Amount Table to help organize the information.

	amount	=	rate	•	base
Descriptions →	cost		price of 1		number
Units → ↓ *Item names* ↓	cents		$\frac{\text{cents}}{\text{item}}$		items
tacos-small meal	❺		❶		2
tacos-big meal	❻		❷		5
drink-small meal	❼		❸		1
drink-big meal	❽		❹		2

• Formulas: if standard formulas will be needed, list them.	• Formulas: amount = rate • base

Step 2 Define unknowns

- Define: assign definitions to *variables*; include units, if appropriate.

- Construct: use the *variables* you just defined to build expressions for other unknowns.

- Amount Table: if you began an Amount Table in Step 1, complete it.

Define: t represents the price of a taco.

Construct: d represents the price of a drink.

Amount Table:

↓ Item names ↓	amount = cost (cents)	rate price of 1 ($\frac{cents}{item}$)	base number (items)
tacos-small meal	$2t$	t	2
tacos-big meal	$5t$	t	5
drink-small meal	$1d$	d	1
drink-big meal	$2d$	d	2

Step 3 Build verbal models   Use English to write relations between the unknowns.

small meal: Total cost = taco cost + drink cost

big meal: Total cost = taco cost + drink cost

Step 4 Build mathematical models   Translate the verbal models into mathematical models. Because the verbal and mathematical models are so closely related we often do Steps 3 and 4 together.

small meal: $390 = 2t + 1d$

big meal: $930 = 5t + 2d$

Step 5 Solve   Solve the mathematical models for the variables.

We use addition to solve. To eliminate d we multiply the small meal equation by 2 to get

$$\begin{cases} -780 = -4t - 2d \\ 930 = 5t + 2d \end{cases} \begin{matrix} \text{small meal} \\ \text{big meal} \end{matrix}$$

Adding the equations gives

$$150 = t$$

Substituting 150 for t into the small meal equation gives

$$390 = 2(\mathbf{150}) + d$$

$90 = d$ The solution is (150,90). We leave the check to you.

Step 6 Values of unknowns   Calculate the values of all unknowns.

❶ and ❷ are t, which is 150

❸ and ❹ are d, which is 90

❺ $2t$ is 300, ❻ $5t$ is 750, ❼ $1d$ is 90, ❽ $2d$ is 180

Step 7 Answer   State the answer to the question.

A taco costs $1.50 and a drink costs $0.90.

Step 8 Check   Using the original words of the problem, check to see that the values make sense and are consistent with the original statement of the problem.

- 2 tacos and 1 drink cost $3.90. Checks because 2 tacos cost $3.00 and 1 drink costs $0.90, which sums to $3.90.

- 5 tacos and 2 drinks cost $9.30. Checks because 5 tacos cost $7.50 and 2 drinks cost $1.80, which sums to $9.30.

Chapter 5 Review Exercises   The answers to all the exercises are in the back of the book.

1. Does (2,0) solve $\begin{cases} x - y = 2 \\ 5x + 2y = 10 \end{cases}$

2. Use graphing to solve: $\begin{cases} -2x + 10y = 8 \\ -5x + 3y = -2 \end{cases}$ **❶** **❷**

Use graphing to solve.

3. $\begin{cases} 3x - y = 2 \\ -6x + 2y = -4 \end{cases}$ **❶** **❷**

4. *Line 1:*

x	y
0	2
1	1

Line 2:

x	y
−1	−7
5	−1

5. $\begin{cases} 2x + y = 0 \\ -4x - 2y = 5 \end{cases}$ **❶** **❷**

Classify this system as consistent or inconsistent and classify the equations of the system as dependent or independent:

6.

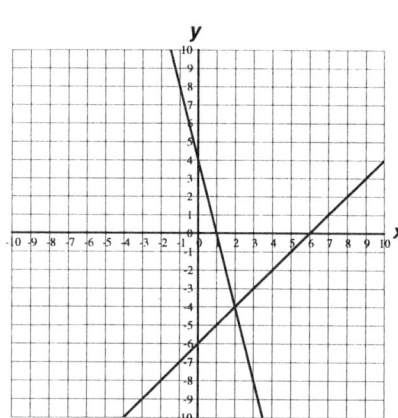

7.

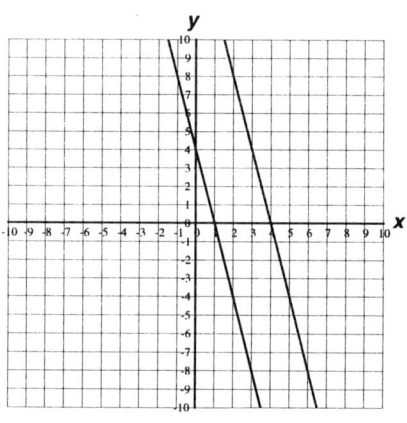

8. A ship springs a leak and water enters at a rate of 100 gallons per minute. Five minutes after the ship begins taking on water the pumps are started and pump out the water at a rate of 150 gallons per minute. A graph of this situation is shown below with minutes since the leak started as the input and gallons of water moved as the output.

   a) Model this situation with a 2 by 2 system of linear equations.
   b) Find the point of intersection and explain what it means.

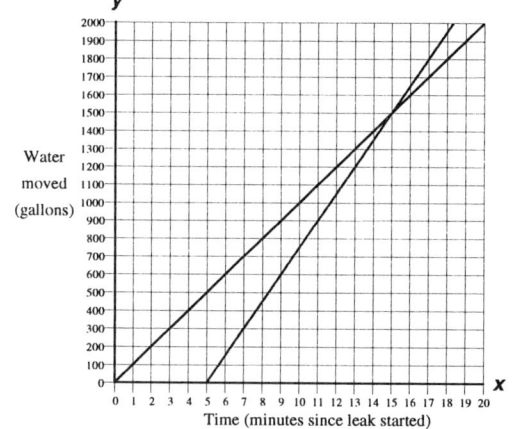

9. Highland Plumbing charges $40 per hour and a trip charge of $60. Penny's Plumbing charges $55 per hour but only charges $15 for the trip.

   a) Model this situation with a 2 by 2 system of linear equations using time as the input and cost as the output.
   b) Graph the system. The input axis should go from 0 to 10 hours in steps of 1 hour; the output axis should go from $0 to $200 in steps of $20.
   c) Find the point of intersection and explain what it means.
   d) Which plumber would cost less money?

10. Use substitution to solve $\begin{cases} 5x - 3y = -5 \\ 7x + y = 19 \end{cases}$ **❶** **❷**

11. Use substitution to solve $\begin{cases} -2x + y = 2 \\ y = 3x \end{cases}$ **❶** **❷**

12. Solve: $\begin{cases} 3 = 5y - 2x \\ 7 = -6x + 15y \end{cases}$ **❶** **❷**

13. Solve: $\begin{cases} 5 + 5y = -4x \\ 8x + 6 = -10y \end{cases}$ **❶** **❷**

14. IntelliCell offers two billing plans for their cell phones. Data from the two plans are given below. Find connect time where the cost of each plan is the same.

x Plan A monthly connect time (minutes)	y Plan A monthly cost (dollars)	x Plan B monthly connect time (minutes)	y Plan B monthly cost (dollars)
0	40	10	32
30	55	30	46

Use addition to solve.

15. $\begin{cases} 3x + y = 4 & ❶ \\ x - y = 8 & ❷ \end{cases}$

16. $\begin{cases} x + 2y = 3 & ❶ \\ -x - 5y = 12 & ❷ \end{cases}$

17. $\begin{cases} 2x - y = 3 & ❶ \\ x + 2y = -11 & ❷ \end{cases}$

18. $\begin{cases} 8x - 3y = 31 & ❶ \\ 2x + y = -1 & ❷ \end{cases}$

19. $\begin{cases} 8x = y + 17 & ❶ \\ x + y = 1 & ❷ \end{cases}$

20. $\begin{cases} 2y = 17 - x & ❶ \\ 5x - 8 = y & ❷ \end{cases}$

Solve.

21. $\begin{cases} 12x + 3y = 15 & ❶ \\ -5x + 2y = -3 & ❷ \end{cases}$

22. $\begin{cases} 9x - 2y = 20 & ❶ \\ 5x + 3y = 7 & ❷ \end{cases}$

23. $\begin{cases} \dfrac{1}{2}x - y = -4 & ❶ \\ x + \dfrac{1}{3}y = -1 & ❷ \end{cases}$

24. $\begin{cases} \dfrac{1}{6}x + y = -1 & ❶ \\ 2x - \dfrac{1}{2}y = 13 & ❷ \end{cases}$

25. A machine that converts $1 and $5 bills into coins returns both quarters and dimes. If you insert a $5 bill and get back 23 coins, how many quarters and how many dimes do you receive?

26. With a tailwind, a plane takes 2 hours to travel 1,600 miles. On the return trip, with a headwind whose speed was equal to that of the tailwind, the plane made the trip in 2.5 hours. Find the speed of the tailwind and the air speed of the plane.

27. A scientist needs to prepare a diet that contains 8 grams of fat and 95 grams of protein. She has available two mixes. The HiPro mix contains 40% protein and 2% fat while the HiFat mix contains 30% protein and 8% fat. How much HiPro and how much HiFat should she combine to obtain the desired mixture?

Chapter 5 Test   The answers to all the exercises are in the back of the book.

1. Does $(-3, -1)$ solve $\begin{cases} 2x - y = -5 & ❶ \\ -3x + 7y = 2 & ❷ \end{cases}$

2. Use graphing to Solve: $\begin{cases} -3x - y = 5 & ❶ \\ 3x - y = -1 & ❷ \end{cases}$

3. Classify this system as consistent or inconsistent and classify the equations of the system as dependent or independent:

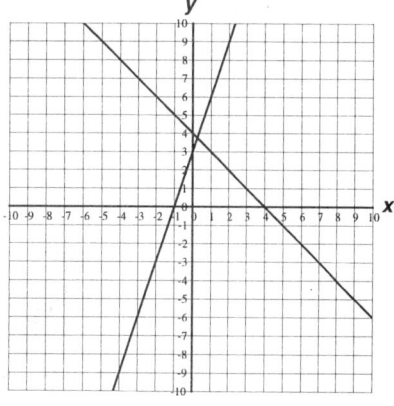

4. A speedboat leaves a marina and heads out towards the open ocean. A quarter of an hour late a police boat starts at the marina and gives chase. The police boat can travel at 60 mph and the speedboat can travel at 50 mph. To be able to stop the speedboat the police boat must catch it before it reaches international waters 100 miles away from the marina. A graph of this situation is shown below with hours since the speedboat left the marina as the input and miles from the marina as the output.

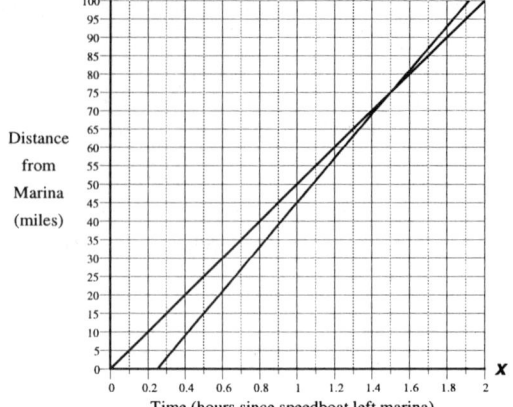

a) Model this situation with a 2 by 2 system of linear equations.
b) Find the point of intersection and explain what it means.

5. Use substitution to solve: $\left\{ \begin{array}{rcl} 2x & - & 4y & = & 14 \\ x & - & y & = & 6 \end{array} \right\}$ ❶ ❷

6. Use addition to solve: $\left\{ \begin{array}{rcl} 8x & - & y & = & 0 \\ -2x & + & y & = & 6 \end{array} \right\}$ ❶ ❷

Solve.

7. $\left\{ \begin{array}{rcl} 20y & = & 8 + 6x \\ 3x + 4 & = & 10y \end{array} \right\}$ ❶ ❷

8. $\left\{ \begin{array}{rcl} 2x + y & = & 5 \\ 2y & = & -4x + 8 \end{array} \right\}$ ❶ ❷

9. $\left\{ \begin{array}{rcl} -5x & + & 2y & = & -9 \\ -3x & - & 3y & = & 3 \end{array} \right\}$ ❶ ❷

10. $\left\{ \begin{array}{rcl} \dfrac{1}{5}x & + & y & = & -4 \\ x & - & \dfrac{2}{3}y & = & 14 \end{array} \right\}$ ❶ ❷

11. Sue rowed 10 miles down the Mississippi River in $2\frac{1}{2}$ hours but it took her 4 hours to return to her starting point. Find the speed of the river's current.

Chapter 6
Exponents and Polynomials

Section 6.1 Properties of Exponents

Topic 6.1 A — Multiplication Property of Exponents

We can use the definition of exponents and the Associative Property of Multiplication to find the product of factors that have the same base. For example, let's find the product of $a^3 a^4$.

$(aaa)(aaaa)$ $\Leftarrow$ Used the definition of an exponent.

$(aaaaaaa)$ $\Leftarrow$ Used the Associative Property of multiplication to group all factors.

a^7 $\Leftarrow$ Used the definition of an exponent.

The exponent of the simplified expression, a^7, is the sum of the exponents of the original expression, $a^3 a^4$. This is true in general.

— *Property* — ***Multiplication Property of Exponents***	English: To multiply two numbers with the same base, add the exponents and keep the base the same. Examples: $3^4 \cdot 3^2 = 3^{4+2} = 3^6$ $x^4 \cdot x^2 \cdot x = x^{4+2+1} = x^7$ Algebra: $a^m a^n = a^{m+n}$

In the property $a^m a^n = a^{m+n}$, m and n can represent any real numbers, including negative numbers, fractions, and decimals. Later, we will discuss expressions that contain exponents such as x^{-2}, and $x^{1/2}$.

Example 6.1.1 Simplify.

a) $3^4 \cdot 3^6$

•SOLUTION• The bases are the same, so we add the exponents to get 3^{4+6}, which simplifies to 3^{10}. Note that this is not 9^{10}. When multiplying factors that have the same base we add the exponents and keep the original base.

— *Note* — ***Convention for simplifying expressions when the base is a constant***	The choice of whether to leave an expression like 3^{10} in exponential form or to calculate it and write 59,049 depends on the situation and personal preference. When we are discussing the properties of exponents we will typically leave constants raised to powers in their exponential form so you can see what we are doing. However, as a general rule of thumb we will calculate such expressions if they turn out to have 3 or fewer digits. For example, we would calculate 3^2 and write 9 but we would leave 3^{10} in exponential form rather than writing 59,049.

b) $x^3 x^8$

•SOLUTION• The bases are the same, so we add the exponents to get x^{3+8}, which simplifies to x^{11}.

c) $b^4 b b$

•SOLUTION• b means b^1. We add the exponents to get b^{4+1+1}, which is b^6.

d) $a^3 b^2$

•SOLUTION• The bases are not the same so the Multiplication Property of Exponents does not apply here. We cannot simplify this expression.

e) $k^6 + k^6$

•SOLUTION• This is the sum of terms, not the product of factors, so the Multiplication Property of Exponents does not apply. However, k^6 and k^6 are like terms so we can add them to get $2k^6$.

Practice 6.1.1 Simplify.

1. $5^6 \cdot 5^4$　　　2. $y^6 y^7$　　　3. $w^3 w^5 w$　　　4. $x^2 y^5$　　　5. $x^3 + x^5$

— *Caution* — *Multiplication is different* *from addition*	The multiplication property of exponents does not extend to addition. That is, when we ADD terms with the same base we may NOT add the exponents. For example, • Since $x^3 x^3$ is a product we may add the exponents to get x^{3+3}, which simplifies to x^6. • Since $x^3 + x^3$ is a sum we may add the coefficients to get $2x^3$ but we may NOT also add the exponents to INCORRECTLY get $2x^6$.

Topic 6.1 B — Using the Multiplication Property to Simplify Expressions

We can combine the multiplication property of exponents with the Associative, Commutative, and Distributive properties to simplify expressions that contain several factors.

Example 6.1.2 Simplify.

a) $2x^4(3x^2)$

•SOLUTION• We use the Associative and Commutative properties to group the coefficients and variables and then simplify.

$$2x^4(3x^2) \quad \Leftarrow \text{ Expression to simplify.}$$

$$(2 \cdot 3)(x^4 x^2) \quad \Leftarrow \text{ Grouped coefficients and grouped variables.}$$

$$(2 \cdot 3)(x^{4+2}) \quad \Leftarrow \text{ Applied multiplication property of exponents.}$$

$$(2 \cdot 3)x^6 \quad \Leftarrow \text{ Added exponents.}$$

$$6x^6 \quad \Leftarrow \text{ Multiplied coefficients.}$$

b) $2x^3(-7x^4)(5x)$

•SOLUTION•

$$\left[(2)(-7)(5)\right]\left[x^3 x^4 x\right] \quad \Leftarrow \text{ Grouped coefficients and grouped variables.}$$

$$[(2)(-7)(5)][x^{3+4+1}] \quad \Leftarrow \text{ Applied multiplication property of exponents.}$$

$$-70x^8 \quad \Leftarrow \text{ Multiplied coefficients and added exponents.}$$

c) $2x^4(7x^3 + 5x^2)$

•SOLUTION• We cannot add the terms inside the parentheses, so we use the Distributive property to remove the parentheses and then simplify.

$$2x^4(7x^3 + 5x^2) \quad \Leftarrow \text{ Expression to simplify.}$$

$$\mathbf{2x^4}(7x^3) + \mathbf{2x^4}(5x^2) \quad \Leftarrow \text{ Distributed } 2x^4 \text{ to each term inside parentheses.}$$

$$(2 \cdot 7)(x^4 x^3) + (2 \cdot 5)(x^4 x^2) \quad \Leftarrow \text{ Grouped coefficients and grouped variables within each term.}$$

$$(2 \cdot 7)(x^{\mathbf{4+3}}) + (2 \cdot 5)(x^{\mathbf{4+2}}) \quad \Leftarrow \text{ Applied multiplication property of exponents.}$$

$$14x^7 + 10x^6 \quad \Leftarrow \text{ Multiplied coefficients and added exponents. Since these are not like terms we cannot combine them.}$$

— *Note* — *Simplifying mentally*	After you have practiced these types of problems you should be able to do them mentally without having to write all the steps.

d) $-y(y^2 - 3)$

•SOLUTION• We distribute $-y$ to each term inside the parentheses to get $-y^3 + 3y$.

Practice 6.1.2 Simplify.

1. $6x^4(3x^3)$

2. $-7m(3m^4)$

3. $\frac{1}{2}t^5(4t^8)$

4. $5x^2(-7x^3)(4x^7)$

5. $\frac{1}{7}x^5(-14x^3)(-3x^5)$

6. $5c^3(6c^5 + 2c)$

7. $4s^2(3s^4 + 8s^3)$

8. $-m(m^2 + 4)$

9   $-4w(-6w^3 + 5)$

Topic 6.1 C — Power Property of Exponents

Exponential expressions with powers raised to powers can be simplified. For example, we can use the definition of exponents as repeated multiplication to simplify $(x^3)^2$.

$$(x^3)^2 \quad \Leftarrow \text{ Expression to simplify.}$$

$$(x^3)(x^3) \quad \Leftarrow \text{ Used definition of exponent as repeated multiplication.}$$

$$x^{3+3} \quad \Leftarrow \text{ Used Multiplication Property of Exponents.}$$

$$x^6 \quad \Leftarrow \text{ Simplified.}$$

Notice the exponent, 6, of the simplified expression is the product of the exponents, $3 \cdot 2$, of the original expression. This observation can be generalized.

— *Property* — *Power of a power property of exponents*	English: To simplify a power raised to a power, multiply the exponents. Examples: $(2^3)^4$ can be written as $2^{3 \cdot 4}$, which simplifies to 2^{12} $(x^2)^5$ can be written as $x^{2 \cdot 5}$, which simplifies to x^{10} Algebra: $(a^m)^n = a^{mn}$

Example 6.1.3   Simplify.

a) $(3^5)^3$

•SOLUTION• We multiply the exponents to get $3^{(5 \cdot 3)}$, which simplifies to 3^{15}.

b) $\left(s^3\right)^2$

•SOLUTION• We multiply the exponents to get $s^{(3 \cdot 2)}$, which simplifies to s^6.

Practice 6.1.3 Simplify.

1. $(2^4)^3$ 2. $\left(d^5\right)^7$ 3. $\left(a^9\right)^4$

Topic 6.1 D — Power of a Product Property of Exponents

When a product is raised to a power, we can use the definition of exponents to simplify. For example,

$(xy)^2$ ⇐ Expression to simplify.

$(xy)(xy)$ ⇐ Used definition of exponents.

$(xx)(yy)$ ⇐ Used Associative and Commutative Properties of Multiplication to group the factors of x and to group the factors of y.

x^2y^2 ⇐ Used definition of exponents.

We see that the exponent of each variable in the simplified expression is the same as the exponent of the product in the original expression. This observation can be generalized.

— *Property* — *Power of a product property of exponents*	English: A product raised to a power can be written as the product of each factor raised to the power. Example: $(5x)^3$ can be written as 5^3x^3 Algebra: $(ab)^m = a^m b^m$

Example 6.1.4   Simplify.

a) $(2w)^4$

•SOLUTION•

$2^4 \cdot w^4$ ⇐ Power of a product so applied exponent to each factor.

$16w^4$ ⇐ Simplified.

b) $(-x)^3$

•SOLUTION• Remember that $(-x)^3$ means $(-1x)^3$.

$(-1)^3(x)^3$ ⇐ Power of a product so applied exponent to each factor.

$-1x^3$ ⇐ Simplified.

$-x^3$ ⇐ Rewrote using shorthand notation.

c) $(-5x^2)^3$

•SOLUTION• When we have exponents inside the parentheses we need to use additional exponent properties.

$(-5x^2)^3$ $\Leftarrow$ Expression to simplify.

$(-5)^3(x^2)^3$ $\Leftarrow$ Power of a product so applied exponent to each factor.

$(-5)^3 x^{2\cdot 3}$ $\Leftarrow$ Applied Power of a Power Property.

$-125x^6$ $\Leftarrow$ Calculated 5^3.

d) $(x^4 y^3)^5$

•SOLUTION•

$(x^4)^5 (y^3)^5$ $\Leftarrow$ Power of a product so applied exponent to each factor.

$(x^{4\cdot 5})(y^{3\cdot 5})$ $\Leftarrow$ Applied Power of a Power Property.

$x^{20} y^{15}$ $\Leftarrow$ Simplified.

Practice 6.1.4   Simplify.

1. $(4s)^2$ 2. $(-v^5)^4$ 3. $(2^3 v^5)^2$ 4. $(-2x^4)^4$ 5. $(m^5 n^2)^7$ 6. $(3xy^3)^2$

— Caution — ***Power of a product does not work with sums***	We cannot use the Power of a Product Property to simplify a term that contains a sum. For example, while $(2w)^4$ simplifies to $2^4 w^4$ $(2 + w)^4$ does NOT simplify to $2^4 + w^4$ We will discuss simplifying sums raised to powers in Section 6.6.

Topic 6.1 E — Division Property of Exponents

We can use the definition of exponents to divide expressions that contain exponents. For example, we can divide x^5 by x^3 as follows:

$\dfrac{x^5}{x^3}$ $\Leftarrow$ Expression to simplify.

$\dfrac{xxxxx}{xxx}$ $\Leftarrow$ Used definition of exponents.

$\left(\dfrac{x}{x}\right)\cdot\left(\dfrac{x}{x}\right)\cdot\left(\dfrac{x}{x}\right)\cdot x \cdot x$ $\Leftarrow$ Used Associative Property of Multiplication.

$1 \cdot 1 \cdot 1 \cdot x \cdot x$ $\Leftarrow$ Reduced each fraction.

x^2 $\Leftarrow$ Used multiplication property of 1 and definition of exponent.

The exponent, 2, of the simplified expression is the difference $(5 - 3)$ of the exponents of the original expressions. We can generalize this observation.

— Property — Division property of exponents	English: To divide factors with the same base, subtract the exponents. Example: $\dfrac{x^5}{x^2}$ can be written as x^{5-2}, which simplifies to x^3 Algebra: If $a \neq 0$ then $\dfrac{a^m}{a^n} = a^{m-n}$

Example 6.1.5   Simplify.

a) $\dfrac{5^6}{5^2}$

> •**SOLUTION**• The bases are the same so we subtract the exponents to get 5^{6-2}, which simplifies to 5^4. We could calculate 5^4 to get 625 but it is easier to leave it in exponential form.

b) $\dfrac{t^8}{t^3}$

> •**SOLUTION**• The bases are the same so we subtract the exponents to get t^{8-3}, which simplifies to t^5.

c) $\dfrac{4c^9}{3c^2}$

> •**SOLUTION**•
>
> $\dfrac{4c^{9-2}}{3}$ $\Leftarrow$ Bases of the variables are the same so we their subtracted exponents.
>
> $\dfrac{4c^7}{3}$ $\Leftarrow$ Subtracted. This also could be written as $\dfrac{4}{3}c^7$.

d) $\dfrac{a^8}{b^4}$

> •**SOLUTION**• The bases are not the same so the Division Property of Exponents does not apply here. We cannot simplify this expression.

Practice 6.1.5   Simplify.

1. $\dfrac{10^7}{10^3}$ 2. $\dfrac{m^6}{m}$ 3. $\dfrac{12x^7}{4x^5}$ 4. $\dfrac{15t^{15}}{-5t^5}$ 5. $\dfrac{x^3y^5}{x^2y^3}$ 6. $\dfrac{m^7n^4}{m^2n}$

— Note — Restrictions on variables because of division by 0	We know that division by 0 is not possible. Therefore, when writing an expression with a variable in the denominator we should also write a restriction that the variable may not take on a value of 0. However, writing something like $x \neq 0$ after every expression gets very cumbersome. Therefore, from now on we will make the blanket assumption that no variables will take on values that will cause expressions in denominators to be 0.

Topic 6.1 F — Power of a Quotient Property of Exponents

When a quotient is raised to a power, we can simplify the expression. For example,

$\left(\dfrac{x}{y}\right)^3$ can be simplified as follows:

$$\left(\dfrac{x}{y}\right)^3 \quad \Leftarrow \text{ Expression to simplify.}$$

$$\left(\dfrac{x}{y}\right)\left(\dfrac{x}{y}\right)\left(\dfrac{x}{y}\right) \quad \Leftarrow \text{ Used definition of exponents.}$$

$$\left(\dfrac{xxx}{yyy}\right) \quad \Leftarrow \text{ Used Associative Property of Multiplication.}$$

$$\dfrac{x^3}{y^3} \quad \Leftarrow \text{ Used definition of exponents.}$$

We see that the exponent of each variable in the simplified expression is the same as the exponent of the quotient in the original expression. This observation can be generalized.

— *Property* — *Power of a quotient property of exponents*	English: A quotient raised to a power can be written as the quotient of the numerator and denominator, each raised to the power. Examples: $\left(\dfrac{2}{3}\right)^5$ can be written as $\dfrac{2^5}{3^5}$ $\left(\dfrac{5}{x}\right)^3$ can be written as $\dfrac{5^3}{x^3}$ Algebra: If $b \neq 0$ then $\left(\dfrac{a}{b}\right)^m = \dfrac{a^m}{b^m}$

Example 6.1.6 Simplify.

a) $\left(\dfrac{3}{w}\right)^4$

•**SOLUTION**• We can apply the exponent to the numerator and the denominator to get $\dfrac{3^4}{w^4}$,

which simplifies to $\dfrac{81}{w^4}$.

b) $\left(\dfrac{5p}{r^2}\right)^3$

•**SOLUTION**• We have to use two properties to simplify this expression.

$$\dfrac{(5p)^3}{(r^2)^3} \quad \Leftarrow \text{ Power of a quotient so applied exponent to numerator and denominator.}$$

$$\dfrac{(5)^3(p)^3}{(r^2)^3} \quad \Leftarrow \text{ Applied Power of a Product Property in numerator.}$$

$$\dfrac{125p^3}{r^6} \quad \Leftarrow \text{ Simplified.}$$

Practice 6.1.6 Simplify.

1. $\left(\dfrac{2}{m}\right)^4$

2. $\left(\dfrac{w}{4}\right)^3$

3. $\left(\dfrac{3w}{y^2}\right)^4$

4. $\left(\dfrac{4d^2}{c^4}\right)^3$

Topic 6.1 G — Simplifying That Requires the Use of Several Exponent Properties

Below is a summary of the exponent properties we have discussed so far.

— *Summary* — *Exponent properties*		
Name	*General Rule*	*Example*
Multiplication Property:	$a^m a^n = a^{m+n}$	$x^2 x^3 = x^{2+3}$, which simplifies to x^5
Power of a Power Property:	$(a^m)^n = a^{m \cdot n}$	$(x^2)^3 = x^{2 \cdot 3}$, which simplifies to x^6
Power of a Product Property:	$(ab)^m = a^m b^m$	$(xy)^2 = x^2 y^2$
Division Property:	$\dfrac{a^m}{a^n} = a^{m-n}, a \neq 0$	$\dfrac{x^5}{x^3} = x^{5-3}$, which simplifies to x^2
Power of a Quotient Property:	$\left(\dfrac{a}{b}\right)^m = \dfrac{a^m}{b^m}, b \neq 0$	$\left(\dfrac{x}{y}\right)^2 = \dfrac{x^2}{y^2}$

When expressions contain several operations involving exponents we must decide which property to use and when to use it. The following procedure works well in most situations:

— *Procedure* — *Simplifying exponential expressions*	*Step 1 Parentheses* Simplify inside parentheses.
	Step 2 Numerator Simplify numerators.
	Step 3 Denominator Simplify denominators.
	Step 4 Fraction Simplify fractions.

Example 6.1.7   Simplify.

a) $\left(x^3\right)^2 x^4$

• SOLUTION •

$x^6 x^4$ $\Leftarrow$ Multiplied exponents using $(a^m)^n = a^{m \cdot n}$

x^{10} $\Leftarrow$ Added exponents using $a^m a^n = a^{m+n}$

b) $\dfrac{y^2 y^3}{y^4}$

• SOLUTION •

$\dfrac{y^5}{y^4}$ $\Leftarrow$ In numerator, added exponents using $a^m a^n = a^{m+n}$

y^1 or just y $\Leftarrow$ Subtracted exponents using $\dfrac{a^m}{a^n} = a^{m-n}$

c) $\dfrac{(x^2y^5)^3}{(y^4)^2}$

•SOLUTION•

$\dfrac{(x^2)^3(y^5)^3}{(y^4)^2}$ ⟸ In numerator, applied exponent to each factor using $(ab)^m = a^m b^m$

$\dfrac{x^6 y^{15}}{(y^4)^2}$ ⟸ In numerator, multiplied exponents using $(a^m)^n = a^{m \cdot n}$

$\dfrac{x^6 y^{15}}{y^8}$ ⟸ In denominator, multiplied exponents using $(a^m)^n = a^{m \cdot n}$

$x^6 y^7$ ⟸ Subtracted exponents using $\dfrac{a^m}{a^n} = a^{m-n}$

d) $\left(\dfrac{2x^3(3y^3)^2}{6x^2y^3} \right)^2$

•SOLUTION• We simplify inside the parentheses first.

$\left(\dfrac{2x^3(9y^6)}{6x^2y^3} \right)^2$ ⟸ In numerator, multiplied exponents using $(a^m)^n = a^{m \cdot n}$.

$\left(\dfrac{18x^3y^6}{6x^2y^3} \right)^2$ ⟸ In numerator, multiplied coefficients (2•9).

$(3xy^3)^2$ ⟸ Reduced coefficients and subtracted exponents using $\dfrac{a^m}{a^n} = a^{m-n}$.

$3^2 x^2 (y^3)^2$ ⟸ Applied exponent to each factor using $(ab)^m = a^m b^m$.

$9x^2y^6$ ⟸ Simplified 3^2 and multiplied exponents using $(a^m)^n = a^{m \cdot n}$ on y.

Practice 6.1.7 Simplify.

1. $m^5(m^2)^6$ 2. $\dfrac{t^7(t^5)^2}{t^3}$ 3. $\dfrac{(x^5x^3)^2}{x^2}$ 4. $\dfrac{(5x^2)^2(2y^8)^3}{10x^2(5y)}$ 5. $\dfrac{16x^6(3y^5)^3}{12x^3(2y^3)^2}$

Exercise Set 6.1   The answers to the odd numbered exercises are at the back of the book.

Simplify.

1. $6^2 6^3$ 2. $8^2 8^5$ 3. $h^3 h$ 4. $j^5 j^7$

5. $y^7 y^6$ 6. $m^2 m^8$ 7. $y^5 z^5$ 8. $t^8 t^4 t^9$

9. $m^4 m^5 m^3$ 10. $t^4 r^7$ 11. $r^2 r^7 r^8$ 12. $ww^5 w^3$

13. $x^2 y^4 x^4 y^5$ 14. $m^5 n^2 m^6 n^5$ 15. $hh^2 jj^3$ 16. $c^8 cg^5 g$

17. $2n^3(3n^5)$ 18. $2y^7(4y^3)$ 19. $-3g^2(-5g^7)$ 20. $-2p^4(-4p^2)$

21. $-3w(2w^4)(4w)$ 22. $-3v^4(4v^3)(5v)$ 23. $8x(-2xy^3)(-3x^3y^8)$ 24. $5xy^2(3x^2)(-2x^2y^5)$

25. $2a^2(3a^3 + 4a)$ 26. $4b^2(5b^5 + 3b^4)$ 27. $5t^2(6t^5 + 4t^4 - 3t)$ 28. $-8r^5(r^5 - 2r^4 - 3r)$

29. $\dfrac{2}{3}w^2(6w^5)(-5w^3)$ 30. $-\dfrac{3}{5}v^3(-10v^7)(2v^{12})$ 31. $\dfrac{2}{3}d^2(6d^5 + 9d^3)$ 32. $\dfrac{3}{4}y^7(4y^3 + 8y)$

33. $(c^5)^6$ 34. $(d^9)^3$ 35. $(k^2)^4$ 36. $(s^9)^5$

37. $(4t)^3$ 38. $(-2m)^3$ 39. $(5w^4)^2$ 40. $(5s^2)^3$

41. $(-4r^3)^4$ 42. $(-5y^6)^2$ 43. $(2x^{12}y^4)^5$ 44. $(-3r^4s)^3$

45. $\dfrac{5^4}{5^2}$ 46. $\dfrac{3^7}{3^3}$ 47. $\dfrac{y^6}{y^2}$ 48. $\dfrac{m^9}{m^3}$

49. $\dfrac{r^8p^4}{rp}$ 50. $\dfrac{w^6z^3}{w^2z^2}$ 51. $\dfrac{s^6t^4}{s^4t^3}$ 52. $\dfrac{x^9y^5}{x^7y^4}$

53. $\dfrac{x^7}{y^5}$ 54. $\dfrac{m^{12}}{n^4}$

55. $\left(\dfrac{8}{x}\right)^2$ 56. $\left(\dfrac{r}{3}\right)^3$ 57. $\left(\dfrac{7}{t^3}\right)^2$ 58. $\left(\dfrac{3}{p^4}\right)^3$ 59. $\left(\dfrac{2x^5}{y^4}\right)^4$ 60. $\left(\dfrac{2r^2}{t^3}\right)^5$

61. $3x^2(2x)^3$ 62. $(2y)^2(3y^2)^3$ 63. $\dfrac{2x^2(4x^3)^2}{16x^5}$ 64. $\dfrac{12m^3(m^2)^5}{(2m)^2}$

65. $\dfrac{8x^3(4xy^2)^2(3x^4)^2}{24x^3(6x^2y^3)}$ 66. $\dfrac{(2x^3)^4(5x)(x^2y^5)^8}{10x^5y(2xy^2)^3}$

Review Exercises   The answers to all of these exercises are at the back of the book.

67. Loretta is driving to her mother's house for the holidays. Due to road construction a part of her trip is driven at a slower speed. If c represents the number of miles she is in the construction zone create an expression using c for the total number of miles she traveled if:

     a)   the part of the trip within the construction zone exceeded the part out of the construction zone by 80 miles.
     b)   the part of the trip outside the construction zone was a quarter as long as the part in the construction zone.
     c)   the distance within the construction was 60 miles less than the distance outside the construction.

68. Two angles are complementary angles if the sum of their measures is 90 degrees. Assume two angles are complementary and the measure of one angle is 10 degrees more than one–fourth the measure of the other. What are the measures of both angles?

69. Calculate the slope of the straight line that passes through (1,1) and (3,–2).   70.   Graph: $9.5 \geq x$

71. Graph the line that passes through (2,–2) with undefined slope.   72.   Find consecutive integers whose sum is –9.

73. Determine whether this graph represents a function: 74.   Determine whether this graph represents a function:

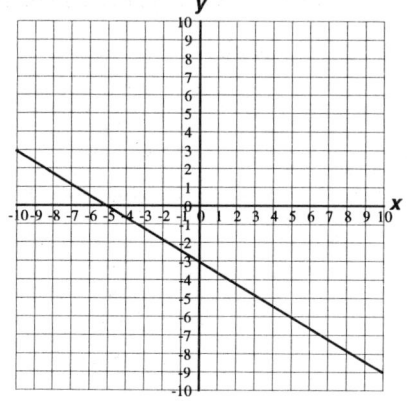

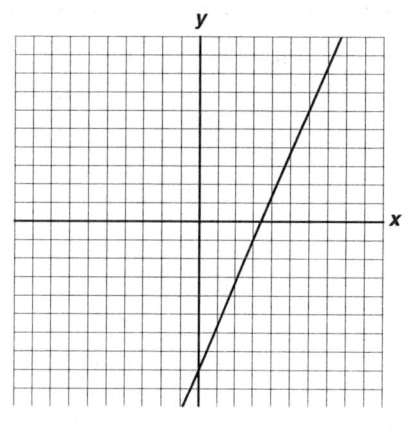

75. Solve: $\begin{cases} x + 7y = -3 & ❶ \\ 2x + 6y = 10 & ❷ \end{cases}$ 76.   Solve: $\begin{cases} 2x - 5y = 7 & ❶ \\ 3x + 2y = 1 & ❷ \end{cases}$

Section 6.2 Exponents Involving 1, 0, and Negative Integers

Topic 6.2 A — The Number 1 as an Exponent

We originally defined exponents as a convenient way to show repeated multiplication. For example, 2^3 means multiply 3 factors of 2, or $2 \cdot 2 \cdot 2$. This works fine for natural number exponents of 2 or greater, but this definition does not make sense when the exponent is 1 because we cannot multiply 1 factor (this would be like trying to clap with one hand). In Chapter 1 we stated that an exponent of 1 can be assumed when no exponent is written. For example, we said that x and x^1 are equivalent. This assumption is consistent with the properties of exponents discussed in the previous section. To show this, let's simplify the expression $\dfrac{x^3}{x^2}$ using two methods we know to be correct and then compare the answers.

- Simplifying $\dfrac{x^3}{x^2}$ using the definition of exponents:

$$\dfrac{xxx}{xx} \quad \Leftarrow \text{ Used definition of exponents.}$$

$$\dfrac{x}{x} \cdot \dfrac{x}{x} \cdot x \quad \Leftarrow \text{ Used Associative Property of Multiplication.}$$

$$1 \cdot 1 \cdot x \quad \Leftarrow \text{ Simplified each fraction.}$$

$$x \quad \Leftarrow \text{ Used Multiplication Property of 1.}$$

- Simplifying $\dfrac{x^3}{x^2}$ using the Division Property of Exponents gives us x^{3-2}, which simplifies to x^1.

We used two different methods to simplify the same expression and we obtained two expressions that look different: x and x^1. This suggests that the simplified expressions represent the same quantity.

— Definition — Exponent of 1	English: An exponent of 1 means a single factor of the base. Examples: 2^1 can be written as 2 y^1 can be written as y Algebra: $a^1 = a$

Topic 6.2 B — 0 as an Exponent

We will use a similar process to determine a meaning for an exponent of 0. We will simplify $\dfrac{x^3}{x^3}$ using two different methods:

- Simplifying $\dfrac{x^3}{x^3}$ using the definition of exponents:

$$\dfrac{xxx}{xxx} \quad \Leftarrow \text{ Used definition of exponents.}$$

$$\dfrac{x}{x} \cdot \dfrac{x}{x} \cdot \dfrac{x}{x} \quad \Leftarrow \text{ Used Associative Property of Multiplication.}$$

$$1 \cdot 1 \cdot 1 \quad \Leftarrow \text{ Simplified each fraction.}$$

$$1 \quad \Leftarrow \text{ Used Multiplication Property of 1.}$$

- Simplifying $\dfrac{x^3}{x^3}$ using the Division Property of Exponents gives us x^{3-3}, which simplifies to x^0.

We used two different methods to simplify the same expression and we obtained two expressions that look different: 1 and x^0. This suggests that the simplified expressions represent the same quantity.

— Definition — *Exponent of 0*	English: An expression with an exponent of 0 is equivalent to the number 1. Examples: $\quad 2^0$ can be written as 1 $\quad y^0$ can be written as 1 (assuming $y \neq 0$) Algebra: If $a \neq 0$, then $a^0 = 1$

In the above definition, we excluded 0 as a possible base because 0^0 is not defined. We will always make this assumption even though we may not explicitly write it.

This result is surprising and may not make intuitive sense. Many people think x^0 should be 0 rather than 1. This is an instance where your intuition might lead you astray. Remember, it's always better to back up intuition with algebra.

One use of this definition is to attach any variable to a term. For example, we may rewrite the constant 5 as an equivalent expression that contains a variable.

$$5 \cdot 1 \quad \Leftarrow \text{We used the Property of Multiplication by 1.}$$

$$5x^0 \quad \Leftarrow \text{We used the definition of a 0 exponent. Note that we must assume that } x \neq 0 \text{ because } 0^0 \text{ is not defined.}$$

It may seem strange to do this. But, we will use this idea later to help us simplify expressions.

Topic 6.2 C — Negative Exponents

Expressions with negative exponents, such as x^{-2}, might not make intuitive sense but we can determine a meaning for them using the same technique we used for exponents of 0 and 1. We will do this by simplifying $\dfrac{x^3}{x^5}$ using two different methods.

- Simplifying $\dfrac{x^3}{x^5}$ using the definition of exponents:

$$\dfrac{x^3}{x^5} \quad \Leftarrow \text{Expression to simplify.}$$

$$\dfrac{xxx}{xxxxx} \quad \Leftarrow \text{Used definition of exponents.}$$

$$\dfrac{x}{x} \cdot \dfrac{x}{x} \cdot \dfrac{x}{x} \cdot \dfrac{1}{xx} \quad \Leftarrow \text{Used Associative Property of Multiplication.}$$

$$1 \cdot 1 \cdot 1 \cdot \dfrac{1}{xx} \quad \Leftarrow \text{Simplified each fraction.}$$

$$\dfrac{1}{x^2} \quad \Leftarrow \text{Used Multiplication Property of 1.}$$

- Simplifying $\dfrac{x^3}{x^5}$ using the Division Property of Exponents gives us x^{3-5}, which simplifies to x^{-2}.

We used two different methods to simplify the same expression and we obtained two answers: $\frac{1}{x^2}$ and x^{-2}. This suggests that the simplified expressions represent the same quantity.

— *Definition* — *Negative Exponents*	English: An expression with a base written on one side of a fraction bar can be written as an equivalent expression with the base on the other side of the fraction bar by changing the sign of the exponent.
	Examples:
	2^{-5} can be written as $\frac{1}{2^5}$
	x^{-3} can be written as $\frac{1}{x^3}$
	$\frac{1}{x^{-4}}$ can be written as x^4
	Algebra: If $a \neq 0$, then $a^{-n} = \frac{1}{a^n}$ and $a^n = \frac{1}{a^{-n}}$

Note that a^{-n} and a^n are reciprocals because their product is 1:

$(a^{-n})(a^n)$ ⇐ Expression to simplify.

$a^{(-n + n)}$ ⇐ Applied multiplication property of exponents.

a^0 ⇐ Added exponents.

1 ⇐ Used definition of 0 exponent.

We can use the idea of negative exponents to "move" factors between the numerator and denominator of a fraction.

— *Procedure* — *Moving factors between numerators and denominators*	• To move a factor from the numerator to the denominator of a fraction, write the factor in the denominator and change the sign of its exponent.
	• To move a factor from the denominator to the numerator of a fraction, write the factor in the numerator and change the sign of its exponent.

For example,

- Moving the x factor from the numerator to the denominator of $\frac{3x^{-2}}{4}$ results in $\frac{3}{4x^2}$.

  Notice the sign of the exponent in each expression is different. We can demonstrate that these two expressions are equivalent as follows:

  $\frac{3x^{-2}}{4}$ ⇐ Expression to simplify.

  $\frac{3}{4}x^{-2}$ ⇐ Split the fraction into two factors.

  $\frac{3}{4} \cdot \frac{1}{x^2}$ ⇐ Used definition of a negative exponent.

  $\frac{3}{4x^2}$ ⇐ Multiplied numerators and multiplied denominators.

- Moving the x factor from the denominator to the numerator of $\frac{2}{5x^{-3}}$ results in $\frac{2x^3}{5}$. We can demonstrate that these two expressions are equivalent as follows:

$$\frac{2}{5x^{-3}} \quad \Leftarrow \text{ Expression to simplify.}$$

$$\frac{2}{5} \cdot \frac{1}{x^{-3}} \quad \Leftarrow \text{ Split the fraction into two factors.}$$

$$\frac{2}{5} \cdot \frac{1}{\frac{1}{x^3}} \quad \Leftarrow \text{ Used definition of a negative exponent.}$$

$$\frac{2}{5} \cdot \frac{x^3}{1} \quad \Leftarrow \text{ Inverted and multiplied second fraction.}$$

$$\frac{2x^3}{5} \quad \Leftarrow \text{ Multiplied numerators and multiplied denominators.}$$

Example 6.2.1   Identify the base of each negative exponent. Then, simplify and write with positive exponents only.

a) 5^{-2}

> •**SOLUTION**• The base is 5. By the definition of a negative exponent, this is equivalent to $\frac{1}{5^2}$, which simplifies to $\frac{1}{25}$.

b) $\dfrac{8}{k^{-2}}$

> •**SOLUTION**• The base of the negative exponent is k. We can move the factors of k to the numerator by changing the sign of the exponent to get $8k^2$.

c) ab^{-3}

> •**SOLUTION**• The base of the negative exponent is b. We can move the factor with the negative exponent to the denominator by changing the sign of its exponent to get $\dfrac{a}{b^3}$.

d) $\dfrac{2^{-3}}{c^{-4}}$

> •**SOLUTION**• The bases of the negative exponents are 2 and c. We can move both factors by changing the signs of their exponents to get $\dfrac{c^4}{2^3}$, or $\dfrac{c^4}{8}$.

e) $\dfrac{1}{2 + y^{-1}}$

> •**SOLUTION**• Base is y; We cannot "move" the y to the numerator because it is not a factor of the denominator (it is added to the 2 rather than multiplied by it). The best we can do at this time is to write $\dfrac{2}{2 + \frac{1}{y}}$. This is called a complex fraction, which we discuss in Section 7.2.

Practice 6.2.1   Identify the base of each negative exponent. Then, simplify and write with positive exponents only.

1. 3^{-4} 2. $5w^{-7}$ 3. $7^{-2}m^{-6}$ 4. $\dfrac{9}{x^{-5}}$

5. $\dfrac{4^{-3}}{d^{-5}}$ 6. $\dfrac{3x^{-2}}{y^{-4}}$ 7. $\dfrac{2 + x^{-1}}{y}$

Example 6.2.2   Simplify.

a) $2^{-1} + 3^{-1}$

•SOLUTION• We must use the definition of negative exponents before combining the terms.

$2^{-1} + 3^{-1}$ ⬅ Expression to simplify.

$\dfrac{1}{2} + \dfrac{1}{3}$ ⬅ Used definition of a negative exponent.

$\dfrac{3}{6} + \dfrac{2}{6}$ ⬅ Converted fractions so they have a common denominator.

$\dfrac{5}{6}$ ⬅ Added numerators.

b) $\dfrac{1}{2^{-2}} - 3^{-1} + 3^{0}$

•SOLUTION•

$2^{2} - \dfrac{1}{3} + 3^{0}$ ⬅ Used definition of a negative exponent.

$4 - \dfrac{1}{3} + 1$ ⬅ Calculated 2^{2} and 4^{0}.

$\dfrac{12}{3} - \dfrac{1}{3} + \dfrac{3}{3}$ ⬅ Converted fractions so they have a common denominator.

$\dfrac{14}{3}$ ⬅ Simplified.

Practice 6.2.2   Simplify.

1. $3^{-1} + 4^{-1}$ 2. $5 + \dfrac{1}{6^{-2}} + 2^{0}$ 3. $4^{-1} + \dfrac{1}{3^{-2}}$ 4. $3 \cdot 2^{-3} - 6^{-1}$

Topic 6.2 D — Simplifying Complicated Exponential Expressions

The following is a summary of the properties of exponents introduced in this Chapter.

— *Summary* — *Exponent definition and properties*		
Name	*General Rule*	*Example*
Multiplication Property:	$a^{m}a^{n} = a^{m+n}$	$x^{2}x^{3} = x^{2+3}$, which simplifies to x^{5}
Power of a Power Property:	$(a^{m})^{n} = a^{m \cdot n}$	$(x^{2})^{3} = x^{2 \cdot 3}$, which simplifies to x^{6}
Power of a Product Property:	$(ab)^{m} = a^{m}b^{m}$	$(xy)^{2} = x^{2}y^{2}$
Division Property:	$\dfrac{a^{m}}{a^{n}} = a^{m-n}, a \neq 0$	$\dfrac{x^{5}}{x^{3}} = x^{5-3}$, which simplifies to x^{2}
Power of a Quotient Property:	$\left(\dfrac{a}{b}\right)^{m} = \dfrac{a^{m}}{b^{m}}, b \neq 0$	$\left(\dfrac{x}{y}\right)^{2} = \dfrac{x^{2}}{y^{2}}$
Exponent of 1 definition:	$a^{1} = a$	$x^{1} = x$
Exponent of 0 definition:	$a^{0} = 1, a \neq 0$	$x^{0} = 1$
Negative Exponent definition:	$a^{-m} = \dfrac{1}{a^{m}}, a \neq 0$	$x^{-2} = \dfrac{1}{x^{2}}$

Example 6.2.3 Simplify and write with positive exponents. Assume all variables represent non–zero real numbers.

a) $\left(x^{-4}x^2\right)^8$

•SOLUTION•

$\left(x^{-2}\right)^8$ $\Leftarrow$ Added exponents using $a^m a^n = a^{m+n}$.

x^{-16} $\Leftarrow$ Multiplied exponents using $(a^m)^n = a^{m \cdot n}$.

$\dfrac{1}{x^{16}}$ $\Leftarrow$ Used definition of a negative exponent, $a^{-m} = \dfrac{1}{a^m}$.

b) $\left(\dfrac{a^{-2}}{-2a^{-4}}\right)^{-3}$

•SOLUTION•

$\left(\dfrac{a^2}{-2}\right)^{-3}$ $\Leftarrow$ Simplified inside parentheses using $\dfrac{a^m}{a^n} = a^{m-n}$. The exponent of a comes from $a^{-2-(-4)}$, which is a^{-2+4}, which is a^2 .

$\dfrac{(a^2)^{-3}}{(-2)^{-3}}$ $\Leftarrow$ Applied exponent of -3 to numerator and denominator using $\left(\dfrac{a}{b}\right)^m = \dfrac{a^m}{b^m}$.

$\dfrac{a^{-6}}{(-2)^{-3}}$ $\Leftarrow$ Multiplied exponents using $(a^m)^n = a^{m \cdot n}$.

$\dfrac{(-2)^3}{a^6}$ $\Leftarrow$ Used definition of a negative exponent, $a^{-m} = \dfrac{1}{a^m}$.

$\dfrac{-8}{a^6}$ $\Leftarrow$ Calculated $(-2)^3$.

Practice 6.2.3 Simplify and write with positive exponents only. Assume all variables represent non–zero real numbers.

1. $\left(z^{-5}z^2\right)^3$ 2. $(b^{-6}b^5)^6$ 3. $\left(\dfrac{c^{-4}}{-4}\right)^{-2}$ 4. $\left(\dfrac{d^{-5}}{-3d^{-2}}\right)^{-4}$ 5. $\dfrac{\left(5m^4\right)^4}{5m^{-2}}$ 6. $\dfrac{\left(6t^{-4}\right)^2}{2t^{-6}}$

Example 6.2.4 Simplify and write with positive exponents only.

a) $\dfrac{(2x^4)^3}{2x^{12}}$

•SOLUTION•

$\dfrac{(2)^3(x^4)^3}{2x^{12}}$ $\Leftarrow$ In numerator, applied exponent of 3 to each factor using $(ab)^m = a^m b^m$

$\dfrac{2^3 x^{12}}{2x^{12}}$ $\Leftarrow$ In numerator, multiplied exponents using $(a^m)^n = a^{m \cdot n}$

$2^2 x^0$ $\Leftarrow$ Subtracted exponents using $\dfrac{a^m}{a^n} = a^{m-n}$

$4x^0$ $\Leftarrow$ Calculated 2^2 .

$4(1)$ $\Leftarrow$ Used Definition of Exponent of Zero.

4 $\Leftarrow$ Multiplied.

b) $\left(\dfrac{m^{-5}m^2}{\left(4n^2\right)^{-1}}\right)^3$

•SOLUTION•

$\left(\dfrac{m^{-3}}{\left(4n^2\right)^{-1}}\right)^3$ $\Leftarrow$ In numerator, added exponents using $a^m a^n = a^{m+n}$

$\left(\dfrac{m^{-3}}{4^{-1}(n^2)^{-1}}\right)^3$ $\Leftarrow$ In denominator, applied exponent of -1 to each factor using
$(ab)^m = a^m b^m$

$\left(\dfrac{m^{-3}}{4^{-1}n^{-2}}\right)^3$ $\Leftarrow$ In denominator, multiplied exponents using $(a^m)^n = a^{m \cdot n}$.

$\left(\dfrac{4n^2}{m^3}\right)^3$ $\Leftarrow$ Used Definition of Negative Exponent to move factors with negative exponents.

$\dfrac{\left(4n^2\right)^3}{\left(m^3\right)^3}$ $\Leftarrow$ Applied exponent of 3 to numerator and denominator using
$\dfrac{a^m}{a^n} = a^{m-n}$.

$\dfrac{(4)^3\left(n^2\right)^3}{\left(m^3\right)^3}$ $\Leftarrow$ Applied exponent of 3 to each factor using $(ab)^m = a^m b^m$.

$\dfrac{64n^6}{m^9}$ $\Leftarrow$ Calculated 4^3 and multiplied exponents using $(a^m)^n = a^{m \cdot n}$.

Practice 6.2.4 Simplify and write with positive exponents only.

1. $\left(\dfrac{x^2 x^5}{(2x^3)^2}\right)^2$

2. $\left(\dfrac{5x^{-2}x^4}{(x^4)^2}\right)^3$

3. $\dfrac{3x^2 \cdot 5x^4}{2x^{-3}(x^{-2})^2}$

4. $\dfrac{2x^5 \cdot 4x^{-3}}{12x^5(x^3)^{-5}}$

Exercise Set 6.2   The answers to the odd numbered exercises are at the back of the book.

Identify the base of each negative exponent. Then, simplify and write with positive exponents only.

1. 3^{-6}
2. 6^{-4}
3. $12z^{-6}$
4. $25r^{-12}$

5. $(7x)^{-5}$
6. $(4w)^{-2}$
7. $\dfrac{2}{d^{-2}}$
8. $\dfrac{6}{p^{-4}}$

9. $4^{-2}v^7$
10. $7^{-3}h^6$
11. $2^{-3}m^{-9}$
12. $5^{-2}w^{-4}$

13. $\dfrac{3^{-2}}{k^{-6}}$
14. $\dfrac{8^{-2}}{c^{-5}}$
15. $\dfrac{3+x^{-1}}{2}$
16. $\dfrac{w^{-2}-3}{7}$

Simplify.

17. $3-2^{-1}$
18. $3^{-1}-2$
19. $2-\dfrac{1}{5^{-1}}-5^0$
20. $8+\dfrac{1}{2^{-1}}-3^0$

21. $5 \cdot 2^{-1} - 3^{-1}$
22. $6^{-1} - 2 \cdot 8^{-1}$

Simplify and write with positive exponents only.

23. $\left(a^{-6}a^4\right)^{-5}$

24. $\left(w^{-9}w^3\right)^{-4}$

25. $\left(\dfrac{d^{-3}}{2^{-3}}\right)^{-2}$

26. $\left(\dfrac{q^{-7}}{3^{-4}}\right)^{-1}$

27. $\left(\dfrac{2x^{-3}}{3x^{-2}}\right)^{-2}$

28. $\left(\dfrac{5x^2}{-2x^{-5}}\right)^{-1}$

29. $\dfrac{(4x^2y^3)^2}{(2xy^4)^{-3}}$

30. $\dfrac{(-3x^2y^3)^2}{(-2xy^{-4})^{-1}}$

31. $\left(\dfrac{12x^4y^{-2}}{18x^{-5}y}\right)^{-2}$

32. $\left(\dfrac{-12x^{-3}y}{9x^5y^{-2}}\right)^{-3}$

33. $\dfrac{(8x^2y^{-5})^4}{(16x^{-3}y^2)^3}$

34. $\dfrac{(2x^{-5}y)^{-3}}{(4x^3y^{-2})^{-1}}$

35. $\left(\dfrac{4^{-2}x^{-3}y^{-1}}{x^{-4}y}\right)\left(\dfrac{8^{-1}x^{-2}y}{x^3y^{-1}}\right)^{-2}$

36. $\left(\dfrac{6xy^{-2}}{x^{-2}y}\right)^{-1}\left(\dfrac{3^{-1}x^{-2}y}{2x^5y^{-3}}\right)^{-2}$

Review Exercises   The answers to all of these exercises are at the back of the book.

37. Simplify: $\dfrac{1}{2} - \dfrac{3}{4}\left(\dfrac{2}{3}\right)^2$

38. Solve: $8x - 3(x - 1) = x + 3$

39. Graph: $x + 2y = 6$

40. Graph: $y \geq -x - 4$

41. Given $f(x) = 2x$

42. Solve: $\left\{\begin{array}{rcrcr} 2x & + & y & = & 7 \\ x & - & 3y & = & -14 \end{array}\right\}$ ❶ ❷

---

43. After a hail storm some of the 85 new cars in a sales lot have been damaged. Of the damaged cars some are still able to be sold and some are not. If x represents the cars which were not damaged, and the number of damaged cars which can still be sold is 15 less than the number not damaged, create a simplified expression to represent the number of damaged cars which cannot be sold.

44. A truck leaves a city at 9:00 am traveling 40 kph (kilometers per hour). How far has the truck traveled after 45 minutes?

45. A machine that converts paper money into coins returns both quarters and dimes. If you insert a $5 bill and get back 29 coins, how many quarters and how many dimes do you receive?

46. Find the slope and y–intercept of the line described by $x + 2y = 3$

47. Find the equation of the line that goes through $(-6,1)$ and $(0,1)$.$+ 7$, find $f(0)$.

---

48. Stalactites are found on the ceilings of caves. Although many factors affect their growth an estimate of 12.5 years for every one inch of growth would not be inappropriate.

   a) Create an equation which will output the number of years the stalactite has been growing given the length of the stalactite in inches.

   b) Complete the following data table:

   c) Plot the points on the following grid:

x stalactite length (inches)	y time growing (years)
3.5	
5	
18	
	125
	100

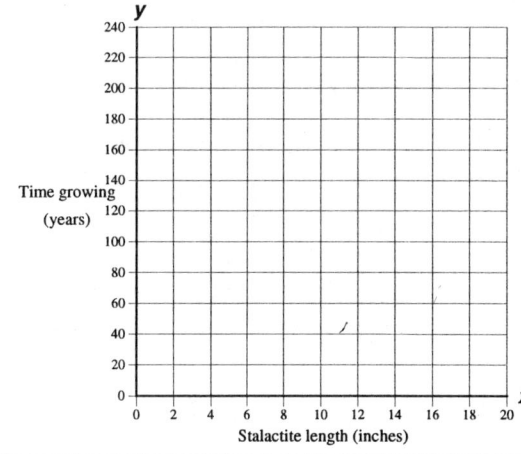

Section 6.3 Scientific Notation

Topic 6.3 A — Definition of Scientific Notation

When we use a scientific calculator to multiply or divide very large or small numbers, the display may switch from regular decimal notation to **scientific notation**. For example, if we wanted to calculate the distance light can travel in a year (a *light year*) we would multiply the speed of light (about 186,000 miles per second) by the number of seconds in a year (about 31,557,600 seconds). Because the product is too large to fit on the display, most calculators would display something like **5.8697136** $^{\times 10^{12}}$ or **5.8697136 12**. This number is written in scientific notation and is equivalent to 5,869,713,600,000 miles.

— *Definition* — *Scientific Notation*	English: A number written in scientific notation has two factors:
	• the first factor is a decimal written with one non–zero digit to the left of the decimal point.
	• the second factor is written as a power of 10.
	Examples:
	2.5×10^{13}
	5.39×10^{-9}
	8×10^{-1}
	Algebra: A number is in scientific notation when it's written in the form $D \times 10^n$
	where n is an integer and D is a decimal number such that
	$1 \le D < 10$ (that is, D is between 1 and 10, including 1 but not including 10).
	or
	$-10 < D \le -1$ (that is, D is between -10 and -1, including -1 but not including -10).

Example 6.3.1   Identify the numbers that are written in scientific notation.

a) 3×10^7

•SOLUTION• Yes.

b) -2.4×10^{-3}

•SOLUTION• Yes.

c) 12.5×10^{-8}

•SOLUTION• No. The decimal number 12.5 is larger than 10.

d) $3\frac{1}{2} \times 10^2$

•SOLUTION• No. Numbers written in scientific notation must be written using decimal notation.

e) 0.5×10^{-4}

•SOLUTION• No. The decimal number 0.5 is not between 1 and 10.

Practice 6.3.1   Identify the numbers that are written in scientific notation.

1. 9.9×10^5, -16.8×10^{-3}, $4\frac{2}{3} \times 10^3$, -4.7×10^9, 0.36×10^{-8}

2. -5.7×10^4, 7.08×10^{-6}, 25.9×10^4, $7\frac{1}{4} \times 10^5$, 0.78×10^2

In some instances, you may see numbers written using **exponent notation** rather than scientific notation. In exponent notation the upper case letter E is written instead of $\times$ 10. For example, some calculators or electronic spreadsheets may display the number 1,530,000,000,000 as 1.53 E12 rather than 1.53×10^{12}.

Topic 6.3 B — Converting from Decimal Notation to Scientific Notation

To convert between scientific and decimal notation we use the fact that moving the decimal point corresponds to multiplying or dividing by a power of 10:

- moving the decimal point one place to the *right* corresponds to *multiplying* by 10.

- moving the decimal point one place to the *left* corresponds to *dividing* by 10.

For example, the number 1234.5 could be written in many different ways depending on where we place the decimal point. All of the following represent the same number:

1234.5	$\Leftarrow$ this is regular decimal notation.
$123.45 \cdot 10^1$	$\Leftarrow$ this is *not* scientific notation because there are 3 non–zero digits to the left of the decimal point.
$12.345 \cdot 10^2$	$\Leftarrow$ this is *not* scientific notation because there are 2 non–zero digits to the left of the decimal point.
$1.2345 \cdot 10^3$	$\Leftarrow$ this *is* scientific notation because there is 1 non–zero digit to the left of the decimal point.
$0.12345 \cdot 10^4$	$\Leftarrow$ this is *not* scientific notation because there are no non–zero digits to the left of the decimal point.

In the above examples, each time we moved the decimal point to the left we had to multiply by 10 to keep the same value. Similarly, we could move the decimal point to the right and divide by 10 to keep the same value. Dividing by 10 is equivalent to multiplying by 10^{-1} because 10^{-1} means $\frac{1}{10}$. All of the following represent the same number:

0.035	$\Leftarrow$ this is regular decimal notation.
$0.35 \cdot 10^{-1}$	$\Leftarrow$ this is *not* scientific notation because there are no non–zero digits to the left of the decimal point.
$3.5 \cdot 10^{-2}$	$\Leftarrow$ this *is* scientific notation because there is 1 non–zero digit to the left of the decimal point.
$35 \cdot 10^{-3}$	$\Leftarrow$ this is *not* scientific notation because there are 2 non–zero digits to the left of the decimal point.

The conversion from decimal notation to scientific notation involves two steps.

— Procedure —	*Step 1*	Move the decimal point until there is one non–zero digit to its left.
Converting from decimal notation to scientific notation	*Step 2*	Multiply by 10 raised to a power whose magnitude is equal to the number of places the decimal was moved. The power is positive if the decimal point was moved to the left (that is, the magnitude of the original decimal was larger than 10). The power is negative if the decimal point was moved to the right.

Example 6.3.2   Convert to scientific notation.

a)   3,500,000,000

•SOLUTION•   Since the decimal point is not explicitly shown it must be on the far right of the number. Move the decimal point until there is one digit to its left.

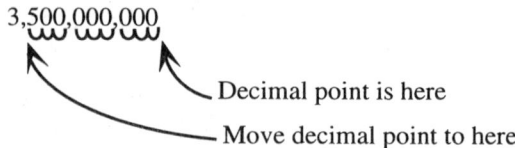

Since we moved the decimal point 9 places to the left the exponent is +9. Thus, 3,500,000,000 written in scientific notation is 3.5×10^9.

b)   −42,340

•SOLUTION•   We move the decimal point 4 places to the left so the exponent is +4. Thus, −42,340 written in scientific notation is -4.234×10^4.

c)   0.000005

•SOLUTION•   We move the decimal point 6 places to the right so the exponent is −6. Thus, 0.000005 written in scientific notation is 5.0×10^{-6}.

d) 32×10^4

•SOLUTION•   This is not scientific notation because there is more than one digit to the left of the decimal point. We move the decimal point one place to the left to get 3.2 and, since we reduced 32 by a factor of 10, we must increase the exponent of 10^4 by 1 to get 10^5. Thus, the answer is 3.2×10^5.

Practice 6.3.2   Convert to scientific notation.

1.   78,023
2. $-6,001 \times 10^3$
3.   0.000256
4. -0.0035×10^{-4}

Topic 6.3 C — Converting from Scientific Notation to Decimal Notation

To convert from scientific notation to decimal notation we reverse the process described above.

— Procedure — *Converting from scientific notation to decimal notation*	Move the decimal point the same number of places as the power of 10. If the power is positive, move the decimal to the right (that is, make the magnitude of the number bigger). If the power is negative, move the decimal to the left (that is, make the magnitude of the decimal smaller).

Example 6.3.3   Convert to decimal notation.

a) 2×10^6

•SOLUTION•

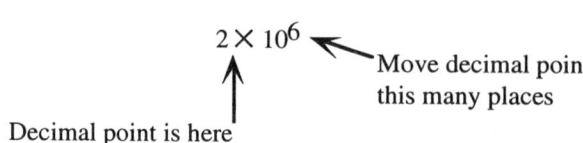

The exponent tells us that we must move the decimal point 6 places to the right. Thus, 2×10^6 can be written as 2000000 or 2,000,000.

b) -4.2 E3

•SOLUTION• This is written using exponent notation. In scientific notation it would be written as -4.2×10^3. To convert it to decimal notation we move the decimal point 3 places to the right. Thus, -4.2×10^3 can be written as -4200. Note that we had to use 0 as a place holder in the tens and units places.

c) 1.3×10^{-7}

•SOLUTION• The exponent is negative, so we must move the decimal point to the left 7 places. Thus, 1.3×10^{-7} can be written as 0.00000013

Practice 6.3.3 Convert to decimal notation.

1. 5×10^5 2. -1.37 E4 3. 5.321×10^{-3} 4. 2.0×10^{-5}

Topic 6.3 D — Using Scientific Notation in Calculations

We may do calculations that involve numbers written in scientific notation using the same methods we used for multiplying and dividing expressions that contain exponents.

Example 6.3.4 Perform the calculations and write the answers in scientific notation.

a) $(4 \times 10^{12})(2 \times 10^6)$

•SOLUTION•

$(4 \cdot 2) \times (10^{12} \cdot 10^6)$ $\Leftarrow$ Used Commutative and Associative properties to group decimal numbers and to group powers of 10. We usually do this step mentally.

$8 \times 10^{12 + 6}$ $\Leftarrow$ Multiplied $4 \cdot 2$ and used Multiplication Property of Exponents.

8×10^{18} $\Leftarrow$ Added exponents.

b) $(0.00035)(92,300,000,000,000)$

•SOLUTION• The first step is to convert the numbers to scientific notation.

$\left(3.5 \times 10^{-4}\right)\left(9.23 \times 10^{13}\right)$ $\Leftarrow$ Converted to scientific notation.

$(3.5 \cdot 9.23) \times (10^{-4} \cdot 10^{13})$ $\Leftarrow$ Used Commutative and Associative properties to group decimal numbers and to group powers of 10.

$32.305 \times 10^{-4 + 13}$ $\Leftarrow$ Multiplied $3.2 \cdot 9.23$ and used Multiplication Property of Exponents.

32.305×10^9 $\Leftarrow$ Added exponents.

3.2305×10^{10} $\Leftarrow$ Converted to scientific notation. Note that since we *decreased* the decimal part by a factor of 10 (we changed 32.305 to 3.2305), we had to *increase* the power of ten part by a factor of 10 (we changed 10^9 to 10^{10}).

c) $\dfrac{(2.3 \times 10^{-18})(9 \times 10^3)}{8.28 \times 10^{-3}}$

•SOLUTION•

$\dfrac{(2.3 \cdot 9) \times (10^{-18} \cdot 10^3)}{8.28 \times 10^{-3}}$ ⇐ In numerator, grouped decimal numbers and grouped powers of 10.

$\dfrac{(2.3 \cdot 9) \times (10^{-18 + 3})}{8.28 \times 10^{-3}}$ ⇐ Used Multiplication Property of Exponents.

$\dfrac{20.7 \times 10^{-15}}{8.28 \times 10^{-3}}$ ⇐ Multiplied 2.3 • 9 and added exponents.

$2.5 \times 10^{-15-(-3)}$ ⇐ Divided decimal numbers and used Division Property of Exponents.

2.5×10^{-12} ⇐ Subtracted exponents.

d) $1.2 \times 10^3 + 4.3 \times 10^2$

•SOLUTION• Just as we cannot add terms such as $1.2x^3 + 4.3x^2$, we cannot directly add $1.2 \times 10^3 + 4.3 \times 10^2$. Before we can add terms written in Scientific Notation we must convert them so that the exponents of 10 are the same. Usually, the easiest way to do this is to convert each number into regular decimal notation.

$1.2 \times 10^3 + 4.3 \times 10^2$ ⇐ Problem to simplify.

$1200 + 430$ ⇐ Converted to regular decimal notation.

1630 ⇐ Added.

— Note — Scientific notation on a calculator	We can enter scientific notation directly on a scientific calculator by using the $\boxed{\textbf{EXP}}$ key or the $\boxed{\textbf{EE}}$ key. For example, to calculate $(4 \times 10^{12})(2 \times 10^4)$ we would tap the following keys: $\boxed{4}\ \boxed{\textbf{EXP}}\ \boxed{1}\ \boxed{2}\ \boxed{\times}\ \boxed{2}\ \boxed{\textbf{EXP}}\ \boxed{4}\ \boxed{=}$ The display will show something like **8.** ^{16} or $8._{\times 10}^{\,16}$ or **8 E 16**.

Practice 6.3.4 Perform the calculations and write the answers in scientific notation.

1. $\left(3 \times 10^5\right)\left(2 \times 10^{-6}\right)$

2. $\left(2 \times 10^{12}\right)\left(4 \times 10^9\right)$

3. $(78{,}000)(1{,}100{,}000{,}000{,}000{,}000)$

4. $(0.000036)\,(25000)$

5. $\left(-5.2 \times 10^6\right)\left(1.4 \times 10^{-8}\right)$

6. $\dfrac{1.21 \times 10^{-8}}{-1.1 \times 10^3}$

7. $\dfrac{9.8 \times 10^4}{(5 \times 10^{-1})(2.5 \times 10^3)}$

8. $\dfrac{(6.3 \times 10^{12})(2.8 \times 10^{-30})}{5.6 \times 10^{-4}}$

9. $8.1 \times 10^{-1} + 2.914 \times 10^2$

Topic 6.3 E — Scientific Notation in Applications

Scientific notation is used in applications where large or small numbers are encountered.

Example 6.3.5 In 1998, the Federal debt (the money owed by the U. S. federal government to its creditors) was approximately \$5,410,000,000,000.

a) Assuming a rate of interest of 7.2% per year, how much interest must the government pay on this debt in 1998?

•**SOLUTION**•

- We must convert the debt into scientific notation because it's too large to fit on a calculator display. We write 5,410,000,000,000 in scientific notation as 5.41×10^{12}.

- We must convert 7.2% into decimal notation by dividing it by 100% to get 0.072. We could write this in scientific notation but this is not necessary since there are not many digits involved.

Now, we multiply the debt by the annual rate of interest as follows:

$$(5.41 \times 10^{12})(0.072) \quad \Leftarrow \text{ Expression to simplify.}$$
$$(5.41 \cdot 0.072) \times 10^{12} \quad \Leftarrow \text{ Used Commutative and Associate Properties.}$$
$$0.38952 \times 10^{12} \quad \Leftarrow \text{ Multiplied}$$

0.38952×10^{12} is the correct answer, but we should convert it into either scientific notation or regular decimal notation to make it easier to understand.

- To convert 0.38952×10^{12} to scientific notation,

  * we move the decimal point 1 place to the right (so there is 1 digit to the left of the decimal point);
  * we decrease the exponent by 1 (because we moved the decimal point 1 place to the right). That is 0.38952×10^{12} can be written using scientific notation as 3.8952×10^{11}.

- To convert 0.38952×10^{12} to regular decimal notation, we move the decimal point 12 places to the right to get 389,520,000,000. This is three hundred eighty–nine billion, five hundred twenty million dollars.

b) If the population of the United States is approximately 260,000,000 people, how much does each person owe as their share of the original debt?

•**SOLUTION**• We must divide the amount owed by the number of people who owe it.

$$\frac{5,410,000,000,000}{260,000,000} \quad \Leftarrow \text{ Expression to simplify.}$$

$$\frac{5.41 \times 10^{12}}{2.6 \times 10^{8}} \quad \Leftarrow \text{ Converted to scientific notation.}$$

$$2.1 \times 10^{12-8} \quad \Leftarrow \text{ Divided decimals and rounded to nearest tenth. Then, we subtracted the exponents.}$$

$$2.1 \times 10^{4} \quad \Leftarrow \text{ Simplified}$$

To convert this to regular decimal notation, we move the decimal point 4 places to the right to get \$21,000. That is, each person's share of the debt is approximately twenty–one thousand dollars!

Practice 6.3.5

1. The earth is approximately 67,250,000 miles from Venus. If a space ship is traveling at approximately 12,500 miles per hour, how long will it take to go from earth to Venus? Express your answer in scientific notation.

2. The earth weighs approximately 13,200,000,000,000,000,000,000 pounds. If a short ton is 2,000 pounds. How many tons does the earth weigh? Express your answer in scientific notation.

Exercise Set 6.3 The answers to the odd numbered exercises are at the back of the book.

Identify the numbers that are written in scientific notation.

1. $2 \times 10^{-3}, 0.2 \times 10^3, 23.2 \times 10^9, 8.01 \times 10^8$

2. $4.01 \times 10^{-1}, 0.65 \times 10^9, 3.82 \times 10^{12}, 12.01 \times 10^7$

Convert to scientific notation.

3. 6,700,000
4. −200,000,000
5. −32,510,000,000
6. −0.000258

7. 0.007003
8. 0.00000006
9. 0.52×10^{-3}
10. 65.2×10^2

11. 273×10^{-3}
12. 0.008×10^5

Convert to decimal notation.

13. 8.995×10^8
14. 7.554×10^4
15. -4.9×10^{-2}
16. -3.22×10^{-4}

17. 8 E15
18. 2.3 E−5

Perform the calculations and write the answers in scientific notation.

19. $(7.5 \times 10^{12})(2.1 \times 10^4)$
20. $(3.2 \times 10^{-6})(5.3 \times 10^3)$
21. $(3.6 \times 10^{-5})(4.5 \times 10^3)$

22. $(6.1 \times 10^{-7})(3.8 \times 10^{-5})$
23. $\dfrac{(-3.6 \times 10^3)(2.3 \times 10^5)}{(1.8 \times 10^{-3})}$
24. $\dfrac{(-4.5 \times 10^7)(7.2 \times 10^{12})}{(5.0 \times 10^{-5})}$

25. $\dfrac{(6,000,000,000,000)(0.000104)}{(520,000,000,000)(0.00002)}$
26. $\dfrac{(820,000)(12,000,000,000)}{300,000}$
27. $2.68 \times 10^{-3} + 5 \times 10^{-2}$

28. $8 \times 10^3 + 2.01 \times 10^5$
29. $5.1 \times 10^4 - 1.5 \times 10^2$
30. $6.152 \times 10^8 - 3 \times 10^6$

31. The mass of an electron is about 9×10^{-28} grams. Find the mass of 11,000 electrons.

32. A proton has a mass about 1,840 times that of an electron (see problem number 1). Find the mass of one proton.

33. One cubic foot of air weighs approximately 0.0778 lb. Find the total weight of air in a balloon with a volume of 32,550 cubic feet.

34. While still afloat, the Titanic displaced approximately 2×10^6 cubic feet of water. If water weighs approximately 62 pounds per cubic foot, find the weight of the water displaced by Titanic.

35. The width of a certain bacterial cell is 0.00005 cm. If 150,000 of these cells were laid in a row, how long would the row be?

36. If the space between the fibers of a pair of surgical gloves is 0.0025 cm, how many of the bacterial cells from the above problem could fit through the space at one time?

Review Exercises The answers to all of these exercises are at the back of the book.

37. Simplify: $2(8 - 3) + [23 - 2(5 + 3)]^2$

38. Evaluate the expression when x is −2 and y is −3: $3x^2 - (x - y)$

39. Simplify: $2(2 - x) - 3(x - 5) - (x - 3)$

40. Solve: $\dfrac{1}{4}x - 10 < 4\left(\dfrac{x}{2} + 1\right)$

41. In a certain area cheap homes tend to sell 3 months sooner than expensive homes. A Realtor sells on average 4.25 cheap homes per month and 1.5 expensive homes per month. After how many months would they have sold 276 homes?

42. How much of $1100 was invested at 8% and how much at 10% if the total amount of simple interest earned by both accounts after 1 year was $97.

43. Solve: $\begin{cases} y + 3 = 6x & \textbf{①} \\ -12x + 6 = -2y & \textbf{②} \end{cases}$

44. Simplify: $\dfrac{1}{2}m^3(2m^3 \cdot 3m)m$

45. Simplify: $\dfrac{1}{5^{-1}} + \dfrac{1}{6^{-1}} + \dfrac{1}{7^{-2}}$

46. Given the following data, find the time when the cost of buying and the cost of leasing are the same and find the cost of buying or leasing for that time.

x Buying time (months)	y Buying cost (dollars)	x Leasing time (months)	y Leasing cost (dollars)
0	1875	0	3000
1	2225	1	3275

47. Find the *x*-intercept and the *y*-intercept:

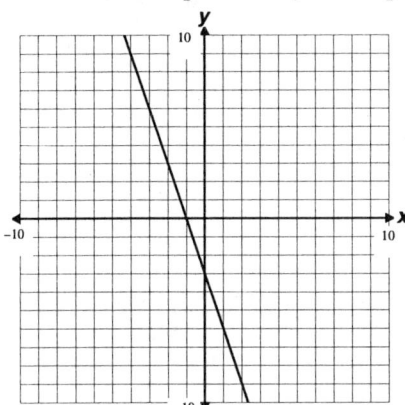

48. Determine whether this graph represents a function:

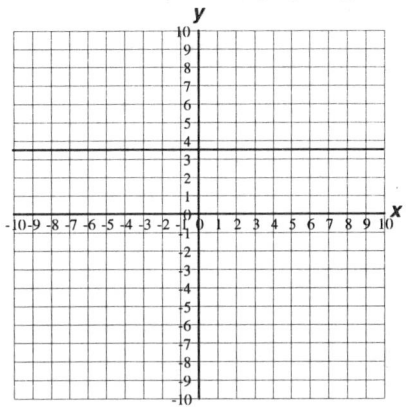

Section 6.4 Introduction to Polynomials

Consider the following data table, which shows the amount of paper that was recovered through recycling between 1970 and 1990.

x time (number of years since 1970)	y paper recovered (millions of tons)
0	13.00
1	12.91
2	12.92
3	13.03
4	13.24
5	13.55
6	13.96

We can plot these points on an *x–y* coordinate system in the usual way:

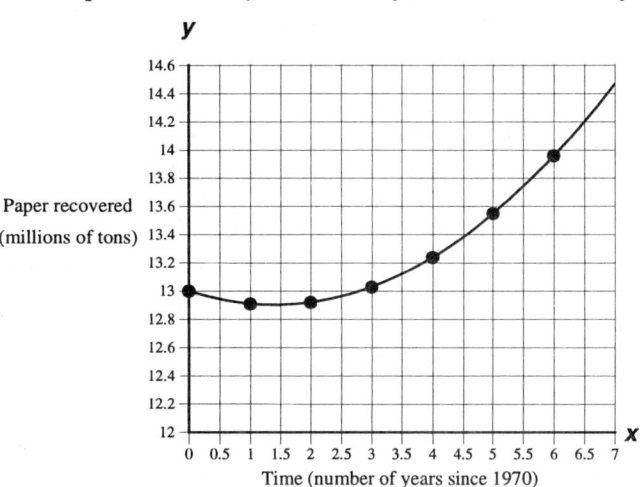

This graph clearly is not a straight line and so we cannot use the linear equation, $y = mx + b$, to model the relation between recovered paper and time. Using techniques you will learn in higher algebra, we can construct the following model for these data: $y = 0.05x^2 - 0.14x + 13$. The expression on the right of this equation is called a **polynomial**.

Topic 6.4 A — Polynomials

We have already discussed polynomials, even though we did not use that word to identify them. Expressions like $\frac{2}{3}x$, $5x^2 + 3x - 2$, and $x^3 - 8$ are considered polynomials.

— Definition — *Polynomial*	**English:** A polynomial is an expression that can be written as a sum of terms where only whole number exponents are allowed and there are no variables in the denominators of fractions. Examples: $3x$ $2x^3 - 5x^2 + x - 3$ **Algebra:** A polynomial in x is a **finite** [†] sum of terms, all of which have the form ax^n where $\quad\quad a$ is a real number, $\quad\quad x$ is a variable, $\quad\quad n$ is a whole number.

We use special words for expressions that have one, two, or three terms:

- a **monomial** has one term (*mono* means 1). For example, $2x^3$ is a monomial.

- a **binomial** has two terms (*bi* means 2). For example, $2x^2 + 3$ is a binomial.

- a **trinomial** has three terms (*tri* means 3). For example, $5x^2 + 2x + 6$ is a trinomial.

If a polynomial has four terms it's called a *four–term polynomial*; if it has five terms it's a *five–term polynomial*; and so on.

Not all expressions are polynomials. For example $4x^{-2} + 1$ does not have a whole number exponent so it's *not* a polynomial. The fraction $\frac{x+2}{x+1}$ has a variable in the denominator so it's *not* a polynomial.

Example 6.4.1 If the expression is a polynomial, state what kind it is. If it's not a polynomial, explain why.

a) $-9x^3$

  •SOLUTION• This is a single term so it is a *monomial*.

b) 5

  •SOLUTION• Even though 5 is written with no variable part we can *attach* a variable using the definition of an exponent of 0.

$$5 \quad \Leftarrow \text{ Expression to modify.}$$

$$5 \cdot 1 \quad \Leftarrow \text{ Used the fact that 1 is the Multiplicative Identity.}$$

$$5 \cdot x^0 \quad \Leftarrow \text{ Used the definition of an exponent of 0. This assumes, of course, that } x \neq 0.$$

Since $5x^0$ is a *monomial*, we can say 5 is a monomial.

[†] Finite (fi–nite) means that the sum does not go on forever (that is, it is not infinite).

c) $g - \dfrac{1}{2g}$

•SOLUTION• This is not a polynomial since there is a variable in the denominator.

d) $\dfrac{1}{2}x^4 - \dfrac{x}{4} + 7$

•SOLUTION• Even though the expression contains fractions, none of the denominators contain variables. This is a *trinomial*.

Practice 6.4.1 If the expression is a polynomial, state what kind it is. If it's not a polynomial explain why.

1. $5n^3$
2. $5n^4 - 3n^2 + 2n - 8$
3. $\dfrac{5}{x}$
4. $3y^{3/4} - 6y + 3$

The coefficient of a term is the number that is multiplied by the variable. For example, in the expression $0.05x^2 - 0.14x + 13$, the number 0.05 is the coefficient of x^2 and -0.14 is the coefficient of x. Since we can write 13 as $13(1)$ or $13x^0$, we may say that 13 is the coefficient of x^0. The 13 is also referred to as the **constant term**. Mathematically we think of $0.05x^2 - 0.14x + 13$ as $0.05x^2 + (-0.14)x^1 + 13x^0$.

In a polynomial like $-x^2 + x$, the coefficient of x^2 is -1 because we may write $-x^2$ as $-1 \cdot x^2$. Likewise, the coefficient of x is 1 because x is a shorthand way of writing $1 \cdot x$.

Example 6.4.2 For each polynomial, identify the terms. For each term, identify the coefficient.

a) $9x^3$

•SOLUTION•
- $9x^3$ is a term.
- 9 is the coefficient of x^3.

b) $2x^2 - 7x + 5$

•SOLUTION• To clearly see the coefficients, let's write this as $2x^2 + (-7x) + 5x^0$.
- $2x^2$, $-7x$, and 5 (or $5x^0$) are terms.
- 2 is the coefficient of x^2.
- -7 is the coefficient of x.
- We can consider 5 the coefficient of x^0; five is also called the constant term.

c) $\dfrac{3x}{4} + 1$

•SOLUTION• We will write this as $\dfrac{3}{4}x + 1x^0$

- $\dfrac{3}{4}x$ and 1 (or $1x^0$) are terms.
- $\dfrac{3}{4}$ is the coefficient of x.
- 1 is the coefficient of x^0; we can say that 1 is the constant term.

Practice 6.4.2 For each polynomial, identify the terms. For each term, identify the coefficient.

1. $7y^2$
2. $5x^2 - 3x + 7$
3. $x^2 - x$
4. $-x + 1$

The expression $0.05x^2 - 0.14x + 13$ is an example of a special class of polynomials called **polynomials in one variable** because it contains only one variable, in this case x. We would refer to this expression as a *polynomial in x*. Likewise, the expression $2y^3 + y$ is a polynomial in y.

A polynomial such as $2l + 2w$ is called a **polynomial in two variables** because it contains two variables, l and w. We would refer to such an expression as a *polynomial in l and w*.

Topic 6.4 B — Degree of Terms and Polynomials

Each term of a polynomial has a characteristic called its **degree.**

— *Definition* — *Degree of a Term*	English: The degree of a term is the number of variable factors in the term. Examples: The degree of $3x^4$ is 4. The degree of $-8xy^2$ is 3. Algebra: If $a \neq 0$ and m and n are whole numbers, then the degree of the term $ax^n y^m$ is $n + m$.

In the polynomial $0.05x^2 - 0.14x + 13$,

- $0.05x^2$ is the *second degree* term because x is squared.

- $-0.14x$ is the *first degree* term because x has an exponent of 1.

- 13 is the *zeroth degree* term because 13 is thought of as $13x^0$. 13 is also called the *constant term* since it has no variable explicitly written.

The **degree of a polynomial** is the degree of the term with the highest degree within the polynomial. For example, $0.05x^2 - 0.14x + 13$ is a second degree polynomial while $5x^7 + 3x^3$ is a seventh degree polynomial.

Example 6.4.3 Decide whether or not each expression is a polynomial. If it's a polynomial, state the degree of each term and the degree of the polynomial. If it's not a polynomial, state the reasons why.

a) $50x + 40$

•SOLUTION• This is a polynomial.

- The degree of $50x$ is 1.

- The degree is 40 is 0.

- Since the term with the highest degree is $50x$, this is a first degree polynomial in x. We could also call it a first degree binomial in x.

b) $-y^5 + 6y^2 - 2y$

•SOLUTION• This is a polynomial.

- The degree of $-y^5$ is 5.

- The degree is $6y^2$ is 2.

- The degree is $-2y$ is 1.

- Since the term with the highest degree is $-y^5$ this is a fifth degree polynomial in y.

c) $3a^7 + 2a^{-3} + 4$

•SOLUTION• This is not a polynomial because the second term does not have a whole number exponent.

d) $2xy^3 + x^2$

> •SOLUTION• This is a polynomial.
>
> - The degree of $2xy^3$ is 4 since x has an exponent of 1 and y^3 has an exponent of 3 and $1 + 3$ is 4.
>
> - The degree is x^2 is 2.
>
> - Since the term with the highest degree is $2xy^3$ this is a fourth degree polynomial in x and y.

Practice 6.4.3   Decide whether or not each expression is a polynomial. If it's a polynomial, state the degree of each term and the degree of the polynomial. If it's not a polynomial, state the reasons why.

1. $x - 17$

2. $-16t^2 + 45t + 25$

3. $7u^{3/2} + 5u^{1/2} - 32$

Topic 6.4 C — Standard Form of a Polynomial

We usually write polynomials in *descending order* of the powers from left to right. Thus, a polynomial with terms 6, $2x$, and $3x^2$ could be written as $2x + 3x^2 + 6$ or $6 + 3x^2 + 2x$, but the standard way of writing it is $3x^2 + 2x + 6$. Polynomials written in this way are said to be in **standard form**.

Example 6.4.4   If the expression is a polynomial, write it in standard form. If it's not a polynomial explain why.

a) $4y^5 - 3y^8 + 9 - y$

> •SOLUTION• Written in standard form, the polynomial is $-3y^8 + 4y^5 - y + 9$.

b) $\dfrac{x^2 - 2x + 1}{x + 5}$

> •SOLUTION• This is not a polynomial because of the variable in the denominator.

c) $\dfrac{3a}{2} + 6a^2 - \dfrac{1}{2}a^3$

> •SOLUTION• Written in standard form, the polynomial is $-\dfrac{1}{2}a^3 + 6a^2 + \dfrac{3}{2}a$. Note that we write the first degree term as $\dfrac{3}{2}a$ rather than $\dfrac{3a}{2}$ because each term of a polynomial in standard form should be in the form ax^n.

Practice 6.4.4   If the expression is a polynomial, write it in standard form. If it's not a polynomial explain why.

1. $12s^5 + 5s^7 - 7s^9 + 23$

2. $\dfrac{3d^2}{4} - \dfrac{d}{9} + \dfrac{5d^4}{9}$

3   $2.75a - 0.3a^2 + 1.9$

4. $-6.7h - 4.9h^2 + 2.6$

5. $\dfrac{x^2 - 1}{x + 2}$

6. $\dfrac{x^3 + x^2 - 8}{x^2 + 7}$

Exercise Set 6.4   The answers to the odd numbered exercises are at the back of the book.

If the expression is a polynomial, state what kind it is. If it's not a polynomial, explain why.

1. -25

2. $-\dfrac{3}{4} + x$

3. $\dfrac{1}{3}t^2 - \dfrac{5}{6}t + 4$

4. $\dfrac{-3}{5}n^2 - \dfrac{4}{9}n$

5. $4y^{-3}$

6. $-15y^{12}$

7. $x^2 + 2x - 1$

8. $\dfrac{4}{3n}$

For each expression, identify the terms. For each term, identify the coefficient.

9. $\frac{2}{3}n^3$

10. $-m^5 + 2m$

11. $3y^3 - 4y^2 + 2y - 4$

12. $\frac{-5t^9}{6}$

13. $\frac{2}{7}d^3 - \frac{3}{7}d + 2$

14. $1.36k^3 - 3.2k^2 + 1.1k - 4.9$

Decide whether or not each expression is a polynomial. If it's a polynomial, state the coefficient of each term and the degree of the polynomial. If it's not a polynomial, state the reasons why.

15. $-15y + 9$

16. $\frac{x-3}{x+2}$

17. $-x + 2$

18. $4m^5 - 3m^3 + 7m + 2$

19. $3m^5 - 7m^{-2} + 9m + 11$

20. $5x^3 - 3x^2 + 7x - 2$

21. $\frac{r+8}{r}$

22. $-9.8t^{-2} + 33t + 18$

If the expression is a polynomial, write it in standard form. If it's not a polynomial, explain why.

23. $5x - 3x^3 + 2x^2 - 7$

24. $2s^3 - 9s + 23 + 3s^5$

25. $\frac{5}{6}y^3 - \frac{2}{3}y^2 + \frac{5}{6}$

26. $\frac{x^2-1}{x+2}$

27. $2w^3 - \frac{4w^7}{3} + 15 + 8w^2$

28. $8w^3 - 6w^2 + 15w + 4$

29. $5m^{-3} + 2m - 12$

30. $\frac{4q}{5} - \frac{q^3}{4} + \frac{6}{7}q^2 + 18$

31. $-6.7h - 4.9h^2 + 2.6$

32. $-6n + 7n^{-3} + 11$

Review Exercises   The answers to all of these exercises are at the back of the book.

33. Simplify: $12 - 5(3 \cdot 2^2/6)$

34. Solve: $\frac{1}{2}(x-4) + \frac{2}{3}x = 5 - \left(\frac{x}{2} - 3\right)$

35. Solve: $0.1(x-3) - (x-0.2) < 1 - 0.9x$

36. Given $f(x) = 5x^2 - 2x + 1$, find $f(-1)$.

37. Solve: $\left\{ \begin{array}{rcrcr} -11x & + & y & = & 8 \\ 8x & - & y & = & -5 \end{array} \right\} \begin{array}{l} ❶ \\ ❷ \end{array}$

38. Simplify $\dfrac{\left(3y^{-1}a^2\right)^2}{\left(6y^{-5}a^{-2}\right)}$

39. Identify each base and its exponent: $3y^5$

40. Is this expression a polynomial? $7 - 4m^2 + 4m^3$

---

41. After paying 15% down on a new house a couple still needs a mortgage of $161,415. How much was the original purchase price for the home?

42. Calculate the slope of the straight line that passes through (8,2) and (−3,2).

43. Find the slope and y–intercept of the line described by $y = x$

44. Determine the domain and range of $\{(0,5),(2,4),(-1,4),(3,1)\}$

45. Use the graph to complete the data table:

x	$f(x)$
−3	
	7
	6
1	
	−9

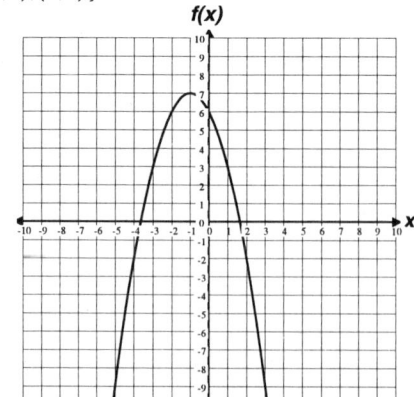

46. The number of paper towels, y, left in a box can be modeled with the equation $y = 200 - 3x$, where x is the number of days since the box was opened. Write an English explanation for this equation.

Section 6.5 Adding and Subtracting Polynomials

Topic 6.5 A — Adding Polynomials

In Section 2.2, we stated that *like terms* (that is, terms which contain identical variable parts) could be combined by adding their coefficients. Thus, $2x + 3x$ can be simplified to a single equivalent term, $5x$. Since polynomials are made up of terms, adding polynomials is accomplished by adding like terms. For example, let's add the trinomials $2x^2 - 5x + 7$ and $4x^2 + 2x - 3$; that is, let's simplify $(2x^2 - 5x + 7) + (4x^2 + 2x - 3)$:

- *Underline* the like terms that have the largest powers (x^2).

$$\underline{2x^2} - 5x + 7 + \underline{4x^2} + 2x - 3$$

- *Combine* the underlined terms into a single term and cross out the underlined terms to show you have used them.

$$2x^2 - 5x + 7 + 4x^2 + 2x - 3$$
$$6x^2$$

- *Underline* the like terms with the next largest powers (x)

$$2x^2 \underline{- 5x} + 7 + 4x^2 \underline{+ 2x} - 3$$

- *Combine* the underlined terms into a single term and cross out the underlined terms.

$$2x^2 - 5x + 7 + 4x^2 + 2x - 3$$
$$6x^2 - 3x$$

- *Underline* the like terms with the next largest powers (these are the constant terms.)

$$2x^2 - 5x \underline{+ 7} + 4x^2 + 2x \underline{- 3}$$

- *Combine* the underlined terms into a single term and cross out the underlined terms (all the terms should have been used).

$$2x^2 - 5x + 7 + 4x^2 + 2x - 3$$
$$6x^2 - 3x + 4$$

The process we illustrated above takes place in only two lines when we write out the solution long hand. It would look like this:

$$2x^2 - 5x + 7 + 4x^2 + 2x - 3$$
$$6x^2 - 3x + 4$$

Example 6.5.1   a)   Add $4x^2 - 3$ and $-x^2 + 7x - 2$

•**SOLUTION**• The first step is to translate the English into mathematics. We will enclose each polynomial in grouping symbols to make it clear what is being added.

$$\left[4x^2 - 3\right] + \left[-x^2 + 7x - 2\right] \quad \Leftarrow \text{ Translated the English into mathematical symbols.}$$

$$3x^2 + 7x - 5 \quad \Leftarrow \text{ Combined like terms; that is}$$

- $4x^2 + (-x^2)$ is $3x^2$
- $7x$ is the only first degree term so it remains unchanged
- $-3 + (-2)$ is -5

b)   Simplify $\left[8x^4 - 3x^2 + 5x - 3\right] + \left[-4x^2 - 3x + 1\right] + \left[-2x^4 + x^3 + x^2\right]$

•**SOLUTION**• We combine like terms to get $6x^4 + x^3 - 6x^2 + 2x - 2$; that is,

- $8x^4 + (-2x^4)$ is **$6x^4$**
- x^3 is the only third degree term so it remains unchanged
- $-3x^2 + (-4x^2) + 1x^2$ is **$-6x^2$**
- $5x + (-3x)$ is **$2x$**
- $-3 + 1$ is **-2**

c) Simplify $\left[3xy^2 + 5x^2y\right] + \left[8xy^2 - 7x^2y\right]$

•SOLUTION• Even though these expressions have two variables we still combine like terms in the usual way. The simplifies expression is $11xy^2 - 2x^2y$.

Practice 6.5.1

1. Add $7t^2 - 8t + 5$ to $-12t^2 + 9$

2. Add $4xw^2 - 12w + 15$ and $5xw^2 - 2$

3. Simplify $[5a^3 - 9a^2 + a - 32] + [-9a^3 + a^2 - 7]$

4. Simplify $(7u^3 - 4u^2 + 3u + 2) + (6u^2 - 12u + 5) + (4u^3 - 9)$

Topic 6.5 B — Adding Polynomials Vertically

Some people like to add polynomials vertically, as we might do with numbers in arithmetic. Using this method, we line up like terms and then add as usual.

Example 6.5.2   Use the vertical method to add $5x^2 - 3x - 5$ and $-7x^2 - 3$.

•SOLUTION• To add vertically, we line up like terms in a column and then add the coefficients. Note that $-7x^2 - 3$ does not have an x term so we will insert the term $0x$ in the middle as a place holder. This helps keep the like terms lined up.

$$
\begin{array}{r}
\ \ 5x^2 - 3x - 5 \\
+\ \ -7x^2 + 0x - 3 \\
\hline
-2x^2 - 3x - 8
\end{array}
$$

Practice 6.5.2   Add using the vertical method.

1. $3r^2 - 5r + 6$ and $-7r^2 + 12$

2. $-12w^3 + 7w^2 - 8w + 5$ and $4w^2 + 9w - 7$

3. $-15s^4 - 8s^3 + 6s^2 - 4$ and $24s^4 + 11s^2 - 15$

Topic 6.5 C — Subtracting Polynomials

Subtracting polynomials is a little more complicated than adding because subtraction causes changes in the signs of the terms being subtracted. Below is the general procedure we will follow for subtracting polynomials:

— Procedure — *Subtracting polynomials*	*Step 1 Insert grouping symbols* If necessary, write the polynomials inside grouping symbols. *Step 2 Remove grouping symbols* Use the Distributive Property to remove grouping symbols, being careful with negative signs. *Step 3 Add like terms* Add the coefficients of like terms and write the resulting expression in decreasing powers of the variable.

Example 6.5.3   a)   Simplify $[8x^2 - 2x + 5] - [x^2 - 6x - 1]$.

•SOLUTION•

Step 1 Insert grouping symbols   The polynomials are already inside brackets.

Step 2 Remove grouping symbols   The first set of grouping symbols are non–essential and may be removed without changing the inside. The second set is removed by changing the sign of the first term and then by changing each addition to subtraction and each subtraction to addition.

$$
y = 8x^2 - 2x + 5 - x^2 + 6x + 1
$$

No changes   Changes

Step 3 Add like terms   Add the coefficients of like terms to get $7x^2 + 4x + 6$.

b)   Subtract $3x^2 - 3x + 2$ from $x - 4$

> •**SOLUTION**• Since the trinomial is to be subtracted **from** the binomial, we must write the problem as (binomial) – (trinomial).
>
> ***Step 1 Insert grouping symbols***
>
> $$[x - 4] - \left[3x^2 - 3x + 2\right]$$
>
> ***Step 2 Remove grouping symbols***   The first set of grouping symbols are non–essential and may be removed without changing the inside. When we remove the second set we must make the changes as shown.
>
>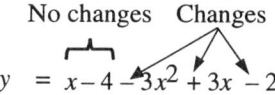
>
> ***Step 3 Add like terms***   Add the coefficients of like terms to get $-3x^2 + 4x - 6$.

Practice 6.5.3

1. Simplify $(4r^3 - 3r^2 + 9) - (6r^3 - 9r + 5)$

2. Subtract $4t^2 - 3t + 12$ from $t^2 - 12t + 7$

3. Subtract $-5s^2 + 7s - 3$ from $6s - 12$

4. Simplify $(15x^2 - 3x + 6) - (-3x^2 + 2x + 5) + (4x^2 - 8)$

5. Simplify $(8x^2y + 5xy^2) - (2x^2y - 7xy^2 + 1)$

Topic 6.5 D — Applications of Polynomials

If you own a car you have *variable costs* such as gas, oil, and tires and *fixed costs* such as depreciation[†], insurance, and registration. The following graphs show how the fixed costs have changed (due to inflation, market forces, etc.) over the period from 1977 to 1993.[†]

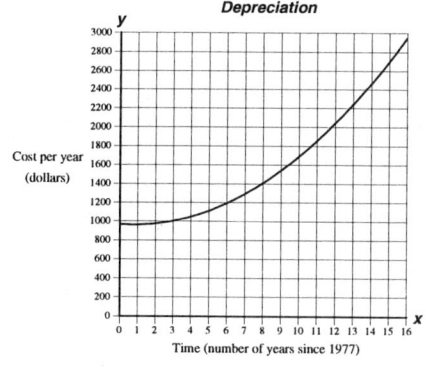

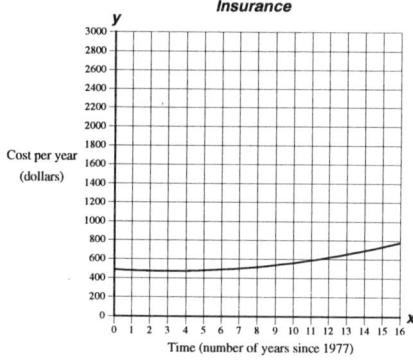

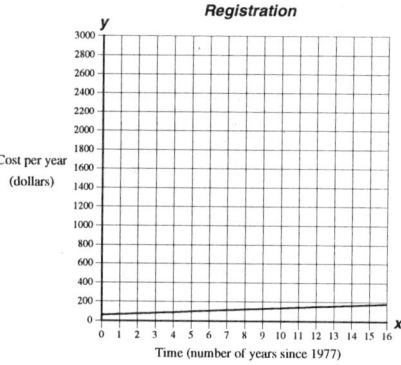

Given these data and a knowledge of statistical techniques (regression) we can build mathematical models that approximate each of these costs. In the following models, the input, x, represents the number of years since 1977, and the output, y, represents the yearly cost in dollars.

- cost of depreciation in dollars: $y = 8.6x^2 - 1.4x + 966$

- cost of insurance in dollars: $y = 1.8x^2 - 1.0x + 488$

- cost of license/registration in dollars: $y = 7.6x + 59$

[†]   Every year you own a car it's resale value is less due to wear and tear. Depreciation estimates this drop in value.

[†]   Source: <u>Statistical Abstract of the United States</u>, 1995, Table No. 1036, Cost of Owning and Operating an Automobile.

Example 6.5.4 a) Given the above three equations for depreciation, insurance, and registration, build a formula that will predict the total fixed cost of owning a car for any given year.

   •**SOLUTION**• We add the three costs to come up with a total fixed cost.

$$\begin{array}{ccccccc} \text{total fixed} \\ \text{costs} \end{array} = \begin{array}{c} \text{depreciation} \\ \text{costs} \end{array} + \begin{array}{c} \text{insurance} \\ \text{costs} \end{array} + \begin{array}{c} \text{registration} \\ \text{costs} \end{array}$$

$$y \qquad = 8.6x^2 - 1.4x + 966 + 1.8x^2 - 1.0x + 488 + 7.6x + 59$$

We can simplify the expression on the right by combining like terms to get $y = 10.4x^2 + 5.2x + 1513$. If we input the number of years since 1977, the output is the total fixed cost of owning a car for that year.

   b) Use $y = 10.4x^2 + 5.2x + 1513$ to calculate the fixed costs for owning a car in 1990.

   •**SOLUTION**• Since x represents the number of years since 1977, we will use 13 as the input value to represent the year 1990.

$$y = 10.4x^2 + 5.2x + 1513 \qquad \Leftarrow \text{Formula to evaluate.}$$

$$y = 10.4(\mathbf{13})^2 + 5.2(\mathbf{13}) + 1513 \quad \Leftarrow \text{Substituted 13 for } x.$$

$$y = 3338.2 \qquad\qquad\qquad \Leftarrow \text{Simplified.}$$

For insurance, depreciation, and registration the formula estimates the average car owner paid $3,338.20 in 1990. The actual average cost was $3,197.00.

As another example, consider landfill management. Although waste paper is both recycled and burned for energy, it still takes up more landfill volume than any other single waste item. Below an example which uses subtraction of polynomials to highlight the problem of landfill space.

Example 6.5.5 The top line of the following graph shows the pounds of waste generated per person in the U. S. each day between 1985 and 1995. The bottom line shows the pounds of waste recovered per person each day. [†]

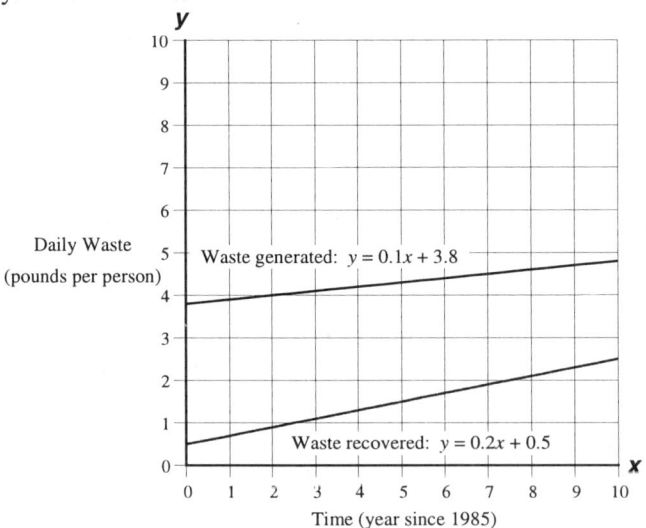

The polynomials that best model these data are:

  • waste generated (pounds per person): $0.1x + 3.8$
  • waste recovered (pounds per person): $0.2x + 0.5$

[†] Source: <u>Statistical Abstract of the United States</u>, 1995, Table No. 372, Municipal Solid Waste Generation, Recovery, and Disposal.

a) Construct a model that represents the daily pounds of waste that goes to the landfill for any given year.

•**SOLUTION**• The waste that goes to the landfill is

$$y = \text{(waste generated)} - \text{(waste recovered)}$$

Therefore, we must subtract the polynomials.

$y = [0.1x + 3.8] - [0.2x + 0.5]$ ⟸ Attached grouping symbols.

$y = 0.1x + 3.8 - \mathbf{0.2x} - \mathbf{0.5}$ ⟸ Removed grouping symbols.

$y = -0.1x + 3.3$ ⟸ Combined like terms.

The model is $y = -0.1x + 3.3$.

b) Calculate the daily pounds of waste that was landfilled per person in 1990.

•**SOLUTION**• Since 1985 is year 0, 1990 is year 5. Therefore, we evaluate $y = -0.1x + 3.3$ when x is 5:

$y = -0.1x + 3.3$ ⟸ Formula to evaluate.

$y = -0.1(\mathbf{5}) + 3.3$ ⟸ Substituted 5 for x.

$y = \mathbf{-0.5} + 3.3$ ⟸ Multiplied.

$y = 2.8$ ⟸ Added.

The model says that every day in 1990 each person in the United States landfilled 2.8 pounds of waste. The actual amount was 2.9 pounds, so the model seems to be a good one.

c) Predict the daily pounds of waste that will be landfilled per person in 2001.

•**SOLUTION**• 2001 is year 16 so we evaluate the polynomial when x is 16:

$y = -0.1x + 3.3$ ⟸ Formula to evaluate.

$y = -0.1(\mathbf{16}) + 3.3$ ⟸ Substituted 16 for x.

$y = \mathbf{-1.6} + 3.3$ ⟸ Multiplied.

$y = 1.7$ ⟸ Added.

The model says that in 2001 each day 1.7 pounds of waste will be landfilled per person in the U. S.

Practice 6.5.5

1. A company sells two types of stamping machines. The profit from one machine is given by $p = x^2 - 3x + 2$ while the profit for the other machine is given by $p = 5x - 3$. Construct a model for the total profit from both machines and then calculate the total profit when x is 12.

2. Acme Golf Cart Corp. has determined that the revenue made from their 'super–deluxe' model can be estimated by the model $y = 0.1x^2 + 2x$, where x represents the number of units sold. They have also determined that the cost of producing this model can be estimated by $y = 7x + 125$.

    a) Construct a model that represents the profit earned for a given number (x) of units sold.
    b) Estimate the profit on 50 units and explain what this means.
    c) Estimate the profit on 100 units and explain what this means.

Exercise Set 6.5   The answers to the odd numbered exercises are at the back of the book.

1. Add $4x^2 - 3x + 7$ to $12x^2 + 11$

2. Add $-5y^2 + 7y - 3$ and $9y^2 + 17$

3. Add $\frac{1}{3}t^2 - \frac{1}{4}t + 7$ and $-\frac{3}{4}t^2 + 9$

4. Add $\frac{4}{9}w^2 - \frac{5}{6}w + 13$ to $\frac{2}{3}w^2 - 15$

Simplify.

5. $[1.7a^3 - 6.3a^2 + 4.5a - 3.2] + [-6.5a^3 + 4.1a^2 - 3.6]$

6. $[-3.2r^3 + 4.1r^2 + 7.5r - 8.5] + [6.2r^3 + 3.8r - 9.1]$

7. $[-9r^3 + 4r^2 + 7r - 5] + [6r^3 + 3r - 9] + [r^3 - 5r^2]$

8. $(-12d^4 + 7d^3 - 9) + (7d^2 - 3) + (4d^4 - 5d^2 + 9)$

Add using the vertical method.

9. $-8r^2 - 6r + 12$ and $-12r^2 + 5$

10. $-14u^2 + 6u - 12$ and $-7u - 13$

11. $-9u^2 + 3u - 8$ and $6u^2 - 5u - 9$

12. $12x^3 - 3x^2 + x - 3$ and $x^3 - 5x - 4$

17. Subtract $-9d^2 - 5d + 11$ from $d^2 - 3d + 4$

18. Subtract $6a^2 + 4a - 7$ from $2a^2 + 4a - 10$

19. Subtract $7s^2 - 4s + 8$ from $6s^2 + 3s - 9$

20. Subtract $-9w^2 + 4w + 1$ from $5w^2 + 2w + 7$

Simplify.

21. $(7w^3 + 4w^2 - 11) - (-2w^3 + 6w^2 - 15)$

22. $(-12c^2 + 2c - 9) - (-5c + 3) + (2c^2 - 4c + 10)$

23. $(4v^2 - 3v + 2) - (5v^3 + 3v^2 + 5v - 8) + (3v^3 - 8v^2 + 3)$

24. $(12b^3 + 4b^2 - 8b + 9) - (6b^3 + 4b^2 - 3b + 2) + (b^2 + 9)$

25. $(4x^2 + 5x + 11) - (3x^2 - 2x + 5) - (2x + 4)$

26. $(3w^2 - 4w + 7) - (5w^2 - 9w - 12) - (5w - 8)$

27. $(1.3r^2 - 2.2r + 6.5) - (-3.1r^2 + 2.4r + 5.1) + (4.2r^2 - 8)$

28. $(-1.3k^2 + 2.2k - 9.2) - (-3.5k + 3.2) + (2.1k^2 - 4.1k + 1.2)$

29. The length of a certain rectangle is given by $l = 2x^2 + 5x + 3$ and its width is given by $w = x^2 + 3$. Construct a model for the perimeter and then calculate the perimeter when x is 4.

30. Southern Timber Company has determined that the revenue made from the sales of high–grade oak boards can be estimated by the following model: $y = 4.5x^2 - 4x$, where x represents the number of boards. They have also determined that the cost of producing the boards can be estimated by the following model $y = 65.5x - 30$.

    a) Construct a model that represents the profit made for a given number of boards sold. The profit is the revenue made minus the cost of producing the boards.
    b) Estimate the profit on 20 units and explain what this means.
    c) Estimate the profit on 15 units and explain what this means.
    d) Estimate the profit on 10 units and explain what this means.

31. From 1960 until 1990 the billions of dollars spent by the federal government for research and development can be modeled using the polynomial $0.21x^2 + (-2x) + 20.8$, where x is the number of years since 1960. During this same period the amount spent for defense and space can be modeled using the polynomial $-0.0005x^4 + 0.031x^3 - 0.554x^2 + 3.51x + 4.83$.

    a) Use the model to predict the amount spent overall for research and development in 1980.
    b) Use the model to predict the amount spent for defense and space in 1980.
    c) Use your answers for parts a) and b) to predict the amount of money not spent in 1980 for defense and space.
    d) Build a model that will predict the amount not spent on defense and space given the number of years since 1960. Then use your model to predict the amount not spent in 1980.

32. The three polynomials below model domestic motor fuel consumption in billions of gallons from 1970 to 1993 (x is the number of years since 1970).

    Total from all sources: $1.81x + 94.95$
    Automobile consumption: $0.0067x^3 - 0.222x^2 + 1.92x + 68.71$
    Truck consumption: $1.7x + 24$

    a) Use the model to predict the total amount of motor fuel consumed in 1985.
    b) Use the model to predict the total amount of automobile fuel used in 1985.
    c) Use the model to predict the total amount of truck fuel used in 1985.
    d) Create a model to predict the total amount of fuel used by both cars and trucks given the number of years since 1970. Then, use the new model to find the total amount used by both cars and trucks in 1985.
    e) If all domestic motor fuel is considered allocated to automobiles, trucks, and buses create a model for the total fuel used by buses. Then, use your model to predict the fuel used by buses in 1985.

Review Exercises   The answers to all of these exercises are at the back of the book.

33. Find the equation of a line perpendicular to $y = x - 1$ and which passes through (6,0).

34. Simplify: $\dfrac{60}{34} \div \dfrac{16}{51}$

35. Identify the sums, terms, products, and factors: $5x + 8y$

36. Prime factor: 225

37. Solve $y = 3x + 8$ for x

38. Determine whether this relation is a function: $y = 3x$

39. Simplify: $t + 3[t - 4(t + 6) + 5]$

40. Solve: $\left\{ \begin{array}{rcrcl} 8x & - & 2y & = & 14 \\ 5x & + & 3y & = & 13 \end{array} \right\} \begin{array}{l} \mathbf{❶} \\ \mathbf{❷} \end{array}$

41. Solve: $\left\{ \begin{array}{rcrcl} 5x & + & 2y & = & 1 \\ 2x & + & y & = & 1 \end{array} \right\} \begin{array}{l} \mathbf{❶} \\ \mathbf{❷} \end{array}$

42. The length of a rectangle is two more than its width. If the width is doubled while the length is increased by 10 the perimeter increases by 26 inches. Find the width of the original rectangle.

43. A theater sold adult tickets for $8.50 per ticket and 204 senior tickets for $2.25 less per ticket than the adult tickets. Find the total amount of money the theater made if they sold one–third as many senior tickets as adult tickets.

44. A biologist has two food mixes on hand. Mix A contains 25% protein and 5% fat while Mix B contains 60% protein and 20% fat. How much of each mix should the biologist combine to obtain a mixture that contains 410 grams of protein and 110 grams of fat?

45. The following data show the number of U. S. citizens visiting Mexico between 1984 and 1994. Graph the data using year since 1984 on the x axis and millions of travelers on the y–axis. The data re not exactly linear so you will have to draw a straight line that you feel best approximates the relationship.

x year	y number of travelers
1984	10,992,000
1988	13,463,000
1991	15,042,000
1994	15,759,000

Section 6.6 Multiplying Polynomials

We found in the last section that the daily amount of waste landfilled per person in the U. S can be modeled by $(-0.1x + 3.3)$, where x is the number of years since 1985. As it turns out, the projected population of the United States (in millions) can be modeled by the polynomial $2.4x + 259$. If we want to find the waste landfilled by everyone in the U. S. we would use the Amount Formula:

amount $=$ rate • base

amount $=$ (daily waste per person) • (number of people)

amount $= (-0.1x + 3.3) • (2.4x + 259)$

To simplify this expression we must multiply the polynomials. That is what we will discuss in this section, beginning with the products of monomials.

Topic 6.6 A — Multiplying Polynomials

As we discussed in Section 6.1, we can multiply monomials by applying the rules of exponents in combination with the Associative and Commutative Properties. The procedure is to multiply the numeric coefficients and then add the exponents of factors that have the same base. For example, $(2w^4)(5w^2)$ simplifies to $(2 • 5)(w^{4+2})$ or $10w^6$.

Multiplying a monomial and a polynomial that has more than one term involves the use of the Distributive Property.

— Procedure — *Multiplying a monomial and a polynomial*	Multiply the monomial by each term of the polynomial and then simplify if possible.

Example 6.6.1   a)   Simplify: $2x^2(3x - 5)$

•SOLUTION• Multiply each term inside the parentheses by $2x^2$ to get $6x^3 - 10x^2$

b)   Simplify: $3y(4y^2 - 5y + 2)$

•SOLUTION• Multiply each term inside the parentheses by $3y$ to get $12y^3 - 15y^2 + 6y$

c)   Simplify: $(a^2 + 2b - ab)(2b)$

•SOLUTION• Because multiplication is commutative , the Distributive Property works whether the monomial is on the left or the right. We multiply each term inside the parentheses by $2b$ to get $2a^2b + 4b^2 - 2ab^2$

Practice 6.6.1   Simplify.

1. $4t(3t - 9)$ 2. $-9m(6m - 5)$ 3. $3m^2(4m^2 - 2m + 9)$

4. $-5b(4b^2 - 7b + 11)$ 5. $(4x^2 - 3x + 5)(3x)$ 6. $(9v^2 - 5v - 10)(-2v)$

If each polynomial has more than one term we can use the Distributive Property several times to find the product. For example, suppose we wish to multiply $(3x + 2)(5x^2 + 2x + 4)$

- Treat the binomial, $3x + 2$, as a single quantity and distribute it to each term of the trinomial.

$$(3x + 2)(5x^2 + 2x + 4)$$
$$(3x + 2)(5x^2) + (3x + 2)(2x) + (3x + 2)(4)$$

- Use the Distributive Property again to distribute the monomials to each term of each binomial.

$$(3x + 2)(5x^2) + (3x + 2)(2x) + (3x + 2)(4)$$
$$(3x)(5x^2) + (2)(5x^2) + (3x)(2x) + (2)(2x) + (3x)(4) + (2)(4)$$

- Multiply the monomials.

$$15x^3 + 10x^2 + 6x^2 + 4x + 12x + 8$$

- Collect like terms.

$$15x^3 + 16x^2 + 16x + 8$$

If you carefully study the above example, you will notice that each term of the binomial was multiplied by each term in the trinomial. Because this will always occur when multiplying two polynomials, we can use the following shortcut procedure:

— Procedure — *Multiplying polynomials*	*Step 1* Multiply each term of the first polynomial by each term of the second polynomial.
	Step 2 Simplify.

Example 6.6.2   a)   Simplify: $(x + 2)(5x - 4)$

•SOLUTION•

$(x)(5x) + (x)(-4) + (2)(5x) + (2)(-4)$ ⇐ Multiplied each term in the first binomial by each term in the second binomial.

$5x^2 + (-4x) + 10x + (-8)$ ⇐ Multiplied.

$5x^2 + 6x - 8$ ⇐ Combined like terms.

b) Simplify. $(2x - y)(4x^2 + 2xy + y^2)$

•SOLUTION•

$(2x)(4x^2) + (2x)(2xy) + (2x)(y^2) + (-y)(4x^2) + (-y)(2xy) + (-y)(y^2)$ ⟸ Multiplied each term in the binomial by each term in the trinomial.

$8x^3 + 4x^2y + 2xy^2 + (-4x^2y) + (-2xy^2) + (-y^3)$ ⟸ Multiplied.

$8x^3 - y^3$ ⟸ Combined like terms.

As you get better at multiplying polynomials (through a LOT of practice), you will be able to do many of the steps mentally.

Practice 6.6.2 Simplify.

1. $(m - 2)(4m - 3)$ 2. $(7d + 3)(3d - 5)$ 3. $(4m - n)(3m^2 - 3mn + 4n^2)$ 4. $(3r - s)(5r^2 + 6rs - 3s^2)$

Keep in mind that when we multiply polynomial expressions we are rewriting the expressions as equivalent expressions. In every case, the original expression represents the same value as the final expression no matter what number we substitute for x. For example, we just found that $(x + 2)(5x - 4)$ and $5x^2 + 6x - 8$ are equivalent expressions. If we substitute any value for x, say 3, both expressions will simplify to the same value:

Value of **Original Expression**	**Value of** **Equivalent Expression**
$(x + 2)(5x - 4)$	$5x^2 + 6x - 8$
$((3) + 2)(5(3) - 4)$	$5(3)^2 + 6(3) - 8$
	$5(9) + 6(3) - 8$
$(3 + 2)(15 - 4)$	$45 + 18 - 8$
$(5)(11)$	$63 - 8$
55	55

We may square or cube polynomials using the same procedure.

Example 6.6.3   a)   Simplify $(3x + 5)^2$

•SOLUTION• The base of the exponent is the binomial $3x + 5$; the base is NOT $3x$ nor is it 5. Therefore, we cannot simply square each term. Instead, we use the definition of exponents to rewrite $(3x + 5)^2$ as the product of two binomials. Then, we multiply each term in the first binomial by each term in the second.

$(3x + 5)^2$ ⟸ Expression to simplify.

$(3x + 5)(3x + 5)$ ⟸ Rewrote "square" as a product.

$(3x)(3x) + (3x)(5) + (5)(3x) + (5)(5)$ ⟸ Multiplied each term in the first binomial by each term in the second.

$9x^2 + 15x + 15x + 25$ ⟸ Multiplied.

$9x^2 + 30x + 25$ ⟸ Combined like terms.

b)   Simplify $(2x + 5)^3$

•SOLUTION• As was the case with squaring a binomial, we may NOT simply cube $2x$ and cube 5 because exponents may not be distributed across a sum or difference. We have to use the definition of exponents to explicitly write the factors of this expression and multiply the factors in the usual way.

$(2x + 5)^3$ ⟸ Expression to simplify.

$(2x + 5)(2x + 5)(2x + 5)$ ⟸ Rewrote "cube" as a product.

$\Big((2x)(2x) + (2x)(5) + (5)(2x) + (5)(5)\Big)(2x + 5)$ ⟸ Multiplied each term in the first binomial by each term in the second binomial. Note that we have not yet involved the third binomial. We leave that until we have completed multiplying and simplifying the first two binomials.

$(4x^2 + 10x + 10x + 25)(2x + 5)$ ⟸ Multiplied.

$(4x^2 + 20x + 25)(2x + 5)$ ⟸ Combined like terms.

$(4x^2)(2x) + (4x^2)(5) + (20x)(2x) + (20x)(5) + (25)(2x) + (25)(5)$ ⟸ Multiplied each term in the trinomial by each term in the binomial.

$8x^3 + 20x^2 + 40x^2 + 100x + 50x + 125$ ⟸ Multiplied.

$8x^3 + 60x^2 + 150x + 125$ ⟸ Combined like terms.

Practice 6.6.3 Simplify.

1. $(4m - 1)^2$ 2. $(5r + s)^2$ 3. $(6y^2 - 3w)^2$ 4. $(4x + 1)^3$ 5. $(2a + 5b)^3$

— Caution — **Squaring a polynomial**	When a polynomial is raised to a power we **cannot** simply raise each individual term to the power. This is because exponents may not be distributed across a sum or difference. For example, $(3x + 5)^2$ is **not** equal to $(3x)^2 + (5)^2$ because the base of the exponent is the binomial $3x + 5$; the base is NOT $3x$ nor is it 5.

Now that we have practiced multiplying polynomials we can work the landfill problem which opened this section.

Example 6.6.4   If x represents the number of years since 1985, then $-0.1x + 3.3$ approximates the pounds of daily waste landfilled per person in the U. S. and $2.4x + 259$ approximates the population (in millions).

a) Construct a mathematical model that outputs the total pounds of waste needed to be landfilled each day in the U. S. when the number of years since 1985 is input.

   •SOLUTION• We use the Amount Formula to construct the model.

waste needed to be landfilled = (daily waste per person)(millions of people in the U. S.)

$$y \qquad = \qquad (-0.1x + 3.3) \qquad (2.4x + 259)$$

Now, we multiply the binomials and simplify.

$y = (-0.1x + 3.3)(2.4x + 259)$ ⟸ Formula to simplify.

$y = (-0.1x)(2.4x) + (-0.1x)(259) + (3.3)(2.4x) + (3.3)(259)$ ⟸ Multiplied each term in the first binomial by each term in the second.

$y = -0.24x^2 + (-25.9x) + 7.92x + 854.7$ ⟸ Multiplied.

$y = -0.24x^2 - 17.98x + 854.7$ ⟸ Combined like terms.

b) Predict the amount of waste that will be landfilled in one day in 2001.

   •SOLUTION• Since x is the number of years since 1985, we will substitute 16 for x.

$y = -0.24x^2 - 17.98x + 854.7$ ⟸ Formula to evaluate.

$y = -0.24(16)^2 - 17.98(16) + 854.7$ ⟸ Substituted 16 for x.

$y = -61.44 - 287.68 + 854.7$ ⟸ Simplified.

$y = 505.58$ ⟸ Simplified.

Every day in 2001 approximately 506 million pounds of waste will need to be landfilled, somewhere.

Practice 6.6.4   Answer the following questions given that x represents the number of years since 1980 and $525x - 25$ approximates the number of automobile accidents involving alcohol per million adults over the age of 21 and $1.5x + 158.7$ approximates the population (in millions) over 21.

1. Construct a model that outputs the total number of alcohol related accidents in a given year when the number of years since 1980 is input.

2. Estimate the number of alcohol related accidents that occurred in 1990. Round to the nearest whole number.

Topic 6.6 B — Multiplying Two Binomials Using FOIL

You probably know several shortcuts for doing arithmetic calculations on numbers or problems that have special forms. For example, to multiply a whole number by 10 we can simply attach a zero onto the end of the number. Thus, 57×10 is 570. "Attaching a zero" is not a mathematical operation, but it works under certain conditions and it saves time. Likewise we can quickly find some polynomial products using shortcuts, such as FOIL.

We use the acronym **FOIL** to remind us of the procedure for multiplying two binomials. The letters in FOIL stand for First, Outer, Inner, Last and represent the products of the terms of the two binomials being multiplied. For example, in the product $(5x + 2)(4x + 3)$

F stands for the product of the First two terms $(5x)(4x)$

O stands for the product of the Outer two terms $(5x)(3)$

I stands for the product of the Inner two terms $(2)(4x)$

L stands for the product of the Last two terms $(2)(3)$

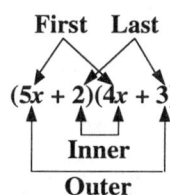

Adding the terms that result from FOIL produces the final result.

Example 6.6.5   Use FOIL to find the product $(5x + 2)(4x + 3)$.

•SOLUTION•

$$(5x)(4x) + (5x)(3) + (2)(4x) + (2)(3) \quad \Leftarrow \text{ Multiplied First, Outer, Inner, and Last terms.}$$

$$20x^2 + 15x + 8x + 6 \quad \Leftarrow \text{ Multiplied.}$$

$$20x^2 + 23x + 6 \quad \Leftarrow \text{ Combined like terms.}$$

Practice 6.6.5   Use FOIL to find the products.

1. $(2x + 1)(3x + 8)$ 2. $(5m + 8)(3m - 2)$ 3. $(2x - y)(3x - y)$ 4. $(3w - y)(4w + y)$

Keep in mind that FOIL is equivalent to the general procedure for multiplying polynomials (multiply each term in one polynomial by each term in the other polynomial). FOIL is easy to remember and use, but it only works when multiplying a pair of binomials.

Topic 6.6 C — Squaring a Binomial

We can use FOIL to square a binomial. For example,

$$(a + b)^2 \quad \Leftarrow \text{ Expression to simplify.}$$

$$(a + b)(a + b) \quad \Leftarrow \text{ Used the definition of squaring.}$$

$$(a)(a) + (a)(b) + (b)(a) + (b)(b) \quad \Leftarrow \text{ Used FOIL.}$$

$$a^2 + ab + ab + b^2 \quad \Leftarrow \text{ Multiplied.}$$

$$a^2 + 2ab + b^2 \quad \Leftarrow \text{ Combined like terms.}$$

If you look carefully at the final expression you might be able to see a pattern:

- the first term, a^2, is the square of the first term of the given binomial.

- the middle term, $2ab$, is twice the product of the terms a and b of the binomial.

- the last term, b^2, is the square of the last term of the binomial.

This gives us a quick and easy way of squaring a binomial.

— Procedure — Squaring a binomial	**Step 1** Square the first term. **Step 2** Add twice the product of the terms. **Step 3** Add the square of the second term. Algebraically, we can write this as $(a + b)^2 = a^2 + 2ab + b^2$

Example 6.6.6   Use the shortcut to square the binomials.

a) $(c - 5)^2$

•SOLUTION•

$$(c)^2 + 2(c)(-5) + (-5)^2 \quad \Leftarrow \text{ Squared first, added twice product, added square of second.}$$
$$ \text{Remember, } c - 5 \text{ means } c + (-5).$$

$$c^2 - 10c + 25 \quad \Leftarrow \text{ Simplified.}$$

b) $(2w - 3z)^2$

•SOLUTION•

$$(2w)^2 + 2(2w)(-3z) + (-3z)^2 \quad \Leftarrow \text{ Squared first, added twice product, added square of second.}$$

$$4w^2 - 12wz + 9z^2 \quad \Leftarrow \text{ Simplified.}$$

Practice 6.6.6   Use the shortcut to square the binomials.

1. $(w + 4)^2$ 2. $(m - 5)^2$ 3. $(w - 9)^2$ 4. $(2r + 11)^2$

Topic 6.6 D — The Product of Conjugates

The expressions $(a + b)$ and $(a - b)$ are called **conjugates** because they are the sum and difference of the same two expressions. When we multiply conjugates, the middle terms of the product (the *OI* in *FOIL*) are opposites and therefore add to zero.

$$(a + b)(a - b) \quad \Leftarrow \text{ Expression to simplify.}$$

$$(a)(a) + (a)(-b) + (b)(a) + (b)(-b) \quad \Leftarrow \text{ Used FOIL.}$$

$$a^2 + (-ab) + ab + (-b^2) \quad \Leftarrow \text{ Simplified.}$$

$$a^2 - b^2 \quad \Leftarrow \text{ Combined like terms.}$$

The product of conjugates is always the difference of the squares of the terms. This gives us a quick and easy way of multiplying conjugates.

— Procedure —	**Step 1**	Square the first term of one of the binomials.
Multiplying conjugates	**Step 2**	Square the second term of one of the binomials.
	Step 3	Subtract the squares.
	Algebraically, we can write this as $(a + b)(a - b) = a^2 - b^2$	

Example 6.6.7   Use the shortcut for multiplying conjugates to find the product $(3x + 2y)(3x - 2y)$.

•**SOLUTION**•

$$(3x)^2 - (2y)^2 \quad \Leftarrow \text{ Wrote the product as the difference of two squares.}$$

$$9x^2 - 4y^2 \quad \Leftarrow \text{ Simplified.}$$

Practice 6.6.7   Use the shortcut to find the products of the conjugates.

1. $(x + 7)(x - 7)$ 2. $(r + 6m)(r - 6m)$ 3. $(4x - 5y)(4x + 5y)$ 4. $(7t - 9s)(7t + 9s)$

Topic 6.6 E — Multiplying Binomials of the Form $(x + m)(x + n)$

A product of binomials that frequently occurs has the form $(x + m)(x + n)$, where x is a variable and m and n are constants. For example, each of the following products has this form:

$(x + 5)(x + 2) \quad \Leftarrow \text{ In this case, } m \text{ is 5 and } n \text{ is 2.}$

$(x + 3)(x - 8) \quad \Leftarrow \text{ We can write this as } (x + 3)(x + (-8)) \text{ so } m \text{ is 3 and } n \text{ is } -8.$

$(9 + x)(x + 7) \quad \Leftarrow \text{ We can write this as } (x + 9)(x + 7) \text{ so } m \text{ is 9 and } n \text{ is 7.}$

We can use FOIL to find the general product.

$$(x + m)(x + n) \quad \Leftarrow \text{ Expression to simplify.}$$

$$(x)(x) + (x)(n) + (m)(x) + (m)(n) \quad \Leftarrow \text{ Used FOIL.}$$

$$x^2 + nx + mx + mn \quad \Leftarrow \text{ Multiplied.}$$

The middle two terms, nx and mx, are not like terms because we do not know the values of m and n. But, we can write the sum $nx + mx$ as a single term if we factor out the x and use parentheses. If we do this, we can write

$$(x + m)(x + n) = x^2 + (m + n)x + mn$$

Carefully examine the structure of the expression on the right:

- it's a trinomial with decreasing powers of x. That is, we have a term with x^2, then a term with x (which is x^1), then a constant term (which is the x^0 term).

- the first term, x^2, has a coefficient of 1.

- the second term, $(m + n)x$, has a coefficient that is the sum of the last terms of the binomials, $m + n$.

- the third term, mn, is the product of the last terms of the binomials.

This gives us a quick way of multiplying binomials that have the form $(x + m)(x + n)$.

— *Procedure* — ***Multiplying binomials of*** ***the form $(x + m)(x + n)$***	When multiplying $(x + m)(x + n)$, the coefficients of the resulting three terms will be • 1 (for the x^2 term) • the sum $m + n$ (for the x term) • the product mn (for the last term). Algebraically we can write this as $(x + m)(x + n) = x^2 + (m + n)x + mn$

Example 6.6.8   Use the shortcut to find the products.

a) $(x + 12)(x - 2)$

    •SOLUTION•

$$x^2 + (12 + (-2))x + (12)(-2) \quad \Leftarrow \text{Used formula using 12 for } m \text{ and 2 for } n.$$

$$x^2 + 10x - 24 \quad \Leftarrow \text{Simplified.}$$

b) $(x + 5)(x - 5)$

    •SOLUTION•

$$x^2 + \left(5 + (-5)\right)x + (5)(-5) \quad \Leftarrow \text{Used formula using 5 for } m \text{ and } -5 \text{ for } n.$$

$$x^2 + (0x) + (-25) \quad \Leftarrow \text{Simplified.}$$

$$x^2 - 25 \quad \Leftarrow \text{Simplified.}$$

Notice the factors were conjugates so we could have used the difference of two squares shortcut instead.

c) $(x + 7)(3x + 2)$

    •SOLUTION• These binomials do not fit the form $(x + m)(x + n)$ because the second x term has a coefficient that is not 1. Therefore, the shortcut will not work. We can still use FOIL on this to get $3x^2 + 23x + 14$.

Practice 6.6.8   Use the shortcut to find the products.

1. $(m + 2)(m + 5)$ 2. $(r - 8)(r - 4)$ 3. $(x + 7)(x - 12)$ 4. $(y - 2)(2y + 1)$

Remember, you can always multiply binomials using the general procedure for polynomials (multiply each term of the first polynomial by each term in the second polynomial) or by using FOIL. However using the shortcuts given in this section will save you a lot of time.

Exercise Set 6.6   The answers to the odd numbered exercises are at the back of the book.

Multiply and simplify.

1. $5t^3(4t - 12)$ 2. $2w^3(4w^5 - 5)$ 3. $-\dfrac{1}{4}m^4(12m^2 - 8)$

4. $-\dfrac{1}{3}p^5(9p^2 - 12)$ 5. $3y^3(4y^4 + 2y^2 - 1)$ 6. $4n^6(5n^4 - 8n^2 - 3)$

7. $6r^2(3r^2 - 5r + 7)$ 8. $-2v(3v^2 - 5v + 12)$ 9. $(2x^2 - 4x + 1)(5x^2)$

10. $(3t^2 - t + 4)(2t^2)$ 11. $(3w^2 - 8w + 7)(4w)$ 12. $(2y^2 - y + 3)(3y)$

13. $(4x + 7)(2x - 8)$ 14. $(5m - 9)(3m + 2)$ 15. $(2x - 8)(3x - 12)$

16. $(3m - 1)(6m + 4)$ 17. $(2x + 1)(4x^2 - 2x + 1)$ 18. $(3y - 2)(9y^2 + 6y + 4)$

19. $(4m + 3n)(5m^2 - 2mn - 6n^2)$ 20. $(5r - 2s)(2r^2 + 5rs - 7s^2)$ 21. $(8x^2 + x - 3)(x^2 + 2x - 6)$

22. $(3y^2 - 2y + 1)(y^2 - y - 2)$

Expand and simplify.

23. $(3x + 7)^2$ 24. $(5y - 2)^2$ 25. $(3m + 5n)^2$ 26. $(6s + 9t)^2$

27. $(3z - w)^2$ 28. $(5x^2 - y)^2$ 29. $(x + 3)^3$ 30. $(3x - 1)^3$

31. $(2x - y)^3$ 32. $(5m - 2n)^3$

Use FOIL to find the products.

33. $(4g - 9)(3g - 7)$ 34. $(5t - 7)(5t - 2)$ 35. $(7s + 3)(4s - 5)$ 36. $(5q + 1)(7q - 5)$

37. $(9r - s)(4r - s)$ 38. $(6m + n)(4m - n)$ 39. $(5m + 3n)(4m + 2n)$ 40. $(3w + 2z)(7w + 5z)$

41. $(9a + 7b)(3a - 4b)$ 42. $(6d + 5c)(7d - 2c)$

Use the shortcut to square the binomials.

43. $(5m - 4)^2$ 44. $(2x - 9)^2$ 45. $(4x - 5y)^2$ 46. $(7j + 3k)^2$

Use the shortcut to find the products of the conjugates.

47. $(x + 1)(x - 1)$ 48. $(m + 2)(m - 2)$ 49. $(a - 4b)(a + 4b)$ 50. $(d - 3c)(d + 3c)$

51. $(3s + 5t)(3s - 5t)$ 52. $(4r + 3w)(4r - 3w)$ 53. $\left(\frac{1}{2}x - \frac{2}{3}\right)\left(\frac{1}{2}x + \frac{2}{3}\right)$ 54. $\left(\frac{1}{4}y - \frac{3}{5}\right)\left(\frac{1}{4}y + \frac{3}{5}\right)$

Use the shortcut to find the products.

55. $(y + 7)(y + 8)$ 56. $(m - 4)(m - 3)$ 57. $(r - 8)(r - 5)$ 58. $(w - 3)(w + 1)$

59. $(w + 7)(w + 5)$ 60. $(t - 2)(t - 9)$

Review Exercises   The answers to all of these exercises are at the back of the book.

61. Name the property that justifies the statement: $3(x + 5) = 3x + 15$

62. Simplify: $\frac{10}{26} \cdot \frac{39}{40}$

63. Graph the straight line that passes through $(0,0)$ with slope $\frac{2}{3}$.

64. Simplify: $-\left(-\frac{1}{2}x^4y^2\right)^4$

65. Use graphing to solve: $\left\{ \begin{array}{rcr} -7x + y &=& 3 \\ x + y &=& -5 \end{array} \right\}$ ❶ ❷

66. Solve: $\left\{ \begin{array}{rcr} 5x - 3y &=& 5 \\ x - y &=& 1 \end{array} \right\}$ ❶ ❷

67. A couple leaves a campground and begins paddling across a lake. A half hour later a second couple leaves the same campground and tries to catch the first couple. The second couple paddles 0.5 mph faster than the first and catches them in 4 hours. How fast was each couple paddling?

68. Decide whether these two straight lines are parallel, perpendicular or neither: *Line 1* passes through the points $(2,8)$ and $(-2,5)$ and *Line 2* passes through the points $(7,9)$ and $(3,6)$.

69. Our galaxy weighs 2.2×10^{41} kilograms. Our sun weighs 2.0×10^{30} kilograms. If we placed our galaxy on one side of a large balance scale how many suns would it take to balance the scale?

70. State the degree of the polynomial: $4pq + 2p + 1$

71. Determine the domain and range of the relation shown in the following graph:

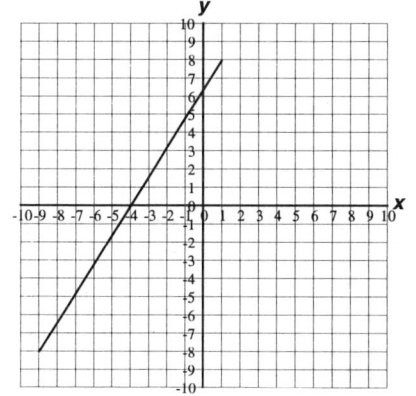

72. Construct the equation of the line shown in the graph.

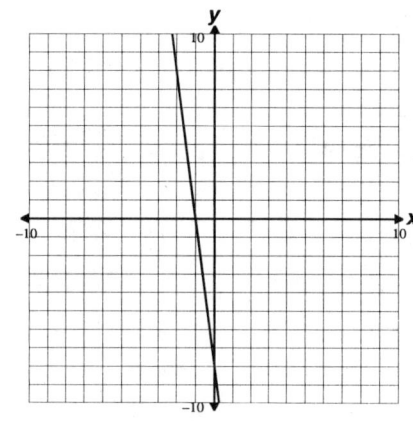

Section 6.7 Dividing Polynomials

Topic 6.7 A — Dividing a Polynomial by a Monomial

From arithmetic, we know that to add fractions which have the same denominator we add the numerators and use the original denominator. For example, $\frac{1}{7} + \frac{3}{7}$ can be written as $\frac{1+3}{7}$. The reverse of this process is also true. If we have a fraction with a sum in the numerator, we can rewrite the fraction as the sum of two separate fractions. Thus, $\frac{1+3}{7}$ can be written as $\frac{1}{7} + \frac{3}{7}$.

This same process holds true for algebraic fractions. Consider the fraction $\frac{x+3}{5}$. We may break up this single fraction into the sum of two fractions and write it as $\frac{x}{5} + \frac{3}{5}$.

Example 6.7.1   Simplify.

a) $\dfrac{6x^2 + 8x}{2x}$

•SOLUTION•

$$\frac{6x^2}{2x} + \frac{8x}{2x} \qquad \Leftarrow \text{ Rewrote as separate fractions.}$$

$$3x + 4 \qquad \Leftarrow \text{ Reduced fractions.}$$

b) $\dfrac{12x^5 + 9x^3 + 15x}{3x^2}$

•SOLUTION•

$$\frac{12x^5}{3x^2} + \frac{9x^3}{3x^2} + \frac{15x}{3x^2} \qquad \Leftarrow \text{ Rewrote as separate fractions.}$$

$$4x^3 + 3x + \frac{5}{x} \qquad \Leftarrow \text{ Reduced fractions.}$$

c) $\dfrac{6x^4 - 5x^2 - 20}{10x^2}$

•SOLUTION•

$$\frac{6x^4}{10x^2} - \frac{5x^2}{10x^2} - \frac{20}{10x^2} \qquad \Leftarrow \text{ Rewrote as separate fractions.}$$

$$\frac{3}{5}x^2 - \frac{1}{2} - \frac{2}{x^2} \qquad \Leftarrow \text{ Reduced fractions.}$$

Practice 6.7.1   Simplify.

1. $\dfrac{14w^8 - 21w^5 + 7w^4}{7w^3}$

2. $\dfrac{-28a^6 + 20a^5 + 12a^3}{4a^2}$

3. $\dfrac{35h^7 - 15b^4 + 10b}{5b^5}$

4. $\dfrac{-36d^9 + 18d^5 + 9d^3}{18d^4}$

— *Caution* — *Canceling*	You may be tempted to simplify expressions such as $\dfrac{x+5}{5}$ by "canceling" the 5's $\dfrac{x + \overset{1}{\cancel{5}}}{\underset{1}{\cancel{5}}}$ to get $x + 1$. This would be like simplifying $\dfrac{15}{5}$ by "canceling" the 5's $\dfrac{1\cancel{5}}{\cancel{5}}$ to get 1. This clearly does not work. We must divide *each* term in the numerator by the denominator. Thus, $\dfrac{x+5}{5}$ may be written as $\dfrac{x}{5} + \dfrac{5}{5}$, which simplifies to $\dfrac{x}{5} + 1$.

Topic 6.7 B — Dividing a Polynomial by a Binomial

One way of dividing a polynomial by a binomial is to use a procedure that is similar to the one you may have used to do long division in arithmetic. To show the process, we will divide side be side $21 \overline{)4853}$ and $2x + 1 \overline{)4x^3 + 8x^2 + 5x + 3}$. Carefully note the similarities in the processes.

Arithmetic	*Algebra*
$21 \overline{)4853}$	$2x + 1 \overline{)4x^3 + 8x^2 + 5x + 3}$
Step 1 Divide Divide 48 by 21 to get 2 $\begin{array}{r} 2 \\ 21 \overline{)4853} \end{array}$	**Step 1 Divide** Divide $4x^3$ by $2x$ to get $2x^2$. Or, ask yourself "What can I multiply $2x$ by to get $4x^3$?" $\begin{array}{r} 2x^2 \\ 2x + 1 \overline{)4x^3 + 8x^2 + 5x + 3} \end{array}$
Step 2 Multiply Multiply 21 by 2 to get 42 $\begin{array}{r} 2 \\ 21 \overline{)4853} \\ 42 \end{array}$	**Step 2 Multiply** Multiply $(2x + 1)$ by $2x^2$ to get $4x^3 + 2x^2$ $\begin{array}{r} 2x^2 \\ 2x + 1 \overline{)4x^3 + 8x^2 + 5x + 3} \\ 4x^3 + 2x^2 \end{array}$
Step 3 Subtract Subtract $48 - 42$ to get 6 $\begin{array}{r} 2 \\ 21 \overline{)4853} \\ \underline{42 } \\ 6 \end{array}$	**Step 3 Subtract** Subtract $(4x^3 + 8x^2) - (4x^3 + 2x^2)$ to get $6x^2$ $\begin{array}{r} 2x^2 \\ 2x + 1 \overline{)4x^3 + 8x^2 + 5x + 3} \\ \underline{4x^3 + 2x^2 } \\ 6x^2 \end{array}$
Step 4 Bring down Bring down the 5 $\begin{array}{r} 2 \\ 21 \overline{)4853} \\ \underline{42 } \\ 65 \end{array}$	**Step 4 Bring down** Bring down the $5x$ $\begin{array}{r} 2x^2 \\ 2x + 1 \overline{)4x^3 + 8x^2 + 5x + 3} \\ \underline{4x^3 + 2x^2 } \\ 6x^2 + 5x \end{array}$

Step 5 Repeat the process

Repeat steps 1 through 4

- divide 65 by 21 to get 3

- multiply 21 by 3 to get 63
- subtract 65 − 63 to get 2
- bring down the 3

$$
\begin{array}{r}
23 \\
21\,\overline{)\,4853} \\
\underline{42} \\
65 \\
\underline{63} \\
23
\end{array}
$$

Step 5 Repeat the process

Repeat steps 1 through 4

- divide $6x^2$ by $2x$ to get $3x$. Or, ask yourself "What can I multiply $2x$ by to get $6x^2$?"

- multiply $2x + 1$ by $3x$ to get $6x^2 + 3x$
- subtract $(6x^2 + 5x) - (6x^2 + 3x)$ to get $2x$
- bring down the 3

$$
\begin{array}{r}
2x^2 + 3x \\
2x + 1\,\overline{)\,4x^3 + 8x^2 + 5x + 3} \\
\underline{4x^3 + 2x^2} \\
6x^2 + 5x \\
\underline{6x^2 + 3x} \\
2x + 3
\end{array}
$$

Step 6 Repeat the process

Repeat steps 1 through 4 until the division is complete.

- divide 23 by 21 to get 1

- multiply 21 by 1 to get 21

- subtract 23 − 21 to get 2

$$
\begin{array}{r}
231 \\
21\,\overline{)\,4853} \\
\underline{42} \\
65 \\
\underline{63} \\
23 \\
\underline{21} \\
2
\end{array}
$$

Step 6 Repeat the process

Repeat steps 1 through 4 until the division is complete.

- divide $2x$ by $2x$ to get 1. Or, ask yourself "What can I multiply $2x$ by to get $2x$?"

- multiply $2x + 1$ by 1 to get $2x + 1$

- subtract $(2x + 3) - (2x + 1)$ to get 2

$$
\begin{array}{r}
2x^2 + 3x + 1 \\
2x + 1\,\overline{)\,4x^3 + 8x^2 + 5x + 3} \\
\underline{4x^3 + 2x^2} \\
6x^2 + 5x \\
\underline{6x^2 + 3x} \\
2x + 3 \\
\underline{2x + 1} \\
2
\end{array}
$$

Step 7 Write the final answer

The quotient is 231 with a remainder of 2. We write this as $231 + \frac{2}{21}$, which we usually write as the mixed number $231\frac{2}{21}$.

Step 7 Write the final answer

The quotient is $2x^2 + 3x + 1$, with a remainder of 2. We write this as $2x^2 + 3x + 1 + \frac{2}{2x + 1}$.

We can check to see if $2x^2 + 3x + 1 + \dfrac{2}{2x + 1}$ is the correct answer by multiplying the divisor by the quotient:

$$(2x + 1)\left(2x^2 + 3x + 1 + \frac{2}{2x + 1}\right) \quad \Leftarrow \text{Expression to simplify.}$$

$$(2x + 1)(2x^2) + (2x + 1)(3x) + (2x + 1)(1) + (2x + 1)\left(\frac{2}{2x + 1}\right) \quad \Leftarrow \text{Multiplied } 2x + 1 \text{ by each term}$$
$$\text{in the second polynomial.}$$

$$4x^3 + 2x^2 + 6x^2 + 3x + 2x + 1 + 2 \quad \Leftarrow \text{Multiplied.}$$

$$4x^3 + 8x^2 + 5x + 3 \quad \Leftarrow \text{Combined like terms.}$$

We see that the product of the quotient and divisor $(2x + 1)\left(2x^2 + 3x + 1 + \frac{2}{2x + 1}\right)$

is the dividend, $4x^3 + 8x^2 + 5x + 3$. Therefore, we are confident that the answer is correct.

Example 6.7.2   Use long division to simplify.

a) $x - 2 \overline{\smash{\big)}\,3x^3 - 7x^2 + 5x - 1}$

•**SOLUTION**• We must be careful when dealing with negative signs, especially when subtracting during the division process.

Subtract here,
$(3x^3 - 7x^2) - (3x^3 - 6x^2)$
$3x^3 - 7x^2 - 3x^3 + 6x^2$
$-7x^2 + 6x^2$
$-1x^2$

$$
\begin{array}{r}
3x^2 - 1x + 3 \\
x - 2 \overline{\smash{\big)}\,3x^3 - 7x^2 + 5x - 1} \\
\underline{3x^3 - 6x^2} \\
-1x^2 + 5x \\
\underline{-1x^2 + 2x} \\
3x - 1 \\
\underline{3x - 6} \\
5
\end{array}
$$

Subtract here,
$(3x - 1) - (3x - 6)$
$3x - 1 - 3x + 6$
5

Subtract here,
$(-1x^2 + 5x) - (-1x^2 + 2x)$
$-1x^2 + 5x + 1x^2 - 2x$
$3x$

The remainder, 5, should be written using the divisor as a denominator. Thus, the final answer is

$$3x^2 - x + 3 + \frac{5}{x - 2}$$

If you have difficulty keeping the signs straight when doing subtractions of negatives, by all means, write the subtraction out along the side of the problem, as we did above.

b) $\dfrac{8x^3 - 1}{2x - 1}$

•**SOLUTION**• The dividend is missing terms with x^2 and x. To make it easier to line up like terms, we will insert the terms $0x^2$ and $0x$ as place holders. Thus, we will be

dividing $\dfrac{8x^3 + 0x^2 + 0x - 1}{2x - 1}$.

Subtract here,
$(8x^3 + 0x^2) - (8x^3 - 4x^2)$
$8x^3 + 0x^2 - 8x^3 + 4x^2$
$4x^2$

$$
\begin{array}{r}
4x^2 + 2x + 1 \\
2x - 1 \overline{\smash{\big)}\,8x^3 + 0x^2 + 0x - 1} \\
\underline{8x^3 - 4x^2} \\
4x^2 + 0x \\
\underline{4x^2 - 2x} \\
2x - 1 \\
\underline{2x - 1} \\
0
\end{array}
$$

Subtract here,
$(4x^2 + 0x) - (4x^2 - 2x)$
$4x^2 + 0x - 4x^2 + 2x$
$2x$

Since the remainder is 0, the final answer is $4x^2 + 2x + 1$.

Practice 6.7.2 Use long division to simplify.

1. $\dfrac{2x^3 + 3x^2 - 17x - 30}{x + 2}$
2. $\dfrac{12y^3 - y^2 - 16y + 7}{y - 1}$
3. $\dfrac{4r^3 + 8r^2 - 17r - 29}{2r + 5}$
4. $\dfrac{8x^3 - 64}{2x - 4}$

Exercise Set 6.7   The answers to the odd numbered exercises are at the back of the book.

Simplify.

1. $\dfrac{4m^2 - 8m}{4m}$
2. $\dfrac{12c^2 - 4c}{2c}$
3. $\dfrac{5y^7 + 10y^4 + 25y}{5y^3}$

4. $\dfrac{12t^6 - 8t^5 + 16t^3 + 4t}{4t^3}$
5. $\dfrac{15m^9 - 10m^7 + 5m^4}{15m^8}$
6. $\dfrac{16a^8 + 12a^4 + 4a^3}{4a^6}$

7. $\dfrac{45t^6 + 3t^4 - 6t^3}{9t^5}$
8. $\dfrac{24w^{12} - 4w^6 + 8w^4}{4w^7}$

Use long division to simplify.

9. $\dfrac{3x^2 - x - 14}{x + 2}$
10. $\dfrac{5w^2 - 24w - 5}{w - 5}$
11. $\dfrac{10y^2 - 28y - 48}{5y + 6}$

12. $\dfrac{6p^2 - p - 15}{2p + 3}$
13. $\dfrac{6x^3 + x^2 - 10x + 3}{x - 1}$
14. $\dfrac{10m^3 + 63m^2 + 14m - 20}{m + 6}$

15. $\dfrac{4x^3 - 12x^2 + 7x - 5}{2x - 5}$
16. $\dfrac{3x^3 + 5x^2 + 7x - 3}{3x - 1}$
17. $\dfrac{5x^3 + 3x^2 - 2x + 7}{x + 2}$

18. $\dfrac{2y^3 + 2y^2 + y + 1}{y - 2}$
19. $\dfrac{2x^4 - 3x^2 + 2}{x - 3}$
20. $\dfrac{y^3 - 125}{y - 5}$

Review Exercises   The answers to all of these exercises are at the back of the book.

21. Graph: $x < 0$
22. Graph: $y - x = 2$
23. Simplify: $-2^{-4} - \dfrac{15}{16} + 3^0$
24. Simplify: $\dfrac{\left(w^4 g\right)^{-2}}{\left(w^{-2}\right)^4 g}$

25. Solve: $\left\{ \begin{array}{rcrcl} 3x & - & y & = & 4 \\ 2x & - & 3y & = & -2 \end{array} \right\} \begin{array}{l} ❶ \\ ❷ \end{array}$

26. Solve by graphing: $\left\{ \begin{array}{rcrcl} 3x & + & y & = & 2 \\ 6x & + & 2y & = & -5 \end{array} \right\} \begin{array}{l} ❶ \\ ❷ \end{array}$

27. Find the slope and y–intercept of the line described by $3 - x = y$
28. Subtract $-8xy + 5x - y$ from $2xy - 3y$

29. Simplify and write the answer in scientific notation: $(6 \times 10^4)(2 \times 10^{-2})$
30. Prime factor: 135

---

31. Two trains leave towns 400 miles apart. If f represents the distance covered by the fast train, use f to create an expression to represent the distance covered by the slow train if:

   a) the slow train covered 115 more miles than the fast train.
   b) the fast train covered twice the distance of the slow train.
   c) the distance covered by the slow train was one–fourth the distance covered by the fast train.
   d) the fast train covered 80 less miles than the slow train.

32. A health club membership is discounted by 15%. If the price after the discount was $340, how much is a membership without a discount?

33. The data table below shows the relation between the number of boxes of cookies a Girl Scout has left and the number of houses she has contacted. Calculate the slope of the line that would connect the given points and discuss the meaning of the slope.

x number of houses contacted	*y* number of boxes of cookies sold
8	168
15	140
30	80
48	8

Chapter 6 Review

Vocabulary to Know

binomial — an expression that has two terms, as in $2x + 3$.

conjugates — a pair of two–termed expressions where one is the sum of the terms and the other is the difference; $(a + b)$ and $(a − b)$ are conjugates.

constant term — a term that contains no variable factors other than those with exponents of 0.

degree of a polynomial — the degree of the term with the highest degree; the degree of $5x^2 + 8$ is 2.

degree — the number of variable factors in a term; the degree of $3x^2y^3$ is 5.

exponent notation — notation for powers of 10 where the letter E is written instead of $\times$; in this notation, 3E5 means 3×10^5.

finite — something that does not go on forever (that is, it's not infinite).

FOIL — acronym for multiplying binomials. It stands for First, Outer, Inner, Last.

monomial — an expression that has only one term; 3 and $2x$ are monomials.

polynomial — an algebraic expression consisting of a sum of terms where only whole number exponents are allowed; $3x^2 + 1$ is a polynomial but $3x^{-2} + 1$ is not.

polynomial in one variable — a polynomial that contains only one variable, as in $x^2 + 3x + 9$.

polynomial in two variables — a polynomial that contains two variables, as in $2l + 2w$.

scientific notation — notation used with very large or small numbers where the number is written as the product of a power of 10 and a decimal number equal to or greater than 1 but less than 10; 5200 written in scientific notation is 5.2×10^3.

standard form of a polynomial — a way of writing polynomials where the terms are written in decreasing powers of the variable; $5x^2 + 2x + 3$ is in standard form but $2x + 5x^2 + 3$ is not.

trinomial — an expression that has three terms as in $2x^3 + x + 2$.

Properties, Definitions, Formulas, and Notation to Understand

Multiplication Property of Exponents — To multiply two numbers with the same base, add the exponents and keep the base the same.	$a^m a^n = a^{m+n}$ $3^4 \cdot 3^2$ can be written as 3^{4+2}, which simplifies to 3^6 $x^4 \cdot x^2 \cdot x$ can be written as x^{4+2+1}, which simplifies to x^7
Power of a power property of exponents — To simplify a power raised to a power, multiply the exponents.	$(a^m)^n = a^{mn}$ $(2^3)^4$ can be written as $2^{3 \cdot 4}$, which simplifies to 2^{12} $(x^2)^5$ can be written as $x^{2 \cdot 5}$, which simplifies to x^{10}
Power of a product property of exponents — A product raised to a power can be written as the product of each factor raised to the power.	$(ab)^m = a^m b^m$ $(5x)^3$ can be written as $5^3 x^3$
Division property of exponents — To divide factors with the same base, subtract the exponents.	If $a \neq 0$ then $\dfrac{a^m}{a^n} = a^{m-n}$ $\dfrac{x^5}{x^2}$ can be written as x^{5-2}, which simplifies to x^3
Power of a quotient property of exponents — A quotient raised to a power can be written as the quotient of the numerator and denominator, each raised to the power.	If $b \neq 0$ then $\left(\dfrac{a}{b}\right)^m = \dfrac{a^m}{b^m}$ $\left(\dfrac{2}{3}\right)^5$ can be written as $\dfrac{2^5}{3^5}$ $\left(\dfrac{5}{x}\right)^3$ can be written as $\dfrac{5^3}{x^3}$

Exponent of 1 — An exponent of 1 means a single factor of the base.	$a^1 = a$ 2^1 can be written as 2 y^1 can be written as y
Exponent of 0 — An expression with an exponent of 0 is equivalent to the number 1.	If $a \neq 0$, then $a^0 = 1$; 0^0 is undefined. 2^0 can be written as 1 y^0 can be written as 1 (assuming $y \neq 0$)
Negative Exponents — An expression with a base written on one side of a fraction bar can be written as an equivalent expression with the base on the other side of the fraction bar by changing the sign of the exponent.	If $a \neq 0$, then $a^{-n} = \dfrac{1}{a^n}$ and $a^n = \dfrac{1}{a^{-n}}$ 2^{-5} can be written as $\dfrac{1}{2^5}$ x^{-3} can be written as $\dfrac{1}{x^3}$ $\dfrac{1}{x^{-4}}$ can be written as x^4
Scientific Notation — A number written in scientific notation has two factors: • the first factor is a decimal written with one non-zero digit to the left of the decimal point. • the second factor is written as a power of 10.	2.5×10^{13} 5.39×10^{-9}

Polynomial — A polynomial is an expression that can be written as a sum of terms where only whole number exponents are allowed and there are no variables in the denominators of fractions.	A polynomial in x is a finite sum of terms, all of which have the form ax^n where
	
	     a is a real number,
	     x is a variable,
	     n is a whole number.
	
	 $3x$ is a one-term polynomial (a monomial).
	
	 $2x^3 - 5x^2 + x - 3$ is a three-term polynomial (a trinomial).

Degree of a Term — The degree of a term is the number of variable factors in the term.	If $a \neq 0$ and m and n are whole numbers, then the degree of the term $ax^n y^m$ is $n + m$. The degree of $3x^4$ is 4. The degree of $-8xy^2$ is 3.

Procedures to Follow

Procedure for simplifying exponential expressions.	Simplify $\dfrac{5x^2(x^{-1}y^3 \cdot 2x)^2}{15x^{-3}(2x)^2}$
Step 1 Parentheses Simplify inside parentheses.	$\dfrac{5x^2(2y^3)^2}{15x^{-3}(2x)^2}$
Step 2 Numerator Simplify numerators.	$\dfrac{20x^2y^6}{15x^{-3}(2x)^2}$
Step 3 Denominator Simplify denominators.	$\dfrac{20x^2y^6}{60x^{-1}}$
Step 4 Fraction Simplify fractions.	$\dfrac{x^3y^6}{3}$

Procedure for moving factors between numerators and denominators.	Move the x factor to the denominator. $\dfrac{2x^{-4}}{3y}$ $\qquad$ $\dfrac{3x^2}{5y^{-3}}$
• To move a factor from the numerator to the denominator of a fraction, write the factor in the denominator and change the sign of its exponent.	$\dfrac{2}{3x^4 y}$ $\qquad$ $\dfrac{3}{5x^{-2}y^{-3}}$
• To move a factor from the denominator to the numerator of a fraction, write the factor in the numerator and change the sign of its exponent.	Move the y factor of $\dfrac{2x^{-4}}{3y}$ $\qquad$ Move the y factor of $\dfrac{3x^2}{5y^{-3}}$ to the denominator. $\qquad$ to the denominator. $\dfrac{2x^{-4}y^{-1}}{3}$ $\qquad\qquad$ $\dfrac{3x^2 y^3}{5}$

Procedure for converting from decimal notation to scientific notation.	Convert to scientific notation: 582,000,000 $\qquad$ 0.0003
Step 1 Move the decimal point until there is one non-zero digit to its left.	5.82 $\qquad\qquad$ 3.
Step 2 Multiply by 10 raised to a power whose magnitude is equal to the number of places the decimal was moved. The power is positive if the decimal was moved to the left and negative if the decimal was moved to the right.	5.82×10^8 $\qquad$ 3×10^{-4}

Procedure for converting from scientific notation to decimal notation.	Convert to decimal notation: 2.3×10^4 $\qquad$ 1.97×10^{-6}
Move the decimal point the same number of places as the power of 10. If the power is positive, move the decimal to the right (that is, make the magnitude of the decimal bigger). If the power is negative, move the decimal to the left (that is, make the magnitude of the decimal smaller).	23,000 $\qquad\qquad$ 0.00000197

Procedure for subtracting polynomials.	Subtract $5x^3 + 2x^2 - 3x + 1$ from $2x^3 - 3x$
Step 1 Insert grouping symbols If necessary, write the polynomials inside grouping symbols.	$(2x^3 - 3x) - (5x^3 + 2x^2 - 3x + 1)$
Step 2 Remove grouping symbols Use the Distributive Property to remove grouping symbols.	$2x^3 - 3x - 5x^3 - 2x^2 + 3x - 1$
Step 3 Add like terms Add the coefficients of like terms and write the resulting expression in decreasing powers of the variable.	$-3x^3 - 2x^2 - 1$

Procedure for multiplying a monomial and a polynomial.	Simplify $-5x^2(3x^2 - 2x + 2)$
Multiply the monomial by each term of the polynomial and then simplify if possible.	$(-5x^2)(3x^2) + (-5x^2)(-2x) + (-5x^2)(2)$ $-15x^4 + 10x^3 - 10x^2$

Procedure for multiplying polynomials.	Simplify $(5x - 2)(3x^2 + x - 6)$
Step 1 Multiply each term of the first polynomial by each term of the second polynomial.	$(5x)(3x^2) + (5x)(x) + (5x)(-6) + (-2)(3x^2) +$ $(-2)(x) + (-2)(-6)$
Step 2 Simplify.	$15x^3 + 5x^2 - 30x - 6x^2 - 2x + 12$ $15x^3 - x^2 - 32x + 12$

Procedure for squaring a binomial.	Simplify $(4x - 3)^2$
Step 1 Square the first term.	$(4x)^2$
Step 2 Add twice the product of the terms.	$(4x)^2 + 2(4x)(-3)$
Step 3 Add the square of the second term.	$(4x)^2 + 2(4x)(-3) + (-3)^2$
We can write this process as $(a + b)^2 = a^2 + 2ab + b^2$	Simplifying we get $16x^2 - 24x + 9$
Procedure for multiplying conjugates.	Simplify $(2x - 7)(2x + 7)$
Step 1 Square the first term of one of the binomials.	$(2x)^2$
Step 2 Square the second term of one of the binomials.	$(7)^2$
Step 3 Subtract the squares.	$(2x)^2 - (7)^2$
We can write this as $(a + b)(a - b) = a^2 - b^2$	Simplifying we get $4x^2 - 49$
Procedure for multiplying binomials of the form $(x + m)(x + n).$	Simplify $(x - 3)(x + 5)$
When multiplying $(x + m)(x + n)$, the coefficients of the resulting three terms will be	
• 1 (for the x^2 term) • the sum $m + n$ (for the x term) • the product mn (for the last term).	$1x^2$ $(-3 + 5)x$ $(-3)(5)$
We can write this as $(x + m)(x + n) = x^2 + (m + n)x + mn$	Adding these gives $1x^2 + (-3 + 5)x + (-3)(5)$, which simplifies to $x^2 + 2x - 15$

Chapter 6 Review Exercises   The answers to all the exercises are in the back of the book.

1. $m^6 n^4 n^3 m$

2. $9b9b^2$

3. $-a(-4a^2 + 3a)$

4. $\frac{1}{5}y^8\left(-15y^2\right)\left(\frac{1}{2}y\right)$

5. $(-x^3 y)^4$

6. $-(a^2 b)^2$

7. $\dfrac{\left(b^2 b^4\right)^2 (2b)^3}{2b^4}$

8. $\dfrac{4x^{-2}}{x^{-5}}$

9. $\dfrac{6xy^{-2}}{3x^{-2}y^{-5}}$

10. $\dfrac{8c^3 y^2}{12c^{-2}y^5}$

11. $\dfrac{4^{-1} + 3^{-2}}{2^{-3}}$

12. $2^{-3} - 3^{-2} + 2^{-2} + 4^0$

13. $\left(\dfrac{-8x^3 y}{2xy^2}\right)^2\left(x^{-2}y^{-2}\right)$

14. $-2t\left(\dfrac{-2t^{-2}}{(-2t)^2}\right)$

15. $\dfrac{8x^{-3}x^{10}}{(2x^2)^2}$

16. $\dfrac{12x^2\left(2xy^3\right)^2}{3x\left(2x^{-3}\right)^2 y}$

Convert to scientific notation.

17. -0.00083

18. 350×10^7

19. 5×10^4

20. -3.02×10^{-3}

Simplify and write the answer in scientific notation.

21. $\dfrac{1.0 \times 10^{-10}}{0.010}$

22. $\dfrac{(7.2 \times 10^{-10})(1 \times 10^{-3})}{1.6 \times 10^{-11}}$

23. A grape weighs about 3×10^{-3} kg while a speck of dust weighs about 6.7×10^{-10} kg. Would a million specks of dust weight more than a grape? If they wouldn't, how many specks of dust would you need?

24. The height of Mt. Everest is about 8.9×10^3 meters while the height of an average person is about 1.7 meters. How many people standing on top of each other would it take to be as high as Mt. Everest?

25. Is this expression a polynomial? $\frac{1}{5}x^2 - \frac{3}{4}x - 2$

26. Is this expression a polynomial? $3a^3 - 2a^{-2} + 4a$

27. State the degree of this polynomial: $3x^2 + 2x - 1$

28. State the degree of this polynomial: $4x^3y^2 + 3x2y$

29. Write in standard polynomial form: $5x - 3x^2 + 2$

30. Add $(-5xy + 2x)$ and $(3x - 2xy)$ and $(4x^2 - xy + 1)$

31. Add $\left(\frac{1}{3}a^3b + \frac{1}{2}ab^3\right)$ and $\left(\frac{3}{4}a^2b^2 + \frac{3}{4}ab^3\right)$ and $\left(\frac{5}{3}a^3b - \frac{1}{2}a^2b^2\right)$

32. Subtract $5x^2 + 7x - 3$ from $-x^2 + 5x$

Simplify.

33. $-4m^2(3m^2 - 2m + 4)$

34. $x(x - y + 1) + y(x - y + 1)$

35. $(4x - y) - (y - 4x) - (x - 4y) - (-4x - y)$

36. $(3m - 1)(6m + 5)$

37. $(k + 1)(k^3 - k^2 - k)$

38. $(2y - 3)^3$

39. $(3a - 2)(5a + 9)$

40. $(z - 8)(2z - 11)$

41. $(5a - 4)^2$

42. $(9d - 5)(9d + 5)$

43. $(6y - 1)(6y + 1)$

Divide and simplify. Use long division where necessary.

44. $\dfrac{-20r^7 - 16r^5 + 8r^3}{-4r^2}$

45. $\dfrac{51d^6 + 57d^5 - 93d^4}{3d^4}$

46. $\dfrac{3y^3 + 15y^2 + 6y - 22}{y + 4}$

47. $\dfrac{12y^4 - 30y^3 - 2y + 4}{2y - 5}$

Chapter 6 Test   The answers to all the exercises are in the back of the book.

Simplify

1. $a^3ba^2b^2$

2. $-\frac{1}{3}z(2z^5)(-6z^{12})$

3. $\dfrac{\left(3x^3y^2\right)^2\left(xy^3\right)^2}{9(xy)^3}$

4. $\dfrac{5m^{-3}}{m^{-6}}$

5. $4 \cdot 3^{-1} - 3 \cdot 4^{-1}$

6. $-\left(\frac{1}{3}\right)^4 + 3^0 + (-3)^{-4}$

7. $\left(\dfrac{3m^{-8}}{(2m)^{-2}}\right)^2$

8. $\dfrac{6x^{-2}\left(3y^3\right)^2}{x^{-3}y(18xy^{-2})}$

9. $3c^2(5c^2 - c - 3)$

10. $(5a - 9)(5a - 4)$

11. $(2r + 3s - t)(r - t)$

12. $(5x - 2)^2$

13. Convert to scientific notation: 49,000,000

14. Convert to decimal notation: 8.01×10^{-4}

15. Simplify and write the answer in scientific notation: $\dfrac{(2.5 \times 10^{-3})(3.2 \times 10^8)}{1.6 \times 10^{-4}}$

16. An elephant weighs about 4,500 kg while an ocean liner weighs about 72,000,000 kg. Would one thousand elephants weigh as much as an ocean liner? If not, how many elephants would it take to equal the weight of an ocean liner?

17. Is this expression a polynomial? $5y + \frac{3}{y} + 11$

18. State the degree of the polynomial: $8x^3 - 3x + 1$

19. Add $5x^2 + 3x - 6$ to $-8x^2 - 3x + 1$

20. Write in standard polynomial form: $12x^3 - x^4 + 7x^2 - 3$

21. Subtract $16c^4 - 14d^2 - 11e$ from $-12c^4 + 11d^2 - 11e$

22. Simplify $(3w^2 - 5w - 2) - (4w^2 - 6w - 5) - (8w^2 + 6w - 1)$

23. Divide and simplify: $\dfrac{15q^3 - 5q^2 - 15q}{5q}$

24. Use long division to simplify: $\dfrac{h^4 + 2h^3 - 23h^2 + 10h + 24}{h + 6}$

Chapter 7
Factoring Algebraic Expressions

Section 7.1 Greatest Common Factor (GCF)

In the last chapter we used the Distributive Property to expand the product of two polynomials into a sum. For example, we multiplied $(x + 2)(x + 3)$ to get $x^2 + 5x + 6$. Much of this chapter will be spent doing the reverse. That is, we will practice converting a sum like $x^2 + 5x + 6$ into a product. Rewriting an expression as a product is known as **factoring** the expression.

Topic 7.1 A — Greatest Common Factor in Arithmetic

In Section 1.2 we defined the **greatest common factor** (**GCF**) of a set of numbers as the largest factor common to all the numbers in the set. For example, the GCF of 10 and 25 is 5 because 5 is the largest factor common to both 10 and 25. It's easy to see the GCF of two numbers like 10 and 25. However, if we need to find the GCF of large numbers, or a set of several numbers, the following procedure is helpful.

— Procedure — *Finding the GCF of a set* *of numbers*	*Step 1 Prime Factor* Prime factor each number and write the factors in exponential form. *Step 2 Greatest Common Power* Multiply the greatest power of each prime factor that is *common* to all the numbers. If a factor does not appear in all the numbers then it will not be a part of the GCF.

Example 7.1.1   a)   Find the GCF of 60 and 48.

•SOLUTION• *Step 1 Prime Factor*   Prime factor 60 and 48.

Prime factor 60	*Prime factor 48*
$\begin{array}{r} 5 \\ 3\ \ 2\,\overline{)15} \\ 2\,\overline{)30} \\ 2\,\overline{)60} \end{array}$	$\begin{array}{r} 3 \\ 2\,\overline{)6} \\ 2\,\overline{)12} \\ 2\,\overline{)24} \\ 2\,\overline{)48} \end{array}$
60 factors as $2^2 3^1 5^1$	48 factors as $2^4 3^1$

Step 2 Greatest Common Power   Since 5 is not a factor of both numbers, only 2 and 3 are *common* factors.

• The largest power of 2 common to both numbers is 2^2.
• The largest power of 3 common to both numbers is 3^1.

The GCF is $2^2 \cdot 3^1$, which simplifies to 12. That is, 12 is the greatest number that is a factor of both 60 and 48.

b)   Find the GCF of 28, 84, and 126.

•SOLUTION• *Step 1 Prime Factor*   Prime factor 28, 84, and 126.

28 factors as $2^2 7^1$
84 factors as $2^2 3^1 7^1$
126 factors as $2^1 3^2 7^1$

Step 2 Greatest Common Power   Since 3 does not appear in all the numbers, only 2 and 7 are *common* factors.

• The largest power of 2 *common* to all three numbers is 2^1.
• The largest power of 7 *common* to all three numbers is 7^1.

The GCF is $2^1 \cdot 7^1$, which simplifies to 14. That is, 14 is the greatest number that is a factor of 28, 84, and 126.

— Note — *Greatest and smallest in a GCF*	The GCF of a set of numbers is the product of the greatest power of each prime factor that is common to all the numbers of the set. You may have noticed that this always turns out to be the smallest power of each common prime factor. How can the greatest be the smallest? The key word here is *common*. For example, if we are given 2^4 and 2^6, the greatest power of 2 is 2^6 but the greatest *common* power of 2 is 2^4. We can see that 2^4 is common to both numbers by writing 2^6 as $2^2 \cdot 2^4$.

Practice 7.1.1 Find the GCF.

1. 28, 42, and 70

2. 72, 96, and 144

3. 36, 72, and 144

Topic 7.1 B — Finding the GCF of Algebraic Expressions

We find the GCF of algebraic expressions using exactly the same procedure as we used for arithmetic numbers. For example, let's find the GCF of $9x^4$ and $48x^7$.

Step 1 Prime Factor

$9x^4$ factors as $3 \cdot 3 \cdot x^4$, which can be written as $3^2 x^4$

$48x^7$ factors as $2 \cdot 2 \cdot 2 \cdot 2 \cdot 3 \cdot x^7$, which can be written as $2^4 3^1 x^7$

Step 2 Greatest Common Power Only 3 and x are common to both terms.

- The greatest power of 3 *common* to both terms is 3^1.

- The greatest power of x *common* to both terms is x^4.

The GCF is $3^1 x^4$, or $3x^4$.

Example 7.1.2

a) Find the GCF of $8x^5$, $12x^5$, and $16x^2$

•SOLUTION• ***Step 1 Prime Factor***

$8x^5$ factors as $2 \cdot 2 \cdot 2 \cdot x^5$ or $2^3 x^5$

$12x^5$ factors as $2 \cdot 2 \cdot 3 \cdot x^5$ or $2^2 3^1 x^5$

$16x^2$ factors as $2 \cdot 2 \cdot 2 \cdot 2 \cdot x^2$ or $2^4 x^2$

Step 2 Greatest Common Power Only 2 and x are common to all terms.

- The greatest power of 2 *common* to all three terms is 2^2.

- The greatest power of x *common* to all three terms is x^2.

The GCF is $(2^2)(x^2)$, which simplifies to $4x^2$.

b) Find the GCF of $10x^5 y^4$ and $15x^3 y^5$.

•SOLUTION• ***Step 1 Prime Factor***

$10x^5 y^4$ factors as $2 \cdot 5 \cdot x^5 y^4$ or $2^1 5^1 x^5 y^4$

$15x^3 y^5$ factors as $3 \cdot 5 \cdot x^3 y^5$ or $3^1 5^1 x^3 y^5$

Step 2 Greatest Common Power Only 5, x, and y are common to all terms.

- The greatest power of 5 *common* to both terms is 5^1.

- The greatest power of x *common* to both terms is x^3.

- The greatest power of y *common* to both terms is y^4.

The GCF is $(5^1)(x^3)(y^4)$, which simplifies to $5x^3 y^4$

Practice 7.1.2   Find the GCF.

1. $4r^4, 16r^2, 24r^5$ 2. $-10t^2, -25t^5, -15t^4$ 3. $-18r^4s^4, -12r^9s^8, -36r^3$ 4. $3a^{11}b^3, 9a^6b, 15a^7b^4$

You can find a GCF even if the factors are not monomials.

Example 7.1.3   a)   Find the GCF of $4x(x + 4)$ and $6y(x + 4)$.

•**SOLUTION**•   **Step 1 Prime Factor**   The first term has factors of 4, x, and the binomial $(x + 4)$; the second term has factors of 6, y, and the binomial $(x + 4)$. These terms prime factor as follows:

$$2^2 x^1 (x + 4)^1$$

$$2^1 3^1 y^1 (x + 4)^1$$

Step 2 Greatest Common Power   Only 2 and $(x + 4)$ are common to both terms.

- The greatest power of 2 *common* to both terms is 2^1.

- The greatest power of $(x + 4)$ *common* to both terms is $(x + 4)^1$.

The GCF is $2^1(x + 4)^1$, which simplifies to $2(x + 4)$. Note that we do not use the Distributive property to write this as $2x + 4$. We want the GCF to be a single term.

b)   Find the GCF of $12(a + 3)^3$ and $18(3 + a)^2$.

•**SOLUTION**•   Since addition is commutative the order of the terms in $(a + 3)$ and $(3 + a)$ doesn't affect the process. To be consistent we will write the terms as $12(a + 3)^3$ and $18(a + 3)^2$.

Step 1 Prime Factor

$12(a + 3)^3$ factors as $2^2 3^1 (a + 3)^3$

$18(a + 3)^2$ factors as $2^1 3^2 (a + 3)^2$

Step 2 Greatest Common Power   All the factors are common to both terms.

- The greatest power of 2 *common* to both terms is 2^1.

- The greatest power of 3 *common* to both terms is 3^1.

- The greatest power of $(a + 3)$ *common* to both terms is $(a + 3)^2$.

The GCF is $2^1 3^1 (a + 3)^2$, which simplifies to $6(a + 3)^2$.

Practice 7.1.3   Find the GCF.

1. $x(x - 8)$ and $y(x - 8)$ 2. $6(x - 1)^3$ and $18(x - 1)^2$ 3. $9(x - 5)^5$ and $12(x - 5)^6$ 4. $21(a + b)^4$ and $14(b + a)^2$

— *Note* —	When finding the GCF of algebraic expressions we will follow two conventions:
Conventions when finding a GCF	• We will only use integers (no fractions) as numeric factors.
	• We usually don't include 1 and −1 in the GCF, even though they are factors of all expressions.

Exercise Set 7.1   The answers to the odd numbered exercises are at the back of the book.

Find the GCF.

1.   36 and 48   2.   45 and 60   3.   24 and 42

4.   18 and 45   5.   27, 54 and 108   6.   54, 72 and 108

7. $2x^3y^5$ and $3x^8y^2$

8. $5s^8t^3$ and $7s^{11}t^9$

9. $5y^4$, $15y^7$, and $10y^3$

10. $8m^4n^3$, $56m^8n^5$, and $24m^5n^2$

11. $9s^6$, $6s^5$ and $15s^4$

12. $12s^7t^{12}$, $18s^4t^6$ and $24s^{13}t^5$

13. $m^3(m+5)^2$ and $m^2(m+5)^8$

14. $x^5(x-3)^5$ and $x^3(x-3)$

15. $m(n+1)^8$ and $5m^3(n+1)^4$

16. $45x^4(w+2)$ and $15x^3(w+2)$

17. $18m^5(m+2)$ and $24m^2(m+2)^3$

18. $10(5-x)^3$ and $15(5-x)^6$

19. $6x^3(3x+7)^3$ and $15x^4(3x+7)^7$

20. $15m^5(6m+5)^2$ and $45(6m+5)$

21. $5t^3(t+12)^2$ and $15(t+12)$

22. $12(r+7)^4$ and $36r(r+7)^3$

Review Exercises   The answers to all of these exercises are at the back of the book.

23. Identify the sums, terms, products, and factors: $3 + 2x$

24. Prime factor: 48

25. Find the equation of the line that goes through $(7,-1)$ and $(7,2)$.

26. Solve: $\frac{1}{2}x - \frac{1}{3}(3-9x) \geq 2x$

27. A company is worried about how fast computers become "obsolete" and so decided to rent two computers. The cost to rent one computer for 18 months turned out to be the same as renting the other computer for 27 months. How much did it cost each month to rent the cheaper computer if it was \$15 more a month than the cheaper computer?

28. The thickness of the metal used to form a pipe is 2 mm. If the circumference of the pipe is 47.1 mm find the inside radius of the pipe. (Use 3.14 for π.)

29. Given $f(x) = x^2 - 5$, find $f(-3)$.

30. Divide and simplify: $\dfrac{18y^4 - 6y^3 - 6y^2}{6y^2}$

31. Solve: $\left\{ \begin{array}{rcrcl} 2x & - & 3y & = & -7 \\ 21 & + & 6x & = & 9y \end{array} \right\}$ ❶ ❷

32. Solve: $\left\{ \begin{array}{rcrcl} 3x & + & 5y & = & -12 \\ -2x & + & 3y & = & -11 \end{array} \right\}$ ❶ ❷

Simplify.

33. $-(-w^2g^5)^2$

34. $-4^{-2} - 2^{-4}$

35. $3x(5x^2 - 4x + 1)$

36. $(x-3)(3x^2 + 2x - 4)$

---

37. Determine whether this relation is a function:

x	y
2	7
−1	4
0	9
−3	7

---

38. How many dollars are in a quarter? Often people will say, "none" but that is not quite right. Assume there is 1/4 dollar in every quarter.

a)   Create an equation which will output the value in dollars given the number of quarters.

b)   Complete the following data table:

x number of quarters	y value (dollars)
8	
18	
30	
	6
	9.5

c)   Plot the points on the following grid:

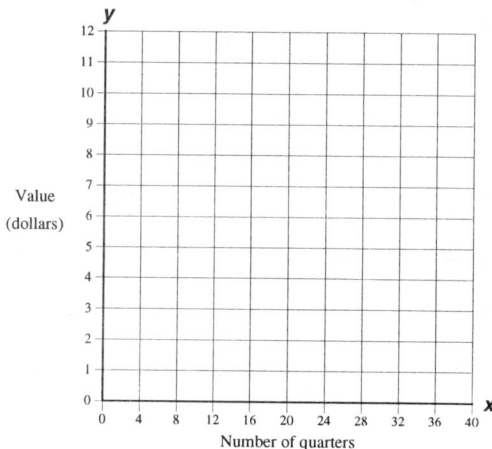

Section 7.2 Factoring the GCF Out of Polynomials

By using the Distributive Property we are able to convert certain sums into products. We referred to this process as factoring. For example, in Section 1.6 we used factoring to justify adding like terms such as $4x + 5x$ (a sum) to get $9x$ (a product).

$4x + 5x \Leftarrow$ Expression to simplify.

$x(4 + 5) \Leftarrow$ Applied the Distributive Property (that is, we factored out the GCF, x).

$x(9) \Leftarrow$ Simplified.

$9x \Leftarrow$ Applied the Commutative Property.

In essence, we found the GCF of $4x$ and $5x$ and then we rewrote the sum $4x + 5x$ as a product. This is an example of factoring the GCF out of a binomial.

We can follow this procedure to rewrite some polynomials as products. For example, let's factor the binomial $3x + 21$:

$3x + 21 \Leftarrow$ Expression to factor.

$(3)(x) + (3)(7) \Leftarrow$ Wrote each term as a product where the GCF is one factor.

$3(x + 7) \Leftarrow$ Used the Distributive Property to factor out the GCF.

We can use the following procedure for most polynomials:

— Procedure —	**Step 1 Find the GCF**
Factoring the GCF out of a polynomial	• We find the numeric part of the GCF by inspection or by prime factoring. By convention, we only consider integer factors, and we include factors of 1 and −1 only in special circumstances. Therefore, when we say "There is no GCF" we really mean that there are no integer factors, other than 1 or −1, which are common to all the terms of the polynomial.
	• For the polynomial factors, select the smallest power of each factor that is common to all terms.
	Step 2 Rewrite the terms Write each term as a product where the GCF is a factor.
	Step 3 Factor Use the Distributive Property to factor out the GCF.

Example 7.2.1   Rewrite as a product by factoring out the GCF.

a) $5x^3 + 10x - 30$

•SOLUTION•   *Step 1 Find the GCF*

• The greatest factor that is common to 5, 10, and 30 is 5.

• There is no variable factor common to all the terms.

The GCF is 5.

Step 2 Rewrite the terms

$5x^3 + 10x - 30 \Leftarrow$ Expression to factor.

$(5)(x^3) + (5)(2x) - (5)(6) \Leftarrow$ Wrote each term as a product where the GCF is a factor.

Step 3 Factor

$5(x^3 + 2x - 6) \Leftarrow$ Used the Distributive Property to factor out the GCF.

b) $51x^3z + 34x^2y$

•**SOLUTION**• *Step 1 Find the GCF*

- Unless you have memorized your 17 times tables, it's difficult to "see" the greatest common factor of 34 and 51. Therefore, we prime factor each number:

$$51 \text{ factors as } 3 \cdot 17$$
$$34 \text{ factors as } 2 \cdot 17$$

Now its easy to see that 17 is the greatest common factor of 34 and 51.

- x is the only variable common to both terms; the smallest power of x is x^2.

The GCF is $17x^2$.

Step 2 Rewrite the terms

$51x^3z + 34x^2y$ ⇐ Expression to factor.

$(\mathbf{17x^2})(3xz) + (\mathbf{17x^2})(2y)$ ⇐ Wrote each term as a product where the GCF is a factor.

Step 3 Factor

$(\mathbf{17x^2})(3xz + 2y)$ ⇐ Used the Distributive Property to factor out the GCF.

We can verify that this is the correct factorization by checking two things:

- The terms of the binomial should have no common variable or integer factors (other than 1 or –1). This is true for $3xz$ and $2y$.

- If we multiply $17x^2$ by each term of the binomial we should get the original expression. If you try it you will see that we do.

c) $20b^3 + 6b^2 + 2b$

•**SOLUTION**• *Step 1 Find the GCF*

- The greatest number that is a factor of 6, 20, and 2 is 2.

- b is common to all terms. The smallest power of b is b^1.

The GCF is $2b$.

Step 2 Rewrite the terms

$20b^3 + 6b^2 + 2b$ ⇐ Expression to factor.

$(\mathbf{2b})(10b^2) + (\mathbf{2b})(3b)+ (\mathbf{2b})(1)$ ⇐ Wrote each term as a product where the GCF is a factor. Notice that we need the factor of 1 in the last term.

Step 3 Factor

$(\mathbf{2b})(10b^2 + 3b + 1)$ ⇐ Used the Distributive Property to factor out the GCF.

We can verify that this is the correct factorization by checking two things:

- The terms of the binomial should have no common variable or integer factors (other than 1 or –1). This is true for $10b^2$, $3b$, and 1.

- If we multiply $2b$ by each term of the binomial we should get the original expression. If you try it you will see that we do.

| — *Caution* —
 A term of 1 may be needed | When the GCF happens to be a term of the polynomial we must have a term of 1 in one of the factors. For example,

 INCORRECT: $20b^3 + 6b^2 + 2b$ does NOT factor as $(2b)(10b^2 + 3b)$

 CORRECT: $20b^3 + 6b^2 + 2b$ factors as $(2b)(10b^2 + 3b + \underline{1})$. |

Practice 7.2.1 Rewrite as a product by factoring out the GCF.

1. $10x + 15$ 2. $3x^6 + 12x^2$ 3. $64r^4 + 16r^2 - 8$ 4. $45a^4b^5 - 15a^3b^2$ 5. $27c^7d^9 - 54c^3d^{10}$

We will use this type of factoring later to reduce algebraic fractions. For example, to reduce the fraction $\dfrac{2x + 6}{x + 3}$ we factor the numerator to get $\dfrac{2(x + 3)}{x + 3}$. Since

$\dfrac{x + 3}{x + 3}$ reduces to 1, the original fraction reduces to $2 \cdot 1$, which is 2.

Topic 7.2 A — Factors of 1 and –1

By convention, we do not include 1 or –1 as a part of the GCF because 1 and –1 are factors of every expression. However, in some of the factoring problems coming up we will have to factor out 1 or –1.

Example 7.2.2   Rewrite as a product by factoring out 1. Then, rewrite as a product by factoring out –1.

a)   5

•SOLUTION• We can write 5 as $1(5)$ or $-1(-5)$.

b) $x + y$

•SOLUTION• We can write $x + y$ as $1(x + y)$ or $-1(-x - y)$. Notice how the signs changed when we factored out –1.

c) $2x^3 - 3x - 7$

•SOLUTION• We can write $2x^3 - 3x - 7$ as $1(2x^3 - 3x - 7)$ or $-1(-2x^3 + 3x + 7)$. Notice how the signs changed when we factored out –1.

— *Caution* — **Be careful with signs**	Be careful with the signs in expressions that involve subtraction. In the last example, we did not write $2x^3 - 3x - 7$ as $2x^3 + (-3x) + (-7)$ because it clutters up the problem and slows down the process of factoring. However, if you are having difficulty keeping the signs straight, continue to write subtractions as additions of the opposite. This is a judgment call you will have to make for yourself.

d) $4 - x^2$

•SOLUTION• We can write $4 - x^2$ as $1(4 - x^2)$ or $-1(-4 + x^2)$. Notice how the signs changed when we factored out –1. Since we usually write terms with variables first, we probably would write this as $-1(x^2 - 4)$.

Practice 7.2.2   Rewrite as a product by factoring out 1. Then, rewrite as a product by factoring out –1.

1. -25 2. 32 3. $2x + y$ 4. $4m - 2n$

5. $7v^4 - 8v^3 + 2$ 6. $-12w^3 + 15w^2 + 11$ 7. $25 - w^2$

When a polynomial has many subtractions, or when the first term has a negative coefficient, we sometimes factor out a negative sign as a part of the GCF.

Example 7.2.3   Rewrite as a product by factoring out a negative sign with the GCF.

a) $-x^3 - 7x$

•SOLUTION• *Step 1 Find the GCF*   We can consider either x or $-x$ as the GCF. We will use $-x$ since we are asked to factor out a negative sign with the GCF.

Step 2 Rewrite the terms $(-x)(x^2) + (-x)(7)$

Step 3 Factor $-x(x^2 + 7)$

b) $-6x^4 + 12x^2 - 9x$

> •SOLUTION• *Step 1 Find the GCF* We can consider either $3x$ or $-3x$ as the GCF. We will use $-3x$ since we are asked to factor out a negative GCF.
>
> *Step 2 Rewrite the terms* $(-3x)(2x^3) + (-3x)(-4x) + (-3x)(3)$
>
> *Step 3 Factor* $-3x(2x^3 + (-4x) + 3)$. We would write the final expression using subtraction as $-3x(2x^3 - 4x + 3)$.

Practice 7.2.3 Rewrite as a product by factoring out a negative sign with the GCF.

1. $-w^7 - 6w$ 　　　　 2. $-6m^5 + 12m$ 　　　　 3. $-4k^4 + 8k^3 - 12k$ 　　　　 4. $-6r^5 + 18r^4 - 12r$

Topic 7.2 B — Polynomial Greatest Common Factors

In some expressions the GCF will be a factor that consists of several terms. For example, in the expression $2x(x + 1) + 3(x + 1)$ the binomial $(x + 1)$ is the GCF. We factor out the GCF in such expressions using the same procedure as before. Study the following example:

Monomial GCF	*Binomial GCF*	
$2xy + 3y$	$2x(x + 1) + 3(x + 1)$	⇐ Expressions to factor.
$2xy + 3y$	$2x(\mathbf{x + 1}) + 3(\mathbf{x + 1})$	⇐ GCF is in **bold**.
$y(2x + 3)$	$(x + 1)(2x + 3)$	⇐ Factored out GCF.

The factorization of $2x(x + 1) + 3(x + 1)$ is $(x + 1)(2x + 3)$. Notice that in its unfactored form, the expression shows $(x + 1)$ in each term. But, in the factored form, $(x + 1)$ appears only once. Sometimes students ask where the other $(x + 1)$ went. It's still there because $(x + 1)(2x + 3)$ implies that $(x + 1)$ is to be multiplied by both $2x$ and 3. We have simply written the expression in a more compact form.

Example 7.2.4 Rewrite as a product by factoring out the GCF.

a) $7x(x - 3) + 5(x - 3)$

> •SOLUTION• *Step 1 Find the GCF* The GCF is $(x - 3)$.
>
> *Step 2 Rewrite the terms* $(7x)(x - 3) + (5)(x - 3)$
>
> *Step 3 Factor* $(x - 3)(7x + 5)$

b) $2y(y + 1) + 8(y + 1)$

> •SOLUTION• *Step 1 Find the GCF* The GCF is $2(y + 1)$.
>
> *Step 2 Rewrite the terms* $y \cdot 2(y + 1) + 4 \cdot 2(y + 1)$
>
> *Step 3 Factor* $2(y + 1)(y + 4)$

Practice 7.2.4 Rewrite as a product by factoring out the GCF.

1. $5z(z + 1) - 8(z + 1)$ 　　　　 2. $2t(t + 9) + 6(t + 9)$ 　　　　 3. $4m^2(n^2 - 8) - 12(n^2 - 8)$

4. $7(x + y)^4 - 14(x + y)^2 + 21(x + y)$ 　　　　 5. $12(m - n)^5 + 8(m - n)^3 - 4(m - n)$

Topic 7.2 C — Factoring by Grouping

Most of the products of binomials we have studied so far have simplified to trinomials. This was because the middle terms turned out to be like terms and so could be combined. This is not always the case. For example,

$$(x + 3)(x + y) \quad \Leftarrow \text{Expression to multiply.}$$

$$x^2 + xy + 3x + 3y \quad \Leftarrow \text{Used FOIL.}$$

We cannot combine the middle terms (the OI in FOIL) because they are not like terms.

When the polynomial we are trying to factor has 4 terms, like the one above, we use a procedure called **factoring by grouping**. This procedure uses the Associative Property to group pairs of terms into binomials which may have a GCF.

— *Procedure* —	*Step 1 Group* Group the four terms into a pair of binomials.
Factoring by grouping	*Step 2 Factor* Factor the GCF out of each binomial.
	Step 3 Factor again Factor the GCF out again.

Example 7.2.5   a)   Factor by grouping $2x^2 - 6x + xy - 3y$.

•SOLUTION• *Step 1 Group*   Group the four terms into a pair of binomials to get
$(2x^2 - 6x) + (xy - 3y)$

Step 2 Factor   Factor the GCF out of each binomial. The GCF of the first binomial is $2x$ and the GCF of the second binomial is y. Factoring out these GCFs we get $\mathbf{2x}(x - 3) + \mathbf{y}(x - 3)$.

Step 3 Factor again   Factor the GCF out again. We factor $x - 3$ out of each term to get $(\mathbf{x - 3})(2x + y)$.

b)   Factor by grouping $2p^2 + p + 6pq + 3q$

•SOLUTION• *Step 1 Group*   Group the four terms into a pair of binomials to get
$(2p^2 + p) + (6pq + 3q)$

Step 2 Factor   Factor the GCF out of each binomial to get
$p(2p + 1) + \mathbf{3q}(2p + 1)$.

Step 3 Factor again   Factor the GCF out again. We factor $(2p + 1)$ out of each term to get $(\mathbf{2p + 1})(p + 3q)$.

Practice 7.2.5 Factor by grouping.

1. $6p^2 + 4p + 3rp + 2r$
2. $10q^2 + 25rq + 14q + 35r$
3. $6x^2 + 2xy + 24xz + 8yz$

When negative signs are present we may have to factor a negative number out of one of the groups. For example, to factor $x^2 - xy - 9x + 9y$ we can factor x out of the first pair. But, for the second pair, do we factor out 9 or –9? Let's try both and see what happens. If we factor 9 out of the second pair we get:

$$x(x - y) + 9(-x + y)$$

The difficulty here is that there is no common factor. The first term has a factor of $(x - y)$ but the second term has a factor of $(-x + y)$. Let's try factoring –9 out of the second pair:

$$x(x - y) - 9(x - y)$$

Now we have a common factor of $(x - y)$. Factoring out the –9 changed the signs of both terms in the second pair. We can factor out the common $(x - y)$ to get $(x - y)(x - 9)$.

— Caution — **Be careful with signs**	If you find the signs confusing, you might try changing the subtractions into additions of the opposites. Doing this will make the signs easier to see. The downside is that the problem gets cluttered up with lots of + and − symbols. The best solution is to practice so much that you become automatic at working with negative signs.

Example 7.2.6   a)   Factor by grouping $p^2 - pw - p + w$.

> •SOLUTION•   ***Step 1 Group***   Group the four terms into a pair of binomial to get $(p^2 - pw) + (-p + w)$. Notice that we changed the subtraction of the third term to addition of the opposite before grouping. We have to do this in order to keep the signs of the second pair correct. It would be INCORRECT to write this as $(p^2 - pw) - (p + w)$ because this says the w term is subtracted, which it is not.
>
> ***Step 2 Factor***   Factor the GCF out of each binomial to get $\boldsymbol{p}(p - w) - \boldsymbol{1}(p - w)$.
>
> ***Step 3 Factor again***   Factor the GCF out again. We factor $(p - w)$ out of each term to get $\boldsymbol{(p - w)}(p - 1)$.

b)   Factor by grouping $15g^2 + 5g - 6gh - 2h$

> •SOLUTION•   ***Step 1 Group***   Group the four terms into a pair of binomials to get $(15g^2 + 5g) + (-6gh - 2h)$. Again, notice that we changed the subtraction of the third term to addition of the opposite before grouping.
>
> ***Step 2 Factor***   Factor the GCF out of each binomial to get $\boldsymbol{5g}(3g + 1) - \boldsymbol{2h}(3g + 1)$
>
> ***Step 3 Factor again***   Factor the GCF out again. We factor $(3g + 1)$ out of each term to get $\boldsymbol{(3g + 1)}(5g - 2h)$.

Practice 7.2.6   Factor by grouping.

1. $y^2 + xy - y - x$ 2. $3a^2 + 6ab - 2a - 4b$ 3. $6x^2 + 14xz - 9xy - 21yz$ 4. $10a^2 + 6ab - 35a - 21b$

Note that not all four–term polynomials can be factored using integers. For example, we cannot factor $x^2 + x + y + 2$.

Exercise Set 7.2   The answers to the odd numbered exercises are at the back of the book.

Rewrite as a product by factoring out the GCF.

1. $12r^4s^3 + 36r^6s^7$ 2. $25m^9n^4 - 10m^6n^2$ 3. $4x^6y^3 - 16x^2y^4 + 2xy$

4. $9s^2t^3 - 18s^7t^8 + 3st$ 5. $12x^5y^7 - 18x^8y^4 + 36x^3y^{12}$ 6. $18m^{11}n^5 - 27m^6n^9 + 81m^2n^7$

7. $15f^5g^{12} - 25f^3g^9 + 45f^{13}g$ 8. $36uv^4 - 24u^5v^3 + 15u^6v$

Rewrite as a product by factoring out 1. Then, rewrite as a product by factoring out −1.

9. $x + 2$ 10. $y - 3$ 11. $5t + u$ 12. $m - 5n$ 13. $4p^4 - 3p^3 + 9$

14. $5w^2 - 2w + 1$ 15. $-5x^3 + 7x - 9$ 16. $-12t^3 + 5t - 8$ 17. $25 - w^3$ 18. $2 - y^3$

Rewrite as a product by factoring out a negative sign with the GCF.

19. $-x^3 + 3x^2 - 5x$ 20. $-t^4 - 5t^3 + 7t^2$ 21. $-2xy^5 + 4xy^3 - 8xy^2 + 6xy$

22. $-3rs^5 + 9rs^7 - 6rs^3 + 12rs$ 23. $-8r^6s^3t^5 + 4r^3s^4t^2 - 12rst$ 24. $-15x^7y^4z^6 + 45x^9y^5z^8 - 25x^2y^3z^6$

Rewrite as a product by factoring out the GCF.

25. $2r(r^2 + 2) - 3(r^2 + 2)$ 26. $5w(w + 1) - 4(w + 1)$

27. $6r(r + 3) - 5r^2(r + 3)$ 28. $5w^2(w + 2) + 3w^3(w + 2)$

29. $16m^3(m + 5)^2 - 12m(m + 5)$

30. $9b^2(b + 2)^3 - 3b(b + 2)^4$

31. $4v(v + 4) - 12(v + 4)$

32. $6c(3c + 2) - 8(3c + 2)$

33. $8m(m + 2n)^5 + 12m(m + 2n)^4 - 16m(m + 2n)^3$

34. $25x(x + 5)^4 - 15x(x + 5)^2 - 35x(x + 5)$

Factor by grouping.

35. $mr + ms + nr + ns$

36. $ac + 2ad + bc + 2bd$

37. $2f^2 - 14f + fg - 7g$

38. $2x^2 + 2xy - 3x - 3y$

39. $15wt - 10t + 3w - 2$

40. $10xy - 6y - 25x + 15$

41. $12xz + 3x + 8yz + 2y$

42. $15mn + 5mt - 3n - t$

Factor by grouping

43. $2r^2 - 2r - 5s + 5sr$

44. $20m^2 - 8mn - 6n + 15m$

45. $ab + 5b - 7a - 35$

46. $mn + 9n - 4m - 36$

47. $8s^2 - 12s - 18rs + 27r$

48. $18rm - 6r - 15m + 5$

49. $8x^2 + 10xy - 4x - 5y$

50. $12b^2 - 3b + 8bc - 2c$

Review Exercises   The answers to all of these exercises are at the back of the book.

51. Simplify: $15 - 2(4 \cdot 3^2/9)$

52. Prime factor: 1287

53. Solve $a = 2(b + c)$ for c

54. Simplify: $(6m + 1)(2m + 3)$

55. Solve: $\begin{cases} x - y = -3 & ❶ \\ 2x + 3y = 4 & ❷ \end{cases}$

56. Graph: $y < -3x + 2$

57. Simplify: $x + 5[4x + 3(x - 2) - 3]$

58. Identify each base and its exponent, and then simplify: -3^6

59. Divide: $\dfrac{8a^4 + 6a^3 + a^2 + 16a + 4}{4a + 1}$

60. State the degree of the polynomial: $2a + 4a^4 + 2a^3 + 4$

---

61. You have a jar of quarters and nickels whose face value totals $28.30. If you have 100 more quarters than nickels, how many nickels do you have?

62. Plot on an x–y coordinate system: $f(2) = 3, f(3) = 9, f(-1) = -4$, and $f(0) = 3$

63. Simplify and write the answer in scientific notation: $\dfrac{(7.4 \times 10^3)(1.5 \times 10^2)}{2.22 \times 10^4}$

---

64. Find the x-intercept and the y-intercept:

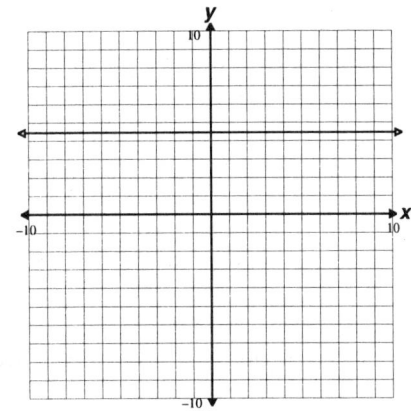

65. The equation $w(d) = -\dfrac{1}{3}d + 218$ models Byron's pounds of weight loss, w, after d days on a weight loss program. Express in English the meaning of each of the following:

   a) $w(60)$

   b) $w(d) = 190$

   c) $w(30) = 208$

   d) $w(d) = w(18) - 7$

Section 7.3 Factoring Trinomials of the Form $x^2 + bx + c$

In Section 7.6, we used FOIL to multiply $(x + m)(x + n)$ to get $x^2 + (m + n)x + mn$. We noted that the product has three important characteristics:

- decreasing powers of x.
- the coefficient of the x term is the sum $m + n$.
- the last term is the product mn.

We can use this formula in reverse to factor a trinomial that has the form $x^2 + bx + c$:

— Procedure — *Product–Sum method of* *factoring a trinomial*	***Step 1 Standard Form*** Write the polynomial in standard form (that is, combine like terms and write all the terms in decreasing powers of the variable). ***Step 2 GCF*** If there is a GCF, factor it out so that the other factor has the form $x^2 + bx + c$. ***Step 3 Identify b and c*** Identify the values of b and c in $x^2 + bx + c$. If the first degree (middle) term is missing, b is 0. ***Step 4 Find Integers*** Find two integers whose product is c and whose sum is b. Think of this as (__)(__) = c and __ + __ = b ***Step 5 Use Integers*** Use the integers found in Step 4 as the constant terms in the binomial factors.

For example, $x^2 + 5x + 6$ has the form $x^2 + bx + c$, where b is 5 and c is 6. To factor this, we find two integers whose product is 6 and whose sum is 5. The numbers 2 and 3 fit this description. Therefore, the factorization is $(x + 2)(x + 3)$.

Example 7.3.1   a)   Factor $x^2 + 8x + 12$

•SOLUTION•   ***Step 1 Standard Form***   The trinomial is already in standard form.

Step 2 GCF   There is no GCF. (Remember, by this we mean that there are no variables or integers, other than 1 or –1, which are common to all the factors of the polynomial.)

Step 3 Identify b and c b is 8 and c is 12.

Step 4 Find Integers   We want to find two integers whose product is 12 and whose sum is 8. That is, (__)(__) = 12 and __ + __ = 8. They are 6 and 2.

Step 5 Use Integers   We use 6 and 2 as the constant terms of the binomial factors. The factored form is $(x + 2)(x + 6)$.

b)   Factor $x^2 + x - 6$

•SOLUTION•   ***Step 1 Standard Form***   The trinomial is already in standard form.

Step 2 GCF   There is no GCF.

Step 3 Identify b and c   Mentally, think of the trinomial as $x^2 + 1x + (-6)$. So, b is 1 and c is –6.

Step 4 Find Integers   We want to find two integers whose product is –6 and whose sum is 1. That is, (__)(__) = –6 and __ + __ = 1. They are –2 and 3.

Step 5 Use Integers   We use –2 and 3 as the constant terms of the binomials. The factored form is $(x + (-2))(x + 3)$. We would write this as $(x - 2)(x + 3)$.

c) Factor $-6x + 15 + x^2 - 2x$

•SOLUTION• **Step 1 Standard Form**   Combine like terms and write the terms in decreasing powers of x to get $x^2 - \mathbf{8}x + 15$.

Step 2 GCF   There is no GCF.

Step 3 Identify b and c b is -8 and c is 15.

Step 4 Find Integers   We want to find two integers whose product is 15 and whose sum is -8. That is, $(\_)(\_) = 15$ and $\_ + \_ = -8$. They are -5 and -3.

Step 5 Use Integers   We use -5 and -3 as the constant terms of the binomials. The factored form is $(x + (-5))(x + (-3))$. We would write this as $(x - 5)(x - 3)$.

Practice 7.3.1   Factor.

1. $x^2 + 8x + 12$

2. $x^2 + 20x + 99$

3. $x^2 - 2x - 15$

4. $x^2 + 2x - 99$

5. $11x + 30 + x^2$

6. $-10x + 72 + x^2 - 7x$

In the next chapter we will use this type of factoring to reduce algebraic fractions like $\dfrac{x^2 + 5x + 6}{x^2 + 7x + 10}$. Factoring the numerator and denominator gives us $\dfrac{(x + 2)(x + 3)}{(x + 2)(x + 5)}$. Since $\dfrac{x + 2}{x + 2}$ reduces to 1 the original fraction reduces to $1 \cdot \dfrac{x + 3}{x + 5}$, which simplifies to $\dfrac{x + 3}{x + 5}$.

If you need to factor an expression such as $x^2 - 5x - 36$, the two integers whose product is -36 and sum is -5 may not be obvious. In these cases, you should write out a list of possibilities. Start the list with $1 \cdot c$, then write $2 \cdot \dfrac{c}{2}$, then write $3 \cdot \dfrac{c}{3}$ and so on until the second factor is smaller than the first. Here is the list when c is 36:

Product		Sum
1 •	-36	-35
2 •	-18	-16
3 •	-12	-9
4 •	-9	-5
5 •		$\Leftarrow$ we don't use 5 because 36 is not divisible by 5
6 •	-6	0
7 •		$\Leftarrow$ we stop here because the second factor now is smaller than the first

Since 4 and -9 multiply to -36 and add to -5 we use those. Thus, $x^2 - 5x - 36$ factors into $(x + 4)(x - 9)$.

— *Note* —	If you have a well developed number sense you may not have to write out the factor pairs as we did in this example. It's usually a good idea to take 10 seconds and see if the integers pop into your head. If they don't, systematically write out the possible factor pairs, using a calculator if necessary. This is easy to do and if the integers exist you will find them.
Try factoring mentally *first*	

Example 7.3.2   a)   Factor $x^2 + 12x - 64$

> •SOLUTION• *Step 1 Standard Form*   The trinomial is already written in standard form.
>
> *Step 2 GCF*   There is no GCF.
>
> *Step 3 Identify b and c*   *b* is 12 and *c* is –64
>
> *Step 4 Find Integers*   We want to find two integers whose product is –64 and whose sum is 12. Since the product is negative, the factors must have different signs. Since the sum is positive, the larger factor must be positive. You may find it difficult to find the numbers without listing the possibilities:
>
> | Product | Sum |
> |---------|-----|
> | –1 • 64 | 63 |
> | –2 • 32 | 30 |
> | –4 • 16 | 12 |
>
> We see that –4 and 16 have a product of –64 and a sum of 12 so we can stop listing factors at this point.
>
> *Step 5 Use Integers*   We use –4 and 16 as the constant terms of the binomials. The factored form is $(x + (-4))(x + 16)$, which we will write as $(x - 4)(x + 16)$.

b)   Factor $x^2 + 5x + 18$

> •SOLUTION• *Step 1 Standard Form*   The trinomial is already written in standard form.
>
> *Step 2 GCF*   There is no GCF.
>
> *Step 3 Identify b and c*   *b* is 5 and *c* is 18
>
> *Step 4 Find Integers*   We need to find two integers whose product is 18 and whose sum is 5. Here is a list of products and sums:
>
> | Product | Sum |
> |---------|-----|
> | 1 • 18 | 19 |
> | 2 • 9 | 11 |
> | 3 • 6 | 9 |
>
> Since there are no two integers whose product is 18 and whose sum is 5 we can say that $x^2 + 5x + 18$ is not factorable using integers.

c)   Factor $x^2 - 25$

> •SOLUTION• *Step 1 Standard Form*   The trinomial is already written in standard form.
>
> *Step 2 GCF*   There is no GCF.
>
> *Step 3 Identify b and c*   Since there is no first degree (middle) term *b* is 0. The value of *c* is –25.
>
> *Step 4 Find Integers*   We need to find two integers whose product is 0 and whose sum is –25. Here is a list of products and sums:
>
> | Product | Sum |
> |---------|-----|
> | –1 • 25 | 24 |
> | –5 • 5 | 0 |
>
> We see that –5 and 5 have a product of –25 and a sum of 0.
>
> *Step 5 Use Integers*   We use –5 and 5 as the constant terms of the binomials. The factored form is $(x + (-5))(x + 5)$, which we write as $(x - 5)(x + 5)$

Practice 7.3.2 Factor.

1. $x^2 - 14x - 72$

2. $y^2 - 49$

3. $r^2 + 17r - 60$

4. $21x + x^2 - 100$

5. $-96 + 29x + x^2$

6. $x^2 + 7x + 15$

So far we have only worked with trinomials where the coefficient of the x^2 term is 1. When this is not the case, we will try to factor out a GCF. If factoring a GCF leaves the x^2 term with a coefficient of 1 we try to factor the remaining trinomial using the product–sum method.

Example 7.3.3   a)   Factor completely $4x^2 + 8x - 60$

•SOLUTION•   *Step 1 Standard Form*   The trinomial is already written in standard form.

Step 2 GCF   The GCF is 4. Factoring this out of the trinomial gives $4(x^2 + 2x - 15)$. Notice the second degree term now has a coefficient of 1.

Step 3 Identify b and c b is 2 and c is -15.

Step 4 Find Integers   Two integers whose product is -15 and whose sum is 2 are 5 and -3.

Step 5 Use Integers   We use 5 and -3 as the constant terms of the binomials. The factored form is $(x + 5)(x + (-3))$. Along with the GCF, we write the final factored form as $4(x + 5)(x - 3)$.

— Note — *Factoring versus* *Factoring Completely*	Note that the direction for this example says *Factor completely* rather than *Factor*. To *factor* means to write as a product; to **factor completely** means to write as a product where each factor cannot be further factored using integers other than 1 or -1. For example, • an acceptable answer to Factor $4x^2 + 8x - 60$ is $4(x^2 + 2x - 15)$. • the answer to Factor completely $4x^2 + 8x - 60$ is $4(x + 5)(x - 3)$. However, be aware that when your teacher asks you to factor something, he or she probably wants you to factor it *completely*.

b)   Factor completely $-x^2 + 10x - 24$

•SOLUTION•   *Step 1 Standard Form*   The trinomial is already written in standard form.

Step 2 GCF   We want the coefficient of x^2 to be 1 so we will factor out a GCF of -1 to get $-1(x^2 - 10x + 24)$. At this point, we have factored the trinomial but we have not factored it completely.

Step 3 Identify b and c b is -10 and c is 24.

Step 4 Find Integers   Two integers whose product is 24 and whose sum is -10 are -4 and -6.

Step 5 Use Integers   We use -4 and -6 as the constant terms of the binomials. The factored form is $(x + (-4))(x + (-6))$. Along with the GCF, we write the final factored form as $-1(x - 4)(x - 6)$.

c)   Factor completely $-70x + 3x^3 - 15x^2 - 2x$

•SOLUTION•   *Step 1 Standard Form*   We combine like terms and then write the trinomial in descending powers of x to get $3x^3 - 15x^2 - 72x$.

Step 2 GCF   We factor out the GCF of $3x$ to get $3x(x^2 - 5x - 24)$. At this point, we have factored the trinomial but we have not factored it completely.

Step 3 Identify b and c b is -5 and c is -24.

Step 4 Find Integers   Two integers whose product is -24 and whose sum is -5 are 3 and -8.

Step 5 Use Integers   We use 3 and -8 as the constant terms of the binomials. The factored form is $(x + 3)(x + (-8))$. Along with the GCF, we write the final factored form as $3x(x + 3)(x - 8)$.

Practice 7.3.3 Factor completely.

1. $5x^2 + 40x + 75$

2. $3x^2 + 27x + 42$

3. $x^2 + 12x - 3$

4. $60 - x^2 - 7x$

5. $4x^3 + 160x + 52x^2$

6. $100x + 5x^3 + 65x^2 + 80x$

Exercise Set 7.3   The answers to the odd numbered exercises are at the back of the book.

Factor completely.

1. $x^2 + 11x + 28$
2. $m^2 + 7m + 10$
3. $x^2 + 13x + 40$
4. $x^2 + 15x + 36$

5. $x^2 + 6x - 27$
6. $x^2 - 3x - 40$
7. $m^2 - 9m - 36$
8. $n^2 - 7n - 44$

9. $x^2 - 18x + 45$
10. $y^2 - 16y + 48$
11. $t^2 - 17t + 72$
12. $p^2 - 19p + 84$

13. $y^2 - 27y + 72$
14. $x^2 + 14x + 40$
15. $m^2 + 11m - 60$
16. $y^2 + 12y + 28$

17. $x^2 - 32x - 144$
18. $x^2 - 73x - 150$
19. $y^2 - 22y + 12$
20. $y^2 - 6y - 432$

21. $t^2 - 28t + 192$
22. $s^2 - 7s - 144$
23. $w^2 - 8w + 19$
24. $u^2 - 144$

25. $6x^2 - 36x - 162$
26. $3x^2 - 6x - 144$
27. $-x^2 + x + 12$
28. $3x^2 - 33x + 90$

29. $5x^2 - 4x + 2$
30. $30x + 4x^2 + 6x - 144$
31. $12 - x^2 + x$
32. $3x^2 + 3x + 2$

33. $-x^2 + 9x + 22$
34. $-30x + 80 - 5x^2$
35. $18x^2 + 165x - 3x^3$
36. $30x^2 + 5x^3 + 20x^2 - 120x$

Review Exercises   The answers to all of these exercises are at the back of the book.

37. The total cost for a cup of pop and a box of popcorn is $4. If the pop cost $1 more than the box of popcorn how much did the box of popcorn cost?

38. A firefighter invested $11,000 of her $15,000 retirement savings in one account and the rest in a second account whose rate of return was 8 percentage points less. Taken together, the value of the accounts increased by $467.50. What was the rate of growth of each account?

39. Find the slope and y-intercept of the line described by $x + y = 0$

40. Computer Depot rents computers for $100 down and $50 per month. Pica Computer Company rents computes for $60 with no money down.

   a) Model this situation with a 2 by 2 system of linear equations using time as the input and cost as the output.
   b) Graph the system. The input axis should go from 0 to 4 months in steps of 0.5 month; the output axis should go from $0 to $300 in steps of $20.
   c) Find the point of intersection and explain what it means.d)Which company would cost less money to rent from?

41. Use graphing to solve: $\begin{cases} x - 2y = -3 \ -4x + 8y = 12 \end{cases} \begin{matrix} ❶ \ ❷ \end{matrix}$

42. Solve: $\begin{cases} \frac{1}{2}x + y = 2 \ \frac{2}{3}x + \frac{1}{2}y = 6 \end{cases} \begin{matrix} ❶ \ ❷ \end{matrix}$

43. Simplify: $\dfrac{64p^7 + 32p^5 - 28p^3}{4p}$

44. Convert to scientific notation: 3100×10^5

45. Find the GCF of $12x^2$ and $18xy$

46. Factor $-2xy^3 - 2y^2$

Simplify.

47. $-3^2 + 3^{-1} - 3(-1)$

48. $\left(\dfrac{a^5}{a^{-3}}\right) \cdot (2a)$

49. $\left(\dfrac{4a^2b^3}{\left(a^3b^2\right)^{-2}}\right)^2$

50. $(4x^2y - 2y) + (3x - y) + (4y + x^2y)$

51. $(2c - 5)(3c - 1)$

52. $(2x - 1)^2$

53. $2n^2(3n)^2$

54. $(4b^2 - 6b + 1) - (2b + 8b^2) - (b - 5)$

55. Calculate the slope of the straight line described by the following data table:

x	y
4	−5
0	−1

56. Graph the following data. Use a scale that has 1992 as year 0 and displays the sales in thousands of dollars.

x year	y Sales (dollars)
1992	6300
1993	8100
1996	13500
1998	17100

Section 7.4 Factoring Trinomials of the Form $ax^2 + bx + c$

Factoring is more complicated when we cannot factor out a GCF to make the coefficient of the second degree term 1. In those cases, we have two tools that are useful. The first, which we call the "Guess and Check" method, is fast if the numbers involved are prime and small; the second tool, which we call the "ac" method, involves several steps that systematically lead to an answer.

Topic 7.4 A — Guess and Check Method of Factoring Trinomials

In the **Guess and Check** method, we make use of FOIL to make an educated guess at the binomial factors, Then, we use FOIL to multiply the binomials and check to see if we end up with the original trinomial.

Example 7.4.1 a) Use "Guess and Check" to factor $3x^2 + 4x + 1$.

•SOLUTION• • Since all the terms have positive coefficients we know that all the constants must be positive.

• FOIL tells us that the first term of the trinomial, $3x^2$, is the product of the first terms of the binomial factors. Therefore, a good guess for the first terms of the binomials would be $(3x)(1x)$.

• FOIL tells us that the last term of the trinomial, 1 , is the product of the last terms of the binomial factors. Therefore, a good guess for the last terms of the binomials would be $(1)(1)$.

Given the above conditions, the only guess we can reasonably make is $(3x + 1)(x + 1)$. We can use FOIL to check this guess.

$$(3x + 1)(x + 1) \quad \Leftarrow \text{Expression to multiply.}$$

$$(3x)(x) + (3x)(1) + (1)(x) + (1)(1) \quad \Leftarrow \text{Used FOIL to expand.}$$

$$3x^2 + 4x + 1 \quad \Leftarrow \text{Simplified.}$$

This shows that the original guess, $(3x + 1)(x + 1)$, is correct.

b) Use "Guess and Check" to factor $2x^2 - x - 6$.

•SOLUTION• • Since some terms have negative coefficients we know that some constants will be negative.

• We want a pair of binomials where the product of the first terms is $2x^2$. Possible integer factors of $2x^2$ are $(1x)(2x)$ and $(-1x)(-2x)$.

• We want a pair of binomials where the product of the last terms is −6. Possible integer factors are $(1)(-6)$, $(2)(-3)$, $(-1)(6)$ and $(-2)(3)$.

Let's try the various possibilities.

Guess	FOILed	Simplified
$(1x + 1)(2x - 6)$	$2x^2 - 6x + 2x - 6$	$2x^2 - 4x - 6$
$(1x + 2)(2x - 3)$	$2x^2 - 3x + 4x - 6$	$2x^2 + 1x - 6$
$(1x - 1)(2x + 6)$	$2x^2 + 6x - 2x - 6$	$2x^2 + 4x - 6$
$(1x - 2)(2x + 3)$	$2x^2 + 3x - 4x - 6$	$2x^2 - x - 6$

We can stop at this point because we have found a pair of binomials with the desired product. The factorization is $(1x - 2)(2x + 3)$.

Practice 7.4.1 Use the "Guess and Check" method to factor.

1. $5x^2 + 6x + 1$
2. $3x^2 - x - 2$
3. $7x^2 - 22x + 3$

If you have a good number sense and if the numbers involved are not too large, guess and check may be a good method to try.

Topic 7.4 B — The "ac" Method of Factoring Trinomials

The **"ac" method of factoring trinomials** is a step by step procedure that uses a variation of factoring by grouping.

— Procedure — *"ac" method of factoring* *a trinomial*	***Step 1 Standard Form*** Write the polynomial in standard form (that is, combine like terms and write all the terms in decreasing powers of the variable). ***Step 2 GCF*** If there is a GCF, factor it out so that the other factor has the form $ax^2 + bx + c$. ***Step 3 Identify a, b, and c*** Identify the values of a, b, and c in $ax^2 + bx + c$. If any term is missing, its coefficient is 0. ***Step 4 Find Integers*** Find two integers whose product is ac and whose sum is b. ***Step 5 Replace middle term*** Replace the middle term, bx, with a sum using the two integers found in step 4. ***Step 6 Factor*** Factor by grouping.

Let's try this method on $2x^2 + 7x + 3$:

Step 1 Standard Form   The trinomial is already in standard form.

Step 2 GCF   There is no GCF.

Step 3 Identify a, b, and c

$$ax^2 + bx + c$$

$$2x^2 + 7x + 3$$

In this example, a is 2, b is 7, and c is 3.

Step 4 Find Integers   We need to find two integers whose product is ac (that is, $2 \cdot 3$, or 6) and whose sum is b (that is, 7). That is, $(\_\_)(\_\_) = 6$ and $\_\_ + \_\_ = 7$. The numbers are 1 and 6.

Step 5 Replace middle term   We replace the middle term, $7x$, with the sum $1x + 6x$. Note that we could also use $6x + 1x$ because addition is commutative.

$$2x^2 + 7x + 3$$

$$2x^2 + \overbrace{1x + 6x} + 3$$

Step 6 Factor

$2x^2 + 1x + 6x + 3$ ⇐ Expression to factor by grouping.

$x(2x + 1) + 3(2x + 1)$ ⇐ Factored x out of the first pair and 3 out of the second pair.

$(2x + 1)(x + 3)$ ⇐ Factored out GCF of $(2x + 1)$.

Thus, $2x^2 + 7x + 3$ factors into $(2x + 1)(x + 3)$. We may check to see that the factorization is correct by multiplying the binomials. We leave the check to you.

This process may seem long, but with a little practice you will be able to do it fairly quickly.

Example 7.4.2   a)   Use the "ac" method to factor $10x^2 + 17x + 3$.

•**SOLUTION**• ***Step 1 Standard Form***   The trinomial is already in standard form.

Step 2 GCF   There is no GCF.

Step 3 Identify a, b, and c a is 10, b is 17, c is 3.

Step 4 Find Integers   We seek two integers whose product is ac, which is 10•3 or 30 and whose sum is b, which is 17. That is, (__)(__) = 30 and __ + __ = 17. The integers are 2 and 15.

Step 5 Replace middle term   We replace the middle term, $17x$, with $2x + 15x$. We could also use $15x + 2x$ because addition is commutative.

Step 6 Factor

$10x^2 + 17x + 3$ ⇐ Expression to factor.

$10x^2 + 2x + 15x + 3$ ⇐ Replaced middle term.

$2x(5x + 1) + 3(5x + 1)$ ⇐ Factored $2x$ out of first pair and 3 out of second pair.

$(5x + 1)(2x + 3)$ ⇐ Factored out GCF of $(5x + 1)$.

b)   Use the "ac" method to factor $5 + 6x^2 - 13x$

•**SOLUTION**• ***Step 1 Standard Form***   We write the terms in descending powers of x to get $6x^2 - 13x + 5$.

Step 2 GCF   There is no GCF.

Step 3 Identify a, b, and c a is 6, b is –13, and c is 5.

Step 4 Find Integers   We seek two integers whose product is ac, which is 6•5 or 30 and whose sum is –13. That is, (__)(__) = 30 and __ + __ = –13. They are –3 and –10.

Step 5 Replace middle term   We replace the middle term, $-13x$, with $-3x - 10x$ or $-10x - 3x$.

Step 6 Factor

$6x^2 - 13x + 5$ ⇐ Expression to factor.

$6x^2 - 3x - 10x + 5$ ⇐ Replaced middle term.

$3x(2x - 1) - 5(2x - 1)$ ⇐ Factored $3x$ out of first pair and –5 out of second pair.

$(2x - 1)(3x - 5)$ ⇐ Factored out GCF of $(2x - 1)$.

c)   Use the "ac" method to factor $32x^2 - 20x - 3$

•SOLUTION•   ***Step 1 Standard Form***   The trinomial is already in standard form.

Step 2 GCF   There is no GCF.

Step 3 Identify a, b, and c a is 32, b is –20, and c is –3.

Step 4 Find Integers   We seek two integers whose product is ac, which is $(32)(-3)$, or –96 and whose sum is –20. This one may be too difficult to do mentally, so we will begin to write out the integer factorizations of –96:

Product	Sum
1 • –96	–95
2 • –48	–46
3 • –32	–29
4 • –24	–20

We can stop here because we have found that 4 and –24 have a product of –96 and a sum of –20.

Step 5 Replace middle term   We replace the middle term, $-20x$, with $4x - 24x$.

Step 6 Factor

$$32x^2 - 20x - 3 \quad \Leftarrow \text{Expression to factor.}$$

$$32x^2 + \mathbf{4x - 24x} - 3 \quad \Leftarrow \text{Replaced middle term.}$$

$$\mathbf{4x}(8x + 1) - \mathbf{3}(8x + 1) \quad \Leftarrow \text{Factored } 4x \text{ out of first pair and } -3 \text{ out of second pair.}$$

$$(\mathbf{8x + 1})(4x - 3) \quad \Leftarrow \text{Factored out GCF of } (8x + 1).$$

d)   Factor completely $12x^2 + 44x - 56$

•SOLUTION•   ***Step 1 Standard Form***   The trinomial is already in standard form.

Step 2 GCF   There is a GCF of 4, so we factor it out to get $4(3x^2 + 11x - 14)$.

Step 3 Identify a, b, and c a is 3, b is 11, and c is –14.

Step 4 Find Integers   We seek two integers whose product is ac, which is $(3)(-14)$ or –42 and whose sum is 11. Because this one may be too difficult to do mentally, we will begin to write out the integer factorizations of –42. Notice the larger number is positive since b is positive:

Product	Sum
–1 • 42	41
–2 • 21	19
–3 • 14	11

We can stop here because we found that –3 and 14 have a product of –42 and a sum of 11.

Step 5 Replace middle term   We replace the middle term, $11x$, with $-3x + 14x$.

Step 6 Factor

$$12x^2 + 44x - 56 \quad \Leftarrow \text{Expression to factor.}$$

$$\mathbf{4}(3x^2 + 11x - 14) \quad \Leftarrow \text{Factored out GCF of 4.}$$

$$4(3x^2 \mathbf{- 3x + 14x} - 14) \quad \Leftarrow \text{Replaced middle term.}$$

$$4(\mathbf{3x}(x - 1) + \mathbf{14}(x - 1)) \quad \Leftarrow \text{Factored } 3x \text{ out of first pair and 14 out of second pair.}$$

$$4((\mathbf{x - 1})(3x + 14)) \quad \Leftarrow \text{Factored out GCF of } (x - 1).$$

e) Factor completely $2x^2 + 5x + 1$

•SOLUTION• *Step 1 Standard Form* The trinomial is already in standard form.

Step 2 GCF There is no GCF.

Step 3 Identify a, b, and c *a* is 2, *b* is 5, and *c* is 1

Step 4 Find Integers We seek two integers whose product is *ac* , which is 2•1 or 2, and whose sum is 5. The possible integer factorizations of 2 are

Product	**Sum**
1 • 2	3
–1 • –2	–3

Since there are no two integers whose product is 2 AND whose sum is 5, the trinomial $2x^2 + 5x + 1$ cannot be factored using integers. [†]

Practice 7.4.2 Use the "ac" method to factor completely.

1. $4w^2 + 9w + 5$
2. $5m^2 + 11m + 6$
3. $3x^2 - 13x - 10$
4. $12w^2 + 17w - 5$

5. $8z^2 + 18z + 10$
6. $12m^2 - 10m + 2$
7. $4x^2 + 7x + 5$

Exercise Set 7.4   The answers to the odd numbered exercises are at the back of the book.

Use the "Guess and Check" method to factor completely.

1. $2x^2 + 5x + 2$
2. $3x^2 + 8x + 4$
3. $35x^2 + 3x - 2$
4. $6x^2 + 5x - 6$

5. $3r^2 - 10r + 7$
6. $7m^2 - 17m + 6$

Use the "ac" method to factor completely.

7. $12z^2 + 20z + 3$
8. $14s^2 + 11s + 2$
9. $6m^2 - 11m + 3$
10. $15z^2 - 16z + 4$

11. $6y^2 - 7y - 20$
12. $18x^2 + 3x - 10$
13. $3r^2 - 2r - 9$
14. $10x^2 - 3x - 4$

15. $12y^2 - 7y - 10$
16. $6m^2 - 5m - 6$
17. $6 + 4w^2 - 11w$
18. $15 + 2t^2 - 11t$

19. $12t^2 - 13t + 3$
20. $15x^2 - 11x + 2$
21. $12x^2 - 5x - 2$
22. $10y^2 + 6y - 3$

23. $50x^2 - 55x + 15$
24. $16w^2 - 28w + 10$
25. $6 + 21m + 15m^2$
26. $12 + 8x^2 - 20x$

27. $8m^2 + 4m - 40$
28. $9x^2 + 39x - 30$
29. $-2 + 3y^2 - 4y$
30. $-y + 5y^2 - 2$

Review Exercises   The answers to all of these exercises are at the back of the book.

31. Simplify: $8x - (7 - x) + 2(3 - 8x)$

32. Solve: $4 - 2(5 - x) = 2x - 3(x - 3)$

33. Solve: $5x - 2 = 3(1 - 2x) - 5$

34. Factor $6x^2y^5 - 3xy^4$

35. Determine the domain and range of $y = 3x - 1$

36. Given $f(x) = 2x^3 - 5x + 7$, find $f(2)$.

37. Solve: $\begin{cases} x - 5y = 3 \\ 3x + y = 9 \end{cases} \begin{matrix} ❶ \\ ❷ \end{matrix}$

38. Solve: $\begin{cases} 8x - 5y = 5 \\ 3x + 3y = -3 \end{cases} \begin{matrix} ❶ \\ ❷ \end{matrix}$

39. Use long division to simplify: $\dfrac{3x^4 + x^3 - x^2 + 2x + 1}{x + 1}$

40. Find the GCF of $4x^2(x + 2)^3$ and $6x^2(x + 2)^4$

41. It takes 1 hour to build a standard VCR circuit board and 30 minutes to install it . To build an extended VCR circuit board takes 2 hours but it takes only 15 minutes to install it. If the machinery to build the boards can run for 140 hours per week and the technicians needed to install the boards can work 40 hours per week, how many of each type of circuit board can be built and installed?

[†]   In a later math class you may learn how to factor this trinomial using non–integer numbers including fractions and square roots.

42. Ned is driving from Hicksville to Renton and notices as he passes the midway point that it has been 1 hour and 15 minutes since he has left Hicksville. On his return trip, the traffic is heavier and so he must reduce his speed by 10. This time, he reaches the midway point 1 hour and 30 minutes after leaving. How fast was he going driving from Hicksville to Renton?

43. Construct the equation of the line shown in the graph.   44. Discuss the meaning of the graph.

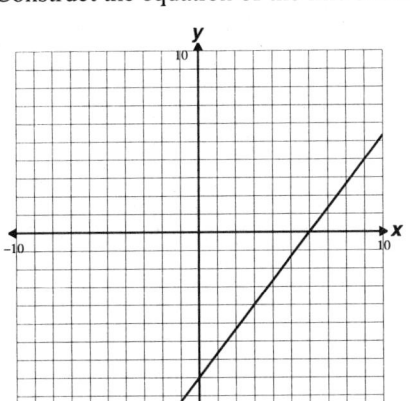

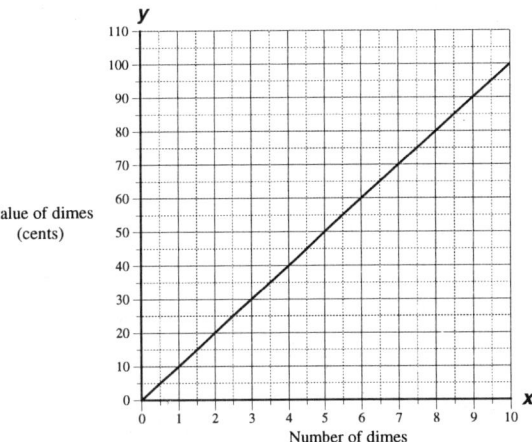

Value of dimes (cents)

Number of dimes

45. Graph the straight line that passes through $(-1, 4)$ with slope 0.

46. Find the equation of a line perpendicular to $y = -2$ and which passes through $(4, 1)$.

47. Decide whether the two straight lines described by the following data tables are parallel, perpendicular or neither:

Line 1 comes from the data table

x	y
3	−1
−2	5

Line 2 comes from the data table

x	y
0	1
8	−3

Simplify.

48. $\dfrac{(c^2 d^{-3})^2}{c^{-3}} \cdot \dfrac{d^4}{c^{-2}}$

49. $\left(\dfrac{6m}{m^3 n7}\right)^2 \left(\dfrac{m^5}{n^{-3}}\right)^2$

50. $8y(6y^3 - 4y^2 + 2)$

51. $(2y - 1)(4y^2 + 2y + 1)$

Section 7.5 Factoring Special Products and General Factoring

Some polynomials have special forms that allow us to factor them quickly by inspection.

Topic 7.5 A — Polynomials That Are the Difference of Two Perfect Squares

Recall that conjugates are the sum and difference of the same two expressions, as in $(2x + 3y)$ and $(2x - 3y)$. Recall what happens when we multiply conjugates:

$$(2x - 3y)(2x + 3y) \quad \Leftarrow \text{ Conjugates to be multiplied.}$$

$$4x^2 + 6xy - 6xy + 9y^2 \quad \Leftarrow \text{ Multiplied using FOIL.}$$

$$4x^2 - 9y^2 \quad \Leftarrow \text{ Combined like terms.}$$

The two middle terms turned out to be opposites and so added to 0. This leaves us with a binomial that consists of the difference of two **perfect squares**, $(2x)^2$ and $(3y)^2$. A perfect square is the square of an integer, the square of a variable, or the square of the product of integers and variables.

Looking at the process in reverse, we can see that the difference of two perfect squares will factor into the product of conjugates. This gives us a shortcut for factoring this special type of polynomial.

— *Procedure* —	**Step 1** Write each term as a perfect square.
Factoring the difference	**Step 2** Write the conjugate factors.
of two perfect squares	

Example 7.5.1   a)   Factor completely $x^2 - 9$

   •**SOLUTION**•

$$(x)^2 - (3)^2 \quad \Leftarrow \text{ Wrote as the difference of two perfect squares.}$$

$$(x - 3)(x + 3) \quad \Leftarrow \text{ Wrote as the product of conjugates.}$$

   b)   Factor completely $16p^2 - 49q^2$

   •**SOLUTION**•

$$(4p)^2 - (7q)^2 \quad \Leftarrow \text{ Wrote as the difference of two perfect squares.}$$

$$(4p - 7q)(4p + 7q) \quad \Leftarrow \text{ Wrote as the product of conjugates.}$$

Practice 7.5.1   Factor completely.

1. $m^2 - 81$ 2. $81w^2 - 4$ 3. $4r^2 - 9s^2$ 4. $36q^2 - 25d^2$

The key to factoring this type of polynomial is to recognize it as the difference of two perfect squares.

Example 7.5.2   If the polynomial is not the difference of two squares, explain why. If possible, factor completely.

   a) $x^2 + 4$

   •**SOLUTION**• This is the sum of two perfect squares, not the difference. We cannot factor the sum of two perfect squares using real numbers.

   b) $x^2 - 5$

   •**SOLUTION**• This is not the difference of two perfect squares because 5 is not a perfect square.

   c) $36x^2 - 1$

   •**SOLUTION**•

$$(6x)^2 - (1)^2 \quad \Leftarrow \text{ Wrote as the difference of two perfect squares.}$$

$$(6x - 1)(6x + 1) \quad \Leftarrow \text{ Wrote as the product of conjugates.}$$

   d) $x^4 - 9$

   •**SOLUTION**• The term x^4 may not look like a perfect square, but it is because we can write it as $(x^2)^2$.

$$(x^2)^2 - (3)^2 \quad \Leftarrow \text{ Wrote as the difference of two perfect squares.}$$

$$(x^2 - 3)(x^2 + 3) \quad \Leftarrow \text{ Wrote as the product of conjugates.}$$

e) $-49x^2 + 4$

•SOLUTION•

$-(49x^2 - 4)$ $\Leftarrow$ Factored out a negative. The expression inside the parentheses is the difference of two perfect squares.

$-\left[(7x)^2 - (2)^2 \right]$ $\Leftarrow$ Wrote as the difference of two perfect squares.

$-(7x - 2)(7x + 2)$ $\Leftarrow$ Wrote as the product of conjugates.

An alternative way of factoring $-49x^2 + 4$ is to use the Commutative Property of Addition to it as $4 - 49x^2$, which is the difference of two perfect squares.

$(2)^2 - (7x)^2$ $\Leftarrow$ Wrote as the difference of two perfect squares.

$(2 - 7x)(2 + 7x)$ $\Leftarrow$ Wrote as the product of conjugates.

Thus, we may write the factored form as $-(7x - 2)(7x + 2)$ or $(2 - 7x)(2 + 7x)$.

Practice 7.5.2   If the polynomial is not the difference of two squares, explain why. If possible, factor completely.

1. $100a^2 - 81b^2$ 2. $9z^2 - 49v^2$ 3. $4x^2 + 49z^2$ 4. $2m^2 - 32$

5. $x^2 + 2x - 1$ 6. $25x^4 - 9$ 7. $-1 + 9x^2$ 8. $-36x^2 + 25$

Topic 7.5 B — Polynomials That Are Perfect Square Trinomials

The square of a binomial is another special product that can be factored quickly by recognizing its structure. To see how this works, let's expand $(2x + 3)^2$.

$(2x + 3)(2x + 3)$ $\Leftarrow$ Used the definition of an exponent.

$4x^2 + 6x + 6x + 9$ $\Leftarrow$ Multiplied using FOIL.

$4x^2 + 12x + 9$ $\Leftarrow$ Combined like terms.

The resulting trinomial has three terms:

- the first term, $4x^2$, is the square of the first term of the original binomial.

- the second term, $12x$, is twice the product of the terms of the original binomial.

- the last term, 9, is the square of the second term of the original binomial.

Because of this structure, we can use the following shortcut for factoring this special type of polynomial.

— *Procedure* —	**Step 1** Write the trinomial as $(x)^2 + 2(x)(y) + (y)^2$
Factoring a perfect	**Step 2** Write the factored form as $(x + y)^2$
square trinomial	

Example 7.5.3   If the polynomial is equivalent to the square of a binomial, factor it. Otherwise, explain why the polynomial is not a perfect square trinomial.

a) $x^2 + 14x + 49$

•SOLUTION•   •   First term can be written as $(x)^2$
- Last term can be written as $(7)^2$
- Middle term can be written as twice the product, $2(x)(7)$

Since this meets the requirements, we can factor this as $(x + 7)^2$

b) $x^2 - 12x + 36$

•SOLUTION•
- First term can be written as $(x)^2$
- Last term can be written as $(-6)^2$
- Middle term can be written as twice the product, $2(x)(-6)$

Since this meets the requirements, we can factor this as $(x - 6)^2$

c) $9x^2 + 6x + 1$

•SOLUTION•
- First term can be written as $(3x)^2$
- Last term can be written as $(1)^2$
- Middle term can be written as twice the product, $2(3x)(1)$

Since this meets the requirements, we can factor this as $(3x + 1)^2$

d) $x^2 + 3x + 9$

•SOLUTION•
- First term can be written as $(x)^2$
- Last term can be written as $(3)^2$
- Middle term IS NOT twice the product, $2(3)(x)$. Therefore, this cannot be factored as the square of a binomial.

Practice 7.5.3   If the polynomial is equivalent to the square of a binomial, factor it. Otherwise, explain why the polynomial is not a perfect square trinomial.

1. $y^2 + 4y + 4$ 2. $z^2 - 2z + 1$ 3. $25m^2 - 10m + 1$ 4. $25x^2 - 40x + 16$ 5. $100c^2 + 40c + 16$

These special products are useful and can save you a lot of time. However, you can always use the ac method if you forget these shortcuts.

Topic 7.5 C — Factoring Polynomials in General

We can put together the factoring techniques we have developed to form a general procedure for factoring polynomials.

| — Procedure —
Factoring polynomials in general | ***Step 1 Standard Form***   Write the polynomial in standard form (that is, combine like terms and write all the terms in decreasing powers of the variable).

Step 2 GCF   If there is a GCF, factor it out.

Step 3 Select   Classify the polynomial and select the appropriate factoring procedure:

• binomial — if it is the difference of two squares, factor it as the product of conjugates. That is, $x^2 - y^2 = (x - y)(x + y)$

• trinomial —

  * if it is a perfect square trinomial, factor it as the square of a binomial. That is, $x^2 + 2xy + y^2 = (x + y)^2$

  * if the lead coefficient is 1, use the product–sum method to factor. That is $x^2 + (m + n)x + mn = (x + m)(x + n)$

  * if the lead coefficient is not 1, use the *ac* or *guess and check* method to factor.

• four term polynomial — factor by grouping.

Step 4 Factor Further   If any factor is not prime, repeat step 3. |

Example 7.5.4   Factor completely.

    a) $3x^2 + 48 + 24x$

       •SOLUTION•   ***Step 1 Standard Form***   We write the terms in descending powers of x. to get
$$3x^2 + 24x + 48$$

            Step 2 GCF   We factor out the GCF of 3 to get $3(x^2 + 8x + 16)$

            Step 3 Select   The first and last terms of the trinomial are perfect squares (x^2 and 4^2) and the middle term is twice the product ($2 \cdot x \cdot 4$) so this is a perfect square trinomial. We now have $3(x + 4)^2$.

            Step 4 Factor Further   The factors are prime so we cannot factor further.

    b) $18 - 2x^2$

       •SOLUTION•   ***Step 1 Standard Form***   We write the terms in descending powers of x to get $-2x^2 + 18$.

            Step 2 GCF   There is a common factor of 2 and since the second degree term has a negative coefficient, we will factor out a -2 to get $-2(x^2 - 9)$.

            Step 3 Select   The binomial factor is the difference of two perfect squares (x^2 and 3^2) so it factors into a pair of conjugates to give us $-2(x - 3)(x + 3)$.

            Step 4 Factor Further   The factors are prime so we cannot factor further.

    c) $6x^2 + 2x - 4 + 3x$

       •SOLUTION•   ***Step 1 Standard Form***   We can combine like terms to get $6x^2 + 5x - 4$.

            Step 2 GCF   There is no GCF. (Remember, by this we mean that there are no variables or integers, other than 1 or -1, which are common to all the factors of the polynomial.)

            Step 3 Select   This is not a special product and the coefficient of the second degree term is not 1. So, we use the ac method, where a is 6 and c is -4. We want two integers whose product is $(6)(-4)$ or -24 and whose sum is 5. The numbers are 8 and -3.

$$6x^2 + 5x - 4 \quad \Leftarrow \text{ Trinomial to factor.}$$

$$6x^2 + 8x - 3x - 4 \quad \Leftarrow \text{ Replaced middle term.}$$

$$\mathbf{2x}(3x + 4) + (\mathbf{-1})(3x + 4) \quad \Leftarrow \text{ Factored by grouping.}$$

$$(\mathbf{3x + 4})(2x - 1) \quad \Leftarrow \text{ Factored out GCF of } 3x + 4.$$

            Step 4 Factor Further   The factors are prime so we cannot factor further.

    d) $2x^2y + x^2 - 2y - 1$

       •SOLUTION•   ***Step 1 Standard Form***   The trinomial is already written using descending powers of x.

            Step 2 GCF   There is no GCF.

            Step 3 Select   This is a four–term polynomial so we will try to factor by grouping.

$$\mathbf{x^2}(2y + 1) + (\mathbf{-1})(2y + 1) \quad \Leftarrow \text{ Factored by grouping.}$$

$$(\mathbf{2y + 1})(x^2 - 1) \quad \Leftarrow \text{ Factored out GCF of } 2y + 1.$$

            Step 4 Factor Further   The second factor is the difference of two perfect squares so we can factor it as the product of conjugates to get $(2y + 1)(\mathbf{x - 1})(\mathbf{x + 1})$.

Practice 7.5.4   Factor completely.

1. $6y^2 + 3y^2 - 4$

2. $2z^2 + 72 - 26z$

3. $36 - m^2$

4. $3m^2 - 12m + 2mn - 8n$

5. $x^4 - 11x^3 + 24x^2$

6. $4x^2 + 7x + 2x^2 - 3$

Exercise Set 7.5   The answers to the odd numbered exercises are at the back of the book.

Factor completely.

1. $c^2 - d^2$

2. $g^2 - k^2$

3. $4t^2 - 121$

4. $64s^2 - 16$

5. $25y^2 - 1$

6. $49n^2 - 4$

7. $16 - 9x^2$

8. $3 - 12x^2$

If the polynomial is not the difference of two squares, explain why. If possible, Factor completely.

9. $25x^2 - 144y^2$

10. $49m^2 - 1$

11. $12n^2 - r^2$

12. $x^2 + y^2$

13. $y^2 - 25z^2$

14. $m^4 - 25$

15. $9x^2 - 5$

16. $t^4 + 81$

17. $36z^4 - 49$

18. $1 - 4x^2$

19. $9 - 100\,d^2$

20. $5y^2 + 10y + 1$

If the polynomial is equivalent to the square of a binomial, factor it. Otherwise, explain why the polynomial is not a perfect square trinomial.

21. $49y^2 + 28y + 4$

22. $9m^2 + 60m + 100$

23. $z^2 + 3z + 1$

24. $9x^2 - 12x + 4$

25. $16m^2 + 40m + 25$

26. $v^2 - 8v + 144$

27. $m^2 + 8m + 16$

28. $k^2 - 24k + 144$

29. $r^2 + 10r + 36$

30. $9x^2 + 6x + 1$

31. $9x^2 - 12x - 4$

32. $49m^2 + 14m + 4$

Factor completely.

33. $4x^2 - 96 + 20x$

34. $5y^2 - 70 - 25y$

35. $2m^2 + 8 + 8m$

36. $9w^2 + 78w + 105$

37. $60r^2 + 10r - 24rs - 4s$

38. $48a^2 - 36a + 40ab - 30b$

39. $y^2 + 75 - 4y^2$

40. $2x^2 + 54 - 8x^2$

41. $3x^2 - 6x + 5x^2 - 5$

42. $y^4 - 3y^3 - 18y^2$

Review Exercises   The answers to all of these exercises are at the back of the book.

43. Solve: $\dfrac{x}{2} + 3(6 - x) = 8 - \dfrac{5}{2}(x - 4)$

44. Solve: $8 - 0.5(x - 2) \geq 5 - 0.3x$

45. Graph: $y = 3$

46. Solve: $\left\{ \begin{array}{rcll} 2x & - & y & = -2 \\ 3x & + & 5y & = -3 \end{array} \right\}$ ❶ ❷

47. Divide and simplify: $\dfrac{-60t^7 - 24t^6 - 12t^4 + 36t^3}{12t^3}$

Simplify.

48. $\left(\dfrac{1}{2}\right)^{-2} - \left(\dfrac{4}{3}\right)^{-2}$

49. $x(4x^{-1})(8x^2)$

50. $\dfrac{(5t)^{-3}}{(5t)^{-1}} \cdot \left(\dfrac{5}{t}\right)^{-2}$

51. $(3a + b)(9a^2 - 3ab + b^2)$

52. $\left(\dfrac{1}{2}m^2 + \dfrac{1}{4}m + \dfrac{1}{8}\right) + \left(\dfrac{3}{2}m^2 - \dfrac{1}{16}\right)$

53. $(0.002x^3 + 0.54x^2 - 3) - (0.2x^4 + 2.5x^3 - 0.07x^2 + 1 + 1)$

Factor completely.

54. $10a^3 + 18ab^3$

55. $x^2 - 14x + 48$

56. $3x^2 + 21x + 30$

57. Factor $2x^2 - 2xy + 3x - 3y$

58. Simplify and write the answer in scientific notation: $(11.8 \times 10^{-7})(2.12 \times 10^{-5})$

59. After riding for some time a biker decides to increase her speed and try to get home before a storm hits. If s represents the speed of the storm create an expression to represent the speed of the biker if:

    a) The biker is moving at one–third the speed of the storm.
    b) The storm is moving 3 mph slower than the biker.
    c) The biker is moving 3 mph slower than the storm.
    d) The storm is gaining on the biker at a rate of 1 mph.

60. After the close of the business day a student who has been selling tickets realizes that he was supposed to have been keeping track of how many of each type of ticket he sold. Luckily, he knew the child ticket numbers began with 111 and the last one he sold was numbered 224. If adult tickets cost $4.75 and child tickets cost $2.50, how many adult tickets did he sell if a total of $1163.75 was collected?

61. The sum of two consecutive odd integers is twice the larger integer. Find the two integers.

62. Calculate the slope of the straight line described by the following data table:

x	y
2	−1
5	0

63. Find the slope and y–intercept of the line described by $2x + 3y + 6 = 0$

64. Find the equation of the line that goes through $(-3,2)$ and $(0,0)$.

65. Determine whether this graph represents a function:

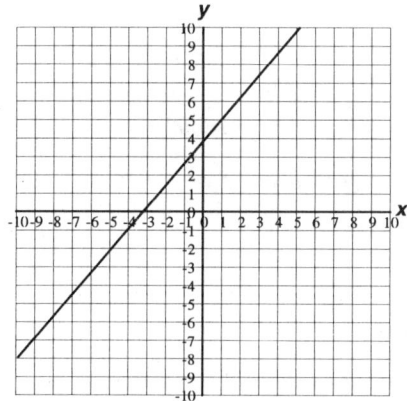

66. The data table below shows the relation between the average life of paper currency and the denomination of the bill. Calculate the slope of the line that would connect the given points and discuss the meaning of the slope.

x denomination of bill (dollars)	y average life (years)
5	2.02
20	4.27
50	8.77

Section 7.6 Solving Polynomial Equations By Factoring

The factoring techniques we just discussed can be a useful tool when solving equations that have variables raised to powers higher than 1. An important class of these equations is called the **quadratic equation in one variable**.

— *Definition* — *Quadratic equation in one variable*	English: A single variable equation where the expressions are polynomials and the highest degree is 2. Examples: $x^2 + 5x + 6 = 0$ $8 - 3x = 5x^2 + 2$ Algebra: If $a \neq 0$, then an equation that can be written in the form $ax^2 + bx + c = 0$ is a quadratic equation in one variable.

Quadratic equations are sometimes called *equations of degree 2* or *second degree equations*.

Topic 7.6 A — Solving Quadratic Equations using the Zero Product Property

Suppose we wish to solve the quadratic equation $x^2 + 5x + 6 = 0$. We would like to combine on one side all terms that contain x and move all the constants to the other side. Unfortunately, we cannot add the terms x^2 and $5x$ because they are not like terms. We could try factoring the trinomial to get $(x + 2)(x + 3) = 0$. Now, we have two numbers, $(x + 2)$ and $(x + 3)$, whose product is 0. The only way we can have 0 as the product of two numbers is if one or both numbers is 0. This fact is called the Zero Product Property.

— *Property* — ***Zero Product Property***	English: If a product is 0, then at least one of its factors is 0. Example: If $(x + 2)(x + 8) = 0$, then $x + 2 = 0$ or $x + 8 = 0$ or both. Algebra: Suppose a and b are real numbers. If $ab = 0$ then $a = 0$, or $b = 0$, or both $a = 0$ and $b = 0$.

Using this property, we can break up the equation $(x + 2)(x + 3) = 0$ into two separate equations, $x + 2 = 0$ and $x + 3 = 0$, each of which is easy to solve. The whole process would look like the following:

$$x^2 + 5x + 6 = 0 \qquad \Leftarrow \text{Equation to solve.}$$

$$(x + 2)(x + 3) = 0 \qquad \Leftarrow \text{Factored the trinomial.}$$

$x + 2 = 0$	$x + 3 = 0$	$\Leftarrow$ Used Zero Product Property to break up the given equation into two equations.
$x = -2$	$x = -3$	$\Leftarrow$ Solved each equation.

The above process shows us that the quadratic equation $x^2 + 5x + 6 = 0$ has two solutions, –2 and –3. We can check to see if these values satisfy the original equation by substituting them for x in the original equation.

Check to see if –2 is a solution:

Left expression	?=?	***Right expression***	
$x^2 + 5x + 6$	=	0	$\Leftarrow$ Equation to check.
$(-2)^2 + 5(-2) + 6$		0	$\Leftarrow$ Substituted –2 for x.
$4 - 10 + 6$		0	$\Leftarrow$ Multiplied.
0		0	$\Leftarrow$ Simplified.

Check to see if –3 is a solution:

Left expression	?=?	***Right expression***	
$x^2 + 5x + 6$	=	0	$\Leftarrow$ Equation to check.
$(-3)^2 + 5(-3) + 6$		0	$\Leftarrow$ Substituted –3 for x.
$9 - 15 + 6$		0	$\Leftarrow$ Multiplied.
0		0	$\Leftarrow$ Simplified.

Example 7.6.1   Find the values of x that make the equation true (that is, solve for x).

a) $(x - 5)(x + 9) = 0$

•SOLUTION• This says the product of two numbers, $x - 5$ and $x + 9$ is 0. By the Zero Product Property, it must be true that $x - 5 = 0$ or $x + 9 = 0$, or both. If we solve each equation we get $x = 5$ and $x = -9$. We leave the check to you.

b) $10x^2 - 15x = 0$

•**SOLUTION**• We cannot combine the terms on the left so we factor to get $5x(2x - 3) = 0$. Then, we use the Zero Product Property to set each factor equal to 0 and solve the resulting equations separately.

First equation:

$$5x \ = \ 0 \qquad \Leftarrow \text{ Equation to solve.}$$

$$x \ = \ 0 \qquad \Leftarrow \text{ Divided both sides by 5.}$$

Second equation:

$$2x - 3 \ = \ 0 \qquad \Leftarrow \text{ Equation to solve.}$$

$$2x \ = \ 3 \qquad \Leftarrow \text{ Added 3 to both sides.}$$

$$x \ = \ \frac{3}{2} \qquad \Leftarrow \text{ Divided both sides by 2.}$$

The solutions to $10x^2 - 15x = 0$ are 0 and $\frac{3}{2}$. We leave the check to you.

c) $4x^2 - 25 = 0$

•**SOLUTION**• We cannot combine the terms on the left so we factor to get $(2x + 5)(2x - 5) = 0$. Then, we use the Zero Product Property to set each factor equal to 0 and solve the resulting equations separately.

First equation:

$$2x + 5 \ = \ 0 \qquad \Leftarrow \text{ Equation to solve.}$$

$$2x \ = \ -5 \qquad \Leftarrow \text{ Subtracted 5 from both sides.}$$

$$x \ = \ \frac{-5}{2} \qquad \Leftarrow \text{ Divided both sides by 2.}$$

Second equation:

$$2x - 5 \ = \ 0 \qquad \Leftarrow \text{ Equation to solve.}$$

$$2x \ = \ 5 \qquad \Leftarrow \text{ Added 5 to both sides.}$$

$$x \ = \ \frac{5}{2} \qquad \Leftarrow \text{ Divided both sides by 2.}$$

The solutions to $4x^2 - 25 = 0$ are $\frac{-5}{2}$ and $\frac{5}{2}$. We leave the check to you.

Practice 7.6.1 Solve.

1. $(y - 12)(y + 5) = 0$ 2. $12x^2 - 36x = 0$ 3. $x^2 - 2x - 15 = 0$ 4. $25t^2 - 36 = 0$

— Caution —	The Zero Product Property is true only for factors whose product is 0. For example, we can break up $(x + 2)(x + 3) = 0$ into $x + 2 = 0$ and $x + 3 = 0$ because we know if a product is 0 then one or both factors must be 0. However, we cannot break up an equation like $(x + 2)(x + 3) = 12$ into $x + 2 = 12$ and $x + 3 = 12$ because we cannot make a similar claim about two numbers whose product is 12. Therefore, before trying to apply the Zero Product Property be sure that one side of the equation is 0.
Zero product property	

The following general procedure is helpful when solving quadratic equations.

— Procedure — *Solving a quadratic equation by factoring*	*Step 1 Standard Form* Write the equation with 0 on one side and a polynomial in standard form on the other.
	Step 2 Factor Factor the polynomial; if it is not factorable, we must use other means to solve the equation (which you will study in intermediate algebra).
	Step 3 Split Set equal to 0 each factor that contains a variable.
	Step 4 Solve Solve each individual equation.
	Step 5 Check Check all solutions.

Example 7.6.2   a)   Solve $3x^2 - 9x = 120$

•SOLUTION•   *Step 1 Standard Form*   Write the equation with 0 on one side and a polynomial in standard form on the other side to get $3x^2 - 9x - 120 = 0$.

Step 2 Factor

$3(x^2 - 3x - 40) = 0$ $\Leftarrow$ Factored out GCF of 3.

$3(x + 5)(x - 8) = 0$ $\Leftarrow$ Found two numbers with product -40 and sum -3.

Step 3 Split   Set equal to 0 each factor that contains a variable. Notice that the factor of 3 has no variable part so we do not set it equal to 0. If we did, we would have the equation $3 = 0$, which has no solution.

$x + 5 = 0$ or $x - 8 = 0$

Step 4   Solve each individual equation.

$x = -5$ or $x = 8$

The solutions are -5 and 8.

Step 5 Check   We leave the check to you.

b)   Solve $14 = (x - 3)(x + 2)$

•SOLUTION•   *Step 1 Standard Form*   First, we move all terms to one side of the equation. Then, we put the polynomial in standard form.

$0 = (x - 3)(x + 2) - \mathbf{14}$ $\Leftarrow$ Subtracted 14 from both sides.

$0 = x^2 + 2x - 3x - 6 - 14$ $\Leftarrow$ Used FOIL to expand the product.

$0 = x^2 - x - 20$ $\Leftarrow$ Simplified.

Step 2 Factor

$0 = (x + 4)(x - 5)$ $\Leftarrow$ Found two numbers whose product is -20 and whose sum is -1.

Step 3 Split   Set equal to 0 each factor that contains a variable.

$x + 4 = 0$ or $x - 5 = 0$

Step 4 Solve   Solve each individual equation.

$x = -4$ or $x = 5$

The solutions are -4 and 5.

Step 5 Check   We leave the check to you.

Practice 7.6.2   Solve.

1. $z^2 - 13z = -36$ 2. $4(x^2 - 24) = 20x$ 3. $6y^2 + y = 4 - y$ 4. $3 - 4a = 6a^2 + 3a$

Topic 7.6 B — Word Problems and Quadratic Equations

It may be necessary to solve quadratic equations when solving some types of word problems.

Example 7.6.3   The product of two consecutive even integers is 224. Find the integers.

•SOLUTION•

Step 1 Organize

Given: • two consecutive even integers.
        • product is 224.

Find: • smaller integer.
       • larger integer.

Sketch: Not helpful.

Rates: None.

Formulas: None needed.

Step 2 Define unknowns

Define: x is the smaller integer.

Construct: $x + 2$ is the next consecutive *even* integer.

Amount Table: None.

Step 3 Verbal models and Step 4 Mathematical Models

Verbal model $\Rightarrow$ product of integers   =   (smaller integer)   •   (larger integer)

Math model $\Rightarrow$ 224   = x • $(x + 2)$

Step 5 Solve

$224 = x(x + 2)$	$\Leftarrow$ Equation to solve.
$224 = x^2 + 2x$	$\Leftarrow$ Multiplied.
$0 = x^2 + 2x - 224$	$\Leftarrow$ Subtracted 224 from both sides.
$0 = (x - 14)(x + 16)$	$\Leftarrow$ Factored.
$x - 14 = 0$ or $x + 16 = 0$	$\Leftarrow$ Set each factor equal to 0.
$x = 14$ or $x = -16$	$\Leftarrow$ Solved each equation.

Step 6 Values of unknowns   We have two values for x and therefore we also have two values for $x + 2$.

• When x is 14, $x + 2$ is 16.
• When x is -16, $x + 2$ is -14.

Step 7 Answer   We have two possible answers to the problem.

• When the smaller integer, x, is 14, the larger integer, $x + 2$, is 16.
• When the smaller integer, x, is -16, the larger integer, $x + 2$, is -14.

Step 8 Check

• consecutive even integers: Both answers check because:
  * 14 and 16 are consecutive even integers
  * -16 and -14 are consecutive even integers.

• the product of the integers is 224: Both answers check:
  * (14)(16) is 224
  * $(-16)(-14)$ is 224

— Caution — *Solutions are not the* *answers*	Note that the equation's solutions, 14 and –16, are NOT the answer to the problem (they are not consecutive even integers and their product is not 224). Just because the original question asked for two numbers and the equation had two solutions does not mean that the solutions are the numbers we were looking for! After we solved the equation we still had more work to do to find the answer to the question.

Practice 7.6.3

1. Find two consecutive integers whose product is 156.

2. One number is one less than twice another. If the product of the numbers is 91, find the numbers.

Finding the solution to some applied problems that involve geometry may require the use of the **Pythagorean Theorem**, which states a relation between the lengths of the sides of a right triangle. (A right triangle is a triangle that has two perpendicular legs — that is, the angle between the legs is 90 degrees, which is a right angle.)

— *Property* — **Pythagorean Theorem**	Given a right triangle with hypotenuse of length c and legs of length a and b, then $c^2 = a^2 + b^2$. 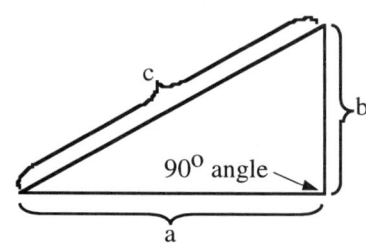

Example 7.6.4   A freight train and a passenger train leave a station at 8 am. The freight train travels east while the passenger travels north and at a rate that is 10 mph faster than the freight train. After one hour the trains are 50 miles apart. What is the speed of the freight train?

•SOLUTION• *Step 1 Organize*

Given: • trains leave at the same time.
   • freight train goes east.
   • passenger train goes north and 10 mph faster.
   • trains are 50 miles apart after 1 hour.

Find: • speed of freight train

Sketch:

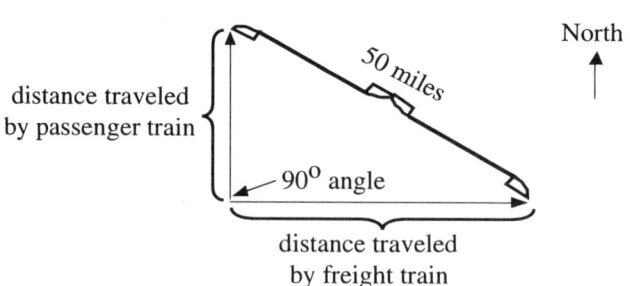

Rates: Rates are involved so we will construct an Amount Table.

	amount	=	rate	•	base
Description →	distance		speed		time
Units →	miles		miles / hour		hours
↓ Item names ↓					
freight train	❸		❶		1
passenger train	❹		❷		1

Formula: amount = rate • base and $a^2 + b^2 = c^2$

Step 2 Define unknowns

Define: x is the speed of the freight train.

Construct: $x + 10$ is the speed of the passenger train.

Amount Table:

	amount	=	rate	•	base
Description →	distance		speed		time
Units → ↓ Item names ↓	miles		$\dfrac{\text{miles}}{\text{hour}}$		hours
freight train	$1x$		x		1
passenger train	$1(x + 10)$		$x + 10$		1

Step 3 Verbal models *and* **Step 4 Mathematical Models** This is just a statement of the Pythagorean Theorem.

$$\textbf{Verbal model} \Rightarrow \quad c^2 \;=\; a^2 \;+\; b^2$$

$$\textbf{Math model} \Rightarrow \quad 50^2 \;=\; (1x)^2 \;+\; [1(x + 10)]^2$$

Step 5 Solve

$50^2 = (1x)^2 + [1(x + 10)]^2$	⇐ Equation to solve.
$2500 = x^2 + x^2 + 20x + 100$	⇐ Squared each term.
$0 = \mathbf{2x^2} + 20x - \mathbf{2400}$	⇐ Combined like terms and subtracted 2500 from both sides.
$0 = 2(x - 30)(x + 40)$	⇐ Factored.
$x - 30 = 0$ or $x + 40 = 0$	⇐ Set equal to 0 each factor that has a variable.
$x = 30$ or $x = -40$	⇐ Solved each equation.

Step 6 Values of unknowns We have two values for x, the speed of the freight train. The negative value, -40, makes no sense for this problem because it would imply that the train was going backwards at 40 mph (possible, but not likely). Therefore, we will only consider 30 as a possible solution.

	amount	=	rate	•	base
Description →	distance		speed		time
Units → ↓ Item names ↓	miles		$\dfrac{\text{miles}}{\text{hour}}$		hours
freight train	$1x$ $1(\mathbf{30})$ **30 miles** **(distance traveled by freight train)**		x **30** **30 mph** **(speed of freight train)**		1
passenger train	$1(x + 10)$ $1(\mathbf{30} + 10)$ **40 miles** **(distance traveled by passenger train)**		$x + 10$ $\mathbf{30} + 10$ **40 mph** **(speed of passenger train)**		1

Step 7 Answer The speed of the freight train, x, is 30 mph.

Step 8 Check

- passenger train goes 10 mph faster: checks because speed of freight train is 30 mph and speed of passenger train is 40 mph.

- distance between the trains is 50 miles: checks because the distance traveled by the freight train is 30 miles and the distance traveled by the passenger train is 40 miles and $30^2 + 40^2$ is 900 + 1600, which simplifies to 2500, which is 50^2.

Practice 7.6.4

1. Two boats start at the same point and travel at right angles away from each other, one going 4 mph faster than the other. If, after one hour, the boats are 20 miles apart, what is the speed of the slow boat?

2. A truck leaves its loading dock and travels east for one hour. It then turns due north and travels 7 mph faster for another hour. If the truck is now 97 miles from its starting place, what was the speed of the truck when it was traveling east?

Topic 7.6 C — Solving Higher Degree Equations using the Zero Product Property

The Zero Product Property may be used to help solve equations that are higher than degree two.

Example 7.6.5   a)   Solve $x^3 + 4x^2 - 5x = 0$

•SOLUTION• ***Step 1 Standard Form*** The polynomial is already in standard form.

Step 2 Factor

$x(x^2 + 4x - 5) = 0$ ⇐ Factored out GCF of x.

$x(x + 5)(x - 1) = 0$ ⇐ Found two numbers with product −5 and sum 4.

Step 3 Split Set equal to 0 each factor that contains a variable.

$x = 0$ and $x + 5 = 0$ and $x - 1 = 0$

Step 4 Solve each equation.

$x = 0$ and $x = -5$ and $x = 1$

The solutions are 0, −5, and 1.

Step 5 Check We leave the check to you.

b)   Solve $x^4 = 9x^2$

•SOLUTION• You may be tempted to divide both sides of the equation by x^2. If you do that, you must stipulate that x cannot be 0. But, 0 turns out to be a solution! We rarely divide both sides of an equation by a variable. So, just follow the steps as before and find the solutions by factoring.

Step 1 Standard Form Write the polynomial in standard form.

$x^4 - 9x^2 = 0$

Step 2 Factor

$x^2(x^2 - 9) = 0$ ⇐ Factored out GCF of x^2.

$x^2(x + 3)(x - 3) = 0$ ⇐ Found two numbers with product −9 and sum 0.

Step 3 Split Set equal to 0 each factor that contains a variable.

$x = 0$ and $x = 0$ and $x + 3 = 0$ and $x - 3 = 0$

Note that we wrote two identical equations, $x = 0$. We did that for completeness, but it is not necessary.

Step 4 Solve   Solve each equation.

$$x = 0 \quad \text{and} \quad x = 0 \quad \text{and} \quad x = -3 \quad \text{and} \quad x = 3$$

The solutions are 0, –3, and 3.

Step 5 Check   We leave the check to you.

Practice 7.6.5   Solve.

1. $a^3 - 7a^2 + 10a = 0$ 2. $m^3 + 4m^2 = 21m$ 3. $4x^3 - 25x = 0$
4. $x^3 + 12x^2 = -2x^2 - 48x$ 5. $x^3 - 2x^2 = 48x$

Exercise Set 7.6   The answers to the odd numbered exercises are at the back of the book.

Solve.

1. $(2x + 3)(x - 1) = 0$ 2. $(3a - 5)(a + 4) = 0$ 3. $(2a - 8)(a - 7) = 0$ 4. $(5x + 1)(x - 3) = 0$

5. $9x(7x - 1) = 0$ 6. $3m(6m + 3) = 0$ 7. $x^2 + 5x + 4 = 0$ 8. $n^2 - 13n - 30 = 0$

9. $x^2 + 4x - 21 = 0$ 10. $w^2 - 2w - 48 = 0$ 11. $x^2 - 144 = 0$ 12. $z^2 - 49 = 0$

13. $8y^2 + 4y = 0$ 14. $6a^2 - 7a = 0$ 15. $8d^2 + 4d = 0$ 16. $3g^2 + 12g = 0$

17. $5t^2 = 15t$ 18. $4c^2 = 3c$ 19. $(x + 6)(x - 4) = -9$ 20. $(x - 1)(x - 3) = 8$

21. $w^2 + 12w = -35$ 22. $x^2 - x = 12$ 23. $a^2 - 5a = 24$ 24. $5y^2 - 25y = 70$

25. $20x^2 + 3x = 2$ 26. $5a^2 + 6 = -13a$ 27. $2x^2 - 72 = 0$ 28. $3y^2 - 27 = 0$

29. $x^2 - 9x = -x - 15$ 30. $c^2 + 4c = 5c + 12$ 31. $2b^2 - 8b = -22b + 88$ 32. $3a^2 + 18a = -27a - 162$

33. $12m^2 - 13m = -10 + 10m$ 34. $8x^2 - 12x + 3 = 10x - 2$

---

35. The length of a rectangle is 8 inches more than its width. The area of the rectangle is 240 square inches. Find the dimensions of the rectangle.

36. Find the dimensions of a rectangle whose area is 84 square meters and whose length is 5 meters more than its width.

37. The product of 3 less than a number and 5 more than that number is 65. Find the number.

38. Nine less than a number is multiplied by 4 more than the number to yield 140. Find the number.

39. A rectangle has a length that is two inches longer than its width. If the length of a diagonal is 10 inches, find the dimensions of the rectangle.

40. The diagonal if a rectangle is one inch longer than the length. If the width of the rectangle is 5 inches, find the length of the diagonal.

41. Ben and Jerry leave camp at the same time. Ben travels west at a rate that is 1 mph faster than Jerry, who walks south. If, after 2 hours, they are 10 miles apart, how fast was Jerry walking?

42. A truck, which is traveling south toward Denver, is 50 miles away from a car, which is traveling west toward Denver and 10 mph faster. If they both arrive at Denver one hour later, how fast was the truck traveling?

Solve.

43. $36x^3 = x$ 44. $4b^3 = b$ 45. $a^3 - 3a^2 - 4a = 0$ 46. $b^3 + 3b^2 = 18b$

47. $b^3 + 4b^2 = 25b^2 + 100b$ 48. $x^3 - 21x^2 = 3x^2 - 63x$ 49. $12w^3 - 10w = 7w^2$ 50. $8t^3 - 5t = 18t^2$

51. $x^3 + 7x^2 = -7x^2 - 49x$ 52. $x^3 + 3x^2 = -3x^2 - 9x$ 53. $4w^4 = 25w^2$ 54. $9t^4 - 4t^2 = 0$

Review Exercises   The answers to all of these exercises are at the back of the book.

55. An instructor gave a 65 question test. 45 questions were worth 5 points each and the rest were worth 10 points each. If a student got one–forth of the 10 point questions right, and 18 more of the 5 point questions then the 10 point questions right, what was their total score for the test?

56. A campground has three types of sites, tent sites, RV sites, and RV sites with electricity. The cost for a tent site is $8 less than an RV site and if you get an RV site with electricity it costs $4 more than an RV site alone. On a Friday night the campground fills it's 27 tent sites and it's 18 RV sites. They only fill 16 of their 22 electrical RV sites. How much do they charge for each type of site if they made $915.50 that Friday night?

57. A mixture contains 2% ammonia and the rest water. After 50 gallons of an 18% ammonia–water mix are added the final mixture is 12% ammonia. How many gallons were in the original container?

58. The earth weighs 6.0×10^{24} kg. while the moon weighs 7.4×10^{22} kg. If we placed our earth on one side of a large balance scale how many moons would it take to balance the scale?

59. Construct the equation of the line shown in the graph.

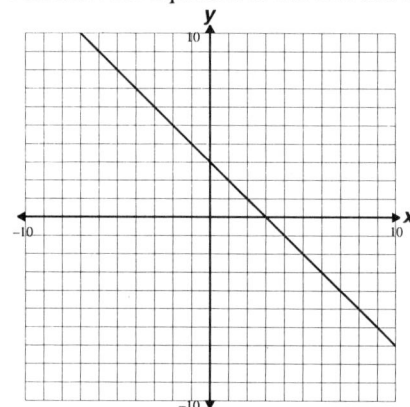

60. Determine the domain and range of

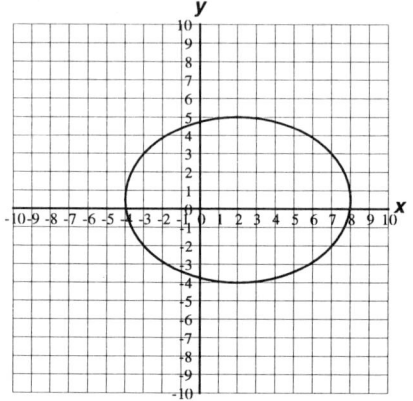

61. Solve: $\left\{ \begin{array}{rcrcr} 4x & - & 6y & = & -1 \\ -8x & + & 12y & = & 2 \end{array} \right\}$ **❶** **❷**

62. Solve: $\left\{ \begin{array}{rcrcr} 3x & - & y & = & -10 \\ x & - & 5y & = & -8 \end{array} \right\}$ **❶** **❷**

63. Use long division to simplify: $\dfrac{2x^3 - 4x^2 - 14x - 9}{x - 4}$

64. Graph: $x < -2.5$

65. Find the GCF of $18x^3(3x - 2)^5$ and $24x^2(3x - 2)^4$

66. Given $g(x) = 10x - 3$, find $g\left(\dfrac{1}{2}\right)$

Factor completely.

67. $x^2 + 2x + xy + 2y$

68. $48 + x^2 - 16x$

69. $5x^2 + 15x - 20$

70. $20x^2 - 3x - 2$

Simplify.

71. $3x(x - 7) - 7x(4 - 2x)$

72. $(8y - 1)(y - 8)$

73. $-x^2(x^3y)^2(-2x)^3$

74. $\dfrac{(c^7d)^2(d^7c)^{-2}}{cd}$

75. $4t + (3t - t^2) + (1 - t - 2t^2 + 3t^3)$

76. In 1975 the average life expectancy for a man was 68.8 years while for a woman it was 76.5 years. In 1990 life expectancy was 71.8 years for a man and 78.7 years for a women. Assume life expectancy for a man is the input and life expectancy for a women is the output.

   a) Build a linear equation for this situation.
   b) Discuss the meaning of the equation.
   c) Predict a women's life expectancy when the life expectancy for a man is 75 years.
   d) Predict what a man's life expectancy will be when the life expectancy for a women is 96 years.

Chapter 7 Review

Vocabulary to Know

ac Method of factoring — a method of factoring a trinomial in which we find two integers whose product and sum are given and then we factor by grouping.

factoring — writing an expression as a product.

factoring completely — writing an expression as a product where each factor cannot be further factored using integers other than 1 or –1.

factoring by grouping — a method of factoring polynomials where the terms are grouped and then the GCF is factored out of each group.

greatest common factor — largest factor common to a set of given numbers; same as GCF; the GCF of $6x^2$ and $8x$ is $2x$.

GCF — same a greatest common factor.

Guess and Check Method of factoring — a method of factoring a trinomial where we guess the binomial factors and then use FOIL to check the correctness of the guess.

perfect square — the square of an integer, the square of a variable, or the square of the product of integers and variables.

Product–Sum Method of factoring — a method of factoring a trinomial in which we find two integers whose product and sum are given.

quadratic equation in one variable — an equation that can be written in the form $ax^2 + bx + c = 0$, where $a \neq 0$.

Properties, Definitions, Formulas, and Notation to Understand

Quadratic equation in one variable — A single variable equation where the expressions are polynomials and the highest degree is 2.	If $a \neq 0$, then an equation that can be written in the form $ax^2 + bx + c = 0$ is a quadratic equation in one variable. $x^2 + 5x + 6 = 0$ $8 - 3x = 5x^2 + 2$
Zero Product Property — If a product is 0, then at least one of its factors is 0.	If $ab = 0$ then $a = 0$, or $b = 0$, or both $a = 0$ and $b = 0$. If $(x + 2)(x + 8) = 0$, then $x + 2 = 0$ or $x + 8 = 0$ or both.
Pythagorean Theorem — The sum of the squares of the lengths of a right triangle equals the square of the length of the hypotenuse.	Given a right triangle with hypotenuse of length c and legs of lengths a and b, then $c^2 = a^2 + b^2$. 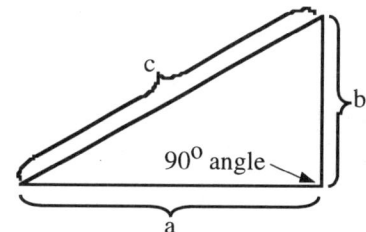

Procedures to Follow

Procedure for finding the GCF of a set of numbers.	Find the GCF of 12, 48, and 30
Step 1 Prime Factor Prime factor each number and write the factors in exponential form.	$12 = 2^2 \cdot 3^1$ $48 = 2^4 \cdot 3^1$ $30 = 2^1 \cdot 3^1 \cdot 5^1$
Step 2 Greatest Common Power Multiply the greatest power of each prime factor that is *common* to all the numbers. If a factor does not appear in all the numbers then it will not be a part of the GCF.	Greatest power of 2 common to all is 2^1 Greatest power of 3 common to all is 3^1 5 is not a factor common to all GCF is $2^1 \cdot 3^1$, which is 6

Procedure for factoring the GCF out of a polynomial.	Factor out the GCF of $6x^3 + 10x^2 - 8x$
Step 1 Find the GCF • We find the numeric part of the GCF by inspection or by prime factoring. • For the polynomial factors, select the smallest power of each factor that is common to all terms.	Largest power of 2 common to all terms is 2^1 Largest power of x common to all terms is x^1 Therefore, the GCF is $2x$
Step 2 Rewrite the terms Write each term as a product where the GCF is a factor.	$\mathbf{2x}(3x^2) + \mathbf{2x}(5x) - \mathbf{2x}(4)$
Step 3 Factor Use the Distributive Property to factor out the GCF.	$\mathbf{2x}(3x^2 + 5x - 4)$

Procedure for factoring by grouping.	Factor $10xy - 4x - 5y + 2$
Step 1 Group Group the four terms into a pair of binomials.	$(10xy - 4x) + (-5y + 2)$
Step 2 Factor Factor the GCF out of each binomial.	$\mathbf{2x}(5y - 2) - \mathbf{1}(5y - 2)$
Step 3 Factor again Factor the GCF out again.	$(\mathbf{2x - 1})(5y - 2)$

Procedure for the product-sum method of factoring a trinomial.	Factor $8x - 144 + 3x^2 - 2x$
Step 1 Standard Form Write the polynomial in standard form (that is, combine like terms and write all the terms in decreasing powers of the variable).	$3x^2 + 6x - 144$
Step 2 GCF If there is a GCF, factor it out so that the other factor has the form $x^2 + bx + c$.	$3(x^2 + 2x - 48)$
Step 3 Identify b and c Identify the values of b and c in $x^2 + bx + c$. If the first degree (middle) term is missing, b is 0.	b is 2 c is -48
Step 4 Find Integers Find two integers whose product is c and whose sum is b. Think of this as $(\_\_)(\_\_) = c$ and $\_\_ + \_\_ = b$	$(\_\_)(\_\_) = -48$ and $\_\_ + \_\_ = 2$ We list the integers whose products are -48 and calculate their sums: Factors Sums of factors $-1(48)$ 47 $-2(24)$ 22 $-3(16)$ 13 $-4(12)$ 8 $-6(8)$ 2 we can stop since the product is -48 and the sum is 2.
Step 5 Use Integers Use the integers found in Step 4 as the constant terms in the binomial factors.	$3(x - \mathbf{6})(x + \mathbf{8})$

Procedure for "ac" method of factoring a trinomial.	Factor $14x + 20x^2 - 12$
Step 1 Standard Form Write the polynomial in standard form (that is, combine like terms and write all the terms in decreasing powers of the variable).	$20x^2 + 14x - 12$
Step 2 GCF If there is a GCF, factor it out so that the other factor has the form $ax^2 + bx + c$.	$2(10x^2 + 7x - 6)$

Step 3 Identify a, b, and c Identify the values of a, b, and c in $ax^2 + bx + c$. If any term is missing, its coefficient is 0.	a is 10 b is 7 c is -6
Step 4 Find Integers Find two integers whose product is ac and whose sum is b.	ac is $(10)(-6)$, which is -60 We list the integers whose products are -60 and calculate their sums: Factors Sums of factors $-1(60)$ 59 $-2(30)$ 28 $-3(20)$ 17 $-4(15)$ 11 $-5(12)$ 7 we can stop since the product is -60 and the sum is 7.
Step 5 Replace middle term Replace the middle term, bx, with a sum using the two integers found in Step 4.	We write $2(10x^2 + 7x - 6)$ as $2(10x^2 - 5x + 12x - 6)$
Step 6 Factor Factor by grouping.	$2[5x(2x - 1) + 6(2x - 1)]$ $2(2x - 1)(5x + 6)$

Procedure for factoring the difference of two perfect squares.	Factor $81x^2 - y^2$
Step 1 Write each term as a perfect square.	$(9x)^2 - (y)^2$
Step 2 Write the conjugate factors.	$(9x - y)(9x + y)$
We can write this procedure as $a^2 - b^2 = (a - b)(a + b)$	

Procedure for factoring a perfect square trinomial.	Factor $9x^2 + 12x + 4$
Step 1 Write the trinomial as $(x)^2 + 2(x)(y) + (y)^2$	$(3x)^2 + 2(3x)(2) + (2)^2$
Step 2 Write the factored form as $(x + y)^2$	$(3x + 2)^2$

Procedure for factoring polynomials in general.	Factor $9 - 10x + 2x^2 + 3$	Factor $9x^2 - 3xy + 18x - 6y$
Step 1 Standard Form Write the polynomial in standard form (that is, combine like terms and write all the terms in decreasing powers of the variable).	$2x^2 - 10x + 12$	$9x^2 + 18x - 3xy - 6y$
Step 2 GCF If there is a GCF, factor it out.	$2(x^2 - 5x + 6)$	$3(3x^2 + 6x - xy - 2y)$

Step 3 Select   Classify the polynomial and select the appropriate factoring procedure:

- binomial — if it's the difference of two squares, factor it as the product of conjugates. That is,
$$x^2 - y^2 = (x - y)(x + y)$$

- trinomial —

  * if it's a perfect square trinomial, factor it as the square of a binomial. That is,
  $$x^2 + 2xy + y^2 = (x + y)^2$$

  * if the lead coefficient is 1, use the product-sum method to factor. That is
  $$x^2 + (m + n)x + mn = (x + m)(x + n)$$

  * if the lead coefficient is not 1, use the *ac* or *guess and check* method to factor.

- four term polynomial — factor by grouping.

We need two numbers whose product is 6 and whose sum is –5. They are –2 and –3.
$$2(x - 2)(x - 3)$$

$$3\big[3x(x + 2) - y(x + 2)\big]$$
$$3(x + 2)(3x - y)$$

Step 4 Factor Further   If any factor is not prime, repeat Step 3.

Procedure for solving a quadratic equation by factoring.

Solve $2x^2 - 6x = 20$

Step 1 Standard Form   Write the equation with 0 on one side and a polynomial in standard form on the other.

$$2x^2 - 6x - 20 = 0$$

Step 2 Factor   Factor the polynomial.

$$2(x^2 - 3x - 10) = 0$$
$$2(x - 5)(x + 2) = 0$$

Step 3 Split   Set equal to 0 each factor that contains a variable.

$$x - 5 = 0 \qquad x + 2 = 0$$

Step 4 Solve   Solve each individual equation.

$$x = 5 \qquad x = -2$$

Step 5 Check   Check all solutions.

Check for 5	Check for –2
$2(5)^2 - 6(5) = 20$	$2(-2)^2 - 6(-2) = 20$
$50 - 30 = 20$	$8 + 12 = 20$
$20 = 20$ Checks	$20 = 20$

Chapter 7 Review Exercises   The answers to all the exercises are in the back of the book.

Find the GCF.

1. 32, 40, 96
2. 36, 72, 108
3. $20x^2y^3$, $40x^4y$, $50x^6y^4$
4. $24x^5y^3z^4$, $30x^3y^4z^2$, $42x^8y^6z^7$
5. $12x^3(5x + 1)^7$, $18x^8(5x + 1)^4$
6. $8x^3y^8(x - 2)^9$, $20x^2y^5(x - 2)^7$

Factor –1 out of the expression.

7. $-9x^2 + 5x + 1$
8. $3x^2 - 2x + 1$
9. $-8x^4 + 3x^2 - 5$

Factor completely.

10. $-24x^3y^2 + 16x^2y^3$
11. $45x^4 - 9x^2$
12. $8x(x - 3) - 4x^2(x - 3)$
13. $12x^2(2x - 1)^3 + 2x(2x - 1)^5$
14. $15x^2 - 6x - 10xy + 4y$
15. $wx + wy - 5hx - 5hy$
16. $x^2 - x - 12$
17. $10 + x^2 - 7x$
18. $x^2 - 7x - 60$

19. $-64 + 12x + x^2$

20. $4x^2 + 12x - 16$

21. $5x^2 - 5x - 60$

22. $18x^2 - 45x - 8$

23. $-15x^2 + 39x - 18$

24. $4x^2 - 9y^2$

25. $25x^2 - 1$

26. $x^2 - 10x + 25$

27. $4x^2 + 28x + 49$

28. $8x^3 + 16x^2 - 8x^2y - 16xy$

29. $75x^3 - 3x$

30. $40x^3 - 45xy^2$

Solve.

31. $10x^2 - 5x = 0$

32. $0 = 2x2 + x - 3$

33. $40 + 6x = 4x2$

34. $9x^2 = -10 - 21x$

35. $x^3 - x^2 = 6x$

36. $x^4 - 4x^2 = 0$

37. Find two consecutive integers whose product is 342

38. One number is 17 more than another and their product is –42. Find the numbers.

39. Two ships are 15 miles apart, one traveling south and the other west, toward the same marker buoy. One ship is traveling 3 mph faster than the other and reaches the buoy in 2 hours. The other ship reaches the buoy in 3 hours. What is the speed of the slower ship?

Chapter 7 Test   The answers to all the exercises are in the back of the book.

1. Find the GCF of $28x^4y^6z^2$, $70y^8z^5$, and $98y^3z^8$

2. Factor –1 out of $4x^6 - 3x^4 - 2x^2 - 3$

Factor completely.

3. $12x^4y^3 - 3xy^2$

4. $6x^4(3x + 2)^5 - 15x^3(3x + 2)^4$

5. $4x^2 - 5x - 8xy + 10y$

6. $x^2 - 5x - 14$

7. $6x^2 - 60x + 126$

8. $5x^2 - 9x - 18$

9. $49x^2 - 9$

10. $7x^2 - 7xy - 21x + 21y$

11. Factor completely: $20 + 8x^2 - 68x + 100$

12. Solve $0 = 15x^2 - 6x$

13. Solve $2x^2 + 12x = -16$

14. The product of two consecutive integers is 5 more than their sum. Find the integers.

Chapter 8
Rational Expressions and Equations

Section 8.1 Reducing Rational Expressions

This chapter will concentrate on **rational expressions**, which are fractions where the numerator and denominator are polynomials.

— *Definition* — *Rational expression*	English: A fraction where the numerator and denominator are polynomials. Examples: $\dfrac{3x}{2}$ and $\dfrac{y^2 - 1}{y + 2}$ Algebra: If P and Q represent polynomials and $Q \neq 0$, then $\dfrac{P}{Q}$ is a rational expression.

Topic 8.1 A — Restricted Values of Rational Expressions

In arithmetic we found that division by 0 was not possible and so we labeled quantities like $\dfrac{6}{0}$ as *undefined*. Likewise, with rational expressions we may not substitute values for variables if this will make the denominator zero. For example, in the expression $\dfrac{y^2 - 1}{y + 2}$ we can't substitute –2 for y because the denominator becomes zero. Values of a variable that would result in division by 0 are called **restricted values**.

Example 8.1.1   Determine the restricted values. That is, find the values of the variable that would make the denominator 0.

a) $\dfrac{5}{x}$

•SOLUTION• 0.

b) $\dfrac{3x}{x + 2}$

•SOLUTION• You may be tempted to say 0 again, but setting x to 0 does not cause any problems since we end up with $\dfrac{3(0)}{(0) + 2}$, which simplifies to $\dfrac{0}{2}$, which is 0. The value of x that makes the denominator 0 is –2 since that gives us $\dfrac{3(-2)}{(-2) + 2}$, which simplifies to $\dfrac{-6}{0}$, which is undefined. Thus, –2 is the restricted value.

c) $\dfrac{8}{5 - x}$

•SOLUTION• Replacing x with 5 causes the denominator to be 0. Thus, 5 is the restricted value.

Practice 8.1.1 Determine the restricted values.

1. $\dfrac{6}{x}$ 2. $\dfrac{y - 8}{2y}$ 3. $\dfrac{4w}{w + 9}$ 4. $\dfrac{3}{1 - w}$

Topic 8.1 B — Fundamental Property of Fractions

A useful tool for working with rational expressions is the Fundamental Property of Fractions.

— *Property* — **Fundamental property of fractions**	English: Multiplying the numerator and denominator of a fraction by the same non-zero number does not change the value of the fraction.

Examples:

$\dfrac{1}{4}$ is equivalent to $\dfrac{1 \cdot 3}{4 \cdot 3}$

$\dfrac{5x}{2}$ is equivalent to $\dfrac{5xy}{2y}$ (assuming $y \neq 0$)

Algebra: If $b \neq 0$ and $k \neq 0$ then $\dfrac{a}{b} = \dfrac{a \cdot k}{b \cdot k}$

This property is a consequence of the fact that when we multiply by 1 we do not change the value of an expression. We can see this by doing the following:

$\dfrac{a}{b}$ ⇐ Original fraction.

$\dfrac{a}{b} \cdot 1$ ⇐ Multiplied by 1, which does not change the value of the fraction.

$\dfrac{a}{b} \cdot \dfrac{k}{k}$ ⇐ Replaced 1 with the equivalent fraction $\dfrac{k}{k}$, assuming $k \neq 0$.

$\dfrac{a \cdot k}{b \cdot k}$ ⇐ Multiplied numerators and denominators.

Starting with $\dfrac{a}{b}$ and rewriting it as the equivalent $\dfrac{a \cdot k}{b \cdot k}$ is called **building up a fraction**.

If we do the reverse, that is, if we begin with $\dfrac{a \cdot k}{b \cdot k}$ and remove the common factor of k, the process is called **reducing the fraction to lowest terms**. [†]

In Section 1.1 we outlined a procedure for reducing arithmetic fractions. That same procedure works well for reducing rational expressions.

— *Procedure* — **Reducing rational expressions**	**Step 1** Prime factor the numerator and the denominator.
	Step 2 Reduce factors common to the numerator and the denominator.

Example 8.1.2   Reduce. That is, remove all factors that are common to the numerator and denominator. Assume no value of a variable will make a denominator 0.

a) $\dfrac{3y - 6}{2y^2 - 4y}$

•SOLUTION•

$\dfrac{3(y - 2)}{2y(y - 2)}$ ⇐ Factored numerator and denominator.

$\dfrac{3}{2y}$ ⇐ Reduced common factor of $y - 2$.

[†]   The word *reducing* may be misleading because we are not making the fraction smaller; we are, however, making the *terms* of the fraction (the numerator and denominator) smaller. Thus, when we reduce $\dfrac{6}{8}$ to $\dfrac{3}{4}$ the 6 was *reduced* to 3 and the 8 was *reduced* to 4 but overall the value of the fraction was not changed.

b) $\dfrac{x^2 - x - 2}{x^2 - 3x - 4}$

•SOLUTION•

$$\dfrac{(x-2)(x+1)}{(x-4)(x+1)} \quad \Leftarrow \text{ Factored numerator and denominator.}$$

$$\dfrac{x-2}{x-4} \quad \Leftarrow \text{ Reduced common factor of } x + 1.$$

— *Caution* — *Canceling*	It's tempting to reduce a fraction like $\dfrac{x-2}{x-4}$ by "canceling" the x's like this: $\dfrac{\cancel{x}-2}{\cancel{x}-4}$ to get $\dfrac{-2}{-4}$. WE CANNOT DO THIS because x is not a *factor* of the numerator and denominator. "Canceling" the x's would be like "canceling" 8 in $\dfrac{18}{28}$ to get $\dfrac{1}{2}$. "Canceling" is not a mathematical operation. To reduce a rational expression we write the numerator and denominator in factored form and then use the multiplication property of 1 to remove the common factors.

Practice 8.1.2 Reduce. That is, remove all factors that are common to the numerator and denominator. Assume no value of a variable will make a denominator 0.

1. $\dfrac{3x + 18}{2x + 12}$

2. $\dfrac{x + 2}{x^2 + 5x + 6}$

3. $\dfrac{m^2 - 7m + 10}{m - 5}$

Remember that when we simplify an expression we are rewriting it in a different way, but the simplified expression is equivalent to the original expression. That is, no matter what number we substitute for the variable (except the restricted values) the original expression and the simplified expression represent the same number. For example, we just reduced $\dfrac{x^2 - x - 2}{x^2 - 3x - 4}$ to get $\dfrac{x-2}{x-4}$. Substituting any value for x, (other than the restricted values of -1 and 4) will give the same answer in both the original and simplified expressions. For example, substituting -5 for x gives us the following:

Original expression	*Simplified expression*
$\dfrac{x^2 - x - 2}{x^2 - 3x - 4}$	$\dfrac{x-2}{x-4}$
$\dfrac{(-5)^2 - (-5) - 2}{(-5)^2 - 3(-5) - 4}$	$\dfrac{(-5) - 2}{(-5) - 4}$
$\dfrac{25 + 5 - 2}{25 + 15 - 4}$	$\dfrac{-7}{-9}$
$\dfrac{28}{36}$	$\dfrac{7}{9}$
$\dfrac{7}{9}$	

The simplified (but equivalent) expression is easier to work with so we prefer to use it in place of the original.

Topic 8.1 C — Factors of –1 in Rational Expressions

In Section 7.1 we mentioned that 1 and –1 are factors of every polynomial and so we usually do not explicitly write them. However, there are times when it is necessary to factor out negative values.

Example 8.1.3 Reduce. Assume no value of a variable will make a denominator 0.

a) $\dfrac{5-x}{x-5}$

• **SOLUTION** • If you look carefully at the numerator and denominator you will notice that they are opposites of each other. Therefore, this fraction must reduce to –1. If you do not notice the special "structure" of the fraction, then you can follow the step by step procedure for reducing rational expressions.

$\dfrac{5-x}{x-5}$ ⇐ Expression to simplify.

$\dfrac{-x+5}{x-5}$ ⇐ Used the Commutative Property to write the numerator in standard form.

$\dfrac{(-1)(x-5)}{(x-5)}$ ⇐ Factored –1 out of the numerator. Note how the signs in the numerator changed.

-1 ⇐ Reduced by the common factor of $(x-5)$.

b) $\dfrac{3-x}{x+3}$

• **SOLUTION** •

$\dfrac{-x+3}{x+3}$ ⇐ Used the Commutative Property to write the numerator in standard form.

$\dfrac{(-1)(x-3)}{x+3}$ ⇐ Factored –1 out of the numerator. Note how the signs in the numerator changed.

The fraction does NOT reduce since there are no factors common to the numerator and denominator. Thus, the original fraction cannot be reduced.

c) $\dfrac{x^2-5x+6}{9-x^2}$

• **SOLUTION** •

$\dfrac{x^2-5x+6}{-x^2+9}$ ⇐ Used the Commutative Property to write the denominator in standard form.

$\dfrac{x^2-5x+6}{(-1)(x^2-9)}$ ⇐ Factored –1 out of the denominator since we prefer the x^2 term to have a positive coefficient.

$\dfrac{(x-2)(x-3)}{(-1)(x-3)(x+3)}$ ⇐ Factored the numerator and the denominator.

$\dfrac{x-2}{(-1)(x+3)}$ ⇐ Reduced by the common factor of $x-3$.

The answer can be left as is but usually we do not leave a factor of –1 in the denominator. That is, we usually write a fraction like $\dfrac{a}{-b}$ as $-\dfrac{a}{b}$ or $\dfrac{-a}{b}$.

Therefore, we would write the final answer as $-\dfrac{x-2}{x+3}$ or $\dfrac{-(x-2)}{x+3}$ or $\dfrac{-x+2}{x+3}$.

Practice 8.1.3 Reduce all common factors. Assume no value of a variable will make a denominator 0.

1. $\dfrac{r-7}{7-r}$
2. $\dfrac{2m-6}{3-m}$
3. $\dfrac{x^2-9x+8}{24+5x-x^2}$
4. $\dfrac{x^2-6x+8}{16-x^2}$
5. $\dfrac{3-x}{x^2-7x+12}$

Topic 8.1 D — An Application

Rational expressions sometimes appear in mathematical models. For example, the following formula predicts the test score s given the time t (hours) a student spent preparing for a particular test: $s = \dfrac{2t^2 + 15t + 7}{t + 7}$ This model was found to be valid for times between 4 and 45 hours.[†]

Example 8.1.4   Predict the score for students who study 23, 35, 42, and 50 hours.

•SOLUTION• We could substitute the given values directly into $s = \dfrac{2t^2 + 15t + 7}{t + 7}$ and then simplify. That would involve a lot of computation. As an alternative, we could simplify the model first and then do the substitutions. We will do the latter.

$$\dfrac{2t^2 + 15t + 7}{t + 7} \quad \Leftarrow \text{ Expression to simplify.}$$

$$\dfrac{(2t + 1)(t + 7)}{t + 7} \quad \Leftarrow \text{ Factored.}$$

$$2t + 1 \quad \Leftarrow \text{ Reduced by the common factor of } t + 7.$$

This simplified version says that to find the scores we just double the preparation time and add one. Therefore,

- an input of 23 gives an output of $2(23) + 1$, which simplifies to 47.

- an input of 35 gives an output of $2(35) + 1$, which simplifies to 71.

- an input of 42 gives an output of $2(42) + 1$, which simplifies to 85.

- we may not use an input of 50 because this is outside the interval where the model is valid (since we said that $4 \le t \le 45$).

The computations using the simplified model are much easier than doing the computations using the original expression.

Practice 8.1.4 Simplify and then find the value of the expression when x takes on the given values. Assume no value of a variable will make a denominator 0.

1. $y = \dfrac{3x^2 + 22x + 7}{x + 7}$ when x is 1 and when x is 3.

2. $y = \dfrac{4x^2 + 3x - 10}{x + 2}$ when x is -3 and when x is 2.

3. $y = \dfrac{5x^2 - 14x - 3}{3 - x}$ when x is -2 and when x is -1.

Exercise Set 8.1   The answers to the odd numbered exercises are at the back of the book.

Determine the restricted values.

1. $\dfrac{12}{m}$

2. $\dfrac{-4}{w}$

3. $\dfrac{5v}{6 + v}$

4. $\dfrac{d}{d + 9}$

5. $\dfrac{-3z}{z + 4}$

6. $\dfrac{9}{8 - x}$

Reduce. Assume no value of a variable will make a denominator 0.

7. $\dfrac{x^2 + 2x - 8}{x^2 + x - 6}$

8. $\dfrac{m^2 + 2m + 1}{m^2 + 7m + 6}$

9. $\dfrac{x^2 - 11x + 18}{x^2 - 6x - 27}$

10. $\dfrac{m^2 + 4m - 5}{m^2 + 8m + 15}$

[†]   This restriction on input values is typical of real–world models. It would be nice if we could develop "one size fits all" models that allow us to use any input but most real world phenomena are too complex to do this.

11. $\dfrac{x^2 - 7x}{x^2 - 49}$ 12. $\dfrac{r^2 + 3r}{r^2 - 9}$ 13. $\dfrac{x - 3}{3 - x}$ 14. $\dfrac{2 - y}{y + 2}$

15. $\dfrac{5 - x}{x - 5}$ 16. $\dfrac{4 + x}{x - 4}$ 17. $\dfrac{x^2 + x - 6}{4 - x^2}$ 18. $\dfrac{x^2 - 6x + 9}{9 - x^2}$

19. $\dfrac{m^2 + 4m - 21}{6 + m - m^2}$ 20. $\dfrac{x^2 - 14x + 49}{49 - x^2}$ 21. $\dfrac{2 - 3x}{6x^2 - x - 2}$ 22. $\dfrac{3 - w}{5w^2 - 14w - 3}$

Simplify and then find the value of the expression when x takes on the given values. Assume no value of a variable will make a denominator 0.

23. $y = \dfrac{x^2 + 14x + 33}{x + 3}$ when x is -1 and when x is 1.   24. $y = \dfrac{x^2 - 15x + 36}{x - 3}$ when x is -2 and when x is 2

25. $y = \dfrac{6x^2 - 11x + 3}{3x - 1}$ when x is -1 and when x is 3   26. $y = \dfrac{8x^2 + 6x - 5}{4x + 5}$ when x is 1 and when x is -3

27. $y = \dfrac{2x^2 + x - 10}{4 - x^2}$ when x is -1 and when x is 4   28. $y = \dfrac{2x^2 + 8x - 42}{18 - 2x^2}$ when x is 1 and when x is 2

Review Exercises   The answers to all of these exercises are at the back of the book.

29. Identify the terms, products, and factors: $6x$ 30. Graph: $y = -2x - 4$ 31. Graph: $x = -4$

32. Simplify: $\dfrac{9}{60} \cdot \dfrac{24}{36}$ 33. Simplify: $\dfrac{\frac{18}{49}}{\frac{30}{56}}$ 34. Simplify: $\dfrac{-4xy^{-2}}{4(xy)^2}$ 35. Simplify: $(4x + 7)(3x - 5)$

36. Divide and simplify: $\dfrac{18w^5 + 9w^4 - 6w^3}{-3w^3}$ 37. Subtract $(417.5x^3 + 200.16x + 15)$ from $(-80.2x^2 + 107x + 155)$

38. Add $4w^4 + 3w^2 + 1$ and $-6w^4 + 6w^3 - w^2 + 7$ 39. Convert to scientific notation: -604×10^{-5}

40. Simplify and write the answer in scientific notation: $\dfrac{(47 \times 10^9)(3.4 \times 10^{11})}{7.99 \times 10^{15}}$ 41. Solve: $x^2 - 16x = 0$

---

42. Write the equation of a line with slope -3 and y–intercept $(0,8)$

43. Data for the supply and demand for a can hand–held electronic game are given in the following table. Find how many games a company should manufacture for supply to equal demand.

x Demand price (dollars)	y Demand quantity (hundreds of units)	x Supply price (dollars)	y Supply quantity (hundreds of units)
2	740	5	113
10	100	12	155

Factor completely.

44. $16x - 5x^3$ 45. $x^2 + x - 6$ 46. $3x^2 - 2x - 5$ 47. $3x^3 + 3x^2 - 18x$

---

48. Brad has decided to split his 650 mile trip into three parts. After driving for 4 hours Brad has covered one–third the distance he will cover on the last part of his trip. If x represents his distance on the last part of his trip, create a simplified expression to represent the distance of the middle part of his trip.

49. A Realtor earned $6,230 on the sale of a house. If the commission rate was 7%, what was the selling price of the house?

50. The thickness of a page of this book is about 0.0004 meters. How many pages laid on top of each other would it take to reach the top of Mt. Everest (8900 meters).

51. Construct the equation of the line shown in the graph.

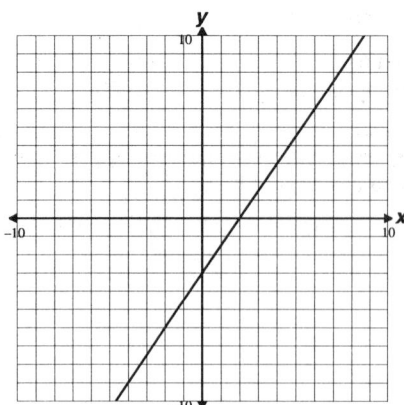

52. The graph shows the relationship between savings bond debt owed by the U. S. government and the number of years since 1985.

   a) If a linear model is appropriate, use two points on the regression line to build the regression equation.
   b) Interpret the meaning of the equation.
   c) What do you estimate the debt was in 1980?
   d) What year do you predict we will double the debt owed in 1990? Use the model to answer the question.

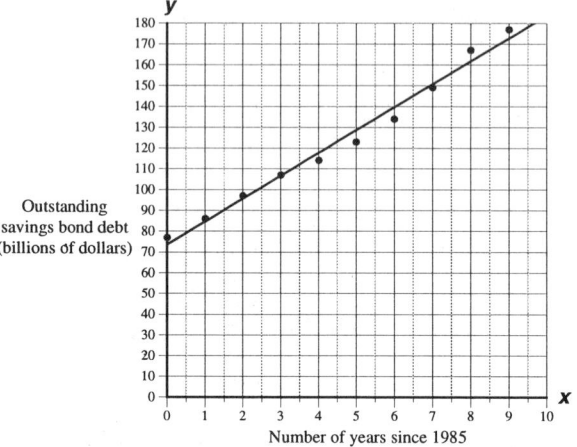

Section 8.2 Multiplying and Dividing Rational Expressions

Topic 8.2 A — Multiplying Rational Expressions

We multiply rational expressions in the same way we multiply rational numbers.

— *Property* — *Multiplying rational expressions*	English: To find the product of two rational expressions, multiply the numerators and multiply the denominators. Examples: $\dfrac{2}{3} \cdot \dfrac{4}{5}$ becomes $\dfrac{2 \cdot 4}{3 \cdot 5}$, which simplifies to $\dfrac{8}{15}$ $\dfrac{2}{x} \cdot \dfrac{y}{3}$ becomes $\dfrac{2 \cdot y}{x \cdot 3}$, which simplifies to $\dfrac{2y}{3x}$ Algebra: If P, Q, R, and S, represent polynomials and $Q \neq 0$ and $S \neq 0$, then $\dfrac{P}{Q} \cdot \dfrac{R}{S} = \dfrac{P \cdot R}{Q \cdot S}$

Because we almost always reduce fractions, it's usually easier to reduce common factors before actually multiplying. For example, to multiply $\dfrac{3}{10} \cdot \dfrac{2}{9}$ we could multiply and then reduce as in

$$\frac{3}{10} \cdot \frac{2}{9} \Rightarrow \frac{3 \cdot 2}{10 \cdot 9} \Rightarrow \frac{6}{90} \Rightarrow \frac{\cancel{6} \cdot 1}{\cancel{6} \cdot 15} \Rightarrow \frac{1}{15}$$

or we could reduce first and then multiply as in

$$\frac{3}{10} \cdot \frac{2}{9} \Rightarrow \frac{3}{2 \cdot 5} \cdot \frac{2}{3 \cdot 3} \Rightarrow \frac{\cancel{3} \cdot \cancel{2}}{\cancel{2} \cdot 5 \cdot \cancel{3} \cdot 3} \Rightarrow \frac{1}{15}$$

The second method is usually faster and it's the one we will use with rational expressions.

— Procedure —	**Step 1 Factor** Factor each numerator and denominator.
Multiplying rational expressions	**Step 2 Reduce** Reduce factors common to the numerator and denominator and then multiply the remaining factors if appropriate.

Example 8.2.1 Simplify. Assume no value of a variable will make a denominator 0.

a) $\dfrac{3}{x^2} \cdot \dfrac{x^5}{6}$

• SOLUTION •

$$\dfrac{3}{x^2} \cdot \dfrac{x^5}{2 \cdot 3} \quad \Leftarrow \text{ Factored.}$$

$$\dfrac{x^3}{2} \quad \Leftarrow \text{ Reduced common factors of 3 and } x^2 \text{ and then multiplied.}$$

We could also write the solution as $\dfrac{1}{2}x^3$.

b) $\dfrac{25w^4}{3q} \cdot 6q \cdot \dfrac{1}{5w}$

• SOLUTION •

$$\dfrac{25w^4}{3q} \cdot \dfrac{6q}{1} \cdot \dfrac{1}{5w} \quad \Leftarrow \text{ Wrote } 6q \text{ with a denominator of 1 to make sure we realize it's in the numerator.}$$

$$\dfrac{5 \cdot 5w^4}{3q} \cdot \dfrac{2 \cdot 3q}{1} \cdot \dfrac{1}{5w} \quad \Leftarrow \text{ Factored.}$$

$$10w^3 \quad \Leftarrow \text{ Reduced common factors of 3, 5, } w \text{ and } q \text{ and then multiplied.}$$

c) $\dfrac{b}{4a} \cdot \dfrac{2c}{b^2} \cdot \dfrac{1}{c}$

• SOLUTION •

$$\dfrac{b}{2 \cdot 2a} \cdot \dfrac{2c}{b^2} \cdot \dfrac{1}{c} \quad \Leftarrow \text{ Factored.}$$

$$\dfrac{1}{2ab} \quad \Leftarrow \text{ Reduced common factors of 2, } b, \text{ and } c \text{ and then multiplied.}$$

Practice 8.2.1 Simplify. Assume no value of a variable will make a denominator 0.

1. $\dfrac{5}{y^5} \cdot \dfrac{y^7}{10}$ 2. $\dfrac{14}{m^8} \cdot \dfrac{m^4}{7}$ 3. $\dfrac{48w^7}{7v} \cdot 14v \cdot \dfrac{1}{24w^{11}}$ 4. $\dfrac{5m}{2n} \cdot \dfrac{4n^4}{3p^2} \cdot \dfrac{2p}{15m^3}$

When the rational expressions contain polynomials the process is the same.

Example 8.2.2   Simplify. Assume no value of a variable will make a denominator 0.

a) $\dfrac{2}{y} \cdot \dfrac{y^2 + 2y}{4}$

• SOLUTION •

$\dfrac{2}{y} \cdot \dfrac{y(y+2)}{2 \cdot 2}$ ⇐ Factored.

$\dfrac{\cancel{2}}{\cancel{y}} \cdot \dfrac{\cancel{y}\,(y+2)}{\cancel{2} \cdot 2}$ ⇐ To help keep track of the common factors, we put a line through factors common to the numerator and the denominator.

$\dfrac{y+2}{2}$ ⇐ Reduced common factors of 2 and y.

You may be tempted to further reduce this by "canceling" the 2's in $\dfrac{y+2}{2}$.
But, the 2 in the numerator is not a factor, it is a term. We may reduce common *factors* but not common *terms*.

b) $\dfrac{x^2 - y^2}{2x+1} \cdot \dfrac{6x^2 + 3x}{x^3 + 2x^2 y + xy^2}$

• SOLUTION •

$\dfrac{(x+y)(x-y)}{2x+1} \cdot \dfrac{3x(2x+1)}{x(x+y)(x+y)}$ ⇐ Factored.

$\dfrac{(\cancel{x+y})(x-y)}{\cancel{2x+1}} \cdot \dfrac{3\cancel{x}(\cancel{2x+1})}{\cancel{x}(\cancel{x+y})(x+y)}$ ⇐ Marked the factors common to the numerator and the denominator.

$\dfrac{3(x-y)}{x+y}$ ⇐ Reduced common factors of x, $(x+y)$ and $(2x+1)$. Note that we left the numerator in its factored form because this is a more useful form if we need to use the expression in a later problem (it also saves us some work).

Practice 8.2.2   Simplify. Assume no value of a variable will make a denominator 0.

1. $\dfrac{5}{x} \cdot \dfrac{x^2 + 9x}{10}$

2. $\dfrac{6}{y+3} \cdot \dfrac{y^2 + 3y}{12}$

3. $\dfrac{m^2 - n^2}{3m+4} \cdot \dfrac{6m+8}{2m^2 + 3mn + n^2}$

4. $\dfrac{a^2 + 2ab + b^2}{a-b} \cdot \dfrac{5a - 5b}{a^2 - b^2}$

5. $\dfrac{3a^2 + 3a}{6a - 6} \cdot \dfrac{a^3 - a}{2a^3 + 3a^2 + a}$

6. $\dfrac{7x + 7y}{14x - 14y} \cdot \dfrac{x^3 - xy^2}{x^3 + 2x^2 y + xy^2}$

Topic 8.2 B — Dividing Rational Expressions

We divide rational expressions in the same way we divide rational numbers.

| — *Property* —
Multiplying rational expressions | English: To find the quotient of two rational expressions, multiply the first by the reciprocal of the second.

Examples:

$\dfrac{2}{3} \div \dfrac{1}{5}$ becomes $\dfrac{2}{3} \cdot \dfrac{5}{1}$, which simplifies to $\dfrac{10}{3}$

$\dfrac{2}{x} \div \dfrac{y}{3}$ becomes $\dfrac{2}{x} \cdot \dfrac{3}{y}$, which simplifies to $\dfrac{6}{xy}$

Algebra: If P, Q, R, and S represent polynomials and $Q \neq 0$, $R \neq 0$, and $S \neq 0$, then

$\dfrac{P}{Q} \div \dfrac{R}{S} = \dfrac{P}{Q} \cdot \dfrac{S}{R}$ |

As was the case with multiplication, it's important to reduce common factors. You may find the following procedure useful.

— Procedure — *Dividing rational* *expressions*	***Step 1 Reciprocal*** Rewrite the problem as the product of the numerator and the *reciprocal* of the denominator. ***Step 2 Factor*** Factor each numerator and denominator. ***Step 3 Reduce*** Reduce factors common to the numerator and the denominator and then multiply the remaining factors if appropriate.

Example 8.2.3   Simplify. Assume no value of a variable will make a denominator 0.

a) $\dfrac{\frac{4}{5}}{\frac{8}{15}}$

•SOLUTION•

$\dfrac{4}{5} \cdot \dfrac{\mathbf{15}}{\mathbf{8}}$ $\Leftarrow$ Changed division to multiplication of the reciprocal.

$\dfrac{2 \cdot 2}{5} \cdot \dfrac{3 \cdot 5}{2 \cdot 2 \cdot 2}$ $\Leftarrow$ Factored.

$\dfrac{3}{2}$ $\Leftarrow$ Reduced common factors of 2^2 and 5.

b) $\dfrac{\frac{x^3}{y}}{\frac{2x}{3y}}$

•SOLUTION•

$\dfrac{x^3}{y} \cdot \dfrac{\mathbf{3y}}{\mathbf{2x}}$ $\Leftarrow$ Changed division to multiplication of the reciprocal.

$\dfrac{3x^2}{2}$ $\Leftarrow$ Reduced common factors of x and y.

c) $\dfrac{\frac{x-2}{2x}}{\frac{x^2-4}{6}}$

•SOLUTION•

$\dfrac{x-2}{2x} \cdot \dfrac{\mathbf{6}}{\mathbf{x^2-4}}$ $\Leftarrow$ Changed division to multiplication of the reciprocal.

$\dfrac{x-2}{2x} \cdot \dfrac{\mathbf{2 \cdot 3}}{\mathbf{(x-2)(x+2)}}$ $\Leftarrow$ Factored.

$\dfrac{\cancel{x-2}}{\cancel{2}\,x} \cdot \dfrac{\cancel{2} \cdot 3}{(\cancel{x-2})(x+2)}$ $\Leftarrow$ Marked common factors.

$\dfrac{3}{x(x+2)}$ $\Leftarrow$ Reduced common factors of 2 and $(x-2)$. We leave the answer in factored form.

d) $\dfrac{x^2 + 5x + 6}{x + 2} \div (x^2 - 9)$

•**SOLUTION**• It is important to recognize that $x^2 - 9$ means $\dfrac{x^2 - 9}{1}$.

$\dfrac{x^2 + 5x + 6}{x + 2} \cdot \dfrac{x^2 - 9}{1}$ ⟸ Explicitly wrote the denominator of the second expression so that we can easily find its reciprocal..

$\dfrac{x^2 + 5x + 6}{x + 2} \cdot \dfrac{1}{x^2 - 9}$ ⟸ Changed division to multiplication of the reciprocal.

$\dfrac{\cancel{(x+2)}(\cancel{x+3})}{\cancel{x+2}} \cdot \dfrac{1}{(x - 3)(\cancel{x+3})}$ ⟸ Factored.

$\dfrac{1}{x - 3}$ ⟸ Reduced common factors of $(x - 3)$ and $(x + 2)$. We leave the answer in factored form.

Practice 8.2.3   Simplify. Assume no value of a variable will make a denominator 0.

1. $\dfrac{\dfrac{5}{6}}{\dfrac{15}{24}}$

2. $\dfrac{\dfrac{x^5}{y^2}}{\dfrac{5x^4}{3y^6}}$

3. $\dfrac{m}{n} \div \dfrac{4m}{4n^2 + 5n}$

4. $\dfrac{6m - 6}{5n} \div \dfrac{9m - 9}{10n}$

5. $\dfrac{y + 4}{2} \div \dfrac{y^2 + 8y + 16}{8}$

6. $\dfrac{x^2 + 4x + 4}{x^2 + 9x + 14} \div x^2 - 4$

Exercise Set 8.2   The answers to the odd numbered exercises are at the back of the book.

Simplify. Assume no value of a variable will make a denominator 0.

1. $\dfrac{45x^7y^6}{4z} \cdot \dfrac{8z^5}{15x^5y^7}$

2. $\dfrac{12a^4b^3}{5c^2} \cdot \dfrac{15c^7}{24ab}$

3. $\dfrac{4a^3b^2}{5c} \cdot \dfrac{15c^3}{10a^2b^4} \cdot \dfrac{25b}{3a^2}$

4. $\dfrac{8m^4}{7n^3r} \cdot \dfrac{21nr^2}{16m^2} \cdot \dfrac{4m^3}{3}$

5. $\dfrac{15x^3}{4y} \cdot 24y \cdot \dfrac{1}{10x}$

6. $\dfrac{36s^5}{5t^4} \cdot 10t \cdot \dfrac{4}{9s^3}$

7. $\dfrac{5r}{7s} \cdot \dfrac{14t}{15r^3} \cdot \dfrac{1}{t}$

8. $\dfrac{3x}{4y} \cdot \dfrac{12y^4}{9x^3z} \cdot \dfrac{x}{y^3z^4}$

9. $\dfrac{6n}{5m} \cdot \dfrac{15m^5}{18n^3} \cdot \dfrac{4n^3}{5m^4}$

10. $\dfrac{5x^2y^5}{7m^3n} \cdot \dfrac{14m^2n^5}{15x^4y} \cdot \dfrac{6xy}{mn^4}$

11. $\dfrac{5}{w} \cdot \dfrac{w^2 + 3w}{15}$

12. $\dfrac{-3}{r} \cdot \dfrac{r^2 - 2r}{-9}$

13. $\dfrac{y - x}{4} \cdot \dfrac{2x + 2y}{x^2 - y^2}$

14. $\dfrac{n - m}{9} \cdot \dfrac{3m + 3n}{m^2 - n^2}$

15. $\dfrac{3m + 4}{2m - 1} \cdot \dfrac{2m^2 + 9m - 5}{3m^2 - 2m - 8}$

16. $\dfrac{4x + 2}{3x + 1} \cdot \dfrac{3x^2 - 14x - 5}{4x^2 + 10x + 4}$

17. $\dfrac{x^2 + 2xy + y^2}{4x + 1} \cdot \dfrac{16x + 4}{2x^2 + 7xy + 5y^2}$

18. $\dfrac{m^2 - 5m + 6}{5m + 1} \cdot \dfrac{10m + 2}{3m^2 - 11m + 6}$

19. $\dfrac{\dfrac{3}{7}}{\dfrac{12}{21}}$

20. $\dfrac{\dfrac{2}{5}}{\dfrac{6}{35}}$

21. $\dfrac{\dfrac{7}{9}}{\dfrac{49}{-36}}$

22. $\dfrac{\dfrac{-3}{4}}{\dfrac{15}{16}}$

23. $\dfrac{\dfrac{6r^9}{5s^3}}{\dfrac{24r^8}{35s^5}}$

24. $\dfrac{\dfrac{9y^5}{8x^5}}{\dfrac{3y^3}{2x^2}}$

25. $\dfrac{\dfrac{9x^3}{11y}}{\dfrac{-27x^2}{22y^4}}$

26. $\dfrac{\dfrac{3m^7}{4n^6}}{\dfrac{m^9}{n^7}}$

27. $\dfrac{\dfrac{4x^4y^7}{9z^3}}{\dfrac{8xy}{3z^6}}$

28. $\dfrac{\dfrac{5m^5n^2}{2p^7}}{\dfrac{15m^3n^2}{8p^5}}$

29. $\dfrac{x}{y} \div \dfrac{2x}{y^2+4y}$

30. $\dfrac{r}{s} \div \dfrac{5r}{s^2-3s}$

31. $\dfrac{r}{s} \div \dfrac{5r}{5s^2-s}$

32. $\dfrac{3y-6}{4x} \div \dfrac{15y-30}{8x}$

33. $\dfrac{3r+6}{5s} \div \dfrac{6r^2+12r}{10s^2-10s}$

34. $\dfrac{8m}{3n^3-27n} \div \dfrac{16m^2+32m}{6n^2-18n}$

Review Exercises   The answers to all of these exercises are at the back of the book.

35. A father wishes to build a rectangular sandbox. If x represents the length of the sandbox, use x to create an expression to represent the width of the sandbox if:

    a) the length is 8 feet greater than the width.
    b) the width if 3 feet more than the length.
    c) the sandbox is a square.
    d) the length is to be twice the width.

36. Find the measures of the three angles of a triangle if the first is 7 degrees less than the second and the third is 7 degrees more than the second. Recall that the sum of the angles of any triangle is 180 degrees.

37. Two city work crews, which are 18 miles apart, are laying sewer pipe in a straight line. It's expected that when they meet both sections of pipe will fit together. The first crew had to work in clay so they moved at 0.5 miles per day. The second crew was working in sandy soil and so could lay 1.5 miles per day. If the second crew worked 6 fewer days than the first, find how many days the first crew worked.

38. Determine the domain and range of $\{(9,3),(-4,5),(9,0),(5,-3)\}$

---

39. Given $f(x) = 3x - 8$, find $f(5)$.

40. Graph: $y \le 5x + 4$

41. Solve: $\left\{ \begin{array}{rcl} 6x & = & 4y + 8 \\ -10 & = & -15x + 10y \end{array} \right\}$ **❶** **❷**

42. Solve: $\left\{ \begin{array}{rcl} 3x + 13y & = & 4 \\ 2x + 9y & = & 3 \end{array} \right\}$ **❶** **❷**

43. Simplify: $(4x + 7 - 3y) - (2y - 3x + 5) + (x - 7y + 12)$

44. Find the GCF of $4x^3y^5$ and $20x^2y^3$

Simplify.

45. $3^{-4} \cdot \left(\dfrac{1}{9}\right)^{-2} + 3^0$

46. $h^3(hg)^{-2}(-h^{-2}g)^2$

47. $\left(\dfrac{5xy^3}{15xy^{-2}}\right)^{-3}$

48. $(x-3)^2$

Factor completely.

49. $30x^4 - 25x^2$

50. $10x^2 - 2xy - 5x + y$

51. $x^2 + 30 - 11x$

52. $x^2 - 12x + 36$

Solve.

53. $(x-2)(3x+5) = 0$

54. $6 = 22x - 20x^2$

55. $3(x-5) + 2x > 5x - 1$

---

56. Find the x-intercept and the y-intercept:

57. Discuss the meaning of the graph.

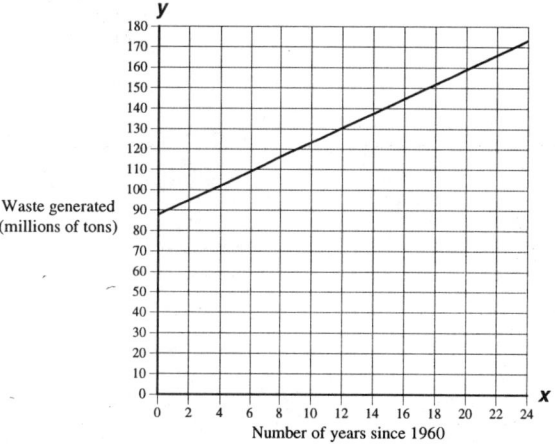

Section 8.3 Prelude to Adding and Subtracting Rational Expressions

In this section we review the ideas needed to successfully add and subtract rational expressions. We know from arithmetic that we can add fractions which have identical denominators by simply adding their numerators. To add fractions with different denominators, such as $\frac{1}{6} + \frac{1}{4}$, we first must convert the fractions into equivalent fractions that have the same denominator. Then, we can add their numerators. We follow the same procedure when adding rational expressions.

Topic 8.3 A — The Least Common Denominator (LCD)

The **Least Common Denominator** or **LCD** of a set of rational expressions is the smallest expression that contains as factors all the denominators of the set. For example, the LCD of $\frac{1}{3x}$ and $\frac{1}{2x^2}$ is $6x^2$. It's easy to see that $6x^2$ has both $3x$ and $2x^2$ as factors because

- $6x^2$ can be written as $(\mathbf{3x})(2x)$.

- $6x^2$ can be written as $(\mathbf{2x^2})(3)$.

In Section 1.2 we presented the following procedure for finding the LCD of a set of arithmetic fractions. This is the method we will use with rational expressions.

— Procedure — *Finding the LCD*	*Step 1 Prime factor* Prime factor each denominator and write it in exponential form.
	Step 2 Select For each prime factor, select the one with the largest power.
	Step 3 Multiply Multiply the selected factors. This is the LCD.

Example 8.3.1 a) Find the least common denominator of $\frac{1}{4}$ and $\frac{1}{18}$.

•SOLUTION• *Step 1 Prime factor*

4 factors as $2 \cdot 2$ or 2^2

18 factors as $2 \cdot 3 \cdot 3$ or $2^1 3^2$

Step 2 Select

largest power of 2 is 2^2

largest power of 3 is 3^2

Step 3 Multiply The LCD is $2^2 3^2$, which is 36. That is, 36 is the smallest number that contains factors of 4 and 18.

b) Find the least common denominator of $\frac{1}{x^2}$ and $\frac{1}{xy^2}$. Assume no value of any variable will make any denominator 0.

•SOLUTION• *Step 1 Prime factor* x^2 and xy^2 are already factored.

Step 2 Select

largest power of x is x^2

largest power of y is y^2

Step 3 Multiply The LCD is $x^2 y^2$

— Note — Try finding the LCD mentally first	If you can see the LCD by inspection that's fine. If you can't see the LCD by inspection, it's important to know how to use the procedure outlined above.

c) Find the least common denominator of $\dfrac{1}{(x-1)^2}$ and $\dfrac{1}{(x-1)(y-1)^2}$. Assume no value of any variable will make any denominator 0.

•SOLUTION• *Step 1 Prime factor* The denominators are already factored.

Step 2 Select

- the largest power of $(x-1)$ is $(x-1)^2$.

- the largest power of $(y-1)$ is $(y-1)^2$.

Step 3 Multiply The LCD is $(x-1)^2(y-1)^2$

— Note — Multiplying out LCDs	When working with LCDs that are made up of monomial factors, we usually multiply the factors to write the LCD as compactly as possible. Thus, we multiplied $2^2 3^2$ to get 36.
	When the LCD has binomial factors we leave it in its factored form because that is usually the most compact and useful way of writing it. Thus, we left $(x-1)^2(y-1)^2$ as is rather than multiplying it out to get $$x^2 y^2 - 2x^2 y - 2xy^2 + x^2 + 4xy + y^2 - 2x - 2y + 1$$

d) Find the least common denominator of $\dfrac{1}{x^2-x}$ and $\dfrac{1}{x^2+x-2}$. Assume no value of any variable will make any denominator 0.

•SOLUTION• *Step 1 Prime factor*

$x^2 - x$ factors as $x(x-1)$

$x^2 + x - 2$ factors as $(x+2)(x-1)$

Step 2 Select

- the largest power of x is x.

- the largest power of $(x-1)$ is $(x-1)$.

- the largest power of $(x+2)$ is $(x+2)$.

Step 3 Multiply The LCD is $x(x-1)(x+2)$.

Practice 8.3.1 Find the least common denominator.

1. $\dfrac{1}{12}$ and $\dfrac{1}{28}$

2. $\dfrac{1}{m^2 n}$ and $\dfrac{1}{mn^2}$

3. $\dfrac{1}{x^3 y}$ and $\dfrac{1}{xy^2}$

4. $\dfrac{1}{3m}$ and $\dfrac{1}{6n}$

5. $\dfrac{1}{x+3}$ and $\dfrac{1}{x-5}$

6. $\dfrac{1}{y+5}$ and $\dfrac{1}{(y+5)^2}$

7. $\dfrac{1}{x^2-36}$ and $\dfrac{1}{x^2-9x+18}$

8. $\dfrac{1}{y^2-49}$ and $\dfrac{1}{y^2-9y+14}$

— Note — Restricted values	From now on, we will not explicitly write the restricted values of rational expressions. We do this in order to save space and to help you focus on the topic under discussion. However, for every rational expression you should assume that no value of any variable will make any denominator 0.

Topic 8.3 B — Building Equivalent Fractions

Once we have found the LCD the next step is to convert each fraction to an equivalent fraction that has the LCD as its denominator. **Equivalent fractions** are fractions that represent the same value. For example $\frac{5}{10}$ and $\frac{2}{4}$ are equivalent fractions because they both are equivalent to 0.5. We will build equivalent fractions using the Fundamental Property of Fractions discussed in the last section (that is, we multiply by another fraction whose value is 1).

— Procedure — *Building equivalent* *fractions*	*Step 1 Find Needed Factors* Determine what the denominator of the original fraction must be multiplied by to transform it into the LCD. *Step 2 Multiply by 1* Multiply BOTH the numerator and the denominator of the original fraction by the factors found in Step 1.

For example, let's build a fraction equivalent to $\frac{3}{4}$ but which has a denominator of 24.

Step 1   The denominator of $\frac{3}{4}$ needs a factor of 6 to transform it into the required denominator of 24.

Step 2   We multiply the original fraction by 1 in the form $\frac{6}{6}$ to build the equivalent fraction: $\frac{3}{4} \Rightarrow \frac{3}{4} \cdot \frac{6}{6} \Rightarrow \frac{18}{24}$.

Example 8.3.2   a)   Build a fraction equivalent to $\frac{3}{x}$ but which has a denominator of $2xy$.

     •SOLUTION•   The denominator of $\frac{3}{x}$ needs a factor of $2y$ to transform it into the required denominator: $\frac{3}{x} \Rightarrow \frac{3}{x} \cdot \frac{2y}{2y} \Rightarrow \frac{6y}{2xy}$

     b)   Build a fraction equivalent to $\frac{a}{3b}$ but which has a denominator of $6b^2$.

     •SOLUTION•   The denominator of $\frac{a}{3b}$ needs a factor of $2b$ to transform it into the required denominator: $\frac{a}{3b} \Rightarrow \frac{a}{3b} \cdot \frac{2b}{2b} \Rightarrow \frac{2ab}{6b^2}$.

Practice 8.3.2   Build a fraction equivalent to the given fraction but which has the given denominator. Assume no value of a variable will make a denominator 0.

1. $\frac{5}{6}$ with a denominator of 24.

2. $\frac{-2}{x}$ with a denominator of x^4

3. $\frac{-9}{r}$ with a denominator of $2qr^5$

4. $\frac{x}{7y}$ with a denominator of $14xy$

We use the same procedure when the denominators contain binomials or trinomials.

Example 8.3.3   a)   Build a fraction equivalent to $\frac{2}{x-2}$ but which has a denominator of $(x-2)(x+2)$.

     •SOLUTION•   The denominator of $\frac{2}{x-2}$ needs a factor of $x+2$ to transform it into the required denominator: $\frac{2}{x-2} \Rightarrow \frac{2}{x-2} \cdot \frac{(x+2)}{(x+2)} \Rightarrow \frac{2(x+2)}{(x-2)(x+2)}$

b) Build a fraction equivalent to $\dfrac{x+1}{x-2}$ but with denominator $x^2 - 4$.

•**SOLUTION**• It may not be obvious what we need to multiply $x - 2$ by in order to get $x^2 - 4$. Therefore, we factor $x^2 - 4$ into $(x - 2)(x + 2)$. Now we see that we need a factor of $x + 2$ to transform it into the required denominator:

$$\frac{x+1}{x-2} \Rightarrow \frac{x+1}{x-2} \cdot \frac{(x+2)}{(x+2)} \Rightarrow \frac{(x+1)(x+2)}{(x-2)(x+2)}$$

Practice 8.3.3 Build a fraction equivalent to the given fraction but which has the given denominator. Assume no value of a variable will make a denominator 0.

1. $\dfrac{5}{w+3}$ with a denominator of $(w+3)(w+3)$

2. $\dfrac{6}{x+2}$ with a denominator of $(x+2)(x-3)$

3. $\dfrac{9}{y+5}$ with a denominator of $y(y+5)$

4. $\dfrac{y+4}{y+3}$ with a denominator of $(y+2)(y+3)$

5. $\dfrac{m+2}{m+5}$ with a denominator of $m^2 - 25$

6. $\dfrac{n+2}{n+7}$ with a denominator of $n^2 + 9n + 14$

Remember, the fractions we build are equivalent to the original fractions. This means that if we substitute any value for the variable, other than the restricted values, the original fraction and the equivalent fraction will simplify to the same number. For example, we found that $\dfrac{2}{x-2}$ and $\dfrac{2(x+2)}{(x-2)(x+2)}$ are equivalent. If we substitute, say, -1 for x into each fraction they will both simplify to the same number.

Topic 8.3 C — Building Equivalent Fractions That Contain the LCD

When we add rational expressions in the next section we will need to combine both procedures just discussed.

— *Procedure* — *Converting to equivalent fractions that contain the LCD*	***Step 1 Build LCD*** Build the LCD of all the fractions by factoring and finding the product of the largest power of each prime factor. ***Step 2 Build equivalent fractions*** Multiply BOTH the numerator and the denominator of each fraction by the factors the denominator needs to become the LCD.

Let's see how this procedure works by converting the fractions $\dfrac{1}{12x^2}$ and $\dfrac{3x}{10}$ into equivalent fractions with their LCD as the denominator.

Step 1 Build LCD

$12x^2$ factors as $2 \cdot 2 \cdot 3 \cdot x^2$ or $2^2 3^1 x^2$

10 factors as $2 \cdot 5$ or $2^1 5^1$

The LCD is $2^2 3^1 5^1 x^2$ or $60x^2$

Step 2 Build equivalent fractions   For each fraction, we multiply BOTH the numerator and the denominator by the factors the denominator needs to become the LCD. An easy way of seeing which factors are needed is to compare the factored form of each denominator with the factored form of the LCD. The factors we need are the ones that are in the LCD but not in the original denominator.

- We compare $12x^2$ to the LCD, $60x^2$:

$$\text{first denominator: } 12x^2 = 2 \cdot 2 \cdot 3 \cdot \qquad x^2$$
$$\updownarrow$$
$$\text{LCD: } 60x^2 = 2 \cdot 2 \cdot 3 \cdot 5 \cdot x^2$$

We see that $12x^2$ is missing a factor of 5 so we multiply $\dfrac{1}{12x^2}$ by 1 in the form $\dfrac{5}{5}$:

$$\frac{1}{12x^2} \Rightarrow \frac{1}{12x^2} \cdot \frac{5}{5} \Rightarrow \frac{5}{60x^2}$$

- We compare 10 to the LCD, $60x^2$:

$$\text{second denominator: } \quad 10 = 2 \cdot \qquad\qquad 5$$
$$\qquad\qquad\qquad\quad \updownarrow \quad \updownarrow \qquad\qquad \updownarrow$$
$$\text{LCD: } 60x^2 = 2 \cdot 2 \cdot 3 \cdot 5 \cdot x^2$$

We see that 10 is missing factors of 2, 3, and x^2 so we multiply $\dfrac{3x}{10}$ by 1 in the form $\dfrac{6x^2}{6x^2}$:

$$\frac{3x}{10} \Rightarrow \frac{3x}{10} \cdot \frac{6x^2}{6x^2} \Rightarrow \frac{18x^3}{60x^2}$$

Example 8.3.4   Find the LCD and then convert each fraction into an equivalent fraction with the LCD as the denominator.

a) $\dfrac{5}{x^2y}$ and $\dfrac{7}{3xy^2}$

    •SOLUTION•   ***Step 1 Build LCD***   The denominators are already in factored form. To find the LCD we multiply the largest power of 3, x, and y to get $3x^2y^2$.

Step 2 Build equivalent fractions

- We compare x^2y to $3x^2y^2$:

$$\text{first denominator: } \quad x^2y = \qquad x \cdot x \cdot y$$
$$\qquad\qquad\qquad\qquad\quad \updownarrow \qquad\qquad\qquad \updownarrow$$
$$\text{LCD: } 3x^2y^2 = 3 \cdot x \cdot x \cdot y \cdot y$$

We see that x^2y is missing factors of 3 and y.

$$\frac{5}{x^2y} \Rightarrow \frac{5}{x^2y} \cdot \frac{3y}{3y} \Rightarrow \frac{15y}{3x^2y^2}$$

- We compare $3xy^2$ to $3x^2y^2$:

$$\text{second denominator: } \quad 3xy^2 = 3 \cdot x \cdot \qquad \cdot y \cdot y$$
$$\qquad\qquad\qquad\qquad\qquad\qquad \updownarrow$$
$$\text{LCD: } 3x^2y^2 = 3 \cdot x \cdot x \cdot y \cdot y$$

We see that $3xy^2$ is missing a factor of x.

$$\frac{7}{3xy^2} \Rightarrow \frac{7}{3xy^2} \cdot \frac{x}{x} \Rightarrow \frac{7x}{3x^2y^2}$$

b) $\dfrac{5}{y-6}$ and $\dfrac{7}{y+3}$

•**SOLUTION**• *Step 1 Build LCD* The denominators are already in factored form. To find the LCD we multiply the largest power of $(y-6)$ and $(y+3)$ to get $(y-6)(y+3)$.

Step 2 Build equivalent fractions

- We compare $y-6$ to the LCD, $(y-6)(y+3)$:

  first denominator: $y - 6 = (y - 6)$

  $\updownarrow$

  LCD: $(y - 6)(y + 3) = (y - 6)(y + 3)$

  We see that $y-6$ is missing a factor of $y+3$.

  $$\frac{5}{y-6} \Rightarrow \frac{5}{y-6} \cdot \frac{y+3}{y+3} \Rightarrow \frac{5(y+3)}{(y-6)(y+3)}$$

- We compare $y+3$ to the LCD, $(y-6)(y+3)$:

  second denominator: $y + 3 = \qquad (y + 3)$

  $\updownarrow$

  LCD: $(y - 6)(y + 3) = (y - 6)(y + 3)$

  We see that $y+3$ is missing a factor of $y-6$.

  $$\frac{7}{y+3} \Rightarrow \frac{7}{y+3} \cdot \frac{y-6}{y-6} \Rightarrow \frac{7(y-6)}{(y-6)(y+3)}$$

c) $\dfrac{6}{x^2-9}$ and $\dfrac{5}{x^2-x-6}$

•**SOLUTION**• *Step 1 Build LCD* Factor the denominators:

$x^2 - 9$ factors as $(x-3)(x+3)$

$x^2 - x - 6$ factors as $(x-3)(x+2)$

Next, write the product of the largest power of each factor to get $(x-3)(x+3)(x+2)$ as the LCD. Remember, because these are binomial factors we don't actually multiply this out. The expanded product would be a mess!

Step 2 Build equivalent fractions

- We compare x^2-9 to the LCD, $(x-3)(x+3)(x+2)$:

  first denominator: $x^2 - 9 = (x - 3)(x + 3)$

  $\updownarrow$

  LCD: $= (x - 3)(x + 3)(x + 2)$

  We see that x^2-9 is missing a factor of $x+2$.

  $$\frac{6}{x^2-9} \Rightarrow \frac{6}{(x-3)(x+3)} \cdot \frac{x+2}{x+2} \Rightarrow \frac{6(x+2)}{(x-3)(x+3)(x+2)}$$

- We compare x^2-x-6 to the LCD, $(x-3)(x+3)(x+2)$:

  second denominator: $x^2 - x - 6 = (x - 3) \qquad (x + 2)$

  $\updownarrow$

  LCD: $= (x - 3)(x + 3)(x + 2)$

  We see that x^2-x-6 is missing a factor of $x+3$.

  $$\frac{5}{x^2-x-6} \Rightarrow \frac{5}{(x-3)(x+2)} \cdot \frac{x+3}{x+3} \Rightarrow \frac{5(x+3)}{(x-3)(x+3)(x+2)}$$

Practice 8.3.4 Find the LCD and then convert each fraction into an equivalent fraction with the LCD as the denominator. Assume no value of a variable will make a denominator 0.

1. $\dfrac{9}{14x}$ and $\dfrac{5}{21x^2}$

2. $\dfrac{5}{y-5}$ and $\dfrac{2}{y+4}$

3. $\dfrac{6}{m+4}$ and $\dfrac{5}{(m+4)^2}$

4. $\dfrac{10}{x^2+2x+1}$ and $\dfrac{3}{x^2+6x+5}$

5. $\dfrac{2}{y^2+6y+8}$ and $\dfrac{5}{y^2+4y+4}$

Exercise Set 8.3   The answers to the odd numbered exercises are at the back of the book.

Find the least common denominator.

1. $\dfrac{1}{14x^2}$ and $\dfrac{1}{21x}$

2. $\dfrac{1}{4x}$ and $\dfrac{1}{6y}$

3. $\dfrac{1}{4x^3y}$ and $\dfrac{1}{6xy^3}$

4. $\dfrac{1}{10m^4n}$ and $\dfrac{1}{15mn^4}$

5. $\dfrac{3}{y+2}$ and $\dfrac{y}{y-3}$

6. $\dfrac{1}{m+5}$ and $\dfrac{2m}{m+1}$

7. $\dfrac{8}{x}$ and $\dfrac{x}{x+1}$

8. $\dfrac{5}{p}$ and $\dfrac{12p}{p-2}$

9. $\dfrac{4}{(m-1)^2}$ and $\dfrac{2}{m+2}$

10. $\dfrac{3}{x-4}$ and $\dfrac{1}{(x-4)^2}$

11. $\dfrac{2}{y^2-3y}$ and $\dfrac{1}{y^2-5y+6}$

12. $\dfrac{1}{x^2+7x}$ and $\dfrac{7}{x^2+8x+7}$

Build a fraction equivalent to the given fraction but which has the given denominator. Assume no value of a variable will make a denominator 0.

13. $\dfrac{5}{8}$ with denominator 32

14. $\dfrac{3}{4}$ with denominator 24

15. $\dfrac{-2}{5}$ with denominator 25

16. $\dfrac{-7}{m}$ with denominator m^3

17. $\dfrac{5}{m}$ with denominator m^2n

18. $\dfrac{3}{y}$ with denominator xy^2

19. $\dfrac{4x}{5y}$ with denominator $20yz$

20. $\dfrac{7m}{8n}$ with denominator $24nt$

21. $\dfrac{5r}{s}$ with denominator $7rs^2$

22. $\dfrac{9h}{g}$ with denominator $15fg^3Y$

23. $\dfrac{7x}{x+4}$ with denominator x^2+4x

24. $\dfrac{5y}{y-3}$ with denominator y^2-3y

25. $\dfrac{2}{t+1}$ with denominator $(t+1)(t+1)$

26. $\dfrac{7}{m-1}$ with denominator $(m-1)(m-5)$

27. $\dfrac{3}{m-9}$ with denominator $m(m-9)$

28. $\dfrac{t+1}{t-3}$ with denominator $(t-3)(t+2)$

29. $\dfrac{w-2}{w-1}$ with denominator w^2-3w+2

30. $\dfrac{x-3}{x-4}$ with denominator $x^2-7x+12$

31. $\dfrac{5}{t+4}$ with denominator $t^2+9t+20$

32. $\dfrac{6}{r-2}$ with denominator $r^2-8r+12$

33. $\dfrac{x-1}{x-6}$ with denominator x^2-36

34. $\dfrac{m+5}{m+6}$ with denominator $m^2+8m+12$

Find the LCD and then convert each fraction into an equivalent fraction with the LCD as the denominator. Assume no value of a variable will make a denominator 0.

35. $\dfrac{3}{5x^3}$ and $\dfrac{2x}{15}$

36. $\dfrac{6}{m^2n}$ and $\dfrac{5}{2mn^3}$

37. $\dfrac{4}{15m}$ and $\dfrac{3}{25m^2}$

38. $\dfrac{7}{2xy^2}$ and $\dfrac{1}{6x^2y^2}$

39. $\dfrac{5}{2m^2n}$ and $\dfrac{7}{3mn^2}$

40. $\dfrac{3}{7rs^2}$ and $\dfrac{4}{5r^2s}$

41. $\dfrac{4}{y^2-3y}$ and $\dfrac{7}{y^2+2y}$

42. $\dfrac{11}{x^2+8x}$ and $\dfrac{12}{x^2-5x}$

43. $\dfrac{-2}{x^2+x-6}$ and $\dfrac{5}{x^2+3x-10}$

44. $\dfrac{-3}{y^2-9y+20}$ and $\dfrac{6}{y^2-7y+12}$

Review Exercises   The answers to all of these exercises are at the back of the book.

45. Evaluate the expression when x is -2 and y is -3: $\dfrac{5x - y}{2x - 3}$ 46. Prime factor: 440   47. Solve: $4x = 3x$

48. Graph the straight line that passes through (7,1) with slope undefined.   49. Graph: $y - 3x \geq 4$

50. Use the graph to complete the data table:

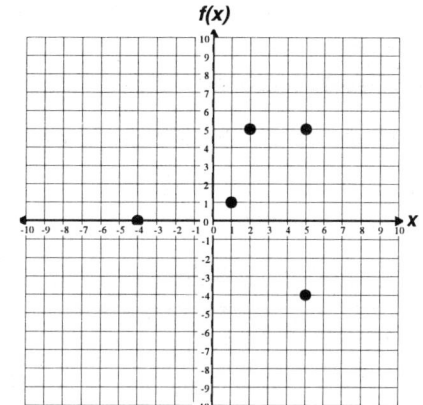

x	$f(x)$
	0
1	
2	
	-4
5	

51. Solve: $\begin{cases} 8 + y = x & \text{❶} \\ 3x + 6y = 6 & \text{❷} \end{cases}$ 52. Solve: $\begin{cases} x + 12y = -4 & \text{❶} \\ 3x - 6y = 30 & \text{❷} \end{cases}$

53. Divide and simplify: $\dfrac{-42y^{17} + 22y^9 - 62y^6}{2y^5}$ 54. Use long division to simplify: $\dfrac{8a^3 + 2a^2 - 7a + 4}{2a - 1}$

Simplify.

55. $9 + 3(9 + 3(9 - 3^2) - 2)$ 56. $\dfrac{7}{8} - \dfrac{5}{9}\left(\dfrac{3}{10}\right)^2$ 57. $\dfrac{24}{x^2} \cdot \dfrac{x^5}{6}$ 58. $\left(\dfrac{(4^{-2}a^3)^2}{(18b^3)^{-1}}\right)\left(\dfrac{b^5}{6a^7}\right)^{-2}$

59. $2(x - 3) - (6 - 2x)$ 60. $(c - 12)(3c + 5)$ 61. $(1.8y^4 - 3.5y^2 + y) + (14.2y^4 - 0.6y^2 + 8)$

Factor completely.

62. $-12x^2 - 18x$ 63. $x^2 - 5x - 15y + 3xy$ 64. $15x^2 - x - 2$ 65. $121 - x^2$

66. $25x^2 - 40x + 16$ 67. $24 - 6x^2$

68. It takes 2 hours to build a model 500 snowboard and 1 hour to custom paint it . The deluxe model 600 takes 3 hours to build and 45 minutes to paint. If Bob plans to spend 32 hours per week building boards and 10 hours per week painting them, how many of each type of board can he make in a week?

69. The data table shows the projected life expectancy for men from 1960 to 2010.

   a) Build a linear equation for this situation.
   b) Discuss the meaning of the equation.
   c) Predict how long the average man will live in 2050.
   d) When will the life expectancy for a man be 85 years?

x number of years since 1960	y life expectancy for men (years)
0	66.8
30	71.6
50	74.8

Section 8.4 Adding and Subtracting Rational Expressions

Topic 8.4 A — Adding Rational Expressions with Like Denominators

To add two arithmetic fractions that have the same denominator we add the numerators and write the sum over the common denominator. For example,

$$\frac{3}{5} + \frac{1}{5} \qquad \Leftarrow \text{Expression to simplify.}$$

$$\frac{3 + 1}{5} \qquad \Leftarrow \text{Added numerators.}$$

$$\frac{4}{5} \qquad \Leftarrow \text{Simplified.}$$

If two rational expressions have the same denominator the process is the same.

Example 8.4.1 Simplify. Assume no value of a variable will make a denominator 0.

a) $\dfrac{4x}{9} + \dfrac{8x}{9}$

•SOLUTION•

$$\frac{4x + 8x}{9} \qquad \Leftarrow \text{Added numerators.}$$

$$\frac{12x}{9} \qquad \Leftarrow \text{Combined like terms.}$$

$$\frac{4x}{3} \qquad \Leftarrow \text{Reduced by common factor of 3.}$$

b) $\dfrac{3}{5x^2} + \dfrac{x + 5}{5x^2}$

•SOLUTION•

$$\frac{3 + x + 5}{5x^2} \qquad \Leftarrow \text{Added numerators.}$$

$$\frac{x + 8}{5x^2} \qquad \Leftarrow \text{Combined like terms. Remember, we cannot reduce the } x \text{ since}$$

it's not a factor of the numerator.

Practice 8.4.1 Simplify. Assume no value of a variable will make a denominator 0.

1. $\dfrac{6m}{5} + \dfrac{4m}{5}$ 2. $\dfrac{y}{9} + \dfrac{2y}{9}$ 3. $\dfrac{2}{7y^2} + \dfrac{y + 4}{7y^2}$ 4. $\dfrac{9}{5t^2} + \dfrac{t - 3}{5t^2}$ 5. $\dfrac{5}{y + 3} + \dfrac{y + 2}{y + 3}$

Topic 8.4 B — Subtracting Rational Expressions with Like Denominators

Subtraction works the same way as addition but we must be careful with the signs. To help keep the signs straight we will convert the subtraction into addition of the opposite as the following example illustrates:

$$\frac{2}{x} - \frac{x - 3}{x} \qquad \Leftarrow \text{Expression to simplify.}$$

$$\frac{2}{x} + (-1)\left(\frac{x - 3}{x}\right) \qquad \Leftarrow \text{Converted subtraction to addition of } -1 \text{ times the fraction.}$$

$$\frac{2}{x} + \left(\frac{-1}{1}\right)\left(\frac{x - 3}{x}\right) \qquad \Leftarrow \text{Explicitly wrote denominator of the factor } -1.$$

$$\frac{2}{x} + \frac{(-1)(x-3)}{x} \quad \Leftarrow \text{Multiplied numerators and denominators in second term.}$$

$$\frac{2 + (-1)(x-3)}{x} \quad \Leftarrow \text{Added numerators.}$$

In practice, we don't write all these steps. We just place the second numerator inside parentheses and go from there. The process would look like this:

$$\frac{2}{x} - \frac{x-3}{x} \Rightarrow \frac{2 - (x-3)}{x}$$

Example 8.4.2 Simplify. Assume no value of a variable will make a denominator 0.

a) $\dfrac{2}{x+1} - \dfrac{x+2}{x+1}$

• SOLUTION •

$$\frac{2 - (x+2)}{x+1} \quad \Leftarrow \text{Subtracted numerators. Note how the second numerator is}$$
placed inside parentheses.

$$\frac{2 - x - 2}{x+1} \quad \Leftarrow \text{Removed parentheses. Note how the signs of terms that were}$$
inside the parentheses changed.

$$\frac{-x}{x+1} \quad \Leftarrow \text{Combined like terms.}$$

— *Caution* — ***Distribute negative sign***	When subtracting rational expressions make sure the subtraction is distributed to every term in the numerator of the expression being subtracted. For example, INCORRECT: $\dfrac{3}{x} - \dfrac{x+2}{x} \neq \dfrac{3 - x + 2}{x}$ CORRECT: $\dfrac{3}{x} - \dfrac{x+2}{x}$ can be written as $\dfrac{3 - (x+2)}{x}$ or $\dfrac{3 - x - 2}{x}$

b) $\dfrac{2x^2 + 3}{x - 1} - \dfrac{x^2 + 4}{x - 1}$

• SOLUTION •

$$\frac{2x^2 + 3 - (x^2 + 4)}{x - 1} \quad \Leftarrow \text{Subtracted numerators. Note how the second numerator is}$$
placed inside parentheses.

$$\frac{2x^2 + 3 - x^2 - 4}{x - 1} \quad \Leftarrow \text{Simplified.}$$

$$\frac{x^2 - 1}{x - 1} \quad \Leftarrow \text{Combined like terms.}$$

$$\frac{(x+1)(x-1)}{x-1} \quad \Leftarrow \text{Factored.}$$

$$x + 1 \quad \Leftarrow \text{Reduced by common factor of } x - 1.$$

c) $\dfrac{6x}{2x-3y} - \dfrac{2x+6y}{2x-3y}$

•SOLUTION•

$\dfrac{6x - (2x + 6y)}{2x - 3y}$ $\Leftarrow$ Subtracted numerators.

$\dfrac{6x - 2x - 6y}{2x - 3y}$ $\Leftarrow$ Distributed negative .

$\dfrac{4x - 6y}{2x - 3y}$ $\Leftarrow$ Combined like terms.

$\dfrac{2(2x - 3y)}{2x - 3y}$ $\Leftarrow$ Factored.

2 $\Leftarrow$ Reduced by common factor of $2x - 3y$.

Practice 8.4.2   Simplify. Assume no value of a variable will make a denominator 0.

1. $\dfrac{3m}{6} - \dfrac{4m}{6}$ 2. $\dfrac{2}{8y^2} - \dfrac{y+4}{8y^2}$ 3. $\dfrac{4y^2+1}{y+2} - \dfrac{2y^2-2}{y+2}$ 4. $\dfrac{3m^2+5}{m+8} - \dfrac{4m^2+6}{m+8}$

Topic 8.4 C — Adding and Subtracting Fractions with Unlike Denominators

When the denominators are not identical we must build equivalent fractions before adding or subtracting, just as we did in arithmetic. For example

$\dfrac{1}{6} + \dfrac{7}{10}$ $\Leftarrow$ Expression to simplify.

$\dfrac{1}{6} \cdot \dfrac{5}{5} + \dfrac{7}{10} \cdot \dfrac{3}{3}$ $\Leftarrow$ Built equivalent fractions that contain the LCD, 30.

$\dfrac{5}{30} + \dfrac{21}{30}$ $\Leftarrow$ Simplified.

$\dfrac{5 + 21}{30}$ $\Leftarrow$ Added numerators.

$\dfrac{26}{30}$ $\Leftarrow$ Simplified.

$\dfrac{13}{15}$ $\Leftarrow$ Reduced by common factor of 2.

The general procedure looks like this:

— Procedure —	
Adding and subtracting rational expressions with unlike denominators	***Step 1 Factor Denominators*** Write all denominators in factored form.
	Step 2 Build LCD Build the Least Common Denominator (LCD) by multiplying the largest power of each prime factor.
	Step 3 Build Equivalent Fractions Build equivalent fractions which contain the LCD.
	Step 4 Combine Numerators Add or subtract the numerators and use the LCD as the denominator.
	Step 5 Reduce Factor the resulting fraction and reduce if possible.

Example 8.4.3   Simplify. Assume no value of a variable will make a denominator 0.

a) $\dfrac{7}{5xy^2} + \dfrac{5}{3x^3y^2}$

•SOLUTION•   ***Step 1 Factor Denominators***   The denominators are already in factored form.

Step 2 Build LCD

Largest power of 3 is 3^1.
Largest power of 5 is 5^1.
Largest power of x is x^3.
Largest power of y is y^2.

LCD is $3^1 \cdot 5^1 \cdot x^3 \cdot y^2$ or $15x^3y^2$.

Step 3 Build Equivalent Fractions

- First term: When we compare $5xy^2$ with the LCD we see that $5xy^2$ needs one factor of 3 and two more factors of x.

$$\frac{7}{5xy^2} \Rightarrow \frac{7}{5xy^2} \cdot \frac{3x^2}{3x^2} \Rightarrow \frac{21x^2}{15x^3y^2}$$

- Second term: When we compare $3x^3y^2$ with the LCD we see that $3x^3y^2$ needs one factor of 5.

$$\frac{5}{3x^3y^2} \Rightarrow \frac{5}{3x^3y^2} \cdot \frac{5}{5} \Rightarrow \frac{25}{15x^3y^2}$$

Step 4 Combine Numerators

$$\frac{21x^2}{15x^3y^2} + \frac{25}{15x^3y^2} \Rightarrow \frac{21x^2 + 25}{15x^3y^2}$$

Step 5 Reduce   We factor numerator and denominator to get $\dfrac{3 \cdot 7x^2 + 5 \cdot 5}{3 \cdot 5 \cdot x^3y^2}$

The numerator and denominator have no common factors so this fraction cannot be reduced. We would write the final answer as $\dfrac{21x^2 + 25}{15x^3y^2}$

b) $\dfrac{1}{10x^2y} + 2$

•SOLUTION•   If you can mentally "see" the LCD, skip to step 3.

Step 1 Factor Denominators

$10x^2y$ factors as $2 \cdot 5 \cdot x^2 \cdot y$ or $2^1 5^1 x^2 y^1$

2 can be written as $\dfrac{2}{1}$. Its denominator, 1, cannot be factored.

Step 2 Build LCD   Note that since factors of 1 do not affect a product we do not include the factor of 1 in the LCD.

Largest power of 2 is 2^1.
Largest power of 5 is 5^1.
Largest power of x is x^2.
Largest power of y is y^1.

LCD is $2^1 \cdot 5^1 \cdot x^2 \cdot y^1$ or $10x^2y$.

Step 3 Build Equivalent Fractions

- First term: When we compare $10x^2y$ with the LCD we see that they match. Therefore, $10x^2y$ does not need any additional factors.

- Second term: We will write 2 as $\frac{2}{1}$ so we can easily see the numerator and denominator. When we compare 1 with the LCD we see that 1 needs one factor of 2, one factor of 5, two factors of x, and one factor of y. In other words, it needs $10x^2y$.

$$2 \Rightarrow \frac{2}{1} \cdot \frac{\mathbf{10x^2y}}{\mathbf{10x^2y}} \Rightarrow \frac{20x^2y}{10x^2y}$$

Step 4 Combine Numerators

$$\frac{1}{10x^2y} + 2 \Rightarrow \frac{1}{10x^2y} + \frac{20x^2y}{10x^2y} \Rightarrow \frac{1 + 20x^2y}{10x^2y}$$

Since we usually write polynomials in standard form we will write this as $\frac{20x^2y + 1}{10x^2y}$.

Step 5 Reduce   We factor to see if there is a GCF.

$$\frac{20x^2y + 1}{10x^2y} \text{ factors as } \frac{2 \cdot 2 \cdot 5x^2y + 1}{2 \cdot 5x^2y}$$

Since the numerator and denominator have no common factors this fraction cannot be reduced. We leave the answer as $\frac{20x^2y + 1}{10x^2y}$.

Practice 8.4.3   Simplify. Assume no value of a variable will make a denominator 0.

1. $\dfrac{4}{15} + \dfrac{3}{10}$ 　　2. $\dfrac{7}{5x} - \dfrac{4}{15x^2}$ 　　3. $\dfrac{3}{7m^2} - \dfrac{5}{14m}$ 　　4. $\dfrac{2}{3s^2t} + \dfrac{5}{6st}$

5. $\dfrac{5}{9a^2b} + \dfrac{5}{6ab^2}$ 　　6. $\dfrac{5}{7mn} + 2$ 　　7. $5 - \dfrac{2}{9xy}$

When the denominators are binomials or trinomials the process is the same but the factoring is more involved.

Example 8.4.4   Simplify. Assume no value of a variable will make a denominator 0.

a) $\dfrac{5}{y - 6} + \dfrac{7}{y + 3}$

•SOLUTION• **Step 1 Factor Denominators**   The denominators are already in factored form.

Step 2 Build LCD

Largest power of $y - 6$ is $(y - 6)^1$.

Largest power of $y + 3$ is $(y + 3)^1$.

LCD is $(y - 6)(y + 3)$. The work that follows will be easier if we leave the LCD in its factored form.

Step 3 Build Equivalent Fractions

Comparing $y - 6$ with the LCD we see that $y - 6$ needs one factor of $y + 3$.

$$\frac{5}{y - 6} \Rightarrow \frac{5}{y - 6} \cdot \frac{\mathbf{y + 3}}{\mathbf{y + 3}} \Rightarrow \frac{5(y + 3)}{(y - 6)(y + 3)}$$

Comparing $y + 3$ with the LCD we see that $y + 3$ needs one factor of $y - 6$.

$$\frac{7}{y + 3} \Rightarrow \frac{7}{y + 3} \cdot \frac{y - 6}{y - 6} \Rightarrow \frac{7(y - 6)}{(y - 6)(y + 3)}$$

Step 4 Combine Numerators

$$\frac{5(y + 3)}{(y - 6)(y + 3)} + \frac{7(y - 6)}{(y - 6)(y + 3)} \qquad \Leftarrow \text{Fractions to add.}$$

$$\frac{5(y + 3) + 7(y - 6)}{(y - 6)(y + 3)} \qquad \Leftarrow \text{Combined over one denominator.}$$

$$\frac{5y + 15 + 7y - 42}{(y - 6)(y + 3)} \qquad \Leftarrow \text{Distributed 5 and 7.}$$

$$\frac{12y - 27}{(y - 6)(y + 3)} \qquad \Leftarrow \text{Combined like terms.}$$

Step 5 Reduce

We factor to get $\dfrac{3(4y - 9)}{(y - 6)(y + 3)}$. There are no common factors so this fraction cannot be reduced.

b) $\dfrac{2x + 4}{x^2 - 3x - 10} + \dfrac{x + 1}{x - 5}$

•SOLUTION•

Step 1 Factor Denominators

$x^2 - 3x - 10$ factors as $(x - 5)^1(x + 2)^1$
$x - 5$ may be written as $(x - 5)^1$

Step 2 Build LCD

Largest power of $x - 5$ is $(x - 5)^1$.
Largest power of $x + 2$ is $(x + 2)^1$.
LCD is $(x - 5)(x + 2)$.

Step 3 Build Equivalent Fractions

$\dfrac{2x + 4}{(x - 5)(x + 2)}$ needs no additional factors (it has the LCD as its denominator).

$\dfrac{x + 1}{x - 5}$ needs one factor of $(x + 2)$

$$\frac{x + 1}{x - 5} \Rightarrow \frac{x + 1}{x - 5} \cdot \frac{(x + 2)}{(x + 2)} \Rightarrow \frac{(x + 1)(x + 2)}{(x - 5)(x + 2)}$$

Step 4 Combine Numerators

$$\frac{2x + 4}{(x - 5)(x + 2)} + \frac{(x + 1)(x + 2)}{(x - 5)(x + 2)} \qquad \Leftarrow \text{Fractions to add.}$$

$$\frac{2x + 4 + (x + 1)(x + 2)}{(x - 5)(x + 2)} \qquad \Leftarrow \text{Combined over one denominator.}$$

$$\frac{2x + 4 + x^2 + 3x + 2}{(x - 5)(x + 2)} \qquad \Leftarrow \text{Multiplied the binomials.}$$

$$\frac{x^2 + 5x + 6}{(x - 5)(x + 2)} \qquad \Leftarrow \text{Combined like terms.}$$

Step 5 Reduce

$$\frac{(x + 2)(x + 3)}{(x - 5)(x + 2)} \Rightarrow \frac{\cancel{(x + 2)}(x + 3)}{(x - 5)\cancel{(x + 2)}} \Rightarrow \frac{x + 3}{x - 5}$$

c) $\dfrac{1}{2x^2 - 4x} - \dfrac{3}{2 - x}$

•SOLUTION• *Step 1 Factor Denominators*

$2x^2 - 4x$ factors as $2^1 x^1 (x - 2)^1$

$2 - x$ should be written in standard form as $-x + 2$. We don't want the x term to have a negative coefficient so we will multiply the fraction by $\dfrac{-1}{-1}$ as shown below:

$\dfrac{3}{2 - x}$ $\Leftarrow$ Original fraction.

$\dfrac{3}{-x + 2}$ $\Leftarrow$ Use Commutative Property to write $2 - x$ as $-x + 2$.

$\dfrac{3}{-x + 2} \cdot \dfrac{-1}{-1}$ $\Leftarrow$ Multiplied numerator and denominator by -1 to remove negative from x term.

$\dfrac{-3}{x - 2}$ $\Leftarrow$ Multiplied.

The denominator of this fraction is $(x - 2)^1$. Note that you should be able to do the above steps mentally.

Step 2 Build LCD

Largest power of 2 is 2^1.
Largest power of x is x^1.
Largest power of $x - 2$ is $(x - 2)^1$.
LCD is $2^1 \cdot x^1 \cdot (x - 2)^1$ or $2x(x - 2)$

Step 3 Build Equivalent Fractions

$\dfrac{1}{2x(x - 2)}$ needs no additional factors (it has the LCD as its denominator).

$\dfrac{-3}{x - 2}$ needs one factor of 2 and one factor of x.

$$\dfrac{-3}{x - 2} \Rightarrow \dfrac{-3}{x - 2} \cdot \dfrac{2x}{2x} \Rightarrow \dfrac{-6x}{2x(x - 2)}$$

Step 4 Combine Numerators

$$\dfrac{1}{2x(x - 2)} - \dfrac{-6x}{2x(x - 2)} \Rightarrow \dfrac{1 - (-6x)}{2x(x - 2)} \Rightarrow \dfrac{1 + 6x}{2x(x - 2)}$$

Step 5 Reduce   There are no common factors so this fraction cannot be reduced.

Practice 8.4.4   Simplify. Assume no value of a variable will make a denominator 0.

1. $\dfrac{7}{m + 4} - \dfrac{5}{m - 2}$

2. $\dfrac{4}{r + 3} + 5$

3. $\dfrac{5}{x^2 - 25} + \dfrac{3}{x + 5}$

4. $\dfrac{7}{y^2 - 9} + \dfrac{2}{y - 3}$

5. $\dfrac{3}{x^2 - 5x + 6} + \dfrac{4}{x - 2}$

6. $\dfrac{7}{x^2 - 5x} + \dfrac{2}{5 - x}$

Exercise Set 8.4 The answers to the odd numbered exercises are at the back of the book.

Simplify. Assume no value of a variable will make a denominator 0.

1. $\dfrac{3x}{7} + \dfrac{11x}{7}$

2. $\dfrac{3w}{8} + \dfrac{w}{8}$

3. $\dfrac{3}{x^2 y} + \dfrac{8}{x^2 y}$

4. $\dfrac{2}{rt} + \dfrac{5}{rt}$

5. $\dfrac{-9}{2x+1} + \dfrac{-x}{2x+1}$

6. $\dfrac{3a}{a-2} + \dfrac{-2}{a-2}$

7. $\dfrac{6}{b+5} + \dfrac{b+3}{b+5}$

8. $\dfrac{11}{y-3} + \dfrac{y+4}{y-3}$

Simplify. Assume no value of a variable will make a denominator 0.

9. $\dfrac{5x}{6} - \dfrac{7x}{6}$

10. $\dfrac{8w}{25} - \dfrac{13w}{25}$

11. $\dfrac{5}{x^2 y^2} - \dfrac{2}{x^2 y^2}$

12. $\dfrac{3}{mn} - \dfrac{5}{mn}$

13. $\dfrac{4}{x+2} - \dfrac{x+3}{x+2}$

14. $\dfrac{7}{y-3} - \dfrac{y+4}{y-3}$

15. $\dfrac{8x^2 + 3x}{5x-3} - \dfrac{3x^2 + 6x}{5x-3}$

16. $\dfrac{5m^2 + 4m}{2m+7} - \dfrac{3m^2 - 3m}{2m+7}$

17. $\dfrac{4x+3y}{2x-3y} - \dfrac{2x-5y}{2x-3y}$

18. $\dfrac{6m-n}{4m+2n} - \dfrac{4n-3m}{4m+2n}$

19. $\dfrac{4m}{3m+2n} - \dfrac{2m-3n}{3m+2n}$

20. $\dfrac{5w}{6w-3s} - \dfrac{3w-2s}{6w-3s}$

Simplify. Assume no value of a variable will make a denominator 0.

21. $\dfrac{3}{4} - \dfrac{5}{6}$

22. $\dfrac{13}{24} + \dfrac{5}{18}$

23. $\dfrac{7}{12xy^3} - \dfrac{9}{8x^3 y}$

24. $\dfrac{4}{9m^2 n} - \dfrac{5}{12mn^4}$

25. $\dfrac{7}{6x} + \dfrac{5}{12y} - \dfrac{4}{9xy}$

26. $\dfrac{3}{10mn} + \dfrac{4}{5m^2} - \dfrac{7}{15n^2}$

27. $\dfrac{5}{6a^2} - 3 + \dfrac{7}{9b}$

28. $\dfrac{2}{5y^2} - 4 + \dfrac{3}{10x^2}$

29. $\dfrac{5}{3m^2 n} + 6 - \dfrac{m}{6n^2}$

30. $\dfrac{7}{4r^2 s} + 5 - \dfrac{3r}{8s^2}$

31. $3 + \dfrac{5}{2m^2 n} - \dfrac{7}{3mn^2}$

32. $\dfrac{5}{4mn} - 2 + \dfrac{4}{3m^2}$

Simplify. Assume no value of a variable will make a denominator 0.

33. $\dfrac{6}{m^2 n} - \dfrac{5}{2mn^3}$

34. $\dfrac{7}{x^2 y} - \dfrac{3}{5x y^2}$

35. $\dfrac{5}{x+2} - \dfrac{3}{x-4}$

36. $\dfrac{6}{m+1} - \dfrac{4}{m+7}$

37. $\dfrac{6}{s-1} - 2$

38. $8 - \dfrac{2x}{2x+3}$

39. $\dfrac{10}{t^2 - 3t} - \dfrac{5}{3-t}$

40. $\dfrac{3}{r^2 - 9r} - \dfrac{4}{9-r}$

41. $\dfrac{3}{x^2 + 6x + 5} + \dfrac{7}{x+5}$

42. $\dfrac{-4}{x^2 - 36} + \dfrac{6}{x+6}$

43. $\dfrac{5x+1}{x^2 - x - 6} + \dfrac{2x+1}{x+2}$

44. $\dfrac{3x-2}{x^2 + 4x - 12} + \dfrac{x+5}{x+6}$

45. $\dfrac{4y-3}{y^2 - 7y} + \dfrac{4}{7-y}$

46. $\dfrac{2m+1}{m^2 - 5m} + \dfrac{5}{5-m}$

47. $\dfrac{2}{x^2 - 7x + 12} + \dfrac{6}{x-4}$

48. $\dfrac{8}{y^2 - 6y} - \dfrac{3}{y-6}$

Review Exercises The answers to all of these exercises are at the back of the book.

49. Identify the sums, terms, products, and factors: $9q + 5$

50. Evaluate $5 - x - xy^2$ when x is -2 and y is -3.

51. The sum of a fraction and its reciprocal is $\dfrac{25}{12}$. If the numerator is 2 less than the denominator, find the fraction.

52. William can lay 40 rolls of sod an hour when the weather is cool but only 30 when the weather is hot. Find the total amount of sod William laid if over an 8 hour period the weather was cool for one fourth of the time and hot three fourths of the time.

53. Find the equation of a line parallel to $x = 3$ and which passes through $(-1,6)$.

54. Determine the domain and range of $x + y = 2$

55. Solve: $3 - \dfrac{1}{2}x = -\dfrac{3x}{2} - (5 - x)$

56. Solve by graphing: $\begin{cases} 4x + y = 2 \\ -2x + y = -4 \end{cases}$ ❶ ❷

57. Solve: $\begin{cases} \dfrac{3}{4}x + y = \dfrac{3}{2} \\ \dfrac{1}{12}x - \dfrac{1}{6}y = -\dfrac{3}{2} \end{cases}$

58. Reduce: $\dfrac{6x^2 + 6x}{18x^2 - 12x}$

59. Factor: $40x^2 + 16x - 5xy - 2y$

60. Factor: $8 + x^2 - 9x$

61. Solve: $0 = (2x - 1)(x - 1)$

Simplify.

62. $(2xy)^{-3}(3x^2y)^{-2}$

63. $\left(\dfrac{a^4}{a^3}\right)^2 \left(\dfrac{a^3}{a^4}\right)$

64. $\dfrac{21}{3}\left(\dfrac{6}{7}\right)^2 - \dfrac{15}{7}$

65. $2x(x - 2) - x(2 - x) - 2(2x - 2)$

66. $\dfrac{8x^3y}{4xy^{-5}}$

67. $-(m^{-2})(-2m^3)$

68. $(4^{-2})\left(\dfrac{1}{2}\right)^{-4} + 4^{-3}\left(\dfrac{1}{2}\right)^{-6} - 2^0$

69. $(18a^2b + 16ab^2 - 4a^2b^2) + (-6ab^2 + 12a^2b^2) - (ab^2 - 12a^2b - 3a^2b^2)$

70. A group of friends is driving to Florida. For the first part of their trip they traveled 268 miles in 4 hours. For the second part they traveled for 6 hours and went 402 miles. For the last part they went 234.5 miles in 3.5 hours. Assuming the distance traveled is the output and the time of travel is the input, calculate the slope of the line that would connect the given points and discuss the meaning of the slope.

Section 8.5 Complex Rational Expressions

A **complex rational expression** is a fraction whose numerator and/or denominator are rational expressions. Examples include:

$\dfrac{\dfrac{2}{3}}{\dfrac{4}{15}}$	$\dfrac{\dfrac{x}{y}}{\dfrac{x^2}{5y}}$	$\dfrac{x - 1}{\dfrac{2}{4}}$	$\dfrac{\dfrac{5}{x^2} + \dfrac{1}{x}}{4 - \dfrac{1}{x}}$

We call the fractions in the numerator or denominator **minor fractions**. For example,

the minor fractions in $\dfrac{\dfrac{5}{x^2} + \dfrac{1}{x}}{4 - \dfrac{1}{x}}$ are $\dfrac{5}{x^2}$, $\dfrac{1}{x}$, and 4 (since 4 can be written as $\dfrac{4}{1}$).

Below we present two common methods for simplifying complex fractions. As you shall see, one method may be better than the other in certain situations.

Topic 8.5 A — Simplifying Complex Rational Expressions Using the Four Basic Operations

This method uses the procedures you just learned for adding, subtracting, multiplying, and dividing rational expressions.

— Procedure —	**Step 1 Simplify** Simplify the numerator and denominator separately.
Simplifying Complex Rational Expressions Using the Four Basic Operations	**Step 2 Invert** Rewrite the problem as the product of the numerator and the *reciprocal* of the denominator.
	Step 3 Reduce Reduce factors common to the numerators and denominators.

Example 8.5.1   Simplify. Assume no value of a variable will make a denominator 0.

a) $\dfrac{\dfrac{x}{y}}{\dfrac{x^2}{5y}}$

•SOLUTION•   *Step 1 Simplify*   The numerator and denominator cannot be simplified.

Step 2 Invert $\dfrac{x}{y} \cdot \dfrac{5y}{x^2}$

Step 3 Reduce   We reduce common factors of x and y to get $\dfrac{5}{x}$.

b) $\dfrac{\dfrac{a-b}{2}}{a-b}$

•SOLUTION•   *Step 1 Simplify*   The numerator and denominator cannot be simplified. But, we can make it easier to identify the minor fractions by writing the denominator as $\dfrac{a-b}{1}$. This gives us $\dfrac{\dfrac{a-b}{2}}{\dfrac{a-b}{1}}$.

Step 2 Invert $\dfrac{a-b}{2} \cdot \dfrac{1}{a-b}$

Step 3 Reduce   We reduce common factor of $a-b$ to get $\dfrac{1}{2}$.

c) $\dfrac{\dfrac{x}{2} - \dfrac{2}{x}}{1 + \dfrac{2}{x}}$

•SOLUTION•   We subtract the fractions in the numerator, then we find the sum of the fractions in the denominator, and finally we do the division.

Step 1 Simplify

$$\dfrac{\dfrac{x}{2} - \dfrac{2}{x}}{1 + \dfrac{2}{x}} \qquad \Leftarrow \text{Expression to simplify.}$$

$$\dfrac{\dfrac{x}{2} \cdot \dfrac{x}{x} - \dfrac{2}{x} \cdot \dfrac{2}{2}}{1 \cdot \dfrac{x}{x} + \dfrac{2}{x}} \qquad \Leftarrow \text{The LCD of the numerator is } 2x \text{ and the LCD of the denominator is } x.$$

$$\dfrac{\dfrac{x^2}{2x} - \dfrac{4}{2x}}{\dfrac{x}{x} + \dfrac{2}{x}} \qquad \Leftarrow \text{Multiplied.}$$

$$\dfrac{\dfrac{x^2 - 4}{2x}}{\dfrac{x+2}{x}} \qquad \Leftarrow \text{Simplified the numerator and simplified the denominator.}$$

Step 2 Invert   We invert the denominator and multiply to get $\dfrac{x^2 - 4}{2x} \cdot \dfrac{x}{x+2}$

Step 3 Reduce

$$\frac{(x-2)(x+2)}{2x} \cdot \frac{x}{x+2} \quad \Leftarrow \text{ Factored.}$$

$$\frac{(x-2)}{2} \quad \Leftarrow \text{ Reduced common factors } x \text{ and } x+2.$$

Practice 8.5.1 Simplify. Assume no value of a variable will make a denominator 0.

1. $\dfrac{\dfrac{2}{3}}{\dfrac{4}{5}}$

2. $\dfrac{\dfrac{6a^5}{b^3}}{\dfrac{3a^2}{4b^5}}$

3. $\dfrac{\dfrac{4w-7z}{15}}{\dfrac{4w-7z}{10}}$

4. $\dfrac{1+\dfrac{4}{m}}{\dfrac{m}{4}-\dfrac{4}{m}}$

Topic 8.5 B — Simplifying Complex Rational Expressions Using the LCD

This second method uses the LCD of all the minor fractions to quickly simplify the given rational expression.

— Procedure —	
Simplifying Complex Rational Expressions Using the LCD	***Step 1 LCD*** Build the LCD of all minor fractions.
	Step 2 Multiply Multiply each minor fraction by the LCD.
	Step 3 Simplify Simplify the numerator and denominator.
	Step 4 Reduce Reduce factors common to the numerators and denominators.

We will use this method on the same fractions as before so that we may compare the two methods.

Example 8.5.2 Simplify. Assume no value of a variable will make a denominator 0.

a) $\dfrac{\dfrac{x}{y}}{\dfrac{x^2}{5y}}$

•SOLUTION•

Step 1 LCD The LCD is $5y$, which we will write as $\dfrac{5y}{1}$.

Step 2 Multiply

$$\frac{\dfrac{x}{y} \cdot \dfrac{5y}{1}}{\dfrac{x^2}{5y} \cdot \dfrac{5y}{1}} \quad \Leftarrow \text{ Multiplied each minor fraction by the LCD.}$$

Step 3 Simplify

$$\frac{5x}{x^2} \quad \Leftarrow \text{ Reduced common factor of } y \text{ in the numerator and reduced}$$
$$\text{common factor of } 5y \text{ in the denominator.}$$

Step 4 Reduce

$$\frac{5}{x} \quad \Leftarrow \text{ Reduced by common factor of } x.$$

b) $\dfrac{\dfrac{a-b}{2}}{a-b}$

•SOLUTION•

Step 1 LCD   The LCD is 2, which we will write as $\dfrac{2}{1}$.

Step 2 Multiply

$$\dfrac{\dfrac{a-b}{2}\cdot\dfrac{\mathbf{2}}{\mathbf{1}}}{\dfrac{a-b}{1}\cdot\dfrac{\mathbf{2}}{\mathbf{1}}} \quad \Leftarrow \text{ Multiplied each minor fraction by the LCD.}$$

Step 3 Simplify

$$\dfrac{a-b}{2(a-b)} \quad \Leftarrow \text{ Reduced the common factor of 2 in the numerator and removed}$$
$$\text{denominator of 1 in the denominator.}$$

Step 4 Reduce

$$\dfrac{1}{2} \quad \Leftarrow \text{ Reduced by common factor of } a-b.$$

c) $\dfrac{\dfrac{x}{2}-\dfrac{2}{x}}{1+\dfrac{2}{x}}$

•SOLUTION•

Step 1 LCD   The LCD is $2x$, which we will write as $\dfrac{2x}{1}$.

Step 2 Multiply

$$\dfrac{\dfrac{x}{2}\cdot\dfrac{\mathbf{2x}}{\mathbf{1}}-\dfrac{2}{x}\cdot\dfrac{\mathbf{2x}}{\mathbf{1}}}{1\cdot\dfrac{\mathbf{2x}}{\mathbf{1}}+\dfrac{2}{x}\cdot\dfrac{\mathbf{2x}}{\mathbf{1}}} \quad \Leftarrow \text{ Multiplied each minor fraction by the LCD.}$$

Step 3 Simplify

$$\dfrac{x^2-4}{2x+4} \quad \Leftarrow \text{ Simplified numerator and denominator.}$$

Step 4 Reduce

$$\dfrac{(x-2)(x+2)}{2(x+2)} \quad \Leftarrow \text{ Factored.}$$

$$\dfrac{x-2}{2} \quad \Leftarrow \text{ Reduced common factor of } x+2.$$

Practice 8.5.2   Simplify. Assume no value of a variable will make a denominator 0.

1. $\dfrac{\dfrac{5}{9}}{\dfrac{10}{27}}$
2. $\dfrac{\dfrac{4ab}{5c^3d}}{\dfrac{8ab}{15cd}}$
3. $\dfrac{\dfrac{5m-10}{3}}{m-2}$
4. $\dfrac{\dfrac{t}{3}-\dfrac{3}{t}}{2+\dfrac{6}{t}}$
5. $\dfrac{3-\dfrac{9}{y}}{\dfrac{y}{3}-\dfrac{3}{y}}$

This last example is a case where both methods take quite a few steps, but the LCD method seems to result in fewer errors by students. As you become more expert, you will be able to do many of the steps in your head.

Example 8.5.3   Simplify $\dfrac{2 - \dfrac{2}{x-1}}{-3x - \dfrac{2}{1-x}}$

• **SOLUTION** •

Step 1 LCD   Although we could use $(x-1)(1-x)$ as the LCD we will put the denominator of $\dfrac{2}{1-x}$ in standard form as follows:

$\dfrac{2}{1-x}$ ⇐ Expression to rewrite.

$\dfrac{2}{-x+1}$ ⇐ Put denominator in standard form.

$\dfrac{2}{(-x+1)} \cdot \dfrac{-1}{-1}$ ⇐ Multiplied numerator and denominator by −1 to make the coefficient of the x term positive.

$\dfrac{-2}{(x-1)}$ ⇐ Multiplied.

Now, we have the complex fraction

$$\dfrac{2 - \dfrac{2}{x-1}}{-3x - \dfrac{-2}{x-1}}$$

The LCD is $x-1$, which we will write as $\dfrac{x-1}{1}$.

Step 2 Multiply

$$\dfrac{\left(\dfrac{x-1}{1}\right) \cdot 2 - \left(\dfrac{x-1}{1}\right) \cdot \left(\dfrac{2}{x-1}\right)}{\left(\dfrac{x-1}{1}\right) \cdot (-3x) - \left(\dfrac{x-1}{1}\right) \cdot \left(\dfrac{-2}{x-1}\right)}$$ ⇐ Multiplied numerator and denominator by the LCD.

Step 3 Simplify

$\dfrac{2(x-1) - 2}{-3x(x-1) + 2}$ ⇐ Reduced minor fractions.

$\dfrac{2x - 2 - 2}{-3x^2 + 3x + 2}$ ⇐ Distributed 2 in the numerator and $-3x$ in the denominator.

$\dfrac{2x - 4}{-3x^2 + 3x + 2}$ ⇐ Combined like terms.

Step 4 Reduce

$\dfrac{2(x-2)}{-3x^2 + 3x + 2}$ ⇐ Factored.

There are no common factors, other than 1 and −1, so we cannot reduce.

Practice 8.5.3   Simplify. Assume no value of a variable will make a denominator 0.

1. $\dfrac{4 - \dfrac{1}{x+3}}{4 + \dfrac{1}{x+3}}$

2. $\dfrac{-\dfrac{3}{t+1} + 1}{\dfrac{3}{t+1} + 1}$

Exercise Set 8.5   The answers to the odd numbered exercises are at the back of the book.

Simplify. Assume no value of a variable will make a denominator 0.

1. $\dfrac{\dfrac{7}{9}}{\dfrac{14}{27}}$

2. $\dfrac{\dfrac{12}{13}}{\dfrac{24}{39}}$

3. $\dfrac{\dfrac{2x}{y}}{\dfrac{4x^3}{5y}}$

4. $\dfrac{\dfrac{3m}{n}}{\dfrac{6m^5}{7n}}$

5. $\dfrac{\dfrac{5m+3}{15}}{\dfrac{5m+3}{10}}$

6. $\dfrac{\dfrac{6w+5}{12}}{\dfrac{6w+5}{14}}$

7. $\dfrac{\dfrac{5x^2y^5}{7a^3b}}{\dfrac{10xy}{21a^2b^3}}$

8. $\dfrac{\dfrac{8r^5s^6}{5p^4q^2}}{\dfrac{4r^2s}{15p^3q}}$

9. $\dfrac{\dfrac{s+5t}{3}}{s+5t}$

10. $\dfrac{\dfrac{3r-5t}{8}}{\dfrac{3r-5t}{12}}$

11. $\dfrac{1+\dfrac{5}{x}}{\dfrac{x}{5}-\dfrac{5}{x}}$

12. $\dfrac{1+\dfrac{7}{d}}{\dfrac{d}{7}-\dfrac{7}{d}}$

13. $\dfrac{\dfrac{a}{5}-\dfrac{5}{a}}{1-\dfrac{5}{a}}$

14. $\dfrac{1+\dfrac{6}{y}}{\dfrac{y}{6}-\dfrac{6}{y}}$

15. $\dfrac{\dfrac{x^2-25}{8}}{x+5}$

16. $\dfrac{\dfrac{m^2-36}{12}}{m-6}$

17. $\dfrac{\dfrac{7x^4}{9y}}{\dfrac{14x}{3y}}$

18. $\dfrac{\dfrac{15m^7}{8n}}{\dfrac{5m}{16n}}$

19. $\dfrac{\dfrac{-5x^4}{y}}{\dfrac{x}{10y^3}}$

20. $\dfrac{\dfrac{-3m^5}{n}}{\dfrac{m}{9n^4}}$

21. $\dfrac{\dfrac{9mn}{8st^5}}{\dfrac{21mn}{12st^6}}$

22. $\dfrac{\dfrac{12xy}{25z}}{\dfrac{42x^2y}{35z^3}}$

23. $\dfrac{\dfrac{x^2-y^2}{5}}{x-y}$

24. $\dfrac{\dfrac{m^2-n^2}{7}}{m+n}$

25. $\dfrac{\dfrac{4x^2-9}{5}}{2x+3}$

26. $\dfrac{\dfrac{5m+1}{7}}{25m^2-1}$

27. $\dfrac{\dfrac{2m+4n}{5}}{2m+4n}$

28. $\dfrac{\dfrac{4x-7y}{12}}{4x-7y}$

29. $\dfrac{1+\dfrac{1}{x}}{1-\dfrac{1}{x}}$

30. $\dfrac{2+\dfrac{1}{n}}{2-\dfrac{1}{n}}$

31. $\dfrac{\dfrac{x}{5}-\dfrac{5}{x}}{3+\dfrac{15}{x}}$

32. $\dfrac{\dfrac{m}{4}-\dfrac{4}{m}}{4-\dfrac{16}{m}}$

33. $\dfrac{2+\dfrac{t+4}{7}}{3+\dfrac{t-3}{14}}$

34. $\dfrac{1+\dfrac{x-3}{2}}{2+\dfrac{x+1}{4}}$

35. $\dfrac{5+\dfrac{1}{x+2}}{5-\dfrac{1}{x+2}}$

36. $\dfrac{4-\dfrac{1}{m+5}}{4+\dfrac{1}{m+5}}$

37. $\dfrac{5+\dfrac{1}{m-2}}{5-\dfrac{1}{m-2}}$

38. $\dfrac{1+\dfrac{2}{r-5}}{1-\dfrac{2}{r-5}}$

Review Exercises   The answers to all of these exercises are at the back of the book.

39. The angles of a triangle must sum to 180 degrees. Assume *a* is the measure of the first angle of a triangle and the measure of the second angle is 2 degrees less than twice the measure of the first. Write a simplified expression using *a* to represent the measure of the third angle.

40. A freight train left town at 8:15 am and traveled at 40 mph. Three–fourths of an hour later a passenger train left the same town on the same track traveling at 45 mph. At what time will the passenger train catch up to the freight train?

41. Determine whether this graph represents a function:

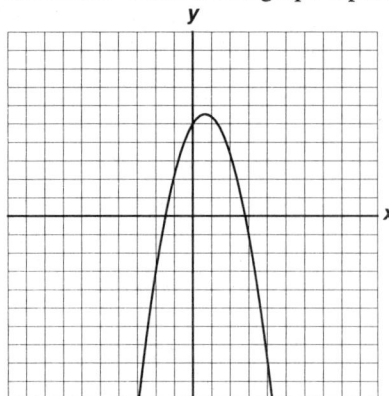

42. Construct the equation of the line shown in the graph.

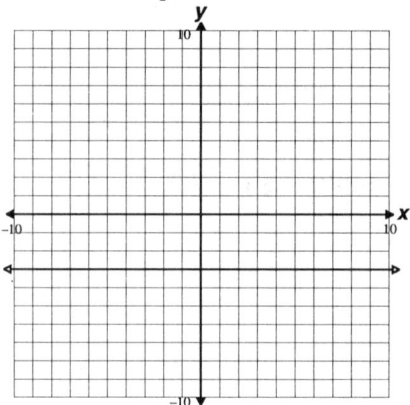

43. Find the slope and y–intercept of the line described by $3y - 2x = 0$

44. Solve: $8x^2 - 16x = 0$

45. Find the equation of the line that goes through (–2,8) and (–2,4).

46. Solve: $18x^2 = 15x - 2$

47. A manufacturer can produce a radio component using two processes. Process A costs $5,000 to set up the machinery and each component costs $15 to produce. Process B costs $2000 to set up and each component costs $25 to produce.

   a) Model this situation with a 2 by 2 system using number of components as the input and cost as the output.
   b) Graph the system. The input axis should go from 0 to 500 hours in steps of 50 components; the output axis should go from $0 to $12,000 in steps of $1,000.
   c) Find the point of intersection and explain what it means.
   d) Which process would cost less money?

48. Solve: $\begin{cases} 4x - y = -6 \quad \textbf{①} \\ 1 - y = x \quad \textbf{②} \end{cases}$

49. Solve: $\dfrac{2}{3}(x + 6) - \dfrac{1}{2}x = -\dfrac{x}{6}$

Simplify.

50. $\dfrac{3a^2b}{2(ab)^2} \cdot \dfrac{4b}{6a^3}$

51. $(m^2)^{-3}(m^2)^3$

52. $(3p^3q^3)^{-2}(-6pq)^3$

53. $5x^{-2}y^{-3}(2x^2y^2)$

54. $(3t + 2)^2$

55. $\dfrac{12}{x^8} \cdot \dfrac{x^{14}}{10}$

56. $\dfrac{3}{4ab} - \dfrac{2}{a^2}$

57. $\dfrac{x^2 + x - 6}{3x} \cdot \dfrac{6x^2 - 30x}{x^2 - 7x + 10}$

58. Simplify and write the answer in scientific notation: $(0.0000101)(0.0008)$

59. Reduce $\dfrac{x^2 + 5x + 6}{2x + 4}$

60. Add $(18ab + 6a + 2b - 1)$ to $(11ab - 5a - b + 2)$

61. Graph: $y = -3$

62. Divide and simplify: $\dfrac{192p^4 - 126p^2}{-8p}$

63. Use long division to simplify: $\dfrac{t^4 + 2t^3 - 3t - 2}{t + 1}$

64. Factor completely: $-8x^2 + 14x - 6$

65. Factor completely: $8x - 18x^3$

66. The equation $e(t) = -1.3t + 1.28$ represents the relationship between millions of tons of carbon monoxide emissions from 1970 to 1995 (e) and the number of years since 1970 (t). Express in English the meaning of each of the following:

   a) $e(5)$
   b) $e(t) = 0.64$
   c) $e(t) = 1/2e(0)$
   d) $e(4) - e(7)$

Section 8.6 Rational Equations

An equation is a statement of equality between two expressions. A **rational equation** is a statement of equality between rational expressions. Examples include:

$\dfrac{x}{2} + 3 = 5$	$\dfrac{3}{x} = \dfrac{5}{x+4}$	$\dfrac{3}{x+1} = \dfrac{5}{2x} + \dfrac{2}{x}$

Topic 8.6 A — Solving Rational Equations

In Chapter 2 we solved equations using the procedure outlined below:

— *Procedure* —	
Solving equations	***Step 1 LCD*** If fractions are present, multiply each term by the LCD.
	Step 2 Simplify Simplify each side of the equation.
	Step 3 Move Terms Use addition or subtraction to move terms with variables to one side of the equation and terms without variables to the other side. Then, simplify each side again.
	Step 4 Divide Divide both sides of the equation by the coefficient of the variable.
	Step 5 Check Check your answer by substituting it into the original equation. If a true statement results, your answer is probably correct. If a false statement results, go back over your work to see if you made an error.

We can use this same procedure to solve rational equations that contain variables in their denominators.

Example 8.6.1   Solve.

a) $\dfrac{3x}{2} = 22 - \dfrac{x}{3}$

•**SOLUTION**•   This rational equation is similar to the ones we solved in Chapter 2. It has no variables in the denominators so we don't have to worry about restricted values.

Step 1 LCD   The LCD of 2 and 3 is 6.

$$6 \cdot \left(\frac{3x}{2}\right) = 6 \cdot \left(\frac{22}{1}\right) - 6 \cdot \left(\frac{x}{3}\right) \quad \Leftarrow \text{Multiplied each term by LCD.}$$

Step 2 Simplify

$$9x = 132 - 2x \qquad \Leftarrow \text{Reduced and then multiplied.}$$

Step 3 Move Terms

$$11x = 132 \qquad \Leftarrow \text{Added } 2x \text{ to both sides.}$$

Step 4 Divide

$$x = 12 \qquad \Leftarrow \text{Divided both sides by 12.}$$

Step 5 Check   The solution is 12. We check this in the usual way:

Left expression *?=?* ***Right expression***

$\dfrac{3x}{2}$	$= 22 - \dfrac{x}{3}$	$\Leftarrow$ Equation to check.
$\dfrac{3(12)}{2}$	$22 - \dfrac{(12)}{3}$	$\Leftarrow$ Substituted 12 for x.
18	18	$\Leftarrow$ Simplified.

Since each expression simplifies to the same number we are confident that 12 is the correct solution.

b) $\dfrac{1}{x} - \dfrac{3}{2} = \dfrac{5}{2x}$

•SOLUTION• *Step 1 LCD* We have denominators of x, 2, and $2x$, so the LCD is $2x$.

$$2x \cdot \left(\dfrac{1}{x}\right) - 2x \cdot \left(\dfrac{3}{2}\right) = 2x \cdot \left(\dfrac{5}{2x}\right) \quad \Leftarrow \text{ Multiplied each term by LCD.}$$

Step 2 Simplify

$$2 - 3x = 5 \qquad \Leftarrow \text{ Reduced and then multiplied.}$$

Step 3 Move Terms

$$-3x = 3 \qquad \Leftarrow \text{ Subtracted 2 from both sides.}$$

Step 4 Divide

$$x = -1 \qquad \Leftarrow \text{ Divided both sides by } -3.$$

Step 5 Check The solution is -1. We check this in the usual way:

Left expression ?=? Right expression

$$\dfrac{1}{x} - \dfrac{3}{2} = \dfrac{5}{2x} \qquad \Leftarrow \text{ Equation to check.}$$

$$\dfrac{1}{(-1)} - \dfrac{3}{2} \quad \Bigg| \quad \dfrac{5}{2(-1)} \qquad \Leftarrow \text{ Substituted } -1 \text{ for } x.$$

$$\dfrac{-5}{2} \quad \Bigg| \quad \dfrac{-5}{2} \qquad \Leftarrow \text{ Simplified.}$$

Since each expression simplifies to the same number we are confident that -1 is the correct solution.

c) $\dfrac{3}{x+1} = \dfrac{5}{2x} + \dfrac{2}{x}$

•SOLUTION• *Step 1 LCD* The denominators are $x + 1$, $2x$, and x so the LCD is $2x(x + 1)$.

$$2x(x+1) \cdot \left(\dfrac{3}{x+1}\right) = 2x(x+1) \cdot \left(\dfrac{5}{2x}\right) + 2x(x+1) \cdot \left(\dfrac{2}{x}\right) \quad \Leftarrow \text{ Multiplied each term by LCD.}$$

Step 2 Simplify

$$2x(3) = 5(x + 1) + 2(x + 1)(2) \qquad \Leftarrow \text{ Reduced and then multiplied.}$$

$$6x = 5x + 5 + 4x + 4 \qquad \Leftarrow \text{ Multiplied.}$$

$$6x = 9x + 9 \qquad \Leftarrow \text{ Combined like terms.}$$

Step 3 Move Terms

$$-3x = 9 \qquad \Leftarrow \text{ Subtracted } 9x \text{ from both sides.}$$

Step 4 Divide

$$x = -3 \qquad \Leftarrow \text{ Divided both sides by } -3.$$

Step 5 Check The solution is -3. We leave the details of the check to you.

d) $\dfrac{x}{x-4} - 1 = \dfrac{8}{x^2 - 16}$

•SOLUTION• *Step 1 LCD* We have denominators of $x - 4$ and, in factored form, $(x - 4)(x + 4)$, so the LCD is $(x - 4)(x + 4)$.

$$(x-4)(x+4)\left(\dfrac{x}{x-4}\right) - (x-4)(x+4)(1) = (x-4)(x+4)\left(\dfrac{8}{(x-4)(x+4)}\right) \quad \Leftarrow \text{ Multiplied by LCD.}$$

Step 2 Simplify

$$(x + 4)(x) - (x - 4)(x + 4)(1) = 8 \quad \Leftarrow \text{Reduced and then multiplied.}$$

$$x^2 + 4x - (x^2 - 16) = 8 \quad \Leftarrow \text{Multiplied}$$

$$x^2 + 4x - x^2 + 16 = 8 \quad \Leftarrow \text{Removed parentheses.}$$

$$4x + 16 = 8 \quad \Leftarrow \text{Combined like terms.}$$

Step 3 Move Terms

$$4x = -8 \quad \Leftarrow \text{Subtracted 16 from both sides.}$$

Step 4 Divide

$$x = -2 \quad \Leftarrow \text{Divided both sides by 4.}$$

Step 5 Check   The solution is –2. We leave the details of the check to you.

— Note —

Three uses for the LCD when working with rational expressions and equations

So far, we have used the LCD in three different ways:

Use 1 Adding and subtracting fractions   We convert each fraction to a given fraction with the LCD in the denominator. For example, the LCD of $\frac{1}{x} + \frac{1}{2}$ is $2x$. We rewrite the fractions as follows:

$$\left(\frac{1}{x}\right) \cdot \left(\frac{2}{2}\right) + \left(\frac{1}{2}\right) \cdot \left(\frac{x}{x}\right) \quad \Leftarrow \text{Converted each fraction into an equivalent}$$

fraction but with the LCD as the denominator.

$$\frac{2}{2x} + \frac{x}{2x} \quad \Leftarrow \text{Simplified.}$$

$$\frac{2 + x}{2x} \quad \Leftarrow \text{Combined into one fraction.}$$

Use 2 Simplifying complex fractions   We multiply the numerator and denominator of the complex fraction by the LCD. For example, the LCD of $\dfrac{1 + \frac{1}{x}}{\frac{1}{2}}$

is $2x$. We rewrite the fractions as follows:

$$\frac{(2x)(1) + (2x)\left(\frac{1}{x}\right)}{(2x)\left(\frac{1}{2}\right)} \quad \Leftarrow \text{Multiplied each minor fraction by the LCD of}$$

all fractions..

$$\frac{2x + 2}{x} \quad \Leftarrow \text{Simplified.}$$

Use 3 Solving rational equations   We multiply each term by the LCD. For example, the LCD of $1 + \frac{1}{x} = \frac{1}{2}$ is $2x$. We rewrite the equation as follows:

$$(2x)(1) + (2x)\left(\frac{1}{x}\right) = (2x)\left(\frac{1}{2}\right) \quad \Leftarrow \text{Multiplied each term by the LCD.}$$

$$2x + 2 = x \quad \Leftarrow \text{Simplified.}$$

$$x = -2 \quad \Leftarrow \text{Solved for } x.$$

Notice that with expressions we use the LCD to build equivalent fractions but with equations we use the LCD to eliminate fractions.

Practice 8.6.1 Solve.

1. $\dfrac{x}{3} = \dfrac{x}{5} + 6$

2. $\dfrac{2}{5x} - \dfrac{1}{2} = \dfrac{7}{5x}$

3. $\dfrac{1}{x} - \dfrac{5}{4} = \dfrac{7}{2x}$

4. $\dfrac{2}{x} = \dfrac{1}{x-1} + \dfrac{3}{2x}$

5. $\dfrac{6}{x} = \dfrac{6}{2x+1} + \dfrac{5}{x}$

6. $\dfrac{x}{x-3} - \dfrac{6}{x^2-9} = 1$

7. $\dfrac{x}{x+7} - \dfrac{14}{x^2-49} = 1$

Topic 8.6 B — Extraneous Roots

Our process for solving equations has always been to rewrite the equation in a simpler but equivalent form until we could find the answer just by looking at the equation. For example, let's solve $\dfrac{x}{x+3} = 2 - \dfrac{3}{x+3}$

Step 1 LCD   The LCD is $x + 3$.

$$(x+3)\left(\dfrac{x}{x+3}\right) = (x+3)(2) - (x+3)\left(\dfrac{3}{x+3}\right) \quad \Leftarrow \text{ Multiplied each term by LCD.}$$

Step 2 Simplify

$$x = (x+3)(2) - 3 \qquad \Leftarrow \text{ Reduced.}$$
$$x = 2x + 6 - 3 \qquad \Leftarrow \text{ Multiplied.}$$
$$x = 2x + 3 \qquad \Leftarrow \text{ Combined like terms.}$$

Step 3 Move Terms

$$-x = 3 \qquad \Leftarrow \text{ Subtracted } 2x \text{ from both sides.}$$

Step 4 Divide

$$x = -3 \qquad \Leftarrow \text{ Divided both sides by } -1.$$

Step 5 Check   In the original equation, we substitute -3 for x and simplify each side.

Left expression   ?=?   Right expression

$$\dfrac{x}{x+3} = 2 - \dfrac{3}{x+3} \qquad \Leftarrow \text{ Equation to check.}$$

$$\dfrac{(-3)}{(-3)+3} \quad \bigg| \quad 2 - \dfrac{3}{(-3)+3} \qquad \Leftarrow \text{ Substituted } -3 \text{ for } x.$$

$$\dfrac{3}{0} \quad \bigg| \quad 2 - \dfrac{3}{0} \qquad \Leftarrow \text{ Simplified.}$$

We found the solution to be -3, but -3 is a restricted value because it causes division by 0 in the original equation. This means we have a contradiction: -3 is the solution but we are not allowed to use -3 for x. Therefore, there is no solution to the equation.

What happened? In Section 2.2 we discussed the Multiplication Property of Equality, which essentially said that if we multiply all terms of an equation by a constant, other than 0, the resulting equation is equivalent to the original equation. We had to exclude multiplication by 0 because that might result in a new equation that is not equivalent to the original.

Multiplication by 0 becomes a possibility with rational equations where we multiply through by a variable. In this example, when we multiplied by $x + 3$ we would be multiplying by 0 if x had a value of -3. Since x is a variable, we cannot be sure that we are not multiplying by 0 at this point. If we are, then the resulting equation may not be equivalent to the original and so the solution may be wrong.

The easiest way to get around this difficulty is to be sure to check the answers we obtain in the original equation. Apparent solutions to an equation that do not satisfy the original equation are called **extraneous roots**. They are not true solutions (roots) to the original equation and are not a part of the solution set; they were introduced when we multiplied through by an LCD which may have been 0.

Example 8.6.2 a) Solve: $\dfrac{3x-1}{x-2} - 1 = \dfrac{5}{x-2}$

•SOLUTION• *Step 1 LCD*   The LCD is $x - 2$.

$$(x-2)\left(\frac{3x-1}{x-2}\right) - (x-2)(1) = (x-2)\left(\frac{5}{x-2}\right) \quad \Leftarrow \text{ Multiplied each term by LCD.}$$

Step 2 Simplify

$$3x - 1 - (x - 2) = 5 \qquad \Leftarrow \text{ Reduced.}$$

$$3x - 1 - x + 2 = 5 \qquad \Leftarrow \text{ Removed parentheses.}$$

$$2x + 1 = 5 \qquad \Leftarrow \text{ Combined like terms.}$$

Step 3 Move Terms

$$2x = 4 \qquad \Leftarrow \text{ Subtracted 1 from both sides.}$$

Step 4 Divide

$$x = 2 \qquad \Leftarrow \text{ Divided both sides by 2.}$$

Step 5 Check

Left expression ?=? Right expression

$$\frac{3x-1}{x-2} - 1 = \frac{5}{x-2} \qquad \Leftarrow \text{ Equation to check.}$$

$$\frac{3(2)-1}{(2)-2} - 1 \quad\Bigg|\quad \frac{5}{(2)-2} \qquad \Leftarrow \text{ Substituted 2 for } x.$$

$$\frac{5}{0} - 1 \quad\Bigg|\quad \frac{5}{0} \qquad \Leftarrow \text{ Simplified.}$$

The solution is 2, but 2 is a restricted value. Therefore, the equation has no solution.

b) $\dfrac{64}{x^2 - 16} = \dfrac{2x}{x-4} - 2$

•SOLUTION• *Step 1 LCD*   We have denominators of $x - 4$ and, in factored form, $(x-4)(x+4)$, so the LCD is $(x-4)(x+4)$.

$$(x-4)(x+4)\left(\frac{64}{(x-4)(x+4)}\right) = (x-4)(x+4)\frac{2x}{x-4} - (x-4)(x+4)(2) \quad \Leftarrow \text{ Multiplied by LCD.}$$

Step 2 Simplify

$$64 = (x+4)(2x) - 2(x-4)(x+4) \qquad \Leftarrow \text{ Reduced.}$$

$$64 = 2x^2 + 8x - 2(x^2 - 16) \qquad \Leftarrow \text{ Multiplied.}$$

$$64 = 2x^2 + 8x - 2x^2 + 32 \qquad \Leftarrow \text{ Removed parentheses.}$$

$$64 = 8x + 32 \qquad \Leftarrow \text{ Simplified each side.}$$

Step 3 Move Terms

$$32 = 8x \qquad \Leftarrow \text{ Subtracted 32 from both sides.}$$

Step 4 Divide

$$4 = x \qquad \Leftarrow \text{ Divided both sides by 8.}$$

Step 5 Check

Left expression ?=? *Right expression*

$$\frac{64}{x^2-16} = \frac{2x}{x-4} - 2 \qquad \Leftarrow \text{Equation to check.}$$

$$\frac{64}{(4)^2-16} \quad \Bigg| \quad \frac{2(4)}{(4)-4} - 2 \qquad \Leftarrow \text{Substituted 2 for } x.$$

$$\frac{64}{0} \quad \Bigg| \quad \frac{8}{0} - 2 \qquad \Leftarrow \text{Simplified.}$$

The solution is 4, but 4 is a restricted value. Therefore, the equation has no solution.

— Note — **Extraneous roots and restricted values**	If the "solution" to a rational equation turns out to be a restricted value of one of the expressions, the "solution" is extraneous and not a true solution.

Practice 8.6.2 Solve.

1. $\dfrac{5x+7}{x+2} - 2 = \dfrac{-3}{x+2}$

2. $\dfrac{x}{2x+1} + 3 = \dfrac{5x+3}{2x+1}$

3. $\dfrac{4x}{x-3} - 4 = \dfrac{72}{x^2-9}$

4. $\dfrac{5x}{x+1} - 5 = \dfrac{10}{x^2-1}$

Topic 8.6 C — Comparing Rational Expressions and Rational Equations

Some students confuse rational *expressions* and rational *equations* because we use many of the same tools when working with each.

— Note — **Rational expressions versus rational equations**	**Rational Expression such as** $\dfrac{1}{x} + \dfrac{1}{2}$	**Rational Equation such as** $\dfrac{1}{x} + \dfrac{1}{2} = 1$
	• Expressions that contain fractions where the numerators and/or denominators are polynomials.	• A statement of equality between two rational expressions.
	• Have NO equal sign.	• Have an equal sign.
	• We *simplify* expressions. That is, we rewrite them as equivalent expressions.	• We *solve* equations. That is, we find the values of the variables that make the equations true.
	• When adding or subtracting we build equivalent fractions that contain the LCD.	• When solving we multiply each term of the equation by the LCD.
	• The simplified expression usually contains the LCD.	• The solution usually contains no variables.

Example 8.6.3 Decide whether to solve or simplify the given equation or expression and then do it.

a) $\dfrac{2}{x} - 1 = \dfrac{-1}{x}$

• **SOLUTION** • The = sign tells us this is an equation so we *solve* it.

$$(x)\left(\frac{2}{x}\right) - (x)(1) = (x)\left(\frac{-1}{x}\right) \quad \Leftarrow \text{ Multiplied each term by the LCD, } x.$$

$$2 - x = -1 \qquad \Leftarrow \text{ Reduced and simplified.}$$

$$-x = -3 \qquad \Leftarrow \text{ Subtracted 2 from both sides.}$$

$$x = 3 \qquad \Leftarrow \text{ Divided both sides by } -1.$$

The solution is 3. We leave the check to you.

b) $\dfrac{2}{x+2} + \dfrac{5}{x} + \dfrac{3}{x(x+2)}$

• **SOLUTION** • This is an expression because there is no =. We will *simplify* the expression by transforming each term so that it has the LCD, $x(x+2)$, in its denominator.

$$\left(\frac{x}{x}\right)\left(\frac{2}{x+2}\right) + \left(\frac{x+2}{x+2}\right)\frac{5}{x} + \frac{3}{x(x+2)} \quad \Leftarrow \text{ Converted each fraction into an}$$

equivalent fraction with the LCD in the denominator.

$$\frac{2x}{x(x+2)} + \frac{5(x+2)}{x(x+2)} + \frac{3}{x(x+2)} \quad \Leftarrow \text{ Reduced and simplified.}$$

$$\frac{2x + 5(x+2) + 3}{x(x+2)} \quad \Leftarrow \text{ Combined fractions over LCD.}$$

$$\frac{7x + 13}{x(x+2)} \quad \Leftarrow \text{ Simplified.}$$

c) $\dfrac{2}{x+2} + \dfrac{5}{x} = \dfrac{3}{x(x+2)}$

• **SOLUTION** • This is an equation because it has an =. We must *solve* the equation. We multiply each term by the LCD, $x(x+2)$, to clear the fractions.

$$x(x+2)\left(\frac{2}{x+2}\right) + x(x+2)\left(\frac{5}{x}\right) = x(x+2)\left(\frac{3}{x(x+2)}\right) \quad \Leftarrow \text{ Multiplied each term by the LCD.}$$

$$2x + 5(x+2) = 3 \qquad \Leftarrow \text{ Reduced.}$$

$$2x + 5x + 10 = 3 \qquad \Leftarrow \text{ Removed parentheses.}$$

$$7x + 10 = 3 \qquad \Leftarrow \text{ Combined like terms.}$$

$$7x = -7 \qquad \Leftarrow \text{ Subtracted 10 from both sides.}$$

$$x = -1 \qquad \Leftarrow \text{ Divided both sides by 7.}$$

Practice 8.6.3 Decide whether to solve or simplify the given equation or expression and then do it.

1. $\dfrac{3}{x} - 5 - \dfrac{4}{5x}$

2. $\dfrac{1}{x+1} + \dfrac{2}{x} - \dfrac{5}{2x}$

3. $\dfrac{1}{x+1} + \dfrac{2}{x} = \dfrac{5}{2x}$

4. $\dfrac{1}{x-3} + \dfrac{2}{7x} + \dfrac{8}{7x}$

Topic 8.6 D — Proportions

You learned in arithmetic that a **ratio** is a fraction which expresses a relation between two quantities. Ratios can be written as fractions, using the word *to*, or using a colon. For example, if a bag of garden fertilizer contains 8 parts nitrogen and 3 parts potash then the ratio of nitrogen to potash can be written as $\frac{8}{3}$, or 8 to 3, or 8:3.

A **proportion** is a statement of equality between two ratios. In other words, a proportion is an equation where the left and right expressions are fractions.

— *Definition* — *Proportion*	English: Statement of equality between two ratios. Example: $$\frac{2}{3} = \frac{x}{6}$$ Algebra: If $b \neq 0$ and $d \neq 0$ then $\frac{a}{b} = \frac{c}{d}$ is a proportion.

Proportions are just special cases of rational equations and so we solve them using the same procedures (that is, multiply both sides by the LCD of all fractions).

Example 8.6.4 Solve the proportions.

a) $\dfrac{51}{x} = \dfrac{17}{7}$

•SOLUTION• **Step 1 LCD**   We have denominators of 7 and x so the LCD is $7x$.

$$7x \cdot \left(\frac{51}{x}\right) = 7x \cdot \left(\frac{17}{7}\right) \quad \Leftarrow \text{ Multiplied each term by LCD.}$$

Step 2 Simplify

$$357 = 17x \qquad \Leftarrow \text{ Reduced and then multiplied.}$$

Step 3 Move Terms   Not needed.

Step 4 Divide

$$21 = x \qquad \Leftarrow \text{ Divided both sides by 17.}$$

Step 5 Check   The solution is 21. We leave the details of the check to you.

b) $\dfrac{3}{x} = \dfrac{5}{x+4}$

•SOLUTION• **Step 1 LCD**   We have denominators of x and $x + 4$ so the LCD is $x(x+4)$.

$$x(x+4) \cdot \left(\frac{3}{x}\right) = x(x+4) \cdot \left(\frac{5}{x+4}\right) \quad \Leftarrow \text{ Multiplied each term by LCD.}$$

Step 2 Simplify

$$3(x+4) = 5x \qquad \Leftarrow \text{ Reduced.}$$

$$3x + 12 = 5x \qquad \Leftarrow \text{ Multiplied.}$$

Step 3 Move Terms

$$12 = 2x \qquad \Leftarrow \text{ Subtracted } 3x \text{ from both sides.}$$

Step 4 Divide

$$6 = x \qquad \Leftarrow \text{ Divided both sides by 2.}$$

Step 5 Check   The solution is 6. We leave the details of the check to you.

Practice 8.6.4 Solve the proportion.

1. $\dfrac{10}{x} = \dfrac{2}{9}$

2. $\dfrac{6}{x} = \dfrac{5}{x+3}$

3. $\dfrac{7}{x+2} = \dfrac{5}{x}$

You may have heard of a shortcut procedure for solving proportions called **cross multiplication**. To see where this comes from, suppose we are given the proportion

$$\frac{a}{b} = \frac{c}{d}$$

If we multiply both sides by the LCD, *bd*, we get an equation that contains no fractions:

$$(bd) \cdot \left(\frac{a}{b}\right) = (bd) \cdot \left(\frac{c}{d}\right)$$

$$ad = bc$$

If you look carefully at the original proportion you will see that the final equation turned out to have expressions that are the product of the numerator of one fraction and the denominator of the other (hence the name "cross" multiplication).

Cross multiplying can be a useful shortcut but it **only** works on proportions. That is, it only works on equations of the form

fraction = fraction

It will not work on any of the following equations because each has more than one term on one or both sides of the equal sign.

$$\frac{51}{x} = \frac{17}{7} + 1$$

$$\frac{1}{x} - \frac{3}{2} = \frac{5}{2x}$$

$$\frac{3}{x+1} = \frac{5}{2x} + \frac{2}{x}$$

Feel free to add cross multiplication to your algebra toolbox. Just be careful to use it ONLY when appropriate.

Topic 8.6 E — Applications Involving Relative Speeds

Typical word problems that may be modeled with rational equations include uniform motion problems which deal with relative speed. In these problems an object can be assigned two different speeds at the same time, depending on the point of reference. For example, what is the speed of a passenger on a train going 60 mph? Since speed is relative, it depends:

- If the passenger is sitting in a seat then he is going 0 mph relative to the train but 60 mph relative to the ground.

- If the passenger is walking at a rate of 3 mph toward the front of the train he is going 3 mph relative to the train but 3 + 60 or 63 mph relative to the ground.

- If the passenger is walking at a rate of 3 mph toward the rear of the train he is again going 3 mph relative to the train but 60 – 3 or 57 mph relative to the ground.

This situation occurs because we often measure speeds relative to something that is also moving. Here are two more examples:

- The airspeed of a plane is the speed of the plane relative to the surrounding air. The airspeed is a function of the power of the engines. If the airspeed of a plane is 500 mph and there is a 50 mph wind blowing in the direction of motion of the plane (a tailwind) then the speed of the plane relative to the ground is 500 + 50 or 550 mph. The tailwind makes the plane go faster relative to the ground.

• If you are on the earth's equator and you walk east at 3 mph your speed would be 1003 mph relative to someone watching you from the moon! This is the result of the 3 mph generated by your legs plus the 1000 mph generated by the earth rotating on its axis (it takes 24 hours for one rotation and the earth's circumference at the equator is 24,000 miles). You don't notice the extra 1000 mph because everything is traveling with you at that speed.

Example 8.6.5

Maren can row at a rate of 5 mph when she practices sculling on a lake. On a nearby river she can row 8 miles downstream in the same time it takes her to row 4.5 miles upstream. What is the speed of the river's current?

•SOLUTION• *Step 1 Organize*

Given: • Maren's rate on a lake is 5 mph.
• she rows 8 miles downstream.
• she rows 4.5 miles upstream.
• time going downstream is the same as time going upstream.

Find: Speed of current in river.

Sketch: Not helpful.

Rates: Rates are involved so we will construct an Amount Table. The items are going upstream and going downstream. We are given a rate of 5 mph but that is not the rate going upstream or downstream, that is the rate when she is rowing on a lake with no current to help or hinder her.

	amount =	*rate* ×	*base*
Descriptions →	distance	rate	time
Units →	miles	$\dfrac{\text{miles}}{1 \text{ hour}}$	hours
↓ *Item names* ↓			
downstream	8	❶	❸
upstream	4.5	❷	❹

Formula: amount = rate × base.

Step 2 Define unknowns

Define: We are asked to find the speed of the current, so we define

c is speed of current.

Construct: When Maren rows downstream the current increases her speed relative to the shore. Her resulting speed is the speed due to her rowing *plus* the speed of the current. Therefore,

$5 + c$ is her speed going downstream.

When she rows upstream the current slows her down so her speed relative to the shore is the speed due to her rowing *minus* the speed of the current. Therefore,

$5 - c$ is her speed going upstream.

	amount =	*rate* ×	*base*
Descriptions →	distance	rate	time
Units →	miles	$\dfrac{\text{miles}}{1 \text{ hour}}$	hours
↓ *Item names* ↓			
downstream	8	$5 + c$	❸
upstream	4.5	$5 - c$	❹

Amount Table: We now complete the Amount Table by calculating ❸, and ❹.

❸ is the time she rows downstream. Since this is the base, we calculate it by dividing the amount by the rate.

$$\frac{8 \text{ miles}}{\dfrac{5 + c \text{ miles}}{1 \text{ hour}}} \Rightarrow (8 \text{ miles}) \cdot \left(\frac{1 \text{ hour}}{5 + c \text{ miles}}\right) \Rightarrow \frac{8}{5 + c} \text{ hours}$$

❹ is the time she rows upstream:

$$\frac{4.5 \text{ miles}}{\dfrac{5 - c \text{ miles}}{1 \text{ hour}}} \Rightarrow (4.5 \text{ miles}) \cdot \left(\frac{1 \text{ hour}}{5 - c \text{ miles}}\right) \Rightarrow \frac{4.5}{5 - c} \text{ hours}$$

	amount	=	*rate*	×	*base*
Descriptions →	distance		rate		time
Units →	miles		$\dfrac{\text{miles}}{1 \text{ hour}}$		hours
↓ *Item names* ↓					
downstream	8		$5 + c$		$\dfrac{8}{5 + c}$
upstream	4.5		$5 - c$		$\dfrac{4.5}{5 - c}$

Step 3 Verbal models and Step 4 Mathematical Models   We are told the time going downstream is the same as the time going upstream.

Verbal model ⇒	time going downstream (hours)	=	time going upstream (hours)
Math model ⇒	$\dfrac{8}{5 + c}$	=	$\dfrac{4.5}{5 - c}$

Step 5 Solve

$$\frac{8}{5 + c} = \frac{4.5}{5 - c} \qquad \Leftarrow \text{ Equation to solve.}$$

$$(5 - c)(5 + c)\left(\frac{8}{5 + c}\right) = (5 - c)(5 + c)\left(\frac{4.5}{5 - c}\right) \qquad \Leftarrow \text{ Multiplied by the LCD, } (5 - c)(5 + c).$$

$$8(5 - c) = 4.5(5 + c) \qquad \Leftarrow \text{ Reduced common factors.}$$

$$40 - 8c = 22.5 + 4.5c \qquad \Leftarrow \text{ Simplified both sides.}$$

$$\mathbf{17.5} - 8c = 4.5c \qquad \Leftarrow \text{ Subtracted 22.5 from both sides.}$$

$$17.5 = \mathbf{12.5c} \qquad \Leftarrow \text{ Added 8c to both sides.}$$

$$1.4 = c \qquad \Leftarrow \text{ Divided both sides by 12.5.}$$

Step 6 Values of unknowns

	amount	=	rate	×	base
Descriptions →	distance		rate		time
Units →	miles		$\dfrac{\text{miles}}{1 \text{ hour}}$		hours
↓ Item names ↓					
downstream	8		$5 + c$ is $5 + 1.4$ **6.4 mph** **(speed going downstream)**		$\dfrac{8}{5 + c}$ is $\dfrac{8}{5 + 1.4}$ **1.25 hours** **(time spent going downstream)**
upstream	4.5		$5 - c$ is $5 - 1.4$ **3.6 mph** **(speed going upstream)**		$\dfrac{4.5}{5 - c}$ is $\dfrac{4.5}{5 - 1.4}$ **1.25 hours** **(time spent going upstream)**

Step 7 Answer   Speed of the current, c, is 1.4 mph.

Step 8 Check

- Time going downstream is the same as time going upstream: checks because each is 1.25 hours.

- Maren travels 8 miles going downstream: checks because she travels 6.4 mph for 1.25 hours, which is $\left(\dfrac{6.4 \text{ miles}}{1 \text{ hour}} \right) \cdot (1.25 \text{ hours})$ or 8 miles.

- Maren travels 4.5 miles going upstream: checks because she travels 3.6 mph for 1.25 hours, which is $\left(\dfrac{3.6 \text{ miles}}{1 \text{ hour}} \right) \cdot (1.25 \text{ hours})$ or 4.5 miles.

— Study Tip — **Study the structure**	It's important to go through a problem like this over and over until you fully understand its structure. It would be a good idea to put this problem on a note card with the problem statement on one side (including the textbook page reference) and an abbreviated solution on the other. Then, when you have a spare moment, read the problem and see if you can construct the solution without looking at the back of the card.

Practice 8.6.5

1. Joe can row 17 miles downstream in the same time that it takes to row 7 miles upstream. If his rate in still water is 6 mph, find the rate of the current.

2. A plane can fly 1125 miles with a tailwind in the same time it can fly 975 miles with a head wind. If the rate of the plane with no wind is 350 mph, find the rate of the wind.

Topic 8.6 F — Applications Involving Work

In general, different workers will do the same job at different rates, depending on their skill, motivation, working conditions, etc. If we assign several workers to a job, and we know how fast each works, we can determine how long it will take the group to finish the job.

Example 8.6.6   Barb and Anne are house painters. Barb can paint an entire room by herself in 6 hours. Anne takes 4 hours to do the same room. If Barb begins painting, and then is joined 1 hour later by Anne, and they complete the job, how long does each painter work?

•SOLUTION•

Step 1 Organize

Given: • Barb can paint a room in 6 hours by herself.

• Anne can paint a room in 4 hours by herself.

Find: The time each painter works.

Sketch: Not helpful.

Rates: Rates are involved so we will construct an Amount Table. Since Barb can paint 1 room in 6 hours her rate is $\dfrac{1 \text{ room}}{6 \text{ hours}}$. Similarly, Anne's rate is $\dfrac{1 \text{ room}}{4 \text{ hours}}$.

	amount =	*rate* ×	*base*
Descriptions →	work done	work done in a given time	time doing work
Units → ↓ *Item names* ↓	rooms	$\dfrac{\text{rooms}}{\text{hours}}$	hours
Barb	❸	$\dfrac{1}{6}$	❶
Anne	❹	$\dfrac{1}{4}$	❷

Formula: amount = rate × base.

Step 2 Define unknowns

Define: We are asked to find the time it take to do the job, which is the same as the time for Barb. We define

t is the time Barb works.

Construct: Since Anne begins 1 hour after Barb, we can write

t − 1 is the time Anne works.

Amount Table: We now complete the Amount Table by determining ❶, ❷, ❸, and ❹.

❶ is the time for Barb, *t*.

❷ is the time for Anne, *t* − 1

❸ is the amount of work Barb does. This is found by multiplying rate × base

$$\frac{1 \text{ room}}{6 \text{ hours}} \cdot (t \text{ hours}) \Rightarrow \frac{1}{6} t \text{ room}$$

❹ is the amount of work Anne does:

$$\frac{1 \text{ room}}{4 \text{ hours}} \cdot (t - 1 \text{ hours}) \Rightarrow \frac{1}{4}(t - 1) \text{ room}$$

	amount	=	*rate*	×	*base*
Descriptions →	work done		work done in a given time		time doing work
Units → ↓ *Item names* ↓	rooms		$\dfrac{\text{rooms}}{\text{hours}}$		hours
Barb	$\dfrac{1}{6}t$		$\dfrac{1}{6}$		t
Anne	$\dfrac{1}{4}(t-1)$		$\dfrac{1}{4}$		$t-1$

Step 3 Verbal models and Step 4 Mathematical Models   The total work done is the work done by Barb plus the work done by Anne. Since each painter completes a fraction of the job, when we add the fraction done by the two workers we get 1 whole job.

Verbal model ⇒   total work done   =   work done by Barb   +   work done by Anne

Math model ⇒ 1 = $\dfrac{1}{6}t$ + $\dfrac{1}{4}(t-1)$

Step 5 Solve

$$1 = \frac{1}{6}t + \frac{1}{4}(t-1) \quad \Leftarrow \text{ Equation to solve.}$$

$$12 = 2t + 3(t-1) \quad \Leftarrow \text{ Multiplied each term by LCD, 12, and reduced.}$$

$$12 = 2t + 3t - 3 \quad \Leftarrow \text{ Distributed the 3.}$$

$$12 = 5t - 3 \quad \Leftarrow \text{ Combined like terms.}$$

$$15 = 5t \quad \Leftarrow \text{ Added 3 to both sides.}$$

$$3 = t \quad \Leftarrow \text{ Divided both sides by 5.}$$

Step 6 Values of unknowns

	amount	=	*rate*	×	*base*
Descriptions →	work done		work done in a given time		time doing work
Units → ↓ *Item names* ↓	rooms		$\dfrac{\text{rooms}}{\text{hours}}$		hours
Barb	$\dfrac{1}{6}t$ is $\dfrac{1}{6}(3)$ $\dfrac{1}{2}$ **(part of the room painted by Barb)**		$\dfrac{1}{6}$		t **is 3 hours** **(time worked by Barb)**
Anne	$\dfrac{1}{4}(t-1)$ is $\dfrac{1}{4}(3-1)$ $\dfrac{1}{2}$ **(part of the room painted by Anne)**		$\dfrac{1}{4}$		$t-1$ **is 2 hours (time worked by Anne)**

Step 7 Answer   Barb works for 3 hours and Anne works for 2 hours.

Step 8 Check

- Barb works one hour more than Anne: checks since Barb works for 3 hours and Anne works for 2 hours.

- They work together to complete one whole job: checks because they each complete $\frac{1}{2}$ of the job.

Practice 8.6.6

1. Using a lawn mower Jane can cut her pasture in 12 hours. Using his riding mower, her brother, Jim, can cut the pasture in 5 hours. How long will it take the two of them working together?

2. Reco figures it will take his worker 8 hours to take inventory in his shop. If he hires someone from a temporary agency he figures it will take that person 15 hours to do the same job. How long should it take the two of them together?

Exercise Set 8.6   The answers to the odd numbered exercises are at the back of the book.

Solve.

1. $\dfrac{x}{3} + 5 = \dfrac{x}{2}$

2. $\dfrac{3x}{4} = \dfrac{x}{2} + 5$

3. $\dfrac{3}{4} + \dfrac{1}{2x} = \dfrac{-5}{2x}$

4. $\dfrac{7}{3} - \dfrac{3}{x} = \dfrac{4}{x}$

5. $\dfrac{1}{x} = \dfrac{2}{9} + \dfrac{5}{3x}$

6. $\dfrac{1}{2} - \dfrac{3}{10x} = \dfrac{6}{5x}$

7. $\dfrac{1}{3x} - \dfrac{3}{4} = \dfrac{11}{6x}$

8. $\dfrac{1}{4x} - \dfrac{1}{6} = \dfrac{7}{12x}$

9. $\dfrac{3}{2x} - \dfrac{1}{3} = \dfrac{5}{2x}$

10. $\dfrac{1}{3x} - \dfrac{1}{4} = \dfrac{-2}{3x}$

11. $\dfrac{5}{2x} = \dfrac{2}{x-5} + \dfrac{3}{x}$

12. $\dfrac{4}{3x} = \dfrac{1}{x+3} - \dfrac{2}{3x}$

13. $\dfrac{1}{x} = \dfrac{1}{x-5} - \dfrac{1}{3x}$

14. $\dfrac{3}{x} = \dfrac{2}{x+4} + \dfrac{1}{2x}$

15. $\dfrac{4}{x} = \dfrac{1}{x-5} + \dfrac{5}{2x}$

16. $\dfrac{2}{x} = \dfrac{1}{x+1} + \dfrac{3}{2x}$

17. $\dfrac{x}{x-3} + \dfrac{9}{x^2 - 2x - 3} = 1$

18. $\dfrac{x}{x+2} - \dfrac{8}{x^2 + 4x + 4} = 1$

19. $\dfrac{x}{x-2} + \dfrac{2}{x^2 - 4} = 1$

20. $\dfrac{x}{x+5} + \dfrac{5}{x^2 - 25} = 1$

21. $\dfrac{6x-1}{x-3} - 5 = \dfrac{17}{x-3}$

22. $\dfrac{3x}{x-5} + 2 = \dfrac{7x-10}{x-5}$

23. $\dfrac{3x+4}{x+1} - 2 = \dfrac{5}{x+1}$

24. $\dfrac{4m+5}{m-1} + 3 = \dfrac{-12}{m-1}$

25. $\dfrac{2x+5}{x+3} - \dfrac{-1}{x+3} = 4$

26. $3 = \dfrac{x-6}{x-2} - \dfrac{x}{x-2}$

27. $\dfrac{4x}{x+4} - 2 = \dfrac{x-8}{x+4}$

28. $\dfrac{2x}{x+5} - 2 = \dfrac{100}{x^2 - 25}$

29. $\dfrac{3x}{x+2} - 3 = \dfrac{18}{x^2 - 4}$

30. $\dfrac{6x}{x-4} - 6 = \dfrac{24}{x^2 - 16}$

Decide whether to solve or simplify the given equation or expression and then do it.

31. $\dfrac{4}{x+3} - \dfrac{2}{x-3}$

32. $\dfrac{3}{x} = \dfrac{4}{x+1}$

33. $\dfrac{1}{2x} + \dfrac{1}{x+3} = \dfrac{3}{4x}$

34. $\dfrac{1}{2x} + \dfrac{1}{x+3} - \dfrac{3}{4x}$

Solve the proportion.

35. $\dfrac{36}{x} = \dfrac{-4}{5}$

36. $\dfrac{14}{x} = \dfrac{-21}{9}$

37. $\dfrac{-12}{x} = \dfrac{16}{9}$

38. $\dfrac{15}{x} = \dfrac{-12}{7}$

39. $\dfrac{9}{x} = \dfrac{3}{7}$

40. $\dfrac{15}{x} = \dfrac{12}{4}$

41. $\dfrac{6}{x} = \dfrac{4}{x+10}$

42. $\dfrac{8}{x} = \dfrac{5}{x-9}$

43. $\dfrac{12}{x+2} = \dfrac{4}{x}$

44. $\dfrac{6}{x-9} = \dfrac{3}{x}$

---

45. Kim drove 300 miles In the same amount of time, Lori drove only 225 miles Find the rate of each if Lori's speed was 15 mph slower than Kim's.

46. A plane flew 900 miles then reduced its speed by 20 mph. In the same amount of time it took to fly 900 miles, it flew only 840 miles at the reduced speed. Find the plane's original speed.

47. A plane can fly 1950 miles with a tailwind in the same time it can fly 1800 miles with a head wind. If the rate of the plane with no wind is 375 mph, find the rate of the wind.

48. Alice can row 11 miles upstream in the same time he can row 26 miles downstream. If the speed of the current is 1.5 mph, find his rate in still water.

49. A small pipe can fill a tank in 5 hours and a larger pipe can fill the same tank in 3 hours. How long will it take the two pipes together to fill the tank?

50. A drain working alone can empty a tank in 15 hours. A larger drain can empty the same tank in 10 hours. How long will it take to drain the tank if both drains are opened?

51. Sue takes 450 hours to make a quilt, while Jane takes 550 hours to do the same job. How long will it take them working together?

52. John can paint a house in 15 hours, while Bill, using a paint sprayer can do the same job in 7 hours. How long will it take the two of them working together?

Review Exercises   The answers to all of these exercises are at the back of the book.

53. Tina buys a waterbed for $735 but pays no money for 9 months. At the end of 9 months she has to pay the total amount of the waterbed along with 12% simple interest. How much does she have to pay?

54. To help the punch at a party last longer Claire adds what's left in a bottle of Juice Aid, which is 10% fruit juice, to the 4 gallon mixture in the punch bowl, which is 60% fruit juice. If the final mixture turns out to be 40% fruit juice how much 10% mixture was in the Juice Aid bottle?

55. The difference of a positive fraction and its reciprocal is $\dfrac{15}{4}$. Find the fraction if the numerator is 9 more than the denominator.

56. Calculate the slope of the straight line described by the following data table:

x	y
3	–6
3	0

57. Determine whether this relation is a function: $\{(-1,2),(4,-2),(8,0),(-5,3),(12,-1)\}$

58. Solve: $\left\{ \begin{array}{l} 20 - 12y = -8x \\ 15 + 6x = 9y \end{array} \right\}$ ❶ ❷

59. Solve: $\left\{ \begin{array}{l} 5x - 2y = 3 \\ 4x + 3y = 7 \end{array} \right\}$ ❶ ❷

Simplify.

60. $\dfrac{5^0}{2^{-1} + 5^{-1}} + (2 + 3)^{-1}$

61. $\left(\dfrac{2x}{y}\right)^2 \left(\dfrac{y^3}{8x}\right)$

62. $\dfrac{\left(\frac{1}{3}k\right)^2}{3k^{-3}}$

63. $\dfrac{\frac{2x-6}{5}}{x-3}$

64. $\left(\dfrac{3m^{-2}n^2}{2m^2n}\right)^{-2} \left(\dfrac{6m^5}{9n^4}\right)$

65. $(m + 4)^2$

66. $3 - \dfrac{5}{2x}$

67. $ab(b + 1) - (b + 1) - b(b - ab - 1)$

68. Simplify and write the answer in scientific notation: $\dfrac{(15000)(18000000)}{4500}$

69. Graph: $2x - y > 5$

70. Subtract $-2xy - 3x^2y + 11xy^2$ from $-2x^2y - 3xy + 7xy^2$

71. Divide and simplify: $\dfrac{-12y^5 + 10y^4 - 8y^3 + 4y}{2y}$

72. Factor completely: $64 - 20x + x^2$

73. Factor completely: $8x^2 - 36x + 16$

74. Factor completely: $4x^2 + 20x + 36 + 4x$

75. Solve: $0 = 6x^2 - 3x$

76. Solve: $15x^2 = 31x - 10$

77. Reduce: $\dfrac{2x^2 - 10x}{x^2 - 25}$

Chapter 8 Review

Vocabulary to Know

building up a fraction — multiplying the numerator and denominator of a rational expression by the same quantity, as in converting $\frac{3}{4}$ to $\frac{6}{8}$.

complex rational expression — a fraction whose numerator and/or denominator are rational expressions.

cross multiplication — in a proportion , multiplying the numerator of one ratio by the denominator of the other ratio, and vice versa.

equivalent fractions — fractions that look different but which represent the same value, as in $\frac{3}{4}$ and $\frac{6}{8}$.

extraneous root — an apparent solution to an equation that does not satisfy the original equation; they are not a part of the solution set.

LCD — same as Least Common Denominator.

least common denominator — the smallest expression that contains as factors all denominators of a set of rational expressions; same as LCD.

minor fraction — a fraction in the numerator or denominator of a complex fraction.

proportion — a statement of equality between two ratios.

ratio — a fraction which expresses a relation between two quantities.

rational equation — an equation that contains at least one rational expression.

rational expression — a fraction where the numerator and denominator are polynomials.

reducing a fraction to lowest terms — removing all common factors from the numerator and denominator of a rational expression.

restricted values — values of a variable which would result in division by 0.

Properties, Definitions, Formulas, and Notation to Understand

Rational expression — A fraction where the numerator and denominator are polynomials.	If P and Q represent polynomials and $Q \neq 0$, then $\frac{P}{Q}$ is a rational expression. Examples include $\frac{3x}{2}$ and $\frac{y^2 - 1}{y + 2}$.
Fundamental property of fractions — Multiplying the numerator and denominator of a fraction by the same non-zero number does not change the value of the fraction.	If $b \neq 0$ and $k \neq 0$ then $\frac{a}{b} = \frac{a \cdot k}{b \cdot k}$ $\frac{1}{4}$ is equivalent to $\frac{1 \cdot 3}{4 \cdot 3}$ $\frac{5x}{2}$ is equivalent to $\frac{5xy}{2y}$ (assuming $y \neq 0$)
Multiplying rational expressions — To find the product of two rational expressions, multiply the numerators and multiply the denominators.	If P, Q, R, and S, represent polynomials and $Q \neq 0$ and $S \neq 0$, then $\frac{P}{Q} \cdot \frac{R}{S} = \frac{P \cdot R}{Q \cdot S}$ $\frac{2}{3} \cdot \frac{4}{5}$ becomes $\frac{2 \cdot 4}{3 \cdot 5}$, which simplifies to $\frac{8}{15}$ $\frac{2}{x} \cdot \frac{y}{3}$ becomes $\frac{2 \cdot y}{x \cdot 3}$, which simplifies to $\frac{2y}{3x}$
Dividing rational expressions — To find the quotient of two rational expressions, multiply the first by the reciprocal of the second.	If P, Q, R, and S represent polynomials and $Q \neq 0$, $R \neq 0$, and $S \neq 0$, then $\frac{P}{Q} \div \frac{R}{S} = \frac{P}{Q} \cdot \frac{S}{R}$ $\frac{2}{3} \div \frac{1}{5}$ becomes $\frac{2}{3} \cdot \frac{5}{1}$, which simplifies to $\frac{10}{3}$ $\frac{2}{x} \div \frac{y}{3}$ becomes $\frac{2}{x} \cdot \frac{3}{y}$, which simplifies to $\frac{6}{xy}$

Uses of the LCD	*Use 1 Adding and subtracting fractions* We convert each fraction to a given fraction with the LCD in the denominator. For example, the LCD of $\frac{1}{x} + \frac{1}{2}$ is $2x$. We rewrite the fractions as $\frac{1}{x} \cdot \frac{2}{2} + \frac{1}{2} \cdot \frac{x}{x}$, which simplifies to $\frac{2}{2x} + \frac{x}{2x}$, or $\frac{2+x}{2x}$.
	Use 2 Simplifying complex fractions We multiply the numerator and denominator of the complex fraction by the LCD. For example, the LCD of $\dfrac{1 + \frac{1}{x}}{\frac{1}{2}}$ is $2x$. We rewrite the fraction as $\dfrac{1 \cdot 2x + \frac{1}{x} \cdot 2x}{\frac{1}{2} \cdot 2x}$, which simplifies to $\frac{2x + 2}{x}$.
	Use 3 Solving rational equations We multiply each term by the LCD. For example, the LCD of $\frac{1}{x} + \frac{1}{2} = 1$ is $2x$. We rewrite the equation as $\frac{1}{x} \cdot 2x + \frac{1}{2} \cdot 2x = 1 \cdot 2x$, which simplifies to $2 + x = 2x$, or $x = 2$.
Proportion — Statement of equality between two ratios.	If $b \neq 0$ and $d \neq 0$ then $\frac{a}{b} = \frac{c}{d}$ is a proportion. $\frac{2}{3} = \frac{x}{6}$

Rational expressions versus rational equations	**Rational Expression such as** $\frac{1}{x} + \frac{1}{2}$	**Rational Equation such as** $\frac{1}{x} + \frac{1}{2} = 1$
	• Expressions that contain fractions where the numerators and/or denominators are polynomials.	• A statement of equality between two rational expressions.
	• Have NO equal sign.	• Have an equal sign.
	• We simplify expressions. That is, we rewrite them as equivalent expressions.	• We solve equations. That is, we find the values of the variables that make the equations true.
	• When adding or subtracting we build equivalent fractions that contain the LCD.	• When solving we multiply each term of the equation by the LCD.
	• The simplified expression usually contains the LCD.	• The solution usually contains no variables.

Procedures to Follow

Procedure for reducing rational expressions.	Simplify $\dfrac{x^2 + 2x + 1}{x^2 - 1}$
Step 1 Prime factor the numerator and the denominator.	$\dfrac{(x+1)(x+1)}{(x-1)(x+1)}$
Step 2 Reduce factors common to the numerator and the denominator.	We reduce the common factor of $(x+1)$ to get $\dfrac{x+1}{x-1}$

Procedure for multiplying rational expressions .	Simplify $\dfrac{x^2 + x - 2}{x^2 + 4x + 4} \cdot \dfrac{x + 3}{x^2 - x}$
Step 1 Factor Factor each numerator and denominator.	$\dfrac{(x + 2)(x - 1)}{(x + 2)(x + 2)} \cdot \dfrac{x + 3}{x(x - 1)}$
Step 2 Reduce Reduce factors common to the numerator and denominator and then multiply the remaining factors if appropriate.	We reduce the common factors of $(x - 1)$ and $(x + 2)$ to get $\dfrac{x + 3}{x(x + 2)}$
Procedure for dividing rational expressions.	Simplify $\dfrac{x^2 + 4x - 21}{x^2 + 7x} \div \dfrac{x^2 - 9}{2x - 8}$
Step 1 Reciprocal Rewrite the problem as the product of the numerator and the *reciprocal* of the denominator.	$\dfrac{x^2 + 4x - 21}{x^2 + 7x} \cdot \dfrac{2x - 8}{x^2 - 9}$
Step 2 Factor Factor each numerator and denominator.	$\dfrac{(x - 3)(x + 7)}{x(x + 7)} \cdot \dfrac{2(x - 4)}{(x - 3)(x + 3)}$
Step 3 Reduce Reduce factors common to the numerator and the denominator and then multiply the remaining factors if appropriate.	We reduce the common factors of $(x - 3)$ and $(x + 7)$ to get $\dfrac{2(x - 4)}{x(x + 3)}$
Procedure for finding the LCD of rational expressions .	Find the LCD of $\dfrac{1}{x^2 - 4}$ and $\dfrac{1}{x^2 + 4x + 4}$
Step 1 Prime factor Prime factor each denominator and write it in exponential form.	$x^2 - 4$ factors as $(x - 2)^1(x + 2)^1$ $x^2 + 4x + 4$ factors as $(x + 2)^2$
Step 2 Select For each prime factor, select the one with the largest power.	Largest power of $x - 2$ is $(x - 2)^1$ Largest power of $x + 2$ is $(x + 2)^2$
Step 3 Multiply Multiply the selected factors. This is the LCD.	LCD $= (x - 2)(x + 2)(x + 2)$
Procedure for adding and subtracting rational expressions with unlike denominators.	Add $\dfrac{1}{x^2 - 4} + \dfrac{1}{x^2 + 4x + 4}$
Step 1 Factor Denominators Write all denominators in factored form.	$\dfrac{1}{(x - 2)(x + 2)} + \dfrac{1}{(x + 2)(x + 2)}$
Step 2 Build LCD Build the Least Common Denominator (LCD) by multiplying the largest power of each prime factor.	LCD $= (x - 2)(x + 2)(x + 2)$
Step 3 Build Equivalent Fractions Build equivalent fractions which contain the LCD.	$\dfrac{1}{(x - 2)(x + 2)} \cdot \dfrac{(x + 2)}{(x + 2)} + \dfrac{1}{(x + 2)(x + 2)} \cdot \dfrac{(x - 2)}{(x - 2)}$
Step 4 Combine Numerators Add or subtract the numerators and use the LCD as the denominator.	$\dfrac{(x + 2) + (x - 2)}{(x - 2)(x + 2)(x + 2)}$ $\dfrac{2x}{(x - 2)(x + 2)(x + 2)}$
Step 5 Reduce Factor the resulting fraction and reduce if possible.	Not possible to reduce.

Procedure for simplifying complex rational expressions using the four basic operations.	Simplify $\dfrac{\dfrac{1}{x} + 2}{3 - \dfrac{1}{x^2}}$
Step 1 Simplify Simplify the numerator and denominator separately.	$\dfrac{\dfrac{1 + 2x}{x}}{\dfrac{3x^2 - 1}{x^2}}$
Step 2 Invert Rewrite the problem as the product of the numerator and the *reciprocal* of the denominator.	$\dfrac{1 + 2x}{x} \cdot \dfrac{x^2}{3x^2 - 1}$
Step 3 Reduce Reduce factors common to the numerators and denominators.	$\dfrac{x(1 + 2x)}{3x^2 - 1}$
Procedure for simplifying complex rational expressions using the LCD.	Simplify $\dfrac{1 - \dfrac{4}{x^2}}{\dfrac{2}{x} + 1}$
Step 1 LCD Build the LCD of all minor fractions.	The LCD is x^2
Step 2 Multiply Multiply each minor fraction by the LCD.	$\dfrac{1 \cdot x^2 - \dfrac{4}{x^2} \cdot x^2}{\dfrac{2}{x} \cdot x^2 + 1 \cdot x^2}$
Step 3 Simplify Simplify the numerator and denominator.	$\dfrac{x^2 - 4}{2x + x^2}$
Step 4 Reduce Reduce factors common to the numerators and denominators.	$\dfrac{(x - 2)\cancel{(x + 2)}}{x\cancel{(2 + x)}}$ $\dfrac{x - 2}{x}$

Procedure for solving equations.	Solve $\dfrac{1}{x} - \dfrac{6}{x^2 + 2x} = \dfrac{10}{2x + 4}$
Step 1 LCD If fractions are present, multiply each term by the LCD.	The denominators factor as x, $x(x + 2)$, $2(x + 2)$ so the LCD is $2x(x + 2)$

$$2x(x + 2)\left(\frac{1}{x}\right) - 2x(x + 2)\left(\frac{6}{x(x + 2)}\right) = 2x(x + 2)\left(\frac{10}{2(x + 2)}\right)$$

Step 2 Simplify Simplify each side of the equation.	$2(x + 2) - 12 = 10x$ $2x + 4 - 12 = 10x$ $2x - 8 = 10x$
Step 3 Move Terms Use addition or subtraction to move terms with variables to one side of the equation and terms without variables to the other side. Then, simplify each side again.	$-8 = 8x$
Step 4 Divide Divide both sides of the equation by the coefficient of the variable.	$-1 = x$

Step 5 Check Check your answer by substituting it into the original equation. If a true statement results, your answer is probably correct. If a false statement results, go back over your work to see if you made an error.

Substituting -1 for x gives

$$\frac{1}{(-1)} - \frac{6}{(-1)^2 + 2(-1)} = \frac{10}{2(-1) + 4}$$

$$-1 + 6 = \frac{10}{2}$$

$$5 = 5 \quad \text{Checks.}$$

Chapter 8 Review Exercises   The answers to all the exercises are in the back of the book.

1. Determine the restricted values of $\dfrac{3x}{x-4}$

2. Determine the restricted values of $\dfrac{w-6}{7w}$

Reduce.

3. $\dfrac{2x+2}{2x^2-2x-4}$

4. $\dfrac{12x^2-18x}{4x^3-2x^2-6x}$

5. $\dfrac{x+3}{9-x^2}$

6. $\dfrac{x^2-8x+15}{3x-x^2+10}$

Simplify.

7. $\dfrac{6x^3}{10y^2} \cdot \dfrac{15y}{x^2}$

8. $\dfrac{8x}{3y^3} \cdot 12y^2 \cdot \dfrac{1}{16x^3}$

9. $\dfrac{6}{x^2} \cdot \dfrac{5x^2-15x}{15x-45}$

10. $\dfrac{x^4-6x^3-7x^2}{x^3-x^2-2x} \cdot \dfrac{x^2+4x-12}{2x^2-14x}$

11. $\dfrac{\dfrac{8x^5}{5y^2}}{\dfrac{4x^3}{10y^8}}$

12. $\dfrac{12-x}{5x} + \dfrac{3}{5x}$

13. $\dfrac{8x+4}{3x^2} + \dfrac{4x-1}{3x^2}$

14. $\dfrac{x^2-9}{2x^2-10x} \div \dfrac{x^2+6x+9}{6x^2-30x}$

15. $\dfrac{y+3}{2y^2} - \dfrac{y-5}{2y^2}$

16. $\dfrac{9x-2}{5x^2} - \dfrac{3-6x}{5x^2}$

17. $\dfrac{21}{6x^2} - \dfrac{5}{2x}$

18. $8 - \dfrac{3}{a^2b}$

19. $\dfrac{5}{3x^2-6x} - \dfrac{x}{2-x}$

20. $\dfrac{\dfrac{1}{k}+3}{1-\dfrac{3}{k}}$

21. $\dfrac{5-\dfrac{1}{y}}{\dfrac{1}{2y}-2}$

22. $\dfrac{1}{x^2-16} + \dfrac{3}{x^2+8x+16}$

Solve.

23. $\dfrac{3}{10x} - \dfrac{1}{6x} = \dfrac{1}{15}$

24. $\dfrac{3}{2x} = 1 - \dfrac{1}{x}$

25. $\dfrac{10}{n^2-25} = \dfrac{n}{n+5} - 1$

26. $\dfrac{x}{x+4} = \dfrac{x+18}{x^2-16} + \dfrac{x}{x-4}$

27. $\dfrac{2}{x} = \dfrac{1}{x+2}$

28. $\dfrac{1}{h-1} = \dfrac{3}{h-5}$

29. $\dfrac{5}{w-5} + 4 = \dfrac{w}{w-5}$

30. $\dfrac{10}{a^2-25} + \dfrac{1}{a+5} = \dfrac{3}{a-5}$

31. Tamika can row 20 miles upstream in the same time she can row 32 miles downstream. If the speed of the current is 1.2 mph, find her rate in still water.

32. Ned can shovel a driveway by himself in 2 hours while it takes Bill 3 hours to do the same job. Working together, how long will it take them to shovel the driveway?

33. An older computer takes 30 minutes to do a job. A second, newer, computer can do the same job in 15 minutes. How long will it take the two computers working together?

Chapter 8 Test   The answers to all the exercises are in the back of the book.

1. Determine the restricted values of $\dfrac{1}{t-5}$

2. Reduce $\dfrac{2x^2+8x}{3x^2+11x-4}$

3. Reduce $\dfrac{9x^3-48x^2-105x}{3x^2-21x}$

Simplify.

4. $\dfrac{y^5}{12x^8} \cdot 18x^3 \cdot \dfrac{2x}{y^3}$

5. $\dfrac{x^2-1}{3x} \cdot \dfrac{6x^2+6x}{x^2+2x+1}$

6. $\dfrac{\dfrac{6x^8}{9y^5}}{\dfrac{x^6}{3y^{12}}}$

7. $\dfrac{\dfrac{1}{x}-\dfrac{2}{y}}{\dfrac{3}{y}-\dfrac{2}{x}}$

8. $\dfrac{4}{3x^2y} - \dfrac{y}{6xy^2}$

9. $\dfrac{2}{x^2-8x+7} + \dfrac{3}{x^2+2x-3}$

10. $\dfrac{x^2+7x-8}{x^2-2x} \div \dfrac{x^2+10x+16}{x^2-4}$

11. $2 - \dfrac{3}{10x} + \dfrac{5}{2x}$

Solve.

12. $\dfrac{2}{x} + 5 = 9$

13. $\dfrac{1}{x^2-1} + \dfrac{x}{x+1} = 1 - \dfrac{2}{x-1}$

14. $\dfrac{b+11}{b^2-5b+4} + \dfrac{3}{b-1} = \dfrac{5}{b-4}$

15. $\dfrac{3}{x-5} = \dfrac{4}{x}$

16. If John can shovel a driveway in 2 hours by himself and Bill can do the same job in 3 hours by himself, how long would it take them working together to do the job?

17. A plane can fly 1240 miles with a tailwind in the same time that it takes to fly 1120 miles with a head wind. If the rate of the wind is 15 mph, find the rate of the plane in still air.

Chapter 9
Radical Expressions and Equations

Section 9.1 Introduction to Radical Expressions

Topic 9.1 A — Definitions and Notation

We use exponents to indicate the number of factors of a base that are to be multiplied. For example, the exponent in 3^2 tells us to multiply two factors of 3 to get 9.

In some situations we may have to do the reverse; that is, we may be given the final product and asked to find the base. For example, we may be asked to find numbers whose square is 16. We call those numbers the **square root** of 16 and we use the **radical symbol** $\sqrt{}$ to indicate the square root. Every positive real number has two square roots which are opposites of each other. The square root that is positive is called the **principal square root**; the square root that is negative is called the **negative square root**. For example

- the *principal square root* of 16 is 4 since 4^2 is 16. We write the principal square root of 16 as $\sqrt{16}$.

- the *negative square root* of 16 is –4 since $(-4)^2$ is 16. We write the negative square root of 16 as $-\sqrt{16}$.

- the *square root* of 16 is 4 or –4. We write the square root of 16 as $\pm\sqrt{16}$, where $\pm$ means *plus or minus*. The symbol $\pm\sqrt{16}$ literally means "$+\sqrt{16}$ or $-\sqrt{16}$ ".

An expression that contains a radical symbol, $\sqrt{}$, is called a **radical expression**, or simply a **radical**. The expression under the radical symbol is called the **radicand**. For example, $\sqrt{16}$ is a radical with a radicand of 16.

A **perfect square** is a number that is the square of a whole number. It's a good idea to memorize the principal square roots of the first 12 perfect squares so you can simplify them mentally, without the aid of a calculator. Here is a list.

$$\sqrt{0} = \sqrt{0^2} = 0 \qquad \sqrt{1} = \sqrt{1^2} = 1 \qquad \sqrt{4} = \sqrt{2^2} = 2$$

$$\sqrt{9} = \sqrt{3^2} = 3 \qquad \sqrt{16} = \sqrt{4^2} = 4 \qquad \sqrt{25} = \sqrt{5^2} = 5$$

$$\sqrt{36} = \sqrt{6^2} = 6 \qquad \sqrt{49} = \sqrt{7^2} = 7 \qquad \sqrt{64} = \sqrt{8^2} = 8$$

$$\sqrt{81} = \sqrt{9^2} = 9 \qquad \sqrt{100} = \sqrt{10^2} = 10 \qquad \sqrt{121} = \sqrt{11^2} = 11$$

Example 9.1.1   Try calculating each square root mentally by asking yourself *what non–negative number squared will result in the radicand?* Use your calculator to check your answer or to find any square roots you are unsure of.

a) $\sqrt{81}$

   •SOLUTION• Since 9^2 is 81, we can say $\sqrt{81}$ is 9.

b) $\sqrt{0.01}$

   •SOLUTION• Since $(0.1)^2$ is 0.01, we can say $\sqrt{0.01}$ is 0.1.

c) $\sqrt{\dfrac{4}{25}}$

   •SOLUTION• Since $\left(\dfrac{2}{5}\right)^2$ is $\dfrac{4}{25}$, we can say $\sqrt{\dfrac{4}{25}}$ is $\dfrac{2}{5}$.

d) $\sqrt{19^2}$

   •SOLUTION• You may be tempted to square 19 to get 361 and then ask yourself what number squared is 361? Instead of doing that, you could do this like the three previous examples and say to yourself 19^2 is 19^2 so $\sqrt{19^2}$ is 19.

Practice 9.1.1 Try calculating each square root mentally. Use your calculator to check your answer or to find any square roots you are unsure of.

1. $\sqrt{16}$ 2. $\sqrt{0}$ 3. $\sqrt{\dfrac{9}{64}}$ 4. $\sqrt{13^2}$ 5. $\sqrt{0.25}$ 6. $\sqrt{0.0049}$

The square roots of some numbers turn out to be irrational (that is, they cannot be written as either repeating or terminating decimals). We refer to such numbers as **irrational numbers**. We can use a calculator to find the approximate square root of an irrational number. We say *approximate* because we can never write the exact decimal representation for an irrational number. For example, to find an approximation of $\sqrt{2}$ we enter 2 into a calculator and then tap the $\boxed{\sqrt{}}$ key [†]. For a calculator that displays 8 digits, the display should read 1.4142136. This is a rounded off version of 1.414213562373... The approximate answer of 1.4142136 is a bit too big. In fact, if you square 1.4142136 you will get 2.0000001 rather than 2.

Depending on the situation, we either leave a value like $\sqrt{2}$ in radical form, or we use a calculator to convert it to decimal notation and round the result. For example, if we needed to cut a piece of lumber to a length of $\sqrt{2}$ meters, we would use a calculator to determine the approximate decimal value and round the result, probably to the hundredths place.

Example 9.1.2 a) Use a calculator to find $\sqrt{17}$ and round your answer to the hundredths place.

• SOLUTION • Evaluating the radical with a calculator gives 4.123105626, which when rounded to the hundredths place gives 4.12.

b) Use a calculator to find $\sqrt{185.8}$ and round your answer to the nearest whole.

• SOLUTION • Evaluating the radical with a calculator gives 13.63084737, which when rounded to the nearest whole gives 14.

Practice 9.1.2 Use a calculator to find the following square roots and then round to the place value indicated.

1. $\sqrt{0.03}$ round to the tenths place 2. $\sqrt{2550}$ round to the ones place.

3. $\sqrt{879.985}$ round to the hundredths place 4. $\sqrt{50.01}$ round to the tenths place.

So far, we have considered only non–negative radicands. What happens if we attempt to find the square root of a negative number? For example, to find $\sqrt{-9}$ we ask *what non–negative number squared gives –9?* We cannot obtain –9 by multiplying a real number by itself because a negative product only results if the two factors have *different* signs. In later math courses you will study a new set of numbers, called **complex numbers**, which are used when some types of expressions have negative radicands.

— *Definition* — *Square root of a negative number*	English: The square root of a negative number is not defined using real numbers.
	Example: $\sqrt{-16}$ is not defined because the radicand, –16, is negative.
	Algebra: If $a < 0$, then $\sqrt{a}$ is not defined using real numbers.

[†] On some calculators, we tap $\sqrt{}$ first and then tap 2 and finally =.

Many calculators work only with real numbers. If you use a calculator to try to find the square root of a negative number the display may show **E, – E –** or **Error**, meaning that the answer is not defined using real numbers. For example, tap $\boxed{9}$, $\boxed{+/-}$, and $\boxed{\sqrt{}}$. An **– E –** should appear on the display. [†]

A negative to the left of the $\sqrt{}$ can be interpreted in several ways:

- as a *negative square* root. For example, $-\sqrt{5}$ can be read as *the negative square root of 5*. This is not the same as the *square root of negative five*, $\sqrt{-5}$, which is not defined using real numbers.

- as the *opposite of* the principal square root. For example, $-\sqrt{25}$ can be read as *the opposite of the principal square root of twenty–five*, which is –5.

- as –1 times the principal square root. For example, $-\sqrt{25}$ can be written as $-1 \cdot \sqrt{25}$, which simplifies to –5.

Example 9.1.3   Decide whether each expression is a real number. If it's a real number, simplify the expression.

a) $\sqrt{-4}$

•**SOLUTION**• This is not defined using real numbers because the radicand is negative.

b) $-\sqrt{4}$

•**SOLUTION**• This is a real number. It's the negative square root of 4, which is –2.

c) $-\sqrt{-9}$

•**SOLUTION**• This is not defined using real numbers because the radicand is negative. The sign to the left of the square root does not affect the sign of the radicand.

Practice 9.1.3 Decide whether each of the following is a real number. If it is a real number, simplify the expression.

1. $-\sqrt{49}$ 2. $\sqrt{-81}$ 3. $-\sqrt{1}$ 4. $-\sqrt{-121}$

Topic 9.1 B — Square Roots and the Order of Operations

Mathematical operations must be carried out in a specified order unless grouping symbols are present or a property, such as the Commutative Property, is applied. In later math courses you will learn that radicals are really another way of writing exponents that happen to be fractions (for example, it turns out that a square root corresponds to an exponent of $\frac{1}{2}$). This means that radicals fall in with exponents in the order of operations:

— Procedure —		
Simplifying expressions using the proper order of operations	*Step 1*	Simplify expressions inside grouping symbols. Grouping symbols include parentheses ( ), brackets [ ], the fraction bar, absolute value bars $\vert \ \vert$, and the radical symbol, $\sqrt{}$.
	Step 2	Simplify absolute values, exponents, and square roots.
	Step 3	Simplify multiplication and division, working left to right.
	Step 4	Simplify addition and subtraction, working left to right.

[†]   On some calculators, entering $\sqrt{-9}$ will display 3*i* or (0,3) rather than –E–. This is the calculator's way of representing complex numbers.

When square roots are involved in multiplication

• we usually omit the multiplication symbol. For example, we will indicate the product of 2 and $\sqrt{3}$ as $2\sqrt{3}$. But, it would also be correct to write this as $(2)\left(\sqrt{3}\right)$ or $2 \cdot \sqrt{3}$.

• we usually write the radical as the right–most numeric factor in a product. For example, we will indicate the product of $\sqrt{3}$ and 2 as $2\sqrt{3}$. But, it would also be correct to write it as $\left(\sqrt{3}\right)(2)$ or $\sqrt{3} \cdot 2$.

Example 9.1.4   Follow the proper order of operations to simplify.

a) $2 - 3\sqrt{64}$

    •**SOLUTION**•

$$2 - 3(\mathbf{8}) \quad \Leftarrow \text{ Simplified square root.}$$
$$2 - \mathbf{24} \quad \Leftarrow \text{ Multiplied.}$$
$$-22 \quad \Leftarrow \text{ Subtracted.}$$

b) $-\sqrt{9} + \sqrt{4}$

    •**SOLUTION**•   Think of the negative sign as -1 times the square root.

$$-\sqrt{9} + \sqrt{4} \quad \Leftarrow \text{ Expression to simplify.}$$
$$(\mathbf{-1})\sqrt{9} + \sqrt{4} \quad \Leftarrow \text{ Wrote negative sign as } -1.$$
$$-1(\mathbf{3}) + \mathbf{2} \quad \Leftarrow \text{ Simplified square roots.}$$
$$\mathbf{-3} + 2 \quad \Leftarrow \text{ Multiplied.}$$
$$-1 \quad \Leftarrow \text{ Added.}$$

c) $5 - 2\sqrt{16} - \sqrt{4} \cdot \sqrt{25}$

    •**SOLUTION**•

$$5 - (\mathbf{2})(\mathbf{4}) - (\mathbf{2})(\mathbf{5}) \quad \Leftarrow \text{ Simplified square roots.}$$
$$5 - \mathbf{8} - \mathbf{10} \quad \Leftarrow \text{ Multiplied.}$$
$$-13 \quad \Leftarrow \text{ Subtracted.}$$

d) $\left(3 + \sqrt{49}\right)\left(5 - \sqrt{100}\right)$

    •**SOLUTION**•

$$(3 + \mathbf{7})(5 - \mathbf{10}) \quad \Leftarrow \text{ Simplified square roots.}$$
$$(\mathbf{10})(\mathbf{-5}) \quad \Leftarrow \text{ Simplified inside parentheses.}$$
$$-50 \quad \Leftarrow \text{ Multiplied.}$$

Note that we could have used FOIL [†] here but it's easier to do the work inside the parentheses first and then multiply the results.

e) $\sqrt{9 + 16}$

    •**SOLUTION**•   Since the radical symbol is also a grouping symbol, we simplify the radicand first to get $\sqrt{25}$, and then calculate the square root to get 5.

[†]   FOIL is the shortcut method of multiplying binomials. FOIL means First, Outer, Inner, Last.

f) $2 + \sqrt{-25}$

•SOLUTION• Simplify the square root first. Because the radicand is negative, the square root is not defined using real numbers; therefore, we say that $2 + \sqrt{-25}$ is undefined.

Practice 9.1.4 Follow the proper order of operations to simplify.

1. $5 + 6\sqrt{49}$

2. $7 - 5\sqrt{25 - 9}$

3. $8 - 12\sqrt{-4}$

4. $-\sqrt{25} + \sqrt{121}$

5. $-\sqrt{16} - \sqrt{81}$

6. $8 - 4 \cdot \sqrt{16} - \sqrt{25} \cdot \sqrt{36}$

7. $12 - 2 \cdot \sqrt{9} + \sqrt{-49} \cdot \sqrt{4}$

8. $\left(5 + \sqrt{81}\right)\left(3 - \sqrt{121}\right)$

9. $\left(2 - \sqrt{16}\right)\left(4 - \sqrt{-64}\right)$

Topic 9.1 C — Applications of Square Roots

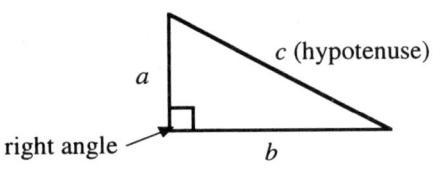

In Section 7.6 we noted that the relation between the lengths of the three sides of a right triangle is described in the Pythagorean Theorem, which can be written symbolically as

$$c^2 = a^2 + b^2$$

where a and b represent the lengths of the legs and c represents the length of the hypotenuse (the side opposite the right angle). Later in this chapter you will find that this formula may be written in the form $c = \sqrt{a^2 + b^2}$.

The Pythagorean Theorem and some other formulas that involve radicals are useful tools when trying to solve certain applied problems.

Example 9.1.5   a) A homeowner wants to place a rectangular fence around his garden. While it is easy for him to accurately measure distances with a tape measure, he has no precise way of measuring angles. If the garden is 60 feet long and 50 feet wide, how long should he make the diagonal to ensure that the sides of the fence are perpendicular?

•SOLUTION• A picture will help us better understand this situation:

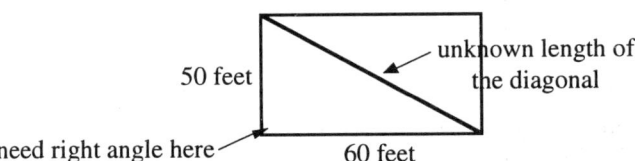

In order for the angle shown to be a right angle, the lengths of the sides of the triangle must obey the Pythagorean Theorem. We will calculate the length of the hypotenuse by substituting 50 for a and 60 for b.

$c = \sqrt{a^2 + b^2}$ ⬅ Pythagorean Theorem.

$c = \sqrt{50^2 + 60^2}$ ⬅ Substituted 50 for a and 60 for b.

$c = \sqrt{2500 + 3600}$ ⬅ Squared.

$c = \sqrt{6100}$ ⬅ Added

$c = 78.10249676$ ⬅ Used a calculator to simplify the square root.

He should adjust the angle between the sides of the fence until the length of the diagonal is about 78.1 feet (about 78 feet $1\frac{1}{4}$ inches).

b) Neglecting air resistance, the speed of a falling object is related to both the distance it has fallen and its acceleration due to gravity. A mathematical model for this relation is $s = \sqrt{2gd}$ where

s represents the speed of the object (feet per second) after it has fallen d feet.

g represents the acceleration due to gravity (about 32 feet/sec^2 on the earth's surface).

d represents the distance the object has fallen (feet).

Suppose a pole vaulter clears a 20 foot high bar. Find his approximate speed as he hits the ground at the end of the jump.

•**SOLUTION**• We substitute 32 for g and 20 for d and then simplify:

$$s = \sqrt{2gd} \qquad \Leftarrow \text{Formula to evaluate.}$$

$$s = \sqrt{2(32)(20)} \qquad \Leftarrow \text{Substituted 32 for } g \text{ and 20 for } d.$$

$$s = \sqrt{1280} \qquad \Leftarrow \text{Multiplied.}$$

$$s = 35.77708764 \qquad \Leftarrow \text{Used a calculator to write the square root as a decimal.}$$

The answer is about 36 feet per second. We can convert this to miles per hour by multiplying by the appropriate conversion factors and reducing:

$$\frac{36 \text{ ft}}{1 \text{ sec}} \cdot \frac{60 \text{ sec}}{1 \text{ min}} \cdot \frac{60 \text{ min}}{1 \text{ hour}} \cdot \frac{1 \text{ mi}}{5280 \text{ ft}}$$

This simplifies to about $\dfrac{25 \text{ mi}}{1 \text{ hour}}$ (rounded). Twenty–five miles per hour is pretty fast so you can see why pole vaulters fall onto thick pads.

Practice 9.1.5

Using the formula $c = \sqrt{a^2 + b^2}$, determine the value of c for the given values of a and b.

1. a is 9.1, b is 1.3; round to the tenths place.
2. a is 408, b is 1700; round to the tens place.

Using the formula $s = \sqrt{2gd}$, determine the value for s for the given values of d and g. Round your answer to the nearest whole number.

3. g is 32 and d is 36
4. g is 32 and d is 49
5. g is 9.8 and d is 15

Exercise Set 9.1   The answers to the odd numbered exercises are at the back of the book.

Try calculating each square root mentally by asking yourself *what non–negative number squared will result in the radicand?* Use your calculator to check your answer or to find any square roots you are unsure of.

1. $\sqrt{64}$
2. $\sqrt{36}$
3. $\sqrt{\dfrac{81}{121}}$
4. $\sqrt{\dfrac{25}{144}}$

5. $\sqrt{0.04}$
6. $\sqrt{0.16}$
7. $\sqrt{1.21}$
8. $\sqrt{1.44}$

Use a calculator to find the following square roots and then round to the place value indicated.

9. $\sqrt{15.4}$ round to the ones place.
10. $\sqrt{80.6}$ round to the tenths place.

11. $\sqrt{254.82}$ round to the tenths place.
12. $\sqrt{3087}$ round to the ones place.

13. $\sqrt{0.05}$ round to the hundredths place.
14. $\sqrt{1.08}$ round to the hundredths place.

Decide whether each of the following is a real number. If it is a real number, simplify the expression.

15. $\sqrt{-25}$ 16. $\sqrt{-49}$ 17. $-\sqrt{169}$ 18. $-\sqrt{144}$ 19. $-\sqrt{-4}$ 20. $-\sqrt{-81}$

Simplify.

21. $15 + 8\sqrt{64}$ 22. $22 + 4\sqrt{36}$ 23. $36 - 2\sqrt{49}$ 24. $40 - 5\sqrt{25}$

25. $-\sqrt{36} + \sqrt{64}$ 26. $-\sqrt{9} + \sqrt{25}$ 27. $3\sqrt{4} - 5\sqrt{16}$ 28. $7\sqrt{81} + 6\sqrt{4}$

29. $8 + 4 \cdot \sqrt{9} + \sqrt{36} \cdot \sqrt{49}$ 30. $10 + 8 \cdot \sqrt{16} + \sqrt{25} \cdot \sqrt{4}$ 31. $(12 - \sqrt{100})(4 + \sqrt{25})$ 32. $(25 - \sqrt{121})(6 + \sqrt{36})$

33. $(4 + 2\sqrt{4})(5 - 3\sqrt{25})$ 34. $(6 + 2\sqrt{36})(2 - 4\sqrt{81})$

Using the formula $c = \sqrt{a^2 + b^2}$, determine the value of c for the given values of a and b.

35. a is 15 and b is 18; round to the ones place.

36. a is 21 and b is 24; round to the ones place.

37. a is 0.0013 and b is 0.056; round to the thousandths place.

38. a is 0.054 and b is 0.009; round to the thousandths place.

39. a is 735 and b is 5010; round to the tens place.

40. a is 4500 and b is 620; round to the tens place.

41. a is 105.050 and b is 60.002; round to the hundredths place.

42. a is 81.07 and b is 313.917; round to the hundredths place.

Under certain conditions, when a car skids to a stop, the length of the skid marks is related to the original speed of the car by the formula $s = 2\sqrt{5d}$,

where

    s represents the speed of the car when the brakes were applied (in mph).

    d represents the length of the skid marks (in feet).

Use the formula to answer the following questions.

43. Calculate the speed of a car whose skid marks are 405 feet long.

44. Calculate the speed of a car whose skid marks are 45 feet long.

45. Calculate the speed of a car whose skid marks are 80 feet long. Round your answer to the nearest one mph.

46. Calculate the speed of a car whose skid marks are 117 feet long. Round your answer to the nearest one mph.

Review Exercises   The answers to all of these exercises are at the back of the book.

47. Prime factor: 117

48. Graph: $x = 6$

49. Simplify: $a(a^2 - ab + b^2) + b(a^2 - ab + b^2)$

50. Divide and simplify: $\dfrac{125h^{11} - 75h^9}{-5h^7}$

51. Use long division to simplify: $\dfrac{6x^3 - 1}{2x - 2}$

Simplify.

52. $\left(\dfrac{(2a)(3ab^2)}{30a^3}\right)^3$

53. $a^2b^3ab^{-2}$

54. $\left(\dfrac{t^3v}{w^{-1}}\right)\left(\dfrac{t^4}{w^3}\right)$

55. $\dfrac{\frac{x^4}{y^2}}{\frac{x^3}{y^5}}$

56. $\dfrac{2m^{-3}}{(2m)^{-3}} \cdot \dfrac{3m^{-2}}{(3m)^{-2}}$

57. $(3s + 2t)(3s - 2t)$

58. $\dfrac{18x^3}{9y^2} \cdot \dfrac{y^5}{2x}$

59. $1 - \dfrac{1}{5x} + \dfrac{3}{2x}$

Factor completely.

60. $45 - 14x + x^2$

61. $12x^2 - 30x + 12$

62. $x^2 + 6x + 9$

63. $10x + 10x^2 - 75x - 35$

Solve.

64. $6(x-1) = \frac{2}{3}(x-9)$ 65. $x^2 - 64 = 0$ 66. $\frac{2}{x-3} - \frac{3}{x+3} = \frac{12}{x^2-9}$ 67. $\left\{ \begin{array}{rrrr} 3x & - & 11y & = & -5 \\ 7x & + & 2y & = & 16 \end{array} \right\}$ ❶ ❷

68. How much would you have to pay back in total if you borrowed $17,000 at 6.5% simple interest for 3 years?

69. On a test a student got 48 right out of 65 questions. His friend, who had a different test of the same material, had 38 questions on his test. How many does the friend have to get right to have the same approximate percentage score as the first student?

70. In 1928 the women's Olympic gold metal distance for the discus was 39.4 meters. Since then the distance has been increasing about 0.51 meters per year. Write a linear equation which will output the current gold metal distance given the year since 1928.

Section 9.2 Simplifying Radical Expressions

Topic 9.2 A — Inverse Operations

The operations of *squaring* a number and *taking its square root* are called **inverse operations** because they "undo" each other. That is, if we begin with a non–negative number, square it, and then take the square root we get back the original number. For example, $\sqrt{7^2}$ simplifies to 7. We can summarize this property as follows:

— *Property* — *Square root of a perfect square*	English: The square root of the square of a number is the absolute value of that number. Example: $\sqrt{5^2}$ simplifies to $\vert 5 \vert$ which is 5. $\sqrt{(-5)^2}$ simplifies to $\vert -5 \vert$ which is 5. Algebra: $\sqrt{x^2} = \vert x \vert$

Notice that $\sqrt{x^2}$ simplifies to the absolute value of x rather than just x. This is because $\sqrt{x^2}$ is the *principle* (positive) square root of the number represented by x. Let's see how this works when x is 4 and when x is -4:

- when x is 4, $\sqrt{x^2} \Rightarrow \sqrt{4^2} \Rightarrow \sqrt{16} \Rightarrow 4$, which is $\vert 4 \vert$

- when x is -4, $\sqrt{x^2} \Rightarrow \sqrt{(-4)^2} \Rightarrow \sqrt{16} \Rightarrow 4$, which is $\vert -4 \vert$

In introductory algebra we usually avoid dealing with the complication of whether or not to use the absolute value by making the assumption that no radicals were formed by squaring negative quantities. Under that assumption, we may write $\sqrt{x^2} = x$.

The square of a square root is similar to the square root of a square.

— *Property* — *Square of a square root*	English: The square of the square root of a non–negative number is that number. Example: $\left(\sqrt{3} \right)^2 = 3$ Algebra: If $x \geq 0$, then $\left(\sqrt{x} \right)^2 = x$

Topic 9.2 B — Product and Quotient Properties of Square Roots

We mentioned in the last section that radicals are really exponents in disguise. A consequence of this is that radicals have many of the same properties as exponents. For example, we know from the Power of a Product Property of exponents that $(ab)^m = a^m b^m$. The corresponding property for square roots is as follows:

— *Property* — *Product property of* *square roots*	English: The square root of a product is the product of the square roots. Example: If $x \geq 0$, $\sqrt{2x}$ may be written as $\sqrt{2}\,\sqrt{x}$ Algebra: For $a \geq 0$ and $b \geq 0$, $\sqrt{ab} = \sqrt{a}\,\sqrt{b}$

We have a similar situation for division. The Power of a Quotient Property of exponents tells us that $\left(\dfrac{a}{b}\right)^m = \dfrac{a^m}{b^m}$. The corresponding property for square roots is as follows:

— *Property* — *Quotient property of* *square roots*	English: The square root of a quotient is the quotient of the square roots. Example: If $x > 0$, $\sqrt{\dfrac{2}{x}}$ may be written as $\dfrac{\sqrt{2}}{\sqrt{x}}$ Algebra: For $a \geq 0$ and $b > 0$, $\sqrt{\dfrac{a}{b}} = \dfrac{\sqrt{a}}{\sqrt{b}}$

Notice that the product and quotient properties of square roots require that the radicands be non–negative numbers. This is because if the radicand is negative the radical expression no longer represents a real number.

We can use these properties to simplify radicals. For example, suppose we want to simplify $\sqrt{49y}$. We can use the Product Property to break up this radical into $\sqrt{49}\,\sqrt{y}$. Since 49 is a perfect square, we can write this as $7\sqrt{y}$.

Example 9.2.1

a) Simplify $\sqrt{25x}$. Assume $x \geq 0$ (we make this assumption because we do not want to have a negative radicand).

• SOLUTION •

$$\sqrt{25}\,\sqrt{x} \qquad \Leftarrow \text{Used Product Property of square roots.}$$

$$5\sqrt{x} \qquad \Leftarrow \text{Simplified square root of perfect square.}$$

b) Rewrite as a product and simplify: $\sqrt{4x^2 y}$. Assume $x \geq 0$ and $y \geq 0$. (We assume $x \geq 0$ so we don't have to use the absolute value when simplifying $\sqrt{x^2}$. We assume $y \geq 0$ because we don't want to have a negative radicand.

• SOLUTION •

$$\sqrt{4}\,\sqrt{x^2}\,\sqrt{y} \qquad \Leftarrow \text{Used Product Property of square roots.}$$

$$2x\sqrt{y} \qquad \Leftarrow \text{Simplified square roots of perfect squares.}$$

c) Simplify $\sqrt{\dfrac{x}{49}}$. Assume $x \geq 0$.

• SOLUTION •

$$\dfrac{\sqrt{x}}{\sqrt{49}} \qquad \Leftarrow \text{Used Quotient Property of square roots.}$$

$$\dfrac{\sqrt{x}}{7} \qquad \Leftarrow \text{Simplified square root of perfect square.}$$

d) Simplify $\sqrt{\dfrac{9x}{y^2}}$. Assume $x \geq 0$ and $y > 0$.

•SOLUTION•

$$\dfrac{\sqrt{9}\ \sqrt{x}}{\sqrt{y^2}} \quad \Leftarrow \text{ Used Product and Quotient Properties of square roots.}$$

$$\dfrac{3\sqrt{x}}{y} \quad \Leftarrow \text{ Simplified square roots of perfect squares.}$$

e) $\sqrt{(8x)^2}$

•SOLUTION• We are taking the square root of a perfect square. We can write the answer using absolute value notation as $\left| 8x \right|$ or we could explicitly write out the two possible answers as follows:

$$\text{if } x \geq 0 \text{ then } \sqrt{(8x)^2} \text{ is } 8x$$

$$\text{if } x < 0 \text{ then } \sqrt{(8x)^2} \text{ is } -8x$$

Practice 9.2.1 Simplify. Assume that no radicands were formed by squaring negative quantities (therefore, you will not have to worry about using the absolute value in your answers).

1. $\sqrt{121w}$ 2. $\sqrt{81xy}$ 3. $\sqrt{\dfrac{16}{t^2}}$ 4. $\sqrt{\dfrac{25y^2t}{z^2}}$ 5. $\sqrt{(3a)^2}$

The product and quotient properties work in both directions. For example, we can multiply $\sqrt{3} \bullet \sqrt{x}$ to get $\sqrt{3x}$.

Example 9.2.2   a) Simplify: $\sqrt{5} \bullet \sqrt{5x}$. Assume $x \geq 0$.

•SOLUTION•

$$\sqrt{25x} \quad \Leftarrow \text{ Used Product Property to combine radicands under one radical symbol.}$$

$$\sqrt{25}\ \sqrt{x} \quad \Leftarrow \text{ Used Product Property to break up into two square roots where one radicand is a perfect square.}$$

$$5\sqrt{x} \quad \Leftarrow \text{ Simplified square root of perfect square.}$$

b) Simplify: $\dfrac{\sqrt{5x}\ \sqrt{2}}{\sqrt{x}}$. Assume $x > 0$.

•SOLUTION•

$$\sqrt{\dfrac{5 \bullet 2x}{x}} \quad \Leftarrow \text{ Used product and quotient properties to combine radicands under one square root symbol.}$$

$$\sqrt{10} \quad \Leftarrow \text{ Reduced common factor of } x \text{ and multiplied.}$$

Practice 9.2.2 Simplify. Assume all variables represent positive real numbers.

1. $\sqrt{2} \bullet \sqrt{2d}$ 2. $\sqrt{18} \bullet \sqrt{8b}$ 3. $\dfrac{\sqrt{5w}\ \sqrt{h}}{\sqrt{w}}$ 4. $\dfrac{\sqrt{48w}\ \sqrt{g}}{\sqrt{3w}}$

| *— Caution —*

Multiplication and
addition of square roots
have different properties | Note that the Product Property of square roots

$$\sqrt{a}\,\sqrt{b} = \sqrt{ab}$$

does not have a counterpart for addition. That is,

$$\sqrt{a} + \sqrt{b} \neq \sqrt{a+b}$$

Using an example, it's easy to show this:

$\sqrt{16} \quad + \quad \sqrt{9} \quad ?=? \quad \sqrt{16+9}$
$\quad 4 \quad + \quad 3 \quad\quad ?=? \quad \sqrt{25}$
$\quad\quad\quad\quad 7 \quad\quad\quad\quad \neq \quad\quad 5$ |

Topic 9.2 C — Simplifying Square Roots

Its easy to simplify square roots whose radicands are perfect squares, like $\sqrt{9}$. But, how do we simplify square roots such as $\sqrt{12}$? And, what does it mean to "simplify" such expressions? We could use a calculator to get an approximate decimal representation of $\sqrt{12}$ but that would not be considered "simplifying" it. Mathematicians have agreed that a simplified square root must meet three conditions:

| *— Definition —*

Simplest radical form | An expression containing a square root is said to be in simplest radical form when

Condition 1 No factors of the radicand are perfect squares.

Condition 2 The radicand contains no fractions.

Condition 3 No fraction has a square root in its denominator. |

The conditions for a simplified square root were set for practical reasons. Before calculators, expressions such as $\sqrt{12}$ were time consuming to calculate by hand. Mathematicians quickly learned that the larger the number under the radical symbol, the more cumbersome were the calculations; also, dividing by an irrational number was more difficult and less precise than dividing by an integer. So, mathematicians simplified radicals in order to make the calculations easier to do. Today, even though we have calculators to convert radicals into decimal notation, we still write square roots in simplest form for three reasons:

- many radicals are irrational numbers and so must be rounded off if written in decimal form. This is a problem if we need an exact answer.

- simplifying can make operations with radicals easier to do.

- simplifying ensures that everyone uses the same standard form when writing and comparing radicals. This makes mathematical communication easier.

Let's consider the first condition: No factors of the radicand are perfect squares. Recall that for whole numbers, the first few perfect squares are 0 (this is 0^2), 1 (this is 1^2), 4 (this is 2^2), 9 (this is 3^2), 16, 25, 36, 49, 64, 81, 100, 121, and 144 (this is 12^2). If we encounter a square root that has a perfect square as a factor of its radicand, we rewrite it so the perfect square is no longer part of the radicand. For example, let's simplify $\sqrt{12}$.

Step 1 Look at 12 and see if you can visualize it as the product of a perfect square and some other number. Think to yourself, *what is the largest perfect square that is a factor of 12?* The answer is 4. Therefore, we rewrite $\sqrt{12}$ as $\sqrt{4 \cdot 3}$.

Step 2 Use the Product Property for square roots to rewrite $\sqrt{4 \cdot 3}$ as $\sqrt{4}\sqrt{3}$.

Step 3 Simplify $\sqrt{4}$, which is 2 and write the simplified expression as $2\sqrt{3}$.

We say that the simplified version of $\sqrt{12}$ is $2\sqrt{3}$. Do $\sqrt{12}$ and $2\sqrt{3}$ represent the same number? We can check this using a calculator:

- Calculate $\sqrt{12}$ by tapping $\boxed{1}$ $\boxed{2}$ $\boxed{\sqrt{}}$. An eight digit calculator should display 3.4641016.

- Calculate $2\sqrt{3}$ by tapping $\boxed{2}$ $\boxed{\times}$ $\boxed{3}$ $\boxed{\sqrt{}}$ $\boxed{=}$. The calculator should again display 3.4641016.

Example 9.2.3 a) Simplify $\sqrt{45}$

•SOLUTION• In the list of perfect squares, find the largest perfect square that is a factor of 45. The answer is 9. So, we will rewrite 45 as the product 9•5 and then simplify:

$\sqrt{45}$ $\Leftarrow$ Expression to simplify.

$\sqrt{9 \cdot 5}$ $\Leftarrow$ Wrote radicand as a product where one factor is a perfect square.

$\sqrt{9}\sqrt{5}$ $\Leftarrow$ Used Product Property of square roots.

$3\sqrt{5}$ $\Leftarrow$ Simplified square root of a perfect square.

b) Simplify $\sqrt{\dfrac{27}{16}}$

•SOLUTION•

$\sqrt{\dfrac{9 \cdot 3}{16}}$ $\Leftarrow$ Wrote numerator of radicand as a product where one factor is a perfect square. The denominator is already a perfect square.

$\dfrac{\sqrt{9}\ \sqrt{3}}{\sqrt{16}}$ $\Leftarrow$ Used Product and Quotient Properties to break into three separate square roots.

$\dfrac{3\sqrt{3}}{4}$ $\Leftarrow$ Simplified square roots of perfect squares.

Practice 9.2.3 Simplify.

1. $\sqrt{8}$ 2. $\sqrt{75}$ 3. $\sqrt{128}$ 4. $\sqrt{\dfrac{24}{49}}$

If the number under the radical symbol is large, you may have difficulty "seeing" perfect square factors. For example, when you look at $\sqrt{98}$ you may not see that 98 has a factor of 49. In those cases, prime factoring can help. For example, to simplify $\sqrt{98}$, prime factor 98 to get 2•7•7. Each time two identical factors appear in the radicand we have a perfect square. In this example the two factors of 7 represent the perfect square 49. Now we can write the square root of the product as the product of square roots and simplify. The whole process looks like this:

$\sqrt{98}$ $\Leftarrow$ Expression to simplify.

$\sqrt{7^2 \cdot 2}$ $\Leftarrow$ Prime factored the radicand.

$\sqrt{7^2} \cdot \sqrt{2}$ $\Leftarrow$ Used Product Property of square roots.

$7\sqrt{2}$ $\Leftarrow$ Simplified square root of a perfect square.

Let's do another example. Suppose we want to simplify $\sqrt{4725}$. First prime factor 4725 to get 3•3•3•5•5•7. Each pair of identical factors represents a perfect square. So, we can write $\sqrt{4725}$ as $\sqrt{3^2 \cdot 3^1 \cdot 5^2 \cdot 7^1}$. Now, since the square root of a product is the product of the square roots we can break this up into several radicals and simplify as follows:

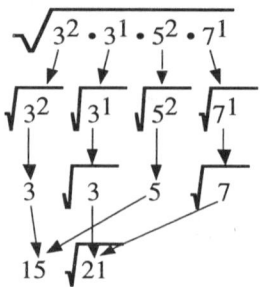

Therefore, $\sqrt{4725}$ simplifies to $15\sqrt{21}$.

If you look carefully at the previous examples you might notice a pattern between the prime factored version of the radicand and the simplified version. If we divide the exponent of each prime factor of the radicand by 2 then the quotient gives us the exponent of the factor which appears outside the radical symbol while the remainder, if there is one, gives us the exponent of the factor inside the radical symbol. This gives us a fairly quick procedure for simplifying square roots.

— Procedure — *Simplifying square roots using the factoring method*	**Step 1** Prime factor the radicand.
	Step 2 For factors with exponents of 2 or more, divide the exponent by 2.
	• the quotient is the power of the factor OUTSIDE the radical symbol.
	• the remainder is the power of the factor INSIDE the radical symbol. If the remainder is 0, the factor does not appear inside the radical symbol. [†]

Let's use this method to simplify $\sqrt{7875}$

Step 1   Prime factoring 7875 gives $3^2 \cdot 5^3 \cdot 7^1$.

Step 2

• The factor 3^2 has an exponent of 2. If we divide the exponent by 2 we get a quotient of 1, so one factor of 3 is outside the radical symbol. The remainder is 0, so no factor of 3 is inside the radical symbol. Now we have $\mathbf{3} \sqrt{5^3 \cdot 7^1}$

• The factor 5^3 has an exponent of 3. If we divide the exponent by 2 we have a quotient of 1, so one factor of 5 is outside the radical symbol. The remainder is 1, so one factor of 5 is inside the radical symbol. Now we have $3 \cdot \mathbf{5} \sqrt{5^1 \cdot 7^1}$.

• Since the factor of 7 has an exponent less than 2, the 7 stays inside the radical symbol. We have $3 \cdot 5 \sqrt{5^1 \cdot 7^1}$, which simplifies to $15\sqrt{35}$.

[†]   This is one of those "convenience" shortcuts like "invert and multiply" to divide fractions or "cross multiply" to solve a proportion. You should spend some time rethinking the two previous examples to see why this shortcut works.

Example 9.2.4   Simplify using the factoring method.

a) $\sqrt{162}$

•**SOLUTION**•

$\sqrt{2^1 \cdot 3^4}$ $\Leftarrow$ Prime factored the radicand.

$3^2\sqrt{2^1}$ $\Leftarrow$ The factor of 2 has an exponent of 1 so 2 stays inside.
The factor of 3 has an exponent of 4 and 4 divided by 2 gives

• a quotient of 2, so 2 is the power of 3 outside,

• a remainder of 0, so no factors of 3 are inside.

$9\sqrt{2}$ $\Leftarrow$ Simplified.

Of course, if you know your 81 times tables you could write $\sqrt{162}$ as $\sqrt{81 \cdot 2}$, which is $\sqrt{81}\sqrt{2}$, which is $9\sqrt{2}$.

b) $\sqrt{864}$

•**SOLUTION**•

$\sqrt{2^5 \cdot 3^3}$ $\Leftarrow$ Prime factored the radicand.

$2^2\sqrt{2^1 \cdot 3^3}$ $\Leftarrow$ The factor of 2 has an exponent of 5 and 5 divided by 2 gives
• a quotient of 2, so 2 is the power of 2 outside,
• a remainder of 1, so 1 is the power of 2 inside.

$2^2 \cdot 3^1\sqrt{2^1 \cdot 3^1}$ $\Leftarrow$ The factor of 3 has an exponent of 3 and 3 divided by 2 gives
• a quotient of 1, so 1 is the power of 3 outside,
• a remainder of 1, so 1 is the power of 3 inside.

$12\sqrt{6}$ $\Leftarrow$ Simplified.

Practice 9.2.4 Simplify using the factoring method.

1. $\sqrt{5000}$ 2. $\sqrt{432}$ 3. $\sqrt{392}$ 4. $\sqrt{968}$

Topic 9.2 D — Simplifying Square Roots that Contain Only Variables

Simplifying square roots that contain variables is often easier than simplifying those with only numbers because the variables usually are presented in factored form. In any case, we follow the same procedure as before. For example, let's simplify $\sqrt{x^3y^6}$.

• divide the exponent of x by 2. The quotient is 1, so 1 is the power of x outside the radical symbol; the remainder is 1, so 1 is the power of x inside the radical symbol. We have $x^1\sqrt{x^1y^6}$

• divide the exponent of y by 2. The quotient is 3, so 3 is the power of y outside the radical symbol; the remainder is 0, so no factors of y remain inside the radical symbol.

The simplified radical is $xy^3\sqrt{x}$.

Example 9.2.5   a) Simplify $\sqrt{y^{21}}$. Assume $y \geq 0$.

•**SOLUTION**• Divide 21 by 2 to get 10 with remainder 1. This gives us 10 as the power of y outside the radical symbol and 1 as the power inside. The simplified expression is $y^{10}\sqrt{y}$.

b) Simplify $\sqrt{a^5 b^3 c}$. Assume all variables represent positive real numbers.

•SOLUTION•

$a^2\sqrt{a^1 b^3 c^1}$ $\Leftarrow$ For a, divided 5 by 2 to get 2 with remainder 1. So 2 is the power of a outside and 1 is the power of a inside.

$a^2 b^1\sqrt{a^1 b^1 c^1}$ $\Leftarrow$ For b, divided 3 by 2 to get 1 with remainder 1. So 1 is the power of b outside and 1 is the power of b inside.

There is only one factor of c, so the c stays inside. The simplified expression is $a^2 b\sqrt{abc}$.

c) Simplify $\sqrt{m^{25}}$. Assume $m \geq 0$.

•SOLUTION• Divide 25 by 2 to get 12 with a remainder of 1. Therefore, 12 is the power of m outside the radical symbol and 1 is the power of m inside. We have $m^{12}\sqrt{m}$. Note that we DO NOT take the square root of 25, even though it is a perfect square; instead, we divide it by 2.

Practice 9.2.5 Simplify. Assume all variables represent positive real numbers.

1. $\sqrt{d^{36}}$ 2. $\sqrt{x^{13}}$ 3. $\sqrt{y^{81} m^{32}}$ 4. $\sqrt{x^5 y^{12} z^9}$ 5. $\sqrt{m^{49} n^4 r^{25}}$

Topic 9.2 E — Simplifying Square Roots that Contain Numbers and Variables

When a square root contains both numbers and variables, we follow the same procedures as outlined above.

Example 9.2.6 a) Simplify $\sqrt{54x^7}$. Assume $x \geq 0$.

•SOLUTION• Try doing this mentally first by asking yourself *what is the largest perfect square that is a factor of $54x^7$?* That is not so easy so you may have to use the factoring method.

$\sqrt{54x^7}$ $\Leftarrow$ Expression to simplify.

$\sqrt{2^1 \cdot 3^3 \cdot x^7}$ $\Leftarrow$ Factored radicand.

$3^1\sqrt{2^1 \cdot 3^1 \cdot x^7}$ $\Leftarrow$ Divided exponent of 3 by 2 to get 1 with remainder 1. So, 1 is the power of 3 outside and 1 is the power of 3 inside.

$3x^3\sqrt{2^1 \cdot 3^1 \cdot x^1}$ $\Leftarrow$ Divided exponent of x by 2 to get 3 with remainder 1. So, 3 is the power of x outside and 1 is the power of x inside.

$3x^3\sqrt{6x}$ $\Leftarrow$ Simplified.

b) Simplify $\sqrt{45x^6 y^{13}}$. Assume all variables represent positive real numbers.

•SOLUTION•

$\sqrt{3^2 \cdot 5^1 \cdot x^6 \cdot y^{13}}$ $\Leftarrow$ Factored radicand.

$3^1 \cdot x^3 \cdot y^6\sqrt{5^1 \cdot y^1}$ $\Leftarrow$ Divided exponent of each factor in radicand by 2. Quotient is power of factor outside radical symbol and remainder is power of factor inside radical symbol.

$3x^3 y^6\sqrt{5y}$ $\Leftarrow$ Simplified.

c) Simplify $\sqrt{16x^{16}y^9}$. Assume all variables represent positive real numbers.

•SOLUTION•

$\sqrt{2^4 \cdot x^{16} \cdot y^9}$ ⟸ Factored radicand.

$2^2 \cdot x^8 \cdot y^4\sqrt{y^1}$ ⟸ Divided exponent of each factor in radicand by 2. Quotient is power of factor outside radical symbol and remainder is power of factor inside radical symbol.

$4x^8y^4\sqrt{y}$ ⟸ Simplified.

— *Note* —	
What you do with a number depends on where it is	When we simplified $\sqrt{16x^{16}y^9}$, notice how the two 16s were treated differently: • we took the square root of the 16 that is the *coefficient* of x. • we divided by 2 the 16 that is the *exponent* of x.

d) Simplify $\sqrt{x^2 + 8x + 16}$.

•SOLUTION• We cannot individually take square roots of sums (we can only do that to products or quotients). Therefore, we first use factoring to write the radicand as a product.

$\sqrt{x^2 + 8x + 16}$ ⟸ Expression to simplify.

$\sqrt{(x+4)(x+4)}$ ⟸ Factored radicand.

$\sqrt{(x+4)^2}$ ⟸ Wrote factors using exponential notation.

$\left| x+4 \right|$ ⟸ Simplified square root of perfect square.

Note the use of absolute value in the last step. We have to do this in order to guarantee that the simplified expression represents a non–negative number since the original expression is the principal (positive) square root. We could get around this requirement by stating the following in the original problem: *Assume that no radicals were formed by squaring negative quantities.* Under that assumption, we could write the answer simply as $x + 4$.

Practice 9.2.6 Simplify. Assume that no radicals were formed by squaring negative quantities.

1. $\sqrt{48a^4b^6}$ 2. $\sqrt{75x^8y^{10}}$ 3. $\sqrt{128r^{15}s^{11}t}$

4. $\sqrt{44x^{17}y^{13}z^8}$ 5. $\sqrt{x^2 - 4x + 4}$ 6. $\sqrt{x^2 + 12x + 36}$

When an expression contains factors that are not under the radical symbol, we simplify the radical part first (remember, the order of operations says to simplify roots before doing multiplication or division).

Example 9.2.7 a) Simplify $2x\sqrt{9x^5}$. Assume $x \geq 0$.

•SOLUTION•

$2x \cdot 3\sqrt{x^5}$ ⟸ The square root of 9 is 3

$2x \cdot 3x^2\sqrt{x}$ ⟸ Divided exponent of x by 2. Quotient is power of factor outside radical symbol and remainder is power of factor inside.

$6x^3\sqrt{x}$ ⟸ Simplified by multiplying $2 \cdot 3$ to get 6 and $x \cdot x^2$ to get x^3.

b) Simplify $\dfrac{3a^2\sqrt{75a^3b^2}}{5b}$. Assume all variables represent positive real numbers.

• SOLUTION •

$$\dfrac{3a^2\sqrt{3 \cdot 5^2 a^3 b^2}}{5b} \quad \Leftarrow \text{ Factored radicand.}$$

$$\dfrac{3a^2 \cdot 5 \cdot a \cdot b\sqrt{3a^1}}{5b} \quad \Leftarrow \text{ Divided exponent of each factor in radicand by 2.}$$

Quotient is power of factor outside radical symbol and remainder is power of factor inside.

$$3a^3\sqrt{3a} \quad \Leftarrow \text{ Reduced common factors of 5 and } b \text{ and combined factors of } a.$$

Practice 9.2.7 Simplify. Assume all variables represent positive real numbers.

1. $7m\sqrt{49m^7}$

2. $2b\sqrt{48b^{25}}$

3. $\dfrac{6x^4\sqrt{72x^3y^5}}{8y}$

4. $\dfrac{2m\sqrt{45m^7n^4}}{3n^4}$

Exercise Set 9.2   The answers to the odd numbered exercises are at the back of the book.

Simplify. Assume that no radicals were formed by squaring negative quantities.

1. $\sqrt{49m}$

2. $\sqrt{64x}$

3. $\sqrt{\dfrac{s}{4}}$

4. $\sqrt{\dfrac{w}{25}}$

5. $\sqrt{\dfrac{10}{y^2}}$

6. $\sqrt{\dfrac{4d}{r^2}}$

7. $\sqrt{(5y)^2}$

8. $\sqrt{(mn)^2}$

9. $\sqrt{4x^2}$

10. $\sqrt{9y^2}$

11. $\sqrt{169x^4}$

12. $\sqrt{144w^4}$

13. $\sqrt{7} \cdot \sqrt{7t}$

14. $\sqrt{32} \cdot \sqrt{2a}$

15. $\dfrac{\sqrt{44}\sqrt{rw}}{\sqrt{11w}}$

16. $\dfrac{\sqrt{32}\sqrt{xy}}{\sqrt{8x}}$

17. $\sqrt{27}$

18. $\sqrt{48}$

19. $\sqrt{50}$

20. $\sqrt{20}$

21. $\sqrt{200}$

22. $\sqrt{162}$

23. $\sqrt{\dfrac{40}{9}}$

24. $\sqrt{\dfrac{90}{49}}$

25. $\sqrt{\dfrac{12}{25}}$

26. $\sqrt{\dfrac{72}{121}}$

27. $\sqrt{150}$

28. $\sqrt{245}$

29. $\sqrt{5832}$

30. $\sqrt{16875}$

31. $\sqrt{x^{15}}$

32. $\sqrt{t^{49}}$

33. $\sqrt{a^{17}}$

34. $\sqrt{g^{27}}$

35. $\sqrt{x^7y^9z^3}$

36. $\sqrt{q^4r^9w^3}$

37. $\sqrt{r^{42}s^{34}t^{16}}$

38. $\sqrt{a^7b^{38}c^{17}}$

39. $\sqrt{12y^5}$

40. $\sqrt{24x^9}$

41. $\sqrt{50a^9b^{15}}$

42. $\sqrt{25x^{25}y^4}$

43. $\sqrt{100w^{25}t^{16}}$

44. $\sqrt{25r^{36}v^{28}}$

45. $\sqrt{80r^2s^5t^3}$

46. $\sqrt{20x^5y^9z^7}$

47. $\sqrt{60x^{20}y^{12}z^{13}}$

48. $\sqrt{12a^{30}b^{19}c^{24}}$

49. $\sqrt{x^2 + 16x + 64}$

50. $\sqrt{x^2 + 2x + 1}$

51. $\sqrt{x^2 - 8x + 16}$

52. $\sqrt{x^2 - 10x + 25}$

53. $\sqrt{4x^2 + 20x + 25}$

54. $\sqrt{9y^2 + 12y + 4}$

55. $4x\sqrt{25x^{10}}$

56. $3y\sqrt{49y^{12}}$

57. $8s\sqrt{36s^{11}}$

58. $5a\sqrt{24a^{15}}$

59. $9wx\sqrt{144w^9x^4}$

60. $6st\sqrt{25s^{12}t^{13}}$

61. $4x^3y^4\sqrt{121x^{11}y^{16}}$ 62. $7m^6n^3\sqrt{200m^3n^7}$ 63. $\dfrac{10a^3\sqrt{27a^{36}b^{10}}}{15b}$ 64. $\dfrac{5t\sqrt{32s^3t^5}}{6s^2}$

65. $\dfrac{8x\sqrt{50x^6y^7}}{5y}$ 66. $\dfrac{9s\sqrt{18s^{11}t^9}}{4t}$

Review Exercises   The answers to all of these exercises are at the back of the book.

67. After playing a video cassette for 1 hour on long play speed (LP) you play the remainder of the tape on standard play (SP), which is twice the speed of long play, for an hour and a half. If a standard T120 VHS video cassette contains 246 meters of tape, what is the LP speed in meters per hour?

68. A train leaves Simm City at 10 am traveling east at 48 mph. One hour later, a train leaves Blue Earth, which is 244 miles east of Simm City, and travels west on a parallel track at 64 mph. At what time do the trains meet?

---

69. Find the slope and y–intercept of the line described by $2y = 6 - 4x$ 70. Factor: $81x^2 - 1$

71. Use the intercept method to graph $x - y = 3$ 72. Reduce $\dfrac{6x - 30}{30x^2 - 15x - 675}$

73. Graph the straight line that passes through $(-2,-8)$ with slope $\dfrac{1}{2}$.   74. Graph: $x + y > 3$

75. Solve: $\dfrac{5}{6}x - \dfrac{1}{2} \geq 2x - 4$ 76. Solve: $5x^2 - 45 = 0$ 77. Solve: $\dfrac{2}{x + 2} + \dfrac{1}{x} = \dfrac{5}{x + 2}$

78. Solve: $\left\{ \begin{array}{rcl} x + y & = & 3 \\ 5x & = & 15 - 5y \end{array} \right\}$ ❶ ❷   79. Solve: $\left\{ \begin{array}{rcl} 4x + 7y & = & -5 \\ 6x + 8y & = & 10 \end{array} \right\}$ ❶ ❷

Simplify.

80. $-(3x + 4) - (8 - 2x)$ 81. $\dfrac{18}{12} \div \dfrac{45}{30}$ 82. $-m^{-1}(-m^3)(-2m^{-2})$ 83. $\left(\dfrac{2a}{3b}\right)^2\left(\dfrac{a}{2b^{-1}}\right)^3\left(\dfrac{a^2}{b}\right)^{-1}$

84. $3.4 \times 10^4 + 5.2 \times 10^3$ 85. $(x - 15p)(x + 15p)$ 86. $\left(\dfrac{3}{5}m^3n + 2\right) + \left(\dfrac{12}{5}m^3n - mn + 2\right) + (6mn - 2)$

---

87. Factor completely: $12x^2 + 44x + 40$ 88. Factor completely: $48x^3 - 104x^2 + 48x$

---

89. In 1932 the winning Olympic time for the women's 100 meter backstroke was 79.4 seconds. Since then the time has been decreasing by approximately 0.33 seconds per year.

   a) Build a linear equation that will predict the women's 100 meter backstroke winning time given the year since 1932.
   b) Use your linear equation to predict the winning time in 2008.
   c) Use your linear equation to predict when the winning time will be less than 1 minute.

Section 9.3 Multiplying and Dividing Radical Expressions

Topic 9.3 A — Multiplying Monomial Radical Expressions

In the last section, we used the fact that $\sqrt{ab} = \sqrt{a}\sqrt{b}$ to simplify square roots (assuming $a \geq 0$ and $b \geq 0$). We can use this property in reverse to multiply square roots. That is $\sqrt{a}\sqrt{b} = \sqrt{ab}$ (assuming $a \geq 0$ and $b \geq 0$).

— *Procedure* —	**Step 1** Multiply the radicands.
Multiplying square roots	**Step 2** Simplify.

Example 9.3.1   a)   Simplify $\sqrt{6}\sqrt{12}$

•SOLUTION• There are two common ways of solving this problem. The first uses the idea of prime factoring:

$\sqrt{6 \cdot 12}$ ⇐ Used Product Property to combine under one radical symbol.

$\sqrt{2 \cdot 3 \cdot 2 \cdot 2 \cdot 3}$ ⇐ Factored radicand.

$\sqrt{2^3 \cdot 3^2}$ ⇐ Wrote factors using exponential notation.

$2^1 \cdot 3^1 \sqrt{2^1}$ ⇐ Divided each exponent by 2. Quotient is power outside radical symbol and remainder is power inside.

$6\sqrt{2}$ ⇐ Simplified.

The second way of solving this problem involves more mental arithmetic:

$\sqrt{6 \cdot 12}$ ⇐ Used Product Property to combine under one radical symbol.

$\sqrt{72}$ ⇐ Multiplied.

$\sqrt{36 \cdot 2}$ ⇐ Wrote the radicand as a product where one factor is the largest perfect square that is a factor of 72. This is where the mental arithmetic comes in.

$\sqrt{36}\sqrt{2}$ ⇐ Used Product Property to break radical into two parts.

$6\sqrt{2}$ ⇐ Simplified.

b)   Simplify $\sqrt{2x}\sqrt{10x^3}$. Assume $x \geq 0$.

•SOLUTION• Again, we can solve this using two common methods.

$\sqrt{2x \cdot 10x^3}$ ⇐ Used Product Property to combine under one radical symbol.

$\sqrt{2x \cdot 2 \cdot 5 \cdot x^3}$ ⇐ Factored radicand.

$\sqrt{2^2 \cdot 5^1 \cdot x^4}$ ⇐ Wrote factors using exponential notation.

$2^1 \cdot x^2 \sqrt{5^1}$ ⇐ Divided each exponent by 2. Quotient is power outside radical symbol and remainder is power inside.

$2x^2\sqrt{5}$ ⇐ Simplified.

The second way of solving this problem involves more mental arithmetic:

$\sqrt{2x \cdot 10x^3}$ ⇐ Used Product Property to combine under one radical symbol.

$\sqrt{20x^4}$ ⇐ Multiplied.

$\sqrt{4x^4 \cdot 5}$ ⇐ Wrote the radicand as a product where one factor is the largest perfect square that is a factor of $20x^4$.

$\sqrt{4x^4}\sqrt{5}$ ⇐ Used Product Property to break radical into two parts.

$2x^2\sqrt{5}$ ⇐ Simplified.

c) Simplify $\sqrt{\dfrac{12}{x^2y^3}}\,\sqrt{\dfrac{15x^4}{y}}$. Assume all variables represent positive real numbers.

•SOLUTION•

$\sqrt{\dfrac{12}{x^2y^3}\cdot\dfrac{15x^4}{y}}$ ⟸ Used Product Property to combine under one radical symbol.

$\sqrt{\dfrac{2^2\cdot3^1}{x^2y^3}\cdot\dfrac{3^1\cdot5^1x^4}{y}}$ ⟸ Factored radicand.

$\sqrt{\dfrac{2^2\cdot3^2\cdot5^1x^4}{x^2y^4}}$ ⟸ Wrote factors using exponential notation.

$\sqrt{\dfrac{2^2\cdot3^2\cdot5^1x^2}{y^4}}$ ⟸ Simplified radicand.

$\dfrac{2^1\cdot3^1\cdot x^1\sqrt{5^1}}{y^2}$ ⟸ Divided each exponent by 2. Quotient is power outside radical symbol and remainder is power inside.

$\dfrac{6x\sqrt{5}}{y^2}$ ⟸ Simplified.

Here is a second way of doing the problem:

$\sqrt{\dfrac{12}{x^2y^3}\cdot\dfrac{15x^4}{y}}$ ⟸ Used Product Property to combine under one radical symbol.

$\sqrt{\dfrac{180x^2}{y^4}}$ ⟸ Simplified radicand.

$\sqrt{\dfrac{36x^2}{y^4}\cdot5}$ ⟸ Wrote the radicand as a product where one factor is the largest perfect square that is a factor of $\dfrac{180x^2}{y^4}$.

$\sqrt{\dfrac{36x^2}{y^4}}\cdot\sqrt{5}$ ⟸ Used Product Property to break radical into two parts.

$\dfrac{6x}{y^2}\sqrt{5}$ ⟸ Simplified.

Practice 9.3.1   Simplify. Assume all variables represent positive real numbers.

1. $\sqrt{32}\sqrt{12}$ 2. $\sqrt{75}\sqrt{27}$ 3. $\sqrt{\dfrac{15x}{7y}}\sqrt{\dfrac{35x}{y^3}}$ 4. $\sqrt{8x^2y^3}\sqrt{12x^4y^5}$ 5. $\sqrt{15m^{11}n^7}\sqrt{75m^5n^{12}}$

Topic 9.3 B — Rationalizing Denominators

In the previous section, we stated that an expression with a square root is in simplest form when:

Condition 1   No factors of the radicand are perfect squares.

Condition 2   The radicand contains no fractions.

Condition 3   No fraction has a square root in its denominator.

We used the Product Property, $\sqrt{ab} = \sqrt{a}\sqrt{b}$ and the Quotient Property $\sqrt{\dfrac{a}{b}} = \dfrac{\sqrt{a}}{\sqrt{b}}$, to simplify square roots so that they met the first two conditions. Now, we will discuss the third condition.

In the previous examples the radicand in the denominator turned out to be a perfect square and so the radical in the denominator simplified to a rational number. This is not always the case. For example, suppose we wish to simplify $\dfrac{1}{\sqrt{2}}$. Our task is to write $\dfrac{1}{\sqrt{2}}$ as an equivalent fraction, but without a radical in the denominator. We ask ourselves, *what number can we multiply $\sqrt{2}$ by to make it a rational number?* Multiplying $\sqrt{2}$ by itself (that is, squaring it) will transform it into a rational number. Therefore, we multiply the given fraction by $\dfrac{\sqrt{2}}{\sqrt{2}}$ as shown below:

$$\dfrac{1}{\sqrt{2}} \cdot \dfrac{\sqrt{2}}{\sqrt{2}} \qquad \Leftarrow \text{ Multiplied by 1 in the form } \dfrac{\sqrt{2}}{\sqrt{2}}.$$

$$\dfrac{\sqrt{2}}{\sqrt{2^2}} \qquad \Leftarrow \text{ Multiplied.}$$

$$\dfrac{\sqrt{2}}{2} \qquad \Leftarrow \text{ Simplified denominator.}$$

This process of converting the denominator from an irrational number, such as $\sqrt{2}$, into a rational number, such as 2, is called **rationalizing the denominator**. We can verify that the original and final expressions represent the same number by using a calculator to convert each fraction into an approximate decimal:

$\dfrac{1}{\sqrt{2}} \quad \Leftarrow$ Expression to simplify.	$\dfrac{\sqrt{2}}{2} \quad \Leftarrow$ Expression to simplify.
$\dfrac{1}{1.4142136} \quad \Leftarrow$ Converted square root to a decimal and rounded.	$\dfrac{1.4142136}{2} \quad \Leftarrow$ Converted square root to a decimal and rounded.
$0.7071068 \quad \Leftarrow$ Divided.	$0.7071067 \quad \Leftarrow$ Divided.

The two values are within one–ten millionth of each other. The slight difference is due to the fact that the calculator rounded off $\sqrt{2}$ to eight decimal places in order to display it.

— *Procedure* — *Rationalizing the denominator*	*Step 1* Multiply numerator and denominator by a radical that will make the radicand in the denominator a perfect square.
	Step 2 Simplify.

Example 9.3.2   a)   Simplify $\dfrac{5}{\sqrt{6}}$

       •SOLUTION•

$$\dfrac{5}{\sqrt{6}} \cdot \dfrac{\sqrt{6}}{\sqrt{6}} \quad \Leftarrow \text{Multiplied by 1 in the form } \dfrac{\sqrt{6}}{\sqrt{6}}. \text{ That is, multiplied numerator}$$

           and denominator by the radical in the denominator.

$$\dfrac{5\sqrt{6}}{\sqrt{6^2}} \quad \Leftarrow \text{Wrote denominator as a perfect square.}$$

$$\dfrac{5\sqrt{6}}{6} \quad \Leftarrow \text{Simplified.}$$

     b)   Simplify $\dfrac{\sqrt{5}}{\sqrt{3x}}$. Assume $x > 0$.

       •SOLUTION•

$$\dfrac{\sqrt{5}}{\sqrt{3x}} \cdot \dfrac{\sqrt{3x}}{\sqrt{3x}} \quad \Leftarrow \text{Multiplied by 1 in the form } \dfrac{\sqrt{3x}}{\sqrt{3x}}. \text{ That is, multiplied numerator}$$

           and denominator by the square root in the denominator.

$$\dfrac{\sqrt{5 \cdot 3x}}{\sqrt{(3x)^2}} \quad \Leftarrow \text{Wrote denominator as a perfect square.}$$

$$\dfrac{\sqrt{15x}}{3x} \quad \Leftarrow \text{Simplified.}$$

     c)   Simplify $\dfrac{2x^2}{\sqrt{6x}}$. Assume $x > 0$.

       •SOLUTION•

$$\dfrac{2x^2}{\sqrt{6x}} \cdot \dfrac{\sqrt{6x}}{\sqrt{6x}} \quad \Leftarrow \text{Multiplied by 1 in the form } \dfrac{\sqrt{6x}}{\sqrt{6x}}. \text{ That is, multiplied numerator}$$

           and denominator by the square root in the denominator.

$$\dfrac{2x^2\sqrt{6x}}{\sqrt{(6x)^2}} \quad \Leftarrow \text{Wrote denominator as a perfect square.}$$

$$\dfrac{2x^2\sqrt{6x}}{6x} \quad \Leftarrow \text{Simplified denominator.}$$

$$\dfrac{x\sqrt{6x}}{3} \quad \Leftarrow \text{Reduced fraction by common factor of } 2x.$$

Practice 9.3.2   Simplify. Assume all variables represent positive real numbers.

1. $\dfrac{3}{\sqrt{11}}$ 2. $\dfrac{7}{\sqrt{12}}$ 3. $\dfrac{10}{\sqrt{5}}$ 4. $\dfrac{\sqrt{11}}{\sqrt{7a}}$ 5. $\dfrac{4m^2}{\sqrt{6m}}$ 6. $\dfrac{12y^2}{\sqrt{8y}}$

Exercise Set 9.3   The answers to the odd numbered exercises are at the back of the book.

Simplify. Assume all variables represent positive real numbers.

1. $\sqrt{48}\sqrt{36}$
2. $\sqrt{75}\sqrt{15}$
3. $\sqrt{10}\sqrt{5}$
4. $\sqrt{50}\sqrt{8}$

5. $\sqrt{20x}\sqrt{10x^9}$
6. $\sqrt{8b}\sqrt{24b^7}$
7. $\sqrt{\dfrac{8t}{q^3}}\sqrt{\dfrac{24t^4}{q}}$
8. $\sqrt{\dfrac{18x^3}{y}}\sqrt{\dfrac{27x^6}{y^5}}$

9. $\sqrt{12m}\sqrt{45m^{11}}$
10. $\sqrt{2b}\sqrt{8b^9}$
11. $\sqrt{54x^2y^3}\sqrt{27x^{10}y^{12}}$
12. $\sqrt{24s^9t^3}\sqrt{15s^{14}t^{15}}$

13. $\sqrt{54x^3y^5}\sqrt{12x^3y^4}$
14. $\sqrt{42a^{13}b^{12}}\sqrt{18a^7b^5}$
15. $\dfrac{4}{\sqrt{6}}$
16. $\dfrac{15}{\sqrt{10}}$

17. $\dfrac{6}{\sqrt{13}}$
18. $\dfrac{5}{\sqrt{8}}$
19. $\dfrac{8}{\sqrt{2}}$
20. $\dfrac{15}{\sqrt{3}}$

21. $\dfrac{\sqrt{7}}{\sqrt{6m}}$
22. $\dfrac{\sqrt{13}}{\sqrt{3x}}$
23. $\dfrac{\sqrt{3}}{\sqrt{5t}}$
24. $\dfrac{\sqrt{6}}{\sqrt{11m}}$

25. $\dfrac{5x^3}{\sqrt{10x}}$
26. $\dfrac{15x^2}{\sqrt{5x}}$
27. $\dfrac{7d^2}{\sqrt{11d}}$
28. $\dfrac{3w^2}{\sqrt{6w}}$

Review Exercises   The answers to all of these exercises are at the back of the book.

29. Evaluate $3(x - xy) - y^2$ when x is -2 and y is -3.

30. Solve: $\begin{cases} 9x + y = 4 \\ 3x - 13 = 2y \end{cases}$ ❶ ❷

---

31. A freight train left town at 8:15 am and traveled at 40 mph. Three–fourths of an hour later a passenger train left the same town on the same track traveling at 45 mph. Will the passenger train have caught up to the freight train by 12:15 pm?

32. Brass is a mixture of copper and zinc. If a 40 gram brass bar, which is 40% zinc and 60% copper, is melted and combined with a second brass bar that is 35 grams the resulting mixture is 53% copper. What percent copper was the second bar?

33. The number of diagonals of a polygon that has n sides is given by the expression $\dfrac{1}{2}n(n - 3)$. Find the number of sides of a polygon that has 14 diagonals.

---

34. Given $f(x) = 2x^2 - 3x + 9$, find $f(-1)$.

35. Use long division to simplify: $\dfrac{3w^4 + 2w^3 - 7w^2 + 6w + 5}{w + 2}$

Simplify.

36. $(10z - 11y)(10z + 11y)$
37. $-p\left(\dfrac{5t}{p}\right)^2\left(\dfrac{2p}{5t}\right)$
38. $(3xy)^3(-2xy)^{-2}$
39. $(4d - 5)(3d + 7)$

40. $\left(\dfrac{-4m^2}{n^{-2}}\right)\left(\dfrac{-4m^{-2}}{n^3}\right)^{-2}\left(\dfrac{m^{-4}}{n}\right)$
41. $\left(\dfrac{b^{-2}}{-2b}\right)^{-3}$
42. $\dfrac{\dfrac{3}{m} - \dfrac{1}{5}}{\dfrac{4}{m}}$
43. $\sqrt{54m^{11}n^3}$

Factor completely.

44. $8x^2 - 20x - 12$
45. $9x^2 - 16$
46. $2x^3 + 32x^2 + 6x^3 - 40x$

Solve.

47. $x^2 - 3x - 10 = 0$
48. $12x^2 + 15x = -3$
49. $\dfrac{27}{x^2 - 5x - 14} + \dfrac{2}{x + 2} = \dfrac{3}{x - 7}$

---

50. The equation $s(t) = 0.3t + 6.7$ predicts the millions of students enrolled in public elementary and secondary schools in the United States between 1869 and 1993. S represents the number of students and t represents the year since 1869. Express in English the meaning of each of the following:

    a) $s(80) = 30.7$
    b) $s(100)$
    c) $s(t) = 40$
    d) $s(t) = 4s(0)$.

Section 9.4 Adding and Subtracting Radical Expressions

Topic 9.4 A — Like Radical Terms

We can only add *like* terms; that is, terms which have identical variable factors. For example, we can add $2x + 3x$ to get $5x$ but we cannot add $2x + 3y$. Adding terms that contain square roots works the same way. We can only add radical terms that are *like* each other. We define **like radical terms** as terms that have the same variable factors *and* the same radical factors.

- $\sqrt{7}$ and $5\sqrt{7}$ are like radical terms because they contain identical variable factors, none, and identical radical factors, $\sqrt{7}$.

- $2x\sqrt{3}$ and $\frac{3}{4}x\sqrt{3}$ are like radical terms because they contain identical variable factors, x, and identical radical factors, $\sqrt{3}$.

- $\frac{\sqrt{5x}}{3}$ and $2\sqrt{5x}$ are like radical terms because they contain identical variable factors, $\sqrt{x}$, and identical radical factors, $\sqrt{5x}$.

We can combine like radical terms in the same way we combined like terms.

— *Procedure* — *Adding (or subtracting)* *like radical terms*	*Step 1* If necessary, simplify each term. *Step 2* Add (or subtract) the numeric coefficients of the like radical terms.

For example, $2\sqrt{3} + 5\sqrt{3}$ simplifies to $7\sqrt{3}$. We can use the Distributive Property to show that this is true:

$$2\sqrt{3} + 5\sqrt{3} \quad \Leftarrow \text{ Expression to simplify.}$$

$$(2 + 5)\sqrt{3} \quad \Leftarrow \text{ Factored } \sqrt{3} \text{ from each term.}$$

$$7\sqrt{3} \quad \Leftarrow \text{ Added numbers inside parentheses.}$$

Example 9.4.1   a) Simplify $3\sqrt{5} + \sqrt{5}$

•**SOLUTION**• The coefficient of $\sqrt{5}$ in the second term is 1. So, adding the coefficients, $3 + 1$, we get $4\sqrt{5}$.

b) Simplify $\sqrt{5y} + \frac{6\sqrt{5y}}{2}$. Assume $y \geq 0$.

•**SOLUTION**•

- The coefficient of $\sqrt{5y}$ is 1.

- We simplify $\frac{6\sqrt{5y}}{2}$ by reducing $\frac{6}{2}$ to get $3\sqrt{5y}$; therefore, its coefficient is 3.

$$\sqrt{5y} + \frac{6\sqrt{5y}}{2} \quad \Leftarrow \text{ Expression to simplify.}$$

$$1\sqrt{5y} + 3\sqrt{5y} \quad \Leftarrow \text{ Simplified each term and wrote coefficients.}$$

$$4\sqrt{5y} \quad \Leftarrow \text{ Added coefficients of like radical terms.}$$

c) Simplify $\dfrac{3\sqrt{2x}}{4} + \dfrac{\sqrt{2x}}{2}$. Assume $x \geq 0$.

•SOLUTION• The coefficient of the first radical is $\dfrac{3}{4}$ and the coefficient of the second

radical is $\dfrac{1}{2}$. To add the terms we first must convert each denominator to the LCD, which is 4.

$$\dfrac{3\sqrt{2x}}{4} + \dfrac{\sqrt{2x}}{2} \quad \Leftarrow \text{Expression to simplify.}$$

$$\dfrac{3\sqrt{2x}}{4} + \dfrac{\mathbf{2}}{\mathbf{2}} \cdot \dfrac{\sqrt{2x}}{2} \quad \Leftarrow \text{LCD is 4, so multiplied numerator and}$$
denominator of second fraction by 2 to convert the fraction into an equivalent fraction with the LCD.

$$\dfrac{5\sqrt{2x}}{4} \quad \Leftarrow \text{Added numerators.}$$

d) Simplify $2\sqrt{7} + 7\sqrt{2}$

•SOLUTION• The radical parts are different so these are not like radical terms. This expression cannot be simplified.

e) Simplify $\sqrt{3}\,x + \sqrt{2}\,x$. Assume $x \geq 0$.

•SOLUTION• These are not like *radical* terms because the radical parts are different.

Practice 9.4.1 Simplify. Assume all variables represent positive real numbers.

1. $12\sqrt{3} + 8\sqrt{3}$

2. $5\sqrt{3} + 2\sqrt{5} + 6\sqrt{3} - \sqrt{5}$

3. $12\sqrt{7} + 4\sqrt{11} - 4\sqrt{7} + 6\sqrt{11}$

4. $22\sqrt{5y} - 8y\sqrt{5y}$

5. $\sqrt{3t} - 8\sqrt{3t}$

6. $12\sqrt{6y} + 8\sqrt{6y} - 7\sqrt{6y}$

7. $\sqrt{7m} + \dfrac{15\sqrt{7m}}{5}$

8. $3\sqrt{8s} - \dfrac{8\sqrt{3s}}{6}$

Topic 9.4 B — Simplifying and Adding Radical Terms

It's important to simplify square roots before attempting to add or subtract them. For example, $3\sqrt{2}$ and $\sqrt{8}$ don't appear to be like radical terms, but if we simplify $\sqrt{8}$ we see that they are.

$$3\sqrt{2} + \sqrt{8} \quad \Leftarrow \text{Expression to simplify.}$$

$$3\sqrt{2} + \sqrt{2^3} \quad \Leftarrow \text{Factored radicand.}$$

$$3\sqrt{2} + 2\sqrt{2} \quad \Leftarrow \text{Simplified } \sqrt{2^3}.$$

$$5\sqrt{2} \quad \Leftarrow \text{Added coefficients of like radical terms.}$$

Example 9.4.2 a) Simplify $8\sqrt{5} + \sqrt{20}$

•SOLUTION•

$$8\sqrt{5} + \sqrt{2^2 \cdot 5^1} \quad \Leftarrow \text{Factored radicand.}$$

$$8\sqrt{5} + 2\sqrt{5} \quad \Leftarrow \text{Simplified } \sqrt{2^2 \cdot 5^1}.$$

$$10\sqrt{5} \quad \Leftarrow \text{Added coefficients of like radical terms.}$$

b) Simplify $7x\sqrt{18x} + 2\sqrt{50x^3}$. Assume $x \geq 0$.

> • **SOLUTION** •

$$7x\sqrt{2 \cdot 3^2 x} + 2\sqrt{2 \cdot 5^2 x^3} \quad \Leftarrow \text{ Factored radicands.}$$

$$7x \cdot 3\sqrt{2x} + 2 \cdot 5x\sqrt{2x} \quad \Leftarrow \text{ Simplified square roots.}$$

$$21x\sqrt{2x} + 10x\sqrt{2x} \quad \Leftarrow \text{ Multiplied.}$$

$$31x\sqrt{2x} \quad \Leftarrow \text{ Added coefficients of like radical terms.}$$

c) Simplify $5\sqrt{4x} - \sqrt{x} + \sqrt{16x}$. Assume $x \geq 0$.

> • **SOLUTION** •

$$5\sqrt{2^2 x} - \sqrt{x} + \sqrt{4^2 x} \quad \Leftarrow \text{ Factored radicands.}$$

$$5 \cdot 2\sqrt{x} - \sqrt{x} + 4\sqrt{x} \quad \Leftarrow \text{ Simplified square roots.}$$

$$10\sqrt{x} - 1\sqrt{x} + 4\sqrt{x} \quad \Leftarrow \text{ Multiplied.}$$

$$13\sqrt{x} \quad \Leftarrow \text{ Added coefficients of like radical terms.}$$

Practice 9.4.2 Simplify. Assume all variables represent positive real numbers.

1. $7\sqrt{75} + \sqrt{12}$

2. $8\sqrt{50} - \sqrt{8}$

3. $4\sqrt{12} + \sqrt{50} - \sqrt{75} + 2\sqrt{8}$

4. $3\sqrt{27} + \sqrt{32} - \sqrt{12} + 4\sqrt{50}$

5. $4m\sqrt{50m} + 6\sqrt{8m^3}$

6. $2\sqrt{49t} + \sqrt{64t} - \sqrt{t}$

Topic 9.4 C — More Complicated Radical Terms

The expressions become more complicated if more operations are included or if there are square roots in the denominators. Below is a procedure that might be helpful:

— Procedure — *Simplifying radical* *expressions*	***Step 1 Simplify inside*** Simplify under the radical symbol.
	Step 2 Multiply Combine radicals that are multiplied.
	Step 3 Simplify radicals Simplify each radical so the radicands contain no perfect squares and combine any like radical terms.
	Step 4 Rationalize Rationalize fractions that contain radicals in their denominators.
	Step 5 Combine Combine like radical terms and simplify.

Example 9.4.3   a) Simplify $\sqrt{6}\left(\sqrt{30} + 5\right) + \sqrt{5}$

> • **SOLUTION** • If the radical expression contains factors that are not monomials we use the Distributive Property to simplify, just as we did with non–radical expressions.

Step 1 Simplify inside There is nothing inside the radical symbols that can be simplified.

Step 2 Multiply

$$\left(\sqrt{6}\right)\left(\sqrt{30}\right) + \left(\sqrt{6}\right)(5) + \sqrt{5} \quad \Leftarrow \text{ Used Distributive Property.}$$

$$\sqrt{6 \cdot 30} + 5\sqrt{6} + \sqrt{5} \quad \Leftarrow \text{ Used Product Property.}$$

Step 3 Simplify radicals

$$\sqrt{2^2 \cdot 3^2 \cdot 5} + 5\sqrt{6} + \sqrt{5} \quad \Leftarrow \text{ Factored radicands.}$$

$$2 \cdot 3\sqrt{5} + 5\sqrt{6} + \sqrt{5} \quad \Leftarrow \text{ Simplified radical.}$$

$$7\sqrt{5} + 5\sqrt{6} \quad \Leftarrow \text{ Combined like radical terms.}$$

Step 4 Rationalize There are no fractions to rationalize.

Step 5 Combine There are no further like radical terms.

b) Simplify $5\sqrt{x}\left(\sqrt{x} - 2\right) - 3\sqrt{4x}$. Assume $x \geq 0$.

•SOLUTION•

Step 1 Simplify inside There is nothing inside the radical symbols that can be simplified.

Step 2 Multiply

$$5\sqrt{x}\left(\sqrt{x}\right) - \left(5\sqrt{x}\right)(2) - 3\sqrt{4x} \quad \Leftarrow \text{ Used Distributive Property to remove parentheses.}$$

$$5\sqrt{x^2} - 10\sqrt{x} - 3\sqrt{4x} \quad \Leftarrow \text{ Used Product Property.}$$

Step 3 Simplify radicals

$$5x - 10\sqrt{x} - 3 \cdot 2\sqrt{x} \quad \Leftarrow \text{ Simplified radicals.}$$

$$5x - 16\sqrt{x} \quad \Leftarrow \text{ Combined like radical terms.}$$

Step 4 Rationalize There are no fractions to rationalize.

Step 5 Combine We cannot subtract because these are not like radical terms.

c) Simplify $\left(\sqrt{3w} + 2\right)^2 + 5\sqrt{3w}$. Assume $w \geq 0$.

•SOLUTION• Be careful! We cannot simply square each term inside the parentheses. We must multiply the binomial by itself $\left(\sqrt{3w} + 2\right)\left(\sqrt{3w} + 2\right)$ using a method such as FOIL.

Step 1 Simplify inside There is nothing inside the radical symbols that can be simplified.

Step 2 Multiply

$$\left(\sqrt{3w} + 2\right) \cdot \left(\sqrt{3w} + 2\right) + 5\sqrt{3w} \quad \Leftarrow \text{ Rewrote using definition of exponent of 2. Notice that we cannot simply square each term in } \left(\sqrt{3w} + 2\right); \text{ we must treat this as a binomial.}$$

$$\left(\sqrt{3w}\right)\left(\sqrt{3w}\right) + 2\left(\sqrt{3w}\right) + 2\left(\sqrt{3w}\right) + (2)(2) + 5\sqrt{3w} \quad \Leftarrow \text{ Used FOIL to remove parentheses.}$$

Step 3 Simplify radicals

$$\sqrt{(3w)^2} + 2\left(\sqrt{3w}\right) + 2\left(\sqrt{3w}\right) + 4 + 5\sqrt{3w} \quad \Leftarrow \text{ Multiplied.}$$

$$3w + 4\sqrt{3w} + 4 + 5\sqrt{3w} \quad \Leftarrow \text{ Simplified.}$$

$$3w + 9\sqrt{3w} + 4 \quad \Leftarrow \text{ Combined like radical terms.}$$

Step 4 Rationalize There are no fractions to rationalize.

Step 5 Combine All like radical terms have been combined.

d) Simplify $3\sqrt{2} + \dfrac{1}{\sqrt{2}}$

•SOLUTION•

Step 1 Simplify inside There is nothing inside the radical symbols that can be simplified.

Step 2 Multiply There is nothing to multiply.

Step 3 Simplify radicals There is nothing to simplify.

Step 4 Rationalize

$$3\sqrt{2} + \frac{1}{\sqrt{2}} \cdot \frac{\sqrt{2}}{\sqrt{2}} \quad \Leftarrow \text{ Rationalized second radical.}$$

$$3\sqrt{2} + \frac{\sqrt{2}}{2} \quad \Leftarrow \text{ Simplified.}$$

Step 5 Combine

$$\frac{6}{2}\sqrt{2} + \frac{1}{2}\sqrt{2} \quad \Leftarrow \text{ Wrote coefficient of each term using LCD of 2.}$$

$$\frac{7}{2}\sqrt{2} \quad \Leftarrow \text{ Added coefficients of like radical terms.}$$

We also could write this as $\dfrac{7\sqrt{2}}{2}$.

Practice 9.4.3 Simplify. Assume all variables represent positive real numbers.

1. $\sqrt{5}\left(\sqrt{20} + 2\right) + \sqrt{45}$ 2. $7\sqrt{y}\left(\sqrt{y} + 4\right) - 6\sqrt{y}$ 3. $3\sqrt{t}\left(\sqrt{t} + 3\right) - 4\sqrt{t}$ 4. $\left(\sqrt{3x} + 3\right)^2 + 2\sqrt{3x}$

5. $5\sqrt{3} + \dfrac{1}{\sqrt{3}}$ 6. $\dfrac{2}{\sqrt{5}} - \sqrt{5}$ 7. $\dfrac{1}{\sqrt{3}} + \dfrac{3}{\sqrt{7}} - \sqrt{7}$

Exercise Set 9.4 The answers to the odd numbered exercises are at the back of the book.

Simplify. Assume all variables represent positive real numbers.

1. $6\sqrt{7} + 15\sqrt{7}$
2. $9\sqrt{11} - 6\sqrt{11}$
3. $5\sqrt{3} + 2\sqrt{3} - 8\sqrt{3}$

4. $6\sqrt{5} + 7\sqrt{5} - 11\sqrt{5}$
5. $8\sqrt{5s} + 9\sqrt{5s} - 3\sqrt{5s}$
6. $5\sqrt{3} + 3\sqrt{5}$

7. $3\sqrt{5} + 2\sqrt{2x} + 7\sqrt{5} - 5\sqrt{2x}$
8. $4\sqrt{2} + 7\sqrt{2w} + 5\sqrt{2} + 6\sqrt{2w}$
9. $\sqrt{6a} - \dfrac{4\sqrt{6a}}{2}$

10. $\sqrt{7w} - \dfrac{15\sqrt{7w}}{5}$
11. $11\sqrt{5c} - \dfrac{12\sqrt{5c}}{4}$
12. $\sqrt{x} + \dfrac{12\sqrt{4x}}{4}$

13. $3\sqrt{5n} + \dfrac{18\sqrt{5n}}{3}$
14. $2\sqrt{6c} + \dfrac{8\sqrt{6c}}{5}$
15. $\sqrt{2}\,t - \sqrt{2}\,w$

16. $\sqrt{3x} - \sqrt{3y}$
17. $\dfrac{5}{2}\sqrt{2s} - \sqrt{2s}$
18. $\dfrac{\sqrt{5d}}{6} - \dfrac{2}{3}\sqrt{5d}$

19. $7\sqrt{8} + \sqrt{32}$
20. $6\sqrt{50} + \sqrt{98}$
21. $4\sqrt{8} - 6\sqrt{50}$

22. $3\sqrt{75} - 5\sqrt{12}$
23. $4\sqrt{36y} + 2\sqrt{49y} - 8\sqrt{25y}$
24. $3\sqrt{64x} + 3\sqrt{81x} - 3\sqrt{100x}$

25. $5\sqrt{12} - 3\sqrt{50} + 2\sqrt{75} - 6\sqrt{8}$
26. $9\sqrt{48} + 4\sqrt{20} - 5\sqrt{75} - 3\sqrt{45}$
27. $5x\sqrt{12x} + 4\sqrt{75x^3}$

28. $6t\sqrt{20t} - 3\sqrt{45t^3}$
29. $3r\sqrt{24r} - 2\sqrt{54r^3} - 2r\sqrt{6r}$
30. $3\sqrt{9w} + \sqrt{w} - \sqrt{25w}$

31. $\sqrt{2}(\sqrt{8}+2)+3\sqrt{32}$

32. $\sqrt{3}(\sqrt{12}+5)+2\sqrt{27}$

33. $\sqrt{6}(\sqrt{6}+3)+\sqrt{28}$

34. $3\sqrt{5}(2\sqrt{5}+2)+\sqrt{20}$

35. $\sqrt{3}(\sqrt{12}-6)+\sqrt{27}$

36. $\sqrt{7}(\sqrt{7}-5)+\sqrt{28}$

37. $6\sqrt{w}(\sqrt{w}+4)+\sqrt{36w}$

38. $4\sqrt{m}(\sqrt{m}-3)+\sqrt{49m}$

39. $(\sqrt{5a}-3)2+2\sqrt{5a}$

40. $(\sqrt{7t}-5)2+3\sqrt{7t}$

41. $(\sqrt{2w}-6)^2-3w\sqrt{2}$

42. $(\sqrt{6r}-4)^2-2\sqrt{6r}$

43. $2\sqrt{5}+\dfrac{1}{\sqrt{5}}$

44. $7\sqrt{5}+\dfrac{3}{\sqrt{5}}$

45. $4\sqrt{3}-\dfrac{2}{\sqrt{3}}$

46. $6\sqrt{5}-\dfrac{3}{\sqrt{5}}$

47. $\dfrac{1}{\sqrt{8}}+\dfrac{1}{\sqrt{12}}$

48. $\dfrac{1}{\sqrt{50}}+\dfrac{1}{\sqrt{18}}$

49. $\dfrac{2}{\sqrt{5}}+\dfrac{5}{\sqrt{6}}-\sqrt{5}$

50. $\dfrac{3}{\sqrt{2}}+\dfrac{3}{\sqrt{7}}-\sqrt{2}$

Review Exercises   The answers to all of these exercises are at the back of the book.

51. Identify each base and its exponent, and then simplify: -5^2

52. The sum of three consecutive integers is 9 more than double the largest. Find the integers.

53. Find the equation of a line perpendicular to $y=5x-1$ and which passes through $(0,0)$.

54. In 1980 the federal government spent \$4,192,000,000 for research and development (R&D) in the life sciences and \$241,000,000 for R&D in mathematics and the computer sciences. In 1990 the federal government spent \$8,838,000,000 for the life sciences and \$841,000,000 for mathematics and computer science. Did the percent spent on mathematics and computer science as compared to the life sciences increase or decrease over this ten year period?

---

55. Factor completely: $-12x^2+14x+6$

56. Reduce: $\dfrac{15x-15}{10x^2-10}$

Simplify.

57. $8(25-2\cdot3^2)$

58. $1-\dfrac{1}{6x}+\dfrac{5}{2x}$

59. $\dfrac{x^2-4}{x-3}\cdot\dfrac{x^2-x-6}{x^2+4x+4}$

60. $\dfrac{\sqrt{648x^2y^2}}{9x^3}$

61. $\left(\sqrt{64}-\sqrt{9}\right)^2$

62. $\sqrt{12a}\,\sqrt{9b^2}$

63. $-(f-2g-3h)-(6h-f)-(-2g+3h)$

Solve.

64. $0=x^2-11x+24$

65. $-30-61x=30x^2$

66. $0.15(12-x)+0.25x=5+0.1x$

67. $\dfrac{7}{m}=\dfrac{3}{4}-\dfrac{m-5}{m}$

---

68. Determine the domain and range of

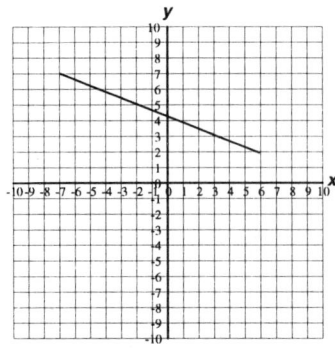

69. In 1960 the average life expectancy in the U. S. was 70.3 years. Since then it has been increasing by 0.16 years each year. Let t represent the year since 1960 and E represent life expectancy in years.

    a) Build a function that shows the average life expectancy as a function of the year since 1960.

    b) Calculate $E(60)$. Interpret the result in English.

    c) Find the year in which $E(t)=60$. Interpret the result in English.

    d) Find the year in which $E(t)=2E(0)$. Interpret the result in English.

Section 9.5 Solving Equations that Involve Square Roots

Topic 9.5 A — Definitions

A **radical equation** is an equation where a variable appears in a radicand. The following are examples of radical equations:

$$\sqrt{x} = 3 \quad \text{or} \quad 2\sqrt{5y + 3} + 8 = \sqrt{y}$$

As was the case with other equations, to solve a radical equation means to find all the values of the variable that make the equation true. We do this using the same techniques we developed in Chapter 3. That is, we transform the equation into simpler and simpler equivalent equations until the variable is isolated on one side.

Since taking a square root and squaring are inverse operations we can "remove" the radical symbol by "undoing" what the radical "does". We can express this property algebraically as follows:

— *Note* —	English: Squaring a number is the inverse of taking it's square root.
Squaring and taking a square root are inverse operations	Example: $\left(\sqrt{5}\right)^2$ simplifies to 5
	Algebra: If $x \geq 0$, then $\left(\sqrt{x}\right)^2 = x$.

For example, let's solve $\sqrt{x} = 3$ by squaring both sides of the equation:

$$\sqrt{x} = 3 \qquad \Leftarrow \text{ Equation to solve.}$$

$$\left(\sqrt{x}\right)^2 = 3^2 \qquad \Leftarrow \text{ Squared both the left and right side.}$$

$$x = 9 \qquad \Leftarrow \text{ Simplified.}$$

We can check the answer in the usual way:

Left expression *?=?* *Right expression*

$\sqrt{x}$ =	3	$\Leftarrow$ Equation to check.
$\sqrt{9}$	3	$\Leftarrow$ Substituted 9 for x.
3	3	$\Leftarrow$ Checks.

The following property can be used to solve radical equations.

— *Property* —	English: If two numbers are equal then their squares are equal.
Squaring property of equality	Examples:
	If $w = 3$ then $w^2 = 3^2$
	If $\sqrt{x} = 5$, then $\left(\sqrt{x}\right)^2 = 5^2$
	Algebra:
	If $a = b$, then $a^2 = b^2$
	Note: Although it's true that if $a = b$ then $a^2 = b^2$ it is *not necessarily* true that if $a^2 = b^2$ then $a = b$. For example $(-2)^2 = (2)^2$ since both are 4 but $-2 \neq 2$.

We can use the Squaring Property of Equality to help us solve equations that contain variables in the radicands of square roots. You may find the following procedure helpful.

— Procedure —	**Step 1 Isolate** Isolate the radical on one side of the equation.
Solving radical equations	**Step 2 Square** Square both sides of the equation.
	Step 3 Solve Solve the resulting equation.
	Step 4 Check Check the solution.

Example 9.5.1   a)   Solve $\sqrt{2y} = 8$

• SOLUTION •

Step 1 Isolate   The radical is already isolated on one side of the equation.

Step 2 Square

$$\left(\sqrt{2y}\right)^2 = 8^2 \qquad \Leftarrow \text{ Squared both sides of the equation.}$$

$$2y = 64 \qquad \Leftarrow \text{ Simplified.}$$

Step 3 Solve

$$y = 32 \qquad \Leftarrow \text{ Divided both sides by 2.}$$

Step 4 Check   The answer, 32, checks since $\sqrt{2(\mathbf{32})}$ is $\sqrt{64}$, which simplifies to 8.

b)   Solve $2\sqrt{y} = 8$

• SOLUTION • This is almost the same problem as in part a) but the constant 2 is outside the radical rather than inside.

Step 1 Isolate

$$\frac{2\sqrt{y}}{2} = \frac{8}{2} \qquad \Leftarrow \text{ Divided both sides by 2}$$

$$\sqrt{y} = 4 \qquad \Leftarrow \text{ Reduced factor of 2.}$$

Step 2 Square

$$\left(\sqrt{y}\right)^2 = 4^2 \qquad \Leftarrow \text{ Squared both sides of the equation.}$$

$$y = 16 \qquad \Leftarrow \text{ Simplified.}$$

Step 3 Solve   There is nothing left to do.

Step 4 Check   The answer, 16, checks since $2\sqrt{\mathbf{16}}$ is 2•4, which simplifies to 8.

Practice 9.5.1 Solve.

1. $\sqrt{5x} = 10$

2. $6 = 3\sqrt{4x}$

3. $\sqrt{x+5} = 3$

4. $5 = \sqrt{x-10}$

Isolating the radical term is critical when the side of the equation containing the radical contains more than one term, as in $\sqrt{x} - 8 = -3$. If we square both sides of this equation we end up with $x - 16\sqrt{x} + 64 = 9$. The squaring process has not cleared the radical and so we are no closer to a solution. If we isolate the radical first the solution is easily found.

Step 1 Isolate $\sqrt{x} = 5$ $\Leftarrow$ Added 8 to both sides to isolate the radical term.

Step 2 Square $x = 25$ $\Leftarrow$ Squared both sides.

Step 3 Solve   There is nothing left to do.

Step 4 Check   The answer, 25, checks since $\sqrt{\mathbf{25}} - 8$ is $5 - 8$, which simplifies to -3.

Example 9.5.2   a)   Solve $\sqrt{3x} - 12 = -6$

• **SOLUTION** •

Step 1 Isolate

$$\sqrt{3x} = 6 \qquad \Leftarrow \text{ Added 12 to both sides.}$$

Step 2 Square

$$3x = 36 \qquad \Leftarrow \text{ Squared both sides.}$$

Step 3 Solve

$$x = 12 \qquad \Leftarrow \text{ Divided both sides by 3.}$$

Step 4 Check   The answer, 12, checks since $\sqrt{3(12)} - 12$ is $\sqrt{36} - 12$, which simplifies to –6.

b)   Solve $3\sqrt{2x-1} = \sqrt{6x+3}$

• **SOLUTION** •

Step 1 Isolate   Even though there are two radicals, one of them is isolated and so we can move on to Step 2 below.

Step 2 Square

$$9(2x-1) = 6x+3 \qquad \Leftarrow \text{ Squared both sides. Be sure you know where the}$$

9 in $9(2x - 1)$ came from. We had to square both 3 and $\sqrt{2x-1}$ in the left expression.

Step 3 Solve

$$18x - 9 = 6x + 3 \qquad \Leftarrow \text{ Distributed the 9.}$$

$$12x - 9 = 3 \qquad \Leftarrow \text{ Subtracted } 6x.$$

$$12x = 12 \qquad \Leftarrow \text{ Added 9.}$$

$$x = 1 \qquad \Leftarrow \text{ Divided by 12.}$$

Step 4 Check   The solution is 1. We must check the solution.

Left expression ?=? ***Right expression***		
$3\sqrt{2x-1}$	$= \quad \sqrt{6x+3}$	$\Leftarrow$ Equation to check.
$3\sqrt{2(1)-1}$	$\sqrt{6(1)+3}$	$\Leftarrow$ Substituted 1 for x.
$3\sqrt{1}$	$\sqrt{9}$	$\Leftarrow$ Simplified radicands.
3	3	$\Leftarrow$ The solution, 1, checks.

Practice 9.5.2   Solve.

1. $\sqrt{5x} - 15 = -5$ 2. $11 - \sqrt{6x} = -1$ 3. $\dfrac{\sqrt{6-x}}{5} + 1 = 2$ 4. $\sqrt{x+6} = \dfrac{\sqrt{25x+6}}{3}$ 5. $5\sqrt{x-7} = \sqrt{3x+1}$

Topic 9.5 B — Extraneous Roots

We need to discuss two situations that might arise when we square both sides of an equation.

Situation One   After squaring both sides of an equation we *may* find that the new equation has more solutions than the original equation. For example, the equation $x = 5$, has one solution, 5. If we square both sides we get $x^2 = 25$, which has two solutions, 5 and –5. The process of squaring introduced another solution!

Situation Two If we begin with a contradiction and square both sides we *may* wind up with a new equation that is not a contradiction. For example, the equation $x + 1 = x$ is a contradiction because it says one more than a number is equal to the number. Watch what happens when we square both sides and solve:

$$x + 1 = x \qquad \Leftarrow \text{ Equation to solve.}$$

$$(x + 1)^2 = x^2 \qquad \Leftarrow \text{ Squared both sides}$$

$$x^2 + 2x + 1 = x^2 \qquad \Leftarrow \text{ Used FOIL to square the binomial.}$$

$$2x + 1 = 0 \qquad \Leftarrow \text{ Subtracted } x^2 \text{ from both sides.}$$

$$2x = -1 \qquad \Leftarrow \text{ Subtracted 1 from both sides.}$$

$$x = \frac{-1}{2} \qquad \Leftarrow \text{ Divided both sides by 2.}$$

The solution, $\frac{-1}{2}$ does not check in the original equation. Once again, we produced an "answer" that doesn't make the original equation true.

As we mentioned in Section 8.6, apparent solutions to an equation that do not satisfy the original equation are called extraneous roots. They are not a part of the solution set.

— Caution — *Checking solutions in radical equations*	Because of the possible introduction of extraneous roots, all the solutions to radical equations *must* be checked in the *original* equation to see if they actually work. (Recall from Chapter 2 that you should be checking solutions in any case to be sure you have not made a mistake while solving the equation.)

Example 9.5.3 a) Solve $\sqrt{2x - 3} = -5$.

•SOLUTION•

Step 1 Isolate Not needed.

Step 2 Square

$$2x - 3 = 25 \qquad \Leftarrow \text{ Squared both sides.}$$

Step 3 Solve

$$2x = 28 \qquad \Leftarrow \text{ Added 3 to both sides.}$$

$$x = 14 \qquad \Leftarrow \text{ Divided both sides by 2.}$$

Step 4 Check We check the solution by substituting 14 for x in the original equation.

Left expression	?=?	*Right expression*	
$\sqrt{2x - 3}$	=	-5	$\Leftarrow$ Equation to check.
$\sqrt{2(\mathbf{14}) - 3}$		-5	$\Leftarrow$ Substituted 14 for x.
$\sqrt{28 - 3}$		-5	$\Leftarrow$ Multiplied.
$\sqrt{25}$		-5	$\Leftarrow$ Subtracted.
5		-5	$\Leftarrow$ Calculated square root.

Since $5 \neq -5$ we see that the solution does not check. This makes sense because the original equation is a contradiction: the expression on the left, $\sqrt{2x - 3}$, is positive (it's the principal square root), but the number on the right is negative. A positive number cannot equal a negative number so we have a contradiction.

b) Solve $\sqrt{x+6} = x$

•SOLUTION•

Step 1 Isolate   Not needed.

Step 2 Square

$$x + 6 = x^2 \qquad \Leftarrow \text{Squared both sides.}$$

Step 3 Solve

$$0 = x^2 - x - 6 \qquad \Leftarrow \text{Because there is a square term, moved all terms to one side of equal sign.}$$

$$0 = (x-3)(x+2) \qquad \Leftarrow \text{Factored.}$$

$x - 3 = 0$	$x + 2 = 0$ $\Leftarrow$ Used Zero Product Property to break up the given equation into two equations.
$x - 3 + 3 = 0 + 3$	$x + 2 - 2 = 0 - 2$ $\Leftarrow$ Solved each equation.
$x = 3$	$x = -2$

Step 4 Check   We check each solution by substituting for x in the original equation.

Check for x = 3

Left expression	***?=?***	***Right expression***	
$\sqrt{x+6}$	$=$	x	$\Leftarrow$ Equation to check.
$\sqrt{(3)+6}$		(3)	$\Leftarrow$ Substituted 3 for x.
$\sqrt{9}$		3	$\Leftarrow$ Added
3		3	$\Leftarrow$ Calculated square root.

Because substituting 3 for x in the original equation resulted in a true statement, 3 is a solution.

Check for x = –2

Left expression	***?=?***	***Right expression***	
$\sqrt{x+6}$	$=$	x	$\Leftarrow$ Equation to check.
$\sqrt{(-2)+6}$		(-2)	$\Leftarrow$ Substituted –2 for x.
$\sqrt{4}$		-2	$\Leftarrow$ Added
2		-2	$\Leftarrow$ Calculated square root.

Because substituting –2 for x in the original equation resulted in a *false* statement, –2 is an extraneous solution. Therefore, the original equation has only one solution, 3.

Practice 9.5.3   Solve.

1. $\sqrt{30-x} = x$ 2. $\sqrt{2x+4} = -6$ 3. $4 = \sqrt{4x} + 7$ 4. $2x = \sqrt{4x+8}$

Topic 9.5 C — Applications

On the surface of the earth you can only see as far as the horizon. The formula for the approximate distance to the horizon is $d = \sqrt{1.5h}$ where

- d is the distance (miles)
- h is the height above the surface (feet)

Example 9.5.4   a)   Calculate the distance to the horizon for a 6 foot tall person.

•SOLUTION•

$$d = \sqrt{1.5h} \qquad \Leftarrow \text{ Formula to evaluate.}$$

$$d = \sqrt{1.5(6)} \qquad \Leftarrow \text{ Substituted 6 for } h.$$

$$d = \sqrt{9} \qquad \Leftarrow \text{ Multiplied.}$$

$$d = 3 \qquad \Leftarrow \text{ Calculated square root.}$$

A 6 foot tall person can see about 3 miles to the horizon.

b)   How many feet above the surface of the water would a submarine periscope have to be to see a distance of 5 miles?

•SOLUTION• We substitute 5 for d and then solve for h.

$$d = \sqrt{1.5h} \qquad \Leftarrow \text{ Equation to solve.}$$

$$5 = \sqrt{1.5h} \qquad \Leftarrow \text{ Substituted 5 for } d.$$

$$25 = 1.5h \qquad \Leftarrow \text{ Squared both sides.}$$

$$16.7 = h \qquad \Leftarrow \text{ Divided both sides by 1.5 and rounded.}$$

We rounded off the answer to the nearest tenth. The periscope would have to be about 16.7 feet above the surface to see 5 miles to the horizon.

Practice 9.5.4 Solve using the formula $d = \sqrt{1.5h}$.

1.   Calculate the distance to the horizon for a 6 foot person on a platform 12 feet high. Round your answer to the nearest tenth of a mile.

2.   A person has an unobstructed view of a school 2.2 miles away. If the school is just at the edge of their view, estimate the height of the person.

3.   How high would an observation platform have to be if a seven foot person wanted to see 8 miles?

Physicists have discovered that the radius of the earth can be approximated by the formula

$$r = \sqrt{\frac{Gm}{g}}$$

where

r = the radius of the earth.

G = Newton's universal gravitation constant.

m = the mass of the earth.

g = the acceleration due to gravity on the earth's surface.

We can use this formula to see how much the earth "weighs".

Example 9.5.5   a)   Solve $r = \sqrt{\dfrac{Gm}{g}}$ for m.

•SOLUTION•

$$(r)^2 = \left(\sqrt{\frac{Gm}{g}} \right)^2 \qquad \Leftarrow \text{ Squared both sides.}$$

$$r^2 = \frac{Gm}{g} \qquad \Leftarrow \text{ Simplified.}$$

$$\frac{gr^2}{G} = m \qquad \Leftarrow \text{ Multiplied by } g \text{ and divided by } G.$$

b) Calculate the weight of the earth given that $r = 6.4 \times 10^6$ meters, $g = 9.8$ meters/sec^2, and $G = 6.7 \times 10^{-11}$ NT–meter^2/kg^2.

•SOLUTION•

$$\frac{gr^2}{G} = m \qquad \Leftarrow \text{ Formula to evaluate.}$$

$$\frac{(9.8)\left(6.4 \times 10^6\right)^2}{6.7 \times 10^{-11}} = m \qquad \Leftarrow \text{ Substituted values.}$$

$$6.0 \times 10^{24} = m \qquad \Leftarrow \text{ Simplified and rounded.}$$

The mass is in kilograms. To convert this to pounds [†] we multiply by the conversion factor 2.2 lbs = 1 kg to get

$$6.0 \times 10^{24} \ \cancel{\text{kg}} \times \frac{2.2 \text{ lbs}}{1 \ \cancel{\text{kg}}}$$

which simplifies to 13.2×10^{24} pounds. This is 13,200,000,000,000,000,000,000,000 pounds.

Practice 9.5.5 The maximum safe speed (in meters per second) for a car rounding a curve can be approximated by the formula $v = \sqrt{\mu g R}$ where μ represents the friction coefficient between the tires and the road, g represents the acceleration of gravity (9.8 meters per second^2), and R represents the radius of the curve.

1. Find the maximum safe speed for a car where $\mu = 0.10$ and R = 200.

2. Find the friction coefficient when v = 12 meters/sec and R = 110.

Exercise Set 9.5 The answers to the odd numbered exercises are at the back of the book.

Solve.

1. $\sqrt{2x} = 10$

2. $\sqrt{6r} = 12$

3. $4 = \sqrt{2w}$

4. $7 = \sqrt{7m}$

5. $5\sqrt{2x} = 15$

6. $7\sqrt{5x} = 28$

7. $4\sqrt{3t} = 24$

8. $12 = 3\sqrt{2x}$

9. $\sqrt{2x + 3} = 5$

10. $\sqrt{3x + 1} = 5$

11. $3 = \sqrt{2x - 5}$

12. $5 = \sqrt{3x - 2}$

13. $\sqrt{y + 4} = 7$

14. $\sqrt{x + 7} = 9$

15. $\sqrt{x^2 - 5} = x$

16. $x = \sqrt{x^2 + 7}$

17. $\sqrt{7x - 2} = 5$

18. $1 = \sqrt{12x - 5}$

19. $8 = \sqrt{3x + 2}$

20. $\sqrt{3m - 9} = -3$

21. $\sqrt{6x - 7} = -1$

22. $\sqrt{5y + 3} = 13$

23. $\sqrt{6x - 5} = 7$

24. $\sqrt{3x - 12} = -3$

25. $\sqrt{2x - 10} = -4$

26. $5 - \sqrt{3x} = -1$

27. $7 - \sqrt{2x} = 1$

28. $7 = \sqrt{3x} - 2$

29. $3 = -2 + \sqrt{5x}$

30. $6 - \sqrt{10y} = -4$

31. $2 = \frac{\sqrt{4 + 6x}}{2} - 3$

32. $\frac{\sqrt{3x + 4}}{5} - 8 = -7$

33. $\frac{\sqrt{4 - 8x}}{3} - 3 = -1$

34. $y - 5 = \frac{\sqrt{3y - 2}}{4}$

35. $\frac{\sqrt{6 - 5x}}{2} - 5 = -2$

36. $\frac{\sqrt{7 - 6x}}{5} - 4 = -3$

37. $5\sqrt{3x - 20} = \sqrt{11x + 12}$

38. $3\sqrt{2x - 1} = \sqrt{16x + 1}$

39. $2\sqrt{10x - 4} = \sqrt{35x + 4}$

40. $3\sqrt{x - 9} = \sqrt{2x - 11}$

41. $\sqrt{y + 72} = y$

42. $\sqrt{72 - t} = t$

43. $5 - \sqrt{2x} = 7$

44. $4 - \sqrt{6x} = 10$

45. $x = \sqrt{20 - x}$

46. $x = \sqrt{3x + 28}$

47. $\sqrt{4 - 16x} = -10$

48. $\sqrt{5x + 8} = 3$

[†] Physicists would leave the answer in kilograms because calculating the "weight" of the earth really makes no sense since the weight of an object is the force of the earth's gravity on the object. But most Americans are used to working with pounds so we do the conversion.

49. $2x = \sqrt{20x + 24}$ 50. $\dfrac{\sqrt{18x + 27}}{3} = x$ 51. $p = \dfrac{6\sqrt{3p + 18}}{5}$ 52. $m = \sqrt{4m + 32}$

53. $-4 = \sqrt{6x} + 2$ 54. $\sqrt{3x} - 2 = -8$

The formula for the approximate distance to the horizon is $d = \sqrt{1.5h}$ where d is the distance (miles) and h is the height above the surface (feet)

55. Calculate the distance to the horizon for a 6 foot person on a platform 16 feet high. Round your answer to the nearest tenth of a mile.

56. Calculate the distance to the horizon for a 5 foot 6 inch person standing on the roof of their house if their roof is 18 feet above the ground. Round your answer to the nearest tenth of a mile.

57. How much higher would a 5 foot tall person have to get to see a distance of 15 miles?

58. How much higher would a 4.25 foot tall person have to get to see a distance of 7.5 miles?

59. How much farther can a 5 foot tall person see than someone 4 feet tall?

60. How much higher can a six foot tall person see standing then sitting if their height is about three feet when sitting?

Review Exercises   The answers to all of these exercises are at the back of the book.

61. Joline is flying a kite and has let out 106 feet of string. Scott is standing directly under the kite and 56 feet from Joline. What is the height of the kite?

62. Calculate the slope of the straight line that passes through (1,1) and (1,0).

63. Write the equation of a line with slope 1 and y–intercept (0,–2)

64. Find the equation of the line that goes through (6,1) and (–8,1).

---

65. Factor completely $4x^2 - 27x + 18$

66. Factor completely $5x^2 - 9x - 18$

67. Find the GCF of $36x^3y^2z$, $24xy^4z^2$, and $40x^2y^3z^6$

68. Solve: $8x^2 - 8x = 0$

69. Solve: $\begin{cases} 3x + y = 2 & \text{\textbf{❶}} \\ 6x + 2y = -5 & \text{\textbf{❷}} \end{cases}$

70. Solve: $2 + \dfrac{1}{b + 1} = \dfrac{b + 2}{b + 1}$

Simplify.

71. $\dfrac{3}{x^2 - 2x} + \dfrac{1}{x^2 - 4}$

72. $\dfrac{\left(2xy^3\right)^4 \left(x^2y\right)^2}{\left(4xy^2\right)^2}$

73. $8 - 3(x - 2) - (5 - 5x)$

74. $5 - \sqrt{9} + 2\left(3 - \sqrt{25}\right)$

75. $\dfrac{\sqrt{18s}\,\sqrt{4s^5}}{6s^7}$

76. $\sqrt{24} + 5\sqrt{54} - 3\sqrt{216}$

77. Find the x-intercept and the y-intercept:

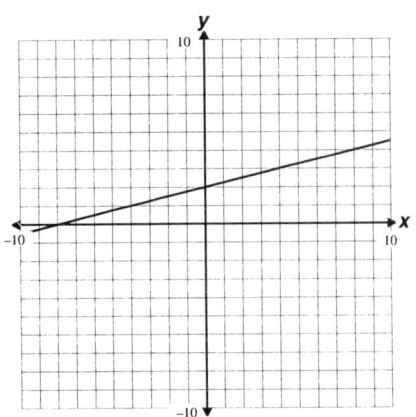

78. Determine whether this graph represents a function

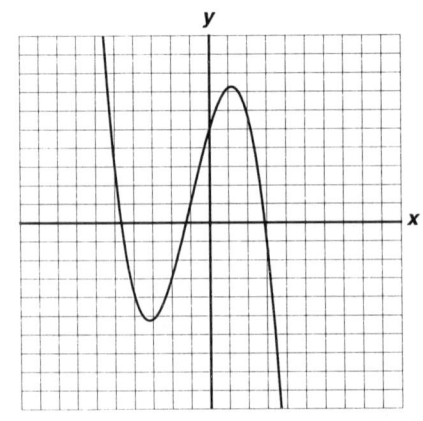

Section 9.6 Quadratic Equations

Topic 9.6 A — Introduction to Quadratic Models

In Chapter 4 we discussed linear relations, which are relations whose graphs are straight lines and which can be modeled by $y = mx + b$.

A second common type of relation is called the **quadratic relation**. These relations can be modeled using $y = ax^2 + bx + c$, where a, b, and c are constants. The graphs of quadratic relations have a distinctive curved shape called a **parabola**. The path taken by a long fly ball at a baseball game has the shape of a parabola.

— Definition — *Quadratic equation in two variables*	English: A two variable equation that can be written in the form $y = ax^2 + bx + c$, where a, b, and c are constants. The graph of a quadratic equation is called a parabola. Example: $y = x^2 - 4$ The graph of this equation is shown below: 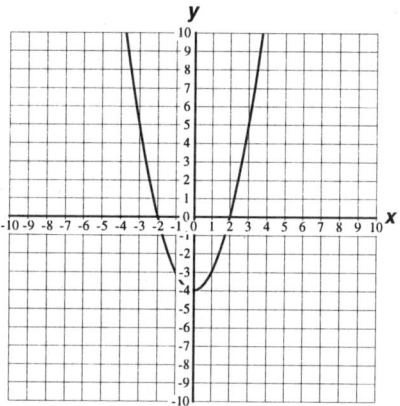 Algebra: If a, b, and c are real numbers and $a \neq 0$, then an equation that can be written in the form $y = ax^2 + bx + c = 0$ is a quadratic equation in two variables.

Quadratic equations may be used to model many applied situations. For example, below is a data table and graph that shows the relation between carbon monoxide released to the air and the number of years since 1940.

x time (number of years since 1940)	y carbon monoxide emissions (millions of tons)
0	93.615
5	98.112
10	102.609
15	106.177
20	109.745
25	118.912
30	128.079
35	115.11
40	115.625
44	114.262
45	114.69
46	109.199
47	108.012
48	115.849
49	103.144
50	100.65
51	97.376
52	94.043
53	94.133
54	98.017

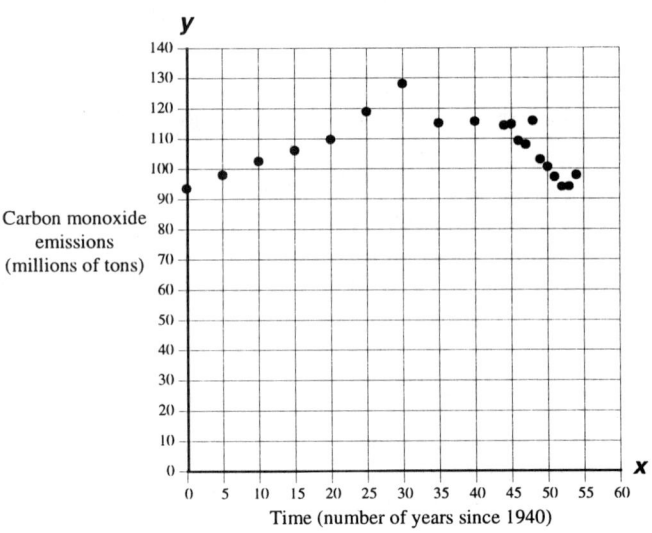

These data would not be modeled very well by a linear equation, $y = mx + b$. However, if we instruct the computer to model the data using a quadratic equation, $y = ax^2 + bx + c$, we get a much better fit as shown below:

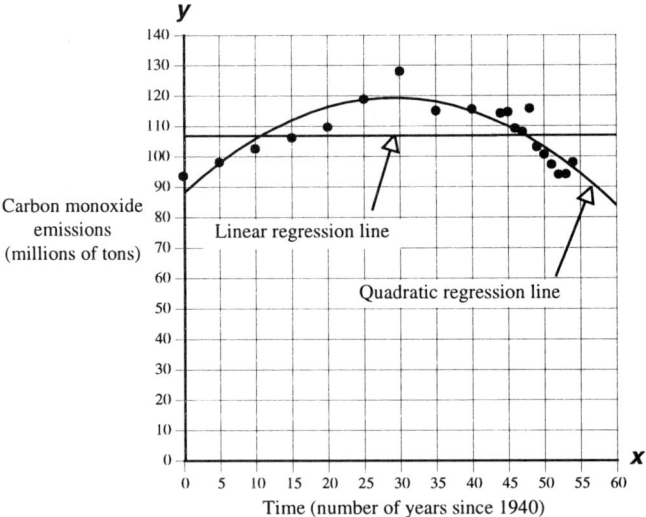

The regression program used by the computer calculates the following values for the constants in the quadratic equation: $a = -0.037$, $b = 2.16$, $c = 88$. Therefore, the equation may be written as $y = -0.037x^2 + 2.16x + 88$

Suppose we wish to use this model to predict when carbon monoxide emissions will drop to 80 million tons. To answer this question we replace y with 80 and attempt to solve for x:

$$y = -0.037x^2 + 2.16x + 88 \quad \Leftarrow \quad \text{Quadratic model.}$$

$$\mathbf{80} = -0.037x^2 + 2.16x + 88 \quad \Leftarrow \quad \text{Substituted 80 for the output.}$$

This is a single variable quadratic equation of the type we solved in Section 7.6. Unfortunately, the technique we used there, which involved factoring and the Zero Product Property, does not work with this quadratic equation since once 80 is subtracted from both sides we cannot factor the resulting quadratic expression using integers. We need a more powerful technique for solving a quadratic equation like this one.

Topic 9.6 B — The Quadratic Formula

The Quadratic Formula allows us to solve any single variable quadratic equation. The derivation of this formula is beyond the scope of this book [†]. However, we can still use the formula to help us solve problems.

— Formula — *The Quadratic Formula*	Algebra: If $a \neq 0$, then a quadratic equation of the form $0 = ax^2 + bx + c$ has the following two solutions: $$x = \frac{-b + \sqrt{b^2 - 4ac}}{2a} \text{ and } x = \frac{-b - \sqrt{b^2 - 4ac}}{2a}$$

Notice that the Quadratic Formula gives us two solutions. This is what we expect because the equation we are solving is a second degree equation. If you look carefully at the two solutions you will note that the formulas are almost identical; the only difference is that in one solution the radical is added and in the other the radical is subtracted. Because of this slight difference we usually write the Quadratic Formula in a more compact form as follows:

$$x = \frac{-b \pm \sqrt{b^2 - 4ac}}{2a}$$

Notice the symbol $\pm$ just to the left of the radical symbol. We introduced this symbol in Section 9.1. It is a combination of the positive sign and the negative sign. The symbol tells us there are two solutions, one that uses the $+$ and the other that uses the $-$.

We may use the following procedure to solve a quadratic equation using the Quadratic Formula:

— Procedure — *Solving a quadratic equation using the Quadratic Formula*	***Step 1 Standard Form*** Write the equation with 0 on one side and a polynomial in standard form on the other, as in $ax^2 + bx + c$.
	Step 2 Substitute Identify the values of a, b, and c and substitute these into the Quadratic Formula, $x = \dfrac{-b \pm \sqrt{b^2 - 4ac}}{2a}$
	Step 3 Simplify Simplify the expression as far as you can without evaluating the $\pm$ symbol.
	Step 4 Split Split the solution at the $\pm$ into two parts, one which uses the $+$ and the other which uses the $-$. Simplify the two expressions.
	Step 5 Check Check all solutions.

[†] The derivation is typically covered in intermediate or college algebra.

Example 9.6.1   Use the Quadratic Formula to solve.

a) $2x^2 + 3x + 1 = 0$ Note that we could use the factoring method to solve this but we want to practice using the Quadratic Formula.

•SOLUTION•

Step 1 Standard form   The equation is given in standard form.

Step 2 Substitute   Substituting 2 for a, 3 for b, and 1 for c gives $x = \dfrac{-3 \pm \sqrt{3^2 - 4(2)(1)}}{2(2)}$

Step 3 Simplify

$$x = \frac{-3 \pm \sqrt{9 - 8}}{4}$$

$$x = \frac{-3 \pm 1}{4}$$

Step 4 Split   We write each solution individually and simplify.

Solution One	***Solution Two***	
$x = \dfrac{-3 + 1}{4}$	$x = \dfrac{-3 - 1}{4}$	⟸ Wrote both solutions.
$x = \dfrac{-2}{4}$	$x = \dfrac{-4}{4}$	⟸ Simplified.
$x = -\dfrac{1}{2}$	$x = -1$	⟸ Simplified.

The solutions are $-\dfrac{1}{2}$ and -1.

Step 5 Check   We leave the check to you.

b) $0 = 3x^2 - 2.5x - 2$

•SOLUTION•

Step 1 Standard form   The equation is given in standard form.

Step 2 Substitute   Substituting 3 for a, -2.5 for b, and -2 for c gives

$$x = \frac{-(-2.5) \pm \sqrt{(-2.5)^2 - 4(3)(-2)}}{2(3)}$$

Step 3 Simplify

$$x = \frac{2.5 \pm \sqrt{6.25 + 24}}{6}$$

$$x = \frac{2.5 \pm 5.5}{6}$$

Step 4 Split   We write each solution individually and simplify.

Solution One	***Solution Two***	
$x = \dfrac{2.5 + 5.5}{6}$	$x = \dfrac{2.5 - 5.5}{6}$	⟸ Wrote both solutions.
$x = \dfrac{4}{3}$	$x = -\dfrac{1}{2}$	⟸ Simplified.

The solutions are $\dfrac{4}{3}$ and $-\dfrac{1}{2}$.

Step 5 Check   We leave the check to you.

Practice 9.6.1 Solve the following using the Quadratic Formula.

1. $3x^2 + 7x - 6 = 0$ 2. $x^2 - 5x + 1 = 0$ 3. $-x^2 + 3x + 7 = 0$ 4. $2.1x^2 - x - 5.4 = 0$

Topic 9.6 C — Using the Quadratic Formula With More Complicated Equations

When the equations become more complicated we may need to use techniques we developed in previous sections, such as rearranging terms or simplifying radical expressions.

Example 9.6.2   Use the Quadratic Formula to solve the following equations.

a) $-11x = 2x^2 - 21$

     •SOLUTION•

     Step 1 Standard form   We move $-11x$ to the right side to get $0 = 2x^2 + 11x - 21$

     Step 2 Substitute   Substituting 2 for a, 11 for b, and -21 for c gives

$$x = \frac{-(11) \pm \sqrt{(11)^2 - 4(2)(-21)}}{2(2)}$$

     Step 3 Simplify

$$x = \frac{-11 \pm \sqrt{121 + 168}}{4}$$

$$x = \frac{-11 \pm \sqrt{289}}{4}$$

$$x = \frac{-11 \pm 17}{4}$$

     Step 4 Solutions

Solution One	***Solution Two***	
$x = \dfrac{-11 + 17}{4}$	$x = \dfrac{-11 - 17}{4}$	⟵ Wrote both solutions.
$x = \dfrac{6}{4}$	$x = \dfrac{-28}{4}$	⟵ Simplified.
$x = \dfrac{3}{2}$	$x = -7$	

     Step 5 Check   We leave the check to you.

b) $x^2 - \frac{1}{6}x = \frac{1}{6}$

     •SOLUTION•   We could first multiply through by the LCD to clear the fractions, or we could simply follow the same procedure as with the other examples. We will do the latter to show how the Quadratic Formula may be used with fractions.

     Step 1 Standard form   We move $\frac{1}{6}$ to the left side to get $x^2 - \frac{1}{6}x - \frac{1}{6} = 0$

     Step 2 Substitute   Substituting 1 for a, $-\frac{1}{6}$ for b, and $-\frac{1}{6}$ for c gives

$$x = \frac{-\left(-\frac{1}{6}\right) \pm \sqrt{\left(\frac{1}{6}\right)^2 - 4(1)\left(-\frac{1}{6}\right)}}{2(1)}$$

Step 3 Simplify

$$x = \frac{\frac{1}{6} \pm \sqrt{\frac{1}{36} + \frac{4}{6}}}{2}$$

$$x = \frac{\frac{1}{6} \pm \sqrt{\frac{25}{36}}}{2}$$

$$x = \frac{\frac{1}{6} \pm \frac{5}{6}}{2}$$

Step 4 Solutions

Solution One	**Solution Two**	
$x = \dfrac{\frac{1}{6} + \frac{5}{6}}{2}$	$x = \dfrac{\frac{1}{6} - \frac{5}{6}}{2}$	⇐ Wrote both solutions.
$x = \dfrac{\frac{6}{6}}{2}$	$x = \dfrac{-\frac{4}{6}}{2}$	⇐ Simplified both solutions.
$x = \dfrac{1}{2}$	$x = -\dfrac{1}{3}$	

Step 5 Check   We leave the check to you.

c) $-2 - x^2 = -4x$

•SOLUTION•

Step 1 Standard form   We move -2 and $-x^2$ to the right to get $0 = x^2 - 4x + 2$

Step 2 Substitute   Substituting 1 for a, -4 for b, and 2 for c gives

$$x = \frac{-(-4) \pm \sqrt{(-4)^2 - 4(1)(2)}}{2(1)}$$

Step 3 Simplify

$$x = \frac{4 \pm \sqrt{16 - 8}}{2}$$

$$x = \frac{4 \pm \sqrt{8}}{2}$$

$$x = \frac{4 \pm 2\sqrt{2}}{2}$$

Step 4 Solutions

Solution One	**Solution Two**	
$x = \dfrac{4 + 2\sqrt{2}}{2}$	$x = \dfrac{4 - 2\sqrt{2}}{2}$	⇐ Wrote both solutions.
$x = \dfrac{4}{2} + \dfrac{2\sqrt{2}}{2}$	$x = \dfrac{4}{2} - \dfrac{2\sqrt{2}}{2}$	⇐ Simplified.
$x = 2 + \sqrt{2}$	$x = 2 - \sqrt{2}$	

Step 5 Check   We leave the check to you.

d) $x^2 + x + 1 = 0$

•SOLUTION•

Step 1 Standard form   The equation is given in standard form.

Step 2 Substitute   Substituting 1 for a, 1 for b, and 1 for c gives

$$x = \frac{-(1) \pm \sqrt{(1)^2 - 4(1)(1)}}{2(1)}$$

Step 3 Simplify

$$x = \frac{-1 \pm \sqrt{1 - 4}}{2}$$

$$x = \frac{-1 \pm \sqrt{-3}}{2}$$

We can stop here because the radicand is negative and so the radical is not a real number. We say the equation has no real solutions. [†]

Practice 9.6.2   Solve the following using the Quadratic Formula:

1. $4x = 2x^2 - 1$ 2. $\frac{1}{4} = \frac{1}{2}x^2 - 3x$ 3. $-2 = 4x^2 - x$ 4. $20 + 2x = 0.1x^2$ 5. $\frac{1}{9}x^2 - \frac{4}{3}x = -4$

Topic 9.6 D — Answering Questions Using the Quadratic Formula

At the beginning of this section we noted that the equation $0 = -0.037x^2 + 2.16x + 8$ predicted the year when carbon monoxide emissions will drop to 80 million tons. We can use the Quadratic Equation to solve this equation.

Step 1 Standard form   The equation is given in standard form.

Step 2 Substitute   Substituting -0.037 for a, -2.16 for b, and 8 for c gives

$$x = \frac{-2.16 \pm \sqrt{(2.16)^2 - 4(-0.037)(8)}}{2(-0.037)}$$

Step 3 Simplify

$$x = \frac{-2.16 \pm \sqrt{4.6656 + 1.184}}{-0.074}$$

$$x = \frac{-2.16 \pm 2.42}{-0.074} \qquad \Leftarrow \text{Simplified the square root and}$$

rounded to two decimal places.

Step 4 Solutions

Solution One	*Solution Two*	
$x = \dfrac{-2.16 + 2.42}{-0.074}$	$x = \dfrac{-2.16 - 2.42}{-0.074}$	$\Leftarrow$ Wrote both solutions.
$x = \dfrac{0.26}{-0.074}$	$x = \dfrac{-4.58}{-0.074}$	$\Leftarrow$ Simplified.
$x = -3.5$	$x = 61.9$	$\Leftarrow$ Divided and rounded to one decimal place.

Step 5 Check   We leave the check to you.

[†]   The equation does have a solution but one that involves a set of numbers which we have not discussed. This new set is called the complex numbers, and you will study it a later mathematics course.

Since x represents the number of years since 1940, the solution -3.5 corresponds to $1936\frac{1}{2}$, which we would interpret as about midway through 1936. The other solution, 61.9, corresponds to about 2002. Since these values involve extrapolation (we are estimating beyond the limits of the data set) we will need to change the scale of the x–axis in order to see the solutions on the graph:

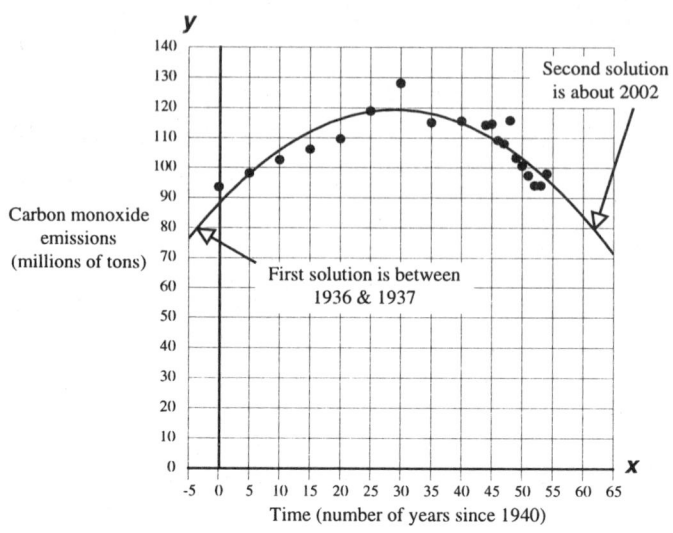

Example 9.6.3   The cost of connecting to the city water system in Saint Paul is based on the size of the water pipe entering a building. The following data show the relation:

x diameter of pipe (inches)	y cost per month (dollars)
0.75	2.4
1	4.2
1.5	8.8
2	20
3	42
4	104
5	150

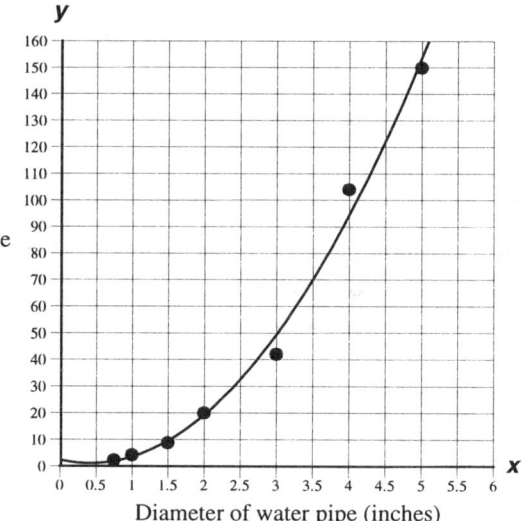

A computer program was used to generate the following quadratic regression equation:

$$y = 10.50x^2 - 20x + 13.60$$

where

    y represents the cost per month (dollars)

    x represents the diameter of the water pipe (inches).

a) A common size of water pipe for a house is 1 inch. Predict the monthly cost for water for one month if the diameter of the pipe is 1 inch.

    •SOLUTION•

$$y = 10.50x^2 - 20x + 13.60 \quad \Leftarrow \text{Quadratic model.}$$

$$y = 10.50(1)^2 - 20(1) + 13.60 \quad \Leftarrow \text{Substituted 1 for the input.}$$

$$y = 10.50 - 20 + 13.60 \quad \Leftarrow \text{Simplified.}$$

$$y = 4.10$$

    We predict it will cost $4.10 per month for water.

b) Predict the largest pipe you could afford if you plan to spend $700 per month for water.

    •SOLUTION•

$$y = 10.50x^2 - 20x + 13.60 \quad \Leftarrow \text{Quadratic model.}$$

$$700 = 10.50x^2 - 20x + 13.60 \quad \Leftarrow \text{Substituted 700 for the output.}$$

$$0 = 10.50x^2 - 20x - 686.40 \quad \Leftarrow \text{Moved all values to the right side.}$$

$$x = -7.20 \text{ or } x = 9.10. \quad \Leftarrow \text{Used Quadratic Formula to solve for } x.$$

    The solution –7.20 inches can not be correct. Since water pipes usually come in full inch or half inch sizes we estimate that the best answer is that we can afford a 9 inch pipe.

c) Find the cost if the pipe is 0 inches in diameter.

    •SOLUTION•

$$y = 10.50x^2 - 20x + 13.60 \quad \Leftarrow \text{Quadratic model.}$$

$$y = 10.50(0)x^2 - 20(0) + 13.60 \quad \Leftarrow \text{Substituted 0 for the input.}$$

$$y = 13.60 \quad \Leftarrow \text{Simplified.}$$

    This implies that if no water can pass through the pipe you must still pay $13.60 a month. The best assumption is that the model should not be used for some minimum input value. (Try using the model for 0.75 inches and reflect on your solution in terms of the answer to question a.)

Example 9.6.4   Wind chill results when the wind blows away the thin layer of insulating air surrounding the skin. The following graph shows the relation between wind speed and wind chill if the air temperature is 30 degrees Fahrenheit.

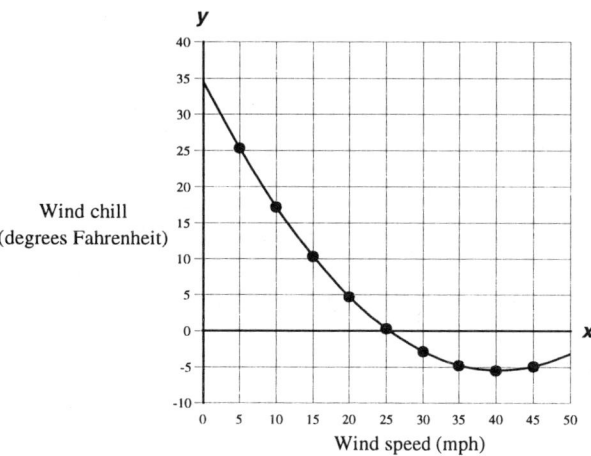

Wind chill
(degrees Fahrenheit)

Wind speed (mph)

Notice that as the wind speed increases the body becomes colder. The quadratic regression equation which approximates this graph is $y = 0.025x^2 - 2x + 34.7$.

a) If the outside temperature is 30 degrees Fahrenheit and we know the wind chill is 0 degrees what is the wind speed?

•SOLUTION• Substituting 0 for y in the wind chill model gives $0 = 0.025x^2 - 2x + 34.7$. We can use the Quadratic Formula to solve this equation.

Step 1 Standard form The equation is given in standard form.

Step 2 Substitute Substituting 0.025 for a, −2 for b, and 34.7 for c gives

$$x = \frac{-(-2) \pm \sqrt{(-2)^2 - 4(0.025)(34.7)}}{2(0.025)}$$

Step 3 Simplify

$$x = \frac{2 \pm \sqrt{4 - 3.47}}{0.05}$$

$$x = \frac{2 \pm \sqrt{0.53}}{0.05}$$

Step 4 Solutions

Solution One	*Solution Two*	
$x = \dfrac{2 + \sqrt{0.53}}{0.05}$	$x = \dfrac{2 - \sqrt{0.53}}{0.05}$	⟸ Wrote both solutions.
$x = \dfrac{2 + 0.73}{0.05}$	$x = \dfrac{2 - 0.73}{0.05}$	⟸ Simplified the square root and rounded to two decimal places.
$x = 54.6$	$x = 25.4$	

Step 5 Check We leave the check to you.

The mathematical model provides us with two possible wind speeds, approximately 55 mph and 25 mph. For reasons you will learn in a biology class, only the 25 mph value is reasonable. This is another case where the results of a mathematical analysis may be used to help guide us to a solution but the final conclusion must be informed by knowledge of other subject areas.

b) Some sources say that a temperature of −20 degrees Fahrenheit can lead to flesh freezing in about 1 minute. Find the wind speed necessary for the wind chill temperature to be −20 degrees Fahrenheit if the air temperature is 30 degrees Fahrenheit.

•SOLUTION• We are given an output value and asked to find the corresponding input values. Substituting −20 for y gives the equation $-20 = 0.025x^2 - 2x + 34.7$, which we may solve using the Quadratic Formula.

Step 1 Standard form We move −20 to the right side to get $0 = 0.025x^2 - 2x + 54.7$

Step 2 Substitute Substituting 0.025 for a, −2 for b, and 54.7 for c gives

$$x = \frac{-(-2) \pm \sqrt{(-2)^2 - 4(0.025)(54.7)}}{2(0.025)}$$

Step 3 Simplify

$$x = \frac{2 \pm \sqrt{4 - 5.47}}{0.05}$$

$$x = \frac{2 \pm \sqrt{-1.47}}{0.05}$$

We can stop here because the radicand is negative and so the equation has no real solutions. Therefore, the formula predicts that at 30 degrees Fahrenheit we will not reach a wind chill temperature of −20 degrees no matter how fast the wind is blowing.

c) Use the quadratic model to find the wind chill if the air temperature is 30 degrees Fahrenheit and the wind speed is 0 mph.

•SOLUTION• Intuition tells us that if there is no wind then the wind chill temperature should be the same as the air temperature. Let's see what happens when we substitute 0 for x in the quadratic model and simplify.

$$y = 0.025x^2 - 2x + 34.7 \quad \Leftarrow \quad \text{Quadratic model.}$$

$$y = 0.025(0)^2 - 2(0) + 34.7 \quad \Leftarrow \quad \text{Substituted 0 for the input.}$$

$$y = 34.7 \quad \Leftarrow \quad \text{Simplified.}$$

This implies that if the wind is not blowing you may feel warmer than the surrounding air. This can happen since the same insulating air the wind blows away to create the wind chill can hold some of the heat of your body close to your skin. This makes your skin warmer than the surrounding air.

Practice 9.6.4

1. The graph shows the relation between the height (feet) of a baseball thrown straight into the air at 60 feet per second (40 mph) and the time (seconds) since it was thrown. The equation which approximates this graph is $y = -16x^2 + 60x + 7$. Use this information to answer the following questions.

    a) Predict the height of the ball after 4 seconds.
    b) Predict when the ball will be at 40 feet.
    c) Predict when the ball will hit the ground.

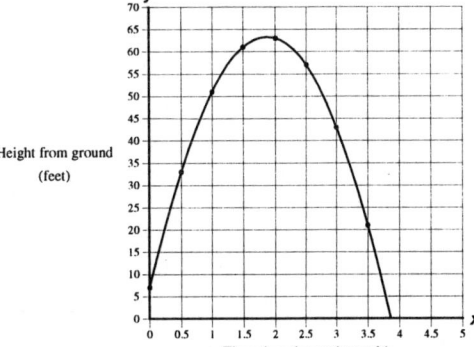

Height from ground (feet)

Time since thrown (seconds)

Exercise Set 9.6 The answers to the odd numbered exercises are at the back of the book.

Solve using the Quadratic Formula.

1. $3x^2 - 2x - 5 = 0$
2. $2x^2 - 19x + 35 = 0$
3. $0 = 2x^2 + 5x + 2$
4. $0 = 3x^2 - 10x + 3$

5. $x^2 - 9x + 4 = 0$
6. $x^2 + 3x + 1 = 0$
7. $-4x^2 + x + 2 = 0$
8. $-3x^2 + 3x + 1 = 0$

9. $9x^2 = 12x - 4$
10. $-28x = -4x^2 - 49$
11. $3x^2 + 3x - \dfrac{21}{4} = 0$
12. $7x^2 - 4x - \dfrac{26}{7} = 0$

13. $-2.07x + 0.05x^2 = 1.15$
14. $7 - 1.01x^2 = 3.19x$
15. $4x^2 + 2x = -2$
16. $9 + 3x^2 = 4x$

17. $-x + 0.5x^2 + 0.5 = 0$
18. $-4 = 2x + \dfrac{1}{4}x^2$

---

19. The graph shows the relation between thousands of births in the U. S., projected to the year 2050, and the number of years since 1985. The equation that approximates this graph is $y = 0.35x^2 + 5.4x + 3877.8$. Use this information to answer the following questions.

    a) Predict the year births will reach 5,000,000.
    b) Predict the number of births in 2075.
    c) Predict when the number of births will be 500,000 more than the number in the year 2000.

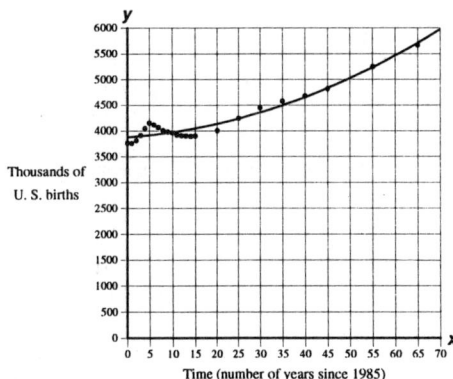

Thousands of U. S. births

Time (number of years since 1985)

20. The graph shows the relation between the number of years since 1980 and the number of reported cases of Hepatitis B in thousands. The equation which approximates this graph is $y = -0.21x^2 + 2.3x + 19.26$. Use this information to answer the following questions.

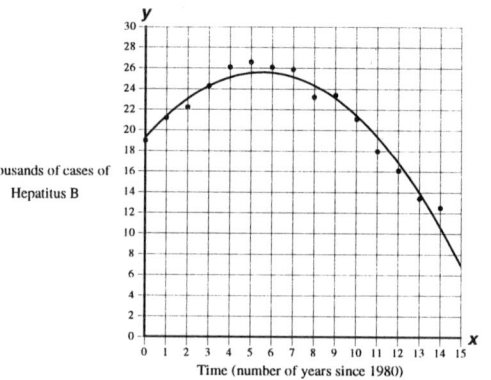

a) Predict the year in which the number of cases of Hepatitis B reached 15,000.
b) Predict the number of expected cases of Hepatitis B in 1995.
c) Predict when the number of cases of Hepatitis B was at the same level as in 1995.

Review Exercises   The answers to all of these exercises are at the back of the book.

21. A swimming pool is filled at the rate of 75 gallons per hour and then drained to empty at the rate of 60 gallons per hour. If it took an hour and a half longer to drain the pool than to fill it how many gallons does the swimming pool hold?

---

22. Graph –0.6 on a number line. Use an interval from –2 to 1 in steps of 0.2.   23. Graph: $2x - 3y \le 6$

24. Use long division to simplify: $\dfrac{2n^5 - 7n^4 - 2n + 7}{2n - 7}$ 25. Factor completely $25x^2 - 4$

Simplify.

26. $(s^6 t^2)^{-2}(s^{-6} t^{-2})$

27. $\dfrac{1.08 \times 10^{-5}}{(0.016)(0.00003)}$

28. $\dfrac{5y^2}{x^2 z} \cdot \dfrac{3x^5}{10y^5} \cdot \dfrac{4z^2}{9}$

29. $\dfrac{2x^2 + 2x}{x - 2} \cdot \dfrac{x^2 - 4}{3x^2 + 3x}$

30. $\dfrac{x^4 - x^3 - 2x^2}{x^2 + x} \div (x - 2)$

31. $\dfrac{1}{x - 1} - 5$

32. $\dfrac{\frac{3}{t} + 1}{\frac{1}{t}}$

33. $\sqrt{\dfrac{4ab^2}{9}}$

34. $\sqrt{150}$

35. $\sqrt{18x^7 y^2}$

36. $\dfrac{4}{\sqrt{6}}$

37. $-8^0 - 64 \div 2^2 \cdot 2(1 - 2)$

Solve.

38. $3x^2 + 5x - 2 = 0$

39. $\dfrac{\sqrt{z - 6}}{2} + 1 = 2$

40. $\begin{cases} 3x - 4y = 2 \\ -6x + 8y = 5 \end{cases} \begin{matrix} ❶ \\ ❷ \end{matrix}$

---

41. Determine whether this graph represents a function:

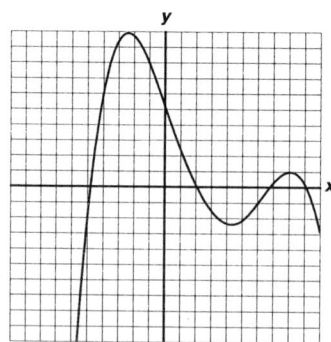

Chapter 9 Review

Vocabulary to Know

complex numbers — a set of numbers that allows values such as $\sqrt{-1}$.

inverse operations — operations that undo each other; addition and subtraction are inverse operations.

irrational numbers — numbers that cannot be written as the ratio of two integers; the decimal representation of an irrational number never repeats or terminates; $\sqrt{2}$ and π are irrational numbers.

like radical terms — terms that have the same variable factors and the same radical factors; $2x\sqrt{3y}$ and $\dfrac{5x\sqrt{3y}}{2}$ are like radical terms.

negative square root — a negative number that, when multiplied by itself, produces the given number; the negative square root of 4, written $-\sqrt{4}$, is -2.

parabola — the curve formed when graphing a quadratic equation.

perfect square — a number that is the square of a whole number; 1, 4, 25, and 81 are perfect squares because they are the squares of 1, 2, 5, and 9, respectively.

principal square root — a positive number that, when multiplied by itself, produces the given number; the principal square root of 4, written $\sqrt{4}$, is 2.

quadratic relation — a relation that can be modeled using $y = ax^2 + bx + c$, where a, b, and c are constants.

radical — same as radical expression.

radical equation — an equation where a variable appears in a radicand.

radical expression — an expression that contains a radical symbol (same as radical).

radical symbol — the symbol $\sqrt{}$ that indicates the principal square root of a number; the principal square root of 5 is written $\sqrt{5}$.

radicand — the expression under the radical symbol; the radicand of $\sqrt{5x}$ is $5x$.

rationalizing the denominator — converting the denominator of a fraction from an irrational to a rational number.

square root — numbers that, when multiplied by themselves, produce the given number; the square root of 4, written $\pm\sqrt{4}$ is ± 2 (that is, $+2$ and -2).

symbols

  $\pm$ plus or minus symbol, indicates two values, one positive and one negative; ±4 represents two numbers, $+4$ and -4.

  $\sqrt{}$ radical symbol or square root symbol, indicates the principal square root of a number; $\sqrt{4}$ is 2.

Properties, Definitions, Formulas, and Notation to Understand

Square root of a negative number — The square root of a negative number is not defined using real numbers.	If $a < 0$, then $\sqrt{a}$ is not defined using real numbers. $\sqrt{-16}$ is not defined because the radicand, -16, is negative.
Square root of a perfect square — The square root of the square of a number is the absolute value of that number.	$\sqrt{x^2} = \lvert x \rvert$. $\sqrt{(2x-3)^2} = \lvert 2x-3 \rvert$
Square of a square root — The square of the square root of a non–negative number is that number.	If $x \geq 0$, then $\left(\sqrt{x}\right)^2 = x$ If $x \geq 0$, then $\left(\sqrt{7x}\right)^2 = 7x$
Product property of square roots — The square root of a product is the product of the square roots.	For $a \geq 0$ and $b \geq 0$, $\sqrt{ab} = \sqrt{a}\,\sqrt{b}$ If $x \geq 0$, $\sqrt{4x}$ may be written as $\sqrt{4}\,\sqrt{x}$
Quotient property of square roots — The square root of a quotient is the quotient of the square roots.	For $a \geq 0$ and $b > 0$, $\sqrt{\dfrac{a}{b}} = \dfrac{\sqrt{a}}{\sqrt{b}}$ If $x > 0$, $\sqrt{\dfrac{2}{x}}$ may be written as $\dfrac{\sqrt{2}}{\sqrt{x}}$

Simplest radical form	An expression containing a square root is said to be in simplest radical form when *Condition 1* No factors of the radicand are perfect squares. *Condition 2* The radicand contains no fractions. *Condition 3* No fraction has a square root in its denominator.
Taking a square root and squaring are inverse operations	$\sqrt{x^2} = \lvert x \rvert$ That is, if $x \geq 0$, then $\sqrt{x^2} = x$ but if $x < 0$, then $\sqrt{x^2} = -x$ $\sqrt{5^2}$ simplifies to 5.
Squaring and taking a square root are inverse operations	If $x \geq 0$, then $\left(\sqrt{x}\right)^2 = x$. $\left(\sqrt{5}\right)^2$ simplifies to 5.
Squaring property of equality — If two numbers are equal then their squares are equal.	If $a = b$, then $a^2 = b^2$ If $w = 3$ then $w^2 = 3^2$ If $\sqrt{x} = 5$, then $\left(\sqrt{x}\right)^2 = 5^2$
Quadratic equation in two variables — A two variable equation that can be written in the form $y = ax^2 + bx + c$, where a, b, and c are constants. The graph of a quadratic equation is called a parabola.	If a, b, and c are real numbers and $a \neq 0$, then an equation that can be written in the form $y = ax^2 + bx + c = 0$ is a quadratic equation in two variables. $y = x^2 - 4$ The graph of this equation is shown below: 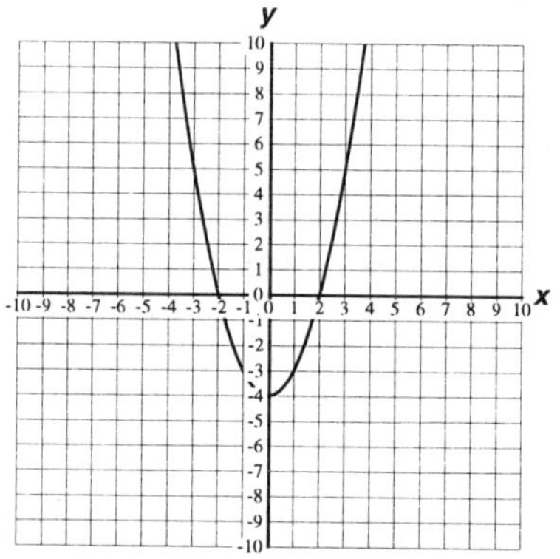
The Quadratic Formula	If $a \neq 0$, then an equation of the form $0 = ax^2 + bx + c$ has the following two solutions (the $\pm$ means use $+$ for one solution and $-$ for the other): $$x = \frac{-b \pm \sqrt{b^2 - 4ac}}{2a}$$ $0 = 2x^2 - 3x - 2$ has solutions $\dfrac{-(-3) \pm \sqrt{(-3)^2 - 4(2)(-2)}}{2(2)}$

Procedures to Follow

Procedure for simplifying expressions using the proper order of operations.	Simplify $6 - \sqrt{16} \cdot \sqrt{25 + 9}$
Step 1 Simplify expressions inside grouping symbols.	$6 - \sqrt{16} \cdot \sqrt{36}$
Step 2 Simplify exponents and radicals.	$6 - 4 \cdot 6$
Step 3 Simplify multiplication and division, working left to right.	$6 - 24$
Step 4 Simplify addition and subtraction, working left to right.	-18

Procedure for simplifying square roots.	Simplify $\sqrt{360x^5}$
Rewrite the radicand as a product where one factor is a perfect square, and then simplify.	Try doing this mentally first by asking yourself *what is the largest perfect square that is a factor of $360x^5$*. If you can "see" $36x^4$, rewrite the radical as a product where this perfect square is a factor and then simplify: $$\sqrt{36x^4}\sqrt{10x}$$ $$6x^2\sqrt{10x}$$
If you have difficulty with the above method, follow the steps given below.	OR, follow the steps given below:
Step 1 Prime factor the radicand.	$\sqrt{360x^5} = \sqrt{2^3 3^2 5^1 x^5}$
Step 2 For factors with exponents of 2 or more, divide the exponent by 2. • the quotient is the power of the factor OUTSIDE the radical symbol. • the remainder is the power of the factor INSIDE the radical symbol. If the remainder is 0, the factor does not appear inside the radical symbol.	• 2^3 has an exponent of 3. If we divide the exponent by 2 we get a quotient of 1, so one factor of 2 is outside the radical symbol. The remainder is 1, so one factor of 2 is inside the radical symbol. Now we have $2^{\mathbf{1}}\sqrt{2^{\mathbf{1}}3^2 5^1 x^5}$ • If we divide the exponent of 3^2 by 2 we get a quotient of 1, so one factor of 3 is outside the radical symbol. The remainder is 0, so no factor of 3 is inside the radical symbol. Now we have $2^1\mathbf{3^1}\sqrt{\mathbf{2^1}5^1 x^5}$ • There is a single factor of 5 under the radical symbol. • If we divide the exponent of x^5 by 2 we get a quotient of 2, so two factors of x are outside the radical symbol. The remainder is 1, so one factor of x is inside the radical symbol. Now we have $2^1 3^1 x^2\sqrt{\mathbf{2^1}5^1 x^1}$ The simplified expression is $6x^2\sqrt{10x}$

Procedure for multiplying square roots.	Simplify $\sqrt{6x^3}\sqrt{18x^5}$
Step 1 Multiply the radicands.	$\sqrt{108x^8}$
Step 2 Simplify.	$\sqrt{2^2 3^3 x^8}$ $2^1 \cdot 3^1 x^4\sqrt{3^1}$ $6x^4\sqrt{3}$

Procedure for rationalizing the denominator.	Simplify $\dfrac{3}{\sqrt{5}}$
Step 1 Multiply numerator and denominator by a radical that will make the radicand in the denominator a perfect square.	$\dfrac{3}{\sqrt{5}} \cdot \dfrac{\sqrt{5}}{\sqrt{5}}$
Step 2 Simplify.	$\dfrac{3\sqrt{5}}{5}$

Procedure for adding (or subtracting) like radical terms.	Simplify $3\sqrt{8x^3} + x\sqrt{18x}$
Step 1 If necessary, simplify each term.	$3 \cdot 2\,x\sqrt{2x} + 3x\sqrt{2x}$ $6x\sqrt{2x} + 3x\sqrt{2x}$
Step 2 Add (or subtract) the numeric coefficients of the like radical terms.	$9x\sqrt{2x}$

Procedure for simplifying radical expressions.	Simplify $\sqrt{2x}\sqrt{12} + 5\sqrt{\dfrac{x^3}{6x^2}}$
Step 1 Simplify inside Simplify under the radical symbol.	$\sqrt{2x}\sqrt{12} + 5\sqrt{\dfrac{x}{6}}$
Step 2 Multiply Combine radicals that are multiplied.	$\sqrt{24x} + 5\sqrt{\dfrac{x}{6}}$
Step 3 Simplify radicals Simplify each radical so the radicands contain no perfect squares and combine any like radical terms.	$2\sqrt{6x} + 5\sqrt{\dfrac{x}{6}}$
Step 4 Rationalize Rationalize fractions that contain radicals in their denominators.	$2\sqrt{6x} + 5\sqrt{\dfrac{x}{6}} \cdot \dfrac{\sqrt{6}}{\sqrt{6}}$ $2\sqrt{6x} + \dfrac{5\sqrt{6x}}{6}$
Step 5 Combine Combine like radical terms and simplify.	$\dfrac{12}{6}\sqrt{6x} + \dfrac{5}{6}\sqrt{6x}$ $\dfrac{17}{6}\sqrt{6x}$ or $\dfrac{17\sqrt{6x}}{6}$

Procedure for solving radical equations .	Solve $3\sqrt{2x} - 12 = 6$
Step 1 Isolate Isolate the radical on one side of the equation.	$3\sqrt{2x} = 18$ $\sqrt{2x} = 6$
Step 2 Square Square both sides of the equation.	$2x = 36$
Step 3 Solve Solve the resulting equation.	$x = 18$

Step 4 Check Check the solution.	Substituting 18 for x gives $$3\sqrt{2(18)} - 12 = 6$$ $$3\sqrt{36} - 12 = 6$$ $$3(6) - 12 = 6$$ $$18 - 12 = 6$$ $$6 = 6 \quad \text{Checks.}$$
Procedure for solving a quadratic equation using the quadratic formula.	Solve $2x^2 + x = 6$
Step 1 Standard Form Write the equation with 0 on one side and a polynomial in standard form on the other, as in $ax^2 + bx + c = 0$	$2x^2 + x - 6 = 0$
Step 2 Substitute Identify the values of a, b, and c and substitute these into the Quadratic Formula, $$x = \frac{-b \pm \sqrt{b^2 - 4ac}}{2a}$$	a is 2, b is 1, and c is -6 $$x = \frac{-(1) \pm \sqrt{(1)^2 - 4(2)(-6)}}{2(2)}$$
Step 3 Simplify Simplify the expression as far as you can without evaluating the $\pm$ symbol.	$$x = \frac{-1 \pm \sqrt{1 + 48}}{4}$$ $$x = \frac{-1 \pm \sqrt{49}}{4}$$ $$x = \frac{-1 \pm 7}{4}$$
Step 4 Split Split the solution at the $\pm$ into two parts, one which uses the + and the other which uses the −. Simplify the two expressions.	$$x = \frac{-1 + 7}{4} \qquad\qquad x = \frac{-1 - 7}{4}$$ $$x = \frac{3}{2} \qquad\qquad\qquad x = -2$$
Step 5 Check Check all solutions.	Substituting $\frac{3}{2}$ for x gives $\qquad$ Substituting -2 for x gives $$2\left(\frac{3}{2}\right)^2 + \left(\frac{3}{2}\right) = 6 \qquad 2(-2)^2 + (-2) = 6$$ $$2\left(\frac{9}{4}\right) + \frac{3}{2} = 6 \qquad\quad 2(4) - 2 = 6$$ $$\frac{9}{2} + \frac{3}{2} = 6 \qquad\qquad\quad 8 - 2 = 6$$ $$\frac{12}{2} = 6 \qquad\qquad\quad 6 = 6 \quad \text{Checks.}$$ $$6 = 6 \quad \text{Checks.}$$

Chapter 9 Review Exercises   The answers to all the exercises are in the back of the book.

1. $-\sqrt{-9}$

2. $-\sqrt{16}$

3. $\left(\sqrt{49} - \sqrt{36}\right)^2$

4. $\dfrac{12\sqrt{4} - 6\sqrt{4}}{4\sqrt{4}}$

5. $\sqrt{49x^2y^2}$

6. $\sqrt{\dfrac{9m^2n}{16}}$

7. $\dfrac{\sqrt{50m}\,\sqrt{6m}}{2\sqrt{3}}$

8. $\sqrt{15x}\,\sqrt{16x}$

9. $\sqrt{72}$

10. $\sqrt{\dfrac{300}{25}}$

11. $\sqrt{1568}$

12. $\sqrt{1440}$

13. $\sqrt{x^2y^4z^{11}}$

14. $4\sqrt{f^3h^9}$

15. $\sqrt{108s^4t^2}$

16. $\sqrt{700m^3n^4p^5}$

17. $\dfrac{3t\sqrt{294s^7}}{21s^4}$

18. $2b^3\sqrt{32b^5cd^8}$

19. $\dfrac{4\sqrt{3a^4}\,\sqrt{6a^3b^2}}{3a^3b}$

20. $\sqrt{\dfrac{5x^3}{2y^5}}\,\sqrt{\dfrac{20x}{y^3}}$

21. $\dfrac{4\sqrt{6}}{\sqrt{12}}$

22. $\dfrac{3y}{\sqrt{18y}}$

23. $\left(2\sqrt{x} + x\right)^2$

24. $6\sqrt{t} + 3\sqrt{t} - 2t\sqrt{t}$

---

25. $14\sqrt{2r} - 2\sqrt{14r} + 3\sqrt{14r} - 11\sqrt{2r}$

26. $\sqrt{12x^3} + \sqrt{27x} - x\sqrt{75x}$

27. $\sqrt{28} - \sqrt{20} - \sqrt{112} + 3\sqrt{5}$

28. $3\sqrt{t}\left(\sqrt{t} - t\right) + 3\left(t + t\sqrt{t}\right)$

Solve.

29. $\sqrt{2y - 4} = 6$

30. $4\sqrt{16t} = 16$

31. $\sqrt{-4t + 6} = 3\sqrt{2t + 8}$

32. $3 - \sqrt{4 - 3x} = -2$

33. $12 = \sqrt{2x} + 14$

34. $\sqrt{x^2 + 5} + 5 = x$

35. $-2x^2 - 3x + 2 = 0$

36. $4y^2 - 8y - 3 = 0$

37. $q^2 - \dfrac{2q}{3} + \dfrac{1}{9} = 0$

38. $b^2 = b + 1.75$

39. Bill is at one corner of a rectangular mall on campus and needs to get to the diagonally opposite corner. He considers walking the diagonal but sees a sign *Please keep off the grass*. If Bill can walk at 3 mph (4.4 feet per second) and the mall is 100 feet by 200 feet, how much time will he lose by walking along the edge rather than across the diagonal?

Chapter 9 Test   The answers to all the exercises are in the back of the book.

Simplify.

1. $\dfrac{16 - 2\sqrt{16} - \left(2\sqrt{25} + 1\right)}{-3}$

2. $\sqrt{\dfrac{36t}{a^2}}$

3. $\dfrac{\sqrt{20y}\,\sqrt{4}}{\sqrt{5y}}$

4. $6\sqrt{5t^3} + t\sqrt{45t} + \sqrt{80t^3}$

5. $\sqrt{588}$

6. $\sqrt{h^{11}j^7}$

7. $\sqrt{15}\left(\sqrt{30} - \sqrt{6}\right) - 3\left(5\sqrt{2} + \sqrt{10}\right)$

8. $\sqrt{9n^2 + 12n + 4}$

9. $\sqrt{\dfrac{15a^2b}{c^3}}\,\sqrt{\dfrac{10b^3}{c}}$

10. $-\dfrac{1}{4}\sqrt{2y} + \dfrac{1}{3}\sqrt{2y} + \dfrac{1}{2}\sqrt{2y}$

11. $\sqrt{6y}\,\sqrt{42y}$

12. $\dfrac{3d}{\sqrt{24}}$

Solve.

13. $3\sqrt{2x} = 12$

14. $\sqrt{x^2 + 4x + 4} - 1 = 2x$

15. $3\sqrt{x + 1} = \sqrt{2x - 5}$

16. $6x^2 + 13x + 6 = 0$

17. $3y^2 - 3y - 1 = 0$

---

18. Jane is building a 4 foot by 7 foot rectangular door for her shed. How long should a cross brace be that reaches from one corner of the door to the diagonally opposite corner?

Answers to Selected Exercises

Chapter 1 The Real Number System

Section 1.1 Introduction to Algebra

Practice 1.1.1

1. 25 + 4 is a sum whose terms are 25 and 4.
2. 5*w* is a product whose factors are 5 and *w*.
3. 6*x* + 2 is a sum whose terms are 6*x* and 2; 6*x* is a product whose factors are 6 and *x*
4. 2*m* + 3*n* is a sum whose terms are 2*m* and 3*n*; 2*m* is a product whose factors are 2 and *m*; 3*n* is a product whose factors are 3 and *n*.
5. 5*abc* is a product whose factors are 5, *a*, *b*, and *c*

Practice 1.1.2

1. 24 **2.** 21 **3.** 9 **4.** 15 **5.** 36 **6.** 6

Practice 1.1.3

1. 4 **2.** 5 **3.** 9 **4.** 7

Practice 1.1.4

1. 31 **2.** 28.26 **3.** 1020 **4.** 48 **5.** 97 **6.** 169

Practice 1.1.5

1.

Variable
Def (pg. 103): A symbol, usually a letter, that is used to represent something, usually an unknown quantity or a quantity that can take on many different values.
Ex: x, y

2.

Simplify
Def (pg. 106): to perform some or all of the allowable operations in an expression.
Ex: The simplified version of 2 + 3 is 5.

3.

Factor
Def (pg. 103): numbers being multiplied.
Ex: In 2x, both 2 and x are factors.

4.

Exponent
Def (pg. 106): in exponential notation, the number of bases being multiplied
Ex: In 3^2, 2 is the exponent. 3^2 means 3 • 3

Exercise Set 1.1

1. 35 + 21 is a sum whose terms are 35 and 21
3. (3.5) • 2 is a product whose factors are 3.5 and 2
5. 4*u* is a product whose factors are 4 and *u*
7. 5*v* + 3*w* is a sum whose terms are 5*v* and 3*w*. 5 *v* is a product whose factors are 5 and *v*; 3*w* is a product whose factors are 3 and *w*.
9. 7 **11.** 8 **13.** 216 **15.** 216 **17.** 11 **19.** 14
21. 14 **23.** 144 **25.** 18 **27.** 8 **29.** 73 **31.** 9
33. 28 **35.** 57 **37.** 121 **39.** 2.25

Section 1.2 Fractions

Practice 1.2.1

1. $2 \cdot 3^2$ **2.** $2 \cdot 3^3$ **3.** 5•7•11 **4.** 7•17 **5.** 3•5•7
6. $2^2 \cdot 3 \cdot 5$

Practice 1.2.2

1. $\frac{9}{10}$ **2.** $\frac{3}{8}$ **3.** $\frac{9}{5}$ **4.** $\frac{2}{3}$ **5.** $\frac{5}{8}$ **6.** $\frac{7}{10}$

Practice 1.2.3

1. $\frac{8}{45}$ **2.** 5 **3.** $\frac{2}{9}$ **4.** $\frac{21}{10}$ **5.** $\frac{8}{3}$ **6.** $\frac{5}{11}$

Practice 1.2.4

1. $\frac{33}{32}$ **2.** $\frac{1}{6}$ **3.** $\frac{x}{1}$ or x **4.** $\frac{1}{y}$

Practice 1.2.5

1. $\frac{1}{3} \cdot 5$ **2.** $\frac{1}{3}x$ **3.** $1w$ **4.** $\frac{5}{4}p$ **5.** $\frac{3}{5}c$

Practice 1.2.6

1. $\frac{4}{3}$ **2.** $\frac{3}{7}$ **3.** $\frac{1}{4}$ **4.** 10 **5.** $\frac{5}{8}$ **6.** $\frac{1}{14}$ **7.** $\frac{5}{98}$ **8.** 2

Practice 1.2.7

1. $\frac{3}{2}$ **2.** $\frac{1}{2}$ **3.** $\frac{1}{5}$ **4.** $\frac{1}{4}$

Practice 1.2.8

1. 36 **2.** 72 **3.** 360 **4.** 510

Practice 1.2.9

1. $\frac{32}{40}$ **2.** $\frac{15}{48}$ **3.** $\frac{72}{18}$ **4.** $\frac{162}{54}$

Practice 1.2.10

1. $\frac{25}{36}$ **2.** $\frac{1}{6}$ **3.** $\frac{31}{18}$ **4.** $\frac{17}{18}$ **5.** $\frac{19}{12}$ **6.** $\frac{1}{2}$

Practice 1.2.11

1. 8 **2.** 2 **3.** $\frac{31}{36}$ **4.** $\frac{5}{6}$ **5.** $\frac{1}{2}$ **6.** $\frac{7}{4}$

Exercise Set 1.2

1. $2^5 \cdot 3$ **3.** $2 \cdot 3^3 \cdot 7$ **5.** $3 \cdot 5^2 \cdot 11$ **7.** $2^2 \cdot 11^2$ **9.** $\frac{3}{10}$

11. $\frac{2}{5}$ **13.** $\frac{15}{14}$ **15.** $\frac{8}{9}$ **17.** $\frac{2}{21}$ **19.** $\frac{7}{3}$ **21.** $\frac{28}{75}$ **23.** $\frac{1}{8}$

25. $\frac{6}{5}$ **27.** $\frac{a}{7}$ **29.** $\frac{1}{t}$ **31.** $\frac{1}{15} \cdot 13$ **33.** $\frac{1}{2}x$ **35.** $\frac{3}{8}t$

37. 1 **39.** $\frac{5}{2}$ **41.** $\frac{1}{28}$ **43.** $\frac{7}{5}$ **45.** 1 **47.** $\frac{4}{5}$ **49.** $\frac{7}{3}$

51. $\frac{1}{6}$ **53.** 56 **55.** 315 **57.** 432 **59.** $\frac{3}{18}$ **61.** $\frac{5}{180}$

63. $\frac{273}{91}$ **65.** $\frac{31}{72}$ **67.** $\frac{3}{5}$ **69.** 4 **71.** $\frac{5}{252}$ **73.** $\frac{77}{144}$

75. $\frac{2}{5}$ **77.** $\frac{10}{3}$ **79.** $\frac{7}{6}$ **81.** $\frac{1}{2}$ **83.** $\frac{16}{21}$

Numbers inside brackets **[ ]** indicate the Chapter, Section, and Example number of a similar worked example.

85. [1.1.1] $-2r + 5s$ is a sum whose terms are $-2r$ and $5s$; $-2r$ is a product whose factors are -2 and r; $5s$ is a product whose factors are 5 and s. **86. [1.1.2]** 21 **87. [1.1.2]** 3

88. [1.1.2] 44 **89. [1.1.2]** 5 **90. [1.1.2]** 2

91. [1.1.3] 22 **92. [1.1.3]** 3 **93. [1.1.4]** 9

94. [1.1.4] 15 **95. [1.1.4]** 48

Section 1.3 Real Numbers

Practice 1.3.1

1.

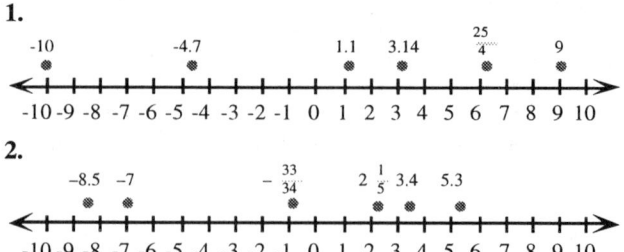

2.

Practice 1.3.2

1.

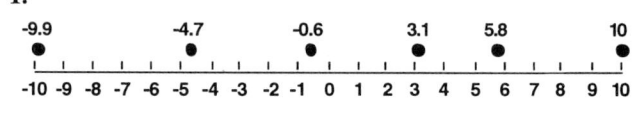

2.

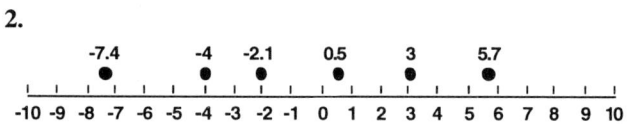

Practice 1.3.3

1.

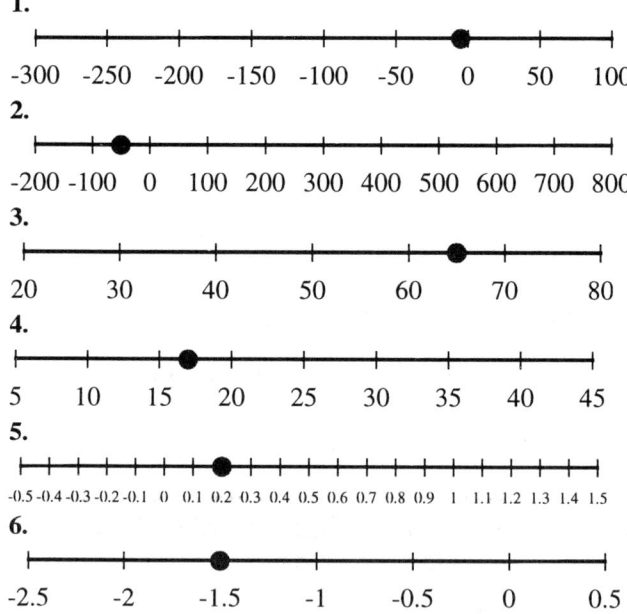

2.

3.

4.

5.

6.

Practice 1.3.4

1. $7 < 12$ **2.** $\frac{1}{2} < \frac{3}{4}$ **3.** $-12 > -22$ **4.** $5 > -12$

Practice 1.3.5

1. -36 **2.** 45 **3.** $2w$ **4.** $-7t$ **5.** 2

Practice 1.3.6

1. 12 **2.** 57 **3.** $\frac{44}{45}$ **4.** -179 **5.** -59

Practice 1.3.7

1. 27 **2.** 18 **3.** 59 **4.** 21 **5.** 9 **6.** 77

Exercise Set 1.3

1.

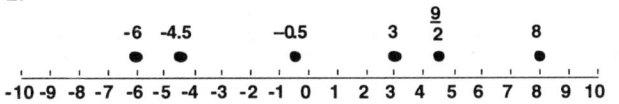

3.

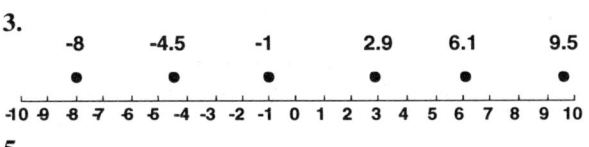

5.

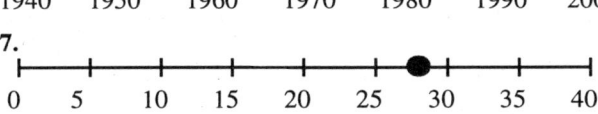

7.

0	5	10	15	20	25	30	35	40

9.

–2	–1.5	–1	–0.5	0	0.5	1	1.5	2

11. $-3 > -4$ **13.** $-\frac{1}{2} < -\frac{1}{3}$ **15.** $5 > -12$ **17.** $0 > -\frac{7}{9}$

19. $\frac{5}{12} > -\frac{6}{7}$ **21.** –12 **23.** $5x$ **25.** –7 **27.** 577

29. –7.5 **31.** 3 **33.** 6 **35.** –14

37. not possible because we get division by 0

Numbers inside brackets **[ ]** indicate the Chapter, Section, and Example number of a similar worked example.

39. [1.1.1] sum: 8 + 2; terms: 8 and 2 **40.** [1.2.2] $\frac{2}{7}$

41. [1.2.4] $\frac{2}{x}$ **42.** [1.2.5] $\frac{1}{3} \cdot 7$ **43.** [1.1.3] 24

44. [1.1.2] 18 **45.** [1.2.11] 5 **46.** [1.2.6] $\frac{28}{15}$

47. [1.2.6] 35 **48.** [1.2.10] $\frac{41}{75}$ **49.** [1.2.10] $\frac{33}{20}$

50. [1.2.11] 6 **51.** [1.2.3] $\frac{5}{16}$

Section 1.4 Multiplying and Dividing Real Numbers

Practice 1.4.1

1. –30 **2.** 54 **3.** –60 **4.** –6 **5.** 8 **6.** –40 **7.** –56

Practice 1.4.2

1. 35 **2.** –25 **3.** –80 **4.** 500 **5.** $\frac{3}{4}$ **6.** 8000

Practice 1.4.3

1. 8 **2.** –9 **3.** 9 **4.** 24

Practice 1.4.4

1. $(-2)(-2)(-2)$ or $(-2)^3$, simplifies to –8.

2. $(5)(-1)$, simplifies to –5.

3. $2(-10)$, simplifies to –20.

4. $\frac{2}{3} \cdot 9$, simplifies to 6.

5. $0.05(8)$, simplifies to 0.4.

6. $15(-4)$, simplifies to –60.

Practice 1.4.5

1. Product of –2 and 4, or –2 multiplied by 4, or –2 times 4.

2. Product of 7 and 8, or 7 multiplied by 8, or 7 times 8.

3. Product of $\frac{1}{2}$ and 5, or $\frac{1}{2}$ multiplied by 5 or $\frac{1}{2}$ of 5.

4. Six to the third power, or six cubed.

Practice 1.4.6

1. –5 **2.** –7 **3.** undefined (we may not divide by 0)

4. 0 (0 divided by any non–zero number is 0, and 0 has no sign) **5.** 5

Practice 1.4.7

1. $-\frac{3}{4}$ **2.** –72 **3.** $\frac{2}{15}$

Practice 1.4.8

1. $\frac{3}{5}$ **2.** $-\frac{1}{12}$ **3.** 180 **4.** $\frac{125}{27}$

Practice 1.4.9

1. $-8/4$, $\frac{-8}{4}$ or $-8 \div 4$, which simplify to –2.

2. $-24/\left(\frac{4}{3}\right)$, $\frac{-24}{\frac{4}{3}}$, or $-24 \div \frac{4}{3}$, which simplify to –18.

3. $\left(\frac{8}{9}\right) / (-3)$, $\frac{\frac{8}{9}}{-3}$ or $\frac{8}{9} \div (-3)$, which simplify to $\frac{-8}{27}$

4. $-36/(-4)$, $\frac{-36}{-4}$ or $-36 \div (-4)$, which simplify to 9.

5. $\frac{1}{2}$ **6.** $\frac{1}{6-4}$, which simplifies to $\frac{1}{2}$.

Practice 1.4.10

1. The quotient of 18 and 4, or 18 divided by 4

2. The quotient of 39 and –5, or 39 divided by –5.

3. The reciprocal of 3 more than 5.

4. The reciprocal of twice –4.

Exercise Set 1.4

1. 75 **3.** –30 **5.** 6 **7.** –1 **9.** –4 **11.** 18 **13.** –6

15. –4 **17.** 144 **19.** –18 **21.** –5

23. $(-5)(-5)$ or $(-5)^2$, which simplifies to 25.

25. $2(-8)$, which simplifies to –16.

27. $\frac{6}{7}(-14)$, which simplifies to -12.

29. $0.12(80)$, which simplifies to 9.6.

31. Product of 6 and 3, or 6 multiplied by 3 or 6 times 3

33. Product of $\frac{3}{4}$ and 12, or $\frac{3}{4}$ multiplied by 12, or $\frac{3}{4}$ of 12.

35. -5 **37.** 9 **39.** -6 **45.** $-\frac{1}{9}$ **47.** $\frac{27}{20}$ **49.** $\frac{3}{5}$

51. $-\frac{8}{3}$ **53.** 12 **55.** $-\frac{1}{4}$ **57.** $\frac{1}{64}$ **59.** $-\frac{2}{3}$

61. $50/(-150)$, $\frac{50}{-150}$ or $50\div(-150)$, simplify to $-\frac{1}{3}$

63. $1/(12-5)$, $\frac{1}{12-5}$ or $1\div(12-5)$, simplify to $\frac{1}{7}$.

65. $\frac{4}{5}/(-2)$, $\frac{\frac{4}{5}}{-2}$ or $\frac{4}{5}\div(-2)$, simplify to $\frac{-2}{5}$.

67. $\left(\frac{6}{15}\right)/\left(\frac{-2}{5}\right)$, $\frac{\frac{6}{15}}{\frac{-2}{5}}$ or $\left(\frac{6}{15}\right)\div\left(\frac{-2}{5}\right)$, simplify to -1

69. The quotient of 64 and -8, or 64 divided by -8.

71. The quotient of 2 and $\frac{1}{3}$, or 2 divided by $\frac{1}{3}$.

73. The quotient of $\frac{1}{3}$ and $\frac{2}{9}$, or $\frac{1}{3}$ divided by $\frac{2}{9}$.

Numbers inside brackets **[ ]** indicate the Chapter, Section, and Example number of a similar worked example.

75. [1.1.1] sum: $12+5$; terms 12 and 5

76. [1.2.1] $2\cdot2\cdot3\cdot5\cdot5$ **77. [1.2.4]** $\frac{8}{3}$ **78. [1.2.5]** $\frac{1}{9}x$

79. [1.2.2] $\frac{4}{5}$ **80. [1.3.5]** -8 **81. [1.3.3]**

-10 -5 0 5 10 15 20 25 30 35 40 45 50 55

82. [1.1.2] 15 **83. [1.1.3]** 1 **84. [1.2.3]** $\frac{1}{10}$

85. [1.2.6] $\frac{15}{4}$ **86. [1.2.10]** $\frac{4}{3}$ **87. [1.2.11]** 3

88. [1.3.7] 7 **89. [1.3.7]** -3

Section 1.5 Adding and Subtracting Real Numbers

Practice 1.5.1

1. -18 **2.** -41 **3.** -27 **4.** 3 **5.** -21 **6.** 11 **7.** -20
8. -51 **9.** $-1\frac{1}{4}$ **10.** $-\frac{1}{12}$ **11.** $-3\frac{7}{8}$ **12.** $-\frac{3}{4}$

Practice 1.5.2

1. -11 **2.** 22 **3.** $\frac{5}{6}$ **4.** -0.002

Practice 1.5.3

1. $-18+10$, which simplifies to -8

2. $-8.6+15$, which simplifies to 6.4

3. $-\frac{5}{7}+\frac{3}{7}$, which simplifies to $-\frac{2}{7}$

4. $-\frac{7}{8}+1$, which simplifies to $\frac{1}{8}$

5. $-16+(-10)$, which simplifies to -26

6. $25+(-36)$, which simplifies to -11

Practice 1.5.4

1. The sum of -12 and 7, or -12 increased by 7, or 7 more than -12, or 7 added to -12.

2. The sum of 15.9 and 25, or 15.9 increased by 25, or 25 more than 15.9, or 25 added to 15.9.

Practice 1.5.5

1. -37 degrees, or 37 degrees below 0

2. -130 feet, or 130 feet below sea level

3. $-\$21$ **4.** $-\$15$ **5.** $\$133.50$ per share

6. $\$251.75$ per share

Practice 1.5.6

1. -18 **2.** -21 **3.** 69 **4.** 40 **5.** -6 **6.** -4 **7.** -9
8. -21 **9.** -29 **10.** -11 **11.** $-\frac{5}{4}$ or $-1\frac{1}{4}$ **12.** $-\frac{39}{40}$

Practice 1.5.7

1. 10 **2.** -13 **3.** 22 **4.** 30 **5.** 48 **6.** -42 **7.** -9
8. -11

Practice 1.5.8

1. -15 **2.** -26 **3.** -39 **4.** -41

Practice 1.5.9

1. $-18-10$, which simplifies to -28

2. $15-9$, which simplifies to 6

3. $-\frac{7}{8}-1$, which simplifies to $-1\frac{7}{8}$

4. $-16-(-10)$, which simplifies to -6

5. $28-(-13)$, which simplifies to 41

6. $5-\left(-\frac{3}{4}\right)$, which simplifies to $5\frac{3}{4}$

Practice 1.5.10

1. The difference of -12 and 7, or -12 decreased by 7, or 7 less than -12, or 7 subtracted from -12.

2. The difference of -1 and -3, or -1 decreased by -3, or -3 less than -1, or -3 subtracted from -1.

Exercise Set 1.5

1. -14 **3.** -2 **5.** 8 **7.** 42 **9.** $\frac{1}{3}$ **11.** $-1\frac{17}{24}$ **13.** -29

15. $\frac{13}{15}$ **17.** 24.9 **19.** $-32 + 24$, which simplifies to -8

21. $-22 + 13$, which simplifies to -9

23. $\frac{5}{8} + \frac{-1}{6}$, which simplifies to $\frac{11}{24}$

25. $5 + \left(\frac{-2}{3}\right)$, which simplifies to $4\frac{1}{3}$

27. The sum of -36 and 11, or -36 increased by 11, or 11 more than -36, or 11 added to -36.

29. The sum of -5 and -2, or -2 added to -5.

31. $\$15$ **33.** $\$24.25$ per share   **35.** -14 **37.** 2

39. -32 **41.** 23 **43.** 30 **45.** $\frac{-15}{14}$ **47.** 16 **49.** 0

51. -19 **53.** -7 **55.** -43 **57.** -5 **59.** -33

61. $-32 - 24$, which simplifies to -56

63. $13 - (-22)$, which simplifies to 35

65. $-\frac{1}{3} - 3$, which simplifies to $-3\frac{1}{3}$

67. $-24 - 5$, which simplifies to -29

69. The difference of -36 and 11, or -36 decreased by 11, or 11 less than -36, or 11 subtracted from -36

71. The difference of $\frac{4}{9}$ and 4, or $\frac{4}{9}$ decreased by 4, or 4 less than $\frac{4}{9}$, or 4 subtracted from $\frac{4}{9}$.

Numbers inside brackets [ ] indicate the Chapter, Section, and Example number of a similar worked example.

73. **[1.2.1]** $2 \cdot 3 \cdot 5 \cdot 5$ **74.** **[1.2.4]** $\frac{1}{x}$

75. **[1.1.1]** product: $4 \cdot 2$; factors 4 and 2 **76.** **[1.2.5]** $\frac{2}{3}t$

77. **[1.3.5]** -0.23 **78.** **[1.2.2]** $\frac{2}{3}$ **79.** **[1.1.2]** 10

80. **[1.2.3]** $\frac{9}{20}$ **81.** **[1.2.10]** $\frac{31}{24}$ **82.** **[1.2.11]** 1

83. **[1.1.3]** 6 **84.** **[1.3.7]** 2 **85.** **[1.4.1]** -3

86. **[1.4.6]** 2.5 **87.** **[1.4.4]** $(-5)(12)$ **88.** **[1.4.9]** $\frac{x}{7}$

Section 1.6 Properties of Real Numbers

Practice 1.6.1

1. 0 **2.** 1 **3.** 7 **4.** -33 **5.** $\frac{3}{2}$ **6.** $\frac{1}{23}$

Practice 1.6.2

1. $22 + 5$ **2.** $-23 + 8$ **3.** $5y^2 + 3y + 2$
4. $5w^3 - 2w^2 + w$ **5.** $-8x - 12$ **6.** $-p^2 + p - 5$
7. $a^2 - 2a - 3$

Practice 1.6.3

1. $5 \cdot 4$ **2.** $5w$ **3.** $12(y - 8)$ **4.** $(d - 3)8$
5. $(x + 2)(x - 12)$ **6.** a^2bc

Practice 1.6.4

1. $\left(\frac{5}{6} + \frac{1}{6}\right) + 9$ **2.** $-10 + \left(\frac{1}{3} + \frac{2}{3}\right)$ **3.** $6 + (7x + 3x)$
4. $(3x + 4x) + 12$

Practice 1.6.5

1. $7 \cdot (3 \cdot 5)$ **2.** $-4 \cdot (2 \cdot (-8))$ **3.** $(4 \cdot 5) \cdot 3$

4. $\left(\frac{2}{3} \cdot 3\right)x$ **5.** $\left(12 \cdot \frac{1}{12}\right)y$ **6.** $\left(8 \cdot \frac{1}{8}\right)z$

Practice 1.6.6

1. $2y + 14$ **2.** $-4x + (-8)$ or $-4x - 8$ **3.** $6r + 18s + 12$
4. $-10x + (-2y) + (-8)$ or $-10x - 2y - 8$ **5.** $-7y + 28$
6. $8x + 12$

Practice 1.6.7

1. $y(2 + 12)$ **2.** $y(5 + 8 + 11)$ **3.** $x(3 - 5)$ **4.** $y(x + w)$
5. $m(n + p)$

Practice 1.6.8

1. Base is 6, exponent is 3; 6^3 simplifies to 216

2. Base is 4, exponent is 5; 4^5 simplifies to 1024

3. Base of first factor is 5, exponent is 1; base of second factor is 4, exponent is 2; $5 \cdot 4^2$ simplifies to 80

4. Base is $3 \cdot 2$, exponent is 3; $(3 \cdot 2)^3$ simplifies to 216

5. Base is -7, exponent is 2; $(-7)^2$ simplifies to 49

6. Base is 8, exponent is 2; $-(8)^2$ simplifies to -64

7. Base is 7, exponent is 2; -7^2 means $-7 \cdot 7$, which simplifies to -49

8. Base is -8 exponent is 2; $-(-8)^2$ means $-(-8) \cdot (-8)$, which simplifies to -64

Practice 1.6.9

1. Base is y, exponent is 7; expanded form is $y \cdot y \cdot y \cdot y \cdot y \cdot y \cdot y$

2. Base of first factor is 8, exponent is 1; base of second factor is z, exponent is 5; expanded form is $8 \cdot z \cdot z \cdot z \cdot z \cdot z$

3. Base is $4y$, exponent is 3; expanded form is $(4y)(4y)(4y)$

4. Base of first factor is 5, exponent is 3; base of second factor is b, exponent is 4; expanded form is $5 \cdot 5 \cdot 5 \cdot b \cdot b \cdot b \cdot b$

Practice 1.6.10

1. Base is 3, exponent is 4; exponential form is 3^4
2. One base is x, exponent is 2; the other base is y, exponent is 5; exponential form is x^2y^5
3. One base is 8, exponent is 3; the other base is x, exponent is 4; exponential form is 8^3x^4
4. One base is 2, exponent is 6; the other base is y, exponent is 1; exponential form is 2^6y
5. One base is -3, exponent is 4; another base is m, exponent is 5; the other base is n, exponent is 3; exponential form is $(-3)^4m^5n^3$

Exercise Set 1.6

1. 0 3. 1 5. 7.7 7. $-\frac{4}{5}$ 9. $\frac{7}{8}$ 11. $-13 + 7$
13. $-2x^2 + 3x + 5$ 15. $26 \cdot 12$ 17. $3r$ 19. $9(x - 5)$
21. $(x + 9)(2x + 3)$ 23. $\left(\frac{1}{16} + \frac{15}{16}\right) + 10$
25. $-2 + (6y + 8y)$ 27. $6 \cdot (2 \cdot 3)$ 29. $\left(\frac{7}{9} \cdot 9\right)y$
31. $12y + 36$ 33. $8x - 16$ 35. $-3y + 15x$
37. $-2x - 2y + 12$ 39. $3y - 6x + 18$ 41. $h(3 + 6)$
43. $x(8 - 2)$ 45. $y(2 - 3 + 5)$ 47. $r(s - t)$
49. Base is 7, exponent is 3; 7^3 simplifies to 343
51. Base of first factor is 2, exponent is 1; base of second factor is 4, exponent is 3; $2 \cdot 4^3$ simplifies to 128
53. Base is $2 \cdot 4$, exponent is 3; $(2 \cdot 4)^3$ simplifies to 512
55. Base is 6, exponent is 2; -6^2 (the opposite of the square of 6) simplifies to -36
57. Base of first factor is 9, exponent is 1; base of second factor is p, exponent is 4; expanded form is $9 \cdot p \cdot p \cdot p \cdot p$
59. Base is $-8r$ exponent is 3; expanded form is $(-8r) \cdot (-8r) \cdot (-8r)$
61. Base of first factor is 4, exponent is 2; base of second factor is q, exponent is 4; expanded form is $4 \cdot 4 \cdot q \cdot q \cdot q \cdot q$
63. Base is r, exponent is 3; expanded form is $-rrr$
65. Base is $-t$, exponent is 4; expanded form is $-(-t)(-t)(-t)(-t)$
67. Base is 11, number of factors is 3; exponential notation is 11^3
69. One base is p, number of factors is 6; the other base is r, number of factors is 3; exponential notation is p^6r^3
71. In the first set of parentheses the base is z and the number of factors is 3. In the second set of parentheses the base is z and the number of factors is 4. Exponential notation is $(z^3)(z^4)$.
73. Base 5, numb. of factors 4; exponential notation $-(5^4)$
75. Base -3, numb. of factors 4; exponential notation $(-3)^4$

Numbers inside brackets [ ] indicate the Chapter, Section, and Example number of a similar worked example.

77. [1.2.4] $-\frac{5}{4}$ 78. [1.2.5] $\frac{5}{2}x$ 79. [1.3.5] $12x$
88. [1.4.9] $\frac{x}{6}$ 81. [1.5.9] $x - 4$ 82. [1.4.4] $2 \cdot \frac{8}{9}$
83. [1.1.3] 32 84. [1.2.3] $\frac{8}{5}$ 85. [1.2.6] $\frac{7}{9}$
86. [1.2.10] $\frac{41}{24}$ 87. [1.3.7] 9 88. [1.5.8] 9
89. [1.2.2] 4 90. [1.3.3]

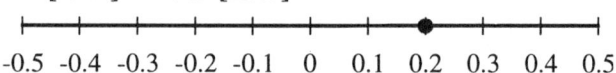

-0.5 -0.4 -0.3 -0.2 -0.1 0 0.1 0.2 0.3 0.4 0.5

Section 1.7 Simplifying Algebraic Expressions

Practice 1.7.1

1. $2x^2$ and $3x^2$ are like terms. 2. $5y$, and $\frac{3}{2}y$ are like terms.
3. $2x^2y^4, \frac{x^2y^4}{\pi}$ and $7x^2y^4$ are like terms; $0.07y^3x^2$ and $-x^2y^3$ are like terms; $-2x^3y^2$ is not like any other term.
4. $5x^5y^2$ and $1.25y^2x^5$ are like terms; $-9x^2y^5$ and $1.1x^2y^5$ are like terms; $10x^3y^2$ is not like any other terms.
5. None are like terms.
6. $7r^2s$ and $8sr^2$ are like terms; $-4rs^2, 15r^2s^2, 6r^2$ and $2s^2$ are not like any other terms.

Practice 1.7.2

1. $9m$ 2. $2r$ 3. $-8w$ 4. $3s$ 5. $-15x^3$ 6. $-22a^2$
7. $7x + y$ 8. $2xy^2 + 5x^2y$

Practice 1.7.3

1. $-6x$ 2. $-11y$ 3. $7w$ 4. $-8m$ 5. 0 6. $19m$
7. $1r$ or r 8. $16n$

Practice 1.7.4

1. $8x + 9$ 2. $9m + 36$ 3. $10y - 21$ 4. $6x$
5. $-36x - 96$ 6. $-20x - 20$ 7. $-6x + 15$ 8. $-9x$
9. $9x + 15$ 10. $2x - 3$

Practice 1.7.5

1. $4x$ 2. $-6m^3 - 3m^2 + 4m$ 3. $3n^2 + 2n - 5$
4. $2x^2 + 3x + 5$

Practice 1.7.6

1. $-13x - 42$ 2. $-26n - 56$ 3. $27r - 60$ 4. $17t - 21$
5. $-2x^2 + x + 6$ 6. $3n^2 + 6n - 5$ 7. $-7x - 2$ 8. $-6y - 7$

Practice 1.7.7

1. $-59x - 6$ 2. $-13x - 12$ 3. $2x + 7$ 4. $-2x - 9$
5. $2x$ 6. $14w$ 7. $-18x + 12$ 8. $-16x + 24$
9. $-25x - 20$ 10. $-8x - 6$

Exercise Set 1.7

1. $3m$ **3.** $-16mn$ **5.** $-2.1xy$ **7.** $\frac{5}{4}x$ **9.** $2x$ **11.** $-3w$

13. $-22x$ **15.** $2x$ **17.** $2m - 42$ **19.** $-16y + 40$

21. $10x + 15$ **23.** $-3x + 1$ **25.** $12y$ **27.** $-x^2 + 3x - 5$

29. $-4w - 6$ **31.** $-2r^2 - 2r + 3$ **33.** $-x + 26$

35. $-2t^2 - 7t + 2$ **37.** $33x - 8$ **39.** $-15x + 34$

41. $x - 42$ **43.** 2

Numbers inside brackets **[ ]** indicate the Chapter, Section, and Example number of a similar worked example.

45. [1.2.4] $-\frac{1}{6}$ **46. [1.2.5]** $\frac{1}{6}w$ **47. [1.3.5]** Since 0

has no sign, it has no opposite. **48. [1.6.7]** $x(7 + 2)$

49. [1.1.2] 16 **50. [1.1.3]** 1 **51. [1.2.3]** $\frac{3}{8}$

52. [1.3.7] 17 **53. [1.4.4]** $0.03 \cdot 9$ **54. [1.4.9]** $\frac{y}{3}$

55. [1.5.9] $x - 5$ **56. [1.5.8]** -3

57. [1.6.11] Commutative Property of Addition
58. [1.6.8] Base of first factor is 3, exponent is 1; base of second factor is w, exponent is 5.

Chapter 1 Review Exercises

Numbers inside brackets **[ ]** indicate the Chapter, Section, and Example number of a similar worked example.
1. [1.1.1] product: $-8rs$; factors -8, r, and s
2. [1.1.1] sum: $3w + 5p$; terms: $3w$ and $5p$; products: within first term $3w$ and within second term $5p$; factors: within first term 3 and w and within second term 5 and p.

3. [1.2.1] $2 \cdot 2 \cdot 2 \cdot 7 \cdot 7$ **4. [1.2.1]** $2 \cdot 2 \cdot 11 \cdot 13$ **5. [1.2.2]** $\frac{16}{39}$

6. [1.2.2] $\frac{59}{24}$ **7. [1.2.4]** $\frac{7}{5}$ **8. [1.2.4]** $-\frac{1}{2}$ **9. [1.2.4]** $\frac{5}{3m}$

10. [1.2.5] $\frac{5}{9}b$ **11. [1.1.2]** 64 **12. [1.1.2]** 40

13. [1.1.3] 2 **14. [1.1.3]** 8 **15. [1.2.3]** $\frac{16}{31}$

16. [1.2.3] $\frac{18}{35}$ **17. [1.2.6]** $\frac{45}{4}$ **18. [1.2.6]** $\frac{25}{24}$

19. [1.2.10] $\frac{47}{48}$ **20. [1.2.10]** $\frac{29}{50}$ **21. [1.2.11]** 4

22. [1.2.11] 2 **23. [1.2.11]** 4 **24. [1.2.11]** 5
25. [1.3.7] 20 **26. [1.3.7]** 14
27. [1.6.8] Base is 3, exponent is 4; simplifies to -81
28. [1.7.3] $7x$ **29. [1.7.2]** $7x^2 - 5x$ **30. [1.7.3]** $-3x$
31. [1.7.6] $x - 2$ **32. [1.7.6]** $-2x - 3$
33. [1.7.7] $-5z - 9$ **34. [1.7.7]** $-19m - 8$ **35. [1.3.1]**

36. [1.3.5] 12.6 **37. [1.3.5]** $-\frac{5}{8}$

38. [1.3.2] $\{-2, -3.5, 0.5, 6\}$
39. [1.3.3]

40. [1.3.3]

41. [1.4.4] $0.08 \cdot 12$ **42. [1.4.4]** $2x$ **43. [1.4.9]** $\frac{x}{-8}$

44. [1.4.9] $\frac{2x}{3}$ **45. [1.5.3]** $x + 5$ **46. [1.5.8]** 7

47. [1.5.8] 0 **48. [1.5.9]** $x - (-2)$ **49. [1.5.9]** $x - 4$

50. [1.5.9] $\frac{1}{2}x + 3$

51. [1.6.11] Commutative Property of Addition
52. [1.6.11] Multiplication Property of 1
53. [1.6.7] $x(2 + 7)$ **54. [1.6.7]** $a(3 - 1)$

Chapter 1 Test

Numbers inside brackets **[ ]** indicate the Chapter, Section, and Example number of a similar worked example.
1. [1.1.1] sum $8 + x$; terms: 8 and x

2. [1.3.1]

3. [1.3.2] $\{-5.5, -2, 0, 3.5\}$
4. [1.3.3]

5. [1.2.4] $-\frac{1}{3}$ **6. [1.3.5]** $-\frac{3}{2}$ **7. [1.2.5]** $\frac{1}{2}w$

8. [1.6.7] $b(5 + 1)$ **9. [1.2.1]** $3 \cdot 3 \cdot 3 \cdot 17$

10. [1.4.4] $2 + \frac{1}{3}x$ **11. [1.2.2]** $\frac{1}{2}$ **12. [1.5.8]** 2

13. [1.6.11] Commutative Property of Addition
14. [1.6.11] Associative Property of Addition

15. [1.1.2] 27 **16. [1.2.3]** $\frac{68}{65}$ **17. [1.2.6]** $\frac{1}{3}$

18. [1.6.8] Base is 2, exponent is 6; simplifies to -64

19. [1.2.10] $\frac{73}{120}$ **20. [1.2.11]** 4 **21. [1.3.7]** 11

22. [1.7.2] $17x^3 - 7x^2$ **23. [1.7.3]** $13x$ **24. [1.7.6]** 6
25. [1.7.7] $-4r + 22$ **26. [1.1.3]** 20

Chapter 2 Equations and Inequalities

Section 2.1 Definitions and Solving Equations

Practice 2.1.1
1. Yes, $-1 = -1$ 2. No, $-1 \neq 12$ 3. Yes, $-18 = -18$
4. No, $9 \neq 6$ 5. Yes, $3 = 3$

Practice 2.1.2
1. $\square = 18$ 2. $\square = 5$ 3. $\square = 11$ 4. $\square = -10$

Practice 2.1.3
1. 25 2. -2 3. -17 4. 11 5. 0 6. 12

Practice 2.1.4
1. -15 2. -9 3. -20

Practice 2.1.5
1. 7 2. -22 3. -1 4. -2 5. 9

Exercise Set 2.1
1. Yes 3. No 5. Yes 7. No 9. Yes 11. No
13. Yes 15. $\square = -5$ 17. $\square = 2.4$ 19. $\square = \frac{5}{4}$ 21. -6
23. 21 25. -24 27. 5 29. -4 31. -7 33. 24
35. 6 37. -3 39. 2 41. 7 43. -9 45. 0 47. 7
49. -10 51. 7

53. [1.2.4] $-\frac{1}{2x}$ 54. [1.2.5] $\frac{5}{8}m$ 55. [1.3.5] $-\frac{1}{3}$
56. [1.1.1] product: $-3y$; factors: -3 and y
57. [1.2.2] $\frac{19}{10}$ 58. [1.6.11] Distributive Property of
Multiplication Over Addition 59. [1.6.7] $y(3 + 1)$
60. [1.7.3] $3x$ 61. [1.2.10] $\frac{11}{10}$ 62. [1.1.2] 36
63. [1.7.2] $-3x + 5y$

Section 2.2 More on Solving Equations

Practice 2.2.1
1. 3 2. 3 3. 5 4. -2 5. 3

Practice 2.2.2
1. 4 2. -1 3. -25 4. -20 5. -22

Practice 2.2.3
1. 3.50 2. 14 3. 8000 4. 1550

Practice 2.2.5
1. All real numbers 2. No solution 3. No solution
4. 0 5. All real numbers 6. 0

Exercise Set 2.2
1. 8 3. $-\frac{35}{13}$ 5. No solution 7. $\frac{8}{7}$ 9. 5 11. -1
13. 2 15. $\frac{1}{2}$ 17. 0 19. $\frac{5}{7}$ 21. $\frac{7}{6}$ 23. $\frac{-1}{4}$ 25. $\frac{1}{2}$
27. 6 29. -1 31. -1 33. -20 35. 15 37. -12
39. $\frac{80}{39}$ 41. -3 43. 2 45. 3 47. -2 49. -1.375
51. 603.75 53. 250 55. 0.75 57. -6.05 59. 20
61. 2 63. All real numbers 65. 0 67. 225

69. [1.2.2] $\frac{64}{45}$

70. [1.3.1]

$$-8\ -7\ -6\ -5\ -4\ -3\ -2\ -1\ 0\ 1\ 2\ 3\ 4\ 5\ 6\ 7\ 8$$

71. [1.3.5] $-\frac{8}{5}$
72. [1.6.11] Commutative Property of Multiplication
73. [1.6.8] Base is 6, exponent is 2; simplifies to 36
74. [2.1.1] Yes. 75. [1.2.1] $2\cdot2\cdot2\cdot3\cdot5$
76. [1.1.3] 222 77. [1.2.6] $\frac{9}{10}$ 78. [1.2.10] $\frac{143}{150}$
79. [1.2.11] 9 80. [1.2.11] $\frac{1}{3}$ 81. [1.2.11] $\frac{47}{5}$
82. [1.7.6] $-6x + 7$

Section 2.3 Literal Equations and Formulas

Practice 2.3.1
1. This is solved for i in terms of p, r and t.
2. This is solved for A in terms of b and h.
3. This is solved for V in terms of r and h. (Remember π has a constant value of about 3.14)
4. This is solved for y in terms of x.

Practice 2.3.2
1. $r = \frac{i}{pt}$ 2. $R = \frac{PV}{nT}$ 3. $b = 2M - a$
4. $t = \frac{a - p}{pr}$ 5. $y = \frac{-ax - c}{b}$ 6. $C = \frac{5(F - 32)}{9}$

Practice 2.3.3
1. $i = p\left(\frac{i}{pt}\right)t; i = i$ 2. $c = 2\pi\left(\frac{c}{2\pi}\right); c = c$
3. $a = p + pr\left(\frac{a - p}{pr}\right); a = p + a - p; a = a$
4. $ax + b\left(\frac{-ax - c}{b}\right) + c = 0; ax + -ax - c + c = 0; 0 = 0$

Exercise Set 2.3

1. Solved for I in terms of E and R.

3. Solved for C in terms of A and B.

5. Solved for A in terms of t.

7. Solved for d in terms of t and n.

9. Solved for t in terms of x.

11. $h = \dfrac{2A}{b}$ **13.** $A = 180 - B - C$ **15.** $b = \dfrac{A}{0.5h} - d$

17. $h = \dfrac{3V}{\pi r^2}$ **19.** $b = 3A - a - c$ **21.** $v = \dfrac{h + 16t^2}{t}$

23. $N = \dfrac{2.5H}{D^2}$ **25.** $h = \dfrac{s - 2\pi r^2}{2\pi r}$ **27.** $M = \dfrac{Fd^2}{km}$

29. $P = \dfrac{LZ}{W}$

31. $C = 180 - A - B$; $C = 180 - A - (180 - A - C)$; $C = 180 - A - 180 + A + C$; $C = C$

33. $s = 180(n - 2)$; $s = 180\left(\dfrac{s}{180} + 2 - 2\right)$; $s = 180\left(\dfrac{s}{180}\right)$; $s = s$

35. $k = \dfrac{1}{2}mv^2$; $k = \dfrac{1}{2}\left(\dfrac{2k}{v^2}\right)v^2$; $k = \left(\dfrac{k}{v^2}\right)v^2$; $k = k$

37. $I = \dfrac{E}{R}$; $I = \dfrac{E}{\dfrac{E}{I}}$; $I = \dfrac{E}{\dfrac{E}{I}}$; $I = \dfrac{E}{1}\left(\dfrac{I}{E}\right)$; $I = I$

39. $A = 3ab + 3bh$; $A = 3\left(\dfrac{A - 3bh}{3b}\right)b + 3bh$; $A = \left(\dfrac{A - 3bh}{b}\right)b + 3bh$; $A = A - 3bh + 3bh$; $A = A$

41. [1.5.3] $x + 12$

42. [1.3.1]

![number line from -8 to 8 with points at -8, -2, -1, 5]

-8 -7 -6 -5 -4 -3 -2 -1 0 1 2 3 4 5 6 7 8

43. [1.3.2] $\{-6.5, 7, -3, 4.5\}$ **44.** [1.1.2] 6

45. [1.2.3] $\dfrac{21}{8}$ **46.** [1.3.7] -1 **47.** [1.7.7] $-13x - 36$

48. [2.2.1] -7 **49.** [2.2.2] 6 **50.** [2.2.3] 114

51. [2.2.4] all real numbers **52.** [2.2.5] 0

53. [2.2.4] no solution

Section 2.4 Properties and Graphs of Inequalities

Practice 2.4.1

1. No $-11 < 10$ **2.** Yes, $-18 < 4$ **3.** Yes, $2 \geq -7$
4. No $3 \leq 5$ **5.** Yes $-6 < -1$ **6.** Yes $2 > -4$

Practice 2.4.2

1. Conditional **2.** Identity **3.** Contradiction
4. Conditional **5.** Contradiction **6.** Identity

Practice 2.4.3

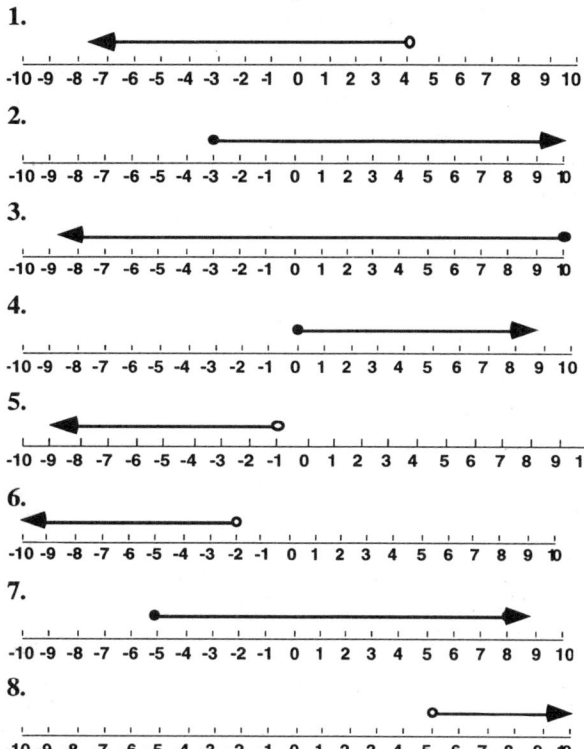

Practice 2.4.4

1. $p \leq 12$ or $12 \geq p$ **2.** $n \geq -3$ or $-3 \leq n$
3. $x < 5$ or $5 > x$ **4.** $f \leq 3$ or $3 \geq f$
5. $c < 7000$ or $7000 > c$ **6.** $m > -4$ or $-4 < m$

Practice 2.4.5

1. y is less than or equal to -6; y is at most -6; the maximum value of y is -6

2. q is greater than 25, q is over 25; q has a value that's above 25.

3. p is greater than or equal to -8; p is at least -8; the minimum value of p is -8.

4. t is less than 9, t is under 9; t is below 9.

Exercise Set 2.4

1. Yes **3.** Yes **5.** No **7.** No **9.** Identity
11. Contradiction **13.** Conditional **15.** Conditional

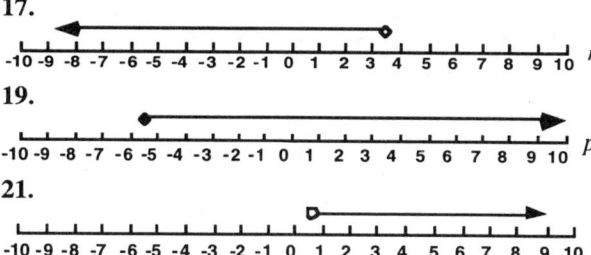

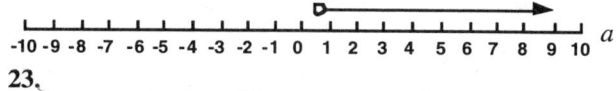

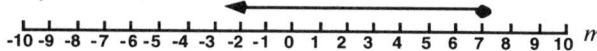

25.

-10 -9 -8 -7 -6 -5 -4 -3 -2 -1 0 1 2 3 4 5 6 7 8 9 10 *r*

27. $p \le \$3.50$ **29.** $h \le 100.75$ **31.** $w > 4\frac{3}{4}$ **33.** $c \le 270$

35. $v > -275$ **37.** $s \ge 90\%$

39. y is greater than or equal to -4.1; y is at least -4.1; the minimum value of y is -4.1

41. s is less than -3.5; s is under -3.5; s is below -3.5

43. y is greater than -1.1; y is above -1.1; y is over -1.1; y is more than -1.1

45. s is greater than -3.5; s is more than -3.5; s is over -3.5; s is above -3.5

47. b is less than or equal to -4; b has a maximum value of -4; b is at most -4.

---

49. [1.3.2] $\{-5, -0.5, 6.5, 8\}$ **50.** [1.6.11] Distributive Property of Multiplication Over Addition

51. [2.3.2] $y = \frac{10 - 3x}{5}$ or $y = 2 - \frac{3}{5}x$ **52.** [1.7.6] $12c + 2$

53. [1.2.11] $\frac{7}{5}$ **54.** [1.7.3] $-2x$ **55.** [1.7.6] $-\frac{7w}{3} + 1$

56. [1.7.7] $33x - 8$ **57.** [1.7.7] $\frac{7}{2}x - 2$

58. [2.2.3] -2.5 **59.** [2.2.2] 4 **60.** [2.2.4] no solution

---

Section 2.5 Solving Inequalities

Practice 2.5.1

1. $x < -10$

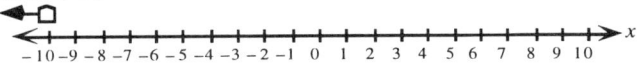
-10 -9 -8 -7 -6 -5 -4 -3 -2 -1 0 1 2 3 4 5 6 7 8 9 10 *x*

2. $x < 2$

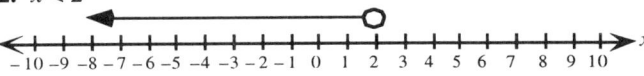
-10 -9 -8 -7 -6 -5 -4 -3 -2 -1 0 1 2 3 4 5 6 7 8 9 10 *x*

3. $x \ge 0$

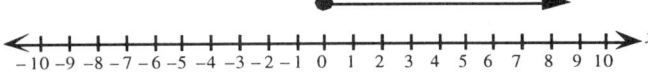
-10 -9 -8 -7 -6 -5 -4 -3 -2 -1 0 1 2 3 4 5 6 7 8 9 10 *x*

4. $y \ge 8$

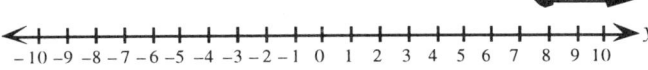
-10 -9 -8 -7 -6 -5 -4 -3 -2 -1 0 1 2 3 4 5 6 7 8 9 10 *y*

5. $x \ge -5$

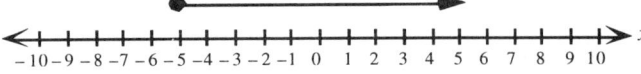
-10 -9 -8 -7 -6 -5 -4 -3 -2 -1 0 1 2 3 4 5 6 7 8 9 10 *x*

6. $x < -2$

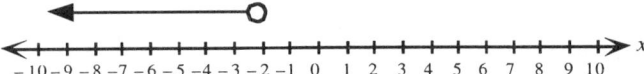
-10 -9 -8 -7 -6 -5 -4 -3 -2 -1 0 1 2 3 4 5 6 7 8 9 10 *x*

Practice 2.5.2

1. $y \ge -4$

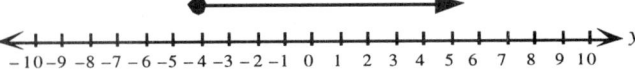
-10 -9 -8 -7 -6 -5 -4 -3 -2 -1 0 1 2 3 4 5 6 7 8 9 10 *y*

2. $x > -4$

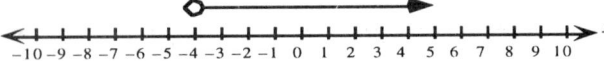
-10 -9 -8 -7 -6 -5 -4 -3 -2 -1 0 1 2 3 4 5 6 7 8 9 10 *x*

3. $x < -9$

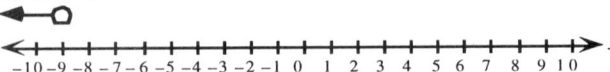
-10 -9 -8 -7 -6 -5 -4 -3 -2 -1 0 1 2 3 4 5 6 7 8 9 10 *x*

4. $t \le -4$

-10 -9 -8 -7 -6 -5 -4 -3 -2 -1 0 1 2 3 4 5 6 7 8 9 10 *t*

5. $x > -9$

-10 -9 -8 -7 -6 -5 -4 -3 -2 -1 0 1 2 3 4 5 6 7 8 9 10 *x*

6. $x > -4$

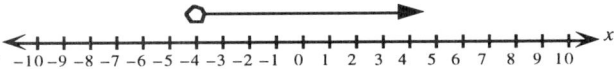
-10 -9 -8 -7 -6 -5 -4 -3 -2 -1 0 1 2 3 4 5 6 7 8 9 10 *x*

Practice 2.5.3

1. $x \le 2$

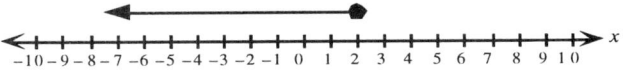
-10 -9 -8 -7 -6 -5 -4 -3 -2 -1 0 1 2 3 4 5 6 7 8 9 10 *x*

2. $x < \frac{-2}{3}$

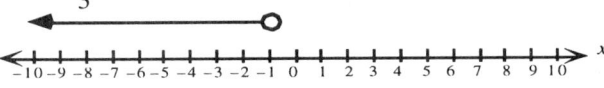
-10 -9 -8 -7 -6 -5 -4 -3 -2 -1 0 1 2 3 4 5 6 7 8 9 10 *x*

3. no solution (note the graph below has no points plotted)

-10 -9 -8 -7 -6 -5 -4 -3 -2 -1 0 1 2 3 4 5 6 7 8 9 10 *m*

4. $y \ge 9$

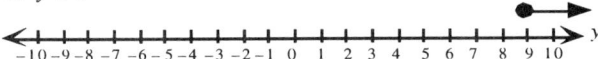
-10 -9 -8 -7 -6 -5 -4 -3 -2 -1 0 1 2 3 4 5 6 7 8 9 10 *y*

5. all real numbers

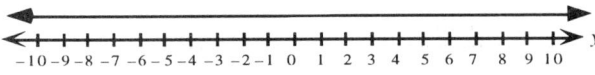
-10 -9 -8 -7 -6 -5 -4 -3 -2 -1 0 1 2 3 4 5 6 7 8 9 10 *y*

6. $y \ge 0$

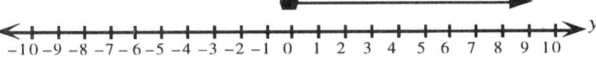
-10 -9 -8 -7 -6 -5 -4 -3 -2 -1 0 1 2 3 4 5 6 7 8 9 10 *y*

7. $x > \frac{-1}{4}$

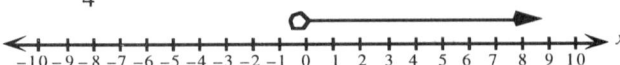
-10 -9 -8 -7 -6 -5 -4 -3 -2 -1 0 1 2 3 4 5 6 7 8 9 10 *x*

8. all real numbers

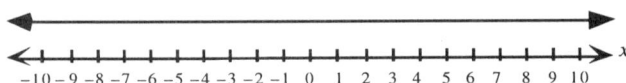
-10 -9 -8 -7 -6 -5 -4 -3 -2 -1 0 1 2 3 4 5 6 7 8 9 10 *x*

9. $x < \dfrac{-29}{10}$

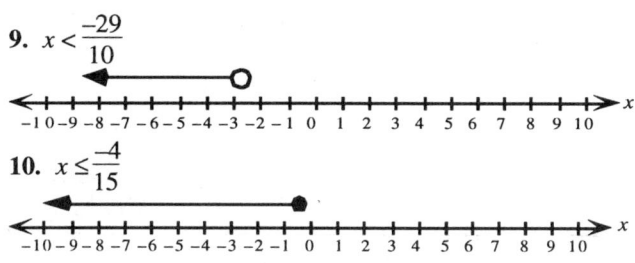

10. $x \le \dfrac{-4}{15}$

Exercise Set 2.5

1. $x \ge 6.8$

3. $y < \dfrac{-1}{4}$

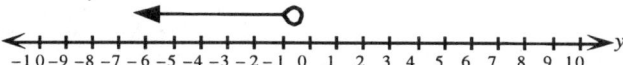

5. $y \ge -6$

7. $x < -3$

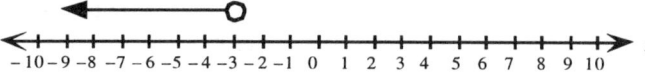

9. $x \le 3$

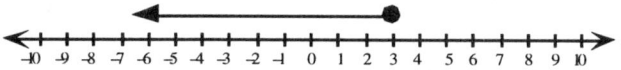

11. $x < 1$

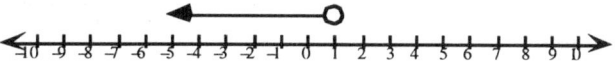

13. $x \ge 1$

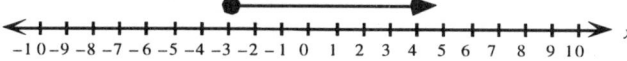

15. $p < 0$

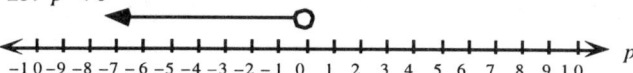

17. all real numbers

19. $x > 0$

21. $x \le 2$

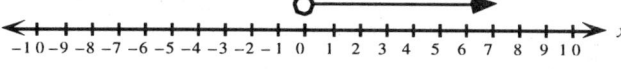

23. [2.4.4] $p \le 12$ **24.** [1.5.3] $3 + x$ **25.** [1.5.9] $x - 8$
26. [2.4.3]

27. [1.7.6] $6c - 18$ **28.** [1.2.11] $\dfrac{7}{6}$ **29.** [1.7.6] $\dfrac{9p}{10} + 2$

30. [2.2.3] 123 **31.** [2.2.5] 0 **32.** [2.2.5] 0
33. [2.2.4] all real numbers
34. [2.3.2] $y = \dfrac{7 - 2x}{3}$ or $y = \dfrac{7}{3} + \dfrac{-2}{3}x$

Chapter 2 Review Exercises

Numbers inside brackets [ ] indicate the Chapter, Section, and Example number of a similar worked example.
1. [2.1.1] Yes. **2.** [2.1.2] -2 **3.** [2.1.2] 5
4. [2.1.4] -15 **5.** [2.1.4] -11 **6.** [2.1.4] -6
7. [2.1.5] 0 **8.** [2.1.5] 3 **9.** [2.2.1] 3 **10.** [2.2.1] -4
11. [2.2.2] -4 **12.** [2.2.2] $\dfrac{1}{2}$ **13.** [2.2.2] 1
14. [2.2.3] 52 **15.** [2.2.4] No solution
16. [2.2.4] All real numbers **17.** [2.2.5] 0
18. [2.2.5] 0
19. [2.3.1] a is solved for in terms of b and c.
20. [2.3.3] $2x + 3\left(-\dfrac{2}{3}x + 2\right) = 6;\ 2x - 2x + 6 = 6;\ 6 = 6$
21. [2.3.2] $y = \dfrac{5}{2}x - \dfrac{3}{2}$
22. [2.3.2] $R = \dfrac{e + E - Inr}{I}$ or $\dfrac{e + E}{I} - nr$
23. [2.3.2] $b = 8a$ **24.** [2.3.2] $p = 2m$
25. [2.4.3]

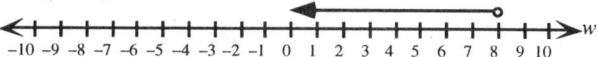

26. [2.4.3]

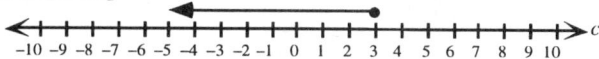

27. [2.4.3]

28. [2.4.3]

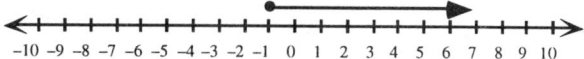

29. [2.4.4] $h \ge 72$ **30.** [2.4.4] $c \le 8$
31. [2.4.4] $x > 20$ **32.** [2.4.4] $w \le 5$
33. [2.4.5] 8 is less than x; x is larger than 8; the value of x is over 8. **34.** [2.4.5] x is greater than or equal to 5; x is at least 5; the minimum value of x is 5.
35. [2.4.5] k is greater than -6; k is more than -6; k is above -6. **36.** [2.5.3] $x > 1$ **37.** [2.5.3] $x \ge 2$
38. [2.5.3] No solution

Chapter 2 Test

Numbers inside brackets [ ] indicate the Chapter, Section, and Example number of a similar worked example.

1. [2.1.1] Yes. **2.** [2.1.2] -2 **3.** [2.1.4] 11
4. [2.1.5] 1 **5.** [2.2.1] 9 **6.** [2.2.2] 3 **7.** [2.2.3] 8.8
8. [2.2.4] All real numbers **9.** [2.2.4] No solution
10. [2.2.5] 0 **11.** [2.3.2] $y = 2x - 4$ **12.** [2.5.3] $x < 2$
13. [2.5.3] $x \geq 8$ **14.** [2.3.2] $a = \dfrac{2V}{w} + c$

15. [2.3.1] w is solved for in terms of d and g.

16. [2.3.3] $w = t + 2s\left(\dfrac{w-t}{2s}\right) \Rightarrow w = t + (w-t); \; w = w.$

17. [2.4.3]

18. [2.4.4] $-5 < t$ or $t > -5$ **19.** [2.4.5] p is less than or equal to 1; p is at most 1; the maximum value of p is 1.

Chapter 3 Using Algebra to Solve Problems

Section 3.1 Modeling with Expressions

Practice 3.1.2

1a) $a + 12$

Items	Number of Tickets	
adult tickets	a	⇐ Given.
child tickets	$a + 12$	⇐ 12 more child tickets.

1b) $a + 12$

Items	Number of Tickets	
adult tickets	a	⇐ Given.
child tickets	$a + 12$	⇐ 12 fewer adult means 12 more child tickets.

1c) a

Items	Number of Tickets	
adult tickets	a	⇐ Given.
child tickets	a	⇐ Same as adult.

1d) $a - 400$

Items	Number of Tickets	
adult tickets	a	⇐ Given.
child tickets	$a - 400$	⇐ 400 more adult means 400 fewer child tickets.

1e) $a + 400$

Items	Number of Tickets	
adult tickets	a	⇐ Given.
child tickets	$a + 400$	⇐ 400 fewer adult means 400 more child tickets.

2a) w

Items	Volume of acid (milliliters)	
weak acid	w	⇐ Given.
strong acid	w	⇐ Same as weak since final mix is half and half.

2b) $w - 75$

Items	Volume of acid (milliliters)	
weak acid	w	⇐ Given.
strong acid	$w - 75$	⇐ 75 ml more of weak means 75 ml less of strong acid.

2c) $w + 10$

Items	Volume of acid (milliliters)	
weak acid	w	$\Leftarrow$ Given.
strong acid	$w + 10$	$\Leftarrow$ 10 ml less weak means 10 ml more of strong acid.

2d) $w - 10$

Items	Volume of acid (milliliters)	
weak acid	w	$\Leftarrow$ Given.
strong acid	$w - 10$	$\Leftarrow$ 10 ml less of strong acid.

3a) $s - 300$

Items	Money invested (dollars)	
savings account	s	$\Leftarrow$ Given.
mutual fund	$s - 300$	$\Leftarrow$ \$300 fewer than savings.

3b) $s + 300$

Items	Money invested (dollars)	
savings account	s	$\Leftarrow$ Given.
mutual fund	$s + 300$	$\Leftarrow$ \$300 more than savings.

3c) $s - 300$

Items	Money invested (dollars)	
savings account	s	$\Leftarrow$ Given.
mutual fund	$s - 300$	$\Leftarrow$ \$300 more in savings.

3d) $s + 300$

Items	Money invested (dollars)	
savings account	s	$\Leftarrow$ Given.
mutual fund	$s + 300$	$\Leftarrow$ \$300 less in savings.

3e) $s + 300$

Items	Money invested (dollars)	
savings account	s	$\Leftarrow$ Given.
mutual fund	$s + 300$	$\Leftarrow$ \$300 more than in savings.

Practice 3.1.3

1.

Items	Number of Tickets	
choice tickets	c	$\Leftarrow$ Given.
bleacher tickets	$385 - c$	$\Leftarrow$ Total number of tickets – number of choice tickets.

2.

Items	Number of Flowers	
roses	r	$\Leftarrow$ Given.
other flowers	$22 - r$	$\Leftarrow$ Total number of flowers – number of roses.

3.

Items	Distance (miles)	
distance already traveled	d	$\Leftarrow$ Given.
distance to shore	$1.75 - d$	$\Leftarrow$ Total distance – distance already traveled.

Practice 3.1.4

1.

Items	Number of Stamps	
All stamps	120	⇐ Given.
28¢ stamps	x	⇐ Given.
20¢ stamps	$x + 12$	⇐ 12 more 20¢ stamps.
32¢ stamps		⇐ To be determined.

$120 - x - (x + 12)$ which simplifies to $108 - 2x$

2.

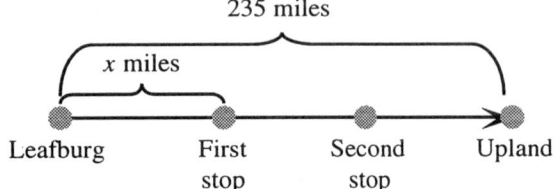

235 miles

x miles

Leafburg   First stop   Second stop   Upland

Items	Distance (miles)	
Leafburg to Upland	235	⇐ Given.
Leafburg to first stop	x	⇐ Given.
First stop to second stop	$\frac{1}{2}x$	⇐ Half distance from Leafburg to first stop.
Second stop to Upland		⇐ To be determined.

$235 - x - \frac{1}{2}x$ which simplifies to $235 - \frac{3}{2}x$

3.

Items	Number of Diapers	
all diapers	45	⇐ Given.
medium diapers	m	⇐ Given.
large diapers	$m - 5$	⇐ Five fewer large than medium.
Small diapers		⇐ To be determined.

$45 - m - (m - 5)$ which simplifies to $50 - 2m$

4.

Items	Volume (gallons)	
all ingredients	2	⇐ Given.
water	w	⇐ Given.
dish soap	$\frac{1}{100}w$	⇐ Five fewer large than medium.
tobacco juice		⇐ To be determined.

$2 - w - \frac{1}{100}w$ which simplifies to $2 - \frac{101}{100}w$

Practice 3.1.5

1.

Items	Length (meters)	
long side	19	⇐ Given.
side B	x	⇐ Given.
side A		⇐ To be determined.

$19 - x$

2.

Items	Length (cm)	
long side	7	⇐ Given.
side A	x	⇐ Given.
side C	$\frac{1}{3}x$	⇐ Given.
side B		⇐ To be determined.

$7 - x - \frac{1}{3}x$ which simplifies to $7 - \frac{4}{3}x$.

3.

Items	Area (sq ft)	
square	$\left(\frac{1}{2}x\right)^2$	⇐ Given.
triangle	$\frac{1}{2}(x)(x)$	⇐ Given.
shaded region		⇐ To be determined.

$\frac{1}{2}x^2 - \left(\frac{1}{2}x\right)^2$ which simplifies to $\frac{1}{4}x^2$

4.

Items	Area (sq ft)	
square	x^2	⇐ Given.
triangle	$\frac{1}{2}(x)(x)$	⇐ Given.
shaded region		⇐ To be determined.

$x^2 - \frac{1}{2}x^2$ which simplifies to $\frac{1}{2}x^2$.

Practice 3.1.6

1a)

Items	Weight (pounds)	
cashews	c	⇐ Given.
peanuts	$c - 6$	⇐ Six less pounds of peanuts.

$c + (c - 6)$ which simplifies to $2c - 6$

1b)

Items	Weight (pounds)	
cashews	c	⇐ Given.
peanuts	$\frac{1}{3}c$	⇐ One–third weight of cashews.

$c + \left(\frac{1}{3}c\right)$ which simplifies to $\frac{4}{3}c$

1c)

Items	Weight (pounds)	
cashews	c	⇐ Given.
peanuts	$\frac{1}{2}c$	⇐ Twice as much cashews so half as many peanuts.

$c + \left(\frac{1}{2}c\right)$ which simplifies to $\frac{3}{2}c$

2a)

Items	Number of Riders	
peak	p	⟸ Given.
non-peak	$\frac{1}{4}p$	⟸ Four peak for every non–peak means one–fourth the number of peak riders.

$p + \left(\frac{1}{4}p\right)$ which simplifies to $\frac{5}{4}p$

2b)

Items	Number of Riders	
peak	p	⟸ Given.
non-peak	$p + 75$	⟸ 75 less peak means 75 more non–peak.

$p + (p + 75)$ which simplifies to $2p + 75$

2c)

Items	Number of Riders	
peak	p	⟸ Given.
non-peak	$p - 75$	⟸ 75 less non–peak.

$p + (p - 75)$ which simplifies to $2p - 75$

Exercise Set 3.1

1a) $a - 50$ **1b)** $a - 40$ **1c)** $4a$ **1d)** $5a$ **1e)** $\frac{1}{2}a$

3a) $t - 1500$ **3b)** $\frac{1}{4}t$ **3c)** t **3d)** $t + 500$ **3e)** $\frac{1}{2}t$

5a) $\frac{1}{15}g$ **5b)** $g - 6$ **5c)** $\frac{1}{3}g$ **5d)** $\frac{1}{12}g$ **5e)** $g - 30$

7. $8500 - b$ **9.** $75 - m$

11.

Items	Time (hours)	
Total time	75	⟸ Given.
weekend	w	⟸ Given.
regular	$5w$	⟸ Five times the weekend hours.
overtime		⟸ To be determined.

$75 - w - 5w$ which simplifies to $75 - 6w$

13.

Items	Time (hours)	
Total spent	285	⟸ Given.
pants	p	⟸ Given.
shoes	$p - 20$	⟸ $20 less on pants.
shirt	$\frac{1}{2}p$	⟸ Half as much as for pants.
jacket		⟸ To be determined.

$285 - p - (p - 20) - \frac{1}{2}p$ which simplifies to $305 - \frac{5}{2}p$

15.

Items	Length (feet)	
height	$10 - 6$	⟸ Whole – Part Concept.
width	$x - 1.5$	⟸ Whole – Part Concept.

area is (height)(width), which is $(10 - 6)(x - 1.5)$, which simplifies to $4x - 6$

17.

Items	Length (feet)	
height of rectangle	x	⇐ Given in sketch.
width of rectangle	x	⇐ Given in sketch.
height of triangle	$\dfrac{x}{2}$	⇐ Given in sketch.
base of triangle	x	⇐ Given in sketch.

area is (area of rectangle) – (area of triangle), which is $(x)(x) - \left(\dfrac{1}{2}\right)(x)\left(\dfrac{x}{2}\right)$ which simplifies to $\dfrac{3x^2}{4}$ or $\dfrac{3}{4}x^2$

19.

Items	Length	
radius of large circle	$\dfrac{x}{2}$	⇐ Radius is half the diameter.
radius of small circle	$\dfrac{1}{2}\left(\dfrac{x}{2}\right)$	⇐ Radius is half the diameter.

area is $\dfrac{1}{2}$ (area of large circle) – (area of small circle, which is $\dfrac{1}{2}\pi\left(\dfrac{x}{2}\right)^2 - \pi\left(\dfrac{x}{4}\right)^2$, which simplifies to $\dfrac{\pi x^2}{16}$

21.

Items	Length (feet)	
base of large triangle	5	⇐ Given in sketch.
height of large triangle	x	⇐ Given in sketch.
base of small triangle	5	⇐ Given in sketch.
height of small triangle	$\dfrac{x}{2}$	⇐ Given in sketch.

area is (area of large triangle) – (area of small triangle), which is $\dfrac{1}{2}(5)(x) - \dfrac{1}{2}(5)\left(\dfrac{x}{2}\right)$, which simplifies to $\dfrac{5x}{4}$

23a)

Items	Distance (miles)	
car	c	⇐ Given.
truck	$c - 200$	⇐ car miles is 200 more than truck miles.

$c + (c - 200)$ which simplifies to $2c - 200$

23b)

Items	Distance (miles)	
car	c	⇐ Given.
truck	$\dfrac{1}{4}c$	⇐ car miles is 4 times truck miles so truck miles is $\dfrac{1}{4}$ car miles.

$c + \dfrac{1}{4}c$ which simplifies to $\dfrac{5}{4}c$

23c)

Items	Distance (miles)	
car	c	⇐ Given.
truck	$c + 2$	⇐ 2 more than car miles.

$c + (c + 2)$ which simplifies to $2c + 2$

25a)

Items	Ingredients (gallons)	
whole milk	w	⇐ Given.
chocolate syrup	$\frac{1}{10}w$	⇐ One–tenth as much syrup as whole milk.
skim milk	$\frac{1}{10}w + 40$	⇐ Forty more gallons of skim milk than of syrup.

$w + \frac{1}{10}w + \left(\frac{1}{10}w + 40\right)$ which simplifies to $\frac{6}{5}x + 40$.

25b)

Items	Ingredients (gallons)	
whole milk	w	⇐ Given.
skim milk	$3w$	⇐ Three times as much skim as whole milk.
chocolate syrup	$3w - 300$	⇐ Three hundred less gallons syrup than skim milk.

$w + 3w + (3w - 300)$ which simplifies to $7w - 300$.

25c)

Items	Ingredients (gallons)	
whole milk	w	⇐ Given.
skim milk	$w - 48$	⇐ Forty–eight gallons more whole than skim.
chocolate syrup	$\frac{1}{4}(w - 48)$	⇐ One–fourth as much syrup as skim milk.

$w + (w - 48) + \frac{1}{4}(w - 48)$ which simplifies to $\frac{9}{4}w - 60$.

27a)

Items	Money invested (dollars)	
money market	m	⇐ Given.
mutual fund	$m - 6000$	⇐ $6000 less in mutual fund as in money market.
bond fund	$5(m - 6000)$	⇐ Five times amount in mutual fund.

$m + (m - 6000) + 5(m - 6000)$ which simplifies to $7m - 36000$.

27b)

Items	Money invested (dollars)	
money market	m	⇐ Given.
mutual fund	$\frac{1}{3}m$	⇐ One–third amount in money market.
bond fund	$\frac{1}{2}\left(\frac{1}{3}m\right)$	⇐ One–half amount in mutual fund.

$m + \frac{1}{3}m + \frac{1}{2}\left(\frac{1}{3}m\right)$ which simplifies to $\frac{3}{2}m$.

27c)

Items	Money invested (dollars)	
money market	m	⇐ Given.
bond fund	$m + 15,000$	⇐ $15,000 more than money market.
mutual fund	$4(m + 15,000)$	⇐ Bond fund is one–fourth mutual fund.

$m + (m + 15,000) + 4(m + 15,000)$ which simplifies to $6m + 75,000$.

---

29. [1.2.2] $\frac{3}{10}$ **30.** [1.4.4] $\frac{1}{2} \cdot 24$ **31.** [1.5.3] $x + 8$ **32.** [1.5.9] $5 - (-3)$ **33.** [1.7.6] $7x - 21$

34. [1.7.6] $-9n + 35$ **35.** [1.7.6] $\frac{m}{6} - 1$ **36.** [2.2.1] -1 **37.** [2.2.3] -11.25 **38.** [2.2.4] no solution

39. [2.5.3] $x < -1$ **40.** [2.5.3] $w > -3$ **41.** [2.5.3] all real numbers

Section 3.2 Applying Algebra to Word Problems

Practice 3.2.1

1. The dinner alone was $38.
2. 325 professors and 1625 students.
3. They drove 175 miles after the break.

Practice 3.2.2

1. 210 gallons unleaded, 70 gallons super–unleaded and 130 gallons premium unleaded.
2. The first prize is $780, the second prize is $280 and the third prize is $140.

Practice 3.2.3

1. The width is 100 feet and the length is 205 feet.
2. The diameter of the lake is about 1.93 miles.

Practice 3.2.4

1. 52 and 54.   2. −34 and −33.   3. −3, −1 and 1.

Exercise Set 3.2

1. dryer $217, washer $298   3. checking $1597, savings $1825   5. pay off $367, transfer $1468   7. 9 feet
9. 10, −2   11. 5, 15, 12   13. savings $500, money market $1000, IRA $1500   15. 94 adult tickets, 134 child tickets,194 senior tickets   17. 55 gallons chocolate syrup, 95 gallons skim milk, 550 gallons whole milk   19. bond fund $19,850, mutual fund $79,400, money market fund $4,750   21. 318 feet.   23. 83 degrees.   25. 58 degrees.
27. Two sides are 9 feet, one side is 3 feet.   27. Each side is 25 inches.   29. −6, −5   31. −15, −13
33. −5, −3, −1

35. [1.2.2] $\frac{33}{17}$ 36. [1.6.7] $r(9-5)$ 37. [2.2.2] $\frac{3}{2}$
38. [1.7.3] $-2x$ 39. [1.2.10] $\frac{26}{63}$ 40. [1.2.11] $\frac{1}{6}$
41. [1.7.6] $16n-26$ 42. [1.7.6] $\frac{25k}{3}-\frac{15}{2}$
43. [1.3.7] −96   44. [2.2.4] all real numbers
45. [2.2.3] 401   46. [2.2.5] 0   47. [1.4.9] $\frac{2x}{5}$

Section 3.3 The Amount Formula: Finding the Amount

Practice 3.3.1

1. $118.75   2. 10 inches

Practice 3.3.2

1. $28.80   2. 32.25 miles

Exercise Set 3.3

1. 170 cents or $1.70   3. $110   5. $2,500   7. The total mixture cost $21.20.   9. John spent $19,125 on the fence.   11. The trains will be 448 miles apart.   13. The company spent $167,952 on computers.   15. The distance for the trip was 207.5 miles.   17. It cost the company $34.40.

19. [1.6.8] Base of first factor is 2, exponent is 1; base of second factor is x, exponent is 5; cannot be simplified.
20. [2.4.3]

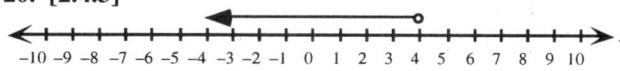

21. [1.7.2] $33x-8y$ 22. [1.7.6] $-m-2$
23. [2.2.2] −2   24. [2.2.3] 6   25. [2.2.4] no solution
26. [2.2.5] 0   27. [2.3.2] $h=\frac{V}{\pi r^2}$ 28. [2.5.3] $x<\frac{3}{2}$

Section 3.4 The Amount Formula: Finding the Rate or Base

Practice 3.4.1

1. 340 regular tickets were sold.
2. They reached approximately 360 houses.
3. The weather was cool for 2 of the 10 hours William worked. Since he can lay 40 rolls an hour while the weather is cool he laid 80 rolls during cool weather.

Practice 3.4.2

1. He has 300 29¢ stamps and 460 15¢ stamps.
2. The experienced solicitor averaged $27 dollars per pledge while the novice solicitor averaged $15 per pledge.

Practice 3.4.3

1. The theater company had 60 premium seats, 120 regular seats and 195 cheap seats.
2. On average they will need to sell 36 $450 drives per hour, 76 $215 drives per hour and 38 $360 drives per hour.

Exercise Set 3.4

1. 55 adult tickets and 125 children tickets   3. 17 ten dollar and 51 five dollar bills   5. It costs 85$ to print a *House and Flower* magazine and 75¢ to print a *Car and Passenger* magazine.   7. She would need to spend 3 minutes per page on the easy reading and 12 minutes per page on the hard reading.   9. The asphalt path would cost $10 per linear foot and the wood path would cost $35 per linear foot.   11. He has 72 fives, 69 tens, and 18 twenties,   13. 198 lower seats, 598 middle seats, and 98 upper seats

15. [1.2.1] $5 \cdot 5 \cdot 7 \cdot 7$ **16.** [2.4.3]

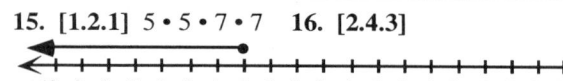

17. [2.5.3] $x > -15$ **18.** [1.3.7] 5
19. [1.7.6] $-2n + 12$ **20.** [1.7.7] $73y + 90$
21. [2.2.1] 0 **22.** [2.2.3] 0.8 **23.** [2.2.4] no solution
24a) [3.1.2] $a + 600$ **24b)** $a - 600$ **24c)** $a - 200$ **24d)** $2a$

Section 3.5 Distance-Rate-Time Word Problems

Practice 3.5.1

1. They pass at 6:30 pm; equation is $78t + 50(t - 2) = 732$
2. The trains will be 885 miles apart at 3:30 pm; equation is $110t + 60(t + 2) = 885$

Practice 3.5.2

1. It would take $1\frac{1}{4}$ hours, or 1 hour 15 minutes.
The equation is $56t = 40\left(t + \frac{1}{2}\right)$
2. The equation is $50t = 450(t - 20)$; t is 22.5 hours; distance is $50(22.5)$ or 1125 miles

Exercise Set 3.5

1. The fire stayed in the swamp for 4 hours.
3. Her speed through the road construction was 45 mph.
5. They can begin communicating at 11:45 am.
7. Dana will need to ride at 18 mph.
9. She traveled at 40 mph back to the office.

---

11. [1.2.1] $3 \cdot 3 \cdot 3 \cdot 11$ **12.** [1.3.5] $\frac{7}{3}$ **13.** [2.4.3]

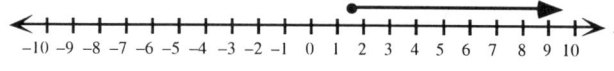

14. [2.5.3] $x \leq \frac{6}{7}$ **15.** [1.6.11] Associative Property of Addition   **16.** [1.7.2] $-1.2x + 11.5y$ **17.** [2.2.3] 1.5
18. [2.2.4] no solution   **19.** [2.5.3] $x \geq -60$

Section 3.6 Word Problems Involving Percent

Practice 3.6.1

1. $9. The equation is $1.35 = 0.15x$
2. 15. The equation is $3 = 20x$
3. $76. The equation is $19 = 0.25x$
4. 92%. The equation is $598 = (598 + 52)x$

Practice 3.6.2

1. $3400   **2.** $2420   **3.** $1500   **4.** $6000   **5.** 10%

Practice 3.6.3

1. $10,500 at 8% and 1,500 at 6%. The equation is $930 = 0.08x + 0.06(12000 - x)$
2. $11,000 at 7% and $2,500 at 18%. The equation is $1220 = 0.07x + 0.18(x - 8500)$

Practice 3.6.4

1. save account 7%; riskier account 17%
2. bonds 6%; stocks 15%.

Practice 3.6.5

1. The whole milk has 1.40 gallons of butterfat while the 2% milk has 1.36 gallons, so the whole milk has more butterfat.   **2.** The 35% solution has 2.1 gallons of antifreeze while the pure antifreeze has 2 gallons. So the 35% solution has more antifreeze.

Practice 3.6.6

1. 40 milliliters of the 25% solution and 60 milliliters of the 30% solution are required. The equation is $0.28(100) = 0.25x + 0.30(100 - x)$
2. 40 tons of 7% ore and 20 tons of 10% ore are required. The equation is $0.08(60) = 0.07x + 0.10(60 - x)$.

Exercise Set 3.6

1. $150   **3.** $12.00   **5.** 35%   **7.** $50,000   **9.** $150
11. $112   **13.** $6,200   **15.** $5720   **17.** 5%
19. $10,000 at 18% and $15,000 at 7%
21. $6000 from his uncle and $2000 from the bank.
23. $20,000 at 18% and $16,000 at 6.5%
25. safe account 6%; riskier account 12%.
27. Pension fund pays 10% and the credit union pays 5%.
29. 25% acid solution has 3.75 liters of acid and 10% acid solution has only 2.5 liters of acid. The 25% solution has the most acid.
31. The 30% solution has 42 milliliters of water and the 25% solution has 56.25 milliliters of water. The 25% solution has the most water.
33. They all have 12 milliliters of acid.
35. 14 gallons
37. 20 liters of the 40% and 10 liters of pure acid.
39. 15 tons of 12% and 45 tons of 4%

---

41. [1.7.7] $-15x + 34$ **42.** [2.4.3]

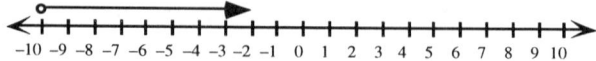

43. [2.2.2] -8 **44.** [2.2.3] -27.5
45. [2.2.4] all real numbers   **46.** [2.5.3] $x > -4$
47. [3.3.1] 45,000 pounds

Chapter 3 Review Exercises

Numbers inside brackets [ ] indicate the Chapter, Section, and Example number of a similar worked example.

1a) [3.1.2] $e - 18$ **1b)** $\frac{1}{2}e$ **1c)** e **1d)** $e - 20$

2a) [3.1.2] $f + 6$ **2b)** $f - 8$ **2c)** $\frac{1}{3}f$ **2d)** $3f$

3. [3.1.3] $40 - p - (p + 8)$, which simplifies to $32 - 2p$

4. [3.1.3] $1200 - s - \frac{1}{2}s$, which simplifies to $1200 - \frac{3}{2}s$

5. [3.1.4] $p + 2p + (2p + 75)$, which simplifies to $5p + 75$

6. [3.1.5] $2\pi(6 - x)$

7a) [3.1.6] $c + \frac{1}{2}c + (c - 200)$, simplifies to $\frac{5}{2}c - 200$

7b) $c + (c - 40) + (c - 40 + 100)$ simplifies to $3c + 20$

7c) $c + (c + 24) + \frac{1}{3}(c + 24)$ simplifies to $\frac{7}{3}c + 32$

8. [3.2.1] $2l + 2\left(\frac{1}{3}l\right) = 68$; 25.5 feet by 8.5 feet.

9. [3.2.2] $c + (c + 16) + \frac{1}{2}c = 41$; $10 for calculator, $26 for backpack $5 for pencil

10. [3.2.3] $2(w) + 2(2w - 13) = (2w - 13) + 67$; width is 20, length is 27

11. [3.2.4] $n + (n + 2) + (n + 4) = 60$; 22

12. [3.3.1] 72 inches **13.** [3.3.2] $64

14. [3.4.1] 12 quarters and 20 dimes

15. [3.4.1] $291.96e + 125.80(e + 16) = 6190.40$; 10 sets of expensive tires.

16. [3.4.2] $3x + 2(x + 15) = 180$; 30 bags per hour.

17. [3.4.3] $5(d - 20) + 10d + 25(d + 17) = 1525$; 30 dimes, 47 quarters, 10 nickels

18. [3.5.1] $20b + 12(b + 0.5) = 14$; 0.25 hour.

19. [3.5.2] $4(s - 30) = 2.5s$; Steve, 80 mph, Lauren, 30 mph **20.** [3.6.1] $p - 0.45p = 385$; $700

21. [3.6.2] $1000 = p(0.06)(5)$; about $3333.33

22. [3.6.3] $2000c + 9500(c + 0.04) = 955$; 5% for checking, 9% for stock

23. [3.6.4] $8000c + 2000(c + 1.2) = 2000$; –0.04; the coffee stand lost 4% of the $8,000 invested in it, but the Internet stock had a positive rate of return of 116%.

24. [3.6.5] Alcohol in 18% solution is $a = 0.18(23)$; 4.14 gallons. Alcohol in 16% solution is $a = 0.16(25)$; 4.00 gallons. 18% solution has more alcohol.

25. [3.6.6] $0.04(210) + 0.30(x - 210) = 0.10(x)$; 273

Chapter 3 Test

Numbers inside brackets [ ] indicate the Chapter, Section, and Example number of a similar worked example.

1a) [3.1.2] $f - 12$ **1b)** $f - 4$ **1c)** $\frac{3}{4}f$ **1d)** $2f$

2. [3.1.3] $\frac{1}{2}(560 - 2w)$, which simplifies to $280 - w$

3. [3.1.4] $3f + f + \frac{1}{2}f$, which simplifies to $\frac{9}{2}f$

4. [3.1.5] $\left(\frac{18 - x}{2}\right) \cdot 8$, which simplifies to $72 - 4x$

5a) [3.1.6] $g + (g + 400) + 2(g + 400)$, which simplifies to $4g + 1200$

5b) $g + (g - 500) + \frac{1}{2}g$, which simplifies to $\frac{5}{2}g - 500$

5c) $g + 2g + g$, which simplifies to $4g$

6. [3.2.1] $38(25) + 27(25 + x) = 2030$; $15

7. [3.2.2] $3(t - 18) + (t - 18) + t = 23$; 19 on third run

8. [3.2.3] $L + (0.5L) + \left(\frac{L + 0.5L}{2} - 1\right) = 2L + 2$; L is 12 so perimeter is 26.

9. [3.2.4] $x + (x + 2) + (x + 4) = x + 2$; –2, 0, 2

10. [3.3.1] $101\frac{1}{4}$ miles

11. [3.4.1] $3x + 5(x + 4) = 180$; There are 20 workers in the smaller group and 24 in the larger group

12. [3.4.2] $9n + 6(n + 10) = 210$; 10 larger cans per shelf

13. [3.4.3] $18(2c + 3) + 21(2c) + 30c = 972$; Special A is $20 per plate, Special B is $17 per plate, Special C is $8.50 per plate

14. [3.5.1] $27(\frac{1}{3}s) + 65s = 296$; s is 4 hours; $65s$ is 260 miles

15. [3.6.2] $i = 500(0.20)(2)$; $200; $i = 200(0.50)(2)$; $200; interest is the same

16. [3.6.4] $2500x + 2500(x + 0.03) = 175$; 2% and 5%

17. [3.6.6] $0.04w + 0.36(1) = 0.20(w + 1)$; 1 gallon

Chapter 4 Mathematical Models

Section 4.1 Introduction to Four Mathematical Models

Practice 4.1.1

1. If we let y represent the distance (miles) and x represent the time (hours) then the equation is $y = 78x$

2. If we let y represent the time (hours) and x represent the number of rooms painted then the equation is $y = \dfrac{10}{3} x$

3. If we let y represent the number of rooms painted and x represent the time (hours) then the equation is $y = \dfrac{3}{10} x$

4. If we let y represent the cost (dollars) and x represent the energy used (kilowatt hours) then the equation is $y = 0.06x$

Practice 4.1.2

1. If we let y represent the distance (feet) and x represent the time (minutes) then the equation is $y = 28x$

x time (minutes)	y distance (feet)
20	560
50	1,400
90	2,520
180	5,040

2. If we let y represent the interest earned (dollars) and x represent the money invested (dollars) then the equation is $y = .085x$

x money invested (dollars)	y interest earned (dollars)
1,000	85
5,000	425
15,000	1,275

3. If we let y represent the cost (dollars) and x represent the number of strings then the equation is $y = 1.25x$

x number of strings	y cost (dollars)
8	10.00
15	18.75
50	62.50
100	125.00

Practice 4.1.3

1.

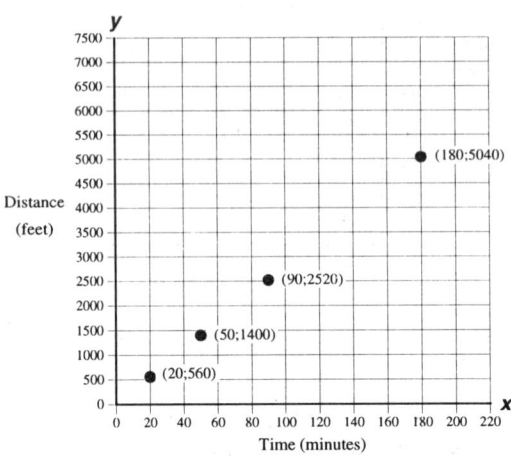

2.

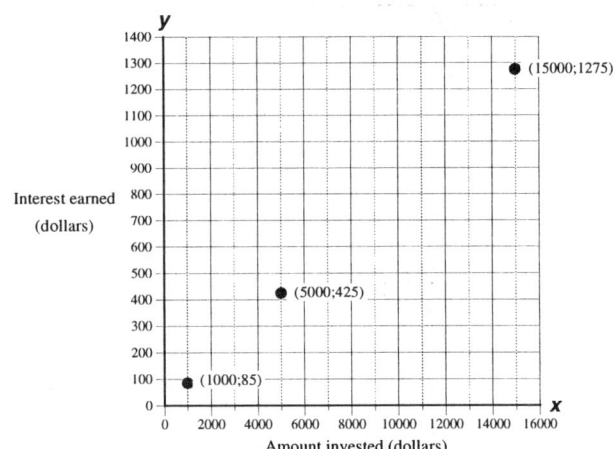

Practice 4.1.4

1a) $y = 13x$

1b)

x time since mall opened (minutes)	y people who have left mall
60	780
120	1,560
180	2,340
230	2,990
308	4,004

1c)

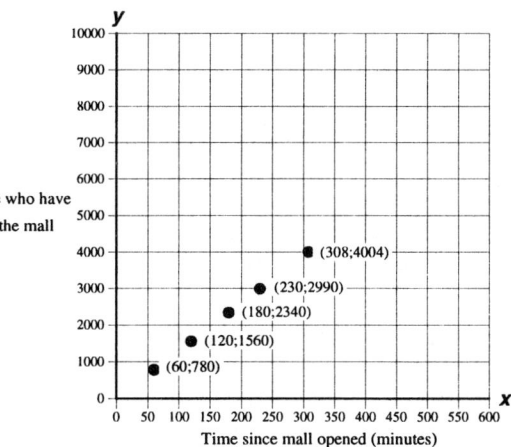

2a) $y = \dfrac{1}{1.5}x$

2b)

x time worked (hours)	y jobs completed
8	5
40	26
120	80
75	50
150	100

2c)

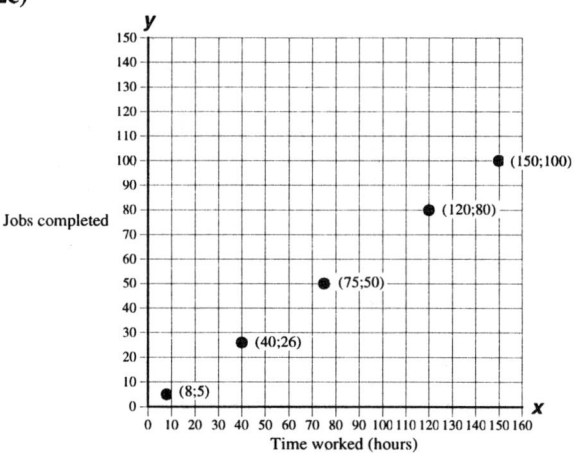

3a) $y = 3.5x$

3b)

x time into program (months)	y weight loss (pounds)
5	17.5
12	42.0
15	52.5
7	24.5
14	49.0

3c)

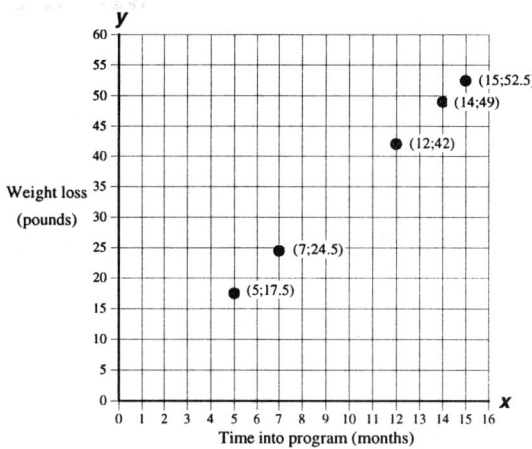

4a) $y = 11.5x$

4b)

x number of booklets made	y cost (dollars)
100	1,150
200	2,300
500	5,750
50	575
434	4,991

4c)

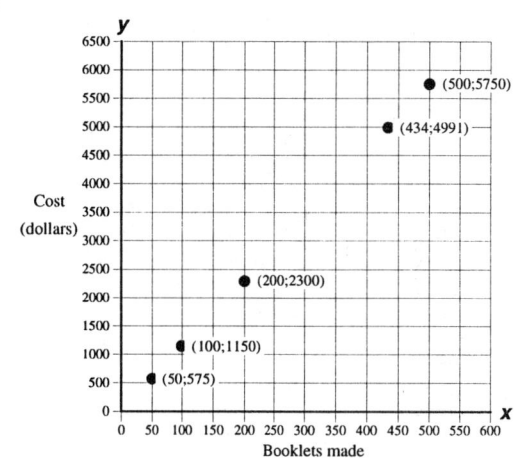

Exercise Set 4.1

1. If we let y represent the size of the oil slick (square miles) and x represent the time since the spill (hours), the equation is $y = 0.25x$

3. If we let y represent the distance covered (miles) and x represent the time since leaving (hours) then the equation is $y = 12x$.

x time since leaving (hours)	y distance covered (miles)
2.5	30
3.75	45
5.5	66

5. If we let y represent the number of roofs completed and x represent the time worked (days) then the equation is
$$y = \frac{1}{4}x$$

x time worked (days)	y number of roofs completed
68	17
92	23
284	71

7.

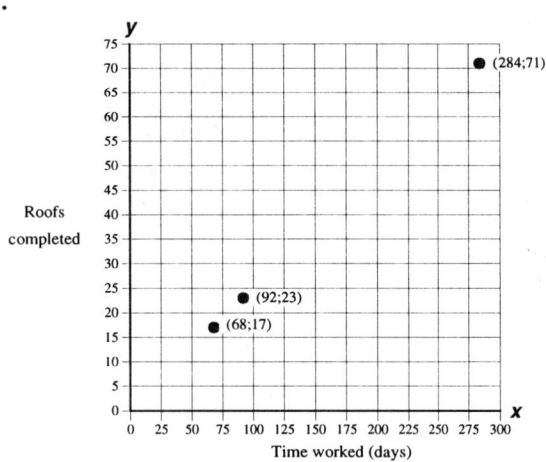

9a) $y = 50.60x$

9b)

x creme purchased (ounces)	y cost (dollars)
4.0	202.40
12.0	607.20
16.0	809.60
5.9	298.54
9.9	500.94

9c)

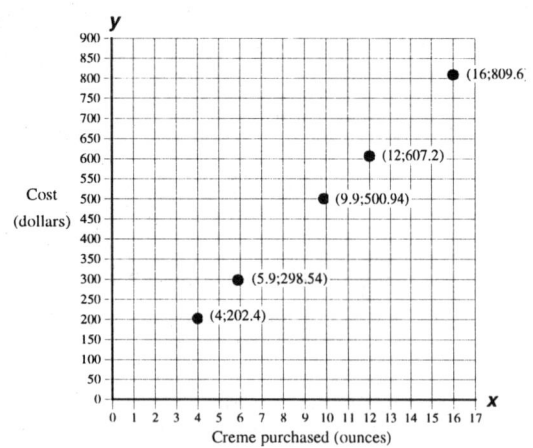

11a) $y = 350x$

11b)

x time since starting child care (weeks)	y cost (dollars)
4	1,400
26	9,100
52	18,200
14	4,900
43	15,050

11c)

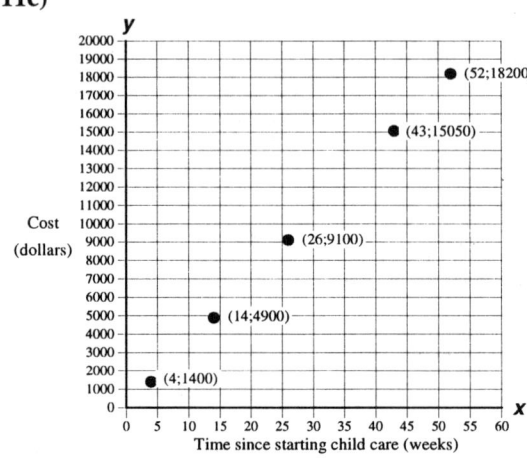

13a) $y = 0.35x$

13b)

x solution in container (milliliters)	y alcohol in container (milliliters)
10	3.5
50	17.5
200	70.0
80	28.0
110	38.5

13c)

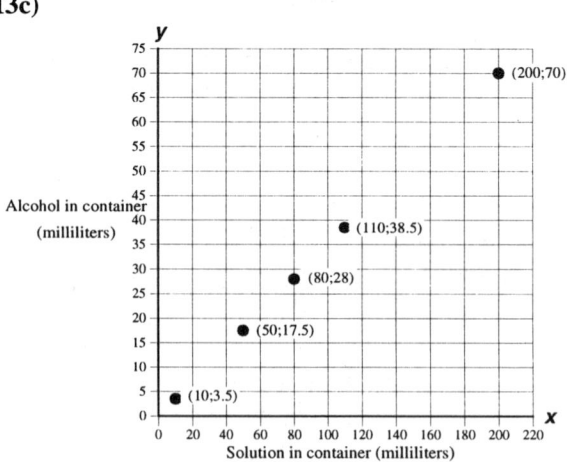

15a) $y = \frac{1}{2}x$

15b)

x counter reading	y number of vehicles
1,400	746
2,116	1,058
1,714	857
800	400
2,000	1,000

15c)

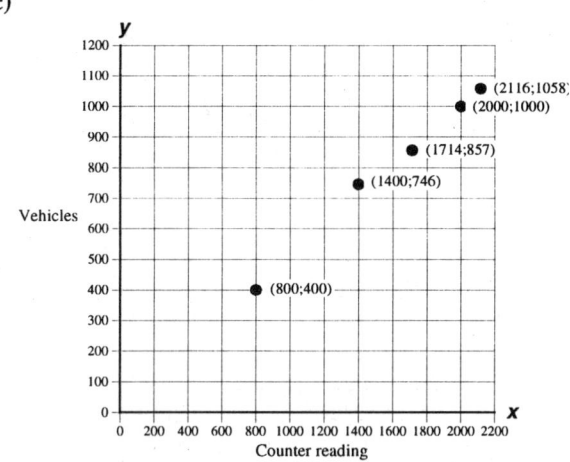

Counter reading

17. [1.1.3] 44   **18.** [1.2.1] $2 \cdot 2 \cdot 7 \cdot 7$

19. [1.2.10] $\frac{16}{49}$ **20.** [1.3.7] 34   **21.** [2.4.3]

22. [1.6.7] $t(5-1)$ **23.** [2.2.2] 0   **24.** [2.5.3] $x < -2$

25. [1.3.3]

-300 -200 -100 0 100 200 300 400 500

26. [3.4.2] $22m + 56(m - 7) = 1168$; 20 miles per gallon

27. [3.6.1] $0.07x = 21.35$; \$305

Section 4.2 The Linear Equation

Practice 4.2.1

1. $y = 2095 + 325x$ **2.** $y = 2$ **3.** $y = 219 - 13x$

4. $y = 25 - \frac{7}{10}x$

Practice 4.2.3

1a) $y = 2700 - 45x$ **1b)** 60 months   **1c)** \$1890

2a) $y = 259 + 2.4x$ **2b)** 2010   **2c)** 273.4 million

3a) $y = 118 + 8.68x$ **3b)** \$248.20   **3c)** 2006

4a) $y = 127 - 3x$ **4b)** 2012   **4c)** 7 million tons

Practice 4.2.4

1. To rent the dining room set you pay \$75 down and \$35 a month.

2. In 1986 7277 thousand tons of lead was emitted in the U. S. From 1986 to 1992 the lead emitted dropped by 400 thousand tons per year

3. A full tank holds 16 gallons and the car can drive 32 miles for every gallon of gas used. In other words the car gets 32 miles per gallon.

Exercise Set 4.2

1. $y = 9000 + 3x$ where y represents the cost to print a book (dollars) and x represents the number of book printed.   **3.** $y = 72 - 4x$ where y represents the height of the pile (inches) and x represents the time since the pile was built (weeks).   **5a)** $y = 1500 + 275x$ **5b)** 23 months   **5c)** 63.6 months.   **7a)** $y = 3.09 + 0.35x$

7b) around 1995   **7c)** \$18.14.   **9a)** $y = 58$ **9b)** 58%

9c) There is no single choice for input.

11a) $y = 27.40 + 2.50x$ **11b)** \$77.40   **11c)** 2014

13a) $y = 54 - 0.39x$ **13b)** 2006   **13c)** 42.3%

15a) $y = \frac{1}{5}x$ (in this case b is 0)

15b) 3.6 miles   **15c)** 8 seconds   **17a)** $y = 125 + 1.50x$

17b) \$297.50   **17c)** 250 items   **19.** When new the computer cost \$2100. Since it was purchased the computer is losing \$35 in value per month.   **21.** In 1970 the average U. S. citizen consumed 25 gallons of whole milk as a beverage. Between 1970 and 1991 consumption dropped 7 gallons every 10 years or $\frac{7}{10}$ gallon every year.

23. In 1940, 91879 thousand tons (or 91,879,000 tons) of carbon monoxide was released. Between 1940 and 1970 the thousands of tons of carbon monoxide increased by 1087 thousand tons (or 1,087,000 tons) per year.

25. The radius of a circle is half its diameter.

27. In 1996 dollars the minimum hourly wage in 1969 was \$6.59. Since 1969 the minimum hourly wage has been declining by about 9¢ a year using 1996 as the base year.

29. Stalactites grow about 1 inch every 12.5 years.

31. [1.2.3] $\frac{27}{10}$ **32.** [2.2.4] no solution   **33.** [2.4.3]

-10 -9 -8 -7 -6 -5 -4 -3 -2 -1 0 1 2 3 4 5 6 7 8 9 10

34. [1.3.2] $\{-3.5, -1, 3, 7\}$ **35.** [1.3.3]

0 150 300 450 600 750 900 1050

36. [1.6.8] Base is 2, exponent is 4; simplifies to -16

37. [3.2.4] $x + (x+1) + (x+2) + (x+3) + (x+4) = 165$; largest integer is 35

Section 4.3 The Graph of a Linear Equation

Practice 4.3.1

1.

x	y
6	−2
4	−1
0	1
−6	4
−2	2
1	**0.5**

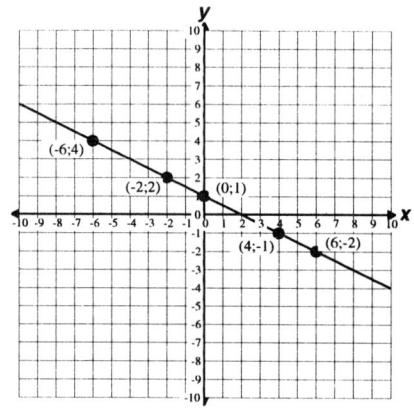

2.

x	y
4	−1
−2	−3
10	−5
	6
−8	
0	

A straight line is not an appropriate model for this graph.

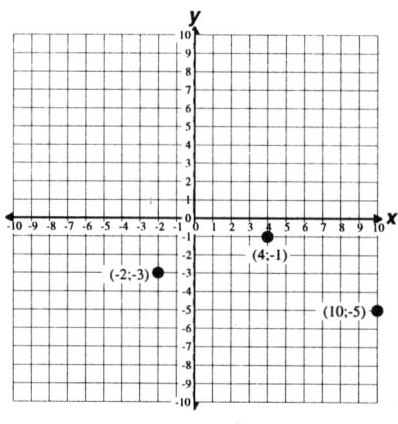

3.

x	y
2	5
−2	−7
1	2
3	**8**
−1	**−4**
−3	−10

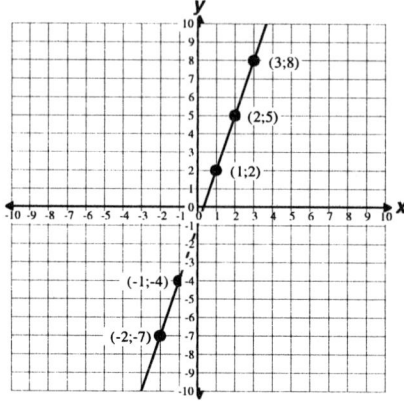

4.

x	y
1	5
−6	5
−2	5
0	**5**
none	−3
8	**5**

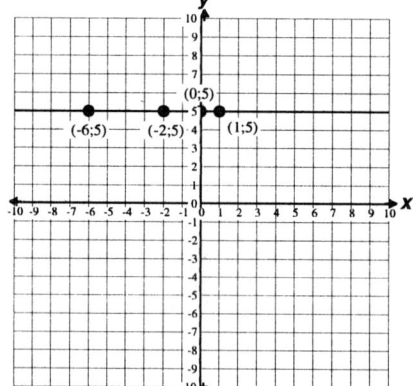

Practice 4.3.2

1. Transformed data table

x time (number of years since 1975)	y number of women earning degrees in dentistry (hundreds)
0	1.70
3	4.61
6	7.52
9	10.43

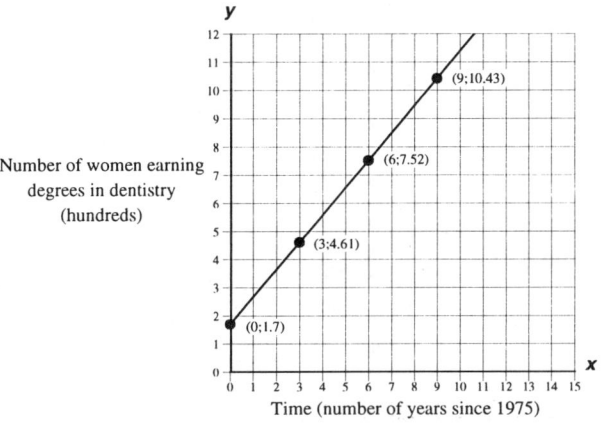

2. Transformed data table

x number of damaged items (hundreds)	y profit (thousands of dollars)
12	68
28	52
6	74
19	61

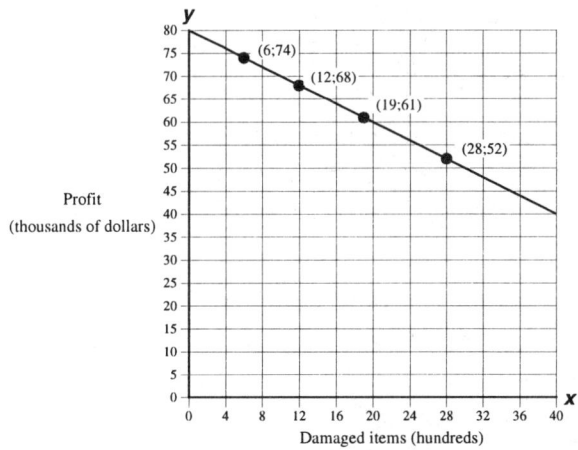

Practice 4.3.3

1a)

x Bars sold (hundreds)	y Profit (dollars)
1	−525
4	−300

1b)

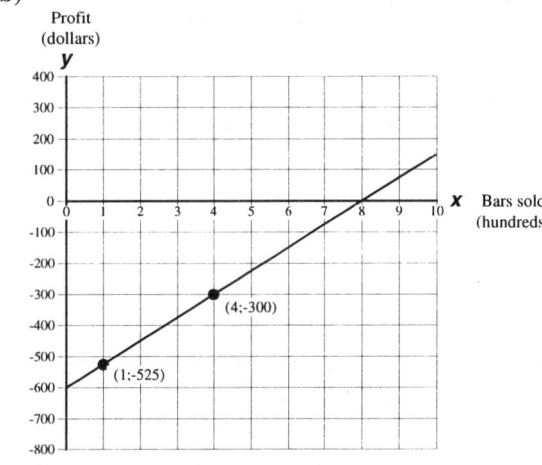

1c) $600

1d) 800 bars

1e) a little over 900 bars

1f) about $150

2a)

x Time since beginning program (days)	y Weight (pounds)
27	209
42	204

2b)

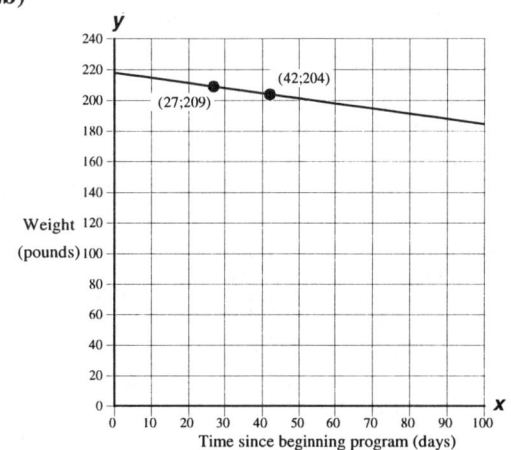

2c) about 85 days

2d) about 198 pounds

2e) about 219 pounds

Practice 4.3.4

1.

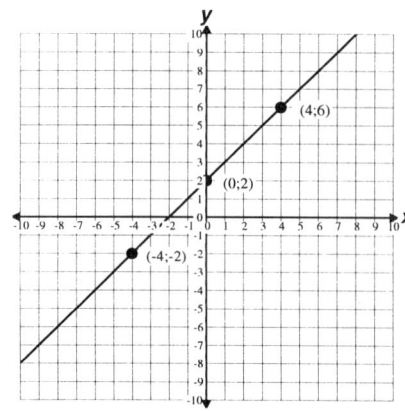

2. Original data table

x	y
0	3000
10	3500
−10	2500

Transformed data table

x	y (thousands)
0	3
10	3.5
−10	2.5

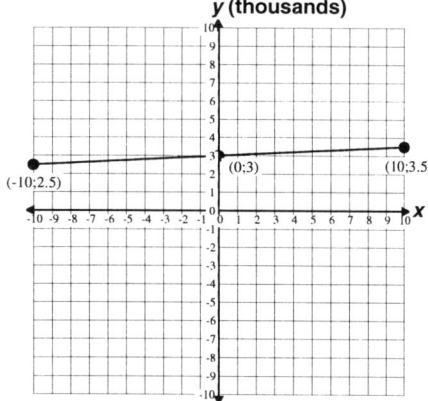

3.

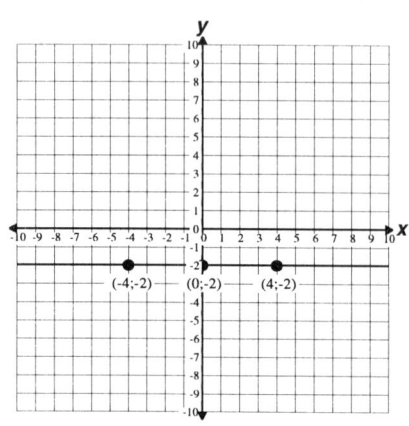

4. Original data table

x	y
0	−10,000
10	0
−10	−20,000

Transformed data table

x	y (thousands)
0	−10
10	0
−5	−15

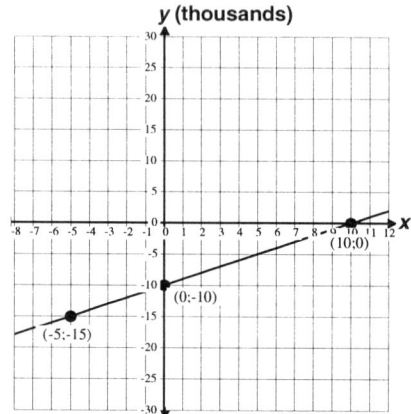

5.

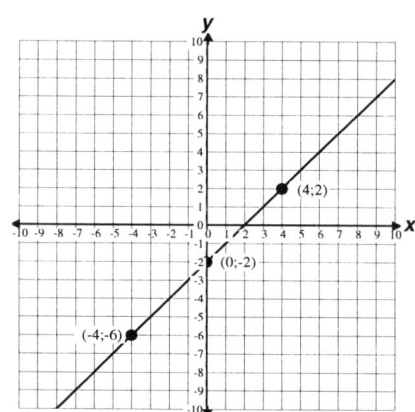

Practice 4.3.5

1. *y*–intercept is (0,4); *x*–intercept is (4,0)

2. *y*–intercept is (0,3); *x*–intercept is (−4,0)

3. *y*–intercept is (0,7); there is no *x*–intercept

4. There is no *y*–intercept; *x*–intercept is (−3.5,0)

5. *y*–intercept is (0,0); *x*–intercept is (0,0)

6. *y*–intercept is (0,0); *x*–intercept is the entire *x*–axis which we may write as (*x*,0)

Practice 4.3.6

1.

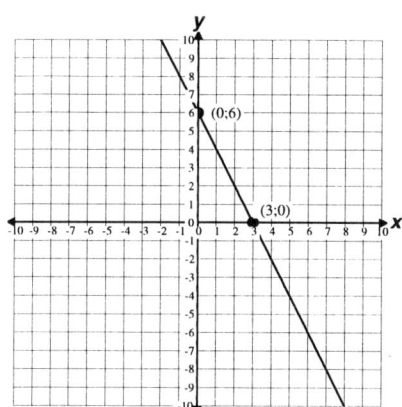

2.

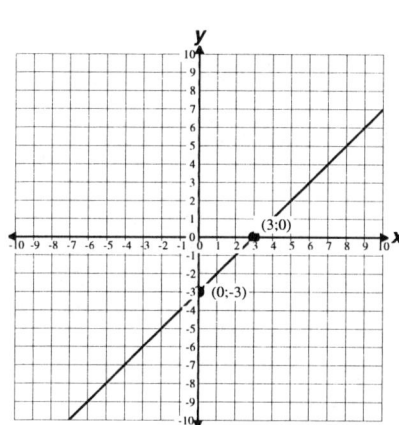

3.

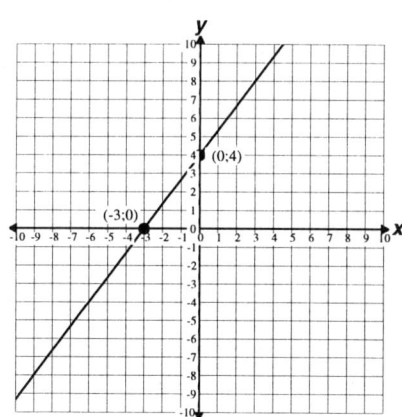

4. Since the x and y intercepts are the same point, we must calculate a second point before graphing. We choose 1 for x and then calculate y to be 2. Therefore, we use (1,2) as the second point.

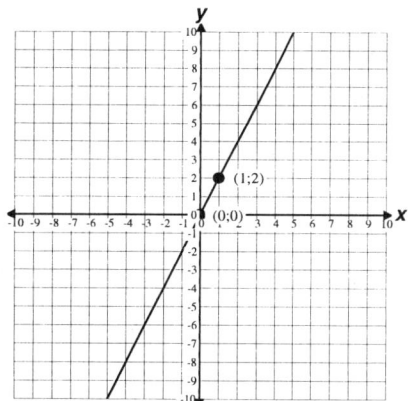

Exercise Set 4.3

1.

x	y
−8	−2
−6	4
2	6
	3
	1
	0

The points do not lie in a straight line. No input values can be determined.

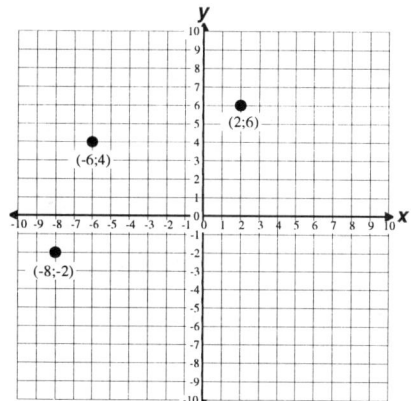

3.

x	y
5	−8
1	−4
−3	0
−7	4
−4	1
−2	−1

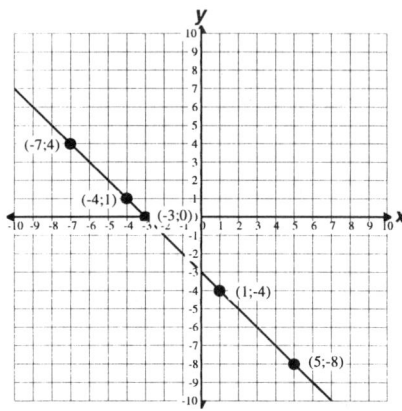

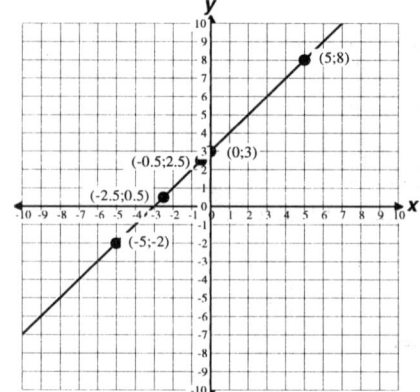

5.

x	y
1	−1
4	8
−1	−7
3	5
0	−4
−2	−10

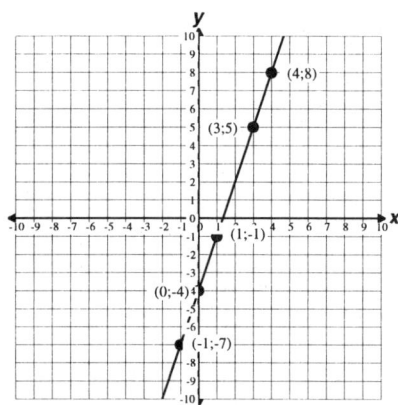

9.

x	y
−3	−2
−3	5
−3	0
−3	1
4	**none**
−3	2

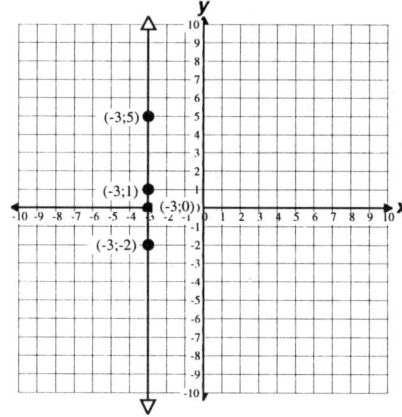

7.

x	y
5	8
0	3
−5	−2
−2.5	0.5
−0.5	2.5
−8.5	**−5.5**

11. Original data table

x	y
0	−500
10	−250
20	0

Transformed data table

x	y (hundreds)
0	−5
10	−2.5
20	0

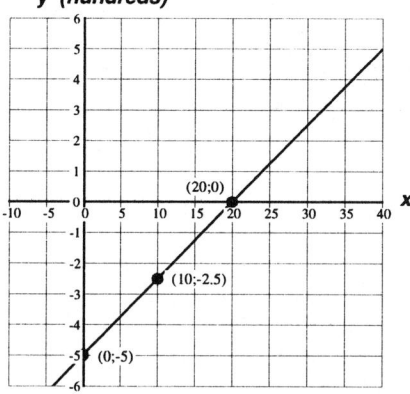

15. Original data table

x	y
0	2,400
−50	1,400
50	3,400

Transformed data table

x	y (hundreds)
0	24
50	14
−50	34

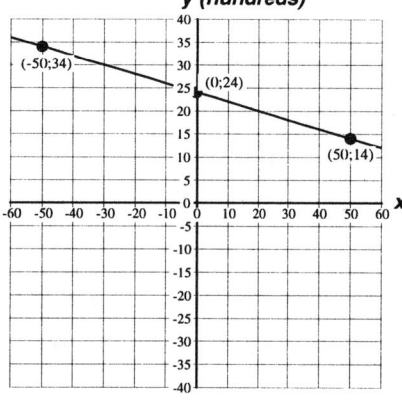

13.

Transformed data table

x number of years since 1960	y number of men earning master's degrees (thousands)
0	47
1	55
8	111
14	159

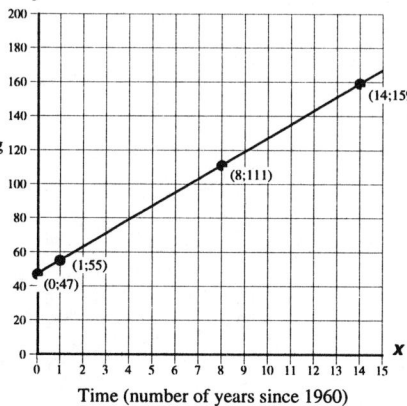

17.

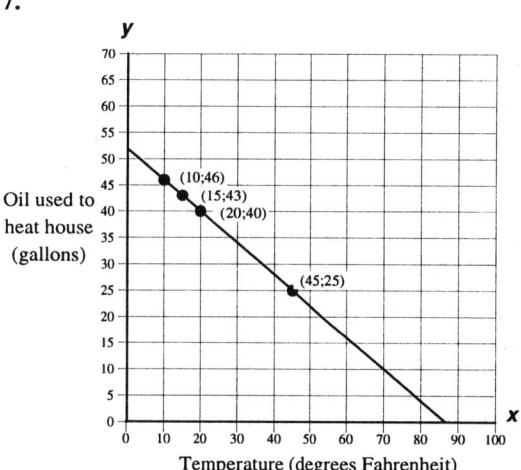

19. Transformed data table

x time (number of years since 1940)	y carbon monoxide emitted (millions of tons)
0	92
10	103
20	114
30	125

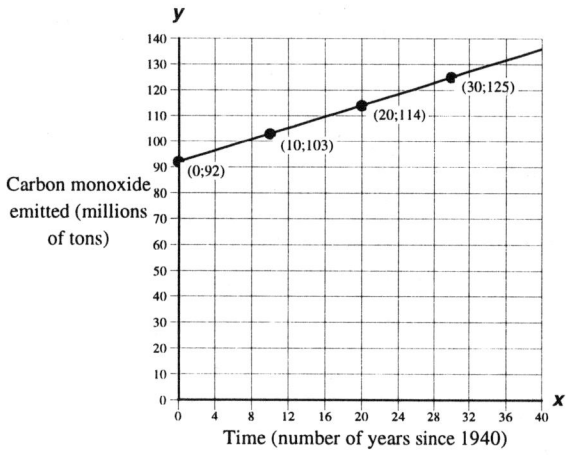

21.

x Temperature (degrees C)	y Temperature (degrees F)
20	68
35	95

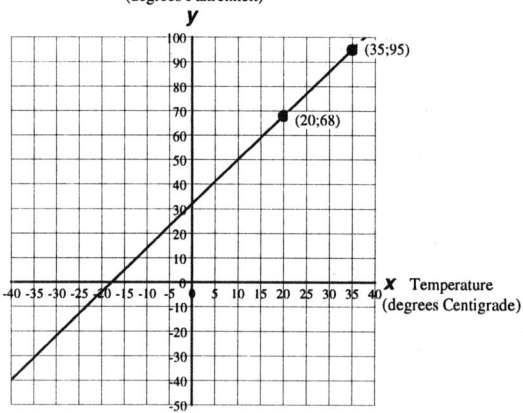

21a) about −20° Fahrenheit.

21b) about 37° Centigrade.

21c) 40° Centigrade corresponds to 104° Fahrenheit, which is a hot day.

21d) about −18 degrees Centigrade.

23.

x Snow depth (inches)	y Time since snow began to fall (hours)
2	8
4	9

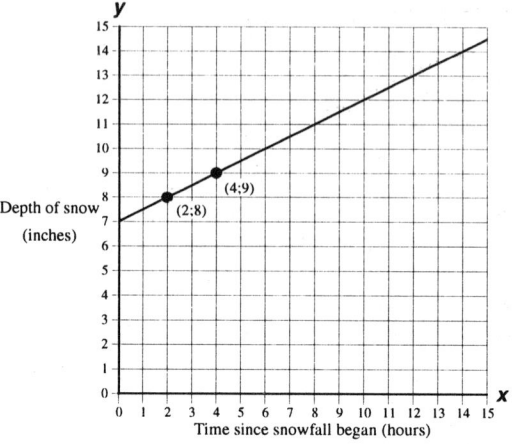

23a) about 7 inches of snow.

23b) about 11 inches.

23c) since there was 7 inches of snow to begin with there would be about 6 inches of new snow.

23d) about 14 hours.

25.

x time (number of years since 1968)	y minimum hourly wage purchasing power (dollars)
6	6.25
14	5.27

25a) about $6.90.

25b) around 1985.

25c) about $4.61.

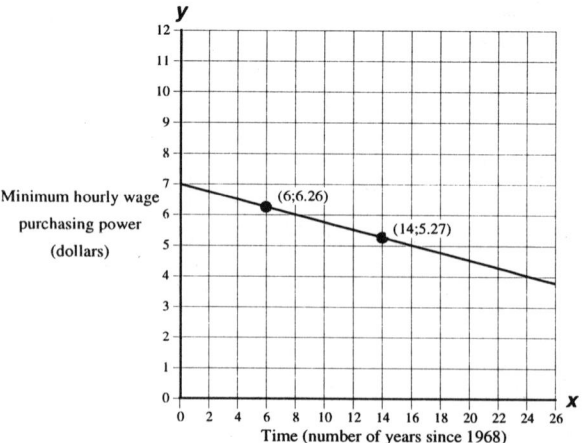

27.

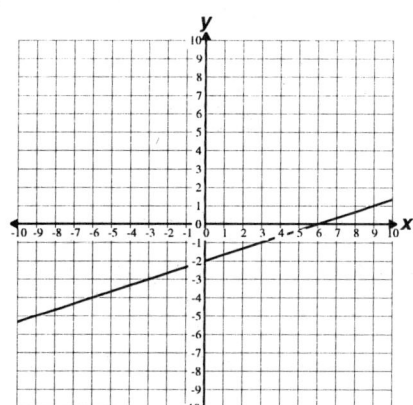

29.

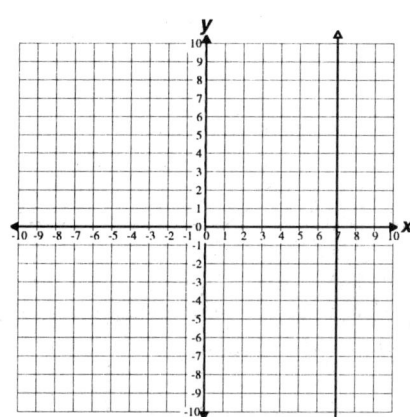

31.

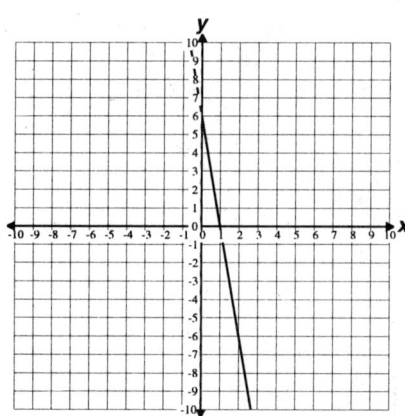

33.

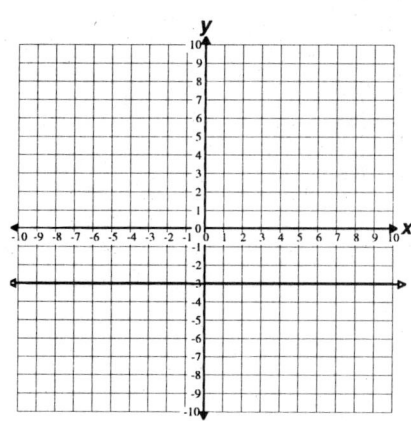

35.

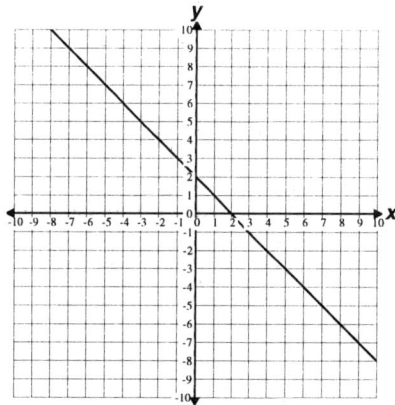

37. y–intercept is $(0,3)$; x–intercept is $(5,0)$

39. y–intercept is $(0,3.5)$; x–intercept is $(-2,0)$

41. y–intercept is $(0,80)$; x–intercept is $(-8,0)$

43. y–intercept is $(0,0)$; x–intercept is $(0,0)$

45.

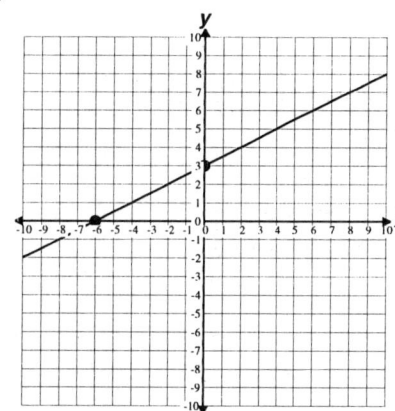

47.

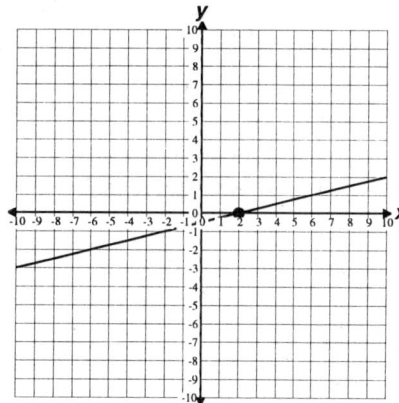

49. Because both intercepts are at (0,0), we had to select a point that is not on either axis for the second point.

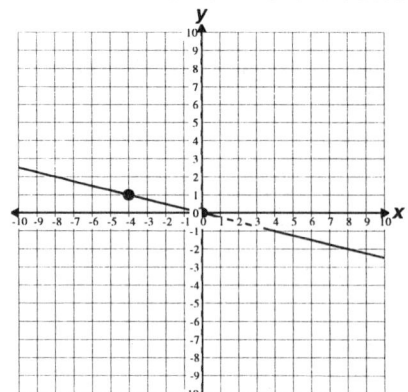

51.

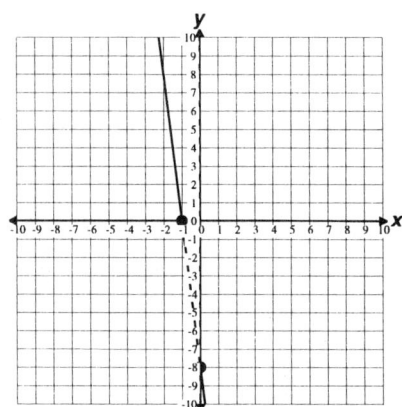

---

53. [1.5.8] 13   **54.** [2.3.2] $p = \dfrac{i}{rt}$ **55.** [1.2.2] $\dfrac{31}{110}$

56. [1.7.7] $-27y + 60$ **57.** [2.2.2] 1   **58.** [1.2.4] $\dfrac{2}{3x}$

59. [3.2.2] $w + (w - 52) + [(w - 52) + 18] = 220$; w is 102 so the bedroom is 50 square feet.

60. [3.4.2] $41x + 55(x + 2) = 446$; 3.5 pounds per brick

Section 4.4 Slope and Some of Its Uses

Practice 4.4.1

1. $y = 1 - 3x$ **2.** $y = 3 + 4x$ **3.** $y = 5 + \dfrac{3}{4}x$

4. $y = -6 - \dfrac{1}{2}x$

Practice 4.4.2

1. Crickets start chirping at 37 degrees Fahrenheit and then increase their chirps by 4 chirps per minute for every 1 degree rise in temperature. (Or by 60 chirps per minute for every 15 degree rise in temperature.)

2. The area of the garden increases by 3 square feet for every 1 foot increase in the length. Notice this also says the width of the garden is 3 feet.

3. Originally the computer cost $2100. Every month since it's purchase it's been losing $35 of it's value.

4. The water level in the pond began at 210 inches and has been decreasing 5 inches every 2 weeks (or 2.5 inches per week) since then.

Practice 4.4.3

1.

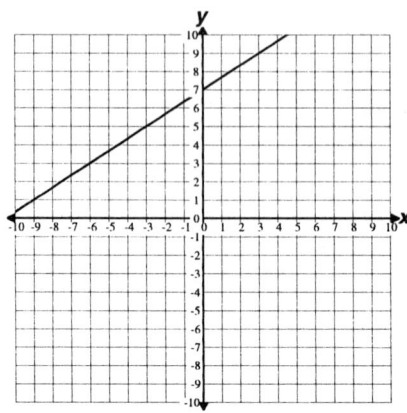

2.

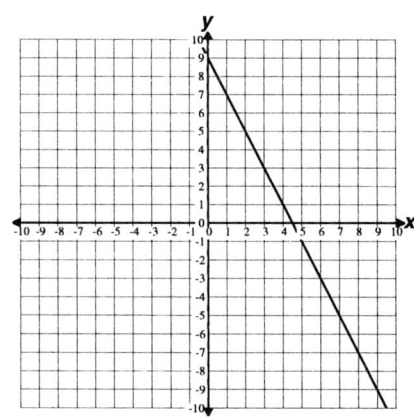

3.

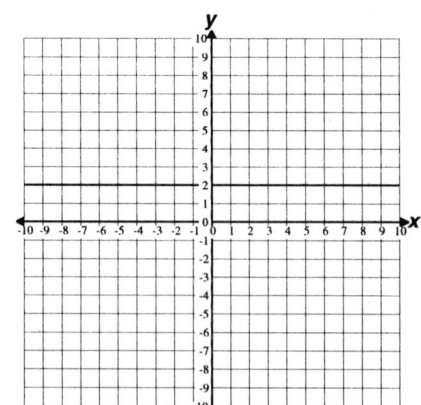

4.

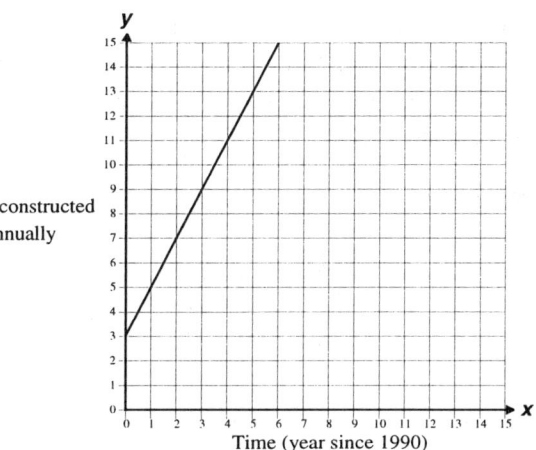

Boats constructed annually / Time (year since 1990)

5.

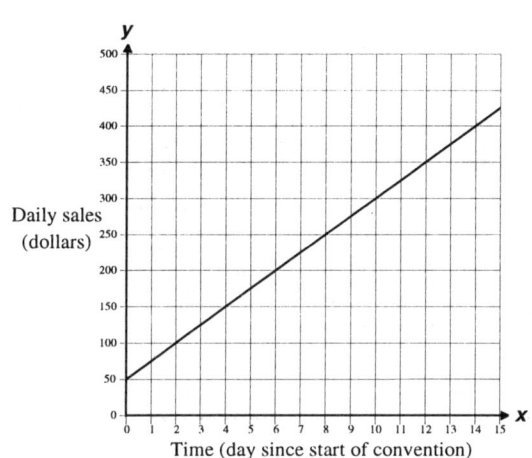

Daily sales (dollars) / Time (day since start of convention)

Practice 4.4.4
1. -9 **2.** $\frac{5}{3}$ **3.** $\frac{1}{2}$ **4.** 15

Practice 4.4.5
1. Slope is $\frac{-2 \text{ pounds}}{1 \text{ year}}$. The slope means beef consumed per person has been dropping 2 pounds every year since 1985.

2. Slope is $\frac{-32,600 \text{ fires}}{1 \text{ year}}$. The slope means structural fires in the U. S. have been dropping by about 32,600 fires each year since 1980.

3. Slope is $\frac{7 \text{ dollars}}{1 \text{ ticket}}$. The slope means the profit increases $7 for every ticket sold.

4. Slope is $\frac{3 \text{ dollars of interest}}{25 \text{ dollars invested}}$ or $\frac{\$.12 \text{ interest}}{\$1 \text{ invested}}$. The slope means you make 12¢ for every dollar you invest. This is a 12% rate of interest. (Notice $\frac{3}{25} = \frac{12}{100}$ which is another way of saying 12%.)

Practice 4.4.6
1. The slope is 0.
2. The slope is undefined. There is no relation between the number of shares purchased and the price per share.
3. The slope is undefined.
4. The slope is 0. There is no relation between the calories consumed and the weight.

Practice 4.4.7
1. Parallel **2.** Perpendicular **3.** Parallel **4.** Parallel

Exercise Set 4.4
1. $y = 7 - \frac{1}{5} x$ **3.** $x = -3$ **5.** $y = -9 + 3x$

7. $y = -6 + \frac{2}{3} x$ **9.** $y = -5$ **11.** Every 5 degree rise in the Centigrade temperature is equivalent to a 9 degree rise in the Fahrenheit temperature. The y–intercept is $(0,32)$, which means that 0 degrees Centigrade corresponds to 32 degrees Fahrenheit.

13. Every 1 inch increase in the diameter of a circle leads to about a 3.1 inch increase in the circumference of the same circle. (Actually it's a 3.14 (or π) inch increase but it's hard to show that precision on this graph.)

15. In 1975 per capita beef consumption was 81 pounds. Since 1975 per capita beef consumption has been dropping about 1 pound per year.

17. 47,000 men earned masters degrees in 1960. Between 1960 and 1975 an additional 8,000 more men earned masters degrees each year.

19.

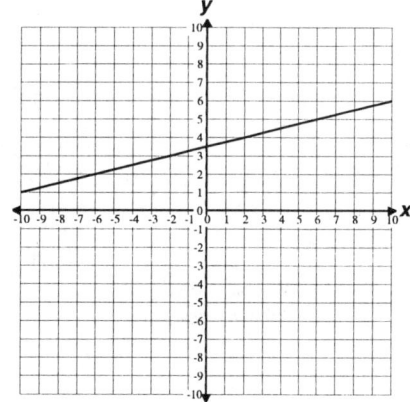

21.

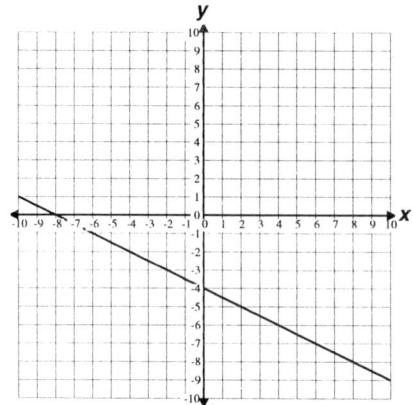

23.

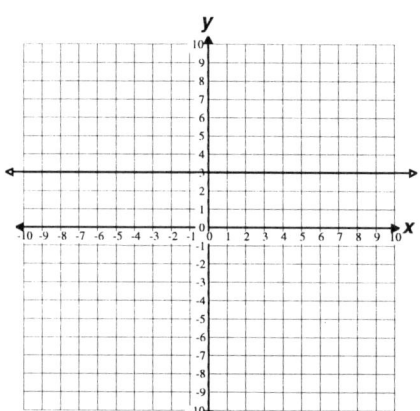

25.

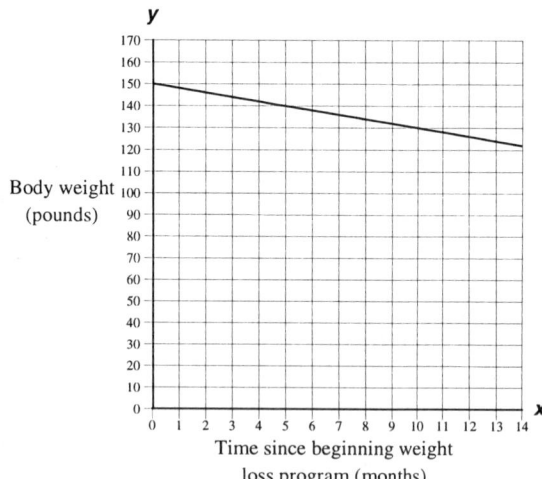

27.

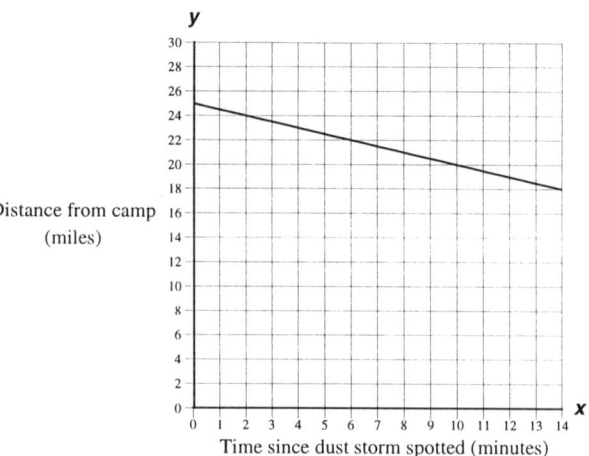

29. 8 **31.** –2 **33.** $\frac{7}{3}$ **35.** $-\frac{1}{4}$

37. Slope is $\frac{11 \text{ gallons of acid}}{50 \text{ gallons of solution}}$ or $\frac{0.22 \text{ gallons of acid}}{1 \text{ gallon of solution}}$.
The slope means you have a 22% acid solution. (Notice that $\frac{11}{50} = \frac{22}{100}$ which is another way of saying 22%.)
or $\frac{0.8 \text{ gallons of orange juice}}{1 \text{ gallon of solution}}$ The slope means you have 80% orange juice in the solution. (Notice that $\frac{4}{5} = \frac{80}{100}$ which also says 80%.)

39. Slope is $\frac{-600 \text{ dollars}}{1 \text{year}}$. The slope means the investment is losing $600 a year.

41. Slope is $\frac{\$13.80}{1 \text{ ounce}}$. The slope means the cost for 1 ounce of dry treatment creme is $13.80.

43. Slope is $\frac{\$0.21}{1 \text{year}}$ or $\frac{21¢}{1 \text{ year}}$. The slope means the average hourly earnings for retail workers was increasing by 21¢ a year.

45. Slope is $\frac{-0.17 \text{ seconds}}{1 \text{ year}}$. The slope means the winning Olympic time in the men's 100 meter freestyle is dropping about 0.17 seconds per year. (Since the Olympics is every four years it may be more appropriate to say it's dropping 0.68 seconds every four years.)

47. Slope is $\frac{63 \text{ miles}}{1 \text{ hour}}$. The slope means that before the road construction the driver was traveling 63 mph.

49. The slope is 0. The pay does not change during the week.

51. The slope is 0.

53. The slope is undefined. Knowing someone paid $25 to park their car does not allow us to estimate their GPA.

55. The slope is undefined.

57. The slope is 0. People did not change their consumption of margarine between 1970 and 1994.

59. Parallel

61. Perpendicular

63. Perpendicular

65. [1.2.6] $\frac{5}{4}$ **66.** [1.2.10] $\frac{17}{36}$ **67.** [1.7.2] $13x^2 - 4x$

68. [2.2.5] 0 **69.** [2.4.3]

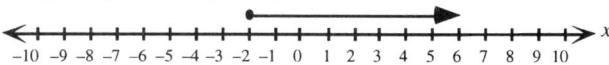

70. [2.5.3] $x \geq -5$

71a) [3.1.4] $t + (t + 3)$, which simplifies to $2t + 3$

71b) $t + (t + 10)$, which simplifies to $2t + 10$

71c) $t + \frac{1}{2}t$, which simplifies to $\frac{3}{2}t$

72. [3.2.1] $x - \frac{1}{3}x = 760$; $1140

73. [3.6.1] $0.80x = 28$; 35 students

Section 4.5 Other Forms of the Linear Equation

Practice 4.5.1

1. $y = -6x + 2$ **2.** slope is 1; y–intercept is (0,–7)

3. slope is –1; y–intercept is (0,8)

4. slope is –2; y–intercept is (0,5)

5. slope is 3; y–intercept is (0,–4)

Practice 4.5.2

1. $y = -2x + 7$ **2.** $y = 8x + 2$ **3.** $y = -2x + 28$

Practice 4.5.3

1. $y = -3x + 4$ **2.** $y = -\frac{1}{3}x - 1$ **3.** $y = -x + 7$

4. $y = 3x + 8$ **5.** $y = \frac{2}{3}x + 4$

Practice 4.5.4

1a) $y = 0.22x + 2.47$.

1b) The linear equation suggests that the average retail worker was making $2.47 an hour in 1970 and that their hourly wage has been increasing 22¢ a year since then.

1c) In 2010 the average retail worker will be making $11.27 an hour in current dollars.

1d) Retail workers will make $20 an hour in 2050.

2a) $y = 0.41x + 5.56$.

2b) The linear equation suggests that the average construction worker made $5.56 an hour in 1970 and that their hourly wage has been increasing 41¢ an hour each year since then.

2c) In 2010 the average construction worker will be making $21.96 an hour in current dollars.

2d) Retail workers will make $20 an hour in 2005.

3a) $y = 0.36x + 2.38$.

3b) The linear equation suggests that the average service worker made $2.38 an hour in 1970 and that their hourly wage has been increasing 36¢ an hour each year since then

3c) In 2010 the average retail worker will be making $16.78 an hour in current dollars.

3d) Retail workers will make $20 an hour in 2019.

Exercise Set 4.5

1. $y = -x - 4$ **3.** $y = 0$ **5.** $y = \frac{3}{7}x - 1$ **7.** $x = -2$

9. $y = -3$ **11.** $y = -x - 1$ **13.** $y = \frac{-4}{9}x$

15. slope is 0 and y–intercept is (0,9)

17. slope is –1 and y–intercept is (0,5)

19. slope is $\frac{1}{3}$ and y–intercept is (0,–10)

21. slope is undefined and y–intercept is (0,0)

23. slope is –4 and y–intercept is $(0,\frac{1}{3})$

25. slope is $-\frac{1}{2}$ and y–intercept is $(0,\frac{1}{4})$

27. slope is –1 and y–intercept is (0,2) **29.** $y = 3x - 15$

31. $y = 0.5x + 6$ or $\frac{1}{2}x + 6$ **33.** $y = -2x + 4$ **35.** $y = 1$

37. $y = 8x$ **39.** $x = 7$ **41.** $y = 0.02x + 6.15$

43. $y = -32.6x + 1044$ **45.** $y = -0.11x + 6.92$

47. $y = 4x + 2$ **49.** $y = -2x - 5$ **51.** $y = \frac{2}{3}x + 1$

53. $y = -1.3x + 151.3$ **55.** $y = 1.25$ **57.** $y = 5x + 12$

59. $y = \frac{1}{2}x$

61a) $y = 199x + 1500$ **61b)** The linear equation suggests that after making the $1500 down payment it costs an additional $199 a month to lease the car. **61c)** about 77 months **61d)** $8664

63a) $y = 1.1x + 92$ **63b)** The linear equation suggests that in 1940 ninety–two million tons of carbon monoxide where released in the United States. After 1940 the amount of carbon monoxide released increased each year by 1.1 million tons. **63c)** 86.5 million tons **63d)** 1993 **63e)** 2024

65a) $y = 52x + 140$ **65b)** The linear equation suggests that the student service fee is $140 and that it costs $52 per credit. **65c)** $556 **65d)** 16 credits (we assume there are no half credits)

67a) $y = 0.4x + 46$. **67b)** The linear equation suggests that in 1970 46% of incoming freshmen were female and that since 1970 the percentage has been increasing by four tenths of one percentage point each year. **67c)** 2005 **67d)** 54% **67e)** 1855 **67f)** 2105

69a) To see if the data are linear, we graph it:

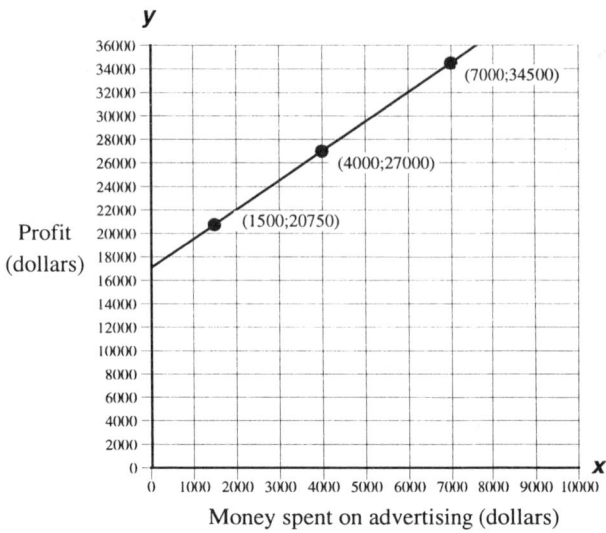

Yes, the graph appears linear.

69b) $y = 2.5x + 17000$ **69c)** $30,750 **69d)** $9,200

71a) $y = -0.7x + 140$ **71b)** about 135 beats per minute

71c) 84 beats per minute **71d)** 34 years old

73. [1.3.7] 14 **74.** [1.7.3] $5x$ **75.** [1.2.11] 14

76. [2.2.4] all real numbers **77.** [2.4.3]

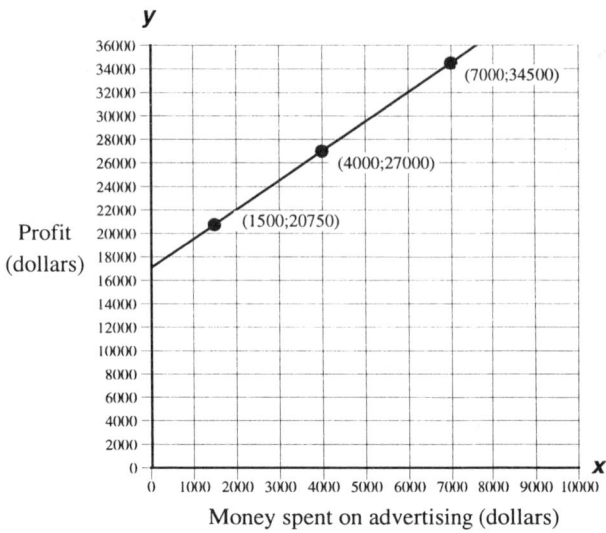

78. [2.5.3] $x > 2$ **79.** [2.4.4] $x \geq 4$

80. [3.5.1] $40(1) + 66(t - 1) = 254.5$; $4\frac{1}{4}$ hours for the trip

so she left at 12:45 pm

Section 4.6 Linear Regression

Practice 4.6.2

1a) The equation suggests that in 1869 there were about 6.7 million students attending public schools and that the number of students has increased by about 0.3 million (300,00) each year since then.

1b) 2013 **1c)** 46 million **1d)** 9 million

2a) The equation suggests that in 1869 students went to public school about 72 days per year and that the number of days attended increased about 1 day per year through 1959. **2b)** 1989 **2c)** The model predicts 1947. The table says this number was reached about 1938.

2d) The predicted rise is 30 days. The actual rise from the table was 36.7 days.

3a) The equation suggests that a job earning 0 points has an annual salary of $12431.71. (This is probably a model which only makes sense for a certain range of positive input values.) The equation also says that annual salary increases by $3.50 for each point earned.

3b) $14,532

3c) The model says a salary of $13,622 should have evaluation points of 705; $14,900 seems a bit high.

3d) $332.50

Practice 4.6.4

1a) This is a case where a linear regression line is not a good model. We should not use the equation to make predictions.

1b) Not appropriate.

1c) Not appropriate.

1d) Not appropriate.

1e) Not appropriate.

2a) Using the points (12,10) and (127,45) leads to the equation $y = 0.3x + 6.4$.

2b) The equation suggests that in 1869 about 6.4 million children were attending public school and since then there has been an increase of 0.3 million children a year

2c) From the graph we estimate the change in enrollment to be 12.5 million. The model estimates the change to be 12 million. The two estimations are pretty close.

2d) From the graph we estimate the change in enrollment to be 21 million. The model estimates the change to be 6 million. Here is a case where the model is not predicting well.

2e) If we estimate the 1969 level to be 46 million then the model says this level will again be reached in 2001.

3a) Using the points (20,40) and (160,90) leads to the equation $y = 0.36x + 32.8$.

3b) This implies that if no people attend it would still take 32.8 minutes to clean the park. (Maybe the model is not very good at small inputs. Maybe the park is dirty before the show.) The slope says that for each person it takes 0.36 minutes to clean or 36 minutes for every 100 people attending.

3c) about 177 minutes

3d) about 256 people

3e) In three hours the normal two–person crew could clean–up for about 409 people. If we expect 600 people that leaves 191 people to clean up for. Since 191 is about half of 409 we should be OK with 1 more person.

Exercise Set 4.6

1a) The equation suggests that a student who scores 0 on the verbal test will get a final math test score of 49% and that for every one point increase in verbal test score the students final exam score will raise by half a point.

1b) The student would need a verbal test score of 102%, an impossibility. This suggests that if the model is valid it is only valid for a limited range of values.

1c) Yes; the equation predicts the final score to be 70.5%.

1d) Under the model a student who receives a verbal test score of 25% is expected to get a final grade of 61.5%. This suggests that verbal ability is not the only variable which has predictive value.

3a) The constant suggests that the dealership adds $1564.00 to the amount they paid for the trade–in before selling the car to someone else. The 1.07 suggests the dealer adds on $1.07 for every $1.00 paid for the trade–in.

3b) $15,150

3c) The equation predicts $16330 while the table says 16300. This is very close considering we are talking about thousands of dollars.

5a) The equation suggests that in 1924 the winning time was about 302 seconds and that the time has been dropping about 1.17 seconds per year (or 4.68 seconds every four years) since then.

5b) Three and a half minutes corresponds to 210 seconds. The equation predicts that this will occur in the year 2002 (so it would first show up in the 2004 Olympics). Notice though that times have not been dropping as fast as expected for the last several Olympic games, which suggests that the model for recent Olympics is different than that for early Olympics.

5c) The difference between 1924 and 1936 is 12 years. Since we predict a drop of 1.17 seconds per year the drop in time over 12 years should be 14.04 seconds. The actual drop is 19.7 seconds, so during these years performance was exceeding the predictions.

5d) The difference between 1984 and 1996 is 12 years. . Since we predict a drop of 1.17 seconds per year the drop in time over 12 years should be 14.04 seconds. The actual drop is only 3.3 seconds. This implies that human performance may be reaching some type of maximum.

7a) The equation suggests that in 1960 165,000 women earned bachelor degrees and since then the number has increased by about 14,000 women per year.

7b) The equation gives us 12 (rounded) for x, which corresponds to 1972.

7c) The equation gives us 26 (rounded) for x, which corresponds to 1986.

9a) Using the points (4,1.6) and (72,2.1) leads to the equation $y = 1.6 + 0.007x$.

9b) The equation suggests that in 1928 the gold medal height was about 1.6 meters and the height has been increasing by about $\frac{7}{1000}$ of a meter per year ($\frac{28}{1000}$ of a meter every four years) since then.

9c) In the year 2014 the gold medal jump should be at 2.2 meters. (This height would first be attained at the 2016 Olympics).

9d) The equation predicts this should have occurred in 1971. This would first show up in the 1972 Olympics (year 44). It seems that the model predicted this rather well.

9e) The equation predicts that the winning height in 1940 would have been 1.68 meters and in 1944 it would have been 1.708 meters.

11a) Using the points (4,0.75) and (43,4.50) leads to the equation $y = 0.10x + 0.35$ when rounded to the cents place.

11b) The equation suggests that in 1950 the minimum hourly wage was 35¢ and it has been rising about 10¢ a year since then.

11c) The equation predicts the federal minimum hourly wage will reach $6.00 in 2006 or 2007.

11d) The predicted value in 1958 is $1.15. Three times this is $3.45. The model predicts this should occur in 1981.

11e) The equation predicts this rise to be 80¢. The graph shows the rise to be closer to $1.70. You can see that the slope of the line from 1969 to 1977 is steeper than the slope of the regression line.

13a) Using the points (1050,2000) and (5250,6750) leads to the equation $y = 1.13x + 813$.

13b) The equation suggests the dealer charges $1.13 for each 1 dollar he pays in trade along with a constant $813.

13c) The predicted retail price is about $7600.

13d) Since the expected retail price is $4881 there is definitely room to negotiate a lower price.

13e) It's sometimes the case that if the car has problems the owner is happy just to get rid of the car. On the other hand sometimes the dealer, who has paint and repair facilities on hand, can fix the car and sell it. So yes this might accurately reflect what happens.

15a) A straight line is not a good model for the given data.

15b) Not appropriate.

15c) Not appropriate.

15d) Not appropriate.

15e) Not appropriate.

17. **[1.6.11]** Multiplication Property of 0   **18.** **[2.2.3]** 65

19. **[1.6.8]** Base is $2x$, exponent is 2; simplifies to $4x^2$

20. **[2.2.4]** no solution   **21.** **[4.4.3]**

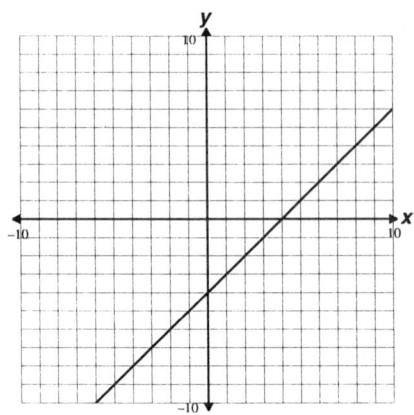

22. **[2.5.3]** $x < -6$ **23.** **[3.2.1]** $\frac{2}{3}(x - 8) + 8 = 32$; 44

Section 4.7   *Building Inequality Models*

Practice 4.7.1

1.

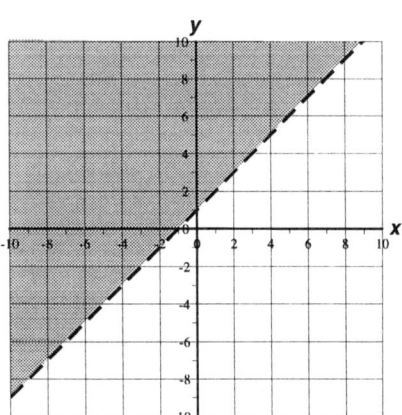

2.

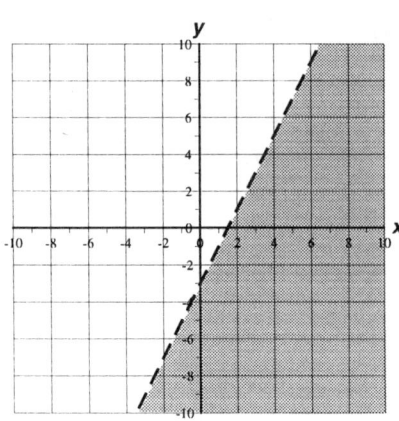

3.

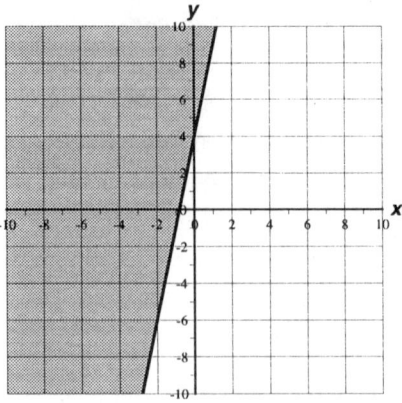

4.

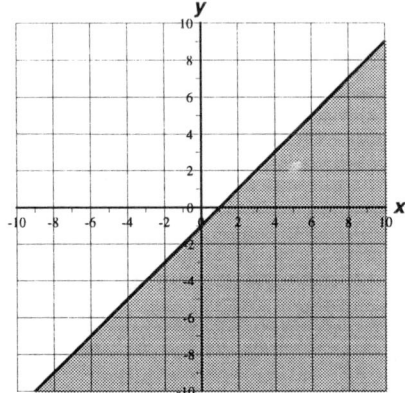

5.

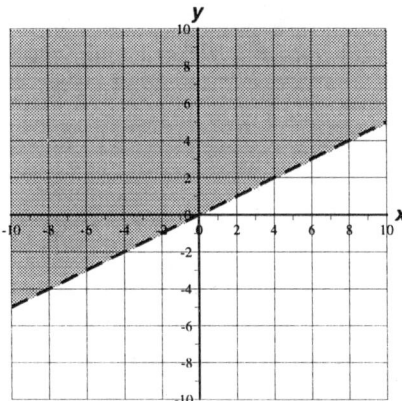

6.

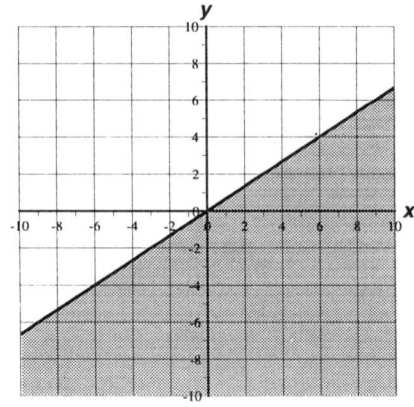

Practice 4.7.2

1a) The problem states that the perimeter must be less than or equal to 175 feet. So we have $2w + 2l \leq 175$

1b)

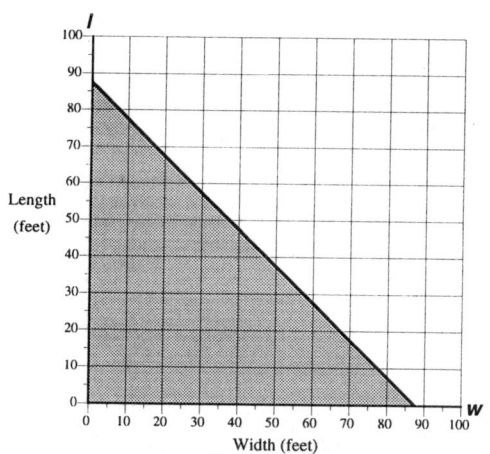

1c) The region above the boundary shows houses that do not meet the standard because they have perimeters that are greater than 175 feet.

1d) The boundary shows houses that just meet the standard because they have perimeters exactly equal to 175.

2a) The problem states that the perimeter must be less than or equal to 100 feet. So we have $2w + 2l \leq 100$

2b)

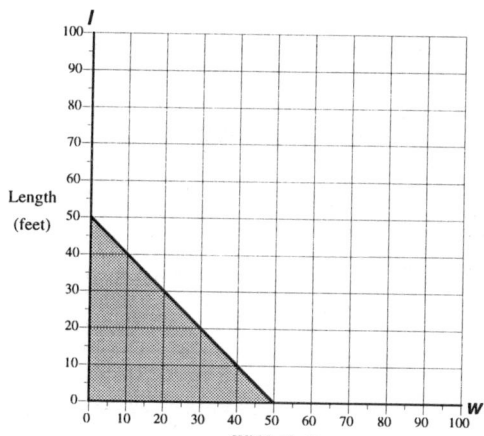

2c) The region below the boundary shows pens that he can make with the 100 feet of fence on hand because they have perimeters less than 100 feet. Because the width and length must be positive, we consider only the region in Quadrant I. All of these pens will have some left over fencing since they use less than 100 feet.

2d) The boundary shows pens that he can make because they have perimeters equal to 100 feet. These pens will have no left over fencing.

Practice 4.7.3

1. $8d + 5s \leq 40$

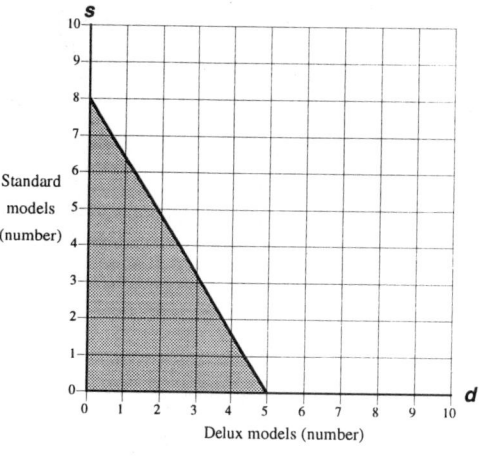

2. $30n + 4m \leq 480$

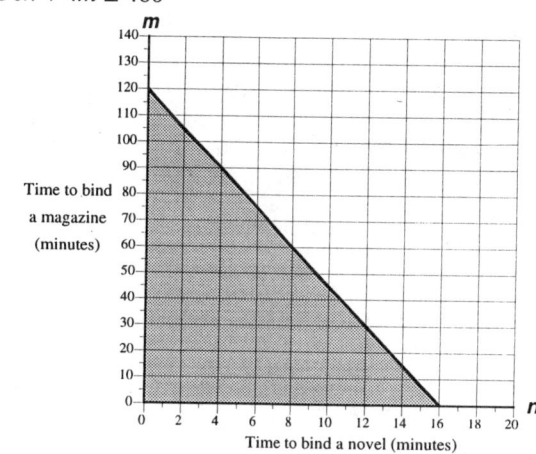

3. $10d + 25q \geq 650$

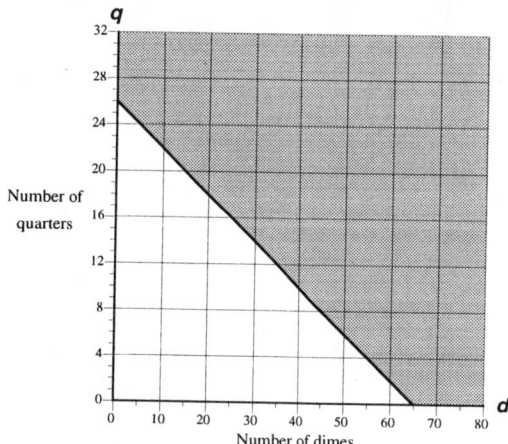

4. $10d + 8c \geq 200$

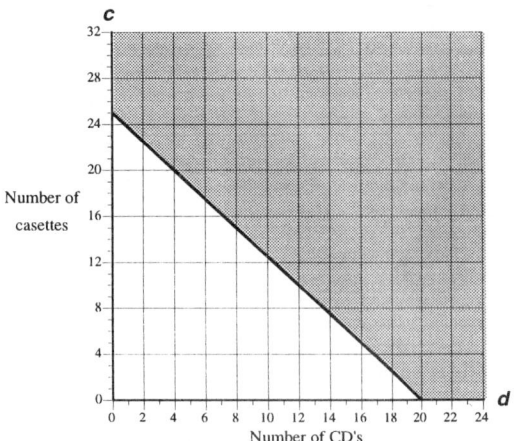

Exercise Set 4.7

1.

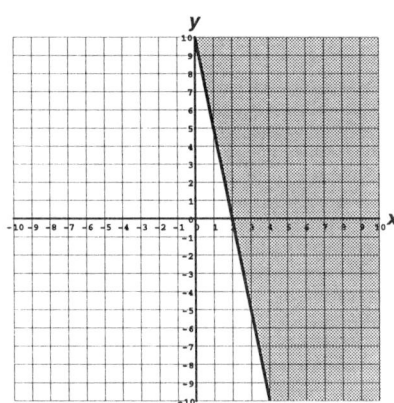

3.

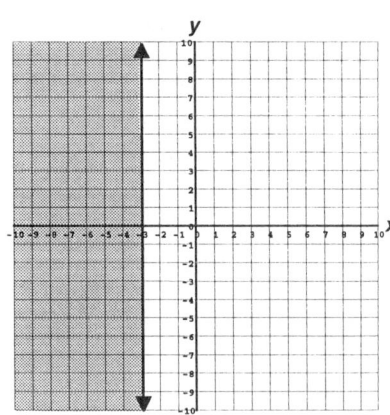

5.

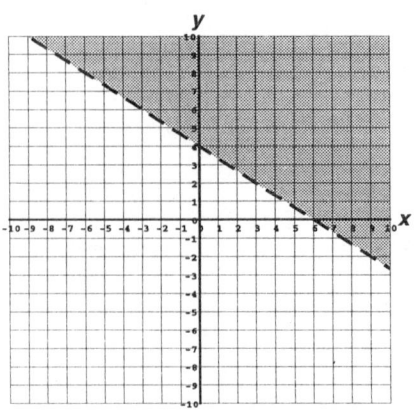

7.

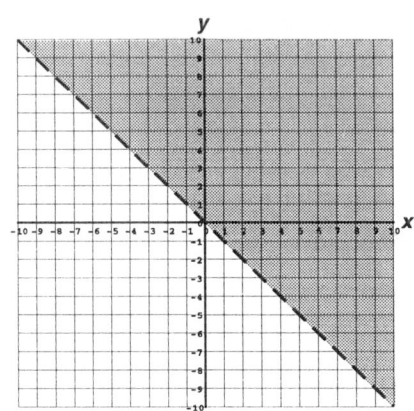

9a) $10r + 5w \leq 10$

9b)

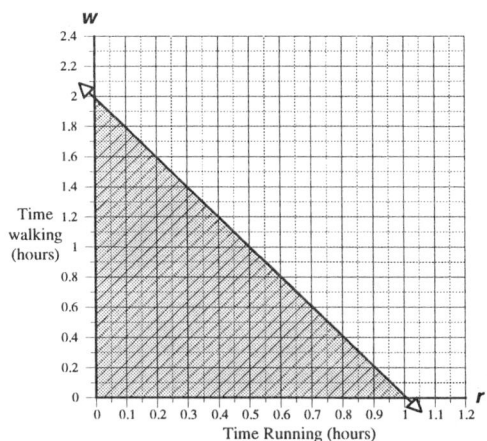

9c) The region below the boundary shows those combinations of running and walking whose sum is less than 10 miles.

9d) The boundary shows those combinations of running and walking whose sum exactly equals a distance of 10 miles.

11a) $1o + 5f \geq 100$

11b)

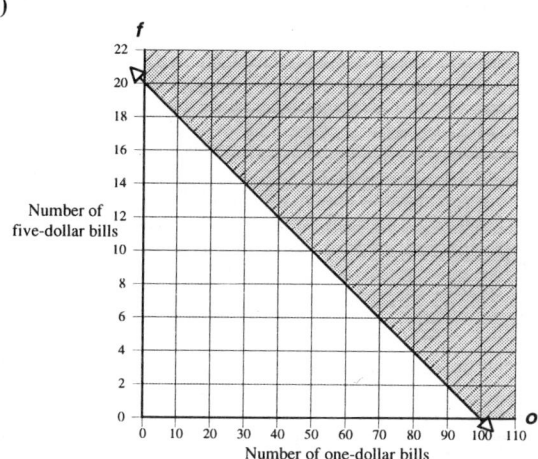

11c) The region above the boundary represents those combinations of bills whose value exceeds $100.

11d) The boundary represents those combinations of bills whose value is exactly $100.

13a) $0.05f + 0.08e \leq 400$

13b)

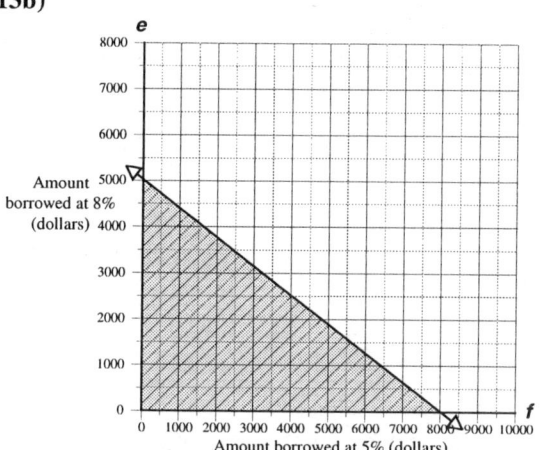

13c) The region below the boundary shows the different dollar amounts she can borrow and keep her interest under $400 a year.

13d) The boundary shows the amounts she can borrow if she want's her interest to be exactly $400 a year.

15. $3x + 4y \leq 240$

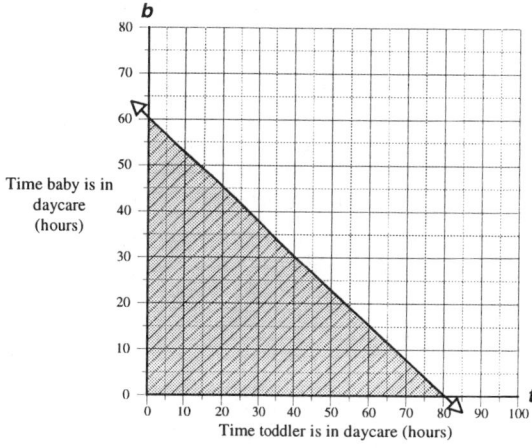

• The point (30,30) represents having both children in daycare 30 hours a week. This appears to be within the region of solutions. A check shows that the total cost would be $3(30) + $4(30) or $210, which is less than the $240 available.

• The point (40,30) represents having the toddler in daycare for 40 hours a week and the baby in daycare for 30 hours a week. The point appears to be on the boundary. A check shows that the cost would be $3(40) + $4(30) or $240, which is equal to the amount the couple is able to spend.

• The point (40,40) represents having both children in daycare for 40 hours a week. The point appears to be outside the range of possibility. A check shows $3(40) + $4(40) is $280, which is more than the couple can afford.

17. $800e + 1000t \geq 80$

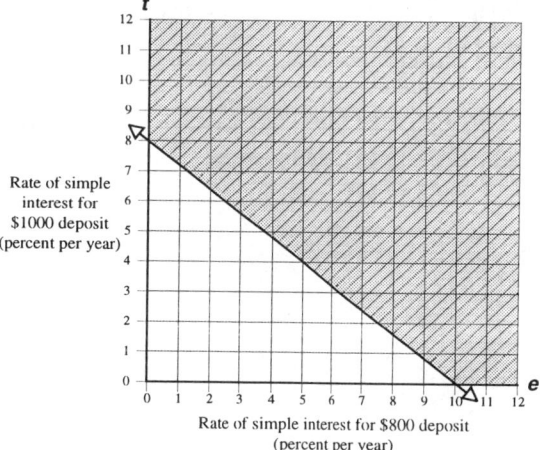

• The point (6,8) represents depositing the $800 in an account that earns 6% simple interest and the $1000 in an account that earns 8% simple interest. The point appears to exceed the minimum of $80 that Justin had set. The total interest earned would be $800(.06) + $1000(.08) or $128.

• The point (4,4.5) represents depositing the $800 in an account that earns 4% simple interest and the $1000 in an account that earns 4.5% simple interest. The point appears to be in the region that represents making less than $80.

• The point (5,4) represents depositing the $800 in an account that earns 5% simple interest and the $1000 in an account that earns 4% simple interest. This appears to be on the boundary and so should represent earning exactly $80 in interest.

19. $50f + 70s > 350$

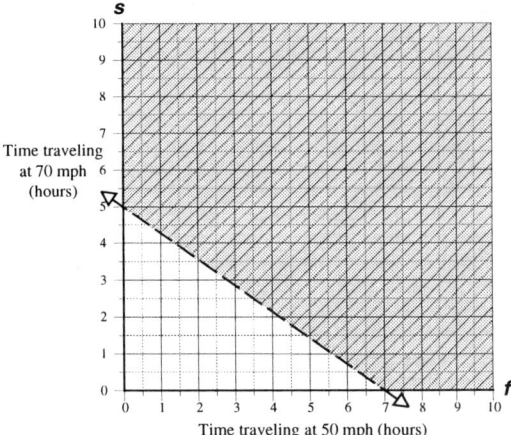

Time traveling at 70 mph (hours) [vertical axis, s]
Time traveling at 50 mph (hours) [horizontal axis, f]

• The point (3.5, 2.5) represents driving for 3.5 hours at 50 mph and 2.5 hours at 70 mph. The point appears to be on the boundary and a check shows this is indeed the case. Notice this is not a solution since Tara wants to exceed 350 miles and so the boundary in this case is a dashed line.

• The point (4,2) represents driving for 4 hours at 50 mph and 2 hours at 70 mph. The point appears to be less than Tara desires and a check shows the total distance would in fact be only 340 miles.

• The point (0, 7) represents driving the entire time (7 hours) at 70 mph for a total distance of 490 miles. This would exceed the 350 miles Tara wishes to travel in one day and the point is clearly in the region of solutions.

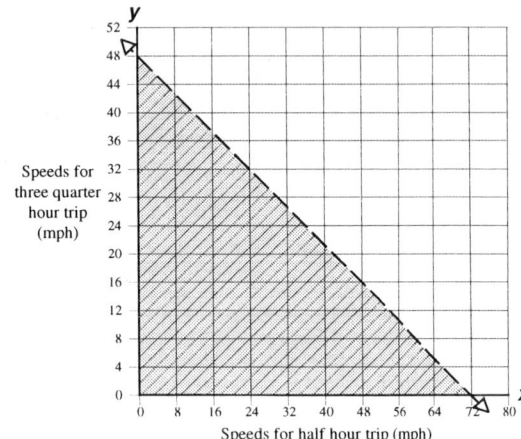

Speeds for three quarter hour trip (mph) [vertical axis, y]
Speeds for half hour trip (mph) [horizontal axis, x]

• The point (35,25) represents driving for half an hour at 35 mph and three–quarters of an hour at 25 mph. The point appears to be near the boundary and a check shows this is indeed the case as the miles traveled would be 36.25, just a bit more than the plumber wanted.

• The point (30,40) represents driving at 30 mph for half an hour and 40 mph for three–quarters of an hour. The point appears to exceed the 36 miles and a check shows this is the case with a total driving distance of 45 miles.

• The point (20,30) represents driving at 20 mph for half an hour and at 30 mph for three–quarters of an hour. The point appears to be within the solution region and a check verifies this with a total distance of 32.5 miles.

21. [1.7.7] $x - 42$ **22.** [2.2.2] 0
23. [4.5.1] Slope is 2; y–intercept is $(0,-1)$.
24. [2.4.3]

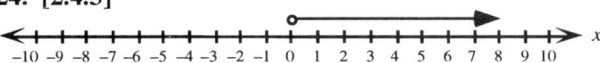

25. [4.5.2] $y = -2x + 9$ **26.** [2.5.3] $x > 7$
27. [3.4.1] $6(45) + 45s = 12.5(45 + s)$; 9 pounds
28. [4.3.5] $(0,-1)$ and $(3,0)$

Section 4.8 Relations and Functions

Practice 4.8.1

1. domain: $\{2,5,6,-1\}$; range$\{3,-20,4,-3\}$
2. domain: $\{0,4,-8\}$; range: $\{0,-5,2\}$
3. This is a non–vertical straight line so x and y can be anything. domain: all real numbers; range: all real numbers.
4. domain: $\{-8,1,3,8\}$; range: $\{-6,1,2,4,9\}$
5. This is a non–vertical straight line segment, starting at $(-5,1)$ and ending at $(4,4)$. So, domain: $\{x \mid -5 \le x \le 4\}$; range: $\{y \mid 1 \le y \le 4\}$.
6. This is an ellipse whose horizontal values lie between -2 and 8 so the domain is $-2 \le x \le 8$. The vertical values lie between -6 and 8 so the range is $\{y \mid -6 \le y \le 8\}$.

Practice 4.8.2

1. Not a function since an in put of 4 gives two different outputs.
2. Not a function since an in put of 1 gives two different outputs.
3. This is a function.
4. This is a function.
5. This is a function.
6. This is not a function since the input value of -2 gives more than one output value.

Practice 4.8.3

1. This is a function.   **2.** This is NOT a function.
3. This is NOT a function.   **4.** This is a function.

Practice 4.8.4

1. –13   **2.** –11   **3.** 28   **4.** –38   **5.** 5

Practice 4.8.5

1. The output is 1 when the input is 5.
2. The output is 4 when the input is –3.
3. The output is 2.4 when the input is –1.6.
4. The output is 5.5 when the input is 3.7.

Practice 4.8.6

1.

x	f(x)
0	1
2	–1
–6	7

2.

x	f(x)
–3	3
–1 or 1	–5
0	–6
2	–2

3. (2,–1), (5,2), (6,3), and (0,–3)
4.

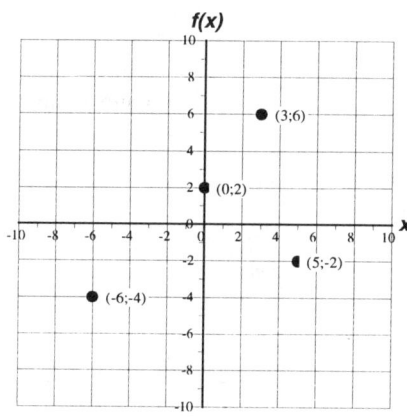

Practice 4.8.7

1. $75   **2.** His taxes on wages of $750
3. The difference in taxes between wages of $450 and
$200   **4.** The taxes on wages of $1000 are $300

Practice 4.8.8

1a) $c(n) = 225n + 1500$ **1b)** 25 units   **1c)** $24000
1d) $35250
2a) $t(n) = 15n + 7500$ **2b)** $t(12) = \$7680$. This is the tax
when there are 12 units left in the warehouse.
2c) The difference in taxes between having 250 units and
125 units stored in the warehouse at the end of the year.
2d) 17 units

Exercise Set 4.8

1. domain: {–1, –2, 2, 8}; range: {4, 8, 5, 32}
3. domain: all real numbers; range: all real numbers
5. domain: $\{x \mid -6 \le x \le 4\}$; range: $\{y \mid -4 \le y \le 5\}$
7. domain: $\{x \mid 2 \le x \le 9\}$; range: $\{y \mid -9 \le y \le -3\}$
9. function   **11.** Not a function since the input 1 has
two different outputs.   **13.** function
15. function   **17.** not a function   **19.** not a function
21. 0   **23.** –16   **25.** 8   **27.** 66   **29.** –3
31.

x	f(x)
–8	2
0	5
8	–4

33.

x	f(x)
–8	–8
0	2
4	4

35.   (–4,3), (0,1), (–1,2), and (6,–4)
37.

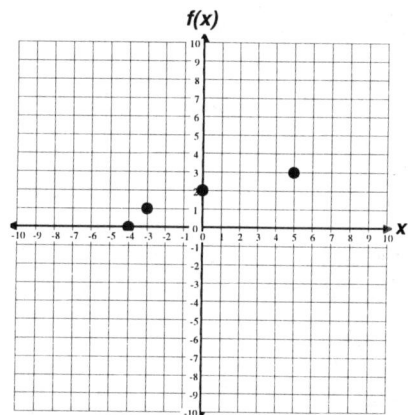

39a) 135 minutes is 2.25 hours; 7(2.25) or 15.75 miles.
39b) The distance the canoe floated in 1.5 hours.
39c) The difference between the distance floated in 1 hour
and 0.75 hour; that is, the distance in 15 minutes.
39d) In $\frac{4}{3}$ hours (1 hour 20 min) canoe floated $9\frac{1}{3}$ miles.
41a) 1980 + 12 is 1992; the number of fires in 1992.
41b) Change in number of fires between 1984 and 1988.
41c) The year the number of structural fires was 700,000.
41d) 1980 + 15 is 1995; in 1995 the number of structural
fires was 555,000.

43. [1.5.8] −2   **44.** [1.7.6] −3x − 10   **45.** [2.2.2] 18

46. [4.3.4]

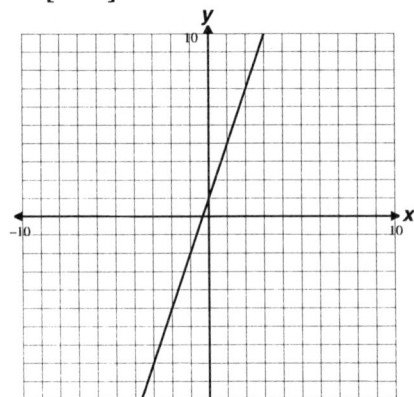

47. [4.3.6]

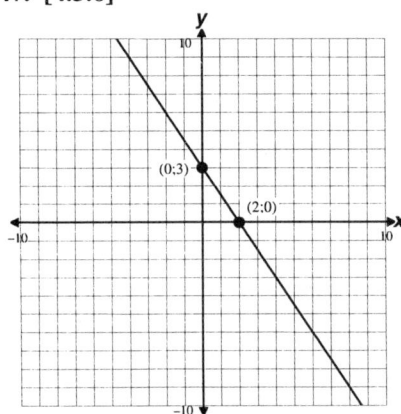

48. [4.5.3] $y = 2x - 5$

49. [4.7.1]

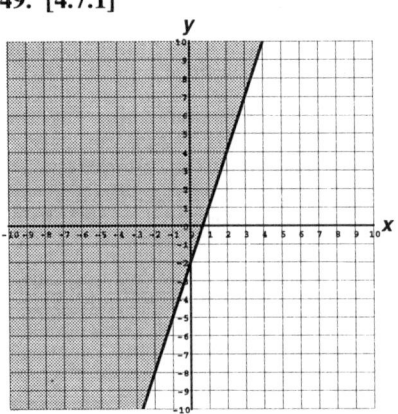

50. [3.2.4] $x + (x + 2) = 122$; 60, 62

51. [4.4.1] $y = \dfrac{6}{5}x + 6$

Chapter 4 Review Exercises

Numbers inside brackets [ ] indicate the Chapter, Section, and Example number of a similar worked example.

1. [4.1.1] If we let y represent the number of envelopes and x represent the time in minutes, then the equation is $y = 80x$.   **2.** [4.1.1] If we let y represent the time in minutes and x represent the number of envelopes, then the equation is $y = \dfrac{x}{80}$.

3. [4.1.2] If we let y represent the height of the plane in feet and x represent the time since takeoff in minutes, then the equation is $y = 500x$.

x time since takeoff (minutes)	y height of plane (feet)
2	1,000
5	2,500
10	5,000

4. [4.1.2] If we let y represent the time since takeoff (minutes) and x represent the height of the plane (feet), then the equation is $y = \dfrac{x}{500}$.

x height of plane (feet)	y time since takeoff (minutes)
100	0.2
800	1.6
1000	2.0

5a) [4.1.4] $y = 2x$.

5b)

x time since started (minutes)	y number of drinks sold
120	**240**
240	**480**
600	**1200**
270	540
435	870

5c)

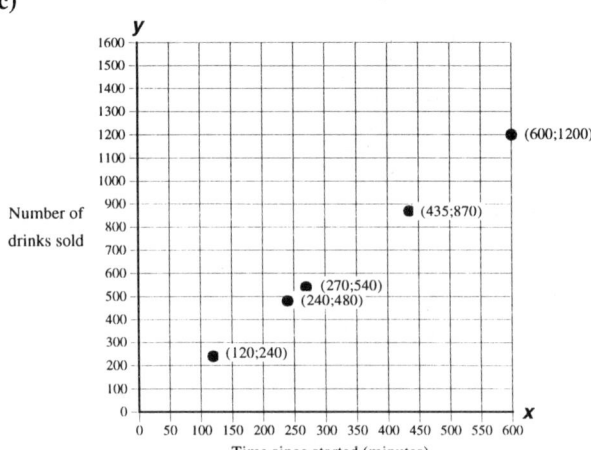

Number of drinks sold

Time since started (minutes)

6a) [4.1.4] $y = 3.80x$.

6b)

x number of lunches bought	y amount spent (dollars)
5	19
20	76
120	456
130	494
200	760

6c)

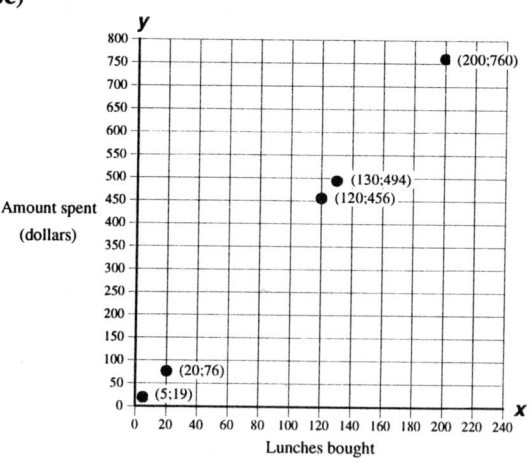

7. [4.2.1] $y = 49.5 + 0.33x$

8. [4.2.1] $y = 13 + 1.64x$

9a) [4.2.3] $y = 12.76 + 3.33x$

9b) 1993

9c) 106 billion dollars

10a) [4.2.3] $y = 420 - 65x$

10b) 208.75 miles

10c) about 2.6 hours. This is 2 hours and 36 minutes.

11. [4.2.4] In 1869 the average length of the school year was 72 days. From 1869 until 1959 the average length increased approximately 1 day per year.

12. [4.3.1]

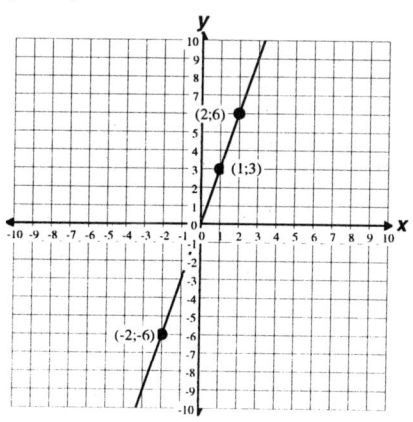

x	y
−2	−6
1	3
2	6
−1	−3
−3	−9

13. [4.3.1]

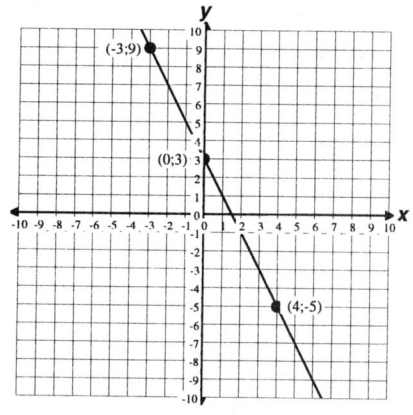

x	y
−3	9
0	3
4	−5
−2	7
1	1

14. [4.3.2] Transformed data table

x mileage (miles per gallon)	y annual fuel cost (thousand of dollars)
9	1.8
11	1.6
12	1.4
13	1.3

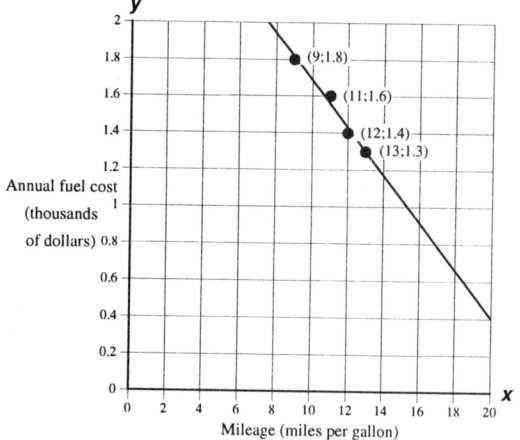

15. [4.4.2] For every $100 invested you earn $8 in interest. This means the money is being invested at a rate of 8%.

16a) [4.3.3]

x gas mileage (miles per gallon)	y annual fuel cost (dollars)
44	392
33	513

16b)

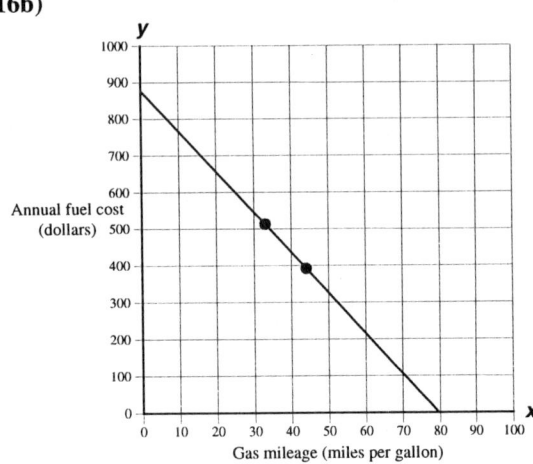

16c) about $550

16d) about 20 mpg

16e) about $770

16f) A car that gets 9 mpg has probably been built for speed, not to conserve fuel. As a consequence it uses more fuel to go the same distance as the smaller, more fuel efficient cars. Using more fuel means the cost will be higher than a model based on fuel efficient cars would predict.

17. [4.3.4]

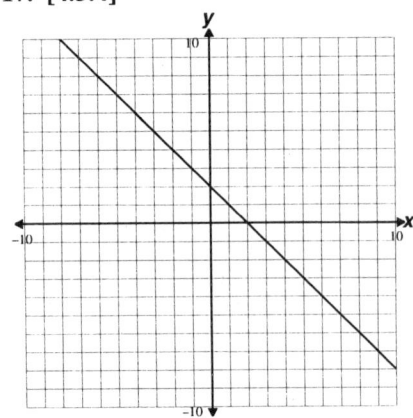

18. [4.3.4]

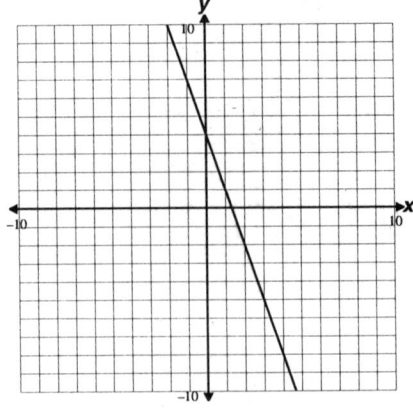

19. [4.3.5] (0,4) and (2,0)

20. [4.3.5] no x–intercept, y–intercept is (0,–2)

21. [4.3.6]

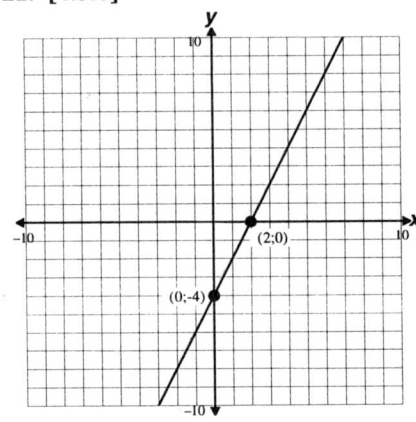

22. [4.3.6]

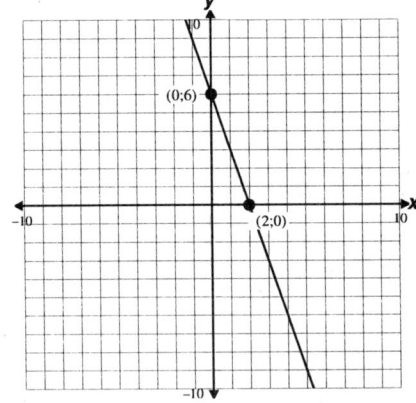

23. [4.4.1] $y = \dfrac{3}{4}x + 3$ **24.** [4.4.1] $y = \dfrac{5}{2}x + 5$

25. [4.4.3]

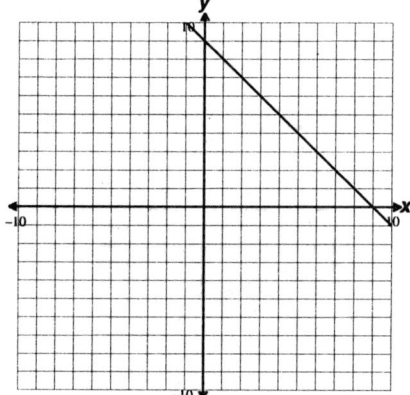

26. [4.4.3]

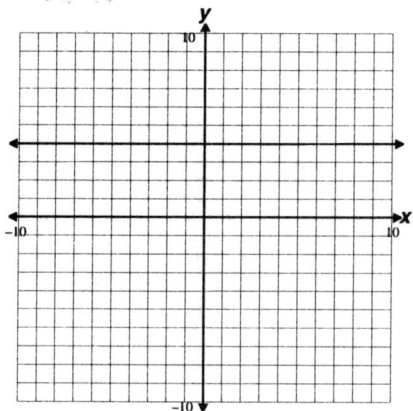

27. [4.4.4] -1 **28. [4.4.4]** $-\dfrac{1}{2}$

29.   [4.4.5] Slope is $\dfrac{1.8 \text{ billion gallons}}{1 \text{ year}}$. The slope means domestic fuel consumption has been rising by 1.8 billion gallons every year since 1970.

30.   [4.4.5] Slope is $\dfrac{0.6 \text{ dollars (female)}}{1 \text{ dollar (male)}}$. The slope implies that women pay about 60¢ for every dollar spent by men for $100,000 of term insurance.

31. [4.4.7] Slope of line 1 is $-\dfrac{1}{5}$; slope of line 2 is 5; lines are perpendicular.

32. [4.4.7] Slope of both lines is $-\dfrac{5}{8}$; lines are parallel.

33. [4.5.1] Slope is 1; y–intercept is $(0,-4)$.

34. [4.5.1] Slope is $\dfrac{1}{3}$; y–intercept is $(0,-2)$.

35. [4.5.1] $y = -2x + 3$ **36. [4.5.2]** $y = -3x + 4$

37. [4.5.2] $y = 3$ **38. [4.5.3]** $y = 3x - 5$

39. [4.5.3] $y = \dfrac{1}{2}x + 6$ **40a) [4.5.4]** $y = 0.14x + 74.5$

40b) In 1960 women lived to an average age of 74.5 years and their life expectancy has been rising by 0.14 years per year since then.   **40c)** 87.1 years   **40d)** 2035

41a) [4.6.2] In 1983 total finances were 118.7 billion dollars and finances have been rising by 14.5 billion dollars each year.

41b) For 1985 the model predicts 147.7 billion while the table shows 147.0. In 1990 the model predicts 220.2 billion while the table shows 218.2 billion. In both cases the equation returns a value close to the one in the table.

41c) The model predicts 147.7 billion in 1985. Twice this is 295.4 billion. Using the model with 295.4 substituted for y we expect finances to double in 1995 (year 12 in the table). You can see the actual rate of growth was a little less than the model would have anticipated.

42a) [4.6.4] Using the points $(0,19000)$ and $(37,40000)$ leads to the equation $y = 568x + 19{,}000$.

42b) In 1947 the mean income was about $19,000 (in 1994 dollars). Since 1947 it has been increasing by about $568 dollars per year in 1994 dollars.   **42c)** $49,104

42d) $11,360   **42e)** Between 2001 and 2002.

43.   [4.6.4] A linear model is not appropriate. Answers to the rest of the questions would be misleading.

44. [4.8.2] no   **45. [4.8.2]** yes   **46. [4.7.1]**

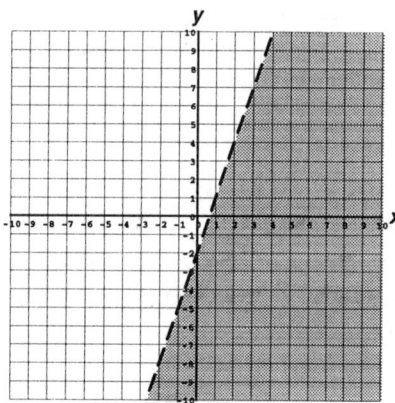

47. [4.8.1] domain: $\{1,3,6,8\}$; range: $\{0,2,5\}$

48. [4.7.1]

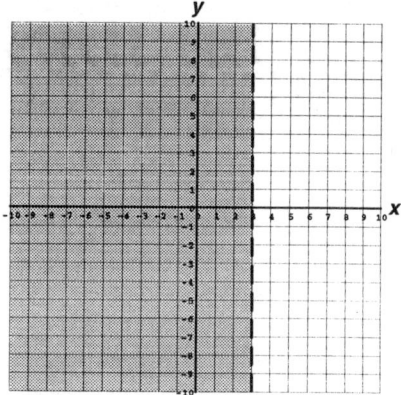

49. [4.8.4] -24 **50. [4.8.4]** 11

51. [4.8.1] domain: $\{x \mid -9 \le x \le 9\}$; range: $\{y \mid -4 \le y \le 7\}$

52. **[4.8.1]** domain: {−7,0,3,9}; range: {−4,0,1,5}

53. **[4.8.3]** yes   **54.** **[4.8.3]** no

55. **[4.8.6]**

x	f(x)
−4	7
1	1
6	−5

56. **[4.8.6]** (1,2), (4,5), (−1,0), and (0,1)

57a) **[4.8.7]** $150

57b) The taxes on wages of $2000.

57c) Difference in taxes between wages of $750 and $225.

57d) The taxes are $150 on wages of $500.

58a) **[4.8.7]** 221.25 miles   **58b)** When simplified this expression represents the distance traveled between the first and second hour of travel.   **58c)** You are 100 miles from home after driving for t hours.   **58d)** After traveling for 1.25 hours you are 318.75 miles from home.

59a) **[4.8.8]** $m(t) = -\frac{2}{5}t + 54$ **59b)** t is 10, which corresponds to 1980. We expect 50% of incoming college freshmen to be male in 1980.   **59c)** $m(-5)$ is 56. This means that in 1965, 56% of incoming freshmen were male. **59d)** $m(10) - m(15)$ is $50 - 48$ or 2. From 1980 to 1985 the number of male freshmen dropped by 2 percentage points.

Chapter 4 Test

Numbers inside brackets [ ] indicate the Chapter, Section, and Example number of a similar worked example.

1. **[4.1.1]** If we let y represent the height of the balloon in feet and x represent the time in minutes, then the equation is $y = 200x$.   **2.** **[4.2.1]** $y = 10.2 + 0.83x$

3a) **[4.2.3]** $y = 1564 + 1.07x$ **3b)** $4988   **3c)** $3490

4. **[4.2.4]** The student began with 400 points and has been losing an average of 7 points per test.

5a) **[4.1.4]** $y = 2x$

5b)

x width (inches)	y length (inches)
2	4
8	16
5	10
6	12
7.5	15

5c)

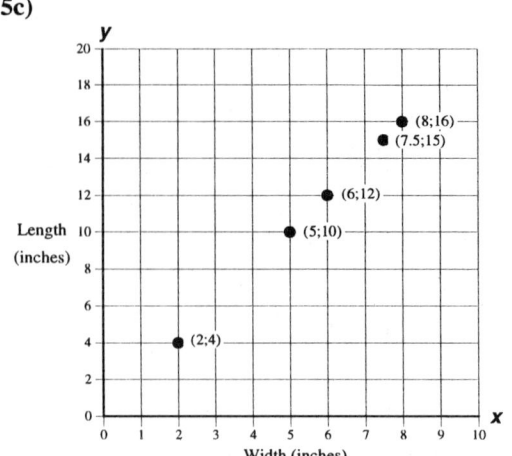

6a) **[4.3.3]**

x number of bills produced produced	y cost to produce (millions of dollars)
3960	14.7
360	1.3

6b)

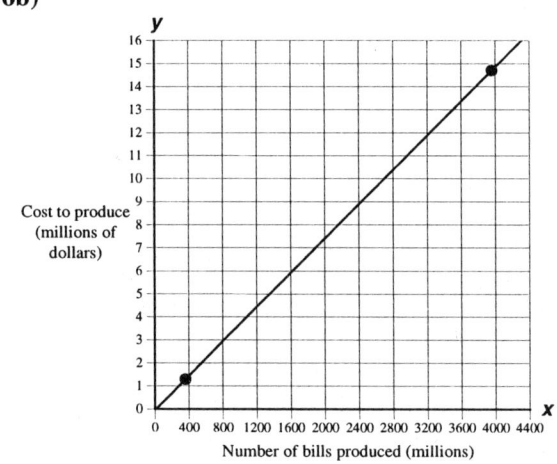

6c) about $4,000,000

6d) about 2,400 million, or 2,400,000,000 bills

6e) The graph only allows us to estimate answers. The graph is good for a quick idea, but not good for exact or close to exact answers. The linear equation is good for exact answers but many people find it less intuitive than a graph.

7. [4.3.1]

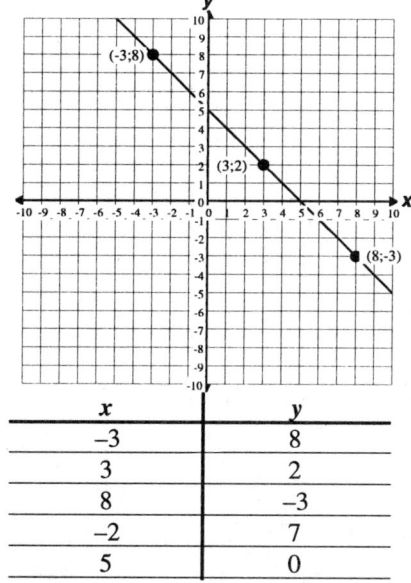

x	y
–3	8
3	2
8	–3
–2	7
5	0

8. [4.3.2]

Transformed data table

x face value (billions of dollars)	y total cost to produce (million of dollars)
3.96	14.65
2.16	7.99
1.08	4.00
0.36	1.33

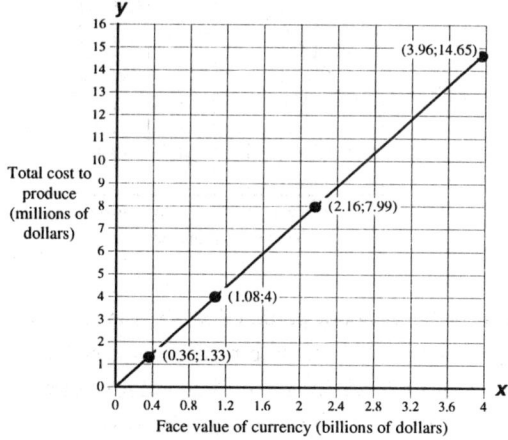

9. [4.3.5] (0,5) and (–5,0)

10. [4.4.1] $y = -x - 8$

11. [4.4.2] In 1947 the median income in 1994 dollars was around 19,000. From then until 1990 it's been rising at about $600 dollars a year, again in 1994 dollars.

12. [4.4.5] Slope is $\dfrac{-\$124}{1 \text{ mpg}}$. The slope means that you will save $124 dollars a year for every 1 mile per gallon increase in your car's gas mileage.

13. [4.4.3]

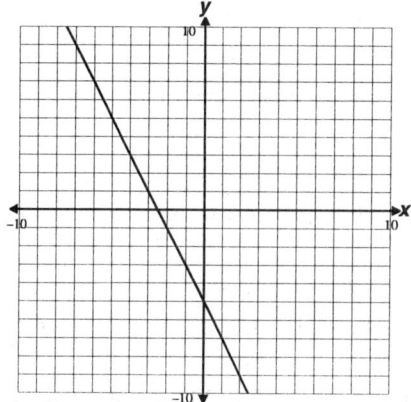

14. [4.4.4] $\dfrac{5}{9}$ **15. [4.4.7]** Slope of line 1 is $-\dfrac{1}{3}$; slope of line 2 is 3; lines are perpendicular.

16. [4.5.1] Slope is –2; y–intercept is (0,5).

17. [4.5.2] $y = -2x - 1$ **18. [4.5.3]** $y = 3x + 4$

19. [4.8.1] domain: {0,1,3,5}; range: {1,3,5}

20. [4.8.2] no

21. a) [4.6.2] In 1974 Americans spent 53.4 billion dollars dining out and the amount spent has been increasing by 10.5 billion each year since then.

21. b) The equation predicts 158.4 billion dollars spent. The table shows 150.0 billion dollars were actually spent. These are fairly close to each other.

21. c) 315.9 billion dollars

22. [4.7.1]

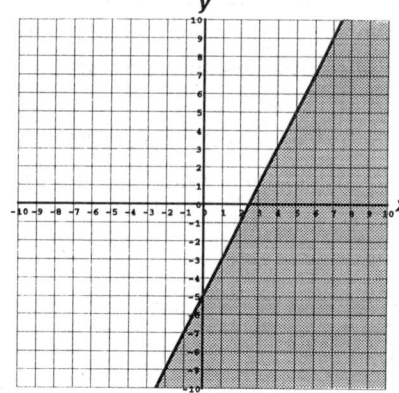

23. [4.3.4]

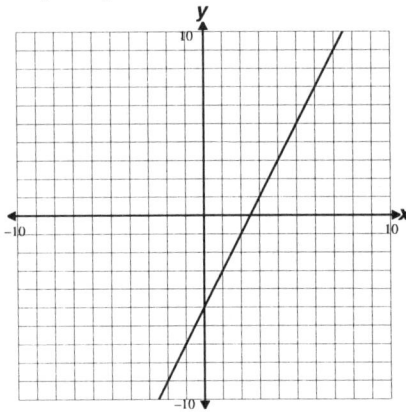

24. [4.8.1] domain: $\{x \mid -9 \le x \le 8\}$;
range: $\{y \mid 1 \le y \le 9\}$
25. [4.8.3] yes
26. [4.8.4] 12

27. [4.8.6]

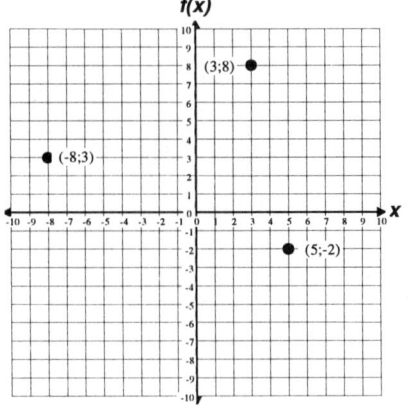

28a) [4.8.8] $A(t) = 6t + 23$
28b) $A(8)$ is 71. In 1988 the number of ATM's was 71,000. **28c)** t is 9.5. There were 80,000 ATM's sometime between 1989 and 1990. **28d)** t is 13.8. About 1993, the number of ATM's was double the number in existence in 1985.

Chapter 5 Systems of Equations

Section 5.1 Solving Systems by Graphing

Practice 5.1.1
1. Yes **2.** No **3.** No **4.** Yes

Practice 5.1.2
1. (0,3) **2.** (2,5) **3.** (3,1)

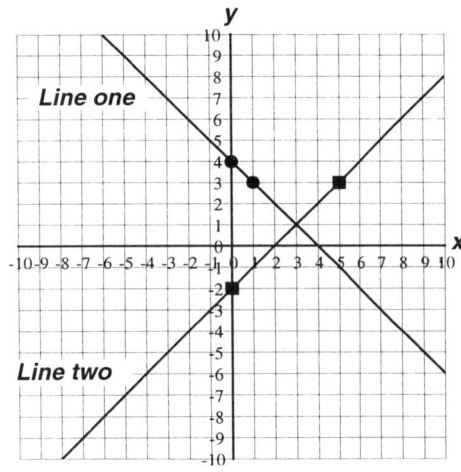

4. (2,–3)

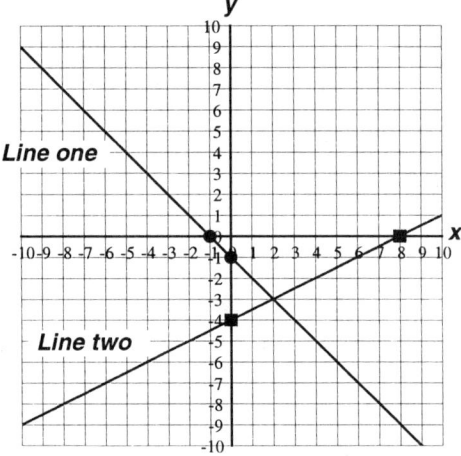

Practice 5.1.3

1. $\left\{ \begin{array}{l} y - x = 1 \\ y = x + 1 \end{array} \right\}$ ❶ ❷

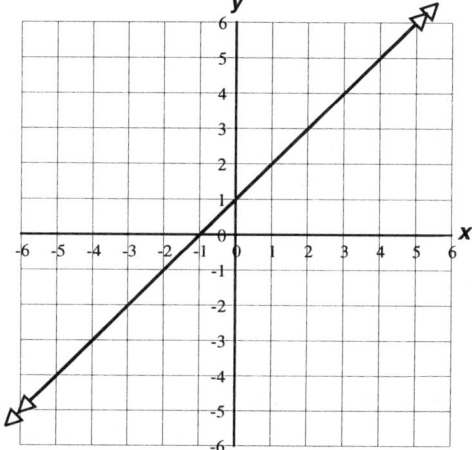

There are an infinite number of solutions, which we will write as $(x, x + 1)$.

2. $\left\{ \begin{array}{l} x + y = 3 \\ x - y = 7 \end{array} \right\}$ ❶ ❷

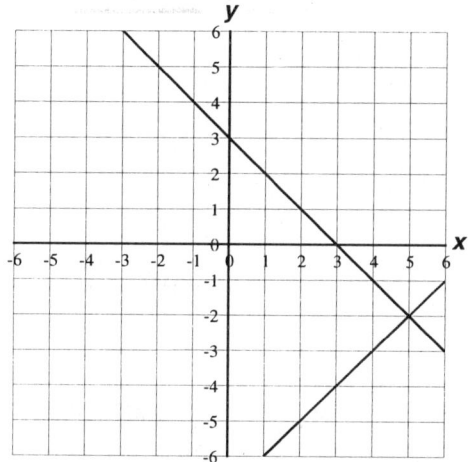

The numbers are 5 and –2.

3. $\left\{ \begin{array}{l} x + y = 4 \\ 2x + 2y = -5 \end{array} \right\}$ ❶ ❷

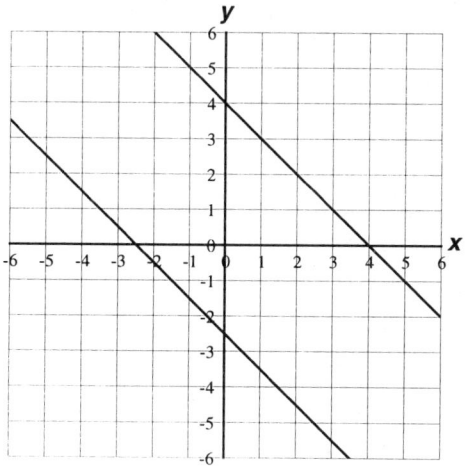

The lines are parallel so there is no solution.

Practice 5.1.4

1. Consistent system, independent equations
2. Consistent system, dependent equations
3. Inconsistent system, independent equations
4. Consistent system, independent equations

Practice 5.1.5

1a) $\left\{ \begin{array}{l} y = 0 + 20x \\ y = 70 + 10x \end{array} \right\}$ ❶ ❷

1b)

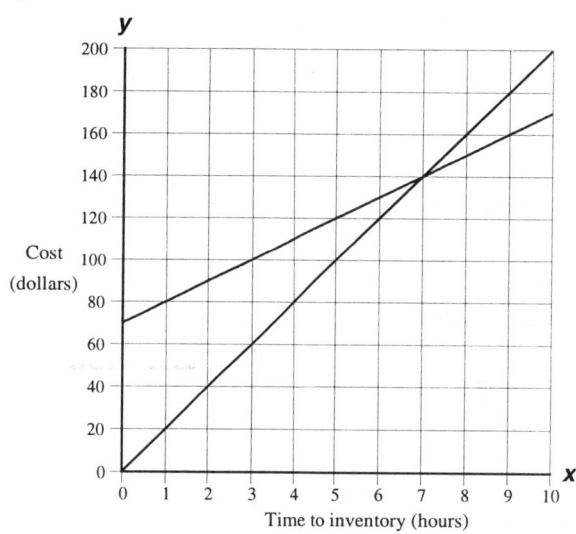

1c) Intersection is at (7,140). The cost will be the same ($140) for the employee and the temporary worker if the inventory takes 7 hours.

1d) If the inventory takes less than 5 hours the employee is cheaper; if it takes over 5 hours the temporary worker is cheaper.

2a) $\left\{ \begin{array}{l} y = 200 + 50x \\ y = 350 + 35x \end{array} \right\}$ ❶ ❷

2b)

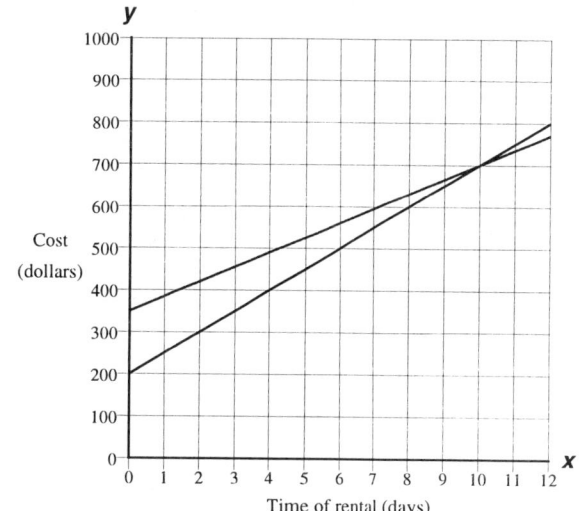

2c) Intersection is at (10,700). The cost is the same ($700) if you rent for 10 days.

2d) If you rent for under 10 days, Company A is cheaper; if you rent for more than 10 days, Company B is cheaper.

3a) The cost is the same where the lines intersect. This is at about $400. The cost of borrowing $400 is about $6 for each card.

3b) If you borrow less than $400, Card 1 is cheaper. If you borrow more than $400, Card 2 is cheaper.

3c) The difference in cost is the vertical distance between the lines when the dollars borrowed is 200. From the graph we see that if $200 is borrowed, Card 2 costs about $4 per month and Card 1 costs about $3 per month. Card 2 is about $1 more expensive.

Exercise Set 5.1

1. Yes   **3.** No   **5.** (0.5,0.5)   **7.** No solution   **9.** (0,1)

11. The lines overlap so there are an infinite number of solutions, including (0,–8), (2,0), and 4,8)

13. $\left\{ \begin{array}{rcrcr} x & + & y & = & 1 \\ 2x & + & 3y & = & -3 \end{array} \right\}$ ❶ ❷

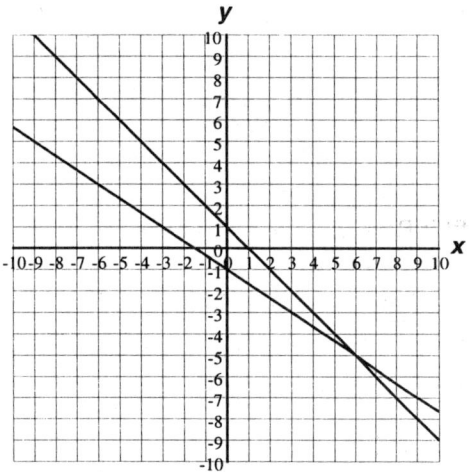

The numbers are 6 and –5.

15. $\left\{ \begin{array}{rcrcr} y & = & 2x & + & 5 \\ 3y & - & 6x & = & 15 \end{array} \right\}$ ❶ ❷

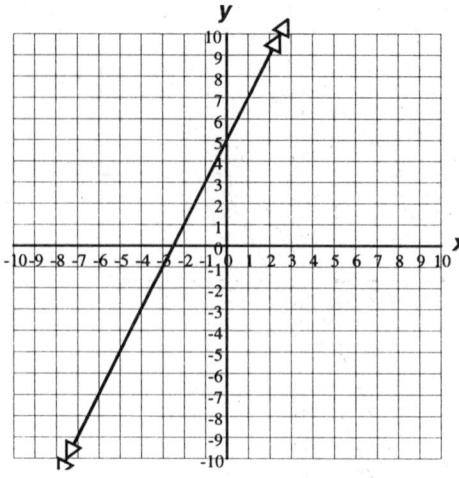

There are an infinite number of solutions, which all satisfy $y = 2x + 5$. We may write the solution as $(x, 2x + 5)$.

17. $\left\{ \begin{array}{rcrcr} x & + & y & = & -6 \\ y & = & 2x & + & 9 \end{array} \right\}$ ❶ ❷

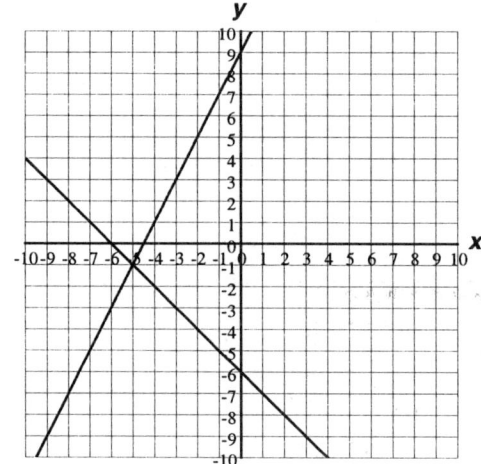

The numbers are –5 and –1

19. $\left\{ \begin{array}{rcrcr} x & + & y & = & -1 \\ 8 & - & x & = & y \end{array} \right\}$ ❶ ❷

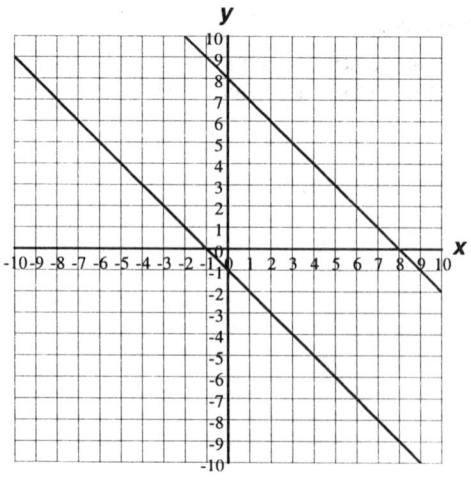

There are no solutions.

21. System is consistent, equations are independent.

23. System is inconsistent, equations are independent.

25a)

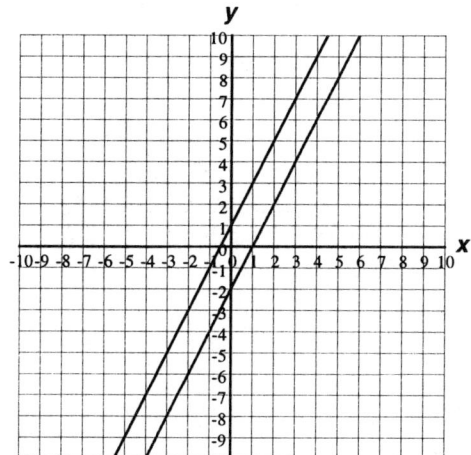

25b) No solution

25c) inconsistent

25d) independent

27a)

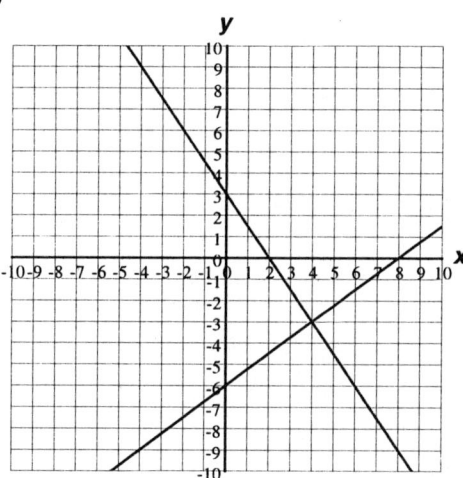

27b) (4,–3)

27c) consistent

27d) independent

29a) $\begin{cases} y = 250x + 4500 \\ y = 500x + 500 \end{cases}$ ❶ ⇐ Company A ❷ ⇐ Company B

29b)

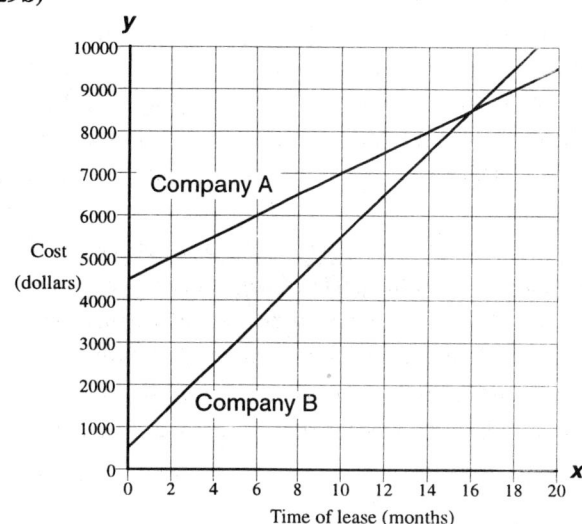

29c) (16,8500) The cost, $8500, will be the same at 16 months.

29d) If owned less than 16 months, buying would be cheaper. If more than 16 months, leasing would be cheaper.

31a) The lines intersect at (20,1100), so the break even point is 20 units sold.

31b) Before the break even point, the cost is higher than the revenue so the business is losing money; after the break even point, the revenue is higher than the cost so the business is making money.

31c) Profit is Revenue – Cost. The revenue for 30 units is almost $1650 while the cost for 30 units is $1400. Therefore, the profit is $1650 – $1400 or $250.

33. [1.3.2] {–8,–4.5,0,1.5}   **34.** [1.6.8] Base is x, exponent is 6; cannot be simplified.   **35.** [4.4.3]

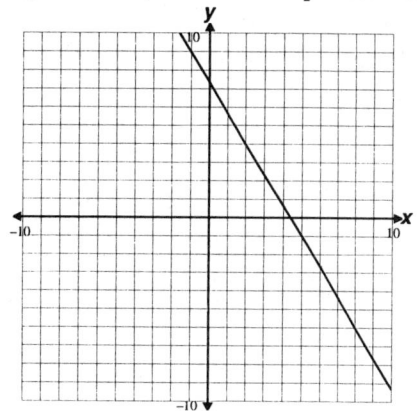

36. [2.2.2] 10   **37.** [2.2.4] no solution

38. [3.1.4] $c + (c + 280) + \frac{1}{4}(c + 280)$, which simplifies to $\frac{9}{4}c + 350$

39. [3.6.3] $0.13(x) + 0.05)(15000 - x) = 1200$; $9375 in 5% account, $5625 in 13% account   **40.** [4.4.4] $-\dfrac{4}{3}$

41. [4.8.6]

x	f(x)
0	4
6	−2
8	−4

Section 5.2   Solving Systems Using Substitution

Practice 5.2.1
1. $(-2,-3)$ **2.** $(4,1)$ **3.** $(2,1)$

Practice 5.2.2
1. $(2,2)$ **2.** $(1,1)$ **3.** $(0,-4)$

Practice 5.2.3
1. $p = 325$ and $t = 75$ **2.** $w = 5$ and $t = 5$

Practice 5.2.4
1. The equations are dependent so there are infinitely many solutions, which we write as $(x, -3x + 7)$
2. There is no solution.
3. There is no solution.
4. The equations are dependent so there are infinitely many solutions, which we write as $\left(x, \dfrac{1}{3}x - \dfrac{7}{3} \right)$.

Practice 5.2.5
1. Cost equation is $y = 35x + 2520$ and revenue equation is $y = 75$. We must sell 63 units for the revenue and cost to each equal $4725.
2. The expensive phone equation is $y = 99 + 11.95x$, where y is the total cost after x months. The cheap phone equation is $y = 2.36 + 17.99x$. After 16 months the total cost for both phones is 290.20.
3. The equation for Joni is $y = -1.3x + 52$ and the equation for Diane is $y = -4x + 80$. The divers will have the same amount of air in their tanks approximately 10.4 minutes into the dive; they will each have 38.5 cf of air at that time.

Exercise Set 5.2
1. $(5,4)$ **3.** $(5,7)$ **5.** $(-3,4)$ **7.** $(2,3)$ **9.** $(2,-1)$
11. $n = 5$ and $d = 20$
13. The equations are dependent so there are infinitely many solutions, which we write as $(x, 4x - 7)$.
15. There is no solution.
17. The equations are dependent so there are infinitely many solutions, which we write as $\left(x, \dfrac{1}{4}x + \dfrac{3}{4} \right)$
19. The equations are dependent so there are infinitely many solutions, which we write as $(x, -2x + 1)$
21. The equation for job 1 is $y = 50x + 250$ and job 2 is $y = 25x + 500$. When 10 cars are sold both jobs will pay $750.
23. They are the same distance (250 miles) from home at 2.4 hours (2 hours and 24 minutes).

25. [1.5.8] -24 **26.** [4.8.4] 2   **27.** [1.7.6] $2w + 3$
28. [1.7.7] $11y + 102$ **29.** [2.4.3]

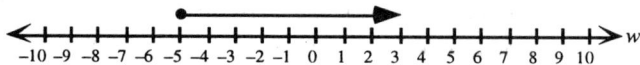

30. [3.3.2] $202.32   **31.** [3.6.1] $x + 0.65x = 42.6$; $40
32. [4.4.3]

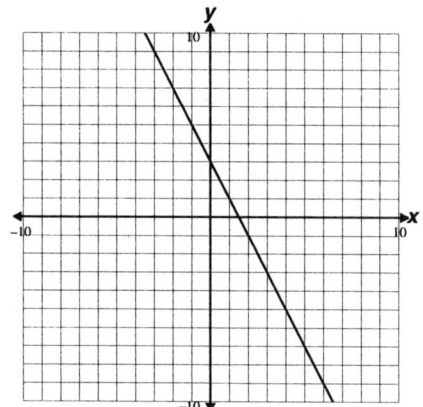

33. [4.8.2] no

Section 5.3   Using the Addition Method to Solve Systems

Practice 5.3.1
1. $(-3,2)$ **2.** $\left(\dfrac{1}{3}, 2 \right)$ **3.** $(4,-5)$

Practice 5.3.2
1. $(3,0)$ **2.** $(0,6)$ **3.** No solution. The lines are parallel.
4. These equations represent the same line. We can solve one equation for y to get $y = 3x + \dfrac{5}{2}$ and then write the solution as $\left(x, 3x + \dfrac{5}{2} \right)$.

Practice 5.3.3

1. $(2,5)$ **2.** $(-1,-1)$ **3.** $(-6,-11)$

Practice 5.3.4

1a) Multiply ❶ by 3 and ❷ by –4 to get

$$\begin{cases} 12x - 15y = -6 \\ -12x + 8y = 20 \end{cases} \begin{matrix} ❶ \\ ❷ \end{matrix}$$

1b) Multiply ❶ by –2 and ❷ by 5 to get

$$\begin{cases} -8x + 10y = 4 \\ 15x - 10y = -25 \end{cases} \begin{matrix} ❶ \\ ❷ \end{matrix}$$

2a) Multiply ❶ by 2 and ❷ by –6 to get

$$\begin{cases} 12x + 10y = 16 \\ -12x - 48y = -6 \end{cases} \begin{matrix} ❶ \\ ❷ \end{matrix}$$

2b) Multiply ❶ by 8 and ❷ by –5 to get

$$\begin{cases} 48x + 40y = 64 \\ -10x - 40y = -5 \end{cases} \begin{matrix} ❶ \\ ❷ \end{matrix}$$

Practice 5.3.5

1. $(-3,-2)$ **2.** $(-2,4)$ **3.** These equations represent the same line. We can solve one equation for y to get $y = 2x - 3$ and then write the solution as $(x, 2x - 3)$.

Practice 5.3.6

1. $(3,1)$ **2.** $(3,3)$ **3.** No solution.

Exercise Set 5.3

1. $(3,4)$ **3.** $(2,1)$ **5.** $(2,3)$ **7.** $(1,3)$ **9.** $(-1,-5)$
11. These equations represent the same line. We can solve one equation for y to get $y = 3x - 7$ and then write the solution as $(x, 3x - 7)$.
13. These equations represent the same line. We can solve one equation for y to get $y = -\frac{1}{5}x + \frac{3}{5}$ and then write the solution as $\left(x, -\frac{1}{5}x + \frac{3}{5}\right)$ **15.** $(5,-1)$ **17.** $(4,1)$
19. These equations represent the same line. We can solve one equation for y to get $y = -\frac{1}{2}x + \frac{3}{2}$ and then write the solution as $\left(x, -\frac{1}{2}x + \frac{3}{2}\right)$. **21.** $\left(\frac{1}{3}, 1\right)$
23. These equations represent the same line. We can solve one equation for y to get $y = \frac{1}{3}x - \frac{2}{3}$ and then write the solution as $\left(x, \frac{1}{3}x - \frac{2}{3}\right)$. **25.** No solution. **27.** $(4,1)$
29. $(2,1)$ **31.** No solution. **33.** $(3,1)$
35. These equations represent the same line. We can solve one equation for y to get $y = -\frac{1}{2}x + 18$ and then write the solution as $\left(x, -\frac{1}{2}x + 18\right)$ **37.** $(0,2)$

39. [1.3.3]

-900 -800 -700 -600 -500 -400 -300 -200 -100   0
40. [2.1.1] Yes. **41.** [2.5.3] $x > 4$
42. [3.1.3] $31 - x - (x + 4)$, which simplifies to $-2x + 27$
43. [4.3.5] $(0,-8)$ and $(-6,0)$ **44.** [4.4.1] $y = -3x + 6$
45. [4.5.2] $y = x - 2$ **46.** [4.5.3] $y = x - 2$
47. [3.2.4] $x + (x + 2) = -14; -8, -6$
48. [3.4.1] $15t + 19(t - 2) = 183$; first worked 6.5 hours, second worked 4.5 hours

Section 5.4 Solving Word Problems Using Systems

Practice 5.4.1

1. $\begin{cases} b + c = 10 \\ 2.75b + 3.50c = 30.50 \end{cases} \begin{matrix} ❶ \\ ❷ \end{matrix};$
weight of butterscotch, $b = 6$ pounds,
weight of chocolate, $c = 4$ pounds.

2. $\begin{cases} n + q = 32 \\ 0.05n + 0.25q = 5.00 \end{cases} \begin{matrix} ❶ \\ ❷ \end{matrix};$
number of nickels, $n = 15$, number of quarters, $q = 17$.

Practice 5.4.2

1. $\begin{cases} f + s = 740 \\ \frac{s}{50} + 2 = \frac{f}{62} \end{cases} \begin{matrix} ❶ \\ ❷ \end{matrix};$

distance at fast speed, $f = 465$ miles, (distance at slow speed, $s = 275$)

2. $\begin{cases} t + 1.5 = n \\ 200t + 240n = 1460 \end{cases} \begin{matrix} ❶ \\ ❷ \end{matrix};$
time going through turbulence, ($t = 2.5$ hours, time going through normal air, $n = 4$ hours)

3. $\begin{cases} p = f - 1 \\ 60p = 50f \end{cases} \begin{matrix} ❶ \\ ❷ \end{matrix};$
time for freight train, f, is 6 hours, which means that train must be moved to a siding before 3 pm. Distance for freight train, $50f$, is 300 miles from the station.

Practice 5.4.3

1. $\begin{cases} 5.5(p + w) = 3575 \\ 6.5(p - w) = 3575 \end{cases} \begin{matrix} ❶ \\ ❷ \end{matrix};$
speed of wind is 50 mph ; speed of plane is 600 mph

2. $\begin{cases} 3(b + c) = 21 \\ 7(b - c) = 21 \end{cases} \begin{matrix} ❶ \\ ❷ \end{matrix}; s$
peed of current is 2 mph; speed of boat is 5 mph

Practice 5.4.4

1. $\left\{ \begin{array}{ccccc} 30s & + & 20e & = & 350 \\ 10s & + & 5e & = & 100 \end{array} \right\}$ ❶ ❷ ;

10 economy cars and 5 sedans

2. $\left\{ \begin{array}{ccccc} 8d & + & 3s & = & 54 \\ 3d & + & 1s & = & 19 \end{array} \right\}$ ❶ ❷ ;

3 deluxe system and 10 standard system

Exercise Set 5.4

1. $\left\{ \begin{array}{ccccc} a & + & b & = & 145 \\ a & = & b & + & 45 \end{array} \right\}$ ❶ ❷ ;

books by Machine A , $a = 50$, books by Machine B, $b = 95$

3. $\left\{ \begin{array}{ccccc} s & + & c & = & 3325 \\ s & = & c & + & 325 \end{array} \right\}$ ❶ ❷ ;

checking, $c = \$1500$, savings, $s = \$1825$

5. $\left\{ \begin{array}{ccccc} 2l & + & 2w & = & 292 \\ l & = & 3w & + & 2 \end{array} \right\}$ ❶ ❷ ;

length, $l = 110$ feet, width, $w = 36$ feet

7. $\left\{ \begin{array}{ccccc} d & + & t & = & 25 \\ 1d & + & 20t & = & 215 \end{array} \right\}$ ❶ ❷ ;

number of one dollar bills, $d = 15$, number of twenty dollar bills, $t = 10$ twenties

9. $\left\{ \begin{array}{ccccc} c & + & e & = & 25 \\ 6.60c & + & 9.50e & = & 199.80 \end{array} \right\}$ ❶ ❷ ;

(weight of cheap coffee, $c = 13$ pounds), weight of expensive coffee, $e = 12$ pounds

11. $\left\{ \begin{array}{ccccc} 110t & + & 60c & = & 985 \\ c & = & t & - & 2 \end{array} \right\}$ ❶ ❷ ;

time for train, $t = 6.5$ hours, time for car, $c = 4.5$ hours, they are 985 miles apart at 5:30 pm

13. $\left\{ \begin{array}{ccccc} 10f & - & 7s & = & 17 \\ s & = & f & - & \frac{1}{2} \end{array} \right\}$ ❶ ❷ ;

time for fast bicyclist, $f = 4.5$ hours, time for slow bicyclist, $s = 4$ hours, they are 17 miles apart at 12:30 pm

15. $\left\{ \begin{array}{ccccc} 3p & + & 3w & = & 1050 \\ 3.5p & - & 3.5w & = & 1050 \end{array} \right\}$ ❶ ❷ ;

speed of plane, $p = 325$ mph, speed of wind, $w = 25$ mph

17. $\left\{ \begin{array}{ccccc} 4m & + & 30n & = & 480 \\ 1m & + & 5n & = & 95 \end{array} \right\}$ ❶ ❷ ;

number of magazines, $m = 45$, number of novels, $n = 10$

19. **[1.6.11]** Commutative Property of Multiplication

20. **[2.4.3]**

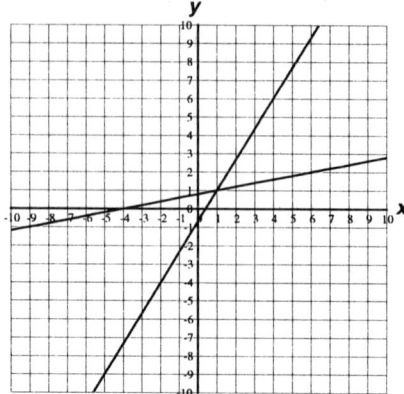

21. **[2.5.3]** $x < -\frac{6}{7}$ 22. **[3.2.3]** $2w + 2(4w) = 130$; 13

23. **[3.3.2]** Yes, with about 1 minute to spare.

24. **[4.4.7]** Slope of line 1 is -1; slope of line 2 is 1; lines are perpendicular.

25. **[4.8.1]** domain: $\{3,5\}$; range: $\{-2,-1,0,2\}$

Chapter 5 Review Exercises

Numbers inside brackets [ ] indicate the Chapter, Section, and Example number of a similar worked example.

1. **[5.1.1]** yes

2. **[5.1.2]** $(1,1)$

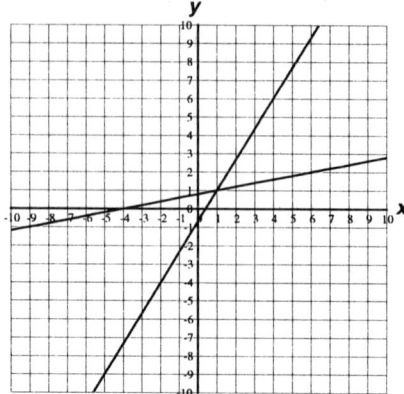

3. **[5.1.3]** $(x, 3x - 2)$

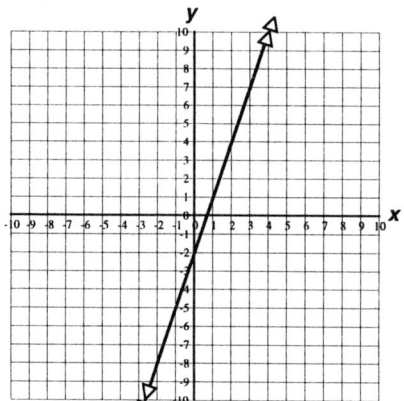

4. **[5.1.2]** $(4,-2)$

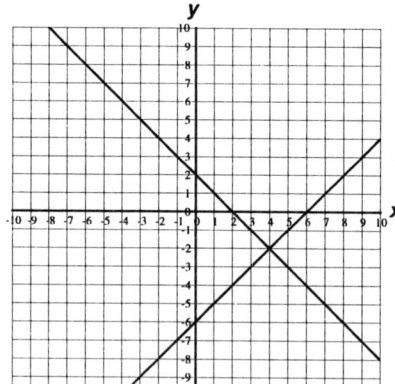

5. **[5.1.3]** No solution

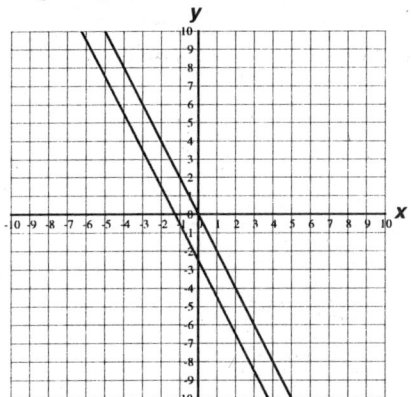

6. **[5.1.4]** System: consistent; Equations: independent

7. **[5.1.4]** System: inconsistent; Equations: independent

8a) **[5.1.5]** Leak: $y = 100x$; Pump: $y = -750 + 150x$

8b) The lines intersect at (15,1500). This means that 15 minutes after the leak started 1500 gallons of water entered the ship through the leak and 1500 gallons was removed by the pumps. That is, the pumps were able to empty all the water 10 minutes after they were started.

9a) **[5.1.5]** Highland: $y = 60 + 40x$; Penny: $y = 50 + 55x$

9b)

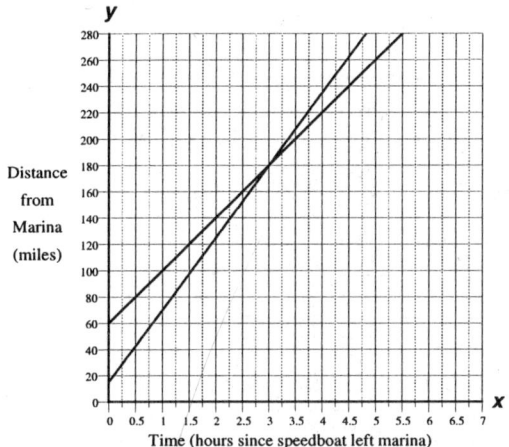

Time (hours since speedboat left marina)

9c) (3,180); If the job takes 3 hours, both plumbers will charge you the same amount, $180.

9d) If the job takes under 3 hours Penny is cheaper; otherwise, Highland is cheaper.

10. **[5.2.1]** (2,5)   **11.** **[5.2.1]** (2,6)

12. **[5.2.4]** No solution   **13.** **[5.2.4]** No solution

14. **[5.2.5]** Plan A: $y = 40 + 0.5x$; Plan B: $y = 25 + 0.7x$; cost is the same when connect time is 75 minutes

15. **[5.2.5]** (1,–7)   **16.** **[5.2.5]** (13,–5)

17. **[5.3.2]** (–1,–5)   **18.** **[5.3.2]** (2,–5)

19. **[5.3.3]** (2,–1)   **20.** **[5.3.3]** (3,7)

21. **[5.3.5]** (1,1)   **22.** **[5.3.5]** (2,–1)   **23.** **[5.3.6]** (–2,3)

24. **[5.3.6]** (6,–2)

25. **[5.4.1]** $\left\{ \begin{array}{rcl} 25q + 10d & = & 500 \\ q + d & = & 23 \end{array} \right\}$ ❶ ❷

Solving gives 18 quarters and 5 dimes.

26. **[5.4.3]** $\left\{ \begin{array}{rcl} 2(p + w) & = & 1600 \\ 2.5(p - w) & = & 1600 \end{array} \right\}$ ❶ ❷

Solving gives 720 mph air speed and 80 mph wind speed.

27. **[5.4.4]** $\left\{ \begin{array}{rcl} 0.02p + 0.08f & = & 8 \\ 0.40p + 0.30f & = & 95 \end{array} \right\}$ ❶ ❷

Solving gives 200 grams HiPro and 50 grams HiFat

Chapter 5 Test

Numbers inside brackets [ ] indicate the Chapter, Section, and Example number of a similar worked example.

1. **[5.1.1]** yes   **2.** **[5.1.2]** (–1,–2)

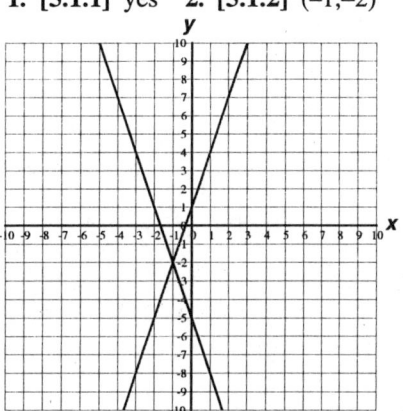

3. **[5.1.4]** System: consistent; Equations: independent

4a) **[5.1.5]** Speedboat: $y = 50x$; Police boat: $y = 60x - 15$

4b) The lines intersect at (1.5,75). This means that 1.5 hours after the speedboat left the marina both it and the police boat had traveled 75 miles. Therefore, the police boat catches the speedboat 1.25 hours after it starts out and before the boat reaches the 100 mile coastal limit.

5. **[5.2.1]** (5,–1)   **6.** **[5.2.5]** (1,8)

7. **[5.2.4]** $(x, \frac{3}{10} x + \frac{2}{5})$ **8.** **[5.2.4]** No solution

9. **[5.3.5]** (1,–2)   **10.** **[5.3.6]** (10,–6)

11. **[5.4.3]** $\left\{ \begin{array}{rcl} 2.5(b + c) & = & 10 \\ 4(b - c) & = & 10 \end{array} \right\}$ ❶ ❷

Solving gives current speed 0.75 mph (and boat speed 3.25 mph).

Chapter 6 Exponents and Polynomials

Section 6.1 Properties of Exponents

Practice 6.1.1
1. 5^{10} 2. y^{13} 3. w^9
4. The bases are not the same so $x^2 y^5$ cannot be simplified.
5. This is the sum of unlike terms so it cannot be simplified.

Practice 6.1.2
1. $18x^7$ 2. $-21m^5$ 3. $2t^{13}$ 4. $-140x^{12}$ 5. $6x^{13}$
6. $30c^8 + 10c^4$ 7. $12s^6 + 32s^5$ 8. $-m^3 - 4m$
9. $24w^4 - 20w$

Practice 6.1.3
1. 2^{12} 2. d^{35} 3. a^{36}

Practice 6.1.4
1. $16s^2$ 2. v^{20} 3. $64v^{10}$ 4. $16x^{16}$ 5. $m^{35}n^{14}$
6. $9x^2y^6$

Practice 6.1.5
1. 10^4 2. m^5 3. $3x^2$ 4. $-3t^{10}$ 5. xy^2 6. m^5n^3

Practice 6.1.6
1. $\dfrac{16}{m^4}$ 2. $\dfrac{w^3}{64}$ 3. $\dfrac{81w^4}{y^8}$ 4. $\dfrac{64d^6}{c^{12}}$

Practice 6.1.7
1. m^{17} 2. t^{14} 3. x^{14} 4. $4x^2y^{23}$ 5. $9x^3y^9$

Exercise Set 6.1
1. 6^5 3. h^4 5. y^{13} 7. Cannot be simplified because the bases are different.   9. m^{12} 11. r^{17} 13. x^6y^9
15. h^3j^4 17. $6n^8$ 19. $15g^9$ 21. $-24w^6$ 23. $48x^5y^{11}$
25. $6a^5 + 8a^3$ 27. $30t^7 + 20t^6 - 15t^3$ 29. $-20w^{10}$
31. $4d^7 + 6d^5$ 33. c^{30} 35. k^8 37. $64t^3$ 39. $25w^8$
41. $256r^{12}$ 43. $32x^{60}y^{20}$ 45. 5^2 47. y^4 49. r^7p^3
51. s^2t 53. Different bases so can't be simplified.
55. $\dfrac{64}{x^2}$ 57. $\dfrac{49}{t^6}$ 59. $\dfrac{16x^{20}}{y^{16}}$ 61. $24x^5$ 63. $2x^3$
65. $8x^8y$

67a) [3.1.4] $c + (c - 80)$, which simplifies to $2c - 80$
67b) $c + \dfrac{1}{4}c$, which simplifies to $\dfrac{5}{4}c$
67c) $c + (c + 60)$, which simplifies to $2c + 60$
68. [3.2.1] $x + (10 + \dfrac{1}{4}x) = 90$; 64 degrees and 26 degrees

69. [4.4.4] $-\dfrac{3}{2}$

70. [2.4.3]

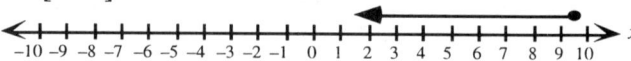

71. [4.4.3]

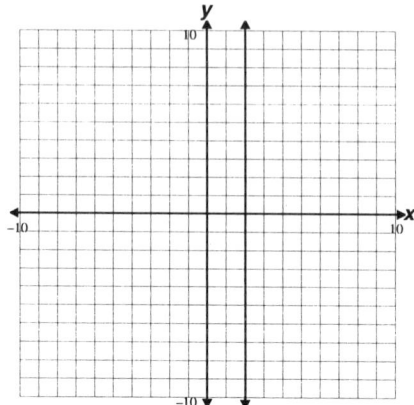

72. [3.2.4] $x + (x + 1) + (x + 2) = -9$; $-4, -3, -2$
73. [4.8.3] yes   74. [4.8.3] yes   75. [5.2.1] $(11, -2)$
76. [5.3.5] $(1, -1)$

Section 6.2 Exponents Involving 1, 0, and Negative Integers

Practice 6.2.1
1. Base is 3; $\dfrac{1}{81}$ 2. Base is w; $\dfrac{5}{w^7}$
3. Bases are 7 and m; $\dfrac{1}{49m^6}$ 4. Base is x; $9x^5$
5. Bases are d and 4; $\dfrac{d^5}{64}$ 6. Bases are x and y; $\dfrac{3y^4}{x^2}$
7. Base is x; We cannot "move" the x to the denominator because it is not a factor of the numerator (it is added to the 2 rather than multiplied by it). The best we can do at this time is to write $\dfrac{2 + \dfrac{1}{x}}{y}$. This is a complex fraction, which we discuss in Section 7.2.

Practice 6.2.2
1. $\dfrac{7}{12}$ 2. 42   3. $9\dfrac{1}{4}$ or $\dfrac{37}{4}$ 4. $\dfrac{5}{24}$

Practice 6.2.3
1. $\dfrac{1}{z^9}$ 2. $\dfrac{1}{b^6}$ 3. $16c^8$ 4. $81d^{12}$ 5. $125m^{18}$ 6. $\dfrac{18}{t^2}$

Practice 6.2.4

1. $\frac{x^2}{16}$ **2.** $\frac{125}{x^{18}}$ **3.** $\frac{15x^{13}}{2}$ **4.** $\frac{2x^{12}}{3}$

Exercise Set 6.2

1. Base is 3; $\frac{1}{3^6}$ **3.** Base is z; $\frac{12}{z^6}$ **5.** Base is $7x$; $\frac{1}{(7x)^5}$

7. Base is d; $2d^2$ **9.** Base is 4; $\frac{v^7}{4^2}$

11. Bases are m and 2; $\frac{1}{2^3 m^9}$ **13.** Bases are k and 3; $\frac{k^6}{3^2}$

15. Base is x; since x is not a factor of the numerator we cannot move it to the denominator. The best we can do is

$$\frac{3 + \frac{1}{x}}{2}$$

17. $\frac{5}{2}$ or $2\frac{1}{2}$ **19.** -4 **21.** $\frac{13}{6}$ or $2\frac{1}{6}$ **23.** a^{10} **25.** $\frac{d^6}{64}$

27. $\frac{9x^2}{4}$ **29.** $128x^7 y^{18}$ **31.** $\frac{9y^6}{4x^{18}}$ **33.** $\frac{x^{17}}{y^{26}}$ **35.** $\frac{4x^{11}}{y^6}$

37. [1.2.11] $\frac{1}{6}$ **38.** [2.2.5] 0

39. [4.3.4]

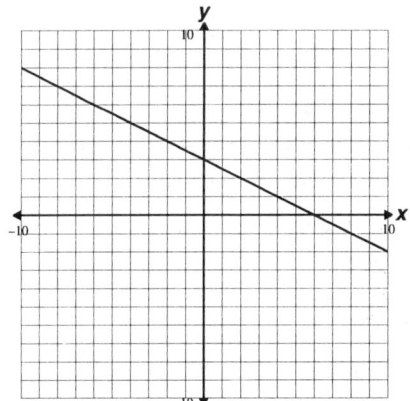

40. [4.7.1]

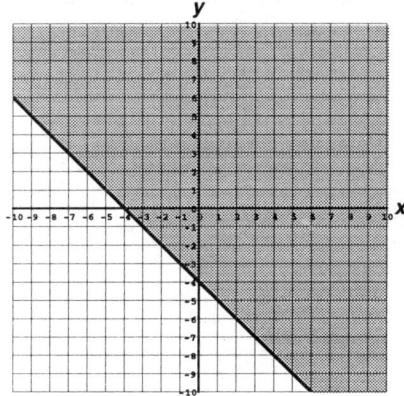

41. [4.8.4] 7 **42.** [5.3.2] (1,5)

43. [3.1.3] $85 - x - (x - 15)$, which simplifies to $100 - 2x$

44. [3.3.1] 30 miles

45. [5.4.1] $\left\{ \begin{array}{rcrcr} 25q & + & 10d & = & 500 \\ q & + & d & = & 29 \end{array} \right\} \begin{array}{l} \mathbf{❶} \\ \mathbf{❷} \end{array}$

Solving gives 14 quarters and 15 dimes.

46. [4.5.1] Slope is $-\frac{1}{2}$; y–intercept is $(0, \frac{3}{2})$.

47. [4.5.2] $y = 1$

48a) [4.1.4] $y = 12.5x$

48b)

x stalactite length (inches)	y time growing (years)
3.5	**43.75**
5	**62.5**
18	**225**
10	125
8	100

48c)

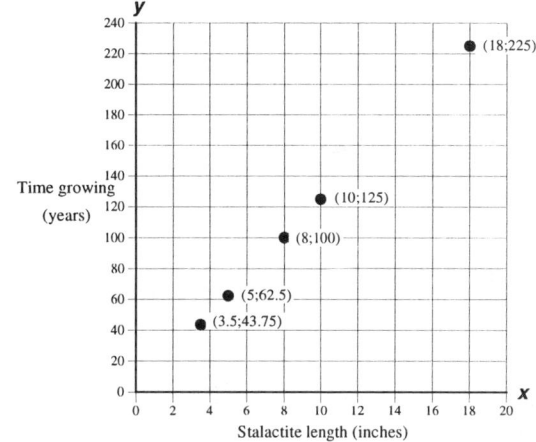

Section 6.3   Scientific Notation

Practice 6.3.1
1. 9.9×10^5, -4.7×10^9 **2.** -5.7×10^4, 7.08×10^{-6}

Practice 6.3.2
1. 7.8023×10^4 **2.** -6.001×10^6 **3.** 2.56×10^{-4}
4. -3.5×10^{-7}

Practice 6.3.3
1. 500,000 **2.** $-13,700$ **3.** 0.005321 **4.** 0.00002

Practice 6.3.4
1. 6×10^{-1} **2.** 8×10^{21} **3.** 8.58×10^{19}
4. 9×10^{-1}
5. -7.28×10^{-2} **6.** -1.1×10^{-11} **7.** 7.84×10^1
8. 3.15×10^{-14} **9.** 2.9221×10^2

Practice 6.3.5
1. 5.380×10^3 hours **2.** 6.6×10^{18} tons

Exercise Set 6.3

1. $2 \times 10^{-3}, 8.01 \times 10^8$ **3.** 6.7×10^6
5. -3.251×10^{10} **7.** 7.003×10^{-3} **9.** 5.2×10^{-4}
11. 2.73×10^{-1} **13.** $899,500,000$ **15.** -0.049
17. $8,000,000,000,000,000$ **19.** 1.575×10^{17}
21. 1.62×10^{-1} **23.** -4.6×10^{11} **25.** 6×10^1
27. 1.34×10^{-4} **29.** 5.085×10^4
31. 9.9×10^{-24} grams **33.** 2.53239×10^3 pounds
35. 7.5 cm

37. [1.1.3] 59 **38.** [1.5.8] 11 **39.** [1.7.6] $-6x + 22$
40. [2.5.3] $x > -8$ **41.** [3.4.1] $4.25c + 1.5c = 276$; 48
months **42.** [3.6.3] $0.08x + 0.10(1100 - x) = 97$; $650 at
8%, $450 at 10%. **43.** [5.2.4] $(x, 6x - 3)$
44. [6.1.2] $3m^9$ **45.** [6.2.2] 60
46. [5.2.5] Equation for buying is $y = 350x + 1875$ and
for leasing is $y = 275x + 3000$. The costs will be the same
($7125) at 15 months.
47. [4.3.5] $(0, -3)$ and $(-1, 0)$ **48.** [4.8.3] yes

Section 6.4 Introduction to Polynomials

Practice 6.4.1

1. Monomial
2. 4 term polynomial
3. Not a polynomial, because the variable is in the
denominator.
4. Not a polynomial, because the exponent on the first
term is not a whole number.

Practice 6.4.2

1. The term is $7y^2$ and the coefficient is 7.
2. The terms are $5x^2$, $-3x$ and 7 (or $7x^0$); the coefficient
of x^2 is 5, the coefficient of x is -3, and the coefficient of
x^0 is 7 (we can also call 7 the constant term).
3. The terms are x^2 and $-x$; the coefficient of x^2 is 1 and
the coefficient of x is -1.
4. The terms are $-x$ and 1 (or $1x^0$); the coefficient of x
is -1 and the coefficient of x^0 is 1 (we can also call 1 the
constant term).

Practice 6.4.3

1. Coefficient of first degree term is 1; coefficient of
zeroth degree term is -17. First degree polynomial in x.
2. Coefficient of second degree term is -16; coefficient of
first degree term is 45; coefficient of zeroth degree term is
25. Second degree polynomial in t.
3. Not a polynomial because the exponents on the first
and second terms are not whole numbers.

Practice 6.4.4

1. $-7s^9 + 5s^7 + 12s^5 + 23$ **2.** $\frac{5}{9}d^4 + \frac{3}{4}d^2 - \frac{1}{9}d$
3. $-0.3a^2 + 2.75a + 1.9$ **4.** $-4.9h^2 - 6.7h + 2.6$
5. Not a polynomial because of the variable in the
denominator.
6. Not a polynomial because of the variable in the
denominator.

Exercise Set 6.4

1. Monomial **3.** Trinomial **5.** This is not a polynomial
since the exponent is not a whole number.
7. Trinomial **9.** The term is $\frac{2}{3}n^3$ and the coefficient is $\frac{2}{3}$.
11. The terms are $3y^3$, $-4y^2$, $2y$, and -4 (or $-4y^0$);
the coefficient of y^3 is 3, the coefficient of y^2 is -4, the
coefficient of y, is 2 and the coefficient of y^0 is -4 (we can
also call 4 the constant term).
13. The terms are $\frac{2}{7}d^3$, $-\frac{3}{7}d$ and 2 (or $2d^0$); the
coefficient of d^3 is $\frac{2}{7}$, the coefficient of d^1 is $-\frac{3}{7}$, and the
coefficient of d^0 is 2.
15. The coefficient of the first degree term is -15 and the
coefficient of the zeroth degree term is 9. First degree
polynomial in y.
17. The coefficient of the first degree term is -1 and the
coefficient of zeroth degree term is 2. First degree
polynomial in x.
19. Not a polynomial because the exponent on the second
term is not a whole number.
21. Not a polynomial because there is a variable in the
denominator. **23.** $-3x^3 + 2x^2 + 5x - 7$
25. Already in standard form.
27. $-\frac{4}{3}w^7 + 2w^3 + 8w^2 + 15$
29. Not a polynomial because the exponent on the first
term is not a whole number. **31.** $-4.9h^2 - 6.7h + 2.6$

33. [1.1.2] 2 **34.** [2.2.2] 6 **35.** [2.5.3] all real
numbers **36.** [4.8.4] 8 **37.** [5.2.1] $(-1, -3)$
38. [6.2.3] $\dfrac{3y^3a^6}{2}$ **39.** [1.6.8] Base of first factor is 3,
exponent is 1; base of second factor is y, base is 5; cannot
be simplified. **40.** [6.4.1] yes **41.** [3.6.1] $x - 0.15x =$
$161,415$; $189,900 **42.** [4.4.4] 0
43. [4.5.1] slope is 1; y–intercept is $(0,0)$
44. [4.8.1] domain: $\{-1,0,2,3\}$; range: $\{1,4,5\}$

45. [4.8.6]

x	f(x)
–3	3
–1	7
–2 or 0	6
1	3
–5 or 3	–9

46. [4.2.4] The box originally had 200 paper towels and they are being used at the rate of 3 towels per day

---

Section 6.5 Adding and Subtracting Polynomials

Practice 6.5.1

1. $-5t^2 - 8t + 14$ **2.** $9xw^2 - 12w + 13$
3. $-4a^3 - 8a^2 + a - 39$ **4.** $11u^3 + 2u^2 - 9u - 2$

Practice 6.5.2

1. $-4r^2 - 5r + 18$ **2.** $-12w^3 + 11w^2 + w - 2$
3. $9s^4 - 8s^3 + 17s^2 - 19$

Practice 6.5.3

1. $-2r^3 - 3r^2 + 9r + 4$ **2.** $-3t^2 - 9t - 5$ **3.** $5s^2 - s - 9$
4. $22x^2 - 5x - 7$ **5.** $6x^2y + 12xy^2 - 1$

Practice 6.5.5

1. $p = (x^2 - 3x + 2) + 5x - 3$, simplifies to
$x^2 + 2x - 1$. When x is 12, p is 167.
2a) $y = 0.1x^2 + 2x - (7x + 125)$, simplifies to
$y = 0.1x^2 - 5x - 125$
2b) $\$-125$. The company is losing money.
2c) $\$375$. The company is making money.

Exercise Set 6.5

1. $16x^2 - 3x + 18$ **3.** $-\frac{5}{12}t^2 - \frac{1}{4}t + 16$
5. $-4.8a^3 - 2.2a^2 + 4.5a - 6.8$ **7.** $-2r^3 - r^2 + 10r - 14$
9. $-20r^2 - 6r + 17$ **11.** $-3u^2 - 2u - 17$
17. $10d^2 + 2d - 7$ **19.** $-s^2 + 7s - 17$
21. $9w^3 - 2w^2 + 4$ **23.** $-2v^3 - 7v^2 - 8v + 13$
25. $x^2 + 5x + 2$ **27.** $8.6r^2 - 4.6r - 6.6$
29. $p = 2(2x^2 + 5x + 3) + 2(x^2 + 3)$, which simplifies to
$p = 6x^2 + 10x + 12$. When x is 4, p is 148.
30a) $y = 4.5x^2 - 4x - (65.5x - 30)$, which simplifies to
$y = 4.5x^2 - 69.5x + 30$ **30b)** $\$440$. The company is
making money. **30c)** $\$0$. The company is breaking even.
30d) $\$-215$. The company is losing money. **31a** $\$64.8$
billion (the actual amount was $\$62.6$ billion).
31b $\$21.43$ billion (the actual amount was $\$18.4$ billion).
31c $\$43.37$ billion.

31d $0.0005x^4 - 0.031x^3 + 0.764x^2 - 5.51x + 15.97$;
predicted amount not spent in 1960 was $\$43.37$ billion (the
actual amount was $\$44.2$ billion). **32a** 122.1 billion gal
(actual amount was 121.3 billion gal)
32b 70.2 billion gal (actual amount was 69.3 billion gal)
32c 49.5 billion gal (actual amount was 51.0 billion gal)
32d $0.0067x^3 - 0.222x^2 + 3.62x + 92.71$; predicted
amount used by automobiles and trucks in 1985 is 119.7
(the same number as the sum for answers b and c)
32e $-0.0067x^3 + 0.222x^2 - 1.81x + 2.24$; predicted
amount used by busses in 1985 is 2.4 billion gallons
(actual amount was 0.8 billion gallons)

---

33. [4.5.3] $y = -x + 6$ **34. [1.2.6]** $\frac{45}{8}$

35. [1.1.1] sum: $5x + 8y$; terms $5x$ and $8y$; products:
within first term $5x$ and within second term $8y$; factors:
within first term 5 and x and within second term 8 and y.

36. [1.2.1] $3 \cdot 3 \cdot 5 \cdot 5$ **37. [2.3.2]** $x = \frac{y-8}{3}$

38. [4.8.2] yes **39. [1.7.7]** $-8t - 57$ **40. [5.3.5]** $(2,1)$
41. [5.3.2] $(-1,3)$ **42. [3.2.3]** $2(w) + 2(w+2) + 26 =$
$2(2w) + 2[(w+2) + 10]$; 3 inches **43. [3.3.2]** $\$6477$

44. [5.4.4] $\begin{cases} 0.25A + 0.60B = 410 \\ 0.05A + 0.20B = 110 \end{cases}$ **❶** **❷**

Solving gives 108 grams of Mix A, 268 grams of Mix B

45. [4.3.2]

Transformed data table

x number of years since 1984	y millions of travelers
0	11.0
4	13.5
7	15.0
10	15.8

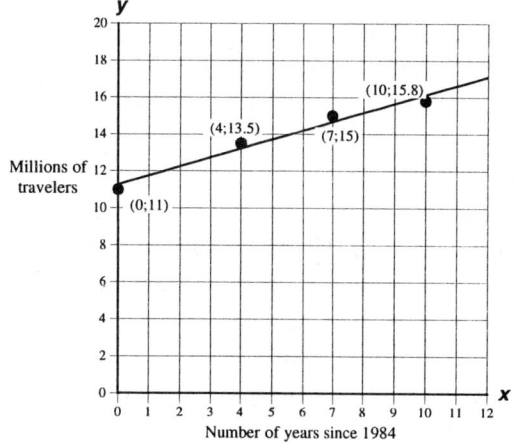

Section 6.6 Multiplying Polynomials

Practice 6.6.1

1. $12t^2 - 36t$ 2. $-54m^2 + 45m$ 3. $12m^4 - 6m^3 + 27m^2$
4. $-20b^3 + 35b^2 - 55b$ 5. $12x^3 - 9x^2 + 15x$
6. $-18v^3 + 10v^2 + 20v$

Practice 6.6.2

1. $4m^2 - 11m + 6$ 2. $21d^2 - 26d - 15$
3. $12m^3 - 15m^2n + 19mn^2 - 4n^3$
4. $15r^3 + 13r^2s - 15rs^2 + 3s^3$

Practice 6.6.3

1. $16m^2 - 8m + 1$ 2. $25r^2 + 10rs + s^2$
3. $36y^4 - 36wy^2 + 9w^2$ 4. $64x^3 + 48x^2 + 12x + 1$
5. $8a^3 + 60a^2b + 150ab^2 + 125b^3$

Practice 6.6.4

1. $787.5x^2 + 83280x - 3967.5$ 2. 907,583 (rounded)

Practice 6.6.5

1. $6x^2 + 19x + 8$ 2. $15m^2 + 14m - 16$
3. $6x^2 - 5xy + y^2$ 4. $12w^2 - wy - y^2$

Practice 6.6.6

1. $w^2 + 8w + 16$ 2. $m^2 - 10m + 25$ 3. $w^2 - 18w + 81$
4. $4r^2 + 44r + 121$

Practice 6.6.7

1. $x^2 - 49$ 2. $r^2 - 36m^2$ 3. $16x^2 - 25y^2$
4. $49t^2 - 81s^2$

Practice 6.6.8

1. $m^2 + 7m + 10$ 2. $r^2 - 12r + 32$ 3. $x^2 - 5x - 84$
4. These binomials do not fit the form $(x + m)(x + n)$ so
the shortcut will not work. We can use FOIL to get
$2y^2 - 3y - 2$.

Exercise Set 6.6

1. $20r^4 - 60t^3$ 3. $-3m^6 + 2m^4$ 5. $12y^7 + 6y^5 - 3y^3$
7. $18r^4 - 30r^3 + 42r^2$ 9. $10x^4 - 20x^3 + 5x^2$
11. $12w^3 - 32w^2 + 28w$ 13. $8x^2 - 18x - 56$
15. $6x^2 - 48x + 96$ 17. $8x^3 + 1$
19. $20m^3 + 7m^2n - 30mn^2 - 18n^3$
21. $8x^4 + 17x^3 - 49x^2 - 12x + 18$ 23. $9x^2 + 42x + 49$
25. $9m^2 + 30mn + 25n^2$ 27. $9z^2 - 6wz + w^2$
29. $x^3 + 9x^2 + 27x + 27$ 31. $8x^3 - 12x^2y + 6xy^2 - y^3$
33. $12g^2 - 55g + 63$ 35. $28s^2 - 23s - 15$
37. $36r^2 - 13rs + s^2$ 39. $20m^2 + 22mn + 6n^2$
41. $27a^2 - 15ab - 28b^2$ 43. $25m^2 - 40m + 16$

45. $16x^2 - 40xy + 25y^2$ 47. $x^2 - 1$ 49. $a^2 - 16b^2$
51. $9s^2 - 25t^2$ 53. $\frac{1}{4}x^2 - \frac{4}{9}$ 55. $y^2 + 15y + 56$
57. $r^2 - 13r + 40$ 59. $w^2 + 12w + 35$

61. **[1.6.11]** Distributive Property of Multiplication Over
Addition 62. **[1.2.3]** $\frac{3}{8}$

63. **[4.4.3]**

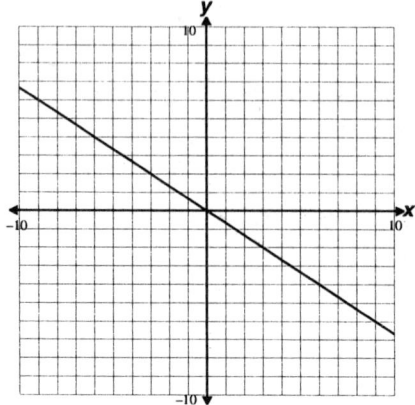

64. **[6.1.4]** $-\frac{1}{16}x^{16}y^8$

65. **[5.1.2]** $(-1,-4)$

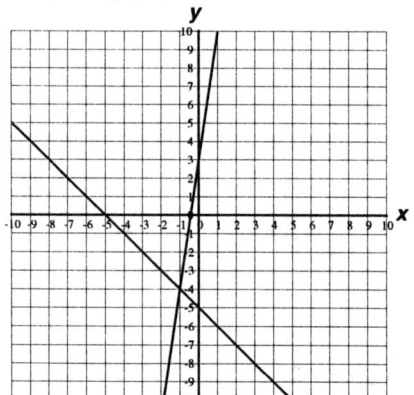

66. **[5.2.1]** $(1,0)$ 67. **[3.5.2]** $4.5f = 4(f + 0.5)$; first couple
4 mph, second couple 4.5 mph

68. **[4.4.7]** Slope of both lines is $\frac{3}{4}$; lines are parallel.

69. **[6.3.5]** 1.1×10^{11} suns (one hundred ten billion
suns). 70. **[6.4.3]** 2

71. **[4.8.1]** domain: $\{x \mid -9 \le x \le 1\}$;
range: $\{y \mid -8 \le y \le 8\}$

72. **[4.4.1]** $y = -8x - 8$

Section 6.7 Dividing Polynomials

Practice 6.7.1

1. $2w^5 - 3w^2 + w$ **2.** $-7a^4 + 5a^3 + 3a$ **3.** $7b^2 - \dfrac{3}{b} + \dfrac{2}{b^4}$

4. $-2d^5 + d + \dfrac{1}{2d}$

Practice 6.7.2

1. $2x^2 - x - 15$ **2.** $12y^2 + 11y - 5 + \dfrac{2}{y-1}$

3. $2r^2 - r - 6 + \dfrac{1}{2r+5}$ **4.** $4x^2 + 8x + 16$

Exercise Set 6.7

1. $m - 2$ **3.** $y^4 + 2y + \dfrac{5}{y^2}$

5. $m - \dfrac{2}{3m} + \dfrac{1}{3m^4}$ **7.** $5t + \dfrac{1}{3t} - \dfrac{2}{3t^2}$ **9.** $3x - 7$

11. $2y - 8$

13. $6x^2 + 7x - 3$ **15.** $2x^2 - x + 1$

17. $5x^2 - 7x + 12 - \dfrac{17}{x+2}$

19. $2x^3 + 6x^2 + 15x + 45 + \dfrac{137}{x-3}$

21. [2.4.3]

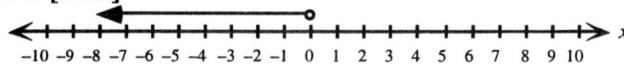

22. [4.3.4]

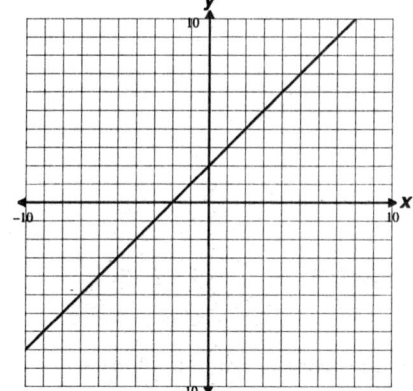

23. [6.2.2] 0 **24.** [6.2.3] $\dfrac{1}{g^3}$ **25.** [5.3.2] $(2,2)$

26. [5.1.3] no solution

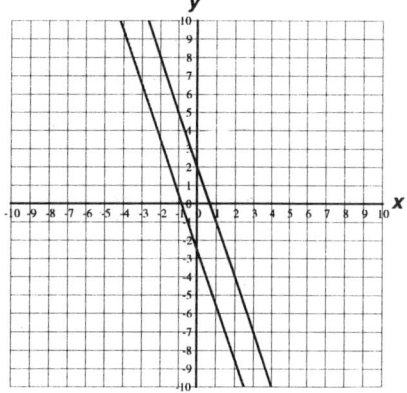

27. [4.5.1] Slope is -1; y–intercept is $(0,3)$.
28. [6.5.3] $10xy - 5x - 2y$ **29.** [6.3.4] 1.2×10^3
30. [1.2.1] $3 \cdot 3 \cdot 3 \cdot 5$ **31a)** [3.1.2] $f + 115$
31b) $\dfrac{1}{2}f$ **31c)** $\dfrac{1}{4}f$ **31d)** $f + 80$
32. [3.6.1] $p - 0.15p = 340$; $\$400$
33. [4.4.5] Slope is $\dfrac{-4 \text{ boxes left}}{1 \text{ house contacted}}$. This implies the
scout is selling 4 boxes of cookies per house.

Chapter 6 Review Exercises

Numbers inside brackets [ ] indicate the Chapter, Section, and Example number of a similar worked example.
1. [6.1.1] m^7n^7 **2.** [6.1.1] $81b^3$ **3.** [6.1.2] $4a^3 - 3a^2$
4. [6.1.2] $\dfrac{-3y^{11}}{2}$ **5.** [6.1.4] $x^{12}y^4$ **6.** [6.1.4] $-a^4b^2$
7. [6.1.7] $4b^{11}$ **8.** [6.2.1] $4x^3$ **9.** [6.2.1] $2x^3y^3$
10. [6.2.2] $\dfrac{2c^5}{3y^3}$ **11.** [6.2.2] $\dfrac{26}{9}$ **12.** [6.2.2] $\dfrac{91}{72}$
13. [6.2.3] $\dfrac{16x^2}{y^4}$ **14.** [6.2.3] $\dfrac{1}{t^3}$ **15.** [6.2.4] $2x^3$
16. [6.2.4] $4x^9y^5$ **17.** [6.3.2] -8.3×10^{-4}
18. [6.3.2] 3.5×10^9 **19.** [6.3.3] $50,000$
20. [6.3.3] -0.00302 **21.** [6.3.4] 1×10^{-8}
22. [6.3.4] 4.5×10^{-2} **23.** [6.3.5] No. You would need
about 4.478 million specks of dust.
24. [6.3.5] 5236 people. **25.** [6.4.1] Yes
26. [6.4.1] No, one of the exponents is negative.
27. [6.4.3] 2 **28.** [6.4.3] 5 **29.** [6.4.4] $-3x^2 + 5x + 2$
30. [6.5.1] $4x^2 + 5x + 1 - 8xy$
31. [6.5.1] $2a^3b + \dfrac{5}{4}ab^3 + \dfrac{1}{4}a^2b^2$
32. [6.5.3] $-6x^2 - 2x + 3$
33. [6.6.1] $-12m^4 + 8m^3 - 16m^2$
34. [6.6.1] $x^2 - y^2 + x + y$ **35.** [6.5.3] $11x + 3y$
36. [6.6.2] $18m^2 + 9m - 5$ **37.** [6.6.2] $k^4 - 2k^2 - k$
38. [6.6.3] $8y^3 - 36y^2 + 54y - 27$

39. [6.6.5] $15a^2 + 17a - 18$ **40.** [6.6.5] $2z^2 - 27z + 88$
41. [6.6.6] $25a^2 - 40a + 16$ **42.** [6.6.7] $81d^2 - 25$
43. [6.6.7] $36y^2 - 1$ **44.** [6.7.1] $5r^5 + 4r^3 - 2r$
45. [6.7.1] $17d^2 + 19d - 31$
46. [6.7.2] $3y^2 + 3y - 6 + \dfrac{2}{y+4}$
47. [6.7.2] $6y^3 - 1 - \dfrac{1}{2y-5}$

Chapter 6 Test

Numbers inside brackets [ ] indicate the Chapter, Section, and Example number of a similar worked example.
1. [6.1.1] a^5b^3 **2.** [6.1.2] $4z^{18}$ **3.** [6.1.7] x^5y^7
4. [6.2.1] $5m^3$ **5.** [6.2.2] $\dfrac{7}{12}$ **6.** [6.2.2] 1

7. [6.2.3] $\dfrac{144}{m^{12}}$ **8.** [6.2.4] $3y^7$
9. [6.6.1] $15c^4 - 3c^3 - 9c^2$ **10.** [6.6.2] $25a^2 - 71a + 36$
11. [6.6.2] $2r^2 + 3rs - 3rt - 35t + t^2$
12. [6.6.3] $25x^2 - 20x + 4$ **13.** [6.3.2] 4.9×10^7
14. [6.3.3] 0.000801 **15.** [6.3.4] 5×10^9
16. [6.3.5] No it would take about 16 thousand elephants.
17. [6.4.1] No, a variable is in the denominator.
18. [6.4.3] 3 **19.** [6.5.1] $-3x^2 - 5$
20. [6.4.4] $-x^4 + 12x^3 + 7x^2 - 3$
21. [6.5.3] $-28c^4 + 25d^2$ **22.** [6.5.3] $-9w^2 - 5w + 4$
23. [6.7.1] $3q^2 - q - 3$ **24.** [6.7.2] $h^3 - 4h^2 + h + 4$

Chapter 7 Factoring Algebraic Expressions

Section 7.1 Greatest Common Factor (GCF)

Practice 7.1.1
1. 14 **2.** 24 **3.** 36

Practice 7.1.2
1. $4r^2$ **2.** $-5t^2$ **3.** $-6r^3$ **4.** $3a^6b$

Practice 7.1.3
1. $x - 8$ **2.** $6(x - 1)^2$ **3.** $3(x - 5)^5$ **4.** $7(a + b)^2$

Exercise Set 7.1
1. 12 **3.** 6 **5.** 27 **7.** x^3y^2 **9.** $5y^3$ **11.** $3s^4$
13. $m^2(m + 5)^2$ **15.** $m(n + 1)^4$ **17.** $6m^2(m + 2)$
19. $3x^3(3x + 7)^3$ **21.** $5(t + 12)$

23. [1.1.1] sum: $3 + 2x$; terms: 3 and $2x$; product: within second term, $2x$; factors: within second term 2 and x
24. [1.2.1] $2 \cdot 2 \cdot 2 \cdot 2 \cdot 3$ **25.** [4.5.2] $x = 7$
26. [2.5.3] $x \geq \dfrac{2}{3}$ **27.** [3.4.2] $18(c + 15) = 27c$; $30 per
month   **28.** [3.2.3] $47.1 = 2(3.14)(r + 2)$; 5.5 mm
29. [4.8.4] 4 **30.** [6.7.1] $3y^2 - y - 1$
31. [5.2.4] $(x, \dfrac{2}{3}x + \dfrac{7}{3})$ **32.** [5.3.5] $(1, -3)$
33. [6.1.4] $-w^4g^{10}$ **34.** [6.2.2] $-\dfrac{1}{8}$
35. [6.6.1] $15x^3 - 12x^2 + 3x$
36. [6.6.2] $3x^3 - 7x^2 - 22x + 12$ **37.** [4.8.2] yes
38a) [4.1.4] $y = \dfrac{1}{4}x$

38b)

x number of quarters	y value (dollars)
8	**2**
18	**4.5**
30	**7.5**
24	6
38	9.5

38c)

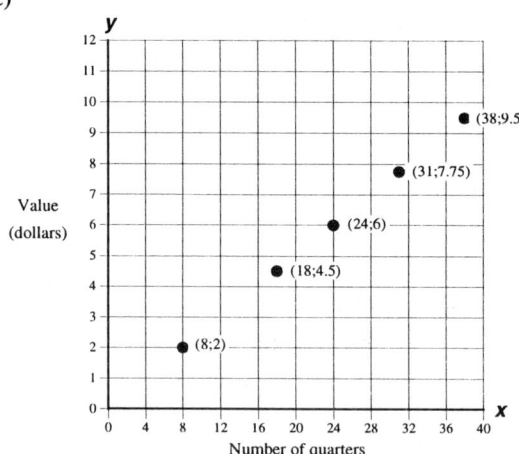

Section 7.2 Factoring the GCF Out of Polynomials

Practice 7.2.1
1. $5(2x + 3)$ 2. $3x^2(x^4 + 4)$ 3. $8(8r^4 + 2r^2 - 1)$
4. $15a^3b^2(3ab^3 - 1)$ 5. $27c^3d^9(c^4 - 2d)$

Practice 7.2.2
1. $1(-25)$ and $-1(25)$ 2. $1(32)$ and $-1(-32)$
3. $1(2x + y)$ and $-1(-2x - y)$
4. $1(4m - 2n)$ and $-1(-4m + 2n)$
5. $1(7v^4 - 8v^3 + 2)$ and $-1(-7v^4 + 8v^3 - 2)$
6. $1(-12w^3 + 15w^2 + 11)$ and $-1(12w^3 - 15w^2 - 11)$
7. $1(25 - w^2)$ and $-1(-25 + w^2)$ or $-1(w^2 - 25)$

Practice 7.2.3
1. $-w(w^6 + 6)$ 2. $-6m(m^4 - 2)$ 3. $-4k(k^3 - 2k^2 + 3)$
4. $-6r(r^4 - 3r^3 + 2)$

Practice 7.2.4
1. $(z + 1)(5z - 8)$ 2. $2(t + 9)(t + 3)$
3. $4(n^2 - 8)(m^2 - 3)$ 4. $7(x + y)[(x + y)^3 - 2(x + y) + 3]$
5. $4(m - n)[3(m - n)^4 + 2(m - n)^2 - 1]$

Practice 7.2.5
1. $(2p + r)(3p + 2)$ 2. $(5q + 7)(2q + 5r)$
3. $(3x + y)(2x + 8z)$

Practice 7.2.6
1. $(y + x)(y - 1)$ 2. $(a + 2b)(3a - 2)$
3. $(3x + 7z)(2x - 3y)$ 4. $(5a + 3b)(2a - 7)$

Exercise Set 7.2
1. $12r^4s^3(1 + 3r^2s^4)$ 3. $2xy(2x^5y^2 - 8xy^3 + 1)$
5. $6x^3y^4(2x^2y^3 - 3x^5 + 6y^8)$
7. $5f^3g(3f^2g^{11} - 5g^8 + 9f^{10})$
9. $1(x + 2)$ and $-1(-x - 2)$
11. $1(5t + u)$ and $-1(-5t - u)$
13. $1(4p^4 - 3p^3 + 9)$ and $-1(-4p^4 + 3p - 9)$
15. $1(-5x^3 + 7x - 9)$ and $-1(5x^3 - 7x + 9)$
17. $1(25 - w^3)$ and $-1(-25 + w^3)$ or $-1(w^3 - 25)$
19. $-x(x^2 - 3x + 5)$ 21. $-2xy(y^4 - 2y^2 + 4y - 3)$
23. $-4rst(2r^5s^2t^4 - r^2s^3t + 3)$ 25. $(r^2 + 2)(2r - 3)$
27. $r(r + 3)(6 - 5r)$
29. $4m(m + 5)[4m^2(m + 5) - 3]$
31. $4(v + 4)(v - 3)$
33. $4m(m + 2n)^3[2(m + 2n)^2 + 3(m + 2n) - 4]$
35. $(m + n)(r + s)$ 37. $(2f + g)(f - 7)$
39. $(3w - 2)(5t + 1)$ 41. $(3x + 2y)(4z + 1)$
43. $(2r + 5s)(r - 1)$ 45. $(a + 5)(b - 7)$
47. $(4s - 9r)(2s - 3)$ 49. $(4x + 5y)(2x - 1)$

51. [1.1.2] 7
52. [1.2.1] $3 \cdot 3 \cdot 11 \cdot 13$ 53. [2.3.2] $c = \frac{1}{2}a - b$
54. [6.6.5] $12m^2 + 20m + 3$ 55. [5.2.1] $(-1, 2)$
56. [4.7.1]

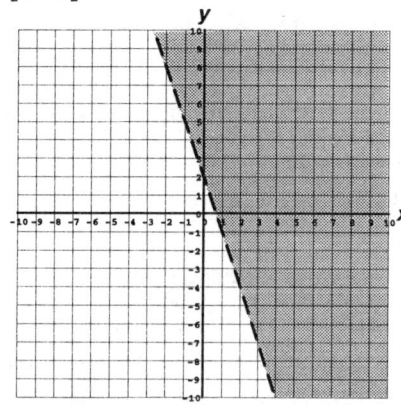

57. [1.7.7] $36x - 45$
58. [1.6.8] Base is 3, exponent is 6; simplifies to -729
59. [6.7.2] $2a^3 + a^2 + 4$
60. [6.4.3] 4
61. [3.4.1] $5n + 25(n + 100) = 2830$; 11 nickels
62. [4.8.6]

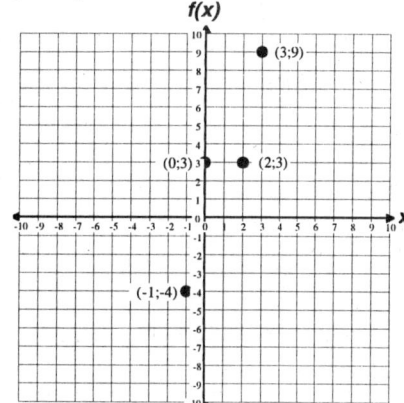

63. [6.3.4] 5×10^1
64. [1.1.1] no x–intercept, y–intercept is $(0, 4.5)$
65a) [4.8.7] Byron's weight 60 days into the program.
65b) [4.8.7] After d days Byron weighs 190 pounds.
65c) [4.8.7] After 30 days Byron's weight is 208 pounds.
65d) [4.8.7] After d days Byron will weigh 7 pounds less than he did the 18th day.

Section 7.3 Factoring Trinomials of the Form $x^2 + bx + c$

Practice 7.3.1
1. $(x + 2)(x + 6)$ **2.** $(x + 9)(x + 11)$ **3.** $(x - 5)(x + 3)$
4. $(x - 9)(x + 11)$ **5.** $(x - 6)(x - 5)$ **6.** $(x - 9)(x - 8)$

Practice 7.3.2
1. $(x - 18)(x + 4)$ **2.** $(y - 7)(y + 7)$ **3.** $(r + 20)(r - 3)$
4. $(x + 25)(x - 4)$ **5.** $(x + 32)(x - 3)$
6. Not factorable using integers because there are no two integers whose product is 15 and whose sum is 7.

Practice 7.3.3
1. $5(x + 3)(x + 5)$ **2.** $3(x + 2)(x + 7)$
3. Not factorable using integers because there are no two integers whose product is –3 and whose sum is 12.
4. $-1(x - 5)(x + 12)$ **5.** $4x(x + 8)(x + 5)$
6. $5x(x + 4)(x + 9)$

Exercise Set 7.3
1. $(x + 7)(x + 4)$ **3.** $(x + 5)(x + 8)$ **5.** $(x + 9)(x - 3)$
7. $(m + 3)(m - 12)$ **9.** $(x - 15)(x - 3)$ **11.** $(t - 8)(t - 9)$
13. $(y - 24)(y - 3)$ **15.** $(m + 15)(m - 4)$
17. $(x - 36)(x + 4)$ **19.** Not factorable using integers because there are no two integers whose product is 12 and whose sum is –22. **21.** $(t - 16)(t - 12)$
23. Not factorable using integers because there are no two integers whose product is 19 and whose sum is –8.
25. $6(x - 9)(x + 3)$ **27.** $-1(x - 4)(x + 3)$
29. Not factorable using integers because there is no GCF and there are no two integers whose product is 10 and whose sum is –4.
31. $-1(x - 4)(x + 3)$ **33.** $-1(x - 11)(x + 2)$
35. $-3x(x - 11)(x + 5)$

---

37. **[3.2.1]** $b + (b + 1) = 4$; $1.50
38. **[3.6.4]** $11000w + 4000(w - 0.08) = 467.5$; 5.25% for first account (she made money) and –2.75% for second account (the negative rate means she lost money in this account).
39. **[4.5.1]** Slope is –1; y–intercept is (0,0).
40a) **[5.1.5]** Computer Depot: $y = 100 + 50x$; Pica Computer Company: $y = 90x$

40b)

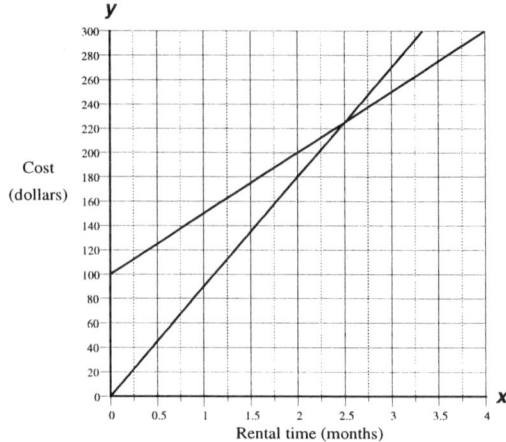

40c) (2.5,225) If you rent for 2.5 months the charge be each company is the same, $225.
40d) If you rent for less than 2.5 months then Pica is cheaper; after 2.5 months Computer Depot is cheaper.
41. **[5.1.3]** ($x, \frac{1}{2}x + \frac{3}{2}$)

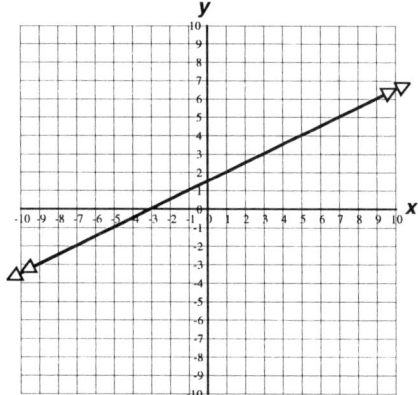

42. **[5.3.6]** (12,–4) **43.** **[6.7.1]** $16p^6 + 8p^4 - 7p^2$
44. **[6.3.2]** 3.1×10^8 **45.** **[7.1.2]** $6x$
46. **[7.2.1]** $-2y^2(xy + 1)$ **47.** **[6.2.2]** $-\frac{17}{3}$
48. **[6.2.3]** $2a^9$ **49.** **[6.2.3]** $16a^{16}b^{14}$
50. **[6.5.1]** $5x^2y + 3x + y$ **51.** **[6.6.2]** $15c^2 - 17c + 5$
52. **[6.6.3]** $4x^2 - 4x + 1$ **53.** **[6.1.7]** $18n^4$
54. **[6.5.3]** $-4b^2 - 9b + 6$ **55.** **[4.4.4]** –1

56. [4.3.2]

Transformed data table

x number of years since 1992	y Sales (thousands of dollars)
0	6.3
1	8.1
4	13.5
6	17.1

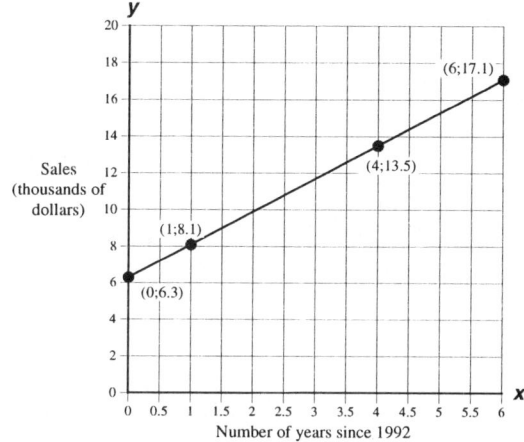

Number of years since 1992

Section 7.4 Factoring Trinomials of the Form $ax^2 + bx + c$

Practice 7.4.1
1. $(5x + 1)(x + 1)$ **2.** $(3x + 2)(x - 1)$ **3.** $(7x - 1)(x - 3)$

Practice 7.4.2
1. $(4w + 5)(w + 1)$ **2.** $(5m + 6)(m + 1)$
3. $(3x + 2)(x - 5)$ **4.** $(4w - 1)(3w + 5)$
5. $2(4z + 5)(z + 1)$ **6.** $2(3m - 1)(2m - 1)$
7. Not factorable using integers because there are no two integers whose product is 20 and whose sum is 7.

Exercise Set 7.4
1. $(2x + 1)(x + 2)$ **3.** $(5x - 1)(7x + 2)$ **5.** $(3r - 7)(r - 1)$
7. $(6z + 1)(2z + 3)$ **9.** $(3m - 1)(2m - 3)$
11. $(3y + 4)(2y - 5)$ **13.** Not factorable using integers because there are no two integers whose product is –27 and whose sum is –2.
15. $(3y + 2)(4y - 5)$ **17.** $(4w - 3)(w - 2)$
19. $(3t - 1)(4t - 3)$ **21.** $(3x - 2)(4x + 1)$
23. $5(5x - 3)(2x - 1)$ **25.** $3(5m + 2)(m + 1)$
27. $4(2m + 5)(m - 2)$ **29.** Not factorable using integers because there are no two integers whose product is –6 and whose sum is –4.

31. [1.7.6] $-7x - 1$ **32.** [2.2.1] 5 **33.** [2.2.5] 0
34. [7.2.1] $3xy^4(2xy - 1)$ **35.** [4.8.1] domain: all real numbers; range: all real numbers **36.** [4.8.4] 13
37. [5.3.2] $(3,0)$ **38.** [5.3.5] $(0,-1)$
39. [6.7.2] $3x^3 - 2x^2 + x + 1$ **40.** [7.1.3] $2x^2(x + 2)^3$
41. [5.4.4] $\begin{cases} 1s + 2e = 140 \\ 0.5s + 0.25e = 40 \end{cases}$ build install
Solving gives 60 standard VCR's and 40 extended VCR's.
42. [3.5.1] $1.25x = 1.5(x - 10)$; 60 mph
43. [4.4.1] $y = \frac{3}{4}x - 8$
44. [4.4.2] The y–intercept is $(0,0)$, which implies there is no value when you have 0 dimes. If we go, for example, from the point $(3,30)$ to $(8,80)$, the change in y is 50 while the change in x is 5. This is a change of $\frac{50 \text{ cents}}{5 \text{ dimes}}$ or 10 cents per dime. Therefore, the value of the dimes changes by 10 cents as the number of dimes changes by 1. The equation is $y = 10x$.
45. [4.4.3]

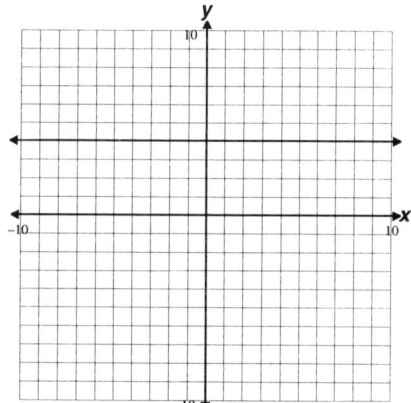

46. [4.5.3] $y = \frac{1}{2}x - 1$ **47.** [4.4.7] Slope of line 1 is $-\frac{6}{5}$; slope of line 2 is $-\frac{1}{2}$; lines are neither parallel nor perpendicular. **48.** [6.2.3] $\frac{c^9}{d^2}$ **49.** [6.2.3] $\frac{36m^6}{n^8}$
50. [6.6.1] $48y^4 - 32y^3 + 16y$ **51.** [6.6.2] $8y^3 - 1$

Section 7.5 Factoring Special Products and General Factoring

Practice 7.5.1

1. $(m-9)(m+9)$ **2.** $(9w-2)(9w+2)$

3. $(2r-3s)(2r+3s)$ **4.** $(6q-5d)(6q+5d)$

Practice 7.5.2

1. $(10a-9b)(10a+9b)$ **2.** $(3z-7v)(3z+7v)$

3. This is the sum, not the *difference* of two squares.

4. $2(m-4)(m+4)$ **5.** This is a trinomial, not a binomial

6. $(5x^2-3)(5x^2+3)$ **7.** $(3x-1)(3x+1)$

8. $(5-6x)(5+6x)$ *or* $-(6x-5)(6x+5)$

Practice 7.5.3

1. $(y+2)^2$ **2.** $(z-1)^2$ **3.** $(5m-1)^2$ **4.** $(5x-4)^2$

5. First term can be written as $(10c)^2$; last term can be written as $(4)^2$; but the middle term is not $2(4)(10c)$. Therefore, this cannot be factored as the square of a binomial.

Practice 7.5.4

1. $(3y-2)(3y+2)$ **2.** $2(z-9)(z-4)$

3. $-1(m+6)(m-6)$ **4.** $(3m+2n)(m-4)$

5. $x^2(x-8)(x-3)$ **6.** $(2x+3)(3x-1)$

Exercise Set 7.5

1. $(c-d)(c+d)$ **3.** $(2t-11)(2t+11)$

5. $(5y-1)(5y+1)$

7. $(4-3x)(4+3x)$ *or* $-(3x-4)(3x+4)$

9. $(5x-12y)(5x+12y)$ **11.** 12 is not a perfect square.

13. $(y-5z)(y+5z)$ **15.** 5 is not a perfect square.

17. $(6z^2-7)(6z^2+7)$

19. $(3-10d)(3+10d)$ *or* $-(10d-3)(10d+3)$

21. $(7y+2)^2$ **23.** First term can be written as $(z)^2$; last term can be written as $(1)^2$; but the middle term is not $2(1)(z)$.   **25.** $(4m+5)^2$ **27.** $(m+4)^2$

29. First term can be written as $(r)^2$; last term can be written as $(6)^2$; but the middle term is not $2(r)(6)$.

31. Last term cannot be written as the square of an integer because it is a negative.

33. $4(x+8)(x-3)$ **35.** $2(x+2)(x+2)$ or $2(x+2)^2$

37. $2(5r-2s)(6r+1)$ **39.** $-3(y+5)(y-5)$

41. $(2x+1)(4x-5)$

43. [2.2.4] all real numbers   **44.** [2.5.3] $x \le 20$

45. [4.3.4]

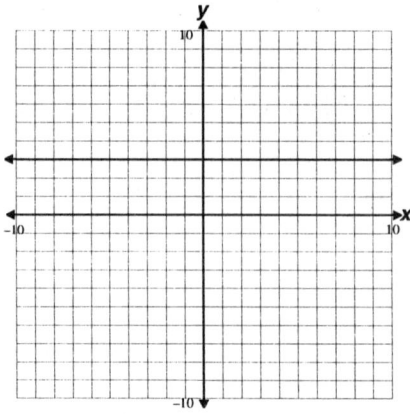

46. [5.2.1] $(-1,0)$ **47.** [6.7.1] $-5t^4 - 2t^3 - t + 3$

48. [6.2.2] $\dfrac{55}{16}$ **49.** [6.2.3] $32x^2$ **50.** [6.2.3] $\dfrac{1}{5^4}$

51. [6.6.2] $27a^3 + b^3$ **52.** [6.5.1] $2m^2 + \dfrac{1}{4}m + \dfrac{1}{16}$

53. [6.5.3] $-0.2x^4 - 2.498x^3 + 0.61x^2 - x - 4$

54. [7.2.1] $2a(5a^2 + 9b^3)$ **55.** [7.3.2] $(x-8)(x-6)$

56. [7.3.3] $3(x+2)(x+5)$ **57.** [7.2.6] $(x-y)(2x+3)$

58. [6.3.4] 2.5016×10^{-11}

59a) [3.1.2] $\dfrac{1}{3}s$ **59b)** $s+3$ **59c)** $s-3$ **59d)** $s-1$

60. [3.4.1] $2.5(114) + 4.75a = 1163.75$; 185 tickets

61. [3.2.4] $(x) + (x+2) = 2(x+2)$; no solution

62. [4.4.4] $\dfrac{1}{3}$

63. [4.5.1] Slope is $-\dfrac{2}{3}$; y–intercept is $(0,-2)$.

64. [4.5.2] $y = -\dfrac{2}{3}x$ **65.** [4.8.3] yes

66. [4.4.5] Slope is $\dfrac{0.15 \text{ years}}{1 \text{ dollar}}$. The slope means that for every dollar increase in the denomination of a bill it's life span increases by 0.15 years. Notice how this model can only speak realistically to certain input values such as 1, 5, 10, and 20 since paper money is printed only in certain denominations.

Section 7.6 Solving Polynomial Equations By Factoring

Practice 7.6.1

1. 12, –5   **2.** 0, 3   **3.** –3, 5   **4.** $\dfrac{6}{5}, \dfrac{-6}{5}$

Practice 7.6.2

1. 9 and 4   **2.** 8 and –3   **3.** –1 and $\dfrac{2}{3}$ **4.** $-\dfrac{3}{2}$ and $\dfrac{1}{3}$

Practice 7.6.3

1. 12 and 13, or –12 and –13   **2.** 7 and 13, or $-\dfrac{13}{2}$ and –14

Practice 7.6.4

1. $s^2 + (s + 4)^2 = 20^2$; 12 mph
2. $x^2 + (x + 7)^2 = 97^2$; 65 mph

Practice 7.6.5

1. 0 and 2 and 5   **2.** 0 and –7 and 3   **3.** 0 and $\dfrac{5}{2}$ and $\dfrac{-5}{2}$

4. 0 and –8 and – 6   **5.** 0 and 8 and –6

Exercise Set 7.6

1. $\dfrac{-3}{2}$ and 1   **3.** 4 and 7   **5.** 0 and $\dfrac{1}{7}$ **7.** –1 and –4

9. 3 and –7   **11.** 12 and –12   **13.** 0 and $-\dfrac{1}{2}$

15. 0 and $-\dfrac{1}{2}$ **17.** 0 and 3   **19.** –5 and 3

21. –7 and –5   **23.** 8 and –3   **25.** $\dfrac{1}{4}$ and $-\dfrac{2}{5}$

27. 6 and –6   **29.** 5 and 3   **31.** –11 and 4

33. $\dfrac{2}{3}$ and $\dfrac{5}{4}$ **35.** 12 in, 20 in   **37.** 8

39. $w^2 + (w + 2)^2 = 10^2$; 6 in by 8 in

41. $(2j)^2 + [2(j + 1)]^2 = 10^2$; 3 mph   **43.** $0, \dfrac{1}{6}, -\dfrac{1}{6}$

45. 0, –1, 4   **47.** 0, 25, –4   **49.** $-\dfrac{2}{3}, \dfrac{5}{4}, 0$ **51.** 0, –7

53. $0, -\dfrac{5}{2}, \dfrac{5}{2}$

---

55. [3.3.2] 165 points
56. [3.4.3] $27(R - 8) + 18R + 16(R + 4) = 915.5$; $9.50
for tent site, $17.50 for RV site, $21.50 for RV site with
electricity   **57.** [3.6.6] $0.02a + 0.18(50) = 0.12(a + 50)$;
30 gallons   **58.** [6.3.5] A little over 81 moons.
59. [4.4.1] $y = -x + 3$
60. [4.8.1] domain: $\{x \mid -4 \le x \le 8\}$;
range: $\{y \mid -4 \le y \le 5\}$ **61.** [5.2.4] $(\dfrac{2}{3}x + \dfrac{1}{6})$

62. [5.3.2] (–3,1)   **63.** [6.7.2] $2x^2 + 4x + 2 - \dfrac{1}{x - 4}$

64. [2.4.3]

65. [7.1.3] $6x^2(3x - 2)^4$ **66.** [4.8.4] 2

67. [7.2.6] $(x + 2)(x + y)$ **68.** [7.3.2] $(x - 12)(x - 4)$
69. [7.3.3] $5(x + 4)(x - 1)$ **70.** [7.4.2] $(4x + 1)(5x - 2)$
71. [6.6.1] $17x^2 - 49x$ **72.** [6.6.5] $8y^2 - 65y + 8$
73. [6.1.7] $8x^{11}y^2$ **74.** [6.2.3] $\dfrac{c^{11}}{d^{13}}$

75. [6.5.1] $3t^3 - 3t^2 + 6t + 1$ **76a)** [4.5.4] $y = 0.72x + 27$
76b) The y–intercept is 27 years. This implies that when
men have a life expectancy of 0 years women have a life
expectancy of 27 years. This is a case where the y–
intercept is so far from the data that is has no meaning for
our questions. The slope, $\dfrac{0.72 \text{ year for women}}{1 \text{ year for men}}$ implies that
the life expectancy for women is increasing by 0.72 year
each time the life expectancy for men increases by 1 year.
76c) 81 years   **76d)** 96 years

Chapter 7 Review Exercises

Numbers inside brackets [ ] indicate the Chapter, Section,
and Example number of a similar worked example.
1. [7.1.1] 8   **2.** [7.1.1] 36   **3.** [7.1.2] $10x^2y$
4. [7.1.2] $6x^3y^3z^2$ **5.** [7.1.3] $6x^3(5x + 1)^4$
6. [7.1.3] $4x^2y^5(x - 2)^7$ **7.** [7.1.3] $1(-9x^2 + 5x + 1)$
8. [7.2.2] $-1(-3x^2 + 2x - 1)$
9. [7.2.2] $-1(8x^4 - 3x^2 + 5)$ **10.** [7.2.1] $-8x^2y^2(3x - 2y)$
11. [7.2.1] $9x^2(5x^2 - 1)$ **12.** [7.2.4] $4x(x - 3)(2 - x)$
13. [7.2.4] $2x(2x - 1)^3(6x + (2x - 1)^2)$
14. [7.2.6] $(3x - 2y)(5x - 2)$ **15.** [7.2.6] $(w - 5h)(x + y)$
16. [7.3.1] $(x - 4)(x + 3)$ **17.** [7.3.1] $(x - 5)(x - 2)$
18. [7.3.2] $(x - 12)(x + 5)$ **19.** [7.3.2] $(x - 4)(x + 16)$
20. [7.3.3] $4(x - 1)(x + 4)$ **21.** [7.3.3] $5(x - 4)(x + 3)$
22. [7.4.2] $(3x - 8)(6x + 1)$
23. [7.4.2] $-3(x - 2)(5x - 3)$
24. [7.5.1] $(2x - 3y)(2x + 3y)$
25. [7.5.1] $(5x - 1)(5x + 1)$
26. [7.5.3] $(x - 5)(x - 5)$ or $(x - 5)^2$
27. [7.5.3] $(2x + 7)(2x + 7)$ or $(2x + 7)^2$
28. [7.5.4] $8x(x - y)(x + 2)$
29. [7.5.4] $3x(5x - 1)(5x + 1)$
30. [7.5.4] $5x(2x - 3y)(2x + 3y)$ **31.** [7.6.1] $0, \dfrac{1}{2}$

32. [7.6.1] $-\dfrac{3}{2}, 1$ **33.** [7.6.2] $-\dfrac{5}{2}, 4$

34. [7.6.2] $-\dfrac{2}{3}, -\dfrac{5}{3}$ **35.** [7.6.5] –2, 0, 3

36. [7.6.5] –2, 0, 2
37. [7.6.3] 18 and 19 or –18 and –19
38. [7.6.3] –14 and 3, or –3 and 14
39. [7.6.4] $(3s)^2 + (2(s + 3))^2 = 15^2$; 3 mph

Chapter 7 Test

Numbers inside brackets [ ] indicate the Chapter, Section, and Example number of a similar worked example.

1. [7.1.2] $14y^3z^2$ **2. [7.2.2]** $-1(-4x^6 + 3x^4 + 2x^2 + 3$
3. [7.2.1] $3xy^2(4x^3y - 1)$
4. [7.2.4] $3x^3(3x + 2)^4(2x(3x + 2) - 5)$

5. [7.2.6] $(x - 2y)(4x - 5)$ **6. [7.3.1]** $(x - 7)(x + 2)$
7. [7.3.3] $6(x - 3)(x - 7)$ **8. [7.4.2]** $(x - 3)(5x + 6)$
9. [7.5.1] $(7x - 3)(7x + 3)$ **10. [7.5.4]** $7(x - 3)(x - y)$
11. [7.5.4] $4(x - 6)(2x - 5)$ **12. [7.6.1]** $0, \dfrac{2}{5}$
13. [7.6.2] $-2, -4$ **14. [7.6.3]** 3 and 4, or -2 and -1

Chapter 8 Rational Expressions and Equations

Section 8.1   Reducing Rational Expressions

Practice 8.1.1
1. 0   **2.** 0   **3.** -9 **4.** 1

Practice 8.1.2
1. $\dfrac{3}{2}$ **2.** $\dfrac{1}{x + 3}$ **3.** $m - 2$

Practice 8.1.3
1. -1 **2.** -2 **3.** $-\dfrac{x - 1}{x + 3}$ **4.** $-\dfrac{x - 2}{x + 4}$ **5.** $\dfrac{-1}{x - 4}$

Practice 8.1.4
1. $3x + 1; 4; 10$ **2.** $4x - 5; -17; 3$ **3.** $-5x - 1; 9; 4$

Exercise Set 8.1
1. 0   **3.** -6 **5.** -4 **7.** $\dfrac{x + 4}{x - 3}$ **9.** $\dfrac{x - 2}{x + 3}$ **11.** $\dfrac{x}{x + 7}$

13. -1 **15.** -1 **17.** $-\dfrac{x + 3}{x + 2}$ **19.** $-\dfrac{m + 7}{m + 2}$ **21.** $\dfrac{-1}{2x + 1}$

23. $x + 11; 10; 12$ **25.** $2x - 3; -5; 3$

27. $-\dfrac{2x + 5}{x + 2}; 3; -\dfrac{13}{6}$

29. [1.1.1] product: $6x$; factors 6 and x

30. [4.3.4]

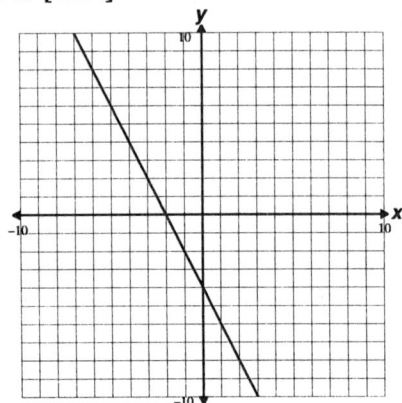

31. [4.3.4]

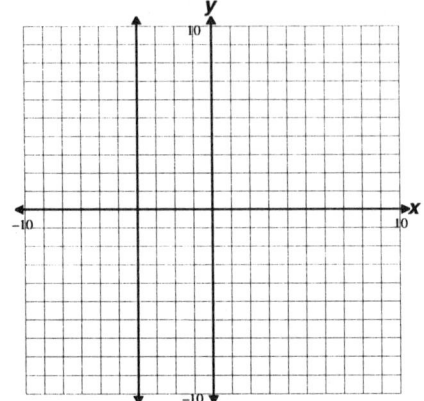

32. [1.2.3] $\dfrac{1}{10}$ **33. [1.2.6]** $\dfrac{24}{35}$ **34. [6.2.3]** $\dfrac{-16x}{y^5}$
35. [6.6.5] $12x^2 + x - 35$ **36. [6.7.1]** $-6w^2 - 3w + 2$
37. [6.5.3] $-417.5x^3 - 80.2x^2 - 93.16x + 140$
38. [6.5.1] $-2w^4 + 6w^3 + 2w^2 + 8$
39. [6.3.2] -6.04×10^{-3} **40. [6.3.4]** 2×10^6
41. [7.6.1] $0, 16$ **42. [4.5.1]** $y = -3x + 8$
43. [5.2.5] Demand: $y = 83 + 6x$; Supply: $y = 900 - 80x$
44. [7.2.1] $x(16 - 5x^2)$ or $-x(5x^2 - 16)$
45. [7.3.1] $(x - 2)(x + 3)$ **46. [7.4.2]** $(x + 1)(3x - 5)$
47. [7.5.4] $3x(x - 2)(x + 3)$

48. [3.1.3] $650 - x - \dfrac{1}{3}x$, which simplifies to $650 - \dfrac{4}{3}x$

49. [3.6.1] $0.07x = 6230$; \$89,000

50. [6.3.5] 22,250,000 pages.   **51. [4.4.1]** $y = \dfrac{3}{2}x - 3$

52 [4.6.4] **52a)** Using the points (0,74) and (6,140) leads to the equation $y = 11x + 74$.

52b) In 1985, 74 billion dollars was owed and since then the amount owed has been increasing by 11 billion dollars each year.   **52c)** Using an input of -5 leads to an output of \$19 billion dollars.

52d) The model predicts that 129 billion dollars will be owed in 1990, . Double this is 258 billion. Using the formula again leads to the year 2002 (approximately).

Section 8.2 Multiplying and Dividing Rational Expressions

Practice 8.2.1

1. $\dfrac{y^2}{2}$ 2. $\dfrac{2}{m^4}$ 3. $\dfrac{4}{w^4}$ 4. $\dfrac{4n^3}{9m^2p}$

Practice 8.2.2

1. $\dfrac{x+9}{2}$ 2. $\dfrac{y}{2}$ 3. $\dfrac{2(m-n)}{2m+n}$) 4. $\dfrac{5(a+b)}{a-b}$

5. $\dfrac{a(a+1)}{2(2a+1)}$ 6. $\dfrac{1}{2}$

Practice 8.2.3

1. $\dfrac{4}{3}$ 2. $\dfrac{3xy^4}{5}$ 3. $\dfrac{4n+5}{4}$ 4. $\dfrac{4}{3}$

5. $\dfrac{4}{y+4}$ 6. $\dfrac{1}{(x-2)(x+7)}$

Exercise Set 8.2

1. $\dfrac{6x^2z^4}{y}$ 3. $\dfrac{10c^2}{ab}$ 5. $9x^2$ 7. $\dfrac{2}{3r^2s}$ 9. $\dfrac{4n}{5}$ 11. $\dfrac{w+3}{3}$

13. $\dfrac{-1}{2}$ 15. $\dfrac{m+5}{m-2}$ 17. $\dfrac{4(x+y)}{2x+5y}$

19. $\dfrac{3}{4}$ 21. $\dfrac{-4}{7}$ 23. $\dfrac{7rs^2}{4}$ 25. $\dfrac{-2xy^3}{3}$

27. $\dfrac{x^3y^6z^3}{6}$ 29. $\dfrac{y+4}{2}$ 31. $\dfrac{5s-1}{5}$ 33. $\dfrac{s-1}{r}$

35a) [3.1.2] $x-8$ 35b) $x+3$ 35c) x 35d) $\dfrac{1}{2}x$

36. [3.2.1] $(s-7)+s+(s+7)=180$;53 degrees, 60 degrees, and 67 degrees

37. [3.5.1] $0.5d+1.5(d-6)=18$; 13.5 days

38. [4.8.1] domain: $\{-4,5,9\}$; range: $\{-3,0,3,5\}$

39. [4.8.4] 7 40. [4.7.1]

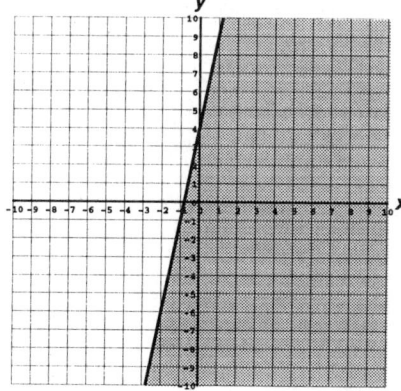

41. [5.2.4] no solution 42. [5.3.5] $(-3,1)$

43. [6.5.3] $8x-12y+14$ 44. [7.1.2] $4x^2y^3$

45. [6.2.2] 2 46. [6.2.3] $\dfrac{1}{h^3}$ 47. [6.2.3] $\dfrac{27}{y^{15}}$

48. [6.6.3] x^2-6x+9 49. [7.2.1] $5x^2(6x^2-5)$

50. [7.2.6] $(2x-1)(5x-y)$ 51. [7.3.1] $(x-5)(x-6)$

52. [7.5.3] $(x-6)(x-6)$ or $(x-6)^2$ 53. [7.6.1] $2, -\dfrac{5}{3}$

54. [7.6.2] $\dfrac{1}{2}, \dfrac{3}{5}$

55. [2.5.3] no solution

56. [1.1.1] x–intercept is $(-4,0)$, no y–intercept.

57. [4.4.2] The y–intercept seems to be around 88 million tons. The slope from the points $(0,88)$ to $(10,123)$ (both are estimated) is $\dfrac{35 \text{ million tons}}{10 \text{ years}}$ or 3.5 million tons per year.

The graph says 88 million tons of waste was generated in 1960 and the amount of waste has been increasing by 3.5 millions a year since then.

Section 8.3 Prelude to Adding and Subtracting Rational Expressions

Practice 8.3.1

1. 84 2. m^2n^2 3. x^3y^2 4. $6mn$ 5. $(x+3)(x-5)$

6. $(y+5)^2$ 7. $(x-6)(x+6)(x-3)$

8. $(y+7)(y-7)(y-2)$

Practice 8.3.2

1. $\dfrac{5\cdot 4}{6\cdot 4}$ which simplifies to $\dfrac{20}{24}$

2. $\dfrac{-2\cdot x^3}{x\cdot x^3}$ which simplifies to $\dfrac{-2x^3}{x^4}$

3. $\dfrac{-9\cdot 2qr^3}{r\cdot 2qr^3}$ which simplifies to $\dfrac{-18qr^4}{2qr^5}$

4. $\dfrac{x\cdot 2y}{7y\cdot 2y}$ which simplifies to $\dfrac{2x^2}{14xy}$

Practice 8.3.3

1. $\dfrac{5(w+3)}{(w+3)(w+3)}$ 2. $\dfrac{6(x-3)}{(x+2)(x-3)}$

3. $\dfrac{9y}{y(y+5)}$ 4. $\dfrac{(y+4)(y+2)}{(y+2)(y+3)}$

5. $\dfrac{(m+2)(m-5)}{m^2-25}$ 6. $\dfrac{(n+2)(n+2)}{n^2+9n+14}$

Practice 8.3.4

1. $\dfrac{27x}{42x^2}$ and $\dfrac{10}{42x^2}$

2. $\dfrac{5(y+4)}{(y-5)(y+4)}$ and $\dfrac{2(y-5)}{(y-5)(y+4)}$

3. $\dfrac{6(m+4)}{(m+4)^2}$ and $\dfrac{5}{(m+4)^2}$

4. $\dfrac{10(x+5)}{(x+1)^2(x+5)}$ and $\dfrac{3(x+1)}{(x+1)^2(x+5)}$

5. $\dfrac{2(y+2)}{(y+2)^2(y+4)}$ and $\dfrac{5(y+4)}{(y+2)^2(y+4)}$

Exercise Set 8.3

1. $42x^2$ **3.** $12x^3y^3$ **5.** $(y+2)(y-3)$ **7.** $x(x+1)$

9. $(m-1)^2(m+2)$ **11.** $y(y-3)(y-2)$ **13.** \f(20,32)

15. $\dfrac{-10}{25}$ **17.** $\dfrac{5mn}{m^2n}$ **19.** $\dfrac{16xz}{20yz}$ **21.** $\dfrac{35r^2s}{7rs^2}$

23. $\dfrac{7x^2}{x^2+4x}$ **25.** $\dfrac{2(t+1)}{(t+1)(t+1)}$

27. $\dfrac{3m}{m(m-9)}$ **29.** $\dfrac{(w-2)(w-2)}{w^2-3w+2}$

31. $\dfrac{5(t+5)}{t^2+9t+20}$ **33.** $\dfrac{(x-1)(x+6)}{x^2-36}$ **35.** $\dfrac{9}{15x^3}$ and $\dfrac{2x^4}{15x^3}$

37. $\dfrac{20m}{75m^2}$ and $\dfrac{9}{75m^2}$ **39.** $\dfrac{15n}{6m^2n^2}$ and $\dfrac{14m}{6m^2n^2}$

41. $\dfrac{4(y+2)}{y(y-3)(y+2)}$ and $\dfrac{7(y-3)}{y(y-3)(y+2)}$

43. $\dfrac{-2(x+5)}{(x+3)(x-2)(x+5)}$ and $\dfrac{5(x+3)}{(x+3)(x-2)(x+5)}$

45. [1.5.8] 1 **46.** [1.2.1] $2\cdot2\cdot2\cdot5\cdot11$ **47.** [2.2.5] 0

48. [4.4.3]

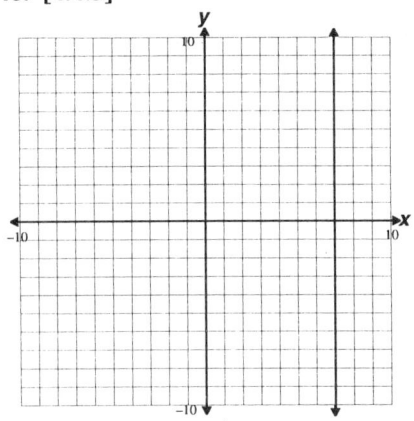

49. [4.7.1]

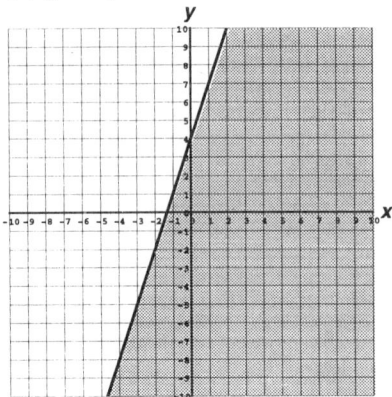

50. [4.8.6]

x	f(x)
–4	0
1	1
2	5
5	–4
5	5

51. [5.2.1] $(6,-2)$ **52.** [5.3.2] $(8,-1)$

53. [6.7.1] $-21y^{12}+11y^4-31y$

54. [6.7.2] $4a^2+3a-2+\dfrac{2}{2a-1}$ **55.** [1.1.3] 30

56. [1.2.11] $\dfrac{33}{40}$ **57.** [8.2.1] $4x^3$ **58.** [6.2.3] $\dfrac{81a^{20}}{2b^7}$

59. [1.7.6] $4x-12$ **60.** [6.6.5] $3c^2-31c-60$

61. [6.5.1] $16y^4-4.1y^2+y+8$ **62.** [7.2.1] $-6x(2x+3)$

63. [7.2.6] $(x-5)(x+3y)$ **64.** [7.4.2] $(3x+1)(5x-2)$

65. [7.5.1] $(11-x)(11+x)$

66. [7.5.3] $(5x-4)(5x-4)$ or $(5x-4)^2$

67. [7.5.4] $-6(x+2)(x-2)$

68. [5.4.4] $\left.\begin{cases} 2f + 3s = 32 \\ 1f + 0.75s = 10 \end{cases}\right\}$ build paint

Solving gives 4 model 500's and 8 model 600's.

69. [4.5.4] **69a)** $y=0.16x+66.8$ **69b)** In 1960 men lived to an average age of 66.8 years and their life expectancy has been rising by 0.16 years per year since then. **69c)** 81.2 years **69d)** 2074

Section 8.4 Adding and Subtracting Rational Expressions

Practice 8.4.1

1. $2m$ **2.** $\dfrac{y}{3}$ **3.** $\dfrac{y+6}{7y^2}$ **4.** $\dfrac{t+6}{5t^2}$ **5.** $\dfrac{y+7}{y+3}$

Practice 8.4.2

1. $\dfrac{-m}{6}$ **2.** $\dfrac{-y-2}{8y^2}$ or $-\dfrac{y+2}{8y^2}$

3. $\dfrac{2y^2+3}{y+2}$ **4.** $\dfrac{-m^2-1}{m+8}$ or $-\dfrac{m^2+1}{m+8}$

Practice 8.4.3

1. $\dfrac{17}{30}$ **2.** $\dfrac{21x-4}{15x^2}$ **3.** $\dfrac{6-5m}{14m^2}$

4. $\dfrac{4+5s}{6s^2t}$ **5.** $\dfrac{10b+15a}{18a^2b^2}$ **6.** $\dfrac{5+14mn}{7mn}$ **7.** $\dfrac{2+45xy}{9xy}$

Practice 8.4.4

1. $\dfrac{2(m-17)}{(m+4)(m-2)}$ **2.** $\dfrac{5r+19}{r+3}$

3. $\dfrac{3x-10}{(x-5)(x+5)}$ **4.** $\dfrac{2y+13}{(y-3)(y+3)}$

5. $\dfrac{4x-9}{(x-2)(x-3)}$ **6.** $\dfrac{-2x+7}{x(x-5)}$

Exercise Set 8.4

1. $2x$ **3.** $\dfrac{11}{x^2y}$ **5.** $\dfrac{-9-x}{2x+1}$ **7.** $\dfrac{b+9}{b+5}$ **9.** $\dfrac{-x}{3}$ or $-\dfrac{1}{3}x$

11. $\dfrac{3}{x^2y^2}$ **13.** $\dfrac{-x+1}{x+2}$ **15.** x

17. $\dfrac{2(x+4y)}{2x-3y}$ **19.** $\dfrac{2m+3n}{3m+2n}$ **21.** $\dfrac{-1}{12}$ **23.** $\dfrac{14x^2-27y^2}{24x^3y^3}$

25. $\dfrac{42y+15x-16}{36xy}$ **27.** $\dfrac{15b-54a^2b+14a^2}{18a^2b}$

29. $\dfrac{10n+36m^2n^2-m^3}{6m^2n^2}$ **31.** $\dfrac{18m^2n^2+15n-14m}{6m^2n^2}$

33. $\dfrac{12n^2-5m}{2m^2n^3}$ **35.** $\dfrac{2(x-13)}{(x+2)(x-4)}$ **37.** $\dfrac{2(4-s)}{s-1}$

39. $\dfrac{10+5t}{t(t-3)}$ **41.** $\dfrac{7x+10}{x^2+6x+5}$

43. $\dfrac{2(x^2-1)}{(x-3)(x+2)}$ **45.** $\dfrac{-3}{y(y-7)}$ **47.** $\dfrac{2(3x-8)}{(x-4)(x-3)}$

49. [1.1.1] sum: $9q+5$; terms: $9q$ and 5; product: $9q$; factors: within first term, 9 and q **50.** [1.5.8] 25

51. [7.6.3] $\dfrac{x}{x+2}+\dfrac{x+2}{x}=\dfrac{25}{12}$; $\dfrac{6}{8}$ or $\dfrac{-8}{-6}$

52. [3.3.2] 260 rolls **53.** [4.5.3] $x=-1$

54. [4.8.1] domain: all real numbers; range: all real numbers **55.** [2.2.4] no solution

56. [5.1.2] $(1,-2)$

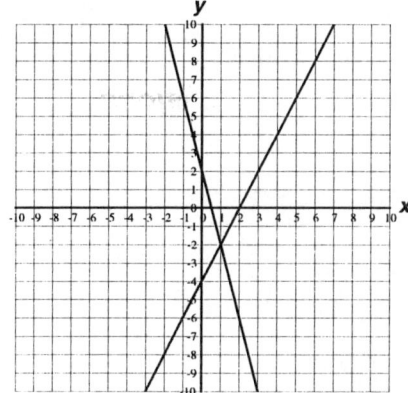

57. [5.3.6] $(-6,6)$ **58.** [8.1.2] $\dfrac{x+1}{3x-2}$

59. [7.2.6] $(8x-y)(5x+2)$ **60.** [7.3.1] $(x-1)(x-8)$

61. [7.6.1] $\dfrac{1}{2},1$ **62.** [6.2.3] $\dfrac{1}{72x^7y^5}$ **63.** [6.1.7] a

64. [1.2.11] 3 **65.** [6.6.1] $3x^2-10x+4$

66. [6.2.1] $2x^2y^6$ **67.** [6.2.3] $2m$ **68.** [6.2.2] 1

69. [6.5.3] $11a^2b^2+30a^2b+9ab^2$

70. [4.4.5] Slope is $\dfrac{67\text{ miles}}{1\text{ hour}}$. This implies they travel at a constant 67 mph..

Section 8.5 Complex Rational Expressions

Practice 8.5.1

1. $\dfrac{5}{6}$ **2.** $8a^3b^2$ **3.** $\dfrac{2}{3}$ **4.** $\dfrac{4}{m-4}$

Practice 8.5.2

1. $\dfrac{3}{2}$ **2.** $\dfrac{3}{2c^2}$ **3.** $\dfrac{5}{3}$ **4.** $\dfrac{t-3}{6}$ **5.** $\dfrac{9}{y+3}$

Practice 8.5.3

1. $\dfrac{4x+11}{4x+13}$ **2.** $\dfrac{t-2}{t+4}$

Exercise Set 8.5

1. $\dfrac{3}{2}$ **3.** $\dfrac{5}{2x^2}$ **5.** $\dfrac{2}{3}$ **7.** $\dfrac{3b^2xy^4}{2a}$ **9.** $\dfrac{1}{3}$ **11.** $\dfrac{5}{x-5}$

13. $\dfrac{a+5}{5}$ **15.** $\dfrac{x-5}{8}$ **17.** $\dfrac{x^3}{6}$ **19.** $-50x^3y^2$ **21.** $\dfrac{9t}{14}$

23. $\dfrac{x+y}{5}$ **25.** $\dfrac{2x-3}{5}$ **27.** $\dfrac{1}{5}$ **29.** $\dfrac{x+1}{x-1}$ **31.** $\dfrac{x-5}{15}$

33. $\dfrac{2t+36}{t+39}$ **35.** $\dfrac{5x+11}{5x+9}$ **37.** $\dfrac{5m-9}{5m-11}$

39. [3.1.3] $180-a-(2a-2)$, simplifies to $182-3a$

40. [3.5.2] $40t=45\left(t-\dfrac{3}{4}\right)$; 3:00 pm **41.** [4.8.3] yes

42. [4.4.1] $y=-3$

43. [4.5.1] Slope is $\dfrac{2}{3}$; y-intercept is $(0,0)$.

44. [7.6.1] 0, 2 **45.** [4.5.2] $x=-2$ **46.** [7.6.2] $\dfrac{1}{6},\dfrac{2}{3}$

47a) [5.1.5] Process A: $y=5000+15x$; Process B: $y=2000+25x$

47b)

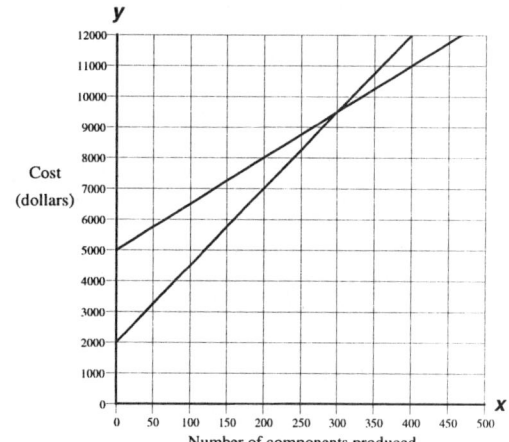

47c) $(300,9500)$; The two processes cost the same, $9,500, when 300 components are manufactured.

47d) If less than 300 components are made, Process B costs less; otherwise, Process A costs less.

48. [5.2.1] $(-1,2)$ **49.** [2.2.2] -12 **50.** [6.1.7] $\dfrac{1}{a^3}$

51. [6.2.3] 1 **52.** [6.2.3] $\dfrac{-24}{p^3 q^3}$ **53.** [6.2.3] $\dfrac{10}{y}$

54. [6.6.6] $9t^2 + 12t + 4$ **55.** [8.2.1] $\dfrac{6}{5} x^6$

56. [8.4.3] $\dfrac{3a - 8b}{4a^2 b}$ **57.** [8.2.2] $2(x + 3)$

58. [6.3.4] 8.08×10^{-9} **59.** [8.1.2] $\dfrac{x + 3}{2}$

60. [6.5.1] $39ab + a + b + 1$ **61.** [4.3.4]

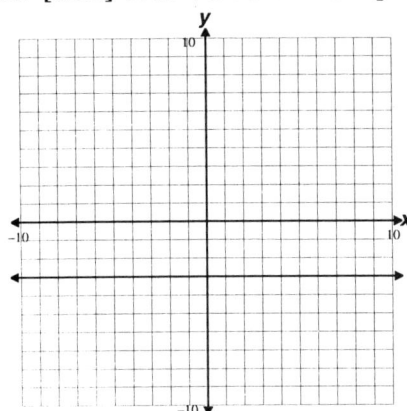

62. [6.7.1] $-24p^3 + 17p$ **63.** [6.7.2] $t^3 + t^2 - t - 2$
64. [7.4.2] $-2(x - 1)(4x - 3)$
65. [7.5.4] $-2x(3x + 2)(3x - 2)$
66a. [4.8.7] Tons of carbon monoxide released in 1975.
66b) 0.64 million tons of carbon monoxide were emitted t years after 1970. **66c)** t years after 1970 emissions were half of what they were in 1970.
66d) The change in emissions between 1974 and 1977.

Section 8.6 Rational Equations

Practice 8.6.1

1. 45 **2.** -2 **3.** -2 **4.** -1 **5.** $\dfrac{1}{4}$ **6.** -1 **7.** 5

Practice 8.6.2

1. No solution **2.** 0 **3.** No solution **4.** No solution

Practice 8.6.3

1. Simplify to get $\dfrac{11 - 25x}{5x}$ **2.** Simplify to get $\dfrac{x - 1}{2x(x + 1)}$

3. Solve to get 1 **4.** Simplify to get $\dfrac{17x - 30}{7x(x - 3)}$

Practice 8.6.4

1. 45 **2.** -18 **3.** 5

Practice 8.6.5

1. 2.5 mph **2.** 25 mph

Practice 8.6.6

1. $3\dfrac{9}{17}$ hours or 3 hours and 32 minutes (rounded)

2. $5\dfrac{5}{23}$ hours or 5 hours and 13 minutes (rounded)

Exercise Set 8.6

1. 30 **3.** -4 **5.** -3 **7.** -2 **9.** 3 **11.** 1 **13.** 20
15. 15 **17.** -4 **19.** -3 **21.** No solution **23.** 3
25. No solution **27.** 0 **29.** -1
31. Simplify to get $\dfrac{2x - 18}{(x - 3)(x + 3)}$ **33.** Solve to get 1
35. -45 **37.** -30 **39.** 21 **41.** -6.75 **43.** 1
45. Kim – 60 mph, Lori – 45 mph **47.** 15 mph
49. $1\dfrac{7}{8}$ hours or 1 hour 53 minutes (rounded)

51. 247.5 hours

53. [3.6.2] $i = 735(0.12)\left(\dfrac{9}{12}\right)$; i is \$66.15 so she pays

\$801.15 **54.** [3.6.6] $0.1t + 0.6(4) = 0.4(t + 4)$; 2 gallons

55. [7.6.3] $\dfrac{x + 9}{x} - \dfrac{x}{x + 9} = \dfrac{15}{4}$; $\dfrac{12}{3}$ **56.** [4.4.4] undefined

57. [4.8.2] yes **58.** [5.2.4] $(x, \dfrac{2}{3} x + \dfrac{5}{3})$

59. [5.3.5] $(1,1)$ **60.** [6.2.2] $\dfrac{7}{5}$ **61.** [6.1.7] $\dfrac{xy}{2}$

62. [6.2.3] $\dfrac{k^5}{27}$ **63.** [8.5.2] $\dfrac{2}{5}$ **64.** [6.2.3] $\dfrac{3}{2m^3 n^2}$

65. [6.6.3] $m^2 + 8m + 16$ **66.** [8.4.3] $\dfrac{6x - 5}{2x}$

67. [6.6.1] $2ab^2 + ab - b^2 - 1$ **68.** [6.3.4] 6×10^7
69. [4.7.1]

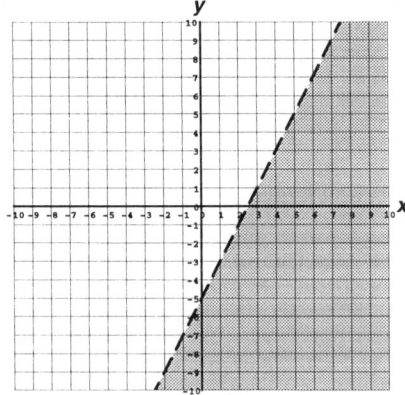

70. [6.5.3] $x^2 y - 4xy^2 - xy$
71. [6.7.1] $-6y^4 + 5y^3 - 4y^2 + 2$
72. [7.3.2] $(x - 16)(x - 4)$ **73.** [7.4.2] $4(x - 4)(2x - 1)$
74. [7.5.4] $4(x + 3)(x + 3)$ **75.** [7.6.1] $0, \dfrac{1}{2}$

76. [7.6.2] $\dfrac{5}{3}, \dfrac{2}{5}$ **77.** [8.1.2] $\dfrac{2x}{x + 5}$

Chapter 8 Review Exercises

Numbers inside brackets [ ] indicate the Chapter, Section, and Example number of a similar worked example.

1. [8.1.1] 4   **2.** [8.1.1] 0   **3.** [8.1.2] $\dfrac{1}{x-2}$

4. [8.1.2] $\dfrac{3}{x+1}$ **5.** [8.1.3] $\dfrac{1}{-x+3}$ or $-\dfrac{1}{x-3}$

6. [8.1.3] $\dfrac{-x+3}{x+2}$ or $-\dfrac{x-3}{x+2}$

7. [8.2.1] $\dfrac{9x}{y}$ **8.** [8.2.1] $\dfrac{2}{x^2 y}$ **9.** [8.2.2] $\dfrac{2}{x}$

10. [8.2.2] $\dfrac{x+6}{2}$ **11.** [8.2.3] $4x^2 y^6$ **12.** [8.4.1] $\dfrac{15-x}{5x}$

13. [8.4.1] $\dfrac{4x+1}{x^2}$

14. [8.2.3] $\dfrac{3(x-3)}{x+3}$ **15.** [8.4.2] $\dfrac{4}{y^2}$ **16.** [8.4.2] $\dfrac{3x-1}{x^2}$

17. [8.4.3] $\dfrac{7-5x}{2x^2}$ **18.** [8.4.3] $\dfrac{8a^2 b-3}{a^2 b}$

19. [8.4.4] $\dfrac{3x^2+5}{3x(x-2)}$ **20.** [8.5.2] $\dfrac{3k+1}{k-3}$

21. [8.5.2] $\dfrac{2(-5y+1)}{4y-1}$ **22.** [8.4.4] $\dfrac{4(x-2)}{(x-4)(x+4)(x+4)}$

23. [8.6.1] 2   **24.** [8.6.1] $\dfrac{5}{2}$ **25.** [8.6.1] 3

26. [8.6.1] −2   **27.** [8.6.4] −4   **28.** [8.6.4] −1

29. [8.6.2] No solution   **30.** [8.6.2] No solution

31. [8.6.5] $\dfrac{20}{b-1.2} = \dfrac{32}{b+1.2}$; 5.2 mph

32. [8.6.6] $\dfrac{1}{2}x + \dfrac{1}{3}x = 1$; $1\dfrac{1}{5}$ hours

33. [8.6.6] $\dfrac{1}{30}x + \dfrac{1}{15}x = 1$; 10 minutes

Chapter 8 Test

Numbers inside brackets [ ] indicate the Chapter, Section, and Example number of a similar worked example.

1. [8.1.1] 5   **2.** [8.1.2] $\dfrac{2x}{3x-1}$ **3.** [8.1.2] $3x+5$

4. [8.2.1] $\dfrac{3y^2}{x^4}$ **5.** [8.2.2] $2(x-1)$ **6.** [8.2.3] $2x^2 y^7$

7. [8.5.2] $\dfrac{-2x+y}{3x-2y}$ **8.** [8.4.3] $\dfrac{-x+8}{6x^2 y}$

9. [8.4.4] $\dfrac{5(x-3)}{(x-7)(x-1)(x+3)}$ **10.** [8.2.3] $\dfrac{x-1}{x}$

11. [8.4.3] $\dfrac{10x+11}{5x}$ **12.** [8.6.4] $\dfrac{1}{2}$ **13.** [8.6.1] −4

14. [8.6.2] No solution   **15.** [8.6.4] 20

16. [8.6.5] $\dfrac{1}{2}x + \dfrac{1}{3}x = 1$; $1\dfrac{1}{5}$ hours

17. [8.6.5] $\dfrac{1240}{x+15} = \dfrac{1120}{x-15}$; 295 mph

---

Chapter 9 Radical Expressions and Equations

Section 9.1 Introduction to Radical Expressions

Practice 9.1.1

1. 4   **2.** 0   **3.** $\dfrac{3}{8}$ **4.** 13   **5.** 0.5   **6.** 0.07

Practice 9.1.2

1. 0.2   **2.** 50   **3.** 29.66   **4.** 7.1

Practice 9.1.3

1. This is a real number. $-\sqrt{49}$ simplifies to −7

2. This is not defined using real numbers.

3. This is a real number. $-\sqrt{1}$ simplifies to −1

4. This is not defined using real numbers.

Practice 9.1.4

1. 47   **2.** −13   **3.** not defined using real numbers

4. 6   **5.** −13   **6.** −38   **7.** not defined using real

numbers   **8.** −112   **9.** not defined using real numbers

Practice 9.1.5

1. 9.2   **2.** 1750   **3.** 48   **4.** 56   **5.** 17   **6.** 29

Exercise Set 9.1

1. 8   **3.** $\dfrac{9}{11}$ **5.** 0.2   **7.** 1.1   **9.** 4   **11.** 16   **13.** 0.22

15. not defined using real numbers.   **17.** −13

19. not defined using real numbers.   **21.** 79   **23.** 22

25. 2   **27.** −14   **29.** 62   **31.** 18   **33.** −80   **35.** 23

37. 0.056   **39.** 5060   **41.** 120.98   **43.** 90 mph

45. 41 mph

47. [1.2.1] $3 \cdot 3 \cdot 13$ **48.** [4.3.4]

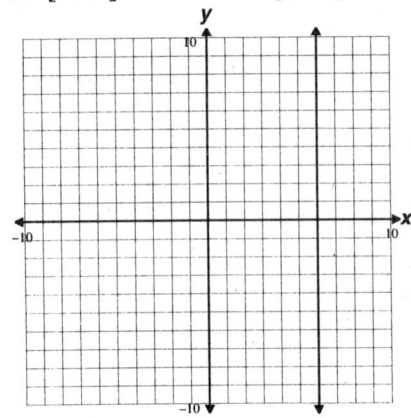

49. [6.6.1] $a^3 + b^3$ **50.** [6.7.1] $-25h^4 + 15h^2$

51. [6.7.2] $3x^2 + 3x + 3 + \dfrac{5}{2x-2}$ **52.** [6.1.7] $\dfrac{b^6}{125a^3}$

53. [6.2.3] a^3b **54.** [6.2.3] $\dfrac{v}{tw^2}$ **55.** [8.2.3] xy^3

56. [6.2.3] 432 **57.** [6.6.7] $9s^2 - 4t^2$ **58.** [8.2.1] x^2y^3

59. [8.4.3] $\dfrac{10x + 13}{10x}$ **60.** [7.3.1] $(x-5)(x-9)$

61. [7.4.2] $6(x-2)(2x-1)$

62. [7.5.3] $(x+3)(x+3)$ or $(x+3)^2$

63. [7.5.4] $5(2x+1)(x-7)$ **64.** [2.2.5] 0

65. [7.6.1] $-8, 8$ **66.** [8.6.2] no solution

67. [5.3.5] $(2,1)$

68. [3.6.2] $i = 17000(0.65)(3)$; \$3,315; payback is
\$20,315   **69.** [8.6.6] $\dfrac{48}{65} = \dfrac{x}{38}$; 28 questions

70. [4.2.1] $y = 39.4 + 0.51x$

Section 9.2   Simplifying Radical Expressions

Practice 9.2.1

1. $11\sqrt{w}$ **2.** $9\sqrt{x}\sqrt{y}$ **3.** $\dfrac{4}{t}$ **4.** $\dfrac{5y\sqrt{t}}{z}$ **5.** $3a$

Practice 9.2.2

1. $2\sqrt{d}$ **2.** $12\sqrt{b}$ **3.** $\sqrt{5h}$ **4.** $4\sqrt{g}$

Practice 9.2.3

1. $2\sqrt{2}$ **2.** $5\sqrt{3}$ **3.** $8\sqrt{2}$ **4.** $\dfrac{2\sqrt{6}}{7}$

Practice 9.2.4

1. $50\sqrt{2}$ **2.** $12\sqrt{3}$ **3.** $14\sqrt{2}$ **4.** $22\sqrt{2}$

Practice 9.2.5

1. d^{18} **2.** $x^6\sqrt{x}$ **3.** $y^{40}m^{16}\sqrt{y}$ **4.** $x^2y^6z^4\sqrt{xz}$
5. $m^{24}n^2r^{12}\sqrt{mr}$

Practice 9.2.6

1. $4a^2b^3\sqrt{3}$ **2.** $5x^4y^5\sqrt{3}$ **3.** $8r^7s^5\sqrt{2rst}$
4. $2x^8y^6z^4\sqrt{11xy}$

5. $x - 2$ If the phrase *Assume that no radicals were formed by squaring negative quantities* were left out of the directions we would write the answer as $|x-2|$

6. $x + 6$ If the phrase *Assume that no radicals were formed by squaring negative quantities* were left out of the directions we would write the answer as $|x+6|$

Practice 9.2.7

1. $49m^4\sqrt{m}$ **2.** $8b^{13}\sqrt{3b}$ **3.** $\dfrac{9x^5y^2\sqrt{2xy}}{2}$ **4.** $\dfrac{2m^4\sqrt{5m}}{n^2}$

Exercise Set 9.2

1. $7\sqrt{m}$ **3.** $\dfrac{\sqrt{s}}{2}$ **5.** $\dfrac{\sqrt{10}}{y}$ **7.** $5y$ **9.** $2x$ **11.** $13x^2$

13. $7\sqrt{t}$ **15.** $2\sqrt{r}$ **17.** $3\sqrt{3}$ **19.** $5\sqrt{2}$ **21.** $10\sqrt{2}$

23. $\dfrac{2\sqrt{10}}{3}$ **25.** $\dfrac{2\sqrt{3}}{5}$ **27.** $5\sqrt{6}$ **29.** $54\sqrt{2}$ **31.** $x^7\sqrt{x}$

33. $a^8\sqrt{a}$ **35.** $x^3y^4z\sqrt{xyz}$ **37.** $r^{21}s^{17}t^8$ **39.** $2y^2\sqrt{3y}$

41. $5a^4b^7\sqrt{2ab}$ **43.** $10w^{12}t^8\sqrt{w}$ **45.** $4rs^2t\sqrt{5st}$

47. $2x^{10}y^6z^6\sqrt{15z}$ **49.** $|x+8|$

51. $|x-4|$ **53.** $|2x+5|$ **55.** $20x^6$ **57.** $48s^6\sqrt{s}$

59. $108w^5x^3\sqrt{w}$ **61.** $44x^8y^{12}\sqrt{x}$ **63.** $2a^{21}b^4\sqrt{3}$

65. $8x^4y^2\sqrt{2y}$

---

67. [3.4.2] $1L + 1.5(2L) = 246$; 61.5 meters per hour

68. [3.5.1] $48s + 64(s-1) = 244$; s is 2.75 hours so they meet at 12:45 pm

69. [4.5.1] Slope is -2; y-intercept is $(0,3)$.

70. [7.5.1] $(9x-1)(9x+1)$

71. [4.3.6]

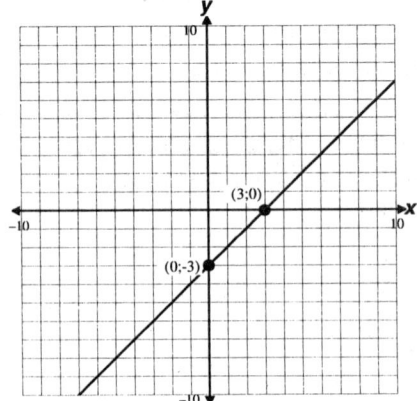

72. [8.1.2] $\dfrac{2}{5(2x+9)}$

73. [4.4.3]

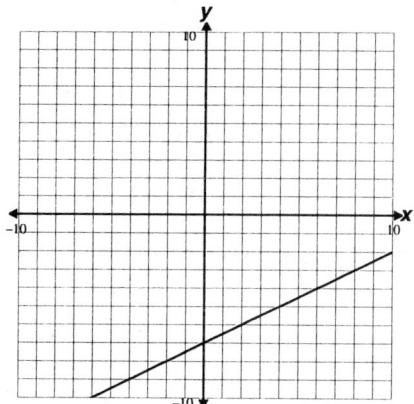

74. [4.7.1]

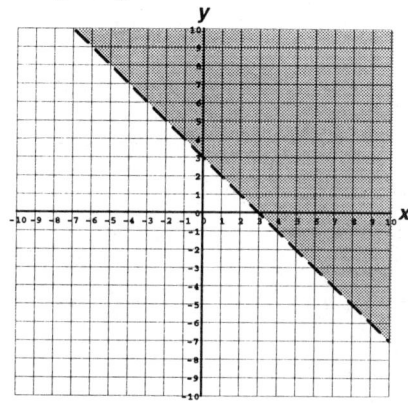

75. [2.5.3] $x < 3$　**76.** [7.6.1] $-3, 3$　**77.** [8.6.1] 1
78. [5.2.4] $(x, -x + 3)$　**79.** [5.3.5] $(11, -7)$
80. [1.7.6] $-x - 12$　**81.** [1.2.6] 1　**82.** [6.2.3] -2
83. [6.2.3] $\dfrac{a^3 b^2}{18}$　**84.** [6.3.4] 3.92×10^4
85. [6.6.7] $x^2 - 225p^2$　**86.** [6.5.1] $3m^3 n + 5mn + 2$
87. [7.4.2] $4(x + 2)(3x + 5)$
88. [7.5.4] $8x(3x - 2)(2x - 3)$
89a) [4.2.3] $y = 79.4 - 0.33x$　**89b)** 54.32 seconds
89c) About 1991. Actually the next Olympics was 1992 and the winning time was 60.68 seconds.

Section 9.3　Multiplying and Dividing Radical Expressions

Practice 9.3.1

1. $8\sqrt{6}$　**2.** 45　**3.** $\dfrac{5x\sqrt{3}}{y^2}$　**4.** $4x^3 y^4 \sqrt{6}$　**5.** $15m^8 n^9 \sqrt{5n}$

Practice 9.3.2

1. $\dfrac{3\sqrt{11}}{11}$　**2.** $\dfrac{7\sqrt{3}}{6}$　**3.** $2\sqrt{5}$　**4.** $\dfrac{\sqrt{77a}}{7a}$　**5.** $\dfrac{2m\sqrt{6m}}{3}$
6. $3y\sqrt{2y}$

Exercise Set 9.3

1. $24\sqrt{3}$　**3.** $5\sqrt{2}$　**5.** $10x^5\sqrt{2}$　**7.** $\dfrac{8t^2\sqrt{3t}}{q^2}$　**9.** $6m^6\sqrt{15}$
11. $27x^6 y^7 \sqrt{2y}$　**13.** $18x^3 y^4 \sqrt{2y}$　**15.** $\dfrac{2\sqrt{6}}{3}$　**17.** $\dfrac{6\sqrt{13}}{13}$
19. $4\sqrt{2}$　**21.** $\dfrac{\sqrt{42m}}{6m}$　**23.** $\dfrac{\sqrt{15t}}{5t}$　**25.** $\dfrac{x^2\sqrt{10x}}{2}$

---

29. [1.5.8] -33　**30.** [5.2.1] $(1, -5)$
31. [3.3.2] 13.75 miles
32. [3.6.6] $0.60(40) + 35x = 0.53(40 + 35)$; 45% copper
33. [7.6.3] $14 = \dfrac{1}{2}n(n - 3)$; 7　**34.** [4.8.4] 14
35. [6.7.2] $3w^3 - 4w^2 + w + 4 - \dfrac{3}{w + 2}$
36. [6.6.7] $100z^2 - 121y^2$　**37.** [6.1.7] $-10t$
38. [6.2.3] $\dfrac{27xy}{4}$　**39.** [6.6.5] $12d^2 + 13d - 35$
40. [6.2.3] $-\dfrac{m^2 n^7}{4}$　**41.** [6.2.3] $-8b^9$
42. [8.5.2] $\dfrac{15 - m}{20}$　**43.** [9.2.6] $3m^5 n\sqrt{6mn}$
44. [7.4.2] $4(x - 3)(2x + 1)$
45. [7.5.1] $(3x - 4)(3x + 4)$　**46.** [7.5.4] $8x(x - 1)(x + 5)$
47. [7.6.1] $-2, 5$　**48.** [7.6.2] $-1, -\dfrac{1}{4}$

49. [8.6.2] no solution
50a) [4.8.7] In 1949 approximately 30.7 million students were enrolled.
50b) The number of students enrolled in 1969.
50c) t years after 1869 40 million students were enrolled.
50d) t years after 1869 the number of students had quadrupled.

Section 9.4　Adding and Subtracting Radical Expressions

Practice 9.4.1

1. $20\sqrt{3}$　**2.** $11\sqrt{3} + \sqrt{5}$　**3.** $8\sqrt{7} + 10\sqrt{11}$
4. not like radical terms　**5.** $-7\sqrt{3t}$　**6.** $13\sqrt{6y}$
7. $4\sqrt{7m}$　**8.** not like radical terms

Practice 9.4.2

1. $37\sqrt{3}$　**2.** $38\sqrt{2}$　**3.** $3\sqrt{3} + 9\sqrt{2}$
4. $7\sqrt{3} + 24\sqrt{2}$　**5.** $32m\sqrt{2m}$　**6.** $21\sqrt{t}$

Practice 9.4.3

1. $10 + 5\sqrt{5}$　**2.** $7y + 22\sqrt{y}$　**3.** $3t + 5\sqrt{t}$
4. $3x + 8\sqrt{3x} + 9$　**5.** $\dfrac{16\sqrt{3}}{3}$　**6.** $-\dfrac{3\sqrt{5}}{5}$　**7.** $\dfrac{\sqrt{3}}{3} - \dfrac{4\sqrt{7}}{7}$

Exercise Set 9.4

1. $21\sqrt{7}$ **3.** $-\sqrt{3}$ **5.** $14\sqrt{5s}$ **7.** $10\sqrt{5} - 3\sqrt{2x}$
9. $-\sqrt{6a}$ **11.** $8\sqrt{5c}$ **13.** $9\sqrt{5n}$

15. These are not like radical terms. **17.** $\frac{3}{2}\sqrt{2s}$

19. $18\sqrt{2}$ **21.** $-22\sqrt{2}$ **23.** $-2\sqrt{y}$
25. $20\sqrt{3} - 27\sqrt{2}$ **27.** $30x\sqrt{3x}$ **29.** $-2r\sqrt{6r}$
31. $4 + 14\sqrt{2}$ **33.** $6 + 3\sqrt{6} + 2\sqrt{7}$ **35.** $6 - 3\sqrt{3}$
37. $6w + 30\sqrt{w}$ **39.** $5a - 4\sqrt{5a} + 9$

41. $2w - 12\sqrt{2w} + 36$ **43.** $\frac{11\sqrt{5}}{5}$ **45.** $\frac{10\sqrt{3}}{3}$

47. $\frac{\sqrt{2}}{4} + \frac{\sqrt{3}}{6}$ **49.** $\frac{-3\sqrt{5}}{5} + \frac{5\sqrt{6}}{6}$

51. [1.6.8] base is 5, exponent is 2; simplifies to -25
52. [3.2.4] $x + (x + 1) + (x + 2) = 2(x + 2) + 9$; 10, 11, 12
53. [4.5.3] $= -\frac{1}{5}x$ **54.** [6.3.5] increased from about
5.7% to about 9.5% **55.** [7.4.2] $-2(2x - 3)(3x + 1)$

56. [8.1.2] $\frac{3}{2(x + 1)}$ **57.** [1.1.2] 56 **58.** [8.4.3] $\frac{3x + 7}{3x}$

59. [8.2.2] $x - 2$ **60.** [9.2.7] $\frac{2\sqrt{2}\,y}{x^2}$ **61.** [9.1.4] 25

62. [9.2.2] $6b\sqrt{3a}$ **63.** [6.5.3] $4g - 6h$
64. [7.6.1] 3, 8 **65.** [7.6.2] $-\frac{6}{5}, -\frac{5}{6}$

66. [2.2.4] no solution **67.** [8.6.1] -8
68. [4.8.1] domain: $\{x \mid -7 \le x \le 6\}$;
range: $\{y \mid 2 \le y \le 7\}$
69a) [4.8.8] $E(t) = 0.16t + 70.3$
69b) $E(60)$ is 79.9. In 2020 the average life expectancy
will be 79.9 years.
69c) t is -64.375. This implies the average life span was
60 years in about 1896. Of course we need to be careful
when extrapolating this far into the past.
69d) t is 439. It will take 439 years, or until 2399, for the
average life expectancy to be double what is was in 1960.
Here's a prediction we certainly should view with
suspicion.

Section 9.5 Solving Equations that Involve Square Roots

Practice 9.5.1

1. 20 **2.** 1 **3.** 4 **4.** 35

Practice 9.5.2

1. 20 **2.** 24 **3.** -19 **4.** 3 **5.** 8

Practice 9.5.3

1. 5 **2.** no solution. **3.** no solution. **4.** 2

Practice 9.5.4

1. $3\sqrt{3}$ miles or 5.2 miles **2.** A little over 3 feet.
3. $35\frac{2}{3}$ feet

Practice 9.5.5

1. 14 meters per second **2.** $\mu = 0.13$

Exercise Set 9.5

1. 50 **3.** 8 **5.** $\frac{9}{2}$ **7.** 12 **9.** 11 **11.** 7 **13.** 45

15. No solution **17.** 7 **19.** 12 **21.** 6 **23.** 24
25. 18 **27.** 18 **29.** 5 **31.** 16 **33.** -4 **35.** -6
37. 8 **39.** 4 **41.** 9 **43.** No solution **45.** 4
47. No solution **49.** 6 **51.** 6 **53.** No solution
55. 5.7 miles **57.** 145 feet **59.** About a third of a mile.

61. [7.6.4] $56^2 + x^2 = 106^2$; 90 feet
62. [4.4.4] Undefined. **63.** [4.5.1] $y = x - 2$
64. [4.5.2] $y = 1$ **65.** [7.4.2] $(x - 6)(4x - 3)$
66. [7.4.2] $(x - 3)(5x + 6)$ **67.** [7.1.2] $4xy^2z$
68. [7.6.1] 0, 1 **69.** [5.2.4] no solution
70. [8.6.2] no solution **71.** [8.4.4] $\frac{2(2x + 3)}{x(x - 2)(x + 2)}$
72. [6.1.7] x^6y^{10} **73.** [1.7.6] $2x + 9$ **74.** [9.1.4] -2

75. [9.3.1] $\frac{\sqrt{2}}{s^4}$ **76.** [9.4.2] $-\sqrt{6}$

77. [4.3.5] (0,2) and (–8,0) **78.** [4.8.3] yes

Section 9.6 Quadratic Equations

Practice 9.6.1

1. $-3, \frac{2}{3}$ **2.** $\frac{5 - \sqrt{21}}{2}, \frac{5 + \sqrt{21}}{2}$ **3.** $\frac{3 - \sqrt{37}}{2}, \frac{3 + \sqrt{37}}{2}$
4. $-1.38305, 1.85924$

Practice 9.6.2

1. $1 - \frac{\sqrt{6}}{2}, 1 + \frac{\sqrt{6}}{2}$ **2.** $3 - \frac{\sqrt{38}}{2}, 3 + \frac{\sqrt{38}}{2}$
3. No solution. (The radicand is a negative value.)
4. $10 + 10\sqrt{3}, 10 - 10\sqrt{3}$ **5.** 6, 6 (This is a double root.
The solution set contains the single number 6.)

Practice 9.6.4

1a) Substituting 4 for x we get –9. This means the ball has already hit the ground so the answer is 0 feet (assuming it did not bounce or fall into a hole).

1b) Substituting 40 for y and solving gives two solutions: $\dfrac{15 - \sqrt{93}}{8}$ and $\dfrac{15 + \sqrt{93}}{8}$. Converting these to approximate decimals we get 0.7 sec and 3.1 sec. That is, the ball will reach 40 feet once on the way up, after about 0.7 seconds, and once on the way down, after about 3.1 seconds.

1c) Substituting 0 for y and solving gives two solutions: $\dfrac{15 - \sqrt{253}}{8}$ and $\dfrac{15 + \sqrt{253}}{8}$. Converting these to approximate decimals we get –0.1 sec and 3.9 sec. The first solution makes no sense since –0.1 sec is before the ball was thrown. The second solution means the ball hits the ground 3.9 sec after it was thrown.

Exercise Set 9.6

1. $-1, \dfrac{5}{3}$ **3.** $-2, -\dfrac{1}{2}$ **5.** $\dfrac{9 - \sqrt{65}}{2}, \dfrac{9 + \sqrt{65}}{2}$

7. $\dfrac{1 - \sqrt{33}}{8}, \dfrac{1 + \sqrt{33}}{8}$

9. $\dfrac{2}{3}, \dfrac{2}{3}$ (The solution is a double root.)

11. $\dfrac{-1 - \sqrt{2}}{2}, \dfrac{-1 + \sqrt{2}}{2}$ **13.** –0.5483, 41.9483

(rounded)   **15.** There is no solution in the real numbers.
17. 1, 1 (The solution is a double root.)
19a) Substituting 5000 for y and solving for x gives 49.4. We add this to 1985 to get 2034.4. Therefore, 5,000,000 births will occur in 2035.
19b) 2075 is 90 years after 1985. Substituting 90 for x and solving for y gives 7198.8, which is 7,198,800 births.
19c) 2000 is 15 years after 1985. Substituting 15 for x and solving for y gives 4037.55, which is 4,037,550 births. 500,000 more than this is 9,037,550. Substituting 9037.55 for y and solving for x gives 113.9, which we round up to 114. 114 years after 1985 is 2099.

21. [8.6.6] $\dfrac{1}{60} g = \dfrac{1}{75} g + 1.5$; 450 gallons

22. [1.3.3]

```
├──┼──┼──┼──●──┼──┼──┼──┼──┤
-2  -1.6  -1.2  -0.8  -0.4   0   0.4   0.8
```

23. [4.7.1]

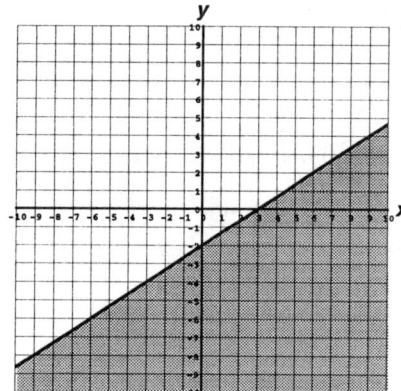

24. [6.7.2] $n^4 - 1$ **25.** [7.5.1] $(5x - 2)(5x + 2)$

26. [6.2.3] $\dfrac{1}{s^{18} t^6}$ **27.** [6.3.4] 2.25×10^1

28. [8.2.1] $\dfrac{2x^3 z}{3y^3}$ **29.** [8.2.2] $\dfrac{2(x + 2)}{3}$ **30.** [8.2.3] x

31. [8.4.4] $\dfrac{-5x + 6}{x - 1}$ **32.** [8.5.2] $t + 3$

33. [9.2.1] $\dfrac{2b\sqrt{a}}{3}$

34. [9.2.3] $5\sqrt{6}$ **35.** [9.2.6] $3x^3 y\sqrt{2x}$

36. [9.3.2] $\dfrac{2\sqrt{6}}{3}$ **37.** [6.2.2] 31   **38.** [7.6.1] $-2, \dfrac{1}{3}$

39. [9.5.2] 10   **40.** [5.2.4] no solution   **41.** [4.8.3] yes

Chapter 9 Review Exercises

Numbers inside brackets [ ] indicate the Chapter, Section, and Example number of a similar worked example.

1. [9.1.3] Not a real number.   **2.** [9.1.3] –4

3. [9.1.4] 1   **4.** [9.1.4] $\dfrac{3}{2}$ **5.** [9.2.1] $7xy$

6. [9.2.1] $\dfrac{3m\sqrt{n}}{4}$ **7.** [9.2.2] $5m$ **8.** [9.2.2] $4x\sqrt{15}$

9. [9.2.3] $6\sqrt{2}$ **10.** [9.2.3] $2\sqrt{3}$ **11.** [9.2.4] $28\sqrt{2}$
12. [9.2.4] $12\sqrt{10}$ **13.** [9.2.5] $xy^2 z^5 \sqrt{z}$
14. [9.2.5] $4fh^4\sqrt{fh}$ **15.** [9.2.6] $6s^2 t\sqrt{3}$

16. [9.2.6] $10mn^2 p^2\sqrt{7mp}$ **17.** [9.2.7] $\dfrac{t\sqrt{6s}}{s}$

18. [9.2.7] $8b^5 d^4\sqrt{2bc}$ **19.** [9.3.1] $4\sqrt{2a}$

20. [9.3.1] $\dfrac{5\sqrt{2}x^2}{y^4}$ **21.** [9.3.2] $2\sqrt{2}$

22. [9.3.2] $\dfrac{\sqrt{2y}}{2}$ **23.** [9.4.3] $x^2 + 4x + 4x\sqrt{x}$

24. [9.4.1] $9\sqrt{t} - 2t\sqrt{t}$ **25.** [9.4.1] $3\sqrt{2r} + \sqrt{14r}$

26. [9.4.2] $-3x\sqrt{3x} + 3\sqrt{3x}$ **27.** [9.4.2] $-2\sqrt{7} + \sqrt{5}$

28. [9.4.3] $6t$ **29.** [9.5.1] 20 **30.** [9.5.1] 1

31. [9.5.2] -3 **32.** [9.5.2] -7 **33.** [9.5.3] No solution. **34.** [9.5.3] No solution. **35.** [9.6.1] $-2, 0.5$

36. [9.6.1] $\dfrac{2 \pm \sqrt{7}}{2}$ **37.** [9.6.2] $\dfrac{1}{3}$

38. [9.6.2] $\dfrac{1 \pm 2\sqrt{2}}{2}$ **39.** [9.1.5] Walking the 224 foot diagonal will take 51 seconds while walking the 300 foot edge will take 68 seconds; Bill will save 17 seconds by walking on the grass.

Chapter 9 Test

Numbers inside brackets [ ] indicate the Chapter, Section, and Example number of a similar worked example.

1. [9.1.4] 1 **2.** [9.2.1] $\dfrac{6\sqrt{t}}{a}$ **3.** [9.2.2] 4

4. [9.4.2] $13r\sqrt{5t}$ **5.** [9.2.4] $14\sqrt{3}$

6. [9.2.5] $h^5 j^3 \sqrt{hj}$ **7.** [9.4.3] $-6\sqrt{10}$

8. [9.2.6] $3n + 2$ **9.** [9.3.1] $\dfrac{5\sqrt{6}\,ab^2}{c^2}$

10. [9.4.1] $\dfrac{7}{12}\sqrt{2y}$ **11.** [9.2.2] $6y\sqrt{7}$

12. [9.3.2] $\dfrac{d\sqrt{6}}{4}$ **13.** [9.5.1] 8 **14.** [9.5.2] 1

15. [9.5.3] No solution.

16. [9.6.1] $-\dfrac{2}{3}, -\dfrac{3}{2}$ **17.** [9.6.1] $\dfrac{3 \pm \sqrt{21}}{6}$

18. [9.1.5] about 8.06 feet (8 feet $\dfrac{3}{4}$ inch)

Index/Glossary

π   Greek letter pi [represents the ratio of the circumference to the diameter of a circle. It's value is about 3.14)], *3*

{ }   braces [indicate a set], *3*

+   addition operator [indicates two numbers are to be added], *3*

+   positive sign [indicates a number is greater than 0], *24*

−   subtraction operator [indicates two numbers are to be subtracted], *4*

−   negative sign [indicates a number is less than 0], *24*

−   opposite sign [indicates the opposite of a number], *28*

−   fraction bar [indicates a fraction or the operation of division (also a grouping symbol)], *4*

÷   division operator [indicates that two numbers are to be divided], *4*

/   division operator [indicates that two numbers are to be divided], *4*

/   slant fraction bar [indicates a fraction], *4*

×   St. Andrew's cross [indicates the operation of multiplication], *4*

•   dot [indicates the operation of multiplication], *4*

=   equal sign [indicates that two expressions represent the same value], *5*

≠   not equal sign [indicates that two expressions do not represent the same value], *5*

( )   parentheses [indicate multiplication as in 3(5)], *4*

( )   parentheses [indicate a grouping as in 2 − (1 + 5)], *6*

[ ]   brackets [indicate a grouping], *6*

>   greater than [indicates the number on the left is larger than the number on the right], *27*

≥   greater than or equal to [indicates the number on the left is larger than the number on the right or is equal to the number on the right], *28*

<   less than, indicates the number on the left is smaller than the number on the right] [*27*

≤   less than or equal to, indicates the number on the left is smaller than the number on the right or is equal to the number on the right], *28*

| | absolute value symbol [indicates the distance of a number from 0 on the number line (also a grouping symbol)], *29*

√   radical symbol [indicates the root of a number], *6, 560*

±   plus or minus [indicates a number and its opposite], *560*

2 by 2 linear system [a system of two unknowns and two equations where the graphs are straight lines], *344*

A

absolute value [the distance of a number from zero on the number line], *29*

"ac" method of factoring trinomials, *475*

Addition Method [a method of solving a system of equations where the equations are added in order to eliminate one of the variables (same as Elimination Method)], *367*

additive identity [the number 0], *54*

additive inverses, *28*

Amount Formula, *143*

Amount Table [a grid used to organize information to be used in an Amount Formula], *146*

amount [in the Amount Formula, the product of the rate and the base], *144*

annual percentage rate [interest rate for a year (APR)], *172*

APR [interest rate for a year (Annual Percentage Rate], *172*

Associative Property of Addition [the way in which terms are grouped does not affect their sum], *57*

Associative Property of Multiplication. The way in which terms are grouped does not affect their product], *57*

B

base [in exponential notation, the number being multiplied — in 32, 3 is the base], *5*

base [in the Amount Formula, the quantity that the rate is based on], *144*

binomial [an expression that has two terms], *426*

building up a fraction [multiplying the numerator and denominator of a rational expression by the same quantity], *503*

C

Caution

  A term of 1 may be needed, *463*

  Adding percents can be tricky, *184*

  Be careful with signs *464, 467*

  Canceling, *98*

  Canceling, *447*

  Canceling, *504*

  Checking solutions in radical equations, *592*

  deciding which way the arrow points, *103*

  Distribute negative sign, *523*

  Distributing a negative, *66*

  Expression versus Equation, *85*

  Identifying numerator and denominator, *11*

  Keep vocabulary straight, *55*

  Multiplication and addition of square roots have different properties, *570*

  Multiplication is different from addition, *401*

  Multiply each term by LCD, *90*

  Power of a product does not work with sums, *404*

  Solutions are not the answers, *490*

  Squaring a polynomial, *440*

  Subtraction and division are not commutative, *56*

  The dash and subtraction, *48*

  The dash and your calculator, *48*

  We cannot distribute over multiplication, *59*

  Working from left to right, *7*

  Zero product property, *487*

cell [the intersection of a row and a column in an Amount Table], *146*

coefficient [a constant multiplied by a variable, *4*

combining like terms [adding and subtracting the coefficients of terms that have identical variable parts], *65*

Commutative Property of Addition [the order in which we add terms does not affect their sum], *55*

Worked Examples

expanded form [the form of a product where repeated bases are written explicitly — x^3 written in expanded form is xxx], *61*

exponent notation [notation for powers of 10 where the letter E is written instead of $\times$], *419*

exponent definition and properties, summary of, *414*

exponent [in exponential notation, the number of bases being multiplied — in 32, 2 is the exponent and tells us to multiply $3 \cdot 3$], *5*

exponential notation, *61*

exponential notation [a shorthand way of writing repeated multiplication, as in 32], *5*

expression [a collection of numbers, operations, letters and grouping symbols such as x, $5y$, and $3x + 2$], *6*

extraneous roots [apparent solutions to an equation that do not satisfy the original equation], *540*

extrapolation [estimating from outside the given data set], *227*

F

factor completely [write as a product where each factor cannot be further factored using integers other than 1 or −1], *472*

factor [a number being multiplied — in $2x$, both 2 and x are factors], *4*

Q

quadratic equation in one variable [an equation that can be written in the form $ax^2 + bx + c = 0$, where a $\neq$ 0], *485*
quadratic relation [a relation that can be modeled using y = $ax^2 + bx + c$, where a, b, and c are constants], *597*
quotient [the combination of two numbers using division], *4*

R

radical equation [an equation where a variable appears in a radicand], *589*
radical expression [an expression that contains a radical symbol (same as radical)], *560*
radical symbol [the symbol $\sqrt{}$ that indicates the principal square root of a number. The principal square root of 5 is written $\sqrt{5}$], *560*
radical [an expression that contains a radical symbol (same as radical expression)], *560*
radicand [the expression under the radical symbol], *560*
range [the set of values allowed for the output variable], *311*
rate of change, *196*
rate [the ratio of two quantities that are measured using different units], *143*

ratio [a fraction which expresses a relation between two quantities], *544*
Rational Numbers [the set of numbers consisting of numbers that can be written as the ratio of two integers, where the denominator is not 0], *24*
rational equation [an equation that contains at least one rational expression], *537*
rational expressions [fractions where the numerator and denominator are polynomials], *502*
rationalizing the denominator [converting the denominator of a fraction from an irrational to a rational number], *580*
Real Numbers [the set of numbers consisting of the rational and irrational numbers], *24*
reciprocals [two numbers whose product is 1 such as \f(3,2) and $\frac{2}{3}$, *14*
reducing the fraction to lowest terms [removing all common factors from the numerator and denominator of a rational expression], *503*
regions [a portion of the x-y plane], *300*
regression line [the line which best fits the data of an input/output relation], *285*
relation [a correspondence between two sets of numbers, such as input and output], *310*
Relative sizes check, *120*
restricted values [values of a variable which would result in division by 0], *502*

rise [the change in output when going from one point on a line to another], *251*
root [a value of a variable that makes an equation or inequality true (same as solution] *78, 100*
run [the change in input when going from one point on a line to another], *251*

S

satisfy [to make an equation true], *78*
satisfy [to make an inequality true], *100*
scale [the size and placement of the numbers on the axes of an x-y graph], *230*
scientific notation [notation used with very large or small numbers where the number is written as the product of a power of 10 and a decimal number equal to or greater than 1 but less than 10, as in 5.2 5 102, *418*
set builder notation, *312*
set builder notation [a notation used to indicate the members of an set by writing a description of the members rather than listing them], *108*
set [a collection of distinct things that are to be treated as an entity], *3*